Remote Sensing Handbook, Volume V (Six Volume Set)

Volume V of the Six Volume *Remote Sensing Handbook*, Second Edition, is focused on the use of remote sensing technologies for studying water resources, including groundwater, floods, snow and ice, and wetlands. It discusses water productivity studies from Earth observation data characterization and modeling, mapping their successes and challenges. Chapters include remote sensing of surface water hydrology; quantitative geomorphology; river basin studies; floods; wetlands, including mangroves and river deltas; groundwater studies; crop water use or actual evapotranspiration modeling and mapping; and snow and ice mapping. This thoroughly revised and updated volume draws on the expertise of a diverse array of leading international authorities in remote sensing and provides an essential resource for researchers at all levels interested in using remote sensing. It integrates discussions of remote sensing principles, data, methods, development, applications, and scientific and social context.

FEATURES

- Provides the most up-to-date comprehensive coverage of remote sensing science for water resources, including wetlands, floods, snow, and ice.
- Provides comprehensive assessments of crop water use and crop water productivity modeling and mapping, including evapotranspiration studies.
- Discusses and analyzes data from old and new generations of satellites and sensors spread across 60 years.
- Includes numerous case studies on advances and applications at local, regional, and global scales.
- Introduces advanced methods in remote sensing, such as machine learning, cloud computing, and artificial intelligence (AI).
- Highlights scientific achievements over the last decade and provides guidance for future developments.

This volume is an excellent resource for the entire remote sensing and GIS community. Academics, researchers, undergraduate and graduate students, as well as practitioners, decision makers, and policymakers, will benefit from the expertise of the professionals featured in this book and their extensive knowledge of new and emerging trends.

Remote Sensing Handbook, Volume V (Six Volume Set)

Water, Hydrology, Floods, Snow and Ice, Wetlands, and Water Productivity

Second Edition

Edited by Prasad S. Thenkabail, PhD

CRC Press is an imprint of the
Taylor & Francis Group, an **informa** business

Designed cover image: © Prasad S. Thenkabail

Second edition published 2025
by CRC Press
2385 NW Executive Center Drive, Suite 320, Boca Raton FL 33431

and by CRC Press
4 Park Square, Milton Park, Abingdon, Oxon, OX14 4RN

CRC Press is an imprint of Taylor & Francis Group, LLC

First edition published by CRC Press 2016

Reasonable efforts have been made to publish reliable data and information, but the author and publisher cannot assume responsibility for the validity of all materials or the consequences of their use. The authors and publishers have attempted to trace the copyright holders of all material reproduced in this publication and apologize to copyright holders if permission to publish in this form has not been obtained. If any copyright material has not been acknowledged, please write and let us know so we may rectify in any future reprint.

Except as permitted under U.S. Copyright Law, no part of this book may be reprinted, reproduced, transmitted, or utilized in any form by any electronic, mechanical, or other means, now known or hereafter invented, including photocopying, microfilming, and recording, or in any information storage or retrieval system, without written permission from the publishers.

For permission to photocopy or use material electronically from this work, access www.copyright.com or contact the Copyright Clearance Center, Inc. (CCC), 222 Rosewood Drive, Danvers, MA 01923, 978–750–8400. For works that are not available on CCC please contact mpkbookspermissions@tandf.co.uk

Trademark notice: Product or corporate names may be trademarks or registered trademarks and are used only for identification and explanation without intent to infringe.

Library of Congress Cataloging-in-Publication Data
Names: Thenkabail, Prasad Srinivasa, 1958- editor.
Title: Remote sensing handbook / edited by Prasad S. Thenkabail ; foreword by Compton J. Tucker.
Description: Second edition. | Boca Raton, FL : CRC Press, 2025. | Includes bibliographical references and index. | Contents: v. 1. Remotely sensed data characterization, classification, and accuracies — v. 2. Image processing, change detection, GIS and spatial data analysis — v. 3. Agriculture, food security, rangelands, vegetation, phenology, and soils — v. 4. Forests, biodiversity, ecology, LULC, and carbon — v. 5. Water, hydrology, floods, snow and ice, wetlands, and water productivity — v. 6. Droughts, disasters, pollution, and urban mapping.
Identifiers: LCCN 2024029377 (print) | LCCN 2024029378 (ebook) | ISBN 9781032890951 (hbk ; v. 1) | ISBN 9781032890968 (pbk ; v. 1) | ISBN 9781032890975 (hbk ; v. 2) | ISBN 9781032890982 (pbk ; v. 2) | ISBN 9781032891019 (hbk ; v. 3) | ISBN 9781032891026 (pbk ; v. 3) | ISBN 9781032891033 (hbk ; v. 4) | ISBN 9781032891040 (pbk ; v. 4) | ISBN 9781032891453 (hbk ; v. 5) | ISBN 9781032891477 (pbk ; v. 5) | ISBN 9781032891484 (hbk ; v. 6) | ISBN 9781032891507 (pbk ; v. 6)
Subjects: LCSH: Remote sensing—Handbooks, manuals, etc.
Classification: LCC G70.4 .R4573 2025 (print) | LCC G70.4 (ebook) | DDC 621.36/780285—dc23/eng/20240722
LC record available at https://lccn.loc.gov/2024029377
LC ebook record available at https://lccn.loc.gov/2024029378

ISBN: 978-1-032-89145-3 (hbk)
ISBN: 978-1-032-89147-7 (pbk)
ISBN: 978-1-003-54140-0 (ebk)

DOI: 10.1201/9781003541400

Typeset in Times
by Apex CoVantage, LLC

Contents

Foreword by Compton J. Tucker xiii
Preface xxi
About the Editor xxix
List of Contributors xxxiii
Acknowledgments xxxvii

PART I Geomorphology

Chapter 1 Geomorphological Studies from Remote Sensing 3

James B. Campbell and Lynn M. Resler

    1.1   Introduction 3
            1.1.1   Historical Perspective 5
    1.2   Alpine and Polar Periglacial Environments 10
            1.2.1   Overview 10
            1.2.2   Change Analysis 11
            1.2.3   Emergent Opportunities and Challenges 12
    1.3   Glacial Geomorphology 12
            1.3.1   Overview 12
            1.3.2   Emergent Opportunities and Challenges 13
    1.4   Mass Wasting 14
            1.4.1   The Role of Synthetic Aperture Radar in Landslide Analysis 15
            1.4.2   Landslide Studies 16
            1.4.3   Summary and Future Outlook 17
    1.5   Fluvial Landforms 18
            1.5.1   Overview 18
            1.5.2   Floodplain Analysis 18
            1.5.3   Channel Migration 20
            1.5.4   Streambank Retreat 21
            1.5.5   Summary 21
    1.6   Coastal Geomorphology 21
            1.6.1   Overview 21
            1.6.2   Multispectral Bathymetry 22
            1.6.3   Airborne LiDAR Bathymetry 22
            1.6.4   Suspended Sediment Concentration (SSC) 23
            1.6.5   Coastal Change and Retreat 23
    1.7   Aeolian Landforms 24
    1.8   Biogeomorphology 26
    1.9   Remote Sensing and Geomorphic Inquiry 27
    References 29

PART II Hydrology and Water Resources

Chapter 2 Remote Sensing Technologies for Multi-Scale Hydrological Studies: Advances and Perspectives 37

Sadiq I. Khan, Ni-Bin Chang, Yang Hong, Xianwu Xue, and Yu Zhang

- 2.1 Introduction 38
- 2.2 Remote Sensing Advances in Hydrology 40
- 2.3 Precipitation Estimation Based on Multi-Sensor Data 41
    - 2.3.1 Physical Principles of Radar Precipitation Estimation 41
    - 2.3.2 Next-Generation Radar (NEXRAD) 43
    - 2.3.3 Global Precipitation Estimation 44
    - 2.3.4 Missions on Precipitation Estimation 45
- 2.4 Evapotranspiration Estimation Using Multi-Sensor Data 46
    - 2.4.1 Physical Principles of Surface Energy Balance (SEB) Method 46
    - 2.4.2 Regional-, Local-, and Field-Scale Studies 47
    - 2.4.3 Global Evapotranspiration Estimation 50
- 2.5 Soil Moisture Estimation Based on Multi-Sensor Data 52
    - 2.5.1 Regional-, Local-, and Field-Scale Studies 52
    - 2.5.2 Global Soil Moisture Estimation 54
    - 2.5.3 Satellite Mission on Soil Moisture Estimation 54
- 2.6 Ground Water Assessment from GRACE 55
- 2.7 Summary and Conclusions 56
- References 57

Chapter 3 Groundwater Targeting Using Remote Sensing 65

Santhosh Kumar Seelan

- 3.1 Introduction 65
- 3.2 Role of Remote Sensing in Groundwater Prospecting 66
- 3.3 Groundwater Information Extraction from Satellite Images 68
    - 3.3.1 Lithology and Lineaments 69
    - 3.3.2 Hydrogeomorphology 70
    - 3.3.3 Soils, Land Use, and Drainage 72
    - 3.3.4 Groundwater Recharge/Discharge 73
    - 3.3.5 Groundwater Use for Irrigation 74
    - 3.3.6 Groundwater Stress 75
    - 3.3.7 The Integrated Approach 77
- 3.4 Indicators and Interpretation Keys in Unconsolidated Rock Terrain 79
    - 3.4.1 Alluvial Aquifers 80
    - 3.4.2 Glacial Terrains 84
    - 3.4.3 Aeolian Deposits 85
    - 3.4.4 Remote Sensing Parameters in Unconsolidated Rock Terrain 85
- 3.5 Indicators and Interpretation Keys in Semi-consolidated to Consolidated Sedimentary Rock Terrain 86
    - 3.5.1 Sandstone–Shale Aquifers 87
    - 3.5.2 Carbonate Rocks 87
    - 3.5.3 Exploration Parameters in Sandstone–Shale and Carbonate Rocks 87
    - 3.5.4 Remote Sensing Parameters in Semi-consolidated to Consolidated Sedimentary Rock Terrain 88

Contents vii

	3.6	Indicators and Interpretation Keys in Volcanic Terrain	89
		3.6.1 Typical Profiles in Volcanic Terrain	90
		3.6.2 Groundwater Occurrence and Exploration Parameters in Volcanic Terrain	92
		3.6.3 Remote Sensing Parameters in Volcanic Terrain	93
	3.7	Indicators and Interpretation Keys in Hard Rock Terrain	94
		3.7.1 Weathered Hard Rocks	95
		3.7.2 Fractured Hard Rocks	96
		3.7.3 Remote Sensing Parameters in Hard Rocks	98
	3.8	Case Studies and Economic Benefits	100
		3.8.1 Case Study I	100
		3.8.2 Case Study II	101
	3.9	Summary	102
	Acknowledgments		103
	References		103

PART III Floods

Chapter 4 Flood Monitoring Using the Integration of Remote Sensing and Complementary Techniques 113

Allan S. Arnesen, Frederico T. Genofre, Marcelo P. Curtarelli, and Matheus Z. Francisco

	4.1	Introduction	114
	4.2	Remote Sensing in Flood Studies	115
		4.2.1 Flood Mapping with SAR and Ancillary Data	116
		4.2.2 Integration of Remote Sensing, GIS, and Hydrological Models	123
		4.2.3 Flood Forecasting Using Remote Sensing–Derived Information	127
	4.3	Conclusions	129
	References		130

Chapter 5 Flood Studies Using Synthetic Aperture Radar Data 135

Sandro Martinis, Claudia Kuenzer, and André Twele

	5.1	Introduction	136
	5.2	Interaction of SAR Signal and Water Bodies	139
		5.2.1 Smooth Open Water	141
		5.2.2 Rough Open Water	143
		5.2.3 Partially Submerged Vegetation	143
		5.2.4 Flooding in Urban Areas	145
	5.3	State of the Art in SAR-Based Water Detection	146
		5.3.1 Visual Interpretation	147
		5.3.2 Thresholding	147
		5.3.3 Change Detection	148
		5.3.4 Contextual Classification	148
		5.3.5 Integration of Auxiliary Data	149
	5.4	Case Studies	150
		5.4.1 Semi-automatic Object-Based Flood Detection (RaMaFlood)	150

		5.4.2	Automatic Pixel-Based Water Detection (WaMaPro)	156
		5.4.3	Fully Automatic Pixel-Based Flood Detection (TerraSAR-X Flood Service)	162
	5.5	Conclusion		168
	References			169

Chapter 6 Remote Sensing of Mangrove Forests 176

Le Wang, Jing Miao, and Ying Lu

6.1	Introduction		176
6.2	Leaf Level		178
6.3	Individual Tree Level		180
6.4	Forest Level (Canopy)		181
6.5	Ecosystem Level		183
	6.5.1	Tidal-Driven Nutrient Exchange	184
	6.5.2	Biodiversity	185
	6.5.3	Carbon Conservation	187
	6.5.4	Coastal Area Protection	187
6.6	Conclusion		188
References			188

PART IV Wetlands

Chapter 7 Remote Sensing of Mangrove Wetlands 195

Chandra Giri

7.1	Introduction	195
7.2	Defining Mangroves for Mapping and Monitoring	197
7.3	Remote Sensing Data	197
7.4	Scale Issues	198
7.5	Methods of Mapping Mangroves Using Remote Sensing	198
7.6	Global Mangrove Mapping	200
7.7	Mangrove Monitoring	201
7.8	Species Discrimination	205
7.9	Impact/Damage Assessment from Natural Disasters	206
7.10	High-Resolution Mangrove Mapping	207
7.11	Conclusions and Recommendations	208
Acknowledgments		210
References		210

Chapter 8 Wetland Mapping Methods and Techniques Using Multi-Sensor, Multi-Resolution Remote Sensing: Successes and Challenges 214

Deepak R. Mishra, X. Yan, Shuvan Ghosh, Christine Hladik, Jessica L. O'Connell, and Hyun Jung ("J.") Cho

8.1	Overview		216
8.2	Evolution of Wetland Remote Sensing		217
	8.2.1	Data	217
	8.2.2	Wetland Remote Sensing Application Scopes and Related Techniques	220

Contents ix

	8.3	The Challenges and Potential Solutions of Wetland Remote Sensing	223
		8.3.1 Proximal Sensing of Wetlands	223
		8.3.2 Multi-Sensor Joint Observation	225
	References		254

Chapter 9 Inland Valley Wetland Cultivation and Preservation for Africa's Green and Blue Revolution Using Multi-Sensor Remote Sensing 267

Murali Krishna Gumma, Pardhasaradhi Teluguntla, Pranay Panjala, Birhanu Zemadim Birhanu, and Pavan Kumar Bellam

9.1	Introduction	268
9.2	Definitions and Study Areas	279
	9.2.1 Definition Used for Mapping Wetlands	279
9.3	Remote Sensing Data for IV Wetland Characterization	280
9.4	Study Area and Ecoregional Approach	281
9.5	Methods of Rapid and Accurate IV Wetland Mapping of WCA	282
	9.5.1 Existing Methods of Wetland Mapping	282
	9.5.2 Automated Methods of Wetland Delineation and Mapping	283
	9.5.3 Semi-Automated Methods of IV Wetland Delineation and Mapping	284
9.6	Characterization and Classification of IV Wetlands	287
9.7	Cloud Computing to Delineate IV Lowlands	294
9.8	Spatial Data Weights Models for Identifying Areas for Agriculture vs. Conservation	295
9.9	Accuracies, Errors, and Uncertainties	302
9.10	Conclusions	302
Acknowledgments		302
References		302

PART V Water Use and Water Productivity

Chapter 10 Remote Sensing of Evapotranspiration from Croplands 313

Trent W. Biggs, Pamela L. Nagler, Anderson Ruhoff, Triantafyllia Petsini, Michael Marshall, Stefanie Kagone, Gabriel B. Senay, George P. Petropoulos, Camila Abe, and Edward P. Glenn

10.1	Introduction	317
10.2	Overview of Methods for ET Calculation Using Remote Sensing	318
	10.2.1 Net Radiation	324
	10.2.2 Vegetation Index–Based Methods for ET Estimation	327
	10.2.3 Radiometric Land Surface Temperature Methods for ET Estimation	335
	10.2.4 Scatterplot-Based Methods for ET Estimation	348
	10.2.5 Land Surface Models and Reanalysis for ET Estimation	353
	10.2.6 Seasonal ET Estimates and Cloud Cover Issues	354
10.3	ET Methods Intercomparison Studies	355
10.4	Special Problems in Cropped Areas	357
	10.4.1 Landscape Heterogeneity and Spatial Disaggregation	357
	10.4.2 Model Complexity, Equifinality, and Sources of Error in ET Models	362

10.5	EO-Based Operational Products Available for ET	363
10.6	Conclusions	363
	Acknowledgments	364
	References	364

Chapter 11 Modeling and Monitoring Water Productivity by Using Geotechnologies: In Some Brazilian Agroecosystems 383

Antônio Heriberto de Castro Teixeira, Janice Freitas Leivas, Celina Maki Takemura, Edson Patto Pacheco, Edlene Aparecida Monteiro Garçon, Inajá Francisco de Sousa, André Quintão de Almeida, Prasad S. Thenkabail, and Ana Flávia Maria Santos

11.1	Introduction	383
11.2	Study Regions and Data Set	386
11.3	Modeling Water Productivity Components	387
11.4	Agricultural Growing Region Scale	392
	11.4.1 Petrolina/Juazeiro Pole	392
	11.4.2 North of Minas Gerais Pole	396
	11.4.3 Fruit Circuit Pole	399
11.5	Crop Scale	401
	11.5.1 Irrigated Crops	401
	11.5.2 Rainfed Crops	406
11.6	Conclusions and Policy Implications	414
	References	415

PART VI Snow and Ice

Chapter 12 Remote Sensing Mapping and Modeling of Snow Cover Parameters and Applications 425

Hongjie Xie, Tiangang Liang, Xianwei Wang, Guoqing Zhang, Xiaodong Huang, and Xiongxin Xiao

12.1	Introduction	426
12.2	Principles of MODIS Snow Cover Mapping and Standard Products	427
	12.2.1 Binary/Fractional Snow Cover Mapping	427
	12.2.2 MODIS Standard Snow Cover Products and Accuracy	429
12.3	Improved Daily and Flexible Multiday Combinations of Snow Cover Mapping	431
	12.3.1 Flexible Multiday Combination of Snow Cover Mapping	432
	12.3.2 Daily Cloud-Free Snow Cover and Snow Water Equivalent Mapping	432
	12.3.3 Improved Daily Cloud-Free Snow Cover Mapping	436
12.4	Snow Cover Phenology Parameters	437
12.5	Snow Cover as a Water Resource for Lake-Level and Watershed Analyses	439
	12.5.1 Snow Cover over Lake Basin for Lake-Level Analysis	440
	12.5.2 Quantitative Water Resource Assessment Using Snowmelt Runoff Model	442
12.6	Scaling Effect in Snow Cover Mapping	446
	12.6.1 Upscaling	446

Contents

		12.6.2	Downscaling	448
	12.7	Snow-Caused Livestock Disasters in Pastoral Area: Risk and Warning		451
		12.7.1	Factors of Snow Disaster Early Warning and Risk Assessment	452
		12.7.2	Models for Early Warning and Qualitative Risk Assessment of Snow Disasters	452
	12.8	Conclusions		456
	Acknowledgments			457
	References			457

PART VII Summary and Synthesis for Volume V

Chapter 13 *Remote Sensing Handbook*, Volume V: Water Resources: Hydrology, Floods, Snow and Ice, Wetlands, and Water Productivity 467

Prasad S. Thenkabail

	13.1	Geomorphological Studies of Remote Sensing Studies	470
	13.2	Hydrological Studies Using Multi-Sensor Remote Sensing	472
		13.2.1 Precipitation	473
		13.2.2 Evapotranspiration	473
		13.2.3 Soil Moisture	473
		13.2.4 Groundwater	473
		13.2.5 Surface Water	474
		13.2.6 Snow, Ice, Glaciers	474
		13.2.7 Basin Characteristics	474
	13.3	Groundwater Studies Using Remote Sensing	475
	13.4	Flood Studies by Integrating Remote Sensing in Hydrological Models and Other Data	477
	13.5	Flood Studies Using SAR Remote Sensing	481
	13.6	Remote Sensing of Mangrove Forests	483
	13.7	Mangrove Wetlands of the World	486
	13.8	Wetland Modeling and Mapping Methods Using Multi-Sensor Remote Sensing	487
	13.9	Inland Valley Wetland Characterization and Mapping	489
	13.10	Actual Evapotranspiration (Water Use) of Croplands from Remote Sensing	490
	13.11	Modeling Water Productivity Studies from Earth Observation Systems	493
	13.12	Remote Sensing of Snow Cover and Its Applications	497
	Acknowledgments	499
	References	500

Index 507

Foreword

Satellite remote sensing has progressed tremendously since the first Landsat was launched on June 23, 1972. Since the 1970s, satellite remote sensing and associated airborne and *in situ* measurements have resulted in geophysical observations for understanding our planet through time. These observations have also led to improvements in numerical simulation models of the coupled atmosphere-land-ocean systems at increasing accuracies and predictive capabilities. This was made possible by data assimilation of satellite geophysical variables into simulation models, to update model variables with more current information. The same observations document the Earth's climate and have driven consensus that *Homo sapiens* are changing our climate through greenhouse gas emissions.

These accomplishments are the work of many scientists from a host of countries and a dedicated cadre of engineers who build and operate the instruments and satellites that collect geophysical observation data from satellites, all working toward the goal of improving our understanding of the Earth. This edition of the *Remote Sensing Handbook* (Second Edition, Vols. I–VI) is a compendium of information for many research areas of the Earth System that have contributed to our substantial progress since the 1970s. The remote sensing community is now using multiple sources of satellite and *in situ* data to advance our studies of Planet Earth. In the following paragraphs, I will illustrate how valuable and pivotal satellite remote sensing has been in climate system study since the 1970s. The chapters in the *Remote Sensing Handbook* provide other specific studies on land, water, and other applications using Earth observation data of the past 60+ years.

The Landsat system of Earth-observing satellites led the way in pioneering sustained observations of our planet. From 1972 to the present, at least one and frequently two Landsat satellites have been in operation (Wulder et al. 2022; Irons et al. 2012). Starting with the launch of the first NOAA-NASA Polar Orbiting Environmental Satellites NOAA-6 in 1978, improved imaging of land, clouds, and oceans and atmospheric soundings of temperature were accomplished. The NOAA system of polar-orbiting meteorological satellites has continued uninterrupted since that time, providing vital observations for numerical weather prediction. These same satellites are also responsible for the remarkable records of sea surface temperature and land vegetation index from the Advanced Very High-Resolution Radiometers (AVHRR) that now span more than 46 years as of 2024, although no one anticipated valuable climate records from these instruments before the launch of NOAA-6 in 1978 (Cracknell 2001). AVHRR instruments are expected to remain in operation on the European MetOps satellites into 2026 and possibly beyond.

The successes of data from the AVHRR led to the MODerate resolution Imaging Spectrometer (MODIS) instruments on NASA's Earth Observing System of satellite platforms that improved substantially upon the AVHRR. The first of the EOS platforms, Terra, was launched in 2000, and the second of these platforms, Aqua, was launched in 2002. Both of these platforms are nearing their operational end of life, and many of the climate data records from MODIS will be continued with the Visible Infrared Imaging Suite (VIIRS) instrument on the Joint Polar Satellite System (JPSS) meteorological satellites of NOAA. The first of these missions, the NPOES Preparation Project was launched in 2012 with the first VIIRS instrument that is operating currently along with similar instruments on JPSS-1 (launched in 2017) and JPSS-2 (launched in 2022). However, unlike the morning/afternoon overpasses of MODIS, the VIIRS instruments are all in an afternoon overpass orbit. One of the strengths of the MODIS observations was morning and afternoon data from identical instruments.

Continuity of observations is crucial for advancing our understanding of the Earth's climate system. Many scientists feel the crucial climate observations provided by remote sensing satellites are among the most important satellite measurements because they contribute to documenting the current state of our climate and how it is evolving. These key satellite observations of our climate are second in importance only to the polar orbiting and geostationary satellites needed for numerical weather prediction that provide natural disaster alerts.

The current state of the art for remote sensing is to combine different satellite observations in a complementary fashion for what is being studied. Climate study is an example of using disparate observations from multiple satellites coupled with *in situ* data to determine if climate change is occurring, where it is occurring, and to identify the various component processes responsible.

1. **Planet warming quantified by satellite radar altimetry**. Remotely sensed climate observations provide the data to understand our planet and identify forces and drivers that influence climate. The primary sea level climate observations come from radar altimetry that started in late 1992 with Topex-Poseidon and has been continued by Jason-1, Jason-2, Jason-3, and Sentinel-6 to provide an uninterrupted record of global sea level. Changes in global sea level provide unequivocal evidence that our planet is warming, cooling, or staying at the same temperature. Radar altimetry from 1992 to date has shown global sea level increases of ~3.5 mm/yr, hence our planet is warming (Figure 0.1). Sea level rise has two components: ocean thermal expansion and ice melt from the ice sheets of Greenland and Antarctica and, to a lesser extent, for glacier concentrations in places like the Gulf of Alaska and Patagonia. The combination of GRACE and GRACE Follow-On gravity measurements quantifies the ice mass losses of Greenland and Antarctica to a high degree of accuracy. Combining the gravity data with the flotilla of almost 4000 Argo floats provides the temperature data with the depth necessary to quantify ocean temperatures and isolate the thermal component of sea level rise.

2. **Our Sun is remarkably stable in total solar irradiance**. Observations of total solar irradiance have been made from satellites since 1979 and show total solar irradiance has varied only ±1 part in 500 over the past 35 years, establishing that our Sun is not to blame for global warming (Figure 0.2).

3. **Determining ice sheet contributions to sea level rise**. Since 2002 gravity observations from the Gravity Recovery and Climate Experiment Satellite, or GRACE, mission and the GRACE Follow-On mission have been measured. GRACE data quantify ice mass changes

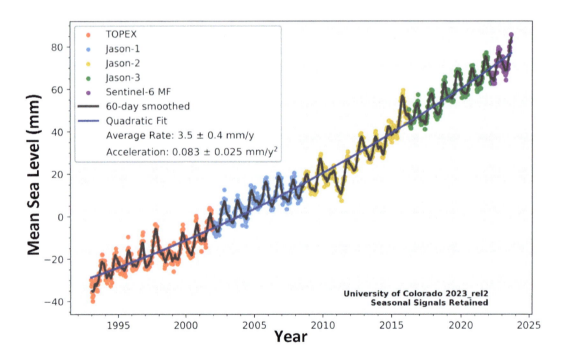

FIGURE 0.1 Seasonal sea level from five satellite radar altimeters from later 1992 to the present. Sea level is the unequivocal indicator of the Earth's climate—when sea level rises, the planet is warming; when sea level falls, the planet is cooling (Nerem et al. 2018 updated to 2023; https://sealevel.colorado.edu/data/total-sea-level-change).

Foreword xv

from the Antarctic and Greenland ice sheets that constitute 98% of the ice mass on land (Luthcke et al. 2013). GRACE data are truly remarkable—their retrieval of variations in the Earth's gravity field is quantitatively and directly linked to mass variations. With GRACE data we are able for the first time to determine the mass balance with time of the Antarctic and Greenland ice sheets and concentrations of glaciers on land. GRACE data show sea level rise is 60% explained by ice sheet mass loss (Figure 0.3). GRACE data have many other uses, such as changes in groundwater storage. See: <www.csr.utexas.edu/grace/>.

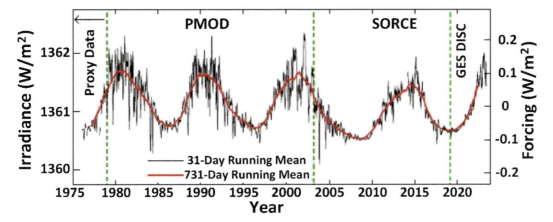

FIGURE 0.2 The Sun is not to blame for global warming, based on total solar irradiance observations from satellites. The few watts per m² solar irradiance variations covary with the sunspot cycle. The luminosity of the Sun varies 0.2% over the course of the 11-year solar and sunspot cycle. The SORCE TSI data set continues these important observations with improved accuracy on the order of ±0.035 (Kopp et al. 2024) and from https://lasp.colorado.edu/sorce/data/tsi-data/.

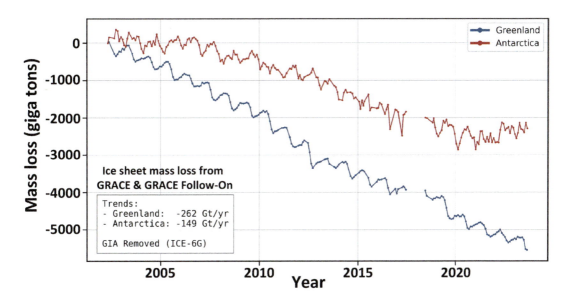

FIGURE 0.3 60% sea level rise explained by mass balance of melting of ice measured by GRACE and GRACE Follow-On satellites. Ice mass variations from 2003 to 2023 for the Antarctic and Greenland ice sheets using gravity data (Croteau et al. 2021 updated to 2023). The Antarctic and Greenland Ice Sheets constitute 98% of the Earth's land ice.

4. **40% sea level rise explained by thermal expansion in the planet's oceans measured by** *in situ* **~ 3700 Argo drifting floats**. The other contributor to sea level rise is the thermal expansion or "steric" component of our planet's oceans. To document this necessitates using diving and drifting floats or buoys in the Argo network to record temperature with depth (Roemmich & the Argo Float Team 2009, Figure 0.4). Argo floats are deployed from ships, then they submerge, descend slowly to 1000 m, recording temperature, pressure, and salinity as they descend. At 1000 m depth, they drift for ten days, continuing their measurements of temperature and salinity. After ten days, they slowly descend to 3000 m and then ascend to the surface, all the time recording their measurements. At the surface, each float transmits all the data collected on the most recent excursion to a geostationary satellite and then descend again to repeat this process.

Argo temperature data show that 40% of sea level rise results from warming and thermal expansion of our oceans. Combining radar altimeter data, GRACE and GRACE Follow-On data, and Argo data provide confirmation of sea level rise and show what is responsible for it and in what proportions. With total solar irradiance being near-constant, what is driving global warming can be determined. Analysis of surface *in situ* air temperature coupled with lower tropospheric air temperature and stratospheric temperature data from remote sensing infrared and microwave sounders show the surface and near-surface is warming while the stratosphere is cooling. This is an unequivocal confirmation that greenhouse gases are warming the planet.

Combining sea level radar altimetry, GRACE and GRACE Follow-On gravity data to quantify ice sheet mass losses, and Argo floats to measure ocean temperatures with depth enables reconciliation of sea level increases with mass loss of ice sheets and ocean thermal expansion. The ice and steric expansion explains 95% of sea level rise (Figure 0.5).

5. **The global carbon cycle**. Many scientists are actively working to study the Earth's carbon cycle, and there are several chapters in this *Remote Sensing Handbook* (Volumes I–VI) on various components under study.

Carbon cycles through reservoirs on the Earth's surface in plants and soils, exists in the atmosphere as gases such as carbon dioxide (CO_2) and exists in ocean water in phytoplankton and in marine sediments. CO_2 is released to the atmosphere from the combustion of fossil fuels, by land cover changes on the Earth's surface, by respiration of green plants, and by decomposition of carbon in dead vegetation and in soils, including carbon in permafrost.

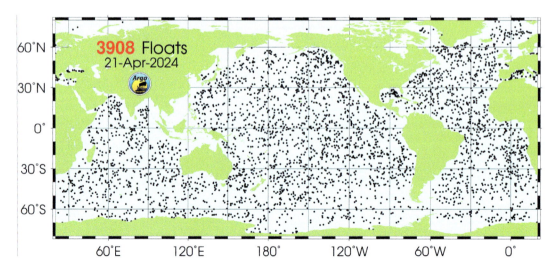

FIGURE 0.4 40% sea level rise explained by thermal expansion in the planet's oceans measured *in situ* by ~3908 drifting floats that were in operation on March 25, 2024. These floats provide the data needed to document thermal expansion of the oceans (Roemmich & the Argo Float Team 2009, updated to 2024 and <www.argo.ucsd.edu/>).

Foreword   xvii

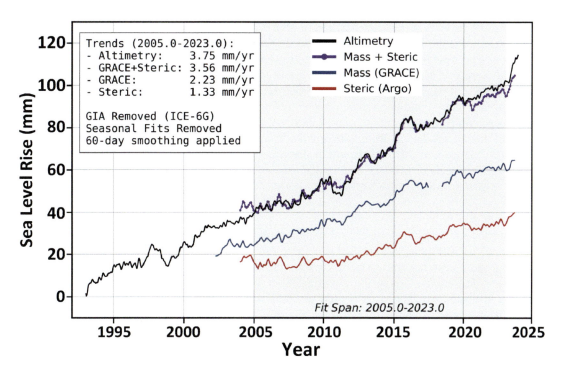

FIGURE 0.5 Sea level rise with the gravity ice mass loss and the Argo thermal expansion quantities added to the plot of global mean sea level. The GRACE and GRACE Follow-On ice sheet gravity term and Argo thermal expansion terms together explain 95% of sea level rise (Croteau et al. 2021 updated to 2023).

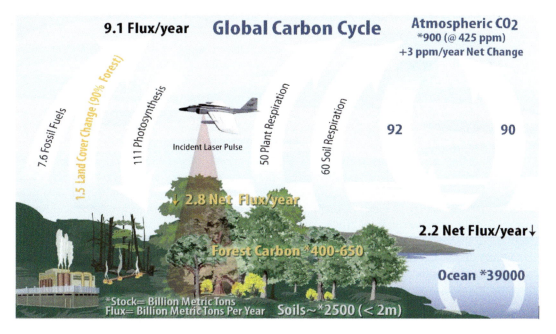

FIGURE 0.6 Global carbon cycle measurements from a multitude of satellite sensors. A representation of the global carbon cycle showing our best estimates of carbon fluxes and carbon reservoirs as of 2024. A series of satellite observations are needed simultaneously to understand the carbon cycle and its role in the Earth's climate system (Ciais et al. 2014 updated to 2023). The major unknowns in the global carbon cycle are fluxes between different reservoirs, oceanic gross primary production, carbon in soils, and the carbon in woody vegetation.

Land gross primary production has been a MODIS product that is extended into the VIIRS era (Running et al. 2004; Román et al. 2024). MODIS data also provide burned area and CO_2 emissions from wildfires (Giglio et al. 2016). Oceanic gross primary production will be provided by the Plankton, Aerosol, Cloud, and ocean Ecosystem, or PACE, satellite that was launched in early 2024 (Gorman et al. 2019). This complements the GPP land portion of the carbon cycle and will enable global gross primary production to be determined by MODIS-VIIRS and PACE.

Furthermore, Harmonized Landsat-8, Landsat-9, and Sentinel-2 30 m data (HLS) provide multispectral time series data at 30 m with a revisit frequency of three days at the equator (Crawford et al. 2023; Masek et al. 2018). This will enable time series improvements in spatial detail to 30 m from the 250 m scale of MODIS. The revisit time of Sentinel-2 with 10 m data is five days at the equator, which is a major improvement from 30 m. Multispectral time series observations are the basis for providing gross primary production estimates on land that are also used for food security (Claverie et al. 2018).

Refinements in satellite multispectral spatial resolution to the 50 cm to 3–4 m scale provided by commercial satellite data have enabled tree carbon to be determined from large areas of trees outside of forests. NASA has started using commercial satellite data to complement MODIS, Landsat, and other observations. One of the uses for Planet 3–4 m and Maxar < 1 m data have been for mapping trees outside of forests (Brandt et al. 2020; Reiner et al. 2023; Tucker et al. 2023). Tucker et al. (2023) mapped 10 billion trees at the 50 cm scale over 10 million km^2 and converted them into carbon at the tree level with allometry. The value of Planet and Maxar (formerly Digital Globe) data allows carbon studies to be extended into areas with discrete trees, and Huang et al. (2024) has successfully mapped one tree species across the entire Sahelian and Sudanian zones of Africa.

The height of trees is an important measurement to determine their carbon content. For areas of contiguous tree crowns, GEDI and ICESat laser altimetry (Magruder et al. 2024) coupled with Landsat and Sentinel-2 observations, enable improved estimates of carbon in these forests (Claverie et al. 2018).

The key to closing several uncertainties in the carbon cycle is to quantify fluxes among the various components. Passive CO_2 retrieval methods from the Greenhouse gases Observing SATellite (GOSAT) (Noël et al. 2021) and the Orbiting Carbon Observatory-2 (OCO-2) (Jacobs et al. 2024) are inadequate to provide this. Passive methods are not possible at night, in all seasons, and require specific Sun-target-sensor viewing perspectives and conditions. A recent development of the Aerosol and Carbon dioxide Detection LiDAR (ACDL) instrument (Dai et al. 2023) by our Chinese colleagues offers a ten-fold coverage improvement in CO_2 retrievals over those provided by OCO-2 and 20-fold coverage improvement over GOSAT. The reported uncertainty of ACDL is in the order of ±0.6 ppm.

Understanding the carbon cycle requires a "full court press" of satellite and *in situ* observations, because all of these observations must be made at the same time. Many of these measurements have been made over the past 30–40 years, but new measurements are needed to quantify carbon storage in vegetation, to quantify CO_2 fluxes, to quantify land respiration, and to improve numerical carbon models. Similar work needs to be performed for the role of clouds and aerosols in climate and to improve our understanding of the global hydrological cycle.

The remote sensing community has made tremendous progress over the last six decades, as captured in various chapters of the *Remote Sensing Handbook* (Second Edition, Vols. I–VI). Handbook chapters provide comprehensive understanding of land and water studies through detailed methods, approaches, algorithms, syntheses, and key references. Every type of remote sensing data obtained from systems such as optical, radar, LiDAR, hyperspectral, and hyperspatial are presented and discussed in different chapters. Chapters in this volume addressing remote sensing data characteristics, within and between sensor calibrations, classification methods, and accuracies taking a wide array of remote sensing data from a wide array of platforms over the last five decades. Volume I also brings in new remote sensing technologies such as radio occultation and reflectometry from the global navigation satellite system or GPS satellites, crowdsourcing, drones, cloud computing,

Foreword

artificial intelligence, machine learning, hyperspectral, radar, and remote sensing law. The chapters in the *Remote Sensing Handbook* are written by leading remote sensing scientists of the world and ably edited by Dr. Prasad S. Thenkabail, Senior Scientist (ST), at the US Geological Survey (USGS) in Flagstaff, Arizona. The importance and the value of the *Remote Sensing Handbook* is clearly demonstrated by the need for a second edition. The *Remote Sensing Handbook* (First Edition, Vols. I–III) was published in 2015, and now after ten years the *Remote Sensing Handbook* (Second Edition, Vols. I–VI), with 91 chapters and nearly 3500 pages, will be published. It is certainly a monumental work in remote sensing science, and for this I want to compliment Dr. Prasad S. Thenkabail. Remote sensing is now important to many scientific disciplines beyond our community, and I recommend the *Remote Sensing Handbook* (Second Edition, six volumes) to not only remote sensers but to the entire scientific community.

We can look forward in the coming decades to improving our quantitative understanding of the global carbon cycle, understanding the interaction of clouds and aerosols in our radiation budget, and understanding the global hydrological cycle.

by Compton J. Tucker
Satellite Remote Sensing Beyond 2025
NASA/Goddard Space Flight Center
Earth Science Division
Greenbelt, Maryland 20771 USA

REFERENCES

Brandt, M., Tucker, C.J., Kariryaa, A., et al. 2020. An unexpectedly large count of trees in the West African Sahara and Sahel. *Nature* 587:78–82. doi: 10.1038/s41586-020-2824-5.

Ciais, P., et al. 2014. Current systematic carbon-cycle observations and the need for implementing a policy-relevant carbon observing system. *Biogeosciences* 11(13):3547–3602.

Claverie, M., Ju, J., Masek, J.G., Dungan, J.L., Vermote, E.F., Roger, J.-C., Skakun, S.V., et al. 2018. The Harmonized Landsat and Sentinel-2 surface reflectance data set. *Remote Sensing of Environment* 219:145–161. doi: 10.1016/j.rse.2018.09.002.

Crawford, C.J., Roy, D.P., Arab, S., Barnes, C., Vermote, E., Hulley, G., et al. 2023. The 50-year Landsat collection 2 archive. *Science of Remote Sensing* 8:100103. ISSN 2666-0172. doi: 10.1016/j.srs.2023.100103. (www.sciencedirect.com/science/article/pii/S2666017223000287).

Cracknell, A. 2001. The exciting and totally unanticipated success of the AVHRR in applications for which it was never intended. *Advances in Space Research* 28:233–240. doi: 10.1016/S0273-1177(01)00349-0.

Croteau, M.J., Sabaka, T.J., and Loomis, B.D. 2021. GRACE fast mascons from spherical harmonics and a regularization design trade study. *Journal of Geophysical Research:Solid Earth* 126:e2021JB022113. doi: 10.1029/2021JB022113.10.1029/2021JB022113.

Dai, G., Wu, S., Sun, K., Long, W., Liu, J., and Chen, W. 2023. Aerosol and Carbon dioxide Detection Lidar (ACDL) overview. Presentation at ESA-JAXA EarthCare Workshop, November.

Giglio, L., Schroeder, W., and Justice, C.O. 2016. The collection 6 MODIS active fire detection algorithm and fire products. *Remote Sensing of Environment* 178:31–41. doi: 10.1016/j.rse.2016.02.054.

Gorman, E.T., Kubalak, D.A., Patel, D., Dress, A., Mott, D.B., Meister, G., and Werdell, P.J. 2019. The NASA Plankton, Aerosol, Cloud, ocean Ecosystem (PACE) mission: An emerging era of global, hyperspectral Earth system remote sensing. *Sensors, Systems, and Next Generation Satellites* 23:11151. doi: 10.1117/12.2537146.

Huang, K., et al. 2024. Mapping every adult baobab (Adansonia digitata L.) across the Sahel to uncover the co-existence with rural livelihoods. *Nature Ecology and Evolution* doi: 10.21203/rs.3.rs-3243009/v1.

Irons, J.R., Dwyer, J.L., and Barsi, J.A. 2012. The next Landsat satellite: The Landsat Data Continuity Mission. *Remote Sensing of Environment* 122:11–21. doi: 10.1016/j.rse.2011.08.026.

Jacobs, N., et al. 2024. The importance of digital elevation model accuracy in X_{CO2} retrievals: Improving the Orbiting Carbon Observatory-2 Atmospheric Carbon Observations from Space version 11 retrieval product. *Atmospheric Measurement Techniques* 17(5):1375–1401. doi: 10.5194/amt-17-1375-2024.

Kopp, G., Nèmec, N.E., and Shapiro, A. 2024. Correlations between total and spectral solar irradiance variations. *Astrophysical Journal* 964(1). doi: 10.3847/1538-4357/ad24e5.

Luthcke, S.B., Sabaka, T.J., Loomis, B.D., Arendt, A.A., McCarthy, J.J., and Camp, J. 2013. Antarctica, Greenland, and Gulf of Alaska land-ice evolution from an iterated GRACE global mascon solution. *Journal of Glaciology* 59(216). doi: 10.3189/2013JoG12J147.

Magruder, L.A., Farrell, S.L., Neuenschwander, A., Duncanson, L., Csatho, B., Kacimi, S., et al. 2024. Monitoring Earth's climate variables with satellite laser altimetry. *Nature Reviews of Earth and Environment* 5(2):120–136. doi: 10.1038/s43017-023-00508-8.

Masek, J., Ju, J., Roger, J.-C., Skakun, S., Claverie, M., and Dungan, J. 2018. Harmonized Landsat/Sentinel-2 products for land monitoring, IGARSS 2018-2018. *IEEE International Geoscience and Remote Sensing Symposium*, Valencia, Spain, pp. 8163–8165. doi: 10.1109/IGARSS.2018.8517760.

Nerem, R.S, Beckley, B.D., Fasullo, J.T., and Mitchum, G.T. 2018. Climate-change—driven accelerated sea-level rise detected in the altimeter era. *Proceeding of the National Academy of Sciences* 115(9):2022–2025. doi: 10.1073/pnas.1717312115.

Noël, S., et al. 2021. XCO_2 retrieval for GOSAT and GOSAT-2 based on the FOCAL algorithm. *Atmospheric Measurement Techniques* 14(5):3837–3869. doi: 10.5194/amt-14-3837-2021.

Reiner, F., et al. 2023. More than one quarter of Africa's tree cover is found outside areas previously classified as forest. *Nature Communications* doi: 10.1038/s41467-023-37880-4.

Roemmich, D., and the Argo Steering Team. 2009. Argo—the challenge of continuing 10 years of progress. *Oceanography* 22(3):46–55.

Román, M., et al. 2024. Continuity between NASA MODIS Collection 6.1 and VIIRS Collection 2 land products. *Remote Sensing of Environment* 302. doi: 10.1016/j.rse.2023.113963.

Running, S.W., Nemani, R.R., Heinsch, F.A., Zhao, M.S., Reeves, M., and Hashimoto, H. 2004. A continuous satellite-derived measure of global terrestrial primary production. *Bioscience* 54(6):547–560. doi: 10.1641/0006-3568.

Tucker, C., Brandt, M., Hiernaux, P., Kariryaa, A., et al. 2023. Sub-continental-scale carbon stocks of individual trees in African drylands. *Nature* 615:80–86. doi: 10.1038/s41586-022-05653-6.

Wulder, M.A., Roy, D.P., Radeloff, V.C., Loveland, T.R., Anderson, M.C., Johnson, D.M., et al. 2022. Fifty years of Landsat science and impacts. *Remote Sensing of Environment* 280:113195. ISSN 0034-4257. doi: 10.1016/j.rse.2022.113195. (www.sciencedirect.com/science/article/pii/S0034425722003054).

Preface

The overarching goal of this six-volume, 91-chapter, about 3500-page, *Remote Sensing Handbook* (Second Edition, Vols. I–VI) was to capture and provide the most comprehensive state of the art in remote sensing science and technology development and advancement in the last 60+ years by clearly demonstrating the: (1) scientific advances, (2) methodological advances, and (3) societal benefits achieved during this period, as well as to provide a vision of what is to come in the years ahead. The book volumes are, to date and to my best knowledge, the most comprehensive documentation of the scientific and methodological advances that have taken place in understanding remote sensing data, methods, and a wide array of applications. Written by 300+ leading global experts in the area, each chapter: (1) focuses on a specific topic (e.g., data, methods, and specific set of applications), (2) reviews existing state-of-the-art knowledge, (3) highlights the advances made, and (4) provides guidance for areas requiring future development. Chapters in the book cover a wide array of subject matters of remote sensing applications. The *Remote Sensing Handbook* (Second Edition, Vosl. I–VI) is planned as a reference material for a broad spectrum of remote sensing scientists to understand the fundamentals as well as the latest advances and the wide array of applications such as for land and water resource practitioners, natural and environmental practitioners, professors, students, and decision-makers.

Special features of the six-volume *Remote Sensing Handbook* (Second Edition) include:

1. Participation of an outstanding group of remote sensing experts, an unparalleled team of writers for such a book project.
2. Exhaustive coverage of a wide array of remote sensing science: data, methods, and applications.
3. Each chapter being led by a luminary and most chapters written by writing teams, which further enriched the chapters.
4. Broadening the scope of the book to make it ideal for expert practitioners as well as students.
5. A global team of writers, global geographic coverage of study areas, a wide array of satellites and sensors; and
6. Plenty of color illustrations.

Chapters in the book have covered remote sensing:

State-of-the-art satellites, sensors, science, technology, and applications
Methods and techniques
A wide array of applications, such as land and water applications, natural resources management, and environmental issues
Scientific achievements and advancements of the aforementioned over the last 60+ years
Societal benefits
Knowledge gaps
Future possibilities in the 21st century

Great advances have taken place over the last 60+ years in the study of Planet Earth from remote sensing, especially using data gathered from a multitude of Earth Observation (EO) satellites launched by various governments as well as private entities. A large part of the initial remote sensing technology was developed and tested during the two world wars. In the 1950s remote sensing slowly began its foray into civilian applications. But during the years of the Cold War civilian and military remote sensing applications increased swiftly. But it was also an age when remote sensing was the domain of very few top experts, often having multiple skills in engineering, science, and computer technology. From the 1960s onwards, there have been many governmental agencies that have initiated civilian remote sensing. The National Aeronautics and Space Administration (NASA) of the USA has been at the forefront of many of these efforts. Others who have provided leadership in civilian remote sensing include, but are not limited to, the European Space Agency (ESA), the Indian Space Research Organization (ISRO), the Centre national d'études spatiales (CNES) of France, the Canadian Space Agency (CSA), the Japan Aerospace Exploration Agency (JAXA), the German Aerospace Center (DLR), the China National Space Administration (CNSA), the United Kingdom Space Agency (UKSA), and the Instituto Nacional de Pesquisas Espaciais (INPE) of Brazil. Many private entities, such as Planet Labs PBC, have launched and operate satellites as well. These government and private agencies and enterprises have launched, and continue to launch and operate, a wide array of satellites and sensors that capture data of the Planet Earth in various regions of the electromagnetic spectrum and in various spatial, radiometric, and temporal resolutions, routinely and repeatedly. However, the real thrust for remote sensing advancement came during the last decade of the 20th century and the beginning of the 21st century. These initiatives included the launch of a series of new-generation EO satellites to gather data more frequently and routinely, the release of pathfinder datasets, web-enabling of the data for free by many agencies (e.g., USGS released the entire Landsat archives as well as real-time acquisitions of the world for free by making them web accessible), and providing processed data ready to users (e.g., the Harmonized Landsat and Sentinel-2, or HLS data, surface reflectance products of MODIS). Other efforts, like Google Earth, made remote sensing more popular and brought in a new platform for easy visualization and navigation of remote sensing data. Advances in computer hardware and software made it possible to handle big data. Crowdsourcing, web access, cloud computing such as in the Google Earth Engine (GEE) platform, machine learning, deep learning, coding, artificial intelligence, mobile apps, and mobile platforms (e.g., drones) added a new dimension to how remote sensing data is used. Integration with global positioning systems (GPS) and global navigation satellite systems (GNSS), as well as the inclusion of digital secondary data (e.g., digital elevation, precipitation, temperature) in analysis has made remote sensing much more powerful. Collectively, these initiatives provided a new vision in making remote sensing data more popular, widely understood, and increasingly used for diverse applications, hitherto considered difficult. Availability of free archival data when combined with more recent acquisitions has also enabled quantitative studies of change over space and time. The *Remote Sensing Handbook* (Vols. I–VI) is targeted to capture these vast advances in data, methods, and applications, so a remote sensing student, scientist, or a professional practitioner will have the most comprehensive, all-encompassing reference material in one place.

Modern-day remote sensing technology, science, and applications are growing exponentially. This growth is as a result of a combination of factors that include: (1) advances and innovations in data capture, access, processing, computing, and delivery (e.g., big data analytics, harmonized and normalized data, inter-sensor relationships, web enabling of data, cloud computing, crowdsourcing, mobile apps, machine learning, deep learning, coding in Python and Java Script, and artificial intelligence); (2) an increasing number of satellites and sensors gathering data of the planet, repeatedly and routinely, in various portions of the electromagnetic spectrum, as well as in an array of spatial, radiometric, and temporal resolutions; (3) efforts at integrating data from multiple satellites and sensors (e.g., Sentinels with Landsat); (4) advances in data normalization, standardization, and harmonization (e.g., delivery of data in surface reflectance, inter-sensor calibration); (5) methods and techniques for handling very large data volumes (e.g., global mosaics); (6) quantum leap in computer hardware and software capabilities (e.g., ability to process several terabytes of data); (7) innovation in methods, approaches, and techniques leading to sophisticated algorithms (e.g., spectral matching techniques, neural network perceptron); and (8) the development of new spectral indices to quantify and study specific land and water parameters (e.g., hyperspectral vegetation indices, or HVIs). As a result of these all-around developments, remote sensing science is today very mature and is widely used in virtually every discipline of

Preface   xxiii

the Earth Sciences for quantifying, mapping, modeling, and monitoring our planet. Such rapid advances are captured in a number of remote sensing and Earth Science journals. However, students, scientists, and practitioners of remote sensing science and applications have significant difficulty in gathering a complete understanding of various developments and advances that have taken place because of their vast spread across the last 60+ years. Thereby, the chapters in the *Remote Sensing Handbook* are designed to give a whole picture of the scientific and technological advances of the last 60+ years.

Today, the science, art, and technology of remote sensing is truly ubiquitous and increasingly part of everyone's everyday life, often without even the user knowing it. Whether looking at your own home or farm (e.g., Figure 0.7), helping you navigate when you drive, visualizing a phenomenon occurring in a distant part of the world (e.g., Figure 0.7), monitoring events such as droughts and floods, reporting weather, detecting and monitoring troop movements or nuclear sites, studying deforestation, assessing biomass carbon, addressing disasters like earthquakes or tsunamis, and a host of other applications (e.g, precision farming, crop productivity, water productivity, deforestation, desertification, water resources management), remote sensing plays a key role. Already, many new innovations are taking place. Companies such as Planet Labs PBC and Skybox are capturing very high spatial resolution imagery and even videos from space using large numbers of microsatellite (CubeSat) constellations. Planet Labs also will soon launch hyperspectral satellites called Tanager. There are others (e.g., Pixxel, India) who have launched and continue

FIGURE 0.7 Google Earth can be used to seamlessly navigate and precisely locate any place on Earth, often with very high spatial resolution data (VHRI; sub-meter to 5 meter) from satellites such as IKONOS, Quickbird, and Geoeye (Note: this image is from one of the VHRI). Here, the editor-in-chief (EiC) (Thenkabail) of this *Remote Sensing Handbook* (Vols. I–VI) located his village home and surroundings that have such land cover as secondary rainforests, lowland paddy farms, areca nut plantations, coconut plantations, minor roads, walking routes, open grazing lands, and minor streams (typically, first and second order) (Note: Land cover is based on the ground knowledge of the EiC). The first primary school attended by the EiC is located precisely. Precise coordinates (1345 39.22 Northern latitude, 75 06 56.03 Eastern longitude) of Thenkabail's village house and the date of image acquisition (March 1, 2014). Google Earth Images are used for visualization as well as for numerous science applications, such as accuracy assessment, reconnaissance, determining land cover, establishing land use, and for various ground surveys.

to launch constellations of hyperspectral or other sensors. China is constantly putting a wide array of satellites into orbit. Just as the smartphone and social media connected the world, remote sensing is making the world our backyard (e.g., Figure 0.7). No place goes unobserved, and no event gets reported without an image. True liberation for any technology and science comes. when it is widely used by common people who often have no idea how it all comes together but understand the information provided intuitively. That is already happening (e.g., how we use smartphones is significantly driven by satellite data–driven maps and GPS-driven locations). These developments make it clear that not only do we need to understand the state of the art, but we must also have a vision of where the future of remote sensing is headed. Thereby, in a nutshell, the goal of the *Remote Sensing Handbook* (Vols. I–VI) is to cover the developments and advancement of six distinct eras (listed next) in terms of data characterization and processing, as well as myriad land and water applications:

Pre-civilian remote sensing era of the pre-1950s: World War I and II when remote sensing was a military tool.
Technology demonstration era of the 1950s and 1960s: The Sputnik-I and NOAA AVHRR era of the 1950s and 1960s.
Landsat era of the 1970s: when the first truly operational land remote sensing satellite (Earth Resources Technology Satellite, or ERTS, later re-named Landsat) was launched and operated.
Earth observation era of the 1980s and 1990s: when a number of space agencies began launching and operating satellites (e.g., Landsat 4,5 by the USA, SPOT-1,2 by France; IRS-1a,1b by India).
Earth observation and new millennium era of the 2000s: when data dissemination to users became as important as launching, operating, and capturing data (e.g., MODIS terra\acqua, Landsat-8, Resourcesat).
Twenty-first-century era starting in the 2010s: when new-generation micro\nano satellites or CubeSats (e.g., Planet Labs PBC, Skybox) and hyperspectral satellite sensors (e.g., Tanager-1, DESIS, PRISMA, EnMAP, upcoming NASA SBG) add to the increasing constellation of multi-agency sensors (e.g., Sentinels, Landsat-8,9, upcoming Landsat-Next).

Motivation to take up editing the six-volume *Remote Sensing Handbook* (Second Edition) wasn't easy. It is a daunting work and requires extraordinary commitment over 2–3 years. After repeated requests from Ms. Irma Shagla-Britton, Manager and Leader for Remote Sensing and GIS books of Taylor and Francis\CRC Press, and considerable thought, I finally agreed to take the challenge in 2022. Having earlier edited the three-volume *Remote Sensing Handbook*, published in 2015, I was pleased that the books were of considerable demand for a second edition. This was enough motivation. Further, I wanted to do something significant at this stage of my career that would make considerable contribution to the global remote sensing community. When I edited the first edition during 2012–2014, I was still recovering from colon cancer surgery and chemotherapy. But this second edition is a celebration of my complete recovery from the dreaded disease. I have not only fully recovered, but I have never felt so completely full of health and vigor. This naturally gave me sufficient energy and the enthusiasm required to back my motivation to edit this monumental six-volume *Remote Sensing Handbook*. At least for me this is the *magnum opus* that I feel proud to have accomplished and feel confident of its immense value for students, scientists, and professional practitioners of remote sensing who are interested in a standard reference on the subject. They will find in this *Remote Sensing Handbook*: "Complete and comprehensive coverage of the state of the art in remote sensing, capturing the advances that have taken place over the last 60+ years, which will set the stage for a vision for the future."

Above all, I am indebted to some 300+ authors and co-authors of the chapters who have spent so much of their creative energy to work on the chapters, deliver them on time, and patiently address

Preface   xxv

all edits and comments. These are amongst the very best remote sensing scientists from around the world. They are extremely busy people, making time for the book project and making outstanding contributions. I went back to everyone who contributed to the *Remote Sensing Handbook* (First Edition, three volumes) published in 2015 and requested that they revise their chapters. Most of the lead authors of the chapters agreed to revise, which was reassuring. However, some were not available, some retired, and some declined for other reasons. In such cases I adopted two strategies: (1) invite a few new chapter authors to make up for this gap or (2) update the chapters myself. I am convinced this strategy worked very well to ensure the integrity of every chapter was maintained. What was also important was to ensure latest advances in remote sensing science were adequately covered. Authors of the chapters amazed me by their commitment and attention to detail. First, the quality of each of the chapters was of the highest standard. Second, with very few exceptions, chapters were delivered on time. Third, edited chapters were revised thoroughly and returned on time. Fourth, all my requests on various formatting and quality enhancements were addressed. My heartfelt gratitude to these great authors for their dedication to quality science. It has been my great honor and privilege to work with these dedicated legends. Indeed, I call them my "heroes" in a true sense. These are highly accomplished, renowned, pioneering scientists of the highest merit in remote sensing science, and I am ever grateful to have their time, effort, enthusiasm, and outstanding intellectual contributions. I am indebted to their kindness and generosity. In the end we had 300+ authors writing 91 chapters.

Overall, the *Remote Sensing Handbook* (Vols. I–VI) took about two years, from the time book chapters and authors were identified to the final publication of the book. The six volumes of the *Remote Sensing Handbook* were designed in such a way that a reader can have all six volumes as a standard reference or have individual volumes to study specific subject areas. The six volumes are:

Remote Sensing Handbook, Second Edition, Vol. I
Volume 1: *Sensors, Data Normalization, Harmonization, Cloud Computing, and Accuracies—9781032890951*

Remote Sensing Handbook, Second Edition, Vol. II
Volume 2: *Image Processing, Change Detection, GIS, and Spatial Data Analysis—9781032890975*

Remote Sensing Handbook; Second Edition, Vol. III
Volume 3: *Agriculture, Food Security, Rangelands, Vegetation, Phenology, and Soils—9781032891019*

Remote Sensing Handbook; Second Edition, Vol. IV
Volume 4: *Forests, Biodiversity, Ecology, LULC, and Carbon—9781032891033*

Remote Sensing Handbook; Second Edition, Vol. V
Volume 5: *Water Resources: Hydrology, Floods, Snow and Ice, Wetlands, and Water Productivity—9781032891453*

Remote Sensing Handbook; Second Edition, Vol. VI
Volume 6: *Droughts, Disasters, Pollution, and Urban Mapping—9781032891484*

There are 18, 17, 17, 12, 13, and 14 chapters, respectively, in the six volumes.

A wide array of topics are covered in the six volumes.

The topics covered in Volume I include: (1) satellites and sensors; (2) global navigation satellite systems (GNSS); (3) remote sensing fundamentals; (4) data normalization, harmonization, and standardization; (5) vegetation indices and their within- and across-sensor calibration; (6) crowdsourcing; (7) cloud computing; (8) Google Earth Engine–supported remote sensing; (9) accuracy assessments; and (10) remote sensing law.

The topics covered in Volume II include: (1) digital image processing fundamentals and advances; (2) digital image classifications for applications such as urban, land use, and land cover; (3) hyperspectral image processing methods and approaches; (4) thermal infrared image processing principles and practices; (5) image segmentation; (6) object-oriented image analysis (OBIA), including geospatial data integration techniques in OBIA; (7) image segmentation in specific applications like land use and land cover; (8) LiDAR digital image processing; (9) change detection; and (10) integrating geographic information systems (GIS) with remote sensing in geoprocessing workflows, democratization of GIS data and tools, fronters of GIScience, and GIS and remote sensing policies.

The topics covered in Volume III include: (1) vegetation and biomass, (2) agricultural croplands, (3) rangelands, (4) phenology and food security, and (5) soils.

The topics covered in Volume IV include: (1) forests, (2) biodiversity, (3) ecology, (4) land use\ land cover, and (5) carbon. Under each of these broad topics, there are one or more chapters.

Volume V focuses on hydrology, water resources, ice, wetlands, and crop water productivity. The chapters are broadly classified into (1) geomorphology, (2) hydrology and water resources, (3) floods, (4) wetlands, (5) crop water use and productivity, and (6) snow and ice.

Volume VI focuses on water resources, disasters, and urban remote sensing. The chapters are broadly classified into (1) droughts and drylands, (2) disasters, (3) volcanoes, (4) fires, and (5) nightlights.

There are many ways to use the *Remote Sensing Handbook* (Second Edition, six volumes). A lot of thought went into organizing the volumes and chapters, so you will see a "flow" from chapter to chapter and from volume to volume. As you read through the chapters, you will see how they are interconnected and how reading all of them provides you with a greater in-depth understanding. You will also realize, as someone deeply interested in one of the topics, that you will have greater interest in a particular volume. Having all six volumes as reference material is ideal for any remote sensing expert, practitioner, or student; however, you can also refer to individual volumes based on your interests. We have also made great attempts to ensure chapters are self-contained so you can focus on a chapter and read it through without having to be overly dependent on other chapters. Taking this perspective, there is a slight amount (~5–10%) of material that may be repeated across chapters. This is done deliberately. For example, when you are reading a chapter on LiDAR or radar, you don't want to go all the way back to another chapter to understand the characteristics of these data. Similarly, certain indices (e.g., vegetation condition index, or VCI, temperature condition index, or TCI) that are defined in one chapter (e.g., on drought) may be repeated in another chapter (also on drought). Such minor overlaps are helpful to readers to avoid going back to another chapter to understand a particular phenomenon or an index or characteristic of a sensor. However, if you want a lot of details of these sensors, indices, or phenomena, then you will have to read the appropriate chapter where there is in-depth coverage of the topic.

Each volume has a summary chapter (the last chapter of each volume). The summary chapter can be read two ways: (1) either as the last chapter to recapture the main points of each of the preceding chapters and\or (2) as an initial overview to first get the feeling for what is in the volume before diving in to read each chapter in detail. I suggest the readers do it both ways: read it first before reading the other chapters in detail to gather an idea on what to expect in each chapter and then read it again at the end to recapture what was read in each of the previous chapters.

It has been a great honor, as well as a humbling experience, to edit the *Remote Sensing Handbook* (Vols. I–VI). I truly enjoyed the effort, albeit I felt overwhelmed at times with never-ending work. What an honor to work with such luminaries in your field of expertise. I learned a lot from them and am very grateful for their support, encouragement, and deep insights. Also, it has been a pleasure working with the outstanding professionals at Taylor & Francis Group\CRC Press. There is no joy greater than being immersed in the pursuit of excellence, knowledge gain, and knowledge capture. At the same time, I am happy it is over. If there will be a third edition a decade or so from now, it will be taken up by someone else (individually or as a team) and certainly not me!

Preface

I expect the book to be a standard reference of immense value to any student, scientist, professional, and practical practitioner of remote sensing. Any book that has the privilege of 300+ truly outstanding and dedicated remote sensing scientists ought to be a *magnum opus* deserving to be a standard reference on the subject.

Dr. Prasad S. Thenkabail, PhD
Editor-in-Chief (EiC)
Remote Sensing Handbook (Second Edition, Vols. I–VI)

Volume 1: Sensors, Data Normalization, Harmonization, Cloud Computing, and Accuracies
Volume 2: Image Processing, Change Detection, GIS, and Spatial Data Analysis
Volume 3: Agriculture, Food Security, Rangelands, Vegetation, Phenology, and Soils
Volume 4: Forests, Biodiversity, Ecology, LULC, and Carbon
Volume 5: Water Resources: Hydrology, Floods, Snow and Ice, Wetlands, and Water Productivity
Volume 6: Droughts, Disasters, Pollution, and Urban Mapping

About the Editor

Dr. Prasad S. Thenkabail, PhD, is a senior scientist with the United States Geological Survey (USGS), specializing in remote sensing science for agriculture, water, and food security. He is a world-recognized expert in remote sensing science with multiple major contributions in the field sustained for 40+ years. Dr. Thenkabail has conducted pioneering research in the hyperspectral remote sensing of vegetation, global croplands mapping for water and food security, and crop water productivity. His work on hyperspectral remote sensing of agriculture and vegetation are widely cited. His papers on hyperspectral remote sensing are the first of their kind and, collectively, they have: (1) determined optimal hyperspectral narrowbands (OHNBs) in the study of agricultural crops, (2) established hyperspectral vegetation indices (HVIs) to model and map crop biophysical and biochemical quantities, (3) created a framework and sample data for the global hyperspectral imaging spectral libraries of crops (GHISA), (4) developed methods and techniques of overcoming Hughes's phenomenon, (5) demonstrated the strengths of hyperspectral narrowband (HNB) data in advancing classification accuracies relative to multispectral broadband (MBB) data, (6) shown the advances one can make in modeling crop biophysical and biochemical quantities using HNB and HVI data relative to MBB data, and (7) created a body of work in understanding, processing, and utilizing HNB and HVI data in agricultural cropland studies. This body of work has become a widely referred reference worldwide. In studies of global croplands for food and water security, Dr. Thenkabail has led the release of the world's first 30-m Landsat Satellite–derived global cropland extent product at 30 m (GCEP30; https://www.usgs.gov/apps/croplands/app/map); (Thenkabail et al. 2021; https://lpdaac.usgs.gov/news/release-of-gfsad-30-meter-cropland-extent-products/) and Landsat-derived global rainfed and irrigated area product at 30 m (LGRIP30; https://lpdaac.usgs.gov/products/lgrip30v001/) (Teluguntla and Thenkabail et al., 2023). Earlier he led production of the world's first global irrigated area map (https://lpdaac.usgs.gov/products/lgrip30v001/; https://lpdaac.usgs.gov/products/gfsad1kcdv001/) using multi-sensor satellite data that led to crop. The global cropland datasets using satellite remote sensing demonstrates a "paradigm shift" in global cropland mapping using remote sensing through big data analytics, machine learning, and petabyte-scale cloud computing on the Google Earth Engine (GEE). The LGRIP30 and GCEP30 products are released through NASA's LP DAAC and published in USGS professional paper 1868 (Thenkabail et al., 2021). He has been principal investigator of many projects over the years, including the NASA-funded global food security support analysis data in the 30-m (GFSAD) project (www.usgs.gov/wgsc/gfsad30).

His career scientific achievements can be gauged by making the list of the world's top 1% of scientists as per the Stanford study ranking the world's scientists from across 22 scientific fields and 176 sub-fields based on deep analysis evaluating about 10 million scientists based on Elsevier SCOPUS data from 1996 to 2023 (Ioannidis, 2023; Ioannidis et al., 2020). Dr. Thenkabail was recognized as Fellow of the American Society of Photogrammetry and Remote Sensing (ASPRS) in 2023. He has published over 150 peer-reviewed scientific papers and edited 15 books. His scientific papers have won several awards over the years, demonstrating world-class quality research. These include: the 2023 Talbert Abrams Grand Award, the highest scientific paper award of the ASPRS (with Itiya Aneece); the 2015 ASPRS ERDAS award for best scientific paper in remote sensing (with Michael Marshall); the 2008 John I. Davidson ASPRS President's Award for practical papers (with Pardha Teluguntla); and the 1994 Autometric Award for outstanding paper in remote sensing (with Dr. Andy Ward).

Dr. Thenkabail's contributions to series of leading edited books places him as a world leader in remote sensing science. There are three seminal book-sets with a total of 13 volumes that he has edited and which have demonstrated his major contributions as an internationally acclaimed remote sensing scientist. These are: (1) *Remote Sensing Handbook* (Second Edition, six-volume book-set, 2024) with 91 chapters and nearly 3000 pages and for which he is the sole editor, (2) *Remote Sensing Handbook* (First Edition, three-volume book-set, 2015) with 82 chapters and 2304 pages and for which he is the sole editor, and (3) *Hyperspectral Remote Sensing of Vegetation* (four-volume book-set, 2018) with 50 chapters and 1632 pages that he edited as the chief editor (co-editors: Prof. John Lyon and Prof. Alfredo Huete).

Dr. Thenkabail is at the center of rendering scientific service to the world's remote sensing community over his long period of service. This includes serving as Editor-in-Chief (2011–present) of *Remote Sensing Open Access Journal*; Associate Editor (2017–present) of *Photogrammetric Engineering and Remote Sensing (PE&RS)*, Editorial Advisory Board Member (2016–present) of the International Society of Photogrammetry and Remote Sensing (ISPRS), and Editorial Board Member (2007–2017) of *Remote Sensing of Environment*.

The USGS and NASA selected him as one of the three international members on the Landsat Science Team (2006–2011). He is an advisory board member of the online library collection to support the United Nations' Sustainable Development Goals (UN SDGs) and is currently a scientist for the NASA and ISRO (Indian Space Research Organization) Professional Engineer and Scientist Exchange Program (PESEP) for 2022–2024. He was the chair of the International Society of Photogrammetry and Remote Sensing (ISPRS) Working Group WG VIII/7 (land cover and its dynamics) from 2013 to 2016; played a vital role for USGS as Global Coordinator, Agricultural Societal Beneficial Area (SBA), Committee for Earth Observation (CEOS) (2010–2013), during which he co-wrote the global food security case study for the CEOS's *Earth Observation Handbook* (EOS), Special Edition for the UN Conference on Sustainable Development, presented in Rio de Janeiro, Brazil; and was the co-lead (2007–2011) of IEEE's "Water for the World" initiative, a nonprofit effort funded by IEEE that worked in coordination with the Group on Earth Observations (GEO) in its GEO Water and GEO Agriculture initiatives.

Dr. Thenkabail worked as a postdoctoral researcher and research faculty at the Center for Earth Observation (YCEO), Yale University (1997–2003) and led remote sensing programs in three international organizations, including:

- International Water Management Institute (IWMI), 2003–2008
- International Center for Integrated Mountain Development (ICIMOD), 1995–1997
- International Institute of Tropical Agriculture (IITA), 1992–1995

He began his scientific career as a scientist (1986–1988) working for the National Remote Sensing Agency (NRSA) (now renamed the National Remote Sensing Center, or NRSC) at the Indian Space Research Organization (ISRO), Department of State, Government of India.

Dr. Thenkabail's work experience spans over 25 countries, including East Asia (China), Southeast Asia (Cambodia, Indonesia, Myanmar, Thailand, and Vietnam), the Middle East (Israel and Syria), North America (United States and Canada), South America (Brazil), Central Asia (Uzbekistan), South Asia (Bangladesh, India, Nepal, and Sri Lanka), West Africa (Republic of Benin, Burkina Faso, Cameroon, Central African Republic, Cote d'Ivoire, Gambia, Ghana, Mali, Nigeria, Senegal, and Togo), and Southern Africa (Mozambique and South Africa). Dr. Thenkabail is regularly invited as keynote speaker or invited speaker at major international conferences and at other important national and international forums every year.

Dr. Thenkabail obtained his PhD in agricultural engineering from the Ohio State University, USA in 1992 and has a master's degree in hydraulics and water resources engineering and a bachelor's degree in civil engineering (both from India). He has 168 publications, including 15 books; 175+ peer-reviewed journal articles, book chapters, and professional papers\monographs; and 15+ significant major global and regional data releases.

REFERENCES

Ioannidis, J.P.A. 2023. October 2023 data-update for "Updated science-wide author databases of standardized citation indicators". *Elsevier Data Repository* V6. https://doi.org/10.17632/btchxktzyw.6

Ioannidis, J.P.A., Boyack, K.W., and Baas, J. 2020. Updated science-wide author databases of standardized citation indicators. *PLoS Biology* 18(10):e3000918. https://doi.org/10.1371/journal.pbio.3000918

About the Editor

SCIENTIFIC PAPERS

https://scholar.google.com/citations?user=9IO5Y7YAAAAJ&hl=en

USGS PROFESSIONAL PAPER, DATA & PRODUCT GATEWAYS, INTERACTIVE VIEWERS

Thenkabail, P.S., Teluguntla, P.G., Xiong, J., Oliphant, A., Congalton, R.G., Ozdogan, M., Gumma, M.K., Tilton, J.C., Giri, C., Milesi, C., Phalke, A., Massey, R., Yadav, K., Sankey, T., Zhong, Y., Aneece, I., and Foley, D., 2021, Global cropland-extent product at 30-m resolution (GCEP30) derived from Landsat satellite time-series data for the year 2015 using multiple machine-learning algorithms on Google Earth Engine cloud: U.S. Geological Survey Professional Paper 1868, 63 p., https://doi.org/10.3133/pp1868 (research paper). https://lpdaac.usgs.gov/news/release-of-gfsad-30-meter-cropland-extent-products/ (download data, documents). www.usgs.gov/apps/croplands/app/map (view data interactively).

P. Teluguntla, P. Thenkabail, A. Oliphant, M. Gumma, I. Aneece, D.Foley and R.McCormick, (2023a). Landsat-derived Global Rainfed and Irrigated-Cropland Product @ 30-m (LGRIP30) of the World (GFSADLGRIP30WORLD). The Land Processes Distributed Active Archive Center (LP DAAC) of NASA and USGS. Pp. 103. https://lpdaac.usgs.gov/news/release-of-lgrip30-data-product/ (download data, documents)

BOOKS

Remote Sensing Handbook (Second Edition, six volumes, 2024)

Thenkabail, Prasad. 2024. *Remote Sensing Handbook (Second Edition, Six Volume Book-set), Volume I: Sensors, Data Normalization, Harmonization, Cloud Computing, and Accuracies*. Taylor and Francis Inc.\CRC Press, Boca Raton, London, New York. 978-1-032-89095-1 — CAT# T132478. Print ISBN: 9781032890951. eBook ISBN: 9781003541141. Pp. 581.

Thenkabail, Prasad. 2024. *Remote Sensing Handbook (Second Edition, Six Volume Book-set), Volume II: Image Processing, Change Detection, GIS, and Spatial Data Analysis*. Taylor and Francis Inc.\CRC Press, Boca Raton, London, New York. 978-1-032-89097-5 — CAT# T133208. Print ISBN: 9781032890975. eBook ISBN: 9781003541158. Pp. 464.

Thenkabail, Prasad. 2024. *Remote Sensing Handbook (Second Edition, Six Volume Book-set), Volume III: Agriculture, Food Security, Rangelands, Vegetation, Phenology, and Soils*. Taylor and Francis Inc.\ CRC Press, Boca Raton, London, New York. 978-1-032-89101-9 — CAT# T133213. Print ISBN: 9781032891019; eBook ISBN: 9781003541165. Pp. 788.

Thenkabail, Prasad. 2024. *Remote Sensing Handbook (Second Edition, Six Volume Book-set), Volume IV: Forests, Biodiversity, Ecology, LULC, and Carbon*. Taylor and Francis Inc.\CRC Press, Boca Raton, London, New York. 978-1-032-89103-3 — CAT# T133215. Print ISBN: 9781032891033. eBook ISBN: 9781003541172. Pp. 501.

Thenkabail, Prasad. 2024. *Remote Sensing Handbook (Second Edition, Six Volume Book-set), Volume V: Water, Hydrology, Floods, Snow and Ice, Wetlands, and Water Productivity*. Taylor and Francis Inc.\CRC Press, Boca Raton, London, New York. 978-1-032-89145-3 — CAT# T133261. Print ISBN: 9781032891453. eBook ISBN: 9781003541400. Pp. 516.

Thenkabail, Prasad. *Remote Sensing Handbook (Second Edition, Six Volume Book-set), Volume VI: Droughts, Disasters, Pollution, and Urban Mapping*. Taylor and Francis Inc.\CRC Press, Boca Raton, London, New York. 978-1-032-89148-4 — CAT# T133267. Print ISBN: 9781032891484; eBook ISBN: 9781003541417. Pp. 467.

Hyperspectral Remote Sensing of Vegetation (First Edition, four volumes, 2018)

Thenkabail, P.S., Lyon, G.J., and Huete, A. (Editors) 2018. Book Title: Hyperspectral Remote Sensing of Vegetation (Second Edition, Four Volume-set).

Volume I Title: Fundamentals, Sensor Systems, Spectral Libraries, and Data Mining for Vegetation. Publisher: CRC Press- Taylor and Francis group, Boca Raton, London, New York. Pp. 449, Hardback ID: 9781138058545; eBook ID: 9781315164151.

Volume II Title: Hyperspectral Indices and Image Classifications for Agriculture and Vegetation. Publisher: CRC Press- Taylor and Francis group, Boca Raton, London, New York. Pp. 296. Hardback ID: 9781138066038; eBook ID: 9781315159331.

Volume III Title: Biophysical and Biochemical Characterization and Plant Species Studies. Publisher: CRC Press- Taylor and Francis group, Boca Raton, London, New York. Pp. 348. Hardback: 9781138364714; eBook ID: 9780429431180.

Volume IV Title: Advanced Applications in Remote Sensing of Agricultural Crops and Natural Vegetation. Publisher: CRC Press- Taylor and Francis group, Boca Raton, London, New York. Pp. 386. Hardback: 9781138364769; eBook ID: 9780429431166.

Remote Sensing Handbook (First Edition, three volumes, 2015)

Thenkabail, P.S., (Editor-in-Chief), 2015. "Remote Sensing Handbook"

Volume I: Remotely Sensed Data Characterization, Classification, and Accuracies. Taylor and Francis Inc.\ CRC Press, Boca Raton, London, New York. ISBN 9781482217865 — CAT# K22125. Print ISBN: 978-1-4822-1786-5; eBook ISBN: 978-1-4822-1787-2. Pp. 678.

Volume II: Land Resources Monitoring, Modeling, and Mapping with Remote Sensing. Taylor and Francis Inc.\CRC Press, Boca Raton, London, New York. ISBN 9781482217957 — CAT# K22130. Pp. 849.

Volume III: Remote Sensing of Water Resources, Disasters, and Urban Studies. Taylor and Francis Inc.\CRC Press, Boca Raton, London, New York. ISBN 9781482217919 — CAT# K22128. Pp. 673.

Hyperspectral Remote Sensing of Vegetation (First Edition, single volume, 2013)

Thenkabail, P.S., Lyon, G.J., and Huete, A. (Editors) 2012. Book entitled: "Hyperspectral Remote Sensing of Vegetation". CRC Press- Taylor and Francis group, Boca Raton, London, New York. Pp. 781 (80+ pages in color). www.crcpress.com/product/isbn/9781439845370

Remote Sensing of Global Croplands for Food Security (First Edition, single volume, 2009)

Thenkabail. P., Lyon, G.J., Turral, H., and Biradar, C.M. (Editors) 2009. Book entitled: "Remote Sensing of Global Croplands for Food Security" (CRC Press- Taylor and Francis group, Boca Raton, London, New York. Pp. 556 (48 pages in color). Published in June, 2009.

FIGURE Snap shots of the Editor-in-Chief's work and life.

Contributors

Camila Abe
Department of Geography
San Diego State University
San Diego, CA, USA

André Quintão de Almeida
Federal University of Sergipe
São Cristóvão, Sergipe, Brazil

Allan S. Arnesen
Department for the Implementation of Research Projects, Development and Innovation
Water and Sanitation Company of the State of São Paulo—SABESP
São Paulo, Brazil

Trent W. Biggs
Department of Geography
San Diego State University
San Diego, CA, USA

Birhanu Zemadim Birhanu
International Water Management Institute
Accra, Ghana

James B. Campbell
Department of Geography
Virginia Tech University
Blacksburg, VA, USA

Ni-Bin Chang
Department of Civil, Environmental and Construction Engineering
University of Central Florida
Orlando, FL, USA

Hyun Jung ("J.") Cho
Department of Integrated Environmental Science
Bethune-Cookman University
Daytona Beach, FL, USA

Marcelo P. Curtarelli
National Institute for Space Research
São José dos Campos
São Paulo, Brazil

Matheus Z. Francisco
Costal Integrated Management Group (LabGerco)
Earth and Sea Technology Centre (CTTMar)
University of Itajaí Valley—UNIVALI
and
Buildings and Municipal Services Department
Itajaí, Brazil

Edlene Aparecida Monteiro Garçon
Embrapa Territory
Campinas, São Paulo, Brazil

Frederico T. Genofre
Environmental Engineer
Santa Catarina, Brazil

Shuvankar Ghosh
Department of Geography
University of Georgia
Athens, GA, USA

Chandra Giri
Earth Resources Observation and Science (EROS) Center
US Geological Survey
Sioux Falls, SD, USA

Edward P. Glenn
Department of Soil, Water and Environmental Science
University of Arizona
Tucson, AZ, USA

Murali Krishna Gumma
International Crops Research Institute for the Semi-Arid Tropics (ICRISAT)
Patancheru, Hyderabad, India

Christine Hladik
Geology and Geography Department
Georgia Southern University
Statesboro, GA, USA

Yang Hong
School of Civil Engineering and Environmental Sciences
Hydrometeorology and Remote Sensing (HyDROS) Laboratory
University of Oklahoma
Norman, OK, USA

Xiaodong Huang
Lanzhou University
Lanzhou City, China

Mangi Lal Jat
International Crops Research Institute for the Semi-Arid Tropics (ICRISAT)
Patancheru, Hyderabad, India

Stefanie Kagone
ASRC Federal Data Solutions
Contractor to US Geological Survey (USGS) Earth Research Observation and Science (EROS) Center
Sioux Falls, SD, US

Sadiq I. Khan
School of Civil Engineering and Environmental Sciences
Hydrometeorology and Remote Sensing (HyDROS) Laboratory
University of Oklahoma
Norman, OK, USA

Claudia Kuenzer
German Remote Sensing Data Center
German Aerospace Center
Oberpfaffenhofen, Germany

Janice Freitas Leivas
Embrapa Territory
Campinas, São Paulo, Brazil

Tiangang Liang
Lanzhou University
Lanzhou City, China

Ying Lu
University at Buffalo
State University of New York
Buffalo, NY, USA

Michael Marshall
University of Twente
Enschede, the Netherlands

Sandro Martinis
German Remote Sensing Data Center
German Aerospace Center
Oberpfaffenhofen, Germany

Jing Miao
University at Buffalo
The State University of New York
Buffalo, NY, USA

Deepak R. Mishra
Geography Department
University of Georgia
Athens, GA, USA

Pamela L. Nagler
Southwest Biological Science Center
Terrestrial Dryland Ecology Branch
US Geological Survey
Tucson, Arizona, USA

Jessica L. O'Connell
Department of Marine Sciences
University of Georgia
Athens, GA, USA

Edson Patto Pacheco
Embrapa Coastal Tablelands
Aracaju, Sergipe, Brazil

Pranay Panjala
International Crops Research Institute for the Semi-Arid Tropics (ICRISAT)
Patancheru, Hyderabad, India

George P. Petropoulos
Department of Geography
Harokopio University of Athens
Athens, Greece

Triantafyllia Petsini
Department of Geography
Harokopio University of Athens
Athens, Greece

Contributors

Lynn M. Resler
Department of Geography
Virginia Tech University
Blacksburg, VA, USA

Anderson Ruhoff
Institute of Hydraulic Research
Universidade Federal do Rio Grande do Sul
Porto Alegre, Rio Grande do Sul, Brazil

Ana Flávia Maria Santos
Federal University of Sergipe
São Cristóvão, Sergipe, Brazil

Santhosh Kumar Seelan
Department of Space Studies, John D. Odegard
 School of Aerospace Sciences
University of North Dakota
Grand Forks, ND, USA

Inajá Francisco de Sousa
Federal University of Sergipe
São Cristóvão, Sergipe, Brazil

Celina Maki Takemura
Embrapa Territory
Campinas, São Paulo, Brazil

Antônio Heriberto de Castro Teixeira
Federal University of Sergipe
São Cristóvão, Sergipe, Brazil

Pardhasaradhi Teluguntla
US Geological Survey (USGS)
Flagstaff, AZ, USA
and
Bay Area Environmental Research Institute
 (BAERI)
West Sonoma, CA, USA

Prasad S. Thenkabail
US Geological Survey (USGS)
Flagstaff, AZ, USA

André Twele
German Remote Sensing Data Center
German Aerospace Center
Oberpfaffenhofen, Germany

Le Wang
University at Buffalo
The State University of New York
Buffalo, NY, USA

Xianwei Wang
Sun Yat-sen University
Guangzhou, China

Xiongxin Xiao
Institute of Geography
University of Bern
Bern, Switzerland

Hongjie Xie
University of Texas at San Antonio
San Antonio, TX, USA

Xianwu Xue
School of Civil Engineering and Environmental
 Sciences
Hydrometeorology and Remote Sensing
 (HyDROS) Laboratory
University of Oklahoma
Norman, OK, USA

Guoqing Zhang
Institute of Tibetan Plateau Research
Chinese Academy of Sciences
Chaoyang District, Beijing, P.R. China

Yu Zhang
School of Civil Engineering
 and Environmental Sciences,
 Hydrometeorology and Remote Sensing
 (HyDROS) Laboratory
University of Oklahoma,
Norman, OK, USA

Acknowledgments

The *Remote Sensing Handbook* (Second Edition, Vosl. I–VI) brought together a galaxy of highly accomplished, renowned remote sensing scientists, professionals, and legends from around the world. I chose the lead authors after careful review of their accomplishments and sustained publication record over the years. The chapters in the second edition were written\revised over a period of two years. All chapters were edited and revised.

Gathering such a galaxy of authors was the biggest challenge. These are all extremely busy people, and committing to a book project that requires a substantial workload is never easy. However, almost all of those whom I requested agreed to write a chapter specific to their area of specialization, and only a few I had to convince to make time. The quality of the chapters should convince readers why these authors are such highly rated professionals and why they are so successful and accomplished in their field of expertise. They not only wrote very high-quality chapters, but delivered them on time, addressed any editorial comments in a timely manner without complaints, and were extremely humble and helpful. Their commitment for quality science is what makes them special. I am truly honored to have worked with such great professionals.

I would like to mention the names of everyone who contributed and made *Remote Sensing Handbook* (Second Edition, Vols. I– VI) possible. In the end, we had 91 chapters, a little over 3000 pages, and a little over 300+ authors. My gratitude goes to each one of them. These are among the well-known "who's who" in the world of remote sensing science. A list of all authors is provided next. The names of the authors are organized chronologically for each volume and the chapters. Each lead author of the chapter is in bold. The names of the 400+ authors who contributed to the six volumes are as follows:

Volume I: Sensors, Data Normalization, Harmonization, Cloud Computing, and Accuracies: 18 chapters written by 53 authors (Editor-in-chief: Prasad S. Thenkabail):

Drs. **Sudhanshu S. Panda**, Mahesh Rao, Prasad S. Thenkabail, Debasmita Misra, and James P. Fitzerald; **Mohinder S. Grewal**; **Kegen Yu**, Chris Rizos, and Andrew Dempster; **D. Myszor**, O. Antemijczuk, M. Grygierek, M. Wierzchanowski, and K.A. Cyran; **Natascha Oppelt** and Arnab Muhuri; **Philippe M. Teillet**; **Philippe M. Teillet** and Gyanesh Chander; **Rudiger Gens** and Jordi Cristóbal Rosselló; **Aolin Jia** and Dongdong Wang; **Tomoaki Miura**, Kenta Obata, Hiroki Yoshioka, and Alfredo Huete; **Michael D. Steven**, Timothy J. Malthus, and Frédéric Baret; **Fabio Dell'Acqua** and Silvio Dell'Acqua; **Ramanathan Sugumaran**, James W. Hegeman, Vivek B. Sardeshmukh, and Marc P. Armstrong; **Lizhe Wang**, Jining Yan, Yan Ma, Xiaohui Huang, Jiabao Li, Sheng Wang, Haixu He, Ao Long, and Xiaohan Zhang; **John E. Bailey** and Josh Williams; **Russell G. Congalton**; **P.J. Blount**; and **Prasad S. Thenkabail**.

Volume II: Image Processing, Change Detection, GIS, and Spatial Data Analysis: 17 chapters written 64 authors (Editor-in-chief: Prasad S. Thenkabail):

Sunil Narumalani and Paul Merani; **Mutlu Ozdogan**; **Soe W. Myint**, Victor Mesev, Dale Quattrochi, and Elizabeth A. Wentz; **Jun Li**, Paolo Gamba, and Antonio Plaza; **Qian Du**, Chiranjibi Shah, Hongjun Su, and Wei Li; **Claudia Kuenzer**, Philipp Reiners, Jianzhong Zhang, and Stefan Dech; **Mohammad D. Hossain** and Dongmei Chen; **Thomas Blaschke**, Maggi Kelly, and Helena Merschdorf; **Stefan Lang** and Dirk Tiede; **James C. Tilton**, Selim Aksoy, and Yuliya Tarabalka; **Shih-Hong Chio**, Tzu-Yi Chuang, Pai-Hui Hsu, Jen-Jer Jaw, Shih-Yuan Lin, Yu-Ching Lin, Tee-Ann Teo, Fuan Tsai, Yi-Hsing Tseng, Cheng-Kai Wang, Chi-Kuei Wang, Miao Wang, and Ming-Der Yang; **Guiying Li**, Mingxing Zhou, Ming Zhang, and Dengsheng Lu; **Jason A. Tullis**, David P. Lanter, Aryabrata Basu, Jackson D. Cothren, Xuan Shi, W. Fredrick Limp, Rachel F. Linck, Sean G. Young, Jason Davis, and Tareefa S. Alsumaiti; **Gaurav Sinha**, Barry J. Kronenfeld, and Jeffrey C.

Brunskill; **May Yuan**; **Stefan Lang**, Stefan Kienberger, Michael Hagenlocher, and Lena Pernkopf; and **Prasad S. Thenkabail**.

Volume III: Agriculture, Food Security, Rangelands, Vegetation, Phenology, and Soils: 17 chapters written by 110 authors (Editor-in-chief: Prasad S. Thenkabail):

Alfredo Huete, Guillermo Ponce-Campos, Yongguang Zhang, Natalia Restrepo-Coupe, and Xuanlong Ma; **Juan Quiros-Vargas**, Bastian Siegmann, Juliane Bendig, Laura Verena Junker-Frohn, Christoph Jedmowski, David Herrera, and Uwe Rascher; **Frédéric Baret**; **Lea Hallik**, Egidijus Šarauskis, Ruchita Ingle, Indrė Bručienė, Vilma Naujokienė, and Kristina Lekavičienė; **Clement Atzberger** and Markus Immitzer; **Agnès Bégué**, Damien Arvor, Camille Lelong, Elodie Vintrou, and Margareth Simoes; **Pardhasaradhi Teluguntla**, Prasad S. Thenkabail, Jun Xiong, Murali Krishna Gumma, Chandra Giri, Cristina Milesi, Mutlu Ozdogan, Russell G. Congalton, James Tilton, Temuulen Tsagaan Sankey, Richard Massey, Aparna Phalke, and Kamini Yadav; **Yuxin Miao**, David J. Mulla, and Yanbo Huang; **Baojuan Zheng**, James B. Campbell, Guy Serbin, Craig S.T. Daughtry, Heather McNairn, and Anna Pacheco; **Prasad S. Thenkabail**, Itiya Aneece, Pardhasaradhi Teluguntla, Richa Upadhyay, Asfa Siddiqui, Justin George Kalambukattu, Suresh Kumar, Murali Krishna Gumma, and Venkateswarlu Dheeravath; **Matthew C. Reeves**, Robert Washington-Allen, Jay Angerer, Raymond Hunt, Wasantha Kulawardhana, Lalit Kumar, Tatiana Loboda, Thomas Loveland, Graciela Metternicht, Douglas Ramsey, Joanne V. Hall, Trenton Benedict, Pedro Millikan, Angus Retallack, Arjan J.H. Meddens, William K. Smith, and Wen Zhang; **E. Raymond Hunt Jr.**, Cuizhen Wang, D. Terrance Booth, Samuel E. Cox, Lalit Kumar, and Matthew C. Reeves; **Lalit Kumar**, Priyakant Sinha, Jesslyn F. Brown, R. Douglas Ramsey, Matthew Rigge, Carson A. Stam, Alexander J. Hernandez, E. Raymond Hunt, Jr., and Matt Reeves; **Molly E. Brown**, Kirsten de Beurs, Kathryn Grace; **José A.M. Demattê**, Cristine L.S. Morgan, Sabine Chabrillat, Rodnei Rizzo, Marston H.D. Franceschini, Fabrício da S. Terra, Gustavo M. Vasques, Johanna Wetterlind, Henrique Bellinaso, Letícia G. Vogel; **E. Ben-Dor**, J.A.M. Demattê; and **Prasad S. Thenkabail**.

Volume IV: Forests, Biodiversity, Ecology, LULC, and Carbon: 12 chapters written by 71 authors (Editor-in-chief: Prasad S. Thenkabail):

E.H. Helmer, Nicholas R. Goodwin, Valéry Gond, Carlos M. Souza Jr., and Gregory P. Asner; **Juha Hyyppä**, Xiaowei Yu, Mika Karjalainen, Xinlian Liang, Anttoni Jaakkola, Mike Wulder, Markus Hollaus, Joanne C. White, Mikko Vastaranta, Jiri Pyörälä, Tuomas Yrttimaa, Ninni Saarinen, Josef Taher, Juho-Pekka Virtanen, Leena Matikainen, Yunsheng Wang, Eetu Puttonen, Mariana Campos, Matti Hyyppä, Kirsi Karila, Harri Kaartinen, Matti Vaaja, Ville Kankare, Antero Kukko, Markus Holopainen, Hannu Hyyppä, Masato Katoh, and Eric Hyyppä; **Gregory P. Asner**, Susan L. Ustin, Philip A. Townsend, and Roberta E. Martin; **Sylvie Durrieu**, Cédric Véga, Marc Bouvier, Frédéric Gosselin, Jean-Pierre Renaud, and Laurent Saint-André; **Thomas W. Gillespie**, Morgan Rogers, Chelsea Robinson, and Duccio Rocchini; **Stefan Lang**, Christina Corbane, Palma Blonda, Kyle Pipkins, and Michael Förster; **Conghe Song**, Jing Ming Chen, Taehee Hwang, Alemu Gonsamo, Holly Croft, Quanfa Zhang, Matthew Dannenberg, Yulong Zhang, Christopher Hakkenberg, and Juxiang Li; **John Rogan** and Nathan Mietkiewicz; **Zhixin Qi**, Anthony Gar-On Yeh, Xia Li, and Qianwen Lv; **R.A. Houghton**; **Wenge Ni-Meister**; and **Prasad S. Thenkabail**.

Volume V: Water Resources: Hydrology, Floods, Snow and Ice, Wetlands, and Water Productivity: 13 chapters written by 60 authors (Editor-in-chief: Prasad S. Thenkabail):

James B. Campbell and Lynn M. Resler; **Sadiq I. Khan**, Ni-Bin Chang, Yang Hong, Xianwu Xue, and Yu Zhang; **Santhosh Kumar Seelan**; **Allan S. Arnesen**, Frederico T. Genofre, Marcelo P. Curtarelli, and Matheus Z. Francisco; **Allan S. Arnesen**, Frederico T. Genofre, Marcelo P. Curtarelli, and Matheus Z. Francisco; **Sandro Martinis**, Claudia Kuenzer, and André Twele; **Le Wang**, Jing Miao, and Ying Lu; **Chandra Giri**; **Deepak R. Mishra**, X. Yan, Shuvan Ghosh,

Acknowledgments

Christine Hladik, J. L. O'Connell, and Hyun Jung ("J.") Cho; **Murali Krishna Gumma**, Prasad S. Thenkabail, Pranay Panjala, Pardhasaradhi Teluguntla, Birhanu Zemadim Birhanu, and Mangi Lal Jat; **Trent W. Biggs**, Pamela L. Nagler, Anderson Ruhoff, Triantafyllia Petsini, Michael Marshall, George P. Petropoulos, Camila Abe, and Edward P. Glenn; **Antônio Heriberto de Castro Teixeira**, Janice Freitas Leivas, Celina Maki Takemura, Edson Patto Pacheco, Edlene Aparecida Monteiro Garçon, Inajá Francisco de Sousa, André Quintão de Almeida, Prasad S. Thenkabail, and Ana Flávia Maria Santos; **Hongjie Xie**, Tiangang Liang, Xianwei Wang, Guoqing Zhang, Xiaodong Huang, and Xiongxin Xiao; and **Prasad. S. Thenkabail**.

Volume VI: Droughts, Disasters, Pollution, and Urban Mapping: 14 chapters written by 53 authors (Editor-in-chief: Prasad S. Thenkabail):

Felix Kogan and Wei Guo; **F. Rembold**, M. Meroni, O. Rojas, C. Atzberger, F. Ham, and E. Fillol; **Brian D.Wardlow**, Martha A. Anderson, Tsegaye Tadesse, Mark S. Svoboda, Brian Fuchs, Chris R. Hain, Wade T. Crow, and Matt Rodell; **Jinyoung Rhee**, Jungho Im, and Seonyoung Park; **Marion Stellmes**, Ruth Sonnenschein, Achim Röder, Thomas Udelhoven, Gabriel del Barrio, and Joachim Hill; **Norman Kerle**; **Stefan Lang**, Petra Füreder, Olaf Kranz, Brittany Card, Shadrock Roberts, and Andreas Papp; **Robert Wright**; **Krishna Prasad Vadrevu** and Kristofer Lasko; **Anupma Prakash**, Claudia Kuenzer, Santosh K. Panda, Anushree Badola, and Christine F. Waigl; **Hasi Bagana**, Chaomin Chena, and Yoshiki Yamagata; **Yoshiki Yamagata**, Daisuke Murakami, Hajime Seya, and Takahiro Yoshida; **Qingling Zhang**, Noam Levin, Christos Chalkias, Husi Letu, and Di Liu; and **Prasad S. Thenkabail**.

The authors not only delivered excellent chapters, they also provided valuable insights and inputs for me in many ways throughout the book project.

I was delighted when **Dr. Compton J. Tucker**, Senior Earth Scientist, Earth Sciences Division, Science and Exploration Directorate, NASA Goddard Space Flight Center (GSFC) agreed to write the foreword for the book. For anyone practicing remote sensing, Dr. Tucker needs no introduction. He has been a "godfather" of remote sensing and has inspired a generation of remote sensing scientists. I have been a student of his without ever really being one. I mean, I have not been his student in the classroom, but I have followed his legendary work throughout my career. I remember reading his highly cited paper (now with citations nearing 7700!):

- Tucker, C.J. (1979) 'Red and Photographic Infrared Linear Combinations for Monitoring Vegetation,' *Remote Sensing of Environment,* 8(2), 127–150.

I first read this paper in 1986 when I had just joined the National Remote Sensing Agency (NRSA, now NRSC) of the Indian Space Research Organization (ISRO). Dr. Tucker's pioneering works have been guiding light for me ever since. After getting his PhD from the Colorado State University in 1975, Dr. Tucker joined NASA GSFC as a postdoctoral fellow and became a full-time NASA employee in 1977. Ever since, he has conducted several path-finding studies. He has used NOAA AVHRR, MODIS, SPOT Vegetation, and Landsat satellite data for studying deforestation, habitat fragmentation, desert boundary determination, ecologically coupled diseases, terrestrial primary production, glacier extent, and how climate affects global vegetation. He has authored or coauthored more than 280 journal articles that have been cited more than 93,000 times, he is an adjunct professor at the University of Maryland and a consulting scholar at the University of Pennsylvania's Museum of Archaeology and Anthropology, and he has appeared in more than 20 radio and TV programs. He is a fellow of the American Geophysical Union and has been awarded several medals and honors, including NASA's Exceptional Scientific Achievement Medal, the Pecora Award from the US Geological Survey, the National Air and Space Museum Trophy, the Henry Shaw Medal from the Missouri Botanical Garden, the Galathea Medal from the Royal Danish Geographical Society, and the Vega Medal from the Swedish Society of Anthropology and Geography. He was the NASA representative to the US Global Change Research Program from 2006 to 2009. He was instrumental in releasing the AVHRR

33-year (1982–2014) Global Inventory Monitoring and Modeling Studies (GIMMS) data. I strongly recommend that everyone reads his excellent foreword before reading the book. In the foreword, Dr. Tucker demonstrates the importance of data from Earth Observation (EO) sensors from orbiting satellites to maintaining a reliable and consistent climate record. Dr. Tucker further highlights the importance of continued measurements of these variables of our planet in the new millennium through new, improved, and innovative EO sensors from sun synchronous and\or geostationary satellites.

I want to acknowledge with thanks for the encouragement and support received by my US Geological Survey (USGS) colleagues. I would like to mention the late Mr. Edwin Pfeifer, Dr. Susan Benjamin (my director at the Western Geographic Science Center), Dr. Dennis Dye, Mr. Larry Gaffney, Mr. David F. Penisten, Ms. Emily A. Yamamoto, Mr. Dario D. Garcia, Mr. Miguel Velasco, Dr. Chandra Giri, Dr. Terrance Slonecker, Dr. Jonathan Smith, Timothy Newman, and Zhouting Wu. Of course, my dear colleagues at USGS, Dr. Pardhasaradhi Teluguntla, Dr. Itiya Aneece, Mr. Adam Oliphant, and Mr. Daniel Foley, have helped me in numerous ways. I am ever grateful for their support and significant contributions to my growth and this body of work. Throughout my career, there have been many postdoctoral-level scientists who have worked with me closely and contributed to my scientific growth in different ways. They include Dr. Murali Krishna Gumma, Head of Remote Sensing at the International Crops Research Institute for the Semi-Arid Tropics; Dr. Jun Xiong, Geo ML ≠ ML with GeoData, Climate Corp.; Dr. Michael Marshall, Associate Professor, University of Twente, Netherlands; Dr. Isabella Mariotto, Former USGS Postdoctoral Researcher; Dr. Chandrashekar Biradar, Country Director, India for World Agroforestry; and numerous others. I am thankful for their contributions. I know I am missing many names: too numerous to mention them all, but my gratitude for them is the same as the names I have mentioned here.

There is a very special person I am very thankful for: the late Dr. Thomas Loveland. I first met Dr. Loveland at USGS, Sioux Falls for an interview to work for him as a scientist in the late 1990s when I was still at Yale University. But even though I was selected, I was not able to join him as I was not a citizen of the United States at that time and working for USGS required that. He has been my mentor and a pillar of strength over two decades, particularly during my Landsat Science Team days (2006–2011) and later once I joined USGS in 2008. I have watched him conduct Landsat Science Team meetings with great professionalism, insight, and creativity. I remember him telling my PhD advisor upon my being hired at USGS: "we don't make mistakes!" During my USGS days, he was someone I could ask for guidance and seek advice, and he would always be there to respond with kindness and understanding. Above all, he would share his helpful insights. It is sad that we lost him too early. I pray for his soul. Thank you, Tom, for your kindness and generosity.

Over the years, there are numerous people who have come into my professional life who have helped me grow. It is a tribute to their guidance, insight, and blessings that I am here today. In this regard I need to mention a few names out of gratitude: (1) Prof. G. Ranganna, my master's thesis advisor in India at the National Institute of Technology (NIT), Surathkal, Karnataka, India. Prof. Ranganna is 92 years old (2024), and I met him a few months back. To this day he is my guiding light on how to conduct oneself with fairness and dignity in professional and personal conduct. Prof. Ranganna's trait of selflessly caring for his students throughout his life is something that influenced me to follow. (2) Prof. E.J. James, former Director of the Center for Water Resources Development and Management (CWRDM), Calicut, Kerala, India. Prof. James was my master's thesis advisor in India who's dynamic personality in professional and personal matters had an influence on me. Dr. James always went out of his way to help his students despite his busy schedule. (3) The late Dr. Andrew Ward, my PhD advisor at the Ohio State University, Columbus, Ohio. He funded my PhD studies in the United States through grants. Through him I learned how to write scientific papers and how to become a thorough professional. He was a tough task maker, your worst critic (to help you grow), but also a perfectionist who helped you grow as a peerless professional and, above all, a very kind human being at the core. He would write you long memos on flaws in your research but then help you out of it by making you work double the time! To make you work harder, he would tell you "You won't get my sympathy." Then when you accomplished your goal, he would tell you "you have paid back for your scholarship many times over!" (4) Dr. John G. Lyon, also my PhD advisor at the Ohio State University, Columbus, Ohio. He was a peerless motivator who encouraged you to believe in yourself. (5) Dr. Thiruvengadachari, Scientist at the National Remote Sensing Agency (NRSA), which is now the National Remote Sensing Center (NRSC), India. He was my first boss at the Indian Space Research Organization (ISRO), and through him I learned initial steps in remote sensing science. I was just 25 years old then and had joined NRSA after my Master of Engineering (hydraulics and water resources) and Bachelor of Engineering (civil engineering) degrees. The first day in the office Dr. Thiruvengadachari asked me how much remote sensing I knew. I told him "zero" and instantly thought he would ask me to leave the room. But his response was "very good!" and he gave me a manual on remote sensing from the Laboratory for Applications of Remote Sensing (LARS), Purdue University to study. Those were the days where there was no formal training in remote sensing in universities. So, my remote sensing lessons began by working practically on projects, and one of our first projects was "drought monitoring for India using NOAA AVHRR data." This was an intense period of learning the fundamentals of remote sensing science for me by practicing on a daily basis. Data came in 9 mm tapes and was read on massive computing systems, image processing was done, mostly working night shifts by booking time on a centralized computer, fieldwork was conducted using false color composite (FCC) outputs and topographic maps (there were no global positioning systems, or GPS), geographic information system (GIS) was in its infancy, a lot of calculations were done using calculators, and we had just started working on IBM 286 computers with floppy disks. So, when I decided to resign my NRSA job and go to the United States to do my PhD, Dr. Thiruvengadachari told me "Prasad, I am losing my right hand, but you can't miss opportunity." Those initial wonderful days of learning from Dr. Thiruvengadachari will remain etched in my memory. I am also thankful to my very good old friend Shri C.J. Jagadeesha, who was my colleague at NRSA\NRSC, ISRO. He was a friend who encouraged me to grow as a remote sensing scientist through our endless rambling discussions over tea in Iranian restaurants outside NRSA those days and elsewhere.

I am ever grateful to my former professors at the Ohio State University, Columbus, Ohio, USA: the late Prof. Carolyn Merry, Dr. Duane Marble, and Dr. Michael Demers. They have taught and\or encouraged, inspired, and given me opportunities at the right time. The opportunity to work for six years at the Yale Center for Earth Observation (YCEO) was incredibly important. I am thankful to Prof. Ronald G. Smith, Director of YCEO for the opportunity, guidance, and kindness. At YCEO I learned and advanced myself as a remote sensing scientist. The opportunities I got working for the International Institute of Tropical Agriculture (IITA), based in Nigeria, and the International Water Management Institute (IWMI), based in Sri Lanka, where I worked on remote sensing science pertaining to a number of applications, such as agriculture, water, wetlands, food security, sustainability, climate, natural resources management, environmental issues, droughts, and biodiversity, were extremely important in my growth as a remote sensing scientist—especially from the point of view of understanding the issues on the ground in real-life situations. Finding solutions, applying one's theoretical understanding to practical problems, and seeing them work has its own nirvana.

As it is clear from the previous paragraphs, it is of great importance to have guiding pillars of light at crucial stages of your education. That is where you become who you will be in the end, where you will grow and make your own contributions. I am so blessed to have had these wonderful guiding lights come into my professional life at the right time of my career (which also influenced me positively in my personal life). From that firm foundation, I could build on what I learned and through the confidence of knowledge and accomplishments pursue my passion for science and do several significant pioneering studies throughout my career.

I mention all of this out of a gratitude for my ability today to edit such a monumental project as the *Remote Sensing Handbook* (Second Edition, Vols. I–VI).

I am very thankful to Ms. Irma Shagla-Britton, Manager and Leader for Remote Sensing and GIS books at Taylor & Francis/CRC Press. Without her consistent encouragement to take on this responsibility of editing the *Remote Sensing Handbook*, especially in trusting me to accomplish this momentous work over so many other renowned experts, I would never have gotten to work on this in the first place. Thank you, Irma. Sometimes you need to ask several times, before one can say yes to something!

I am very grateful to my wife (Sharmila Prasad), my daughter (Spandana Thenkabail), and my son-in-law (Tejas Mayekar) for their usual unconditional understanding, love, and support. My wife and daughter have always been pillars of my life, now joined by my equally loving son-in-law. I learned the values of hard work and dedication from my revered parents. This work wouldn't come through without their life of sacrifices to educate their children and their silent blessings. My father's vision in putting emphasis on education and sending me to the best of places to study despite our family's very modest income and my mother's endless hard work are my guiding light and inspiration. Of course, there are many, many others to be thankful for, but there are too many to mention here. Finally, it must be noted that a work of this magnitude, editing the monumental *Remote Sensing Handbook* (Second Edition, Vols. I–VI), continuing from the three-volume first edition, requires blessings of the almighty. I firmly believe nothing happens without the powers of the universe blessing you and providing needed energy, strength, health, and intelligence. To that infinite power my humble submission of everlasting gratefulness.

It has been my deep honor and great privilege to have edited the *Remote Sensing Handbook* (Second Edition, Vols. I–VI) after having edited the three-volume first edition that was published in 2015. Now after ten years, we will have the six-volume second edition in 2024. A huge thanks to all the authors, the publisher, family members, friends, and everyone else who made this huge task possible.

Dr. Prasad S. Thenkabail, PhD
Editor-in-Chief
Remote Sensing Handbook **(Second Edition, Vols. I–VI)**

Volume 1: Sensors, Data Normalization, Harmonization, Cloud Computing, and Accuracies
Volume 2: Image Processing, Change Detection, GIS, and Spatial Data Analysis
Volume 3: Agriculture, Food Security, Rangelands, Vegetation, Phenology, and Soils
Volume 4: Forests, Biodiversity, Ecology, LULC, and Carbon
Volume 5: Water Resources: Hydrology, Floods, Snow and Ice, Wetlands, and Water Productivity
Volume 6: Droughts, Disasters, Pollution, and Urban Mapping

Part I

Geomorphology

1 Geomorphological Studies from Remote Sensing

James B. Campbell and Lynn M. Resler

ACRONYMS AND DEFINITIONS

ALB	Airborne LiDAR Bathymetry
ASTER	Advanced Spaceborne Thermal Emission and Reflection Radiometer
CSC	Coastal Service Center
DEMs	Digital Elevation Models
DOQQs	Digital Orthophoto Quarter Quads
DTMs	Digital Terrain Models
ERS	European Remote Sensing Satellites
ETM+	Enhanced Thematic Mapper Plus
GIS	Geographic Information Systems
GPS	Global Positioning Systems
InSAR	Interferometric SAR
IRS	Indian Remote Sensing Satellites
LiDAR	Light Detection and Ranging
MLS	Mobile Laser Scanning
MSS	Multi-Spectral Scanner
NASA	National Aeronautics and Space Administration
NDVI	Normalized Difference Vegetation Index
NOAA	National Oceanic and Atmospheric Administration
OBIA	Object-Based Image Analysis
PDTD	photogrammetrically derived topographic data
PSI	Persistent Scattering Interferometry
RADARSAT	RADAR Satellite
RGAs	rapid geomorphic assessments
RS	Remote Sensing
SARs	Synthetic Aperture Radars
SONAR	Sound Navigation and Ranging
SR	Streambank Retreat
SRTM	Shuttle Radar Topography Mission
SSC	Suspended Sediment Concentration
TLS	Terrestrial Laser Surveying
TM	Thematic Mapper
TS	Total Station

1.1 INTRODUCTION

Remote sensing analysis has provided important insights into geomorphic systems and landforms through its analytical capabilities and data integration capacities (Torre et al., 2024; Zangana et al., 2023; Misra, 2022; Conforti et al., 2021; Lausch et al., 2020; Langat et al., 2019; Nie et al., 2018;

Dikpal et al., 2017; Rajaveni et al., 2017). This chapter reviews the role of remotely sensed imagery in geomorphic inquiry, initially through a historical overview of contributions of aerial imagery to understanding geomorphological systems at varied spatial and temporal scales.

With historical background as context, the following discussion documents the significance of remote sensing's role in developing current geomorphic understanding through (1) enhancing the ability to conduct inquiries on established topics and (2) developing technologies that open new lines of inquiry that were previously unavailable. Thus, the remaining portions of the chapter will address the following questions as they apply to a selection of topics (Table 1.1), with a focus on selected subfields of geomorphology that represent periglacial, glacial, mass wasting, fluvial, coastal, aeolian, and biogeomorphologic processes:

- In what ways has remote sensing been used in various subfields of geomorphology?
- How has remote sensing analysis provided unique perspectives and advanced understanding in each subfield?

TABLE 1.1
Summary of Selected Geomorphic Topics

Selected Topics	Selected References
Alpine & Periglacial active layer thickness thermokarst soil organic layer ground ice	French and Thorn (2006); Kääb et al. (2008); Polar Research Board, NAS (2014)
Glacial Landforms glacial extent changes in glacial extent ice movement	Kääb (2005)
Mass Wasting landslides debris flows soil creep	Metternicht et al. (2005); Scaioni (2013)
Fluvial Landforms channel migration flooding and floodplain analysis streambank retreat	Carbonneau and Piegay (2012)
Coastal Landforms coastal form and change suspended sediment concentrations bathymetry	Allen and Wang (2010); Gao (2011)
Aeolian Landforms dune patterns rates of movment dune volume	Mitasova et al. (2005); Livingston et al. (2007)
Biogeomorphology feedback between process and pattern treeline position and dynamics channel migration	Walsh et al. (1998); Bryant and Gilvear (1999); Bryant et al. (1999); Hudson et al. (2006)

Although applications of remote sensing already form a significant dimension to geomorphic inquiry, their impact remains largely dispersed among various subfields and is seldom discussed in an overview. We intend our sketch of these developments to form the basis for further efforts to examine their contributions.

1.1.1 Historical Perspective

The practice of remote sensing has advanced the field of geomorphology for almost 100 years (Muzirafuti, 2024; Zangana et al., 2023; Darwish et al., 2017; Vitek et al., 1996). The history of remote sensing in geomorphology begins with the role of aerial imagery in providing insights into landscape evolution at local to global scales.

Aerial photography played a pivotal role in the early adoption of remote sensing technologies for the examination of physiography and, later, of geomorphic systems. Aerial photography, as the prevailing aerial survey technology of the day, already had a long history by the beginning of World War I. During that conflict aerial photography began its transformation from its status as a curiosity into a practical and scientific tool for examining soils, geological features, coastlines, and river systems. Previously, the airplane and camera existed as two separate, largely incompatible, technologies that could be used together only with great inconvenience. By the time of the 1918 armistice, the groundwork was set for the design of photographic systems that could systematically collect aerial imagery to support analytical objectives, together with organizational structures to interpret and report results of aerial survey, which form forerunners of today's practices.

After military demobilization, in the 1920s, such developments formed foundations for the institutionalization of aerial survey in both military and civil enterprises. Lee's (1922) survey of aerial photography's potential for civil applications outlines its future role in urban planning, land use planning, and examination of geomorphic features, such as coastal landforms, overwash fans, tidal deltas, coastal marshes, and spits.

Progress in development of aerial cameras, aircraft, and supporting technologies during the inter-war decades increased interest in applications of aerial photography. By 1939 (i.e., Melton, 1939), aerial photographs were used as teaching resources for introductory geology courses, especially for illustrating the basics of landforms shaped by steams, wind, waves, and glaciers.

Although aerial photography contributed to geomorphological inquiry during the inter-war years, such knowledge was fragmented. By the beginning of World War II, however, methods and techniques were organized systematically into military handbooks, field manuals, and training programs. Examples include the detailed knowledge of coastal landforms, especially beach profiles, initially in the Pacific Theater, but later in European campaigns. What formerly was considered esoteric geographic knowledge of remote locations became essential for planning successful operations, since many of World War II's key campaigns were conducted in locations lacking systematic geomorphological knowledge.

In this context, aerial photography, as the prevailing form of remote sensing imagery of the time, formed an essential component of geomorphological knowledge supporting: (1) combat intelligence, supporting studies of terrain, trafficability (assessment of terrain suitability for motorized vehicles), relief, and intervisibility (assessing ability to observe landscapes from a given location), and (2) engineering intelligence, supporting construction of airfields and roads in remote, unmapped, regions. Successful military operations required development, often without systematic field data, of detailed understanding of (for example) subsurface beach configurations, river systems, and wetlands. In this context, the role of aerial photography grew from a source of qualitative data to form a more precise quantitative tool for assessing, for example, topographic slope, and depths of offshore topography. World War II saw some of the first attempts to integrate earlier piecemeal practices into systematized photointerpretation to support geologic inquiry (Smith, 1942, 1943).

In the years after World War II, many of these capabilities found applications in civil society, and simultaneously in the national security domain, as the Cold War era began. Many of the successful uses of aerial photography in World War II confirmed and validated its use for civil applications in government and commerce, which in turn contributed to a growing archive of imagery and to investments in new sensor technologies.

For the earth sciences, many of these initial applications can be seen as providing descriptive content, basically recording locations of labeled features—an important mapping contribution, but one that did little to advance the cause of geomorphology's primary mission—investigation of geomorphic processes (Quesada-Román and Peralta-Reyes, 2023; Lausch et al., 2020; Ahmed et al., 2017). During the post–World War II era, aerial imagery contributed to the emerging paradigm of landscape quantification; analysis of aerial imagery became an increasingly important tool for extraction of quantitative landscape metrics (Chorley, 2008). The "mid-century revolution" in fluvial geomorphology (Chorley, 2008) sought to replace a descriptive, uni-dimensional approach to geomorphology with a quantitative/analytical strategy, attributed in part to the impact of World War II, especially in facilitating application of concepts derived from engineering and the physical sciences. For example, Horton (1945) and Strahler (1957) proposed drainage density as a strategy for quantitative assessment of watersheds and drainage systems, and more generally for introducing quantitative analysis in the earth sciences. Such measures, initially proposed in the context of map analysis, were later applied to aerial photography and other remotely sensed imagery (Lausch et al., 2020; Nie et al., 2018; Rudraiah et al., 2008). Thus, although the mid-century revolution began before the field of remote sensing was formalized, it set the context for uses of aerial imagery as a source of quantitative analysis.

Ray's (1960) *Aerial Photographs in Geologic Interpretation and Mapping*, a compilation of annotated aerial photographs illustrating landforms and geologic structures, in addition promoted aerial imagery as a source of quantitative information to support geomorphic and geological analysis. This work was later reinforced by Scheidegger's (1961) survey of quantitative strategies for geomorphological analysis.

Parallels between the advancement of remote sensing techniques, and in geomorphic knowledge, have emerged more recently. Increasingly sophisticated remote sensing data acquisition and analysis have extended our understanding of geomorphic processes and patterns at varied spatial and temporal scales (e.g., Walsh et al., 1998). For example, by the 1960s and 1970s, the availability of new remote sensing systems, such as imaging radars, and later Landsat imagery, provided the ability to examine geomorphic features at broader scales. Early applications of imaging radars focused on its ability to provide a synoptic view, with clear, crisp, and clearly defined representations of relief, drainage, open water, and coastlines (Simpson, 1966).

By 1972, the availability of Landsat imagery provided additional capabilities for broader-scale examination of geomorphic features (Short and Blair, 1986). Beginning in the 1980s and continuing into subsequent decades, Landsat's systematic multispectral coverage has provided a comprehensive sequential record of the earth's surface and changes to topography, vegetation, and land use.

Remote sensing technologies increased to include LiDAR and specialized applications of imaging radars, to be discussed later. In a geomorphological context, their value has been to provide data that extends imagery's capabilities beyond providing a basic map-like view of the terrain to providing metric data that record precise detail and changes in terrain elevation and motion. These innovations have moved the field of geomorphology beyond description and labeling of landforms to providing an improved understanding of process through modeling and forecasting.

Current geomorphological research has employed several forms of remotely sensed imagery, including aerial photography, satellite imagery, imaging radars, and LiDAR imagery. Aircraft and satellite systems contribute to geomorphological analysis through their ability to provide data with map-like views of the terrain surface and representing landscape features in relation to drainage, relief, and land-use patterns. LiDAR and imaging radars are *active systems* that illuminate terrain using their own energy, then compare transmitted and received signals to derive detailed information about terrain surfaces. Aerial photography collects imagery typically using panchromatic (a single channel in the visible spectrum) imagery, natural color (separate channels for the blue, green, and red primaries), and color infrared (using green, red, and near infrared channels). Such imagery often provides spatial detail at about 1 m or finer, but coarser detail if acquired higher altitudes. Aerial photography is widely used for a range of geomorphic topics but often is limited in function to providing the spatial context for field reconnaissance and planning fieldwork.

Satellite imagery provides systematic coverage using standardized formats and technical specifications. For present purposes we can think of satellite imagery as either (1) fine-scaled, detailed data (at a resolution of perhaps 1 m or less), with three to four spectral channels, (in the optical region of the spectrum), typically acquired by commercial organizations, or (2) broad-scale data with coarser levels of detail (perhaps ≥ 30 m).

Commercial satellite corporations collect imagery at a variety of resolutions and spectral regions; the most common specifications and formats resemble those of aerial photography—a section of panchromatic, natural color, and color infrared, with spatial detail at several meters, or at sub-meter detail for some systems (Kant et al., 2023; Conforti et al., 2021; Ahmed et al., 2017).

For regional coverage, land resource satellites provide coarse detail for broader regions using multispectral imagery (perhaps eight to 12 spectral channels) at spatial detail in the range of about 10–30 m). Such systems employ systematic coverage, consistent in format and coverage pattern. For example, there is special value in the ability of satellite imagery, especially Landsat imagery, to provide sequential coverage. Landsat has an image archive of many decades, and imagery with consistent calibration permits reliable change detection, including coastline changes, fluvial systems, and mass wasting events. Sequential imagery provides the ability to assess before/after changes for events, including landslides, hurricane beach erosion, or flooding to permit diagnostic analysis. Equally valuable is the capability provided by sequential imagery to examine incremental processes, such as coastal erosion, fluvial processes, and retreating periglacial landforms. In each instance, it is the ability of remote sensing to illuminate an understanding of geomorphic process that forms its value to geomorphic inquiry.

Two families of remote sensing systems that have special significance for geomorphic inquiry, LiDAR and synthetic aperture radars (SARs), are increasingly finding important roles in geomorphic inquiry. Although other remote sensing systems will continue to be valuable, we can expect LiDAR and SAR imagery to form the core of remote sensing analysis through their ability to acquire accurate and detailed terrain data, as outlined next.

In the 1980s, LiDAR imagery began to find a role in observing landscapes using instruments carried by aircraft to systematically observe a swath of terrain at fine levels of detail. LiDARs observe the earth's surface by transmitting pulses of light at wavelengths in or near the visible spectrum (approximately, within the interval 0.25 μm [250 nm] to 10 μm [10,000 nm], depending on the specific instrument) toward the earth, and recording the time delay of the returned radiation. Although basic LiDAR technology had been developed in earlier decades, by the mid-1980s it was integrated with navigation systems and GPS to form scanning systems that observe the earth at fine detail and high accuracy (Zangana et al., 2023; Ewertowski et al., 2019; Langat et al., 2019; Dikpal et al., 2017). Such systems can detect differences in time delays of returned pulses to separate those reflected from forest canopy and those that have penetrated the canopy to reach the terrain below. Thus, LiDAR offers geomorphologists the ability to observe the terrain surface free of vegetation cover at levels of detail and accuracy not previously available. Although LiDAR imagery is often thought of as a resource best suited to observe relatively small regions at fine spatial detail, we note that, in the United States, states such as Iowa and North Carolina have collected state-wide LiDAR coverage, and that North Carolina is now covered by sequential LiDAR coverage—a likely precursor to future developments that will routinely provide broader coverage of the fine-scale LiDAR data. Because of these capabilities, LiDAR is rapidly becoming one of geomorphology's principal research tools, a growth that will likely accelerate as acquisitions create larger archives that can be applied for sequential analyses of landscapes and landscape changes.

Likewise, synthetic aperture radar (SAR), developed over several decades beginning in the 1970s, plays a significant, and growing, role in role in geomorphologic, geophysical, and geologic inquiry. SAR, a specific form of imaging radar, collects imagery by broadcasting a beam of microwave energy (wavelengths in the range of approximately 1 cm to 1 m, as selected for specific systems), then recording the energy backscattered from the earth's surface to form images. As active systems, SAR can measure the time delay between transmission of the original signal and receipt of its echo, as well as differences in phase and polarization. These differences provide a basis for characterization of terrain surfaces to include roughness, moisture status, subsurface features, vegetation cover and structure, and drainage systems.

Special techniques based upon comparison of signal phase, known as SAR *interferometry* (InSAR), can assess changes in surface elevation using dual views of the same terrain from different tracks to acquire two views of the same region (Klees and Massonnet, 1999). (Figure 1.1). A related implementation, known as *along-track interferometry* examines two views illuminated at different times as the satellite or aircraft follows its trajectory along a single path. If the echo of this signal is generated by a moving object, phase differences between transmitted and received signals will reveal the velocity of the lateral motion (in ocean currents, volcanic flows, or displacement of glacial ice, for example). Another InSAR variation, *Persistent Scattering Interferometry (PSI)* (described later) can assess a collection of sequential interferograms to detect lateral movement of surface debris, which may signal increasing danger of mass wasting events.

InSAR, therefore, forms a valuable tool for assessing subsidence, displacement, and lateral motion of terrain surfaces associated with geomorphologic analysis. Reported accuracies of InSAR analyses vary with specifics of terrain, sensor systems (including wavelength), atmospheric conditions, and validation strategies, among others. The European Space Agency's TanDEM-X SAR system is specifically designed for terrain assessment at levels of detail and accuracies that support assessment of mass wasting hazards (Bamler et al., 2009; Zebker et al., 2010). (In general, tabulation of summary

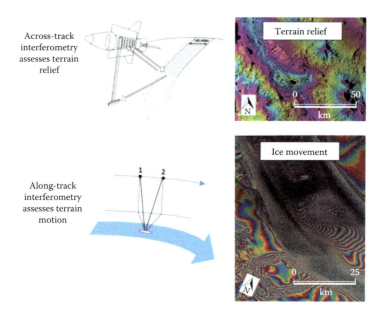

FIGURE 1.1 Schematic diagrams illustrating uses of SAR interferometry to illuminate the terrain from two separate views (shown here with the two antennas of the Shuttle Imaging Radar Topography Mission [SRTM]), or alternatively from two passes of the same instrument, known as *cross-track interferometry*, and analysis of phase differences in the returned signal to derive terrain elevation. *Along-track interferometry* broadcasts microwave signals from differing positions to illuminate a surface and then analyzes phase differences in the reflected signals to assess surface movement, such as glacial ice (shown here), ocean currents, or lava flows. The colored contours (known as *banded color*, or as *wrapped interferograms*) indicate elevation differences, terrain subsidence, uplift, or lateral motion, depending upon the application. Terrain relief image credit: USGS. Ice movement image credit: Jet Propulsion Laboratory, California Institute of Technology. SRTM diagram credit: J. B. Campbell and R. H. Wynne, *Introduction to Remote Sensing*, 2011, copyright Guilford Press. Reprinted with permission of The Guilford Press.

Geomorphological Studies from Remote Sensing

information regarding accuracies of geomorphic remote sensing applications represent so many different measures, methodologies, and variations in experimental design that the data are often not comparable.)

LiDAR and SAR therefore play important roles in current progress in geomorphological remote sensing for their ability to provide cogent geomorphological data and their inherent capability to provide metric data at high positional detail and accuracy. Their significance will increase in future decades, as analytical capabilities advance, numbers of systems increase, and as the scope of archives expand. Although these systems may be seen as displacing conventional imagery, it seems likely that future applications will use multiple forms of imagery together to exploit their synergistic capabilities (Rosario González-Moradas and Viveen, 2020; Bozzano et al., 2017; Darwish et al., 2017).

Despite active investigation of remote sensing applications to geomorphology, relatively few such applications are employed operationally, in routine usage, to meet societal needs. One such example of operational usage is North Carolina's application of LiDAR imagery for flood zone mapping. Such applications require an accurate understanding of the accuracy and precision in mapping flood plain topography, for its use in meeting societal needs in land-use planning and insurance underwriting. However, basic geomorphic research is usually conducted in a much different context, to investigate specific research objectives with findings that may be difficult to generalize across topics or disciplines.

Figure 1.2 illustrates relationships between spatial and temporal scales as they apply to uses of remotely sensed data for geomorphic analysis (Smith and Pain, 2009). This diagram, based upon

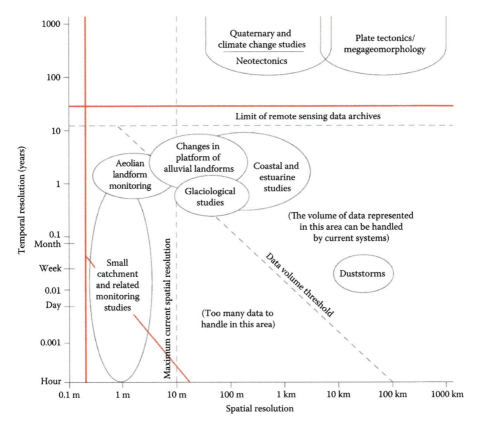

FIGURE 1.2 Schematic representation of relationships between spatial and temporal scales as they apply to uses of remotely sensed data for geomorphic analysis. Dashed lines represent thresholds as reported in 1987; red lines signify current thresholds, indicating expansion of the capabilities of archives, improved resolutions, and improved data handling capabilities. Credit: M. J. Smith and C. F. Paine. 2009. Applications of remote sensing in geomorphology. *Progress in Physical Geography*. Vol. 33, pp. 568–582. Used with permission.

a previous analysis by Millington and Townshend (1987), depicts a data space defined by spatial resolution (X-axis) and temporal resolution (Y-axis). The dashed lines illustrate the analytical space for selected geomorphic analyses, defined by the capabilities of archives and the limits of spatial detail, as available at the time of the 1987 study. The red lines represent current thresholds, which define a larger space created by improvements in spatial detail of sensor systems and the increased capabilities of archive systems and sensor systems and data processing capabilities (Smith and Pain, 2009). This increased analytical space offers opportunities for future remote sensing to contribute to geomorphic inquiry.

1.2 ALPINE AND POLAR PERIGLACIAL ENVIRONMENTS

In both alpine and polar periglacial environments, geomorphological processes are tied to seasonal or perennial freezing and thermal changes in the ground. Furthermore, such processes are influenced by climatic variation that may occur over decadal or millennial temporal scales (Zangana et al., 2023; Misra, 2022; Langat et al., 2019). Many processes in arctic permafrost regions have impacts of global relevance, such as the carbon cycle of wet tundra lowlands (Semiletov et al., 1996; Grosse et al., 2006) and their rapid responses to climate change (ACIA, 2004).

In periglacial geomorphology, field measurements have a long and rich history. Manual measurements prevailed prior to the 1980s (Humlum, 2008). Arctic periglacial geomorphology has been associated with the use of high-resolution aerial imagery (Cabot, 1947; Frost et al., 1966), but spatial and temporal coverage of such imagery is limiting. Furthermore, political (in addition to physical) remoteness prevails in regions of the Russian Arctic, where aerial imagery and detailed topographic maps are still classified or unavailable.

Adoption of remote sensing techniques in remote arctic and mountain environments has increased our ability to analyze locations that are difficult to access and lack detailed topographic data. Accessibility is often cited as an obstacle to obtaining understanding of slope instability processes on periglacial high-mountain landscapes (Fisher et al., 2011). Applications of remote sensing data for the monitoring of these remote and extensive areas are cost-effective and, therefore, a fast-growing research area in the field of periglacial geomorphology (Conforti et al., 2021; Chandler et al., 2018; Dikpal et al., 2017).

1.2.1 Overview

Periglacial geomorphologists often combine a historical perspective coupled with interests in contemporary processes and patterns (Zangana et al., 2023; Ewertowski et al., 2019; Bozzano et al., 2017). In recent years, remote sensing techniques have been employed to detect surface change (Grosse et al., 2005) and map and classify general permafrost properties (Lausch et al., 2020; Bozzano et al., 2017; Rajaveni et al., 2017; Morrissey et al., 1986; Leverington and Duguay, 1997; Plug et al., 2008) and distributions (Ulrich et al., 2009) (Figure 1.3). These techniques often combine data from multiple sensors to enhance the landscape view, analytical power, and temporal perspective. Kääb (2008) provides an excellent review of spaceborne, airborne, and ground-based remote sensing techniques applied in periglacial studies, especially as applied to periglacial hazards.

Degradation of permafrost is a widespread, striking, and important process interrelated with global change dynamics. The value of remote sensing for permafrost studies has been demonstrated (e.g., Morrissey et al., 1986) (Figure 1.3) Analysis of permafrost degradation has involved both identification of periglacial terrain and change detection, and in recent years has integrated remote sensing methods that optimize the spatial, temporal, and spectral resolution of different sensors, using them in combination. For example, Grosse et al. (2005) present an integrated use of optical remote sensing DEM- and GIS-based image stratification for the identification of periglacial terrain surfaces with a focus on thermokarst terrain.

Geomorphological Studies from Remote Sensing

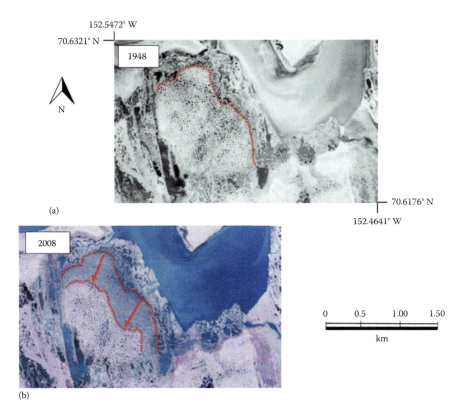

FIGURE 1.3 Changes in Patterned Ground, 1948–2008, Garry Creek, Alaska. Sequential aerial photography, 1948 (top) and 2008 (bottom), illustrating subsiding ice wedge polygonal ground near the mouth of Garry Creek, North Slope of Alaska. Arrows in the 2008 image indicate retreat of the polygonal ground during the interval 1948–2008. Field observations indicate that this and similar near-coastal areas of subsided tundra have been covered with brackish sediments and colonized by saline vegetation. Source: USGS (2011).

In periglacial environments, practical obstacles to acquiring airborne data have led to reliance upon satellite imagery, which provide economical coverage of remote regions. As a result, Landsat-7 ETM+ image, DEM data, and CORONA images offer options for coverage, and in some instances may be useful as reference data to supplement field data. Hjort and Luoto's (2006) integrative approach incorporating topographical, ground, and remote sensing in predictive geomorphological mapping used generalized additive modeling. Plug et al. (2008) used Landsat scenes from 1978–2001 to classify and determine changes in lake coverage on the Tuktoyaktuk Peninsula in northwestern Canada.

1.2.2 CHANGE ANALYSIS

Change in mountainous periglacial geomorphic features has also incorporated integrated remote sensing techniques (Ewertowski et al., 2019; Langat et al., 2019; Dikpal et al., 2017; Walsh et al., 2003; Fischer et al., 2011). For example, Fischer et al. (2011) developed a time series of high-resolution digital terrain models (DTMs) with a 2-m resolution from digital aerial photogrammetry for 1956, 1988, and 2001 and from airborne LiDAR for 2005 and 2007 to characterize topographic change in steep periglacial high-mountain faces in the Swiss Alps associated with increased rock and ice avalanche activity. Their study necessitated spatial resolution and accuracy assessment at a meter scale, and therefore used digital photogrammetry based on aerial images and combined LiDAR for multi-temporal topographic analysis of fine-scale change in periglacial processes. A time series of high-resolution DTMs from high-precision digital aerial photogrammetry and airborne LiDAR was developed to study topographic changes in steep periglacial rock walls in detail over a long time. Walsh et al. (2003) used remotely sensed imagery in concert with spatial analysis to investigate geomorphic elements hypothesized to be direct or indirect influences on alpine treeline in Glacier National Park. Relict solifluction terraces were investigated and mapped using data fusion techniques combining airborne ADAR data from 1999 with a spatial resolution of 50 cm–3 m and Landsat TM. In their study, ADAR data provided greater spatial resolution and Landsat greater spectral resolution (seven spectral channels). Walsh et al. (2003) suggested that broad-scale imagery, such as Landsat TM can be used to describe signatures of periglacial processes or patterns, whereas finer resolution data can be used to assess the nature of the process or pattern.

1.2.3 Emergent Opportunities and Challenges

Many remote sensing applications in periglacial geomorphology, despite availability of high-resolution data, remain focused on characterization (description) of periglacial features, and analytical research to date still remains largely focused on change detection analysis. Relatively little emphasis has been placed on analysis and prediction at this point (Muzirafuti, 2024; Torre et al., 2024; Conforti et al., 2021; Langat et al., 2019; Darwish et al., 2017). Furthermore, in dynamic environments, such as rapidly occurring processes and hazards, airborne and spaceborne remote sensing may offer support for assessing hazard susceptibility (Kääb, 2008).

However, the use of remote sensing imagery into studies of periglacial processes is likely to enable analysis of spatial patterns as a more fundamental part of analyses in the development, parameterization, and validation of spatially explicit models (Walsh et al., 2003). Humlum (2008) claims that a practical priority for periglacial geomorphology should be to improve understanding of climatic controls on geomorphic processes and to predict hazards and hazard risks in the context of climate change. Additionally, modeling landscape evolution in cold climates (French and Thorn, 2006) will likely be aided as they increasingly integrate RS into their development and validation.

1.3 GLACIAL GEOMORPHOLOGY

Glacial geomorphology is the study of geomorphological processes and resultant landforms and relief tied to glacier dynamics (Muzirafuti, 2024; Conforti et al., 2021; Langat et al., 2019; Chandler et al., 2018; Rajaveni et al., 2017). Glacial dynamics have global relevance through their rapid response to climate change and resultant geomorphic impacts. Mountain glaciers are especially sensitive to climate variation, thus are considered to be a high-priority climate indicator (Kääb et al., 2005). Glacial landforms are often studied as proxies for past glacial activities (Smith et al., 2006) (Figure 1.3).

1.3.1 Overview

Remote sensing techniques have supported analysis of glacial landforms and sediment in remote locations lacking detailed topographic data (Lausch et al., 2020; Chandler et al., 2018; Darwish et al., 2017). Mapping of previously glaciated terrain inherently involves application of a historical perspective to glacial landscapes (e.g., Humlum, 2000 for rock glaciers), given that glaciated landforms are landscape manifestations of past processes. Reconstruction of glacier dynamics through landforms originally involved contour mapping derived from field observations. Analysis using remotely sensed data, digital elevation models (DEMs; e.g., Clark and Meehan, 2001), and/or aerial photography, to a large extent, has replaced, or corroborates, this methodology (Smith et al., 2006).

The study of glaciated terrain from the air has its roots in, and still carries a strong tradition in, aerial photography. Aerial photography analysis remains a valuable methodology because it enables fine-resolution mapping of glacial landforms that may not be assessed with coarser scale data (Smith et al., 2006). Humlum (2000), for example, estimated weathering rates, debris volume, and Holocene rockwall retreat rates at 400 locations on Disko Island, Greenland, using a mixed-methods approach incorporating aerial photography.

Digital procedures to extract terrain models from aerial photography have enhanced the collection of topographic data. Use of digital elevation models to outline topography and to study surface and erosional processes is often used in conjunction with, or to corroborate other analyses. For example, Schiefer and Gilbert (2007) prepared high-resolution (submeter) DEMs from archived photography to map changes since the mid-1900s in proglacial terrain. Their study exemplifies a unique application of historical aerial photography for generation of historical DEMS that provide information on landscape response to environmental change.

Satellite imagery, such as low- to moderate-resolution Landsat and ASTER data, has broader aerial coverage and relatively low costs compared to aerial photography and has been used effectively to uncover course-scale patterns created by scale glacier processes (Smith et al., 2006). For example, Clark (1993) used Landsat images to unveil previously unsuspected large-scale pattern of streamlining within drift that is assumed to reflect former phases of ice flow. Use of Landsat, as opposed to aerial photography, was essential in documenting a previously undocumented pattern of an ice-molded landscape. They have been applied to document the form and topology of features such as glacial lineations (large-scale ice-molded landform assemblages), drumlins, glacial megaflutes, and previously undocumented ice-molded landforms. These features have such broad scales that they are not effectively examined using the narrower coverage of airborne sensors. In general, a pattern cannot be recognized as such unless a large enough area is viewed, to reveal its continuity or the repetition of its internal elements.

1.3.2 Emergent Opportunities and Challenges

Despite the availability of high-resolution data, remote sensing applications in glacial geomorphology still emphasize description of glacial features (Quesada-Román and Peralta-Reyes, 2023; Zangana et al., 2023; Ewertowski et al., 2019; Bozzano et al., 2017). Analytical research to date largely remains focused on analysis of change over time, with little emphasis on predictive modeling. However, with the recognition of the importance of surface erosion in mountain building (Raymo and Ruddiman, 1992; Pinter and Brandon, 2005), one promising application of remote sensing technologies for glacial geomorphologists is in mapping and assessing glacial erosion and feedback with climate as they relate to mountain building. Bishop and Shroder (2000) and Bishop et al. (2003) used satellite imagery and digital elevation data to examine erosion and depositional features at Nanga Parbat to identify geomorphic events and to better understand the role of surface processes in the denudation cascade. Furthermore, in dynamic glacial environments, airborne and spaceborne remote sensing may offer increased support for assessing hazard susceptibility, such as glacial lake outbursts (Huggel et al., 2002). For example, Liu et al. (2013) examined movement of rock glaciers in California's Sierra Nevada mountains, reporting local flow rates of 48 cm yr^{-1} (Figure 1.4), finding that their motion exhibited high seasonal variation and high year-to-year variation, indicating the significance of high frequency of observations over long time intervals to understand the dynamics of rock glaciers.

Important limitations to our understanding of mountains through remote sensing are imposed by the challenges associated with complex terrain and environmental factors such as atmosphere and land cover, which control irradiant and radiant flux. Landforms with forested cover may be represented inaccurately in locations with forest cover, even using LiDAR imagery (Smith et al., 2006). Reliable information depends upon reducing spectral variation caused by topography and land cover,

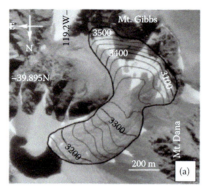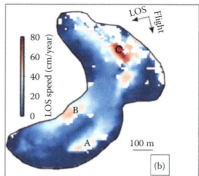

FIGURE 1.4 Rock glacier movement as assessed by InSAR. Liu et al. (2013) applied InSAR analysis to examine the movement of rock glaciers in California's Sierra Nevada mountains. A: Aerial photography, September 25, 1993, showing the Mt. Gibbs rock glacier outlined by the heavy black line, and 20 m contours as narrower black lines. B: InSAR analysis of rock glacier motion (along the LOS axis), with letters and red tones indicating zones of rapid motion (maximum = 48 cm/yr.). A & B: from Liu et al. (2013): used with permission. C: Aerial photography, September 14, 2013, accessed from Google Earth, June 23, 2014.

often by application of band ratios to reduce atmosphere and topographic effects. Such constraints lead to the development of new and innovative approaches in the combined area of remote sensing and geomorphometry, as shown in Bishop and Shroder (2004). Furthermore, more advanced forms of modeling may be needed to accurately apply remote sensing to investigate landform development.

1.4 MASS WASTING

Concisely defined, mass wasting is the bulk transport of regolith downslope, chiefly under the force of gravity, often facilitated by saturation and/or tectonic and weather events. Typical mass wasting processes include landslides, soil creep, debris flows, and solifluction. Contributing factors often include interactions between local slope, surface materials, geologic structure, hydrology, freeze-thaw, shrink-swell, and vegetative cover, which, in specific circumstances can assume dominant or causative roles. Mass wasting events can be incremental or episodic—in some cases, discrete events are preceded by long intervals of incremental processes that set the stage for events often observed as distinct, unexpected, events.

The study of mass wasting includes, among others, geotechnical engineering, hydrology, and geography. These disciplines all include, to varying degrees, impacts of including human behavior, especially prevailing and historic land-use practices, providing the historical context (e.g., previous events that signal the prevailing dynamics), as well as an understanding of local lithology and geological structure, climate, vegetation, and hydrology.

1.4.1 THE ROLE OF SYNTHETIC APERTURE RADAR IN LANDSLIDE ANALYSIS

Research has long used aerial photography, optical satellite imagery, and supporting topographic, geologic, and vegetation data to assess landslide and mass movement risk (Rosario González-Moradas and Viveen, 2020; Langat et al., 2019; Darwish et al., 2017). These analyses basically target risk by identifying the geomorphic context for increased hazard. However, in recent decades, development of analytical strategies using synthetic aperture radar (SAR) imagery increases opportunities to extract more direct information specifying the locations and nature of landslides and the risks they may present. Because these techniques offer the potential to extract immediate, site-specific information about landslide risk, they form an important resource for research addressing mass wasting processes.

As an active remote sensing system, SAR illuminates terrain with microwave energy of known wavelengths and polarizations, so as it receives backscatter, it can detect changes in wavelength, phase, and polarization. Analysis of phase change provides the basis for the detection of terrain displacement, and therefore the monitoring of subsidence and assessing hazard.

A variation on the basic InSAR strategy, known as *persistent scatterer interferometry* (or *permanent scatter interferometry*) (PSI), relies upon analysis of SAR pixels that maintain stable (known as "coherent") backscatter over a sequence of SAR images of the same region. A critical, and often difficult, step in implementation of PSI is identification of these stable pixels (i.e., those easily identifiable in sequential images) within a sequence of SAR images of the same region. Tantianuparp et al. (2013) and Del Ventisette et al. (2013) offer specifics.

Whereas basic InSAR extracts changes in vertical terrain displacement, PSI seeks to define horizontal displacement, which often forms a precursor to landslide events. Because SAR satellites observe terrain from an (almost) overhead perspective, assessment of vertical displacement is easier than assessment of horizontal displacement (Figure 1.5). Nonetheless, using both processing algorithms and ground-based networks of ground observations, projects are underway in Europe to monitor lateral motion to provide warnings of developing landslide hazards (e.g., Ghuffar et al., 2013). Because urban regions usually have larger and more reliable populations of stable backscatters, they are well-positioned to benefit from such systems. In contrast, for rural regions, the number of radar benchmarks can be low, limiting application of the PSI strategy.

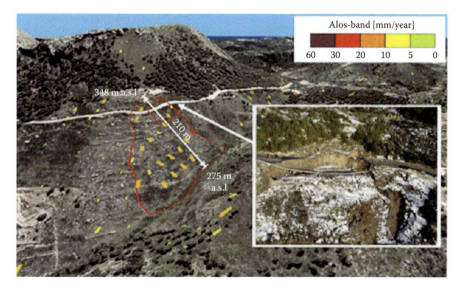

FIGURE 1.5 Mapped landslide in the Estellencs, Italy, area, showing terrain displacement as assessed by PSI (colored arrows), with an inset illustrating a subsequent rotational landslide event, March 2010. Credit: Bianchini et al. (2013). Used with permission.

1.4.2 Landslide Studies

> Landslides are like people: Every one is different. They occur in settings ranging from jungles to deserts. The materials involved range from mud to rock to ice, including mixtures of all three. Some slides are wet; others are dry. Some roar down steep mountainsides; others creep along at barely perceptible rates. Some are so rigid that they are truly "slides"; others, so fluid that they are best described as "flows." But the motion of landslides, small or large, is always governed by the conservation laws of physics. Kieffer (2014, p. 298).

As is clear from Kiefer's statement, landslides are among the most dangerous of geomorphic phenomena and among the most difficult to study. Remote sensing, in partnership with other research methods, provides an important resource for understanding landslides and their behavior. Figure 1.6 illustrates the value of the detail provided by LiDAR imagery to record the configuration of landslide events and permit the examination of changes. Scaioni (2013) identified several themes in applications of remote sensing to landslide studies: (1) inventories of past landslide events; (2) site-by-site examination and monitoring of potential mass wasting sites, largely using SAR imagery; and (3) application of ground-based lasers to examine active sites and monitor stability.

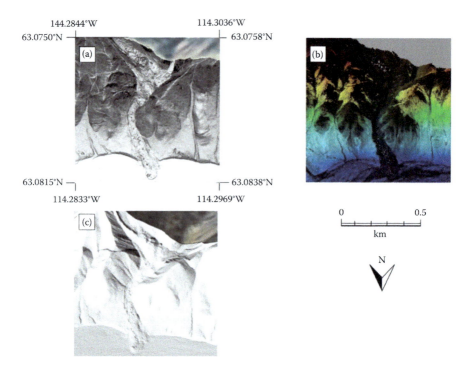

FIGURE 1.6 LiDAR image, landslide, Alaska Range along the Denali-Totschunda Fault System, Alaska. Pictured here are oblique views of a landslide adjacent to the Denali Fault. The last major rupture of this fault system was a magnitude 7.9 earthquake on November 3, 2002, which could have caused this landslide. Image A shows a slope-shade model, which darkens steeper slopes and illuminates gentle slopes. Notice that there are dark, round spots in the landslide patch, which are likely large boulders displaced by the event. In image B, the lineament of a fault is visible in the blue shading extending from higher (light blue) to lower (dark blue) elevations. Image C shows a hillshaded relief display of the same region. Image credit: Emily Kleber, OpenTopography. Dataset: EarthScope (2008). EarthScope Alaska Denali Totschunda LiDAR Project. Plate Boundary Observatory & National Center for Airborne Laser Mapping. Distributed by OpenTopography. http://dx.doi.org/10.5069/G9QN64NF, date accessed: February 25, 2014. Used with permission. Google Earth imagery accessed May 15, 2014.

1. Inventory: The inventory strategy seeks to apply geospatial data to identify areas of increased risk through convergence of contributing factors. Sankar and Kanungo (2004) developed integrated remote sensing and GIS strategy through their studies of a region of the Darjeeling Himalaya. They used IRS satellite data, field data, and topographic maps, integrated using GIS, to assess topographic, hydrologic, and geological conditions within their study area. Their numerical rating strategy, which defined four susceptibility classes, provided close agreement with existing field instability reports and with occurrence of observed landslides.

Strozzi et al. (2013) combined interpretations of aerial photographs and surface displacement derived from SAR interferometry to examine landslides and their intensities. Such approaches provide valuable contributions toward analysis of landslide hazards in areas where traditional monitoring techniques are sparse or unavailable.

2. Monitoring critical sites: Othman and Gloaguen (2013) examined satellite imagery of river channels at the Iraq/Iran borderlands to examine displacement of river channels by landslides that have blocked river channels, thereby diverting river channels. They quantified river offsets using two geomorphic indices that assess the displacement with respect to basin midline; and with respect to the river's principal orientation. Their analysis used these indices to assess intensities of landslides and to assess risks of future events.

Tofani et al. (2013) examined landslides in the Northern Apennines, Italy, to assess surficial displacement as a measure of surface behavior and its relationships with triggering factors. They applied PSI to estimate the yearly deformation velocity of the Santo Stefano d'Aveto landslide. Their study compared PSI results with data from *in situ* data to confirm the value of the PSI strategy.

Tantianuparp et al. (2013) applied InSAR and PSI to examine multiple SAR data sets of the Three Gorges area in China to study slow-moving landslides over long periods. Ghuffar et al. (2013) examined an active landslide in Doren, Austria, using multitemporal airborne and terrestrial laser scanning, 2003–2012. They applied 3D motion vectors for the time series to identify displacements up to 10 m, whereas other regions often did not change for several years. Akbarimehr et al. (2013) applied observations from both InSAR (2004–2006) and GPS (2010–2012) to assess slope stability of the Sarcheshmeh landslide in northeast Iran.

3. Applying ground-based laser scans to examine active sites, and monitor stability: At local scales, remote sensing can assess site-specific risks, warn of impending hazards, and provide diagnostic analysis of historic events. At broader perspectives, remote sensing can identify the significance of latent causative factors and links to contributing factors (e.g., biogeography, hydrology, geology, tectonics, terrain, and land use).

Band et al. (2012) linked landslides in mountainous terrain to forest regulation of hydrologic dynamics at the Coweeta Hydrological Laboratory, a US Forest Service experimental watershed in North Carolina. Their ecohydrological approach examined impacts of hydrologic and canopy patterns upon slope stability and development of landslide risk in steep forested catchments. LiDAR data provided the basis for deriving the flowpath structure, and the pattern of pore pressures, in relation to elevation, slope, and aspect. This strategy revealed ecological and hydrological dimensions of mass wasting processes at this site.

1.4.3 SUMMARY AND FUTURE OUTLOOK

Recent developments in remote sensing technologies and analytical strategies have advanced studies of mass wasting events from one that basically focuses upon descriptive before and after observations of landforms to one that can deploy remote sensing systems to observe dynamic processes in progress, both to examine geomorphic processes and to assess risk with temporal and spatial detail not previously feasible. Key technologies include LiDARs, InSAR, and PSI as mentioned earlier. Also significant is the application of these technologies with established image analytical strategies, including multi-temporal LiDAR and Object-Based Image Analysis (OBIA), to provide the detail and accuracy necessary for volumetric analyses.

Future applications will see increased applications of LiDAR, especially sequential LiDAR analyses, as the archive of repeat coverage increases as acquisitions increase. Research to apply remote sensing mass wasting will apply multi-source remote sensing; continued use to integrate remote sensing with other geospatial data, especially DEMs; increased use of in situ data, including in situ GPS; and applications of fine-resolution InSAR to study geomorphological dynamics and features at finer scales. We note also that some of these applications require systematic and sequential coverage at resolutions, revisit times, and spatial coverage not always routinely available using current systems.

1.5 FLUVIAL LANDFORMS

1.5.1 Overview

Fluvial landforms are formed by channelized flow within rivers and streams, as well as associated deposits and landforms, including terraces, sedimentary deposits, erosional features, stream beds, and river valleys that accompany channelized flow. Fluvial systems are characterized by their dynamic behavior (flooding, channel migration, alternate episodes of deposition and erosion) and their multi-scale character (processes that operate simultaneously at varied spatial and temporal scales). Temporal variation in stream systems combine with incremental processes, and those that act in an episodic manner. Although these characteristics present numerous obstacles for systematic inquiry, the field of remote sensing, with its capabilities for multi-temporal (sequential) and multi-scale imagery provides insight into many of these dimensions of fluvial systems that otherwise would be less well understood (Muzirafuti, 2024; Torre et al., 2024; Zangana et al., 2023; Langat et al., 2019).

Aerial imagery of fluvial landscape were among the earliest images used for geomorphic analysis—Lee (1922), for example, provides representations of complex landscapes that were previously difficult to depict from ground-level views. Likewise, early imaging radar imagery (Simpson, 1966) and, later, Landsat images (Short and Blair, 1986; El-Baz et al., 2000; Sarkar et al., 2012) provided synoptic views of landscapes to highlight regional variations in drainage and fluvial systems and in active geomorphological processes. (Figure 1.7 provides a more recent example.) Currently, LiDAR and sequential aerial photography form two of the most valuable resources for the study of fluvial systems. The detail and accuracy of LiDAR (Hohenthal et al., 2011) forms an especially valuable resource for study of fluvial systems and the length of the aerial photography archive, especially in the United States. The ability of SAR imagery, because of its all-weather capabilities and its ability to detect flooded areas, forms an important resource for understanding flood events and flood plain delineation, especially when cloud cover prevents observation using other sensors. For broad-scale inquiry, the length and consistency of the Landsat archive provides an invaluable resource for geomorphic inquiry. NAS (2007) provides a useful overview of floodplain mapping technologies. Overall, with respect to remote sensing of fluvial systems, Marcus and Fonstad (2010, p. 1868) note that the "recent proliferation of review papers on the topic is one indicator that this is a rapidly emerging field..., as are science agency reports advocating increased use of remote sensing to systematically monitor, manage, and understand river systems."

1.5.2 Floodplain Analysis

Defining and understanding the nature and dynamics of floodplain forms an important dimension of fluvial geomorphology, to include understanding sediment budgets, fluvial erosion, and deposition, with significant implications for land-use policy, hazardous waste, property values, and insurance policy. Remote sensing technologies offer capabilities to examine the fundamental characteristics of fluvial systems (e.g., Figure 1.7).

Geomorphological Studies from Remote Sensing 19

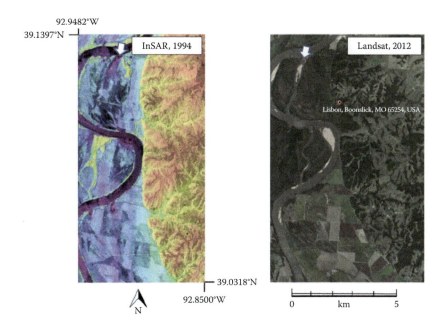

FIGURE 1.7 Missouri River floodplain, August 1994, following major flooding during the summer of 1993, observed using airborne InSAR (TOPOSAR). Here colors represent elevations, with the darkest blues indicating low elevations created by recent scour channels as flood waters cut incipient channels. The right-hand image depicts the same region as imaged by Landsat in 2012. The two arrows identify the location of the incipient scour channel identifiable in the 1994 image as a linear blue streak, but in 2012 present as a major diversion of the river flow. Courtesy of NASA/JPL-Caltech/ PIA01806.

Administrative floodplain mapping applied conventional methods based on analysis of topographic maps and aerial photography to define floodplain limits and to inventory floodplain land use. In recent decades, despite limitations in LiDAR's horizontal accuracy, its superior vertical accuracy, fine spatial detail (Chen, 2007), and ability to measure ground elevation in vegetated areas has demonstrated its advantages relative to conventional aerial survey (Sasaki et al., 2008). As a result, LiDAR data forms a principal source of topographic information for hydrologic/hydraulic analysis and floodplain delineation (Deshpande, 2013).

Analytical and research investigations often require broader spatial scope and a broader range of spectral observations, so a range of remote sensing technologies still form significant resources for floodplain analyses. For example, Ghoshal et al. (2010) examined a 100-year record of stream bank erosion in Yuba River, California, an interval that encompasses a history of hydraulic mining, 1853–1884. The study focused on the interval 1906–1999, recording a period of recovery from the disruption from mining, using historical maps, aerial photographs, digital orthophoto quarter quads (DOQQs), SONAR, and photogrammetrically derived topographic data (PDTD). Their analysis, using aerial photographs for the period 1947–2006, documented flooding, erosion, sedimentation, reworking and redistribution of sediments, the calculation of volumes, and reworking of sediment throughout the 100-year interval, 1906–2006. Their study reveals the long period of hydraulic adjustment to episodic sedimentation events—a long but declining reaction to a major disturbance.

Townsend and Foster (2002) employed synthetic aperture radar (SAR) imagery as part of a simulation of flood extent and duration for the Roanoke River floodplain (United States). Their results suggest that the intermediate sector of the hydrologic gradient has been compressed to emphasize extremes of either wetter or drier conditions. Their model represents a simple method to simulate hydroperiod regimes at landscape scales in situations where data necessary for more complex physical models are not available.

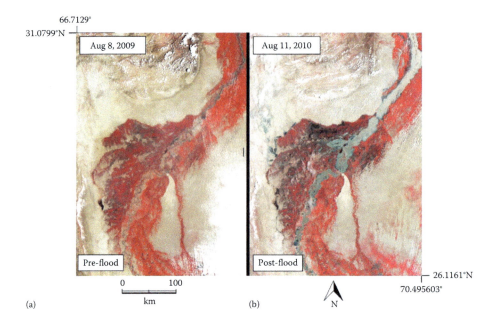

FIGURE 1.8 Flooding of the Indus River as recorded by sequential satellite imagery, Sindh Province, South Pakistan, August 2010 (left: August 8, right: August 11). The Indus River crosses the image from the upper right to the lower left. In late July 2010, heavy monsoon rains caused flooding in several regions of Pakistan, including the Khyber Pakhtunkhwa, Sindh, Punjab, and parts of Baluchistan. These false-color views display the near-infrared, red and green regions of the spectrum in a combination that highlights contrasts between water and vegetation on the riverbanks. Such imagery provides the capability to record the extent, duration, and impacts of such broad-scale flood events with precision and accuracy. Image credit: NASA/GSFC/LaRC/JPL, MISR Team. PIA13337.

1.5.3 Channel Migration

Channel migration describes the lateral movement of a river channel across the alluvial floodplain, through incremental erosion and deposition, and especially through *avulsion*—the sudden translocation of a river channel, typically in response to flood events. Understanding channel migration is a critical dimension of addressing geomorphologic and river management problems. Because of the large magnitudes and episodic and rapid rates of change, special surveillance systems are needed to efficiently measure and monitor channel migration. Because channel migration must be observed at rather broad scales, at infrequent intervals, remote sensing forms an important tool for recording and analyzing channel migration.

Yang et al. (1999) examined sequential Landsat imagery of the Yellow River Delta in China, encompassing an interval of approximately 19 years. They used GIS to map channel position and systematically examined changes of riverbanks and channel centerlines, to relate these to natural and human processes. Likewise, Bhaskar and Kumar (2011) examined satellite imagery to study geomorphic processes that control river channel migration in the Thengapatnam coastal tract bordering the Arabian Sea in the Kanyakumari District, Tamil Nadu, Southern India. Their strategies permitted examination of geomorphic units, recording positions of river channels and, over time, documenting migration of the Kuzhithura river channel in coordination with field data revealing entrenching of the channel, possibly relative uplift relative to mean sea level.

Sequential multispectral satellite imagery of alluvial rivers provides specifics of channel characteristics and positions over time and provides the geomorphic and land-use context for understanding the nature of changes and interrelationships with human occupation of the landscape of the Amazon floodplain (Leal et al., 2016).

1.5.4 STREAMBANK RETREAT

While considerable effort has been directed toward reducing erosion from agricultural and urban lands, the significance of streambank degradation has only recently been acknowledged. Studies have shown that sediment loss from streambank retreat (SR) can account for as much as 85% of watershed sediment yields and bank retreat rates as great as 1.5 m–1100 m/year have been documented (Simon et al., 2000). In addition to water-quality impairment, streambank retreat impacts floodplain residents, riparian ecosystems, bridges, and other structures (ASCE, 1998).

Streambank retreat typically occurs by a combination of subaerial erosion, fluvial erosion, and streambank failure, all related to local soil conditions, land use, streamflow regime, and drainage basin hydrology. Heeren et al. (2012) describe their procedure for evaluating their rapid geomorphic assessments (RGAs) for assessing streambank retreat using sequential aerial photography, 2003–2008.

Recent research has developed strategies for applying ground-based LiDAR. Resop and Hession (2010) surveyed an 11-m streambank at Stroubles Creek, Blacksburg, Virginia, six times over a two-year period using both conventional surveying with total station (TS), and terrestrial laser surveying (TLS). Wang et al. (2013) employed mobile laser scanning (MLS) to record high precision data (including grain size, sphericity, and orientation) of coarse sediment in a gravel bar of a sub-arctic stream, Utsjok River, northern Finland.

1.5.5 SUMMARY

Relative to other topics examined here, remote sensing applications to examine fluvial processes have employed a wide variety of technologies over a broad range of spatial scales and temporal scales (Ewertowski et al., 2019; Langat et al., 2019; Chandler et al. 2018; Nie et al., 2018; Ahmed et al., 2017). Because of the broad spatial and temporal scales required to monitor channel migration, satellite systems have often been employed to examine channel migration. At finer scales, LiDAR imagery provides the spatial detail, and the ability to record bare-earth terrain data necessary for modern floodplain maps. Streambank erosion research has employed sequential aerial photography and LiDAR imagery but has also used sequential ground-based LiDAR, with other instruments, to examine local changes in streambanks. It seems likely that LiDAR will continue to increase in its significance but that this mix of different technologies will continue to be important in fluvial studies.

1.6 COASTAL GEOMORPHOLOGY

1.6.1 OVERVIEW

Coastal geomorphology encompasses the dynamic interface between oceans and land surfaces—a zone subject to constant wave action that forms the principal driver among the numerous processes that erode, transport, and deposit sediments. Such processes are significant now and in the past, but they have special significance in the current context of rising sea levels and increases in populations inhabiting coastal zones (Figure 1.9).

Remote sensing's capabilities to support understanding of the dynamic interplay of processes at local and regional scales, and over time, forms an important resource for the geomorphic study of coastal environments (Zangana et al., 2023; Ewertowski et al., 2019; Bozzano et al., 2017). Remote sensing is especially valuable for the study of dynamic phenomena, such as dunes, coastal erosion, and deposition. Geomorphic dimensions of current concerns, such as changing sea levels, protection of coastlines from wave action, formation of land-use policies in the littoral zone, and nearshore bathymetry, all can benefit from applications of remote-sensed imagery. Allen and Wang (2010) provide a review of remote sensing contributions to studies of coastal regions.

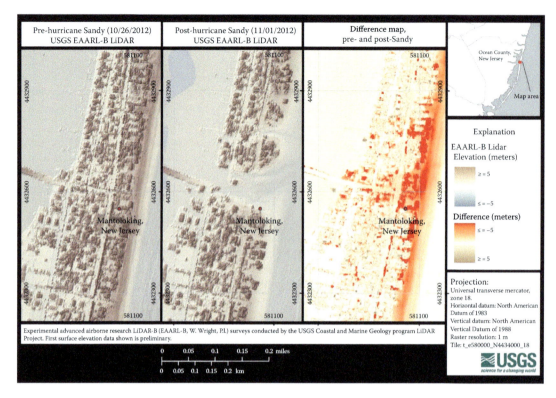

FIGURE 1.9 Before and after LiDAR surveys of coastal elevations at Mantoloking, New Jersey, documenting impacts of Hurricane Sandy (left, October 26, and center, November 1, 2012). At right, the difference map shows, in red, losses in elevation of as much as 5 m (16 ft). Sequential imagery with the high levels of precision and detail of LiDAR form valuable geomorphological information in dynamic coastal environments. Figure by Cindy Thatcher, USGS. Credit: US Geological Survey.

1.6.2 Multispectral Bathymetry

Water depth and subsurface topography are important dimensions of coastal geomorphology, especially in monitoring changes in coastal processes and landforms. Multispectral bathymetry techniques examine brightnesses in several spectral channels to assess differences in brightness against a reference brightness, in the same channels, recorded in a region known to have deep water (e.g., deeper than 50 m), assuming a clear, calm, water body. Thus, for example, values of the brightnesses in the blue and green regions of the spectrum are observed to increase in shallower water due to the scattering of solar radiation from the subsurface terrain. From such observations, water depth can be estimated from reference data that records maximum penetration of the several spectral channels (again, assuming a clear, calm, water surface). As the number of spectral channels available increases, and radiometric resolution increases, depth estimates can attain greater detail and reliability. For example, Gao (2011) derived bathymetric information over clear waters at depths up to 70 m. Gao's (2011) review of remote sensing–based bathymetric methods reports that detectable depths are usually limited to about 20 m, with accuracy decreasing with increasing depth, especially at depths below 12 m. Subsurface topography includes the same irregularities in surface materials, configuration, and vegetation cover, observed in terrestrial remote sensing, which all further contribute to inaccuracies.

1.6.3 Airborne LiDAR Bathymetry

Airborne LiDAR Bathymetry (ALB) uses LiDAR technology to assess water depth directly by illuminating the water surface with a LiDAR beam (Irish, 2000). As each LiDAR pulse reaches the air/water interface, a portion of the transmitted energy is returned from the surface back to the sensor (surface return). The remaining portion of the energy propagates through the water column, to be reflected from the subsurface topography (bottom return), assuming water depth is shallow enough for the pulse to reach the bottom; the time delay between the two returns permits assessment of water depth. An important capability for bathymetric LiDARs is the ability to use a single instrument to collect, in a single pass, high-resolution topographic data, as well as reliable water surface elevation, a capability especially useful at the interface between land and shallow water. Bathymetric LiDAR instruments specifically tailored for bathymetric surveys include green bands designed for penetration of the water body to permit subsurface mapping at varied depths.

1.6.4 SUSPENDED SEDIMENT CONCENTRATION (SSC)

Remote sensing has been applied to observe sediment transport, provide information for sediment budgets, monitor sediment sources, and identify sources of pollution. Multispectral imagery, and in some instances panchromatic imagery, provides the ability to observe distinctions between clear and turbid water. For example, Brakel (1984) examined Landsat MSS data to track sediment plumes from rivers discharging into the Indian Ocean. They found seasonal patterns related to seasonal rain events that carry sediment to the coastline, alternately northward and southward along the coastline, as monsoon winds and currents change seasonally.

Remote sensing has also been applied to derive quantitative estimates of SSC using both airborne and spaceborne sensors. Such surveys offer synoptic overviews of large water bodies that can, in principle, be coordinated with simultaneous, or near-simultaneous, on-site data collection. These methods exploit the positive associations between spectral radiance measured at the instrument and SSC as assessed on-site, usually by boat survey.

Curran and Novo (1988) reviewed the ability of multispectral observation to estimate SSC. Their review highlights difficulties of defining robust relationships between remote sensing observations and on-site SSC because of effects of atmosphere, solar angle, sensor altitude, and wave height. They conclude that future research should focus on improved field sampling of SSC and compensation of environmental influences upon observed radiances.

1.6.5 COASTAL CHANGE AND RETREAT

The ability of sequential imagery to record changes in positions and configurations of coastal landforms provide the opportunity to observe and analyze changes at varied scales (Figure 1.10). Moore and Griggs (2002) applied aerial photography, GIS, and soft-copy photogrammetry to examine long-term coastal cliff retreat and erosion (1953–1994), estimating an average rate of retreat of 7–15 cm/year, but identified episodic hot spots with rates as high as 20–63 cm/yr. Gibson et al. (2009) examined sequential aerial photography of Gulf Shores, Alabama, from 1955 to 2006. At coarser scales, they only identified small coastline changes, but at finer scales, their analysis revealed a variety of effects arising from interactions between residential development, hurricane overwash fans, and artificially engineered tidal inlets.

Several investigators have devised image analysis strategies to track shoreline changes. Shu et al. (2010) devised thresholding algorithms for semi-automated analyses of RADARSAT-2 for shoreline extraction. Marghany et al. (2010) examined multitemporal SAR (ERS-1 and RADARSAT-1) to examine spit migration along the Malaysian coastline, applying filters to define the spit and to track migration of the spit boundary, 1993–2003. Their analysis concluded that their filtering strategy was able to accurately track accretionary processes over this interval. Brock et al. (2002) described applications of LiDAR surveys to examine bare earth topography, vegetation structure, coastal dunes, barrier islands, shoreline change, landslide coastal cliffs, subsidence, storm surge, and tsunami inundation.

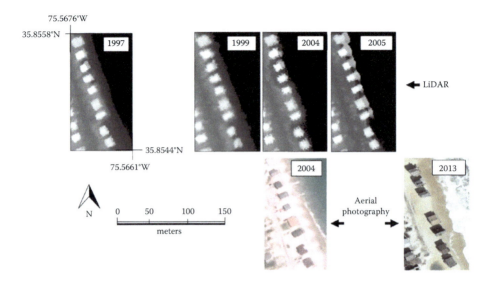

FIGURE 1.10 Coastal erosion as recorded by sequential LiDAR, North Carolina Coast, 1997–2005. Top: Sequential LiDAR imagery, 1997–2005. Here black represents open water, the white rectangular shapes are beachfront homes, and the gray background represents the sand surfaces of the beach. Progress of coastal retreat is revealed by the advance of the coastline toward the residential structures, 1997–2005. Below, see aerial photography of the same region, with the 2013 image depicting the loss of one of the structures visible in the earlier imagery. LiDAR point data collected 1997–2005 by NASA/NOAA coastal mapping programs, hosted by the Coastal Service Center (CSC) of NOAA, Charleston, South Carolina, and processed at East Carolina University. Ground reference data were collected between 2006 and 2008. See White and Wang (2003) and Allen and Wang (2010). Used with permission. Google Earth imagery accessed May 15, 2014.

Kawakuboa et al. (2011) examined TM data of coastlines in southeastern Brazil, applying spectral linear mixing and contextual classification using a segmentation technique, resolved differences between vegetated zones, clear water (a proxy for shade), and soil. Comparison of these classification results revealed a transformation process active in coastal environments and that erosive and depositional features are highly dynamic over short periods of time.

1.7 AEOLIAN LANDFORMS

Aeolian landforms are formed by the action of wind, either by erosion or the deposition of sand, silt, or clay-sized particles. Erosional features include lag deposits formed by selective erosion of finer particles by wind, leaving behind coarser sediments as distinctive geomorphic surfaces, and deflation hollows (blowouts). Such features are typically formed in arid or semi-arid regions where local vegetation cover is disturbed, allowing wind to remove finer particles to create shallow depressions. Here our discussion focuses chiefly upon depositional forms, including both coastal and desert landscapes. Aeolian entrainment occurs by rolling or sliding (*creep*), the Bernoulli Effect of winds (*lift*), bouncing (*saltation*), and the impact of one particle upon another. Entrained particles ultimately form dunes, notable for their distinctive forms and their mobility. Dunes are commonly associated with sand, although notable dunes have formed from gypsum (White Sands National Park, USA) and loess deposits throughout the world from silt-sized sediments.

Dune formation depends upon nearby sources of sediment, such as playas, washes, and fans—some dunes are active and others are relict from previous episodes when conditions favored dune formation. Dunes can be stabilized by natural vegetation, perhaps associated with changes in local climate, or by stabilization programs managed to prevent or minimize dune movement, especially in

Geomorphological Studies from Remote Sensing

coastal zones. Because of the dynamic character of active dunes, and their potential threats to agriculture in some regions and their role in protecting coastal zones, remote sensing has become a valuable tool for recording the movement of dunes and assessing their volume and changes in volume.

Livingston et al. (2007) report that "in the past many of the single-dune studies described a largely inductive approach: that is, they collected field data, generally about wind flow and sand flux, and then tried to make sense of those data" (p. 254). Remote sensing imagery, especially LiDAR imagery, provides the ability to support analysis of volumetric data and, in sequential studies, to examine dune movement, provides a new paradigm opening new arenas for inquiry (Figure 1.11).

Aerial photography has long been a resource for assessing dunes and their movement (Brown and Arbogast, 1999; Mitasova et al., 2005). Other technologies have provided the synoptic view to understand dune systems in their broader geologic and ecological contexts. Blumberg (2006) examined dunes using Shuttle Radar Topography Mission (SRTM), and Short and Blair (1986) discuss applications of Landsat imagery to the study of dune fields. Al-Masrahy and Mountney (2013) examined the broad scale context for dune fields of Saudi Arabia. However, at fine scales, the detail and accuracy of LiDAR imagery establish it as the premier sensor system for remote sensing of aeolian landforms.

Hugenholtz et al. (2012) reviewed applications of remote sensing to examine aeolian dunes, focusing on: (1) dune activity, (2) dune patterns and hierarchies, and (3) extra-terrestrial dunes. LiDAR's

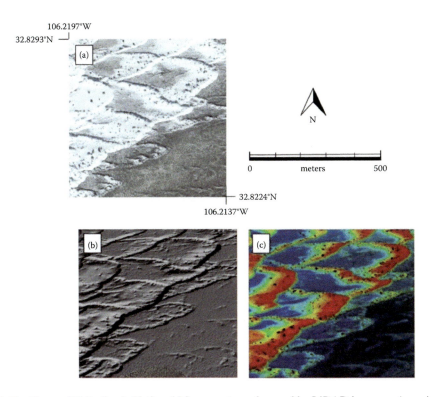

FIGURE 1.11 Dunes, White Sands National Monument, as observed by LiDAR imagery. A: aerial photography (Google Earth) of the crescent-shaped gypsum barchan dunes formed by SW-NE winds. B: the same region as represented by LiDAR imagery, shown as a hillshaded relief display; C: color-coded relief display of LiDAR data (red represents the highest elevations). LiDAR's detail and accuracy permit calculation of sand volume and, in the case of sequential imagery, measurement of rates of movement. Image credit: Emily Kleber, OpenTopography. Dataset: Ewing, R. (2010). White Sands National Monument, NM: LiDAR Survey of Dune Fields. National Center for Airborne Laser Mapping. Distributed by OpenTopography. Google Earth imagery accessed May 2014. http://dx.doi.org/10.5069/G97D2S2D date accessed: February 25, 2014. Used with permission. Google Earth imagery accessed May 15, 2014.

ability to assess dune volume has been studied under varied circumstances (Woolard and Colby, 2002; Grohmann and Sawakuchi, 2013): integration of RS data with field-based measurements of vegetation cover, structure, and aeolian transport rate in order to develop predictive models of dune field activity and expanding observational evidence of dune evolution at temporal and spatial scales that can support validation and refinement of simulation models.

1.8 BIOGEOMORPHOLOGY

Biogeomorphology is the subfield of geomorphology that explicitly recognizes interdependencies between ecologic and geomorphic systems (Viles, 2020; Stallins, 2006). Different methodological and theoretical approaches employed by both biogeographers and geomorphologists can confound formation of unified theories in biogeomorphic research (Fonstad, 2006); however, remote sensing may make important contributions in this area. Through synthesis at different scales, analysis of remotely sensed imagery may activate scale-independent perspectives of biogeomorphic analyses to integrate the process-centered focus of geomorphology with the pattern-centered emphasis of biogeography. Furthermore, remote sensing analysis may contribute to a fundamental goal in biogeomorphic research to determine the types and geographic extent of landscape patterns that emerge from interactions among vegetation dynamics and geomorphic processes (Phillips, 1999).

Biogeomorphological inquiries that may profit from remote sensing analysis include fluvial geomorphology and hydrology, as well as corresponding vegetation cover and/or land use and land cover change (e.g., Mertes et al., 1995; Bryant and Gilvear, 1999; Hudson et al., 2006). Mertes et al. (1995) used Landsat TM data to investigate relationships between hydrogeomorphology and vegetation communities in the Amazon. Using Landsat images and semivariogram analysis, they identified important constraints on the transfer and storage of water, sediment, and other materials during high flood levels, as revealed on three Landsat images. Similar questions have also been examined through the use of SAR (e.g., Martinez and Le Toan, 2007).

Characterization of pattern gives important clues to process, especially if examining patterns over time. Archival aerial photography provides the best temporal coverage for such analyses and are still applied in innovative ways to examine pattern-process relationships over time. For example, Meitzen (2009) measured lateral channel migrations and their effects on structure and function of the riparian forests of the Congaree River, South Carolina, using aerial photography (1938–2006) to measure migration rates. She found that lateral migration of channels produced a directional control on riparian forests, with forest types and successional stage dependent upon controls related to frequency of floods, elevation, and migration rates. They note that recent advances in geospatial technologies have provided "efficient and accurate" methods for examining such dynamics using temporally sequenced aerial photographs.

Innovative uses of remote sensing technologies have also been applied to biogeographic studies in mountain environments at multiple spatial scales (e.g., Butler et al., 2003; Crowley et al., 2003; Walsh et al., 2003). Walsh et al. (2003), for example, studied how biogeographic and geomorphic processes are constrained at alpine treelines by geomorphic features. They used a combined GISc approach that incorporated multi-resolution remote sensing systems, geospatial information, and treeline simulations. Data fusion linking ADAR and Landsat TM data enhanced their visualizations and analytical power by linking data sets with "different biological sensitivities." ADAR provided greater spatial resolution and thus could discern smaller objects and fine-scale landscape pattern. Landscape TM corroborates the ADAR data by providing greater spectral resolution. Using this integrated strategy, they discerned previously undetected vegetation patterns related to fine-scale periglacial features.

Emerging areas include development of simulation models for uncovering feedback between process and pattern. Fonstad (2006), for example, argues that cellular automata models that use remotely sensed data in raster grid formats are highly flexible, enabling them to link ecology and geomorphology that involve competing schema and different process rates of change. Such an approach enables detection of feedback and emergent phenomena. Geodiversity mapping—an emerging area of research that examines the close relationship between geomorphic and topographic diversity with biological diversity—also forms a promising and emerging research area where remote sensing is likely to make important contributions (e.g., Hjort and Luoto, 2012).

1.9 REMOTE SENSING AND GEOMORPHIC INQUIRY

Remote sensing's contributions to the field of geomorphology may be perceived to result primarily from enhancement of data acquisition capabilities with advances in imaging technologies (Muzirafuti, 2024; Torre et al., 2024; Hasan et al., 2023; Kant et al., 2023; Quesada-Román and Peralta-Reyes, 2023; Zangana et al., 2023; Misra, 2022; Conforti et al., 2021; Lausch et al., 2020; Rosario González-Moradas and Viveen, 2020; Ewertowski et al., 2019; Langat et al., 2019; Chandler et al. 2018; Nie et al., 2018; Ahmed et al., 2017; Bozzano et al., 2017; Darwish et al., 2017; Dikpal et al., 2017; Rajaveni et al., 2017). However, our examples here clearly show that these imaging capabilities have contributed to the development of the discipline and its epistemology in profound ways. Geomorphic knowledge has been expanded to encompass increasingly broader ranges of landscapes by enabling extraction of information with detail and precision not previously feasible. Examples mentioned previously include LiDAR imagery and interferometric SAR, which each convey data of a nature, quality, and precision that were beyond any reasonable expectations just a few decades ago. Applications of these newly emerging technologies contribute to an increasing focus on the role of temporal and spatial change within geomorphic inquiry and to advance the role of analytical strategies in relation to descriptive methodologies.

However, these capabilities form only a part of the broader contributions of remote sensing to studies of geomorphology, which also include:

1. Archives and data resources. Remotely sensed data, as well as related data in cognate fields, is often available at low cost, in convenient formats, with minimal administrative overhead.
2. Ancillary benefits of such archives include access to sequential data, available often with the quality, continuity, and consistency to provide the ability to examine changes over time with the confidence to know that observed changes indicate changes within the landscape rather than changes in the sensor systems.
3. Capabilities within the broader field of remote sensing have been applied to good effect to support research in geomorphology. For example, Band et al. (2012) applied the NDVI ratio, originally developed in agriculture and ecology, to examine forest canopies to evaluate hydrologic processes and their contributions to mass wasting.
4. Applications of remote sensing open a gateway into the broader realm of geospatial data. Figure 1.12 represents (as a schematic approximation) each of the three principal components of geospatial data (GIS, remote sensing, and GIS), initially developed within their own realms largely isolated from external interactions, as they have increasingly interacted with each other to form an integrated system. (In Figure 1.12, status as an independent technology is represented by the apexes of the triangle.) Remote sensing is devoted to the acquisition of the informational (geomorphic, in this context) and temporal dimensions of an analysis, whereas GPS provides real-time positional data and GIS contributes a locational framework to permit analysis in a common spatial context. Thus, much of the power of remote sensing derives from the ability of these technologies to integrate informational, temporal, positional, and contextual dimensions of a geomorphic problem into a common framework.
5. Development of new subfields of geomorphology that rely primarily on remotely sensed data, such as the emerging field of planetary geomorphology. In the last decade, improvements in spatial, spectral, and temporal resolution of remotely sensed data has allowed

for the identification of a range of active and relict geomorphic processes, many of which are similar to those on earth, including impact cratering, volcanism, aeolian, fluvial, mass wasting, rock breakdown, periglacial, coastal, and meteoritic gardening. This has sparked renewed activity within this subfield, as evidenced through new publications (e.g., Melosh, 2011; Greeley, 2013) and the emergence of professional groups. Earth field analogs are often used to better understand geomorphic processes on other planets and new ways to examine equifinality (e.g., Burr et al., 2009).
6. Initially, such technologies develop in their own realms, largely isolated from each other. As they mature, these technologies interact with each other through improved design, data formats, and instrumentation. For example, the capabilities of LiDAR systems reside in LiDAR's ability to integrate its sensor systems with the real-time GPS data and orientation data from the aircraft's navigation system.

Figure 1.12 illustrates transitions of geospatial systems as they evolve from an initial narrowly defined technological focus (represented by 1 in Figure 1.12) toward a broader integration with cognate technologies (2 and 3 in Figure 1.12). Such integration and interactions exploit mutual synergies represented by positions occupying the central regions of the diagram (2 and 3) distant from the apex positions at the periphery of the diagram.

Likewise, we can envision the applications of remote sensing to geomorphic inquiry within this diagram, some following a trajectory from the edges toward more central positions. Geomorphic applications of imaging radars (SAR), considered over the decades, can be seen to have made this transition as it changed from use as a static image to a multi-temporal resource using InSAR and PSI capabilities that provide quantitative data at fine spatial and temporal detail (Figure 1.5). Likewise, LiDAR systems, operating near the central portion of Figure 1.12, acquire accurate site-specific data, which when deployed to acquire sequential imagery, provide precise information that documents geomorphic processes at a detail not previously feasible.

As such technologies find their niches within geographic inquiry, they offer the opportunity to expand and strengthen our methodologies and increase the caliber and robustness of its contributions to scientific inquiry, to informing the public of geomorphic and environmental hazards, and to addressing current and emerging societal problems.

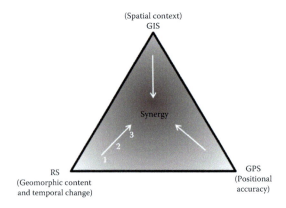

FIGURE 1.12 Schematic diagram depicting interrelationships between the three key geospatial technologies, illustrating evolution over time from independent, isolated, development toward mutually supportive, synergistic contributions. The three dimensions are intended to approximate the contributions of remote sensing, GPS, and GIS, in providing thematic content, precise location, and spatial context, respectively, to the geospatial enterprise. The arrows, and the numbers, indicate progressive degrees of interactions among the three dimensions. Within this framework, we can visualize how geomorphic analysis using these technologies has migrated from the periphery of this diagram toward the central zone of multifaceted applications of geospatial technologies.

REFERENCES

ACIA. 2004. Impacts of a Warming Arctic: Arctic Climate Impact Assessment. Cambridge: Cambridge University Press.

Ahmed, A. A., M. Abdelkareem, A. M. Asran, et al. 2017. Geomorphic and lithologic characteristics of Wadi Feiran basin, southern Sinai, Egypt, using remote sensing and field investigations. J Earth Syst Sci 126: 85. https://doi.org/10.1007/s12040-017-0861-8

Akbarimehr, M., M. Motagh, and M. Haghshenas-Haghighi. 2013. Slope stability assessment of the Sarcheshmeh Landslide, Northeast Iran, investigated using InSAR and GPS observations. Remote Sens 5: 3681–3700.

Allen, T. R., and Y. Wang. 2010. Selected scientific analyses and practical applications of remote sensing: Examples from the coast. In: Manual of Geospatial Science and Technology, 2nd ed., eds. J. D. Bossler, J. B. Campbell, R. B. McMaster, and C. Rizos. London: Taylor & Francis: 467–485.

Al-Masrahy, M. A., and N. P. Mountney. 2013. Remote sensing of spatial variability in aeolian dune and interdune morphology in the Rub' Al-Khali, Saudi Arabia. Aeolian Research 1: 155–170. https://doi.org/10.1016/j.aeolia.2013.06.004

Bamler, R., M. Eineder, N. Adam, X. Zhu, and S. Gerhardt. 2009. Interferometric potential of high-resolution spaceborne SAR. Photogrammtrie, Fernerkundung, Geoinformation 5: 407–419.

Band, L. E., T. Hwang, T. C. Hales, J. Vose, and C. Ford. 2012. Ecosystem processes at the watershed scale: Mapping and ecohydrological control of landslides. Geomorphology 137: 159–167.

Bhaskar, A. S., and R. B. Kumar. 2011. Remote sensing of coastal geomorphology to understand river migration in the Thengapatnam area, southern India. Intl J Remote Sens 32: 5287–5301.

Bianchini, S., G. Herrera, R. M. Mateos, D. Notti, I. Garcia, O. Mora, and S. Moretti. 2013. Landslide activity maps generation by means of persistent scatterer interferometry. Remote Sens 5: 6198–6222. (www.mdpi.com/2072-4292/5/12/6198)

Bishop, M. P., and J. F. Shroder. 2000. Remote sensing and geomorphometric assessment of topographic complexity and erosion dynamics in the Nanga Parbat massif. Geol Soc London Spec Publ 170: 181–200.

Bishop, M. P., and J. F. Shroder, Jr., eds. 2004. Geographic Information Science and Mountain Geomorphology. Berlin: Springer-Praxis.

Bishop, M. P., J. F. Shroder, Jr., and J. D. Colby. 2003. Remote sensing and geomorphometry for studying relief production in high mountains. Geomorphology 55: 345–361.

Blumberg, D. G. 2006. Analysis of large aeolian (wind-blown) bedforms using the Shuttle Radar Topography Mission (SRTM) digital elevation data. Remote Sens Environ 100: 179–189. https://doi.org/10.1016/j.rse.2005.10.011.

Bozzano, F., P. Mazzanti, D. Perissin, A. Rocca, P. De Pari, and M. E. Discenza. 2017. Basin scale assessment of landslides geomorphological setting by advanced InSAR analysis. Remote Sens 9(3): 267. https://doi.org/10.3390/rs9030267

Brakel, W. H. 1984. Seasonal dynamics of suspended-sediment plumes from the Tana and the Sabaki Rivers, Kenya: Analysis of Landsat data. Remote Sens Environ 165–173.

Brock, J. C., C. W. Wright, A. H. Salenger, and W. N. Swift. 2002. Basis and methods of NASA Airborne Topographic Mapper LiDAR surveys for coastal studies. J Coastal Res 18: 1–13.

Brown, D. G., and A. F. Arbogast. 1999. Digital photogrammetric change analysis as applied to active coastal dunes in Michigan. Photogramm Eng Remote Sens 65: 467–474.

Bryant, R. G., and D. J. Gilvear. 1999. Quantifying geomorphic and riparian land cover changes either side of a large flood event using airborne remote sensing: River Tay, Scotland. Geomorphology 29: 307–321.

Burr, D. M., K. L. Tanaka, and K. Yoshikawa. 2009. Pingos on Earth and Mars. Planet Space Sci 57: 541–555.

Butler, D. R., G. P. Malanson, M. F. Bekker, and L. M. Resler. 2003. Lithologic, structural, and geomorphic controls on ribbon forest patterns in a glaciated mountain environment. Geomorphology 55: 203–217.

Cabot, E. C. 1947. The northern Alaskan coastal plain interpreted from aerial photographs. Geogr Rev 639–648.

Chandler, B. M. P., H. Lovell, C. M. Boston, S. Lukas, L. D. Barr, I. O. Benediktsson, D. I. Benn, C. D. Clark, C. M. Darvill, D. J. A. Evans, M. W. Ewertowski, D. Loibl, M. Margold, J. Otto, D. H. Roberts, C. R. Stokes, R. D. Storrar, and A. P. Stroeven. 2018. Glacial geomorphological mapping: A review of approaches and frameworks for best practice. Earth Sci Rev 185: 806–846, ISSN 0012-8252, https://doi.org/10.1016/j.earscirev.2018.07.015. (www.sciencedirect.com/science/article/pii/S0012825217305986)

Chen, Q. 2007. Airborne LiDAR data processing and information extraction. Photogramm Eng Remote Sens 73: 109–112.

Chorley, R. J. 2008. The mid-century revolution in fluvial geomorphology: Chapter 19. In: The History of the Study of Landforms, eds. T. P. Burt, R. J. Chorley, D. Brunsden, and A. S. Goudie. London: The Geological Society: 925–960.

Clark, C. D. 1993. Mega-scale glacial lineations and cross-cutting ice-flow landforms. Earth Surf Proc Land 18: 1–29.

Clark, C. D., and R. T. Meehan. 2001. Subglacial bedform geomorphology of the Irish Ice Sheet reveals major configuration changes during growth and decay. J Quatern Sci 16: 483–496.

Conforti, M., M. Mercuri, and L. Borrelli. 2021. Morphological changes detection of a large earthflow using archived images, LiDAR-derived DTM, and UAV-based remote sensing. Remote Sens 13(1): 120. https://doi.org/10.3390/rs13010120

Crowley, J. K., B. E. Hubbard, and J. C. Mars. 2003. Analysis of potential debris flow source areas on Mount Shasta, California, by using airborne and satellite remote sensing data. Remote Sens Environ 87: 345–358.

Curran, P. J., and E. M. M. Novo. 1988. The relationship between suspended sediment concentration and remotely sensed spectral radiance: A review. J Coast Res 4: 351–368.

Darwish, K., S. E. Smith, M. Torab, H. Monsef, and O. Hussein. 2017. Geomorphological changes along the Nile Delta coastline between 1945 and 2015 detected using satellite remote sensing and GIS. https://doi.org/10.2112/JCOASTRES-D-16-00056.1

Del Ventisette, C., A. Ciampalini, M. Manunta, F. Calò, L. Paglia, F. Ardizzone, A. C. Mondini, P. Reichenbach, R. M. Mateos, S. Bianchini, I. Garcia, B. Füsi, Z. V. Deák, K. Rádi, M. Graniczny, Z. Kowalski, A. Piatkowska, M. Przylucka, H. Retzo, T. Strozzi, D. Colombo, O. Mora, F. Sánchez, G. Herrera, S. Moretti, N. Casagli, and F. Guzzetti. 2013. Exploitation of large archives of ERS and ENVISAT C-Band SAR data to characterize ground deformations. Remote Sens 5: 3896–3917. https://doi.org/10.3390/rs5083896

Deshpande, S. S. 2013. Improved floodplain delineation method using high-density LiDAR data. Comput-Aided Civ Infrastruct Eng 28: 68–79.

Dikpal, R. L., T. J. Renuka Prasad, and K. Satish. 2017. Evaluation of morphometric parameters derived from Cartosat-1 DEM using remote sensing and GIS techniques for Budigere Amanikere watershed, Dakshina Pinakini basin, Karnataka, India. Appl Water Sci 7: 4399–4414. https://doi.org/10.1007/s13201-017-0585-6

El-Baz, F., M. Maingue, and C. Robinson. 2000. Fluvio-aeolian dynamics in the north-eastern Sahara: the relationship between fluvial/aeolian systems and ground-water concentration. J Arid Environ 44(2): 173–183.

Ewertowski, M. W., A. M. Tomczyk, D. J. A. Evans, D. H. Roberts, and W. Ewertowski. 2019. Operational framework for rapid, very-high resolution mapping of glacial geomorphology using low-cost unmanned aerial vehicles and structure-from-motion approach. Remote Sens 11(1): 65. https://doi.org/10.3390/rs11010065

Fischer, L., H. Eisenbeiss, A. Kääb, C. Huggel, and W. Haeberli. 2011. Monitoring topographic changes in a periglacial high-mountain face using high-resolution DTMs, Monte Rosa East Face, Italian Alps. Permafr Periglac Process 22: 140–152.

Fonstad, M. A. 2006. Cellular automata as analysis and synthesis engines at the geomorphology—ecology interface. Geomorphology 77: 217–234.

French, H., and C. E. Thorn. 2006. The changing nature of periglacial geomorphology. Géomorphologie: Relief Process Environ 3.

Frost, R. E., J. H. Mclerran, and R. D. Leighty. 1966. Photo-interpretation in the arctic and subarctic. Proceedings, First International Conference on Permafrost. National Academy of Sciences-National Research Council of Canada, Publication. 1287: 343–348.

Gao, J. 2011. Bathymetric mapping by means of remote sensing: Methods, accuracy and limitations. Prog Phys Geogr 35: 782–809.

Ghoshal, S., L. A. James, M. B. Singer, and R. Aalto. 2010. Channel and floodplain change analysis over a 100-year period: Lower Yuba River, California. Remote Sens 2: 1797–1825.

Ghuffar, S., B. Székely, A. Roncat, and N. Pfeifer. 2013. Landslide displacement monitoring using 3D range flow on airborne and terrestrial LiDAR data. Remote Sens 5: 2720–2745.

Gibson, G., J. B. Campbell, and L. Kennedy. 2009. Fifty-one years of shoreline change at Little Lagoon, Alabama. Southeast Geog 49: 67–83.

Greeley, R. 2013. Introduction to Planetary Geomorphology. Cambridge: Cambridge University Press.

Grohmann, C. H., and A. O. Sawakuchi. 2013. Influence of cell size on volume calculation using digital terrain models: A case of coastal dune fields. Geomorphology 180–181: 130–136.

Grosse, G., L. Schirrmeister, and T. J. Malthus. 2006. Application of Landsat-7 satellite data and a DEM for the quantification of thermokarst-affected terrain types in the periglacial Lena—Anabar coastal lowland. Polar Res 25: 51–67.

Grosse, G., L. Schirrmeister, V. V. Kunitsky, and H. W. Hubberten. 2005. The use of CORONA images in remote sensing of periglacial geomorphology: An illustration from the NE Siberian coast. Permafr Periglac Process 16: 163–172.

Hasan, M. A., A. N. Mayeesha, and M. Z. A. Razzak. 2023. Evaluating geomorphological changes and coastal flood vulnerability of the Nijhum Dwip Island using remote sensing techniques. Remote Sens Appl Soc Environ 32: 101028, ISSN 2352-9385, https://doi.org/10.1016/j.rsase.2023.101028. (www.sciencedirect.com/science/article/pii/S2352938523001106)

Heeren, D. M., A. R. Mittelstet, G. A. Fox, D. E. Storm, A. T. Al-Madhhachi, T. L. Midgley, A. F. Stringer, K. B. Stunkel, and R. D. Tejral. 2012. Using rapid geomorphic assessments to assess streambank stability in Oklahoma Ozark Streams. Trans Am Soc Agric Biol Eng 55: 957–968.

Hjort, J., and M. Luoto. 2012. Can geodiversity be predicted from space? Geomorphology 153: 74–80.

Hohenthal, J., P. Alho, J. Hyyppä, and H. Hyyppä. 2011. Laser scanning applications in fluvial studies. Prog Phys Geog 35: 782–809.

Horton, R. E. 1945. Erosional development of streams and their drainage basins; hydrophysical approach to quantitative morphology. Bull Geol Soc Am 56: 275–370.

Hudson, P. F., R. R. Colditz, and M. Aguilar-Robledo. 2006. Spatial relations between floodplain environments and land use—land cover of a large lowland tropical river valley: Panuco basin, Mexico. Environ Manag 38: 487–503.

Hugenholtz, C. H., N. Levin, T. E. Barchyn, and M. C. Baddock. 2012. Remote sensing and spatial analysis of aeolian sand dunes: A review and outlook. Earth Sci Rev 111: 319–334.

Huggel, C., A. Kääb, W. Haeberli, P. Teysseire, and F. Paul. 2002. Remote sensing based assessment of hazards from glacier lake outbursts: A case study in the Swiss Alps. Can Geotech J 39: 316–330.

Humlum, O. 2000. The geomorphic significance of rock glaciers: Estimates of rock glacier debris volumes and headwall recession rates in West Greenland. Geomorphology 35: 41–67.

Humlum, O. 2008. Alpine and polar periglacial processes: The current state of knowledge. Plenary Paper, Ninth International Conference on Permafrost. 743–7 59.

Irish, J. L. 2000. An Introduction to Coastal Zone Mapping with Airborne LiDAR: The SHOALS System. Technical Report: Corp of Engineers, Mobile, AL. Joint Airborne LiDAR Bathymetry Technical Center of Expertise.

Kääb, A. 2008. Remote sensing of permafrost-related problems and hazards. Permafr Periglac Process 19: 107–136.

Kääb, A., C. Huggel, L. Fischer, S. Guex, F. Paul, I. Roer, N. Salzmann, S. Schlaefli, K. Schmutz, D. Schneider, T. Strozzi, and Y. Weidmann. 2005. Remote sensing of glacier- and permafrost-related hazards in high mountains: An overview. Nat Hazards Earth Syst Sci 5: 527–554.

Kant, C., G. Kumar, and R. S. Meena. 2023. Modeling morphometric and geomorphological parameters of mountainous river basin for water resource management using remote sensing and GIS approach. Model. Earth Syst Environ 9: 2151–2163. https://doi.org/10.1007/s40808-022-01614-0

Kawakuboa, F. S., R. G. Moratoa, R. S. Naderb, and A. Luchiari. 2011. Mapping changes in coastline geomorphic features using Landsat TM and ETM+ imagery: Examples in southeastern Brazil. Int J Rem Sens 32: 2547–2562. https://doi.org/10.1080/01431161003698419

Kieffer, S. W. 2014. The deadly dynamics of landslides. Am Sci 102: 298–303.

Klees, R., and D. Massonnet. 1999. Deformation measurements using SAR interferometry: Potential and limitations. Geologie en Mijinbouw 77: 161–176.

Langat, P. K., L. Kumar, and R. Koech. 2019. Monitoring river channel dynamics using remote sensing and GIS techniques. Geomorphology 325: 92–102, ISSN 0169-555X, https://doi.org/10.1016/j.geomorph.2018.10.007. (www.sciencedirect.com/science/article/pii/S0169555X18304070)

Lausch, A., M. E. Schaepman, A. K. Skidmore, S. C. Truckenbrodt, J. M. Hacker, J. Baade, L. Bannehr, E. Borg, J. Bumberger, P. Dietrich, et al. 2020. Linking the Remote Sensing of Geodiversity and Traits Relevant to Biodiversity—Part II: Geomorphology, Terrain and Surfaces. Remote Sens 12(22): 3690. https://doi.org/10.3390/rs12223690

Leal, C. G., P. S. Pompeu, T. A. Gardner, R. P. Leitão, R. M. Hughes, P. R. Kaufmann, J. Zuanon, F.R. de Paula, S.F. Ferraz, J.R. Thomson, and R. Mac Nally. 2016. Multi-scale assessment of human-induced changes to Amazonian instream habitats. Land Ecol 31: 1725–1745.

Lee, W. T. 1922. The Face of the Earth as Seen from the Air. Special Publication No. 4. New York: American Geographical Society.

Liu, L., C. I. Millar, R. D. Westfall, and H. A. Zebker. 2013. Surface motion of active rock glaciers in the Sierra Nevada, California, USA: Inventory and a case study using InSAR. The Cryosphere 7(4): 1109–1119.

Livingston, I., G. F. S. Wiggs, and C. M. Weaver. 2007. Geomorphology of desert sand dunes: A review of recent progress. Earth Sci Rev 80: 239–257.

Marcus, A., and M. A. Fonstad. 2010. Remote sensing of rivers: The emergence of a subdiscpline in the river sciences. Earth Surf Proc Land 35: 1867–1872.

Marghany, M., Z. Sabu, and M. Hashim. 2010. Mapping coastal geomorphology changes using synthetic aperture radar data. Int J Phys Sci 5: 1890–1896.

Martinez, J.-M., and T. Le Toan. 2007. Mapping of flood dynamics and spatial distribution of vegetation in the Amazon floodplain using multitemporal SAR data. Remote Sens Environ 108: 209–223.

Meitzen, K. M. 2009. Lateral channel migration effects on riparian forest structure and composition, Congaree River, South Carolina, USA. Wetlands 29: 465–475.

Melosh, H. J. 2011. Planetary Surface Processes (No. 13). Cambridge: Cambridge University Press.

Melton, F. A. 1939. Aerial photographs and the first course in geology. Photogramm Eng 5: 74-7 7.

Mertes, L. A. K., D. L. Daniel, J. M. Melack, B. Nelson, L. A. Martinelli, and B. R. Forsberg. 1995. Spatial patterns of hydrology, geomorphology, and vegetation on the floodplain of the Amazon River in Brazil from a remote sensing perspective. Geomorphology 13: 215–232.

Metternicht, G., L. Hurni, and R. Gogu. 2005. Remote sensing of landslides: An analysis of the potential contribution to geo-spatial systems for hazard assessment in mountainous environments. Rem Sens Environ 98: 284–303.

Millington, A. C., and J. R. G. Townshend. 1987. The potential of satellite remote sensing for geomorphological investigations: An overview. In: International Geomorphology, ed. V. Gardiner. Chichester: Wiley: 331–342.

Misra, A. A. 2022. Fluvial Geomorphology in a part of the Spiti River basin, Himachal Pradesh, India. In: Atlas of Structural Geological and Geomorphological Interpretation of Remote Sensing Images. 93–110. https://doi.org/10.1002/9781119813392.ch9

Mitasova, H., M. Overton, and R. S. Harmon. 2005. Geospatial analysis of a coastal sand dune field evolution: Jockey's Ridge, North Carolina. Geomorphology 72: 204–221.

Moore, L. J., and G. B. Griggs. 2002. Long-term cliff retreat and erosion hotspots along the central shores of the Monterey Bay National Marine Sanctuary. Mar Geol 181: 265–283.

Morrissey, L. A., L. Strong, and D. H. Card. 1986. Mapping permafrost in the boreal forest with thematic mapper satellite data. Photogramm Eng Remote Sens 52: 1513–1520.

Muzirafuti, A. 2024. Remotely sensed and field data for geomorphological analysis of water springs: A case study of Ain Maarrouf. Geosciences 14(2): 51. https://doi.org/10.3390/geosciences14020051

Nie, Y., Q. Liu, J. Wang, Y. Zhang, Y. Sheng, and S. Liu. 2018. An inventory of historical glacial lake outburst floods in the Himalayas based on remote sensing observations and geomorphological analysis. Geomorphology 308: 91–106, ISSN 0169-555X, https://doi.org/10.1016/j.geomorph.2018.02.002. (www.sciencedirect.com/science/article/pii/S0169555X1830045X)

Othman, A., and R. Gloaguen. 2013. River courses affected by landslides and implications for hazard assessment: A high resolution remote sensing case study in NE Iraq—W Iran. Remote Sens 5: 1024–1044. https://doi.org/10.3390/rs5031024

Phillips, J. D. 1999. Divergence, convergence, and self-organization in landscapes. Ann Assoc Am Geogr 89: 466–4 88.

Pinter, N., and M. T. Brandon. 2005. How erosion builds mountains. Sci Am 15: 74–81.

Plug, L. J., C. Walls, and B. M. Scott. 2008. Tundra lake changes from 1978 to 2001 on the Tuktoyaktuk Peninsula, western Canadian Arctic. Geophys Res Lett 35(3).

Quesada-Román, A., and M. Peralta-Reyes. 2023. Geomorphological mapping global trends and applications. Geographies 3(3): 610–621. https://doi.org/10.3390/geographies3030032

Rajaveni, S. P., K. Brindha, and L. Elango. 2017. Geological and geomorphological controls on groundwater occurrence in a hard rock region. Appl Water Sci 7: 1377–1389. https://doi.org/10.1007/s13201-015-0327-6

Ray, R. G. 1960. Aerial Photographs in Geologic Interpretation and Mapping. USGS Professional Paper. Washington, D.C.: GPO.

Raymo, M. E., and W. F. Ruddiman. 1992. Tectonic forcing of late Cenozoic climate. Nature 359: 117–122.

Resop, J. P., and W. C. Hession. 2010. Terrestrial laser scanning for monitoring streambank retreat: Comparison with traditional surveying techniques. J Hydraulic Eng 136: 794–798.

Rosario González-Moradas, M. D., and W. Viveen. 2020. Evaluation of ASTER GDEM2, SRTMv3.0, ALOS AW3D30 and TanDEM-X DEMs for the Peruvian Andes against highly accurate GNSS ground control points and geomorphological-hydrological metrics. Remote Sens Environ 237: 111509, ISSN 0034-4257, https://doi.org/10.1016/j.rse.2019.111509. (www.sciencedirect.com/science/article/pii/S0034425719305280)

Rudraiah, M., S. Govindaiah, and S. Srinivas Vittala. 2008. Morphometry using remote sensing and GIS techniques in the sub-basins of Kagna River basin, Gulburga District, Karnataka, India. J India Soc Remote Sens 36: 351–360.

Sankar, S., and D. P. Kanungo. 2004. An integrated approach for Landslide mapping using remote sensing and GIS. Photogramm Eng Remote Sens 70: 617–625.

Sarkar, A., R. D. Garg, and N. Sharma. 2012. RS-GIS based assessment of river dynamics of Brahmaputra River in India. J Water Resource Prot 4: 63–72. https://doi.org/10.4236/jwarp.2012.42008. Published Online February 2012 (www.SciRP.org/journal/jwarp)

Sasaki, T., J. Imanishi, K. Ioki, Y. Morimoto, and K. Kitada. 2008. Estimation of leaf area index and canopy openness in broad-leaved forest sing airborne laser scanner in comparison with high-resolution near-infrared digital photography. Land Ecol Eng 4: 47–55.

Scaioni, M. 2013. Remote sensing for landslide investigations: From research into practice. Remote Sens 5: 5488–54 92.

Scheidegger, A. E. 1961. Theor Geomorphology. New York: Springer: 333.

Schiefer, E., and R. Gilbert. 2007. Reconstructing morphometric change in a proglacial landscape using historical aerial photography and automated DEM generation. Geomorphology 88: 167–178.

Short, N., and R. W. Blair, eds. 1986. Geomorphology from Space: A Global Overview of Regional Landforms. NASA SP-486. Washington DC: GPO.

Shu, Y., J. Li, and G. Gomes. 2010. Shoreline extraction from RADARSAT-2 intensity imagery using a narrow band level set segmentation approach. Mar Geod 33: 187–203. https://doi.org/10.1080/01490419.2010.496681

Simon, A., A. Curini, S. E. Darby, and E. J. Langendoen. 2000. Bank and near-bank processes in an incised channel. Geomorphology 35(3–4): 193–217.

Simpson, R. 1966. Radar, geographic tools. Ann Assoc Am Geog 80–96. https://doi.org/10.1111/j.1467-8306.1966.tb00545.x

Smith, H. T. U. 1942. Aerial photographs in geomorphic studies. Photogramm Eng 8: 129–155.

Smith, H. T. U. 1943. Aerial Photographs and Their Applications. New York: Century Crofts, Inc.

Smith, M. J., and C. F. Pain. 2009. Applications of remote sensing in geomorphology. Prog Phys Geogr 33: 568–582. https://doi.org/10.1177/0309133309346648

Smith, M. J., J. Rose, and S. Booth. 2006. Geomorphological mapping of glacial landforms from remotely sensed data: An evaluation of the principal data sources and an assessment of their quality. Geomorphology 76: 148–165.

Stallins, J. A. 2006. Geomorphology and ecology: Unifying themes for complex systems in biogeomorphology. Geomorphology 77: 207–2 16.

Strahler, A. N. 1957. Quantitative analysis of watershed geomorphology. Trans Am Geophys Union 38: 913–920.

Strozzi, T., C. Ambrosi, and H. Raetzo. 2013. Interpretation of aerial photographs and satellite SAR interferometry for the inventory of landslides. Remote Sens 5: 2554–2570.

Tantianuparp, P., X. Shi, L. Zhang, T. Balz, and M. Liao. 2013. Characterization of landslide deformations in Three Gorges area using multiple InSAR data stacks. Remote Sens 5: 2704–2719.

Tofani, V., F. Raspini, F. Catani, and N. Casagli. 2013. Persistent Scatter Interferometry (PSI) technique for landslide characterization and monitoring. Remote Sens 5: 1045–1065.

Torre, D., J. P. Galve, C. Reyes-Carmona, et al. 2024. Geomorphological assessment as basic complement of InSAR analysis for landslide processes understanding. Landslides. https://doi.org/10.1007/s10346-024-02216-w

Townsend, P. A., and J. R. Foster. 2002. A synthetic aperture radar—based model to assess historical changes in lowland floodplain hydroperiod. Water Resour Res 38: 20–21.

Ulrich, M., G. Grosse, S. Chabrillat, and L. Schirrmeister. 2009. Spectral characterization of periglacial surfaces and geomorphological units in the Arctic Lena Delta using field spectrometry and remote sensing. Remote Sens Environ 113: 1220–1235.

Viles, H. (2020). Biogeomorphology: Past, present and future. Geomorphology 366: 106809.

Vitek, J. D., J. R. Giardino, and J. W. Fitzgerald. 1996. Mapping geomorphology: A journey from paper maps, through computer mapping to GIS and virtual reality. Geomorphology 16: 233–249.

Walsh, S. J., D. R. Butler, and G. P. Malanson. 1998. An overview of scale, pattern, process relationships in geomorphology: A remote sensing and GIS perspective. Geomorphology 21: 183–205.

Walsh, S. J., D. R. Butler, G. P. Malanson, K. A. Crews-Meyer, J. P. Messina, and N. Xiao. 2003. Mapping, modeling, and visualization of the influences of geomorphic processes on the alpine treeline ecotone, Glacier National Park, MT, USA. Geomorphology 53: 129–145.

Wang, Y., X. Liang, C. Flener, A. Kukko, H. Kaartinen, M. Kurkela, M. Vaaja, H. Hyyppä, and P. Alho. 2013. 3D modeling of coarse fluvial sediments based on mobile laser scanning data. Remote Sens 5: 4571–4592.

White, S. A., and Y. Wang. 2003. Utilizing DEMs derived from LIDAR data to analyze morphologic change in the North Carolina coastline. Remote Sens Environ 113: 39–47.

Woolard, J. W., and J. D. Colby. 2002. Spatial characterization, resolution, and volumetric change of coastal Dunes using Airborne LIDAR: Cape Hatteras, North Carolina. Geomorphology 48: 269–287.

Yang, X., C. J. Michiel, R. A. Damen, and R. A. van Zuidam. 1999. Satellite remote sensing and GIS for the analysis of channel migration changes in the active Yellow River Delta, China. Int J Appl Earth Obs Geoinf 1: 146–157. https://doi.org/10.1016/S0303-2434(99)85007-7

Zangana, I., J. Otto, R. Mäusbacher, and L. Schrott. 2023. Efficient geomorphological mapping based on geographic information systems and remote sensing data: An example from Jena, Germany. Journal of Maps 19: 1. https://doi.org/10.1080/17445647.2023.2172468

Part II

Hydrology and Water Resources

2 Remote Sensing Technologies for Multi-Scale Hydrological Studies
Advances and Perspectives

Sadiq I. Khan, Ni-Bin Chang, Yang Hong, Xianwu Xue, and Yu Zhang

ACRONYMS AND DEFINITIONS

ABI	Advanced Baseline Imager
AIRS	Atmospheric Infrared Sounder
AMSR-E	Advanced Microwave Scanning Radiometer—Earth Observing System
ARM	Atmospheric Radiation Measurements
ATMS	Advanced Technology Microwave Sounder
AVHRR	Advanced Very High Resolution Radiometer
CERES	Clouds and the Earth's Radiant Energy System
CMORPH	Climate Prediction Center morphing algorithm
CNES	The Centre national d'études spatiales, or the National Center of Space Studies
CrIS	Cross-track Infrared Sounder
DESDynI	Deformation, Ecosystem Structure, and Dynamics of Ice
DLR	German Aerospace Center
DMSP	Defense Meteorological Satellite Program
DPR	Dual-frequency Precipitation Radar
EDRs	Environmental Data Records
EF	Evaporation Fraction
EOS	Earth Observing System
EROS	Earth Resources Observation and Science
ET	Evapotranspiration
ETMA	ET Mapping Algorithm
EXIS	Extreme Ultra-Violet and X-Ray Irradiance Sensors
GEO	Geostationary Earth Orbital
GLM	Geostationary Lightning Mapper
GMI	GPM Microwave Imager
GOES	Geostationary Operational Environmental Satellite
GPM	Global Precipitation Measurement
GRACE	Gravity recovery and Climate Experiment
HyspIRI	Hyperspectral Infrared Imager
ICESat	Instrument aboard the Ice, Cloud, and land Elevation
iGOS	integrated Grassland Observing Site
IR	Infrared
LEO	Low Earth Orbital

DOI: 10.1201/9781003541400-4

LiDAR	Light Detection and Ranging
LP	DAAC Land Processes Distributed Active Archive Center
LST	Land Surface Temperature
LULC	Land Use Land Cover
MAG	Magnetometer
METRIC	Mapping Evapotranspiration with Internalized Calibration
MIRAS	Microwave Imaging Radiometer with Aperture Synthesis
MOD16	MODIS Global Evapotranspiration Dataset
MODIS	Moderate-Resolution Imaging Spectroradiometer
MOISST	Marena Oklahoma In Situ Sensor Testbed
NASA	National Aeronautics and Space Administration
NESDIS	National Environmental Satellite, Data, and Information Service
NIR	Near-Infrared
NOAA	National Oceanic and Atmospheric Administration
NPOESS	National Polar-Orbiting Operational Environmental Satellite System
NPP	NPOESS Preparatory Project
NTSG	Numerical Terradynamic Simulation Group
OMPS	Ozone Mapping and Profiler Suite
PDSI	Palmer Drought Severity Index
PERSIANN	Precipitation Estimation from Remotely Sensed Information using Artificial Neural Networks
PERSIANN-CCS	PERSIANN Cloud Classification System
POES	Polar Operational Environmental Satellites
PSU	Practical Salinity Unit
RADAR	Radio Detection and Ranging
SAR	Synthetic Aperture Radar
SEB	Surface Energy Balance
SEBAL	Surface Energy Balance Algorithm
SEBS	Surface Energy Balance System
SEISS	Space Environment In Situ Suite
SEM	Space Environment Monitor
SGP	Southern Great Plains
SMAP	Soil Moisture, Active and Passive
S-SEBI	Simplified Surface Energy Balance Index
SSMI/S	Special Sensor Microwave Imager
SUVI	Solar Ultraviolet Imager
SWIR	shortwave Infrared
SWOT	Surface Water and Ocean Topography
TIR	Thermal Infrared
TMI	TRMM Microwave Imager
TRMM	Tropical Rainfall Measuring Mission
TSIS	Total Solar Irradiance Sensor
TWS	Terrestrial Water Storage
VI	Vegetation Indices
VIIRS	Visible/Infrared Imager Radiometer Suite
VIS	Visible
VSM	Volumetric Soil Moisture

2.1 INTRODUCTION

Remote sensing measurements of hydrologic variables and processes represent one of the most challenging research problems in Earth system sciences (Saha and Pal 2024; Duan et al. 2021; Jiao et al.

2021; Kandus et al. 2018; Dang and Kumar 2017). Hydrologic processes in the nexus of energy and water cycles include precipitation, runoff, infiltration and soil water contents, evapotranspiration, and groundwater movement. Remote sensing estimates provide water budget information of the temporal and spatial distribution of water fluxes, often difficult to capture by conventional ground-based sparse sensor network with point-measurement instruments. Many hydrological processes can be observed from satellite to ground-based remote sensing technologies individually. However, the integrated approach to measure all relevant hydrological processes while completely addressing the changing hydrological states during the cycling of water remains a scientific challenge. Scale linkage of water fluxes integrating hydrological processes observed from a small scale to regional and continental scales has been a contemporary challenge for decades.

The terrestrial water budget, commonly defined by the water balance, is the overall change of water storage (ΔS) and the difference between the incoming amount of precipitation (P) and subtracted amount of water in the form of evapotranspiration (ET) and runoff (R) with respect to time:

$$\Delta S = P - ET - R \qquad (1)$$

The water budget variables in equation (1) have different spatiotemporal footprints. For example, precipitation follows a different time scale than groundwater recharge. However, all these processes are interconnected and several satellite sensors discussed throughout this chapter can be used to estimate these processes at different spatiotemporal scales. To accomplish each independent task, satellite and/or airborne sensors are designed to utilize a wide range of electromagnetic spectrum that can provide unique information of radiative reflectance regarding the targeted hydrologic processes. Besides the electromagnetic sensors, microgravity sensors are used to measure the spatiotemporal variations in the terrestrial water storage (Tapley et al. 2004; Syed et al. 2008).

At both regional and global scales, the real-time operational hydrological prediction can be increasingly appreciated and supported by the current and future Earth observation missions. Most of the processes in the terrestrial water budget such as precipitation, evapotranspiration, soil moisture, and water storage changes can be observed at different spatiotemporal resolutions and accuracy through remote sensing (Saha and Pal 2024; Duan et al. 2021; Ha et al. 2018; Zeng et al. 2017). The current Earth Observing System (EOS) with coarse spatiotemporal resolution can be a limiting factor for small watershed scale studies; however, it should be noted that there are satellite missions with advanced sensor technologies (some missions discussed in the following subsections), and this limitation is expected to become less important in the near future. The new satellite missions are anticipated to provide better precipitation and soil moisture data in terms of coverage, accuracy, and resolution. The upcoming soil moisture mission is expected to significantly enhance our understanding of wet and dry conditions of land surface and therefore may improve numerical modeling efforts. Some of the important satellite missions that will provide better estimates of Earth's water cycle are discussed in this chapter (Bhaga et al. 2020; Herman et al. 2018; Gleason et al. 2017).

This chapter introduces advances and perspectives of multi-scale hydrological studies from spaceborne to airborne and to ground-based remote sensing observations (Govender et al. 2022; Kabeja et al. 2022; Jiang and Wang 2019; Dang and Kumar 2017; Thakur et al. 2017a; Thakur et al. 2017b). The detailed discussion in remote sensing hydrology provides an overview of various orbital satellite platforms/sensors focusing on the Earth's hydrological cycle within major global satellite missions associated with some local and regional applications in the world. Soil moisture, precipitation, and evapotranspiration are emphasized in different sections. The principles of physics associated with each of these three important hydrologic variables are entailed individually with an emphasis on the use of remote sensing data products and their availability without involving sensor design details and specific instrumentation. Particular emphasis is also given to science and technology used for spaceborne data estimation, validation, and its applications.

Some of the current remote sensing sensors that are employed to study the water cycle are described throughout the chapter. Along this line, Section 2.2 details the evolution of different quantitative precipitation estimation methods; Section 2.3 is related to evapotranspiration estimation using different surface energy balance algorithms; and Section 2.4 is linked to satellite remote sensing methods for soil moisture estimation. At the end of this chapter, the focus is placed on how remote sensing estimates can be utilized to concatenate the fundamental hydrologic processes such as precipitation, evapotranspiration, and soil moisture and how mission-oriented remote sensing programs are angled to close the hydrological cycle. The last section encompasses perspectives and conclusions.

2.2 REMOTE SENSING ADVANCES IN HYDROLOGY

Advances in hydrology have been made possible with the introduction of research satellite platforms such as the Landsat program, Terra and Aqua missions, and operational platforms such as the Tropical Rainfall Measuring Mission (TRMM) and the Geostationary Operational Environmental Satellite (GOES) series, Gravity Recovery and Climate Experiment (GRACE), Surface Water and Ocean Topography (SWOT) as well as many others. Satellite remote sensing offers a framework to complement and provide better understanding of regional- as well as global-scale hydrosystem processes (Jiao et al. 2021; Herman et al. 2018; Thakur et al. 2017a; Thakur et al. 2017b; Zeng et al. 2017). Many hydrological state variables and fluxes can be evaluated through satellite remote sensing. Optical sensors, as well as both active and passive sensors, are used as a source of observations, particularly in regions where ground-based observation is sparsely available or sometimes unavailable. For instance, remote sensors are used to acquire data on topography that can be represented in the form of three-dimensional data, referred to as the Digital Elevation Model. Multispectral and hyperspectral remote sensing detect reflected or emitted energy from an object in a number of different spectral bands of the electromagnetic spectrum. In remote sensing this spectral signature is the most diagnostic tool in remotely identifying the composition of an object. Generally, there is a trade-off between the spectral resolution, spectral coverage, radiometric resolution, and temporal resolution.

Precipitation is one of the major drivers of the hydrologic cycle and requires accurate measurement in order to help access the overland runoff fluxes and other hydrological state variables. With scarcity and poor quality of ground-based observations, satellite remote sensing estimates can be a viable source for precipitation estimation. Microwave techniques are used to directly observe the rain rates, as microwave radiation relates strongly with different hydrometeors. Some of the well-researched remote sensing precipitation products include CMORPH (Joyce et al. 2004), Precipitation Estimation from Remotely Sensed Information Using Artificial Neural Networks-PERSIANN (Sorooshian et al. 2000), and Tropical Rainfall Measuring Mission (TRMM) Multisatellite Precipitation Analysis (Huffman et al. 2007). The TRMM satellite with TRMM Microwave Imager (TMI) on board and the Advanced Microwave Scanning Radiometer-EOS (AMSR-E) on board the EOS Aqua satellite, and the Global Precipitation Measurement Mission (GPM) are providing precipitation estimates.

ET is one of the most important components of water balance analysis (Amiri et al. 2024; Duan et al. 2021; Herman et al. 2018; Kandus et al. 2018; Dang and Kumar 2017; Gleason et al. 2017; Roy et al. 2017; Thakur et al. 2017a; Thakur et al. 2017b; Zeng et al. 2017). ET estimation using ground network-based methods such as pan-measurement, Bowen ratio, eddy correlation, weighing lysimeter, and scintillometer, are complex techniques that require a lot of data. In comparison to conventional point-based measurements, remote sensing techniques have the following advantages: (1) reasonable spatial and temporal coverage, (2) economic viability, and (3) functional in complex terrains and ungauged areas. Numerous ET models have been used for more than three decades to make use of visible, near infrared (NIR), shortwave infrared (SWIR), and most importantly, thermal data acquired by sensors on airborne and satellite platforms (Figure 2.1).

It is difficult to measure ET directly from remote sensing methods. However, a number of algorithms are available to estimate ET from remotely sensed estimates of surface energy fluxes (Figure 2.1). These models include the Simplified Surface Energy Balance Index (S-SEBI; Roerink et al. 2000), the Surface Energy Balance System (SEBS; Su 2002), the Surface Energy Balance Algorithm (SEBAL) (Bastiaanssen et al. 1998a, 1998b), the ET Mapping Algorithm (ETMA; Loheide and

Multi-Scale Hydrological Studies

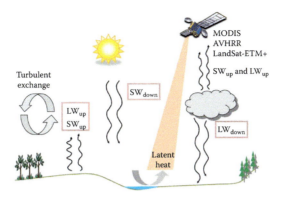

FIGURE 2.1 Surface energy budget remotely sensed components.

Gorelick 2005), and the Mapping Evapotranspiration with Internalized Calibration (METRIC; Allen et al. 2007). There are ongoing research efforts to improve the spatial resolution of ET estimation methods to integrate these products into hydrologic models.

Soil moisture is another important hydrologic variable for understanding water cycle. The distributed information of soil moisture is of great importance in the hydrology and climate studies due to its high spatial and temporal variability. Microwave remote sensing provides a unique capability for direct observation of soil moisture. Remote measurements from space afford the possibility of obtaining frequent, global sampling of soil moisture over a large fraction of the Earth's land surface. Integrating estimates from both passive and active microwave sensors promotes potential for improved surface soil moisture estimates. Such methods have been evaluated with in situ observations and validated in several field experiments (O'Neill et al. 1996; Chauhan 1997; Njoku et al. 2002; Bolten et al. 2003). The Advanced Microwave Scanning Radiometer (AMSR-E) sensor on the Aqua satellite (Njoku et al. 2003) and the Soil and the Moisture and Ocean Salinity (SMOS) mission launched by the European Space Agency (ESA) (Kerr et al. 2001) are providing low frequency–based soil moisture estimates. The spatial resolution of these sensors is 56 and 37 km, respectively.

Understanding how all these variables interact within the geosphere is critical to link the physical drivers of water variability to the health and sustainability of the ecosystem. Furthermore, by coupling unconventional datasets and robust computer models, new contributions can be made in understanding the relationships between precipitation, surface water movement, groundwater storage, aquifer recharge and drawdown, and ecosystem services. Some of the short- and long-term satellite missions are listed in Table 2.1.

2.3 PRECIPITATION ESTIMATION BASED ON MULTI-SENSOR DATA

2.3.1 Physical Principles of Radar Precipitation Estimation

Microwave remote sensing that is a part of electromagnetic sensing has opened new frontiers in hydrologic science. Radio detection and ranging (radar), with its active sensing capabilities, allows data acquisition during both day and night in all weather conditions. Radar sensors are able to provide high-resolution data on the horizontal (1 m) and vertical (10 cm) movement of the Earth's surface (Tronin 2006). Radar consists of basic components, such as the transmitter that generates electromagnetic (EM) radiation as a pulse or continuous wave. Weather radars transmit EM energy in the microwave spectrum that travel at the speed of light in a vacuum at 3×108 meters per second (m/s). The relationship between radio frequency (f), wavelength (λ), and velocity at the speed of light (c) is shown by Eq. 1:

$$c = f\lambda \quad \text{(Eq. 1)}$$

TABLE 2.1
Some of NASA's Missions, Hydrologic Observations, and Societal Benefits[1]

Missions	Launch Date	Partner	Observed Quantities	Societal Benefit
LDCM	May 2013	USGS	Land cover/land use, thermal and IR surface properties	Land cover, ecosystem composition, algal blooms, and waterborne and zoonotic disease
GPM-Core	Jul 2013	JAXA	Precipitation	Heat stress and drought, precipitation, all-weather temperature and humidity, surface water and ocean topography
SAGE III/ISS	May 2014	SOMD	Vertical profiles of ozone, water vapor atmosphere	Ozone processes (health), clouds and aerosols (climate)
SMAP	Nov 2014		Soil moisture, freeze-thaw state	Algal blooms and waterborne infectious disease, soil moisture, surface water and ocean topography
ICESat 2	Oct 2015		Ice sheet thickness, vegetation height	Clouds, aerosols, ice, carbon, glacier surface elevation, glacier retreat
GRACE II	2016	DLR	Time variable gravity, including mass of water/ice	Ocean circulation, heat storage, and climate forcing, groundwater storage, ice sheet mass balance, ocean mass
DESDynI radar and LiDAR	Nov 2017	DLR	Earth surface deformation; vegetation height, canopy volume	Ice dynamics, ecosystem structure, biomass, and biodiversity, heat stress and drought, glacier velocity
SWOT	2020	CNES/USGS	Lake levels (3 million lakes), ocean topography	Surface water and ocean topography, ocean circulation and heat storage, river discharges
HysPIRI	2020	USGS	Aquatic ecosystems, volcanic hazards	Heat stress and drought, vector-borne and zoonotic diseases (health), surface composition and thermal properties

where c is 3x10⁸ m/s, f is in cycles per second, or hertz (Hz), and λ is in m. Table 2.2 shows the most common bands, frequencies, and associated wavelengths that correspond to radars that have hydrologic applications. Note that the typical values for radar microwave frequencies are on the order of 10⁷–10¹¹ Hz, thus it is convenient to use Mega (10⁶) and Giga (10⁹) prefixes, or MHz and GHz. The corresponding radar wavelengths span a few millimeters (mm) up to m. The radar wavelength and diameter (d) of the parabolic dish dictate the angular width of the radar beam, or beamwidth (θ), as follows in Eq. 2:

$$\theta = \frac{73\lambda}{d} \quad \text{(Eq. 2)}$$

where λ and d are both in the same distance units and θ is in degrees. In the case of the WSR-88D radar, it operates at an approximate 10.7-cm wavelength and has an 8.5-m diameter dish. This corresponds to a beamwidth of approximately 0.92° (in both azimuth and elevation directions). Targets with horizontal cross-sections (for a horizontally polarized wave) less than λ/16, or approximately 7 mm for the WSR-88D, are Rayleigh scatters and thus have predictable radar signatures for different-sized raindrops. The targets are assumed to produce scattering equal in all directions,

TABLE 2.2
List of Most Common Radar Bands, Frequencies, and Associated Wavelengths with Their Hydrologic Applications

Band	Frequency	Wavelength	Hydrologic Applications
W	75–110 GHz	2.7–4.0 mm	Detection of cloud droplets
Ka	24–40 GHz	0.8–1.1 cm	Precipitation estimation from spaceborne radar
Ku	12–18 GHz	1.7–2.5 cm	Precipitation estimation from spaceborne radar
X	8–12 GHz	2.5–3.8 cm	High-resolution precipitation and microphysical studies
C	4–8 GHz	3.8–7.5 cm	Precipitation estimation from operational systems
S	2–4 GHz	7.5–15 cm	Precipitation estimation from operational systems
L	1–2 GHz	15–30 cm	Top-layer soil moisture
UHF	300–1000 MHz	0.3–1 m	Ground-penetrating radar for water table
VHF	30–300 MHz	1–10 m	Ground-penetrating radar for water table

called isotropic scattering. The radar detects the component of scattering that comes back to the radar (backscatter). Shorter wavelength radars at X-band and shorter have a lower upper limit on the diameter of targets that cause Rayleigh scattering. But these smaller wavelength radars do not require such large dishes to maintain the small beamwidth desirable for high-resolution precipitation measurements, and thus are more amenable to spaceborne, transportable, and mobile radar platforms. Table 2.2 lists the most common radar bands, frequencies, and associated wavelengths with their hydrologic applications.

2.3.2 Next-Generation Radar (NEXRAD)

While weather radar applications are diverse and far-reaching, the focus in this section is on the hydrologic use of weather radar. In many cases, examples and typical values for the variables will be provided for the Weather Surveillance Radar—1988 Doppler (WSR-88D) (REF); the radar that constitutes the Next Generation Radar (NEXRAD) network (REF) in operation across the United States.

In conventional single-polarization radar detection, the raindrops scatter the incident wave back to the radar antenna. It uses a pulse that is typically polarized about the horizontal plane (H), and the primary measurement used for QPE is radar reflectivity, Z. Single-polarization radar can receive one polarization horizontal (H) or Vertical (V) wave, while a polarimetric radar can transmit/receive waves in both polarizations, hence obtaining more information by which the target medium can be better characterized. In addition to challenges with data quality (i.e., contamination by non-weather scatterers), many studies have shown that Z alone is insufficient to reveal the natural variability of precipitation (Battan 1973; Rosenfeld and Ulbrich 2003). The drop size distribution (DSD) exhibits variability and thus cannot be adequately described using a single reflectivity-to-rainfall rate (Z-R) relation. Polarimetric radar variables are signatures of electromagnetic waves scattering from a targeted medium. For hydrometeors, the radar echo depends on their size, shape, orientation, and density. If the scattering amplitude of each hydrometeor and the particle size distribution (PSD) are known, polarimetric radar variables can be calculated by the integration of scattering amplitude over all the sizes. Depending on their measurement effects and physical meanings, polarimetric radar variables have different applications. Reflectivity factor and differential reflectivity, related to the power of the radar echo, are widely used in various polarimetric radar algorithms. The correlation coefficient is a primary parameter for radar data quality control. Specific differential phase can be used for rain estimation and attenuation correction. Specific attenuation and specific differential attenuation are usually applied by algorithms of attenuation correction.

FIGURE 2.2 Map showing NEXRAD network stations over the contiguous United States.

With the development of dual-polarization radar (also called polarimetric radar), the accuracy of QPE has been improved through the use of polarimetric variables (Bringi and Chandrasekar 2001). The US Next-Generation Radar (NEXRAD) network (Figure 2.2) has been upgraded with dual-polarization technology, and similar upgrades are being conducted or are planned in many other countries.

2.3.3 GLOBAL PRECIPITATION ESTIMATION

Precipitation estimates have been successful using microwave remote sensing. Microwave remote sensing can be categorized into active remote sensing, when a signal is first emitted from aircraft or satellites and then received by onboard sensors, or passive remote sensing when reflective radiation information from sunlight is merely recorded by sensors onboard aircraft or satellites.

Precipitation estimates at varying spatial and temporal scales is vital for climatic and hydrologic studies. The underlying principle in global precipitation estimation from remote sensing is to observe the backscatter from different hydrometeor types (rain, hail, snow, ice crystals) in the atmosphere. Satellite-based precipitation estimation started with the use of visible (VIS) and infrared (IR) instruments by looking at the cloud top temperatures (Petty 1995; Ba and Gruber 2001; Kuligowski 2002; Bellerby 2004; Yan and Yang 2007; Thies et al. 2008). Since the late 1970s infrared satellite remote sensing techniques were first used for precipitation estimation (Arkin and Meisner 1987). The majority of algorithms attempt to correlate the surface rain rate with IR cloud-top brightness temperatures (Tb) using the information obtained from IR imagery. The algorithms developed to date may be classified into three groups depending on the level of information extracted from the IR cloud images: cloud-pixel-based, cloud-window-based, and cloud-patch-based (Yang et al. 2011). Several examples of these algorithms may clarify this classification further.

Advancement in global precipitation products through improved accuracy, coverage, and spatio-temporal resolution was materialized by combining data from GEO VIS/IR and LEO MW sensors.

TABLE 2.3
Summary of Satellite-Based Global Rainfall Products for Hydrology, Meteorology, and Climate Studies. Source (Yang et al. 2011)

Product Name	Agency/Country	Scale	Period	Developer
GPCP	NASA/USA	2.5°/monthly	1979~	Adler et al. 2003
CMAP	NOAA/USA	2.5°/5-day	1979~	Xie et al. 2003
GPCP IDD	NASA/USA	1°/daily	1998~	Huffman et al. 2001
TMPA	NASA-GSFC/USA	25 km/3-hourly	1998~	Huffman et al. 2007
CMORPH	NOAA-CPC/USA	25 km/3-hourly	2002~	Joyce et al. 2004
PERSIANN	University of Arizona	25 km/6-hourly	2002~	Sorooshian et al. 2000
NRL-Blend	Naval Research Lab/USA	10 km/3-hourly	2003~	Turk and Miller 2005
GSMAP	JAXA/Japan	10 km/hourly	2005~	sharaku.eor.jaxa.jp
UBham	University of Birmingham/UK	10 km/hourly	2002~	Kidd et al. 2003
PERSIANN-CCS	University of California Irvine/USA	4 km/half-hourly	2006~	Hong et al. 2004
HE	NOAA/NESDIS	4 km/half-hourly	2004~	Scofield and Kuligowski 2003

The first such merging algorithm was performed at a relatively coarse scale to ensure reasonable error characteristics. For instance, the first such multi-sensor blending algorithm is the Global Precipitation Climatology Project (GPCP), a multi-sensor combination is computed on a monthly 2.5° latitude-longitude grid (Adler et al. 2003) and at 1° daily (Huffman et al. 2001).

In the past decade, a number of quasi-global scale estimates have been developed, including the TRMM-based Multi-satellite Precipitation Analysis (TMPA; Huffman et al. 2007), the Naval Research Laboratory Global Blended-Statistical Precipitation Analysis (Turk and Miller 2005), Climate Prediction Center morphing algorithm (CMORPH; Joyce et al. 2004), PERSIANN Cloud Classification System (PERSIANN-CCS) (Hong et al. 2004, 2005), and UC Irvine PERSIANN (Hsu et al. 1997; Sorooshian et al. 2000). To date, the most commonly available satellite global rain products are summarized in Table 2.3. These quasi-global precipitation products are discussed in terms of accuracy and earth science applications throughout scientific literature. Some of these products are mentioned in Table 2.3.

2.3.4 MISSIONS ON PRECIPITATION ESTIMATION

2.3.4.1 The Global Precipitation Mission

The Global Precipitation Measurement (GPM) is an international initiative, as a follow up of the TRMM. According to the GPM mission documentation and information published online at http://pmm.nasa.gov/GPM, the two main sensors in the GPM Core are the GPM Microwave Imager (GMI), and the Dual-frequency Precipitation Radar (DPR). The GMI is a conical-scan, nine-channel passive microwave radiometer. The configuration of this instrument provides a broad measurement swath (850 km), and like the TMI, maintains a constant earth incidence angle of 52.8° and a constant footprint size for each measurement channel regardless of scan position. The GMI has a 1.2-m diameter main reflector, which is twice the diameter of the TMI. The DPR will provide high-resolution (approximately 4-km), high precision measurements of rainfall, rainfall processes, and cloud dynamics.

The DPR is essentially two radars, the Ku-PR and the Ka-PR. The Ku-PR operates at 13.6 GHz and is of similar design to the TRMM-PR. The Ka-PR is based on the PR's design but operates at 35.55 GHz, and the size of its antenna has been selected such that its measurement footprint matches the footprint size of the Ku-PR. Each radar uses a phased array, slotted wave guide antenna. The Ka-PR also has a selectable high-sensitivity mode, which provides an interlacing scan with a swath width of 120 km; this high-sensitivity mode will aid in the measurement of light rain and snow. The two phased array antennas assures better precipitation estimation in terms of higher spatial resolution (Flaming 2005).

In addition to GPM, the Geostationary Operational Environmental Satellite-R Series (GOES-R) will be an important program of the NOAA operations. The first launch of the GOES-R series satellite was conducted in 2016. The GOES-R mission is playing a key role in weather monitoring, warning, and forecast. The GOES-R satellites consists of the following instruments:

1. The 16-channel Advanced Baseline Imager (ABI) for viewing Earth's clouds, atmosphere, and surface
2. The Geostationary Lightning Mapper (GLM) for monitoring hemispheric lightning flashes
3. The Extreme Ultra-Violet and X-Ray Irradiance Sensors (EXIS) for measuring solar particles
4. the Solar Ultraviolet Imager (SUVI) for imaging the Sun
5. The space environment monitoring suite that includes the Space Environment In Situ Suite (SEISS) and Magnetometer (MAG) for monitoring Earth's space environment and geomagnetic storms

These instruments on the GOES-R satellite series produces more than 50 times the information provided by the current GOES system and offer a wide variety of unique observations of the environment, with particular emphasis on hazardous weather in the Western Hemisphere and space weather impacts. Also GOES-R's improved communication systems with higher data rates ensure a continuous and reliable flow of remote sensing products and relay of other environmental and emergency services information critical to a broad range of users and interests.

2.4 EVAPOTRANSPIRATION ESTIMATION USING MULTI-SENSOR DATA

2.4.1 Physical Principles of Surface Energy Balance (SEB) Method

An overview of remote sensing principles and technologies useful for hydrological studies are presented, with additional reviews of precipitation retrieval models and case studies on a regional and field scale (Amiri et al. 2024; Duan et al. 2021; Dong 2018; Gleason et al. 2017; Roy et al. 2017; Thakur et al. 2017a; Thakur et al. 2017b; Zeng et al. 2017). The remote sensing fundamental of ET retrieval is to obtain the latent heat flux (i.e., ET) as a residual of the Surface Energy Balance (SEB) budget by solving the main components of the SEB equation (Eq. 3), which includes the sensible heat flux (H) and ground heat flux (G).

The latent heat accompanying ET is λ, where λ is the latent heat of vaporization. The ET process requires a source of heat energy to convert water from the liquid to the vapor state. This is ultimately supplied by net radiation (Rn), the amount of incident solar radiation (Rs) that is absorbed at the Earth's surface; a simplified equation for the Surface Energy Balance (SEB) is

$$\lambda ET = Rn - H - G \qquad \text{(Eq. 3)}$$

where the available net radiant energy Rn (Wm−2) is combined between the soil heat flux G and the atmospheric convective fluxes (sensible heat flux H) and latent heat flux LE, which is readily converted to ET. The Rn and other variables (H and G) of the equation can be solved using remotely sensed data of surface characteristics such as vegetation cover, surface temperature albedo, and leaf area index. Rn is energy coming from the Sun less any radiation that gets reflected (or emitted as thermal infrared radiation) back to the atmosphere. Some of that Rn is the sensible heat flux (H), some of it is stored in the soil, and other objects such as woody material and the rest of the energy is absorbed by water, which can be converted to water vapor for ET (Figure 2.3).

FIGURE 2.3 Energy flux process with radiation budget components.

A certain amount of energy per mass of water is required to vaporize moisture, and this is called the latent heat of vaporization. Energy coming from the sun less any radiation that gets reflected (or emitted as thermal infrared radiation) back to the atmosphere—or, net radiation (Rn)—is energy available for AET. Most of these remote sensing ET estimation models use remotely sensed data of surface radiation, temperature, and vegetation properties, which can be retrieved from multi-spectral sensors from visible to thermal infrared (TIR) bands and are listed in Table 2.4.

2.4.2 Regional-, Local-, and Field-Scale Studies

A regional-scale modified form of the SEB method is applied to estimate actual ET (Amiri et al. 2024; Saha and Pal 2024; Jiao et al. 2021; Herman et al. 2018; Roy et al. 2017). Energy balance algorithms are developed by applying satellite-derived surface radiation, meteorological parameters, and vegetation characteristics. The fundamentals of SEB are used to calculate ET as the residual of the energy balance. This method calculates the actual ET by assimilating the MODIS daily dataset and meteorological observations from the Oklahoma Mesonet network. The algorithm is named the MOD/METRIC (thereinafter M/M-ET) and was developed and evaluated for southern Plains in the United States, particularly for the state of Oklahoma. The M/M-ET measurements are assessed on daily, eight-day, and regular bases at both the field and watershed scales. Two distinctive field sources explained next are used to compare the assessed results: one with meteorological towers for latent heat flux observation and another with the Mesonet sites for crop ET. Latent heat flux observations from the two Atmospheric Radiation Measurements (ARM) (www.arm.gov) AmeriFlux eddy covariance tower sites were used for the comparison. These sites are located at the ARM SGP extended facilities in Lamont and El Reno, Oklahoma. The two Mesonet sites at El Reno and Medford with Crop ET data are also selected for the evaluation. The estimated ET is also compared with the crop ET at the selected sites during wheat-growing season for multiple years. Figure 2.4a,b shows the daily time series and scatter plots for El Reno sites for the years 2005 and 2006, respectively.

The estimated ET is in good agreement with the observed crop ET, with an underestimation of −7% and overestimation with 6% for 2005 and 2006, respectively. Correspondingly, the correlation coefficient values indicate that ET values form the model related strongly with values of 0.86 and 0.75 for 2005 and 2006 observations. The correlation coefficient values also indicate that the ET estimates correlate measurements at the El Reno site relatively agreed with AmeriFlux observations. The M/M-ET estimation algorithm is implemented for the entire state of Oklahoma (Figure 2.5). Khan et al. 2010 provides a detailed evaluation of these results.

TABLE 2.4
List of Satellite Precipitation Products and Their Characteristics (Modified from Tapiador et al. 2012)

Precipitation Product	Satellite Sensors	Spatiotemporal Resolution	Areal Coverage/Start Date	Producer URL
GPI	GEO-IR, LEO-IR	2.5/monthly	Global-40 N-S/1986–Feb. 2004	NOAA/NWS CPC [1]
	GEO-, LEO-IR	2.5/pentad	Global-40 N-S/1986–Nov. 2004	NOAA/NWS CPC [2]
	GEO-, LEO-IR	1/daily	Global-40 N-S/Oct. 1996	NOAA/NWS CPC [3]
GPROF2004	AMSR-E	0.5 orbits	Global-70 N-S/June 2002	NSIDC [4]
GPROF2010	AMSR-E	0.25/daily 0.25/monthly	Global-70 N-S/June 2002	Colo. State Univ. [5]
GPROF2010	SSM/I	0.25/daily 0.25/monthly	Global-70 N-S/July 1987–Nov. 2009	Colo. State Univ. [5]
	Level 2 (swath/pixel)/orbit	Global-70 N-S/July 1987–Nov. 2009	-	Colo. State Univ. [6]
GPROF2010	SSMIS	0.25/daily 0.25/monthly	Global-70 N-S/Oct. 2003	Colo. State Univ. [5]
GPROF2010	TMI	0.25/daily 0.25/monthly	Global-40 N-S/Dec. 1997	Colo. State Univ. [5]
GPROF2010 (3G68)	TMI	0.5/hourly	0.1/hourly land	NASA/GSFC PPS [7]
Hydro-Estimator	GEO-IR	4 km/hourly	Global-60 N-S/March 2007	NOAA/NESDIS/STAR [8]
METH	SSM/I, SSMIS	2.5/monthly	Global ocean-60 N-S/July 1987–2010	George Mason Univ. [9]
METH (3A11)	TMI	5/monthly	Global ocean-40 N-S/Jan. 1998	NASA/GSFC PPS [7]
MiRS	AMSU/MHS, SSMIS	Swath	Global/Aug. 2007	NOAA OSDPD [10]
NESDIS/FNMOC Scattering Index	SSM/I	1.0/monthly 2.5/pentad, monthly	Global/July 1987–Nov. 2009	NESDIS/STAR [11]
NESDIS High Frequency	AMSU/MHS	0.25/daily 1.0/pentad, monthly 2.5/pentad, monthly	Global/2000	NESDIS/STAR [12]
OPI	AVHRR	2.5/daily	Global/1979	NOAA/NWS CPC [13]
RSS	TMI, AMSRE, SSM/I, SSMIS, QSCAT	0.25/1-, 3-, 7-day	Monthly	RSS [14]
TRMM PR Precip (3 G68)	PR	0.5/hourly	Global-37 N-S/Dec. 1997	NASA/GSFC PPS [7]
AIRS	AIRS sounding retrievals	Swath/orbit segments	Global/May 2002	NASA/GSFC 610 [15]
CMORPH	TMI, AMSR-E, SSM/I, AMSU, IR vectors	8 km/30-min	50° N-S /1998	NOAA/CPC [16]
GSMaP NRT	TMI, AMSR-E, SSM/I, SSMIS, AMSU, IR vectors	0.1°/hourly	60° N-S /Oct. 2007	JAXA [17]

TABLE 2.4 (*Continued*)
List of Satellite Precipitation Products and Their Characteristics (Modified from Tapiador et al. 2012)

Precipitation Product	Satellite Sensors	Spatiotemporal Resolution	Areal Coverage/Start Date	Producer URL
GSMaP MWR	TMI, AMSR-E, AMSR, SSM/I, IR vectors	0.25°/hourly, daily, monthly	60° N-S /1998–2006	JAXA [18]
GSMaP MVK +	TMI, AMSR-E, AMSR, SSM/I, AMSU, IR vectors	0.1°/hourly	60° N-S /2003–2006	JAXA [18]
NRL Real TIme	SSM/I-cal PMM (IR)	0.25°/hourly	Global 40° N-S /July 2000	NRL Monterey [19]
TCI (3G68)	PR, TMI	0.5°/hourly	Global 37° N-S/Dec. 1997	NASA/GSFC PPS [20]
TOVS	HIRS, MSU	1°/daily	Global/1979–April 2005	NASA/GSFC 610 [15]
TRMM Real-Time HQ (3B40RT)	TMI, TMI-SSM/I, TMI-AMSU	0.25°/3-hourly	Global 70° N-S/Feb. 2005	NASA/GSFC PPS [21]
TRMM Real-Time VAR (3B41RT)	MW-VAR	0.25°/hourly	Global 50° N-S/Feb. 2005	NASA/GSFC PPS [22]
TRMM Real-Time HQVAR (3B42RT)	HQ, MW-VAR	0.25°/3-hourly	Global 50° N-S/Feb. 2005	NASA/GSFC PPS [23]

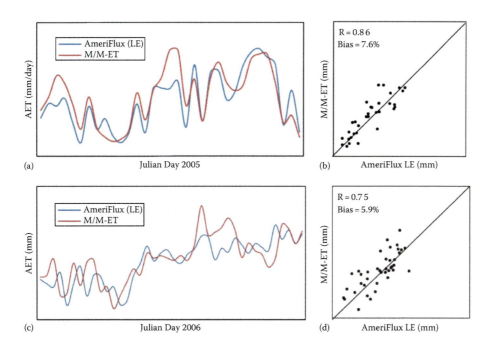

FIGURE 2.4 Comparisons of actual ET from AmeriFlux tower observations and SEB-based M/M-ET estimates at ARM SGP El Reno site. Panels (a) and (b) show the daily time series and scatter plot comparison for 2005; (c) and (d) are for 2006. Source: (Khan et al. 2010).

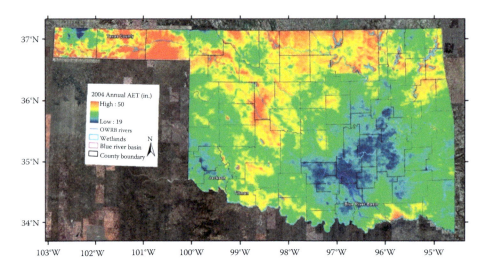

FIGURE 2.5 Annual AET over the entire state of Oklahoma for the year 2004.

Numerous studies have examined the SEB approach at the local scale to retrieve remotely sensed ET (Bastiaanssen et al. 1998a; Bastiaanssen and Harshadeep 2005; Su 2002; Allen et al. 2007). Liu et al. (2010) evaluated the spatiotemporal variations of actual ET for four types of LULC in urban settings in Oklahoma. Landsat 5 data and Oklahoma Mesonet (www.mesonet.org/) data were used for actual ET estimation. A meteorological gauged river basin in central Oklahoma, near Lovell, with an area of approximately 1033.59 km² was studied. The catchment includes very diverse land use land cover (LULC) from agriculture (Garfield County) to urban areas (Oklahoma County). The annual precipitation is around 870 mm, and average high and low temperatures are 21°C and 8 °C, respectively. The study watershed is in a semi-arid region where agriculture activities were mainly sustained by irrigation. This study evaluates the possible closure of the heat balance equation using unique environmental monitoring network; and to estimate actual ET and determine the variation with regard to varying types of LULC in urban settings (Figure 2.6). Overall, wetlands have the highest ET, wetlands and forests present a higher rate of ET than grass and agricultural lands, and the highly urbanized areas have the lowest ET (Figure 2.6).

2.4.3 Global Evapotranspiration Estimation

Until recently, long-term changes in ET have been evaluated by studying reference evaporation using measurements of pan evaporation. A research work published earlier showed that, on average, pan evaporation had decreased over North America, Europe, and Asia from the beginning of 1950 to the 1990s (Hobbins et al. 2004; Peterson et al. 1995). Recent studies (Amiri et al. 2024; Bouizrou et al. 2023; Govender et al. 2022; Jiao et al. 2021; Bhaga et al. 2020; Dang and Kumar 2017) have reiterated this to be an overall trend throughout the Northern latitudes. For instance, over the last half-century decreases in pan evaporation have been reported in India (Chattopadhyay and Hulme 1997), China (Thomas 2000), and parts of Europe (Moonen et al. 2002), albeit some mixed trends have also been reported, e.g., East Asia (Xu 2001) and a similar anomaly in the Middle East (Cohen et al. 2002; Eslamian et al. 2011). Another comprehensive study on the decrease in pan evaporation over the conterminous United States for the past half-century is presented by Hobbins et al. (2004). One of the critical points in most of these studies is that mean observations are used over a wide area, with some sites showing a decreasing trend while others show increasing anomalies.

Multi-Scale Hydrological Studies

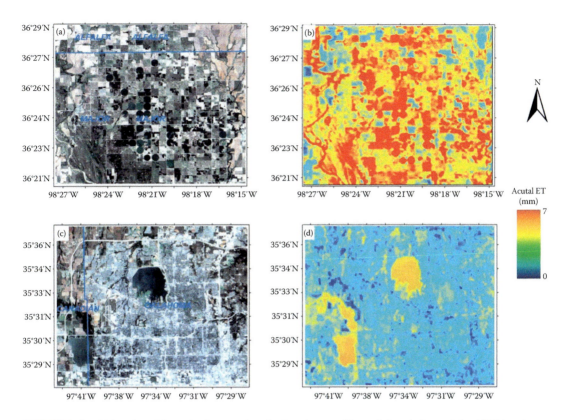

FIGURE 2.6 (a) Landsat false color image of agricultural areas, (b) spatial variations of the AET (mm) over the study area on July 31, 2005, (c) Landsat false color image of urban areas and nearby water body, and (d) distributed AET at urban areas and nearby water body. Source: modified from (Liu et al. 2010).

The advent of remote sensing estimation of land cover, surface temperature (Ts) and reflectance, vegetation indices (VI), emissivity, and surface albedo lead to development of ET algorithms based on a surface energy balance approach (Su 2002; Bastiaanssen and Harshadeep 2005; Allen et al. 2007). The three main remote sensing–based ET estimation methods can be categorized as: (1) empirical methods that integrate remotely sensed vegetation indices with measured ET (Glenn et al. 2008a; Glenn et al. 2008b; Jung et al. 2010; Nagler et al. 2005); (2) integrated remote sensing and classical Penman–Monteith approach (Monteith 1965) (Cleugh et al. 2007; Mu et al. 2007); and (3) physically based SEB models (Allen et al. 2007; Bastiaanssen and Harshadeep 2005; Kustas and Anderson 2009; Overgaard et al. 2006). For physically based SEB ET models, thermal-IR-based land surface temperature (LST) is a critical remote sensing variable (Bastiaanssen et al. 1998a, 1998b; Nishida et al. 2003; Su 2002).

Global terrestrial ET retrieval at a fine spatial scale was never achieved before until the satellite-driven estimation of the terrestrial ET, using the Moderate Resolution Imaging Spectroradiometer (MODIS) satellite sensor on board the Aqua satellite launched on May 4, 2002, with 36 spectral bands, 20 reflective solar bands, and 16 thermal emissive bands. MODIS provides exceptional data for observing vegetation and surface energy (Justice et al. 2002), which is utilized to develop a remotely sensed ET model (Mu et al. 2007). The moderated resolution global terrestrial ET algorithm is developed and refined by the Numerical Terradynamic Simulation Group (NTSG) is a research laboratory at the University of Montana in Missoula. The data product derived is known as the "MOD16" and is defined as an "evaporation fraction" (EF), the energy budget equivalent of an index of actual to potential ET, over the global land surface with 1-km resolution every eight days (Mu 2007; Mu et al. 2011). MODIS data related to ET retrieval are land surface temperature and emissivity (MOD11), surface reflectance products (MOD09), vegetation index (MOD13), and albedo (MOD43B3) obtained by assimilating bi-hemispherical reflectance data modeled. These data sets were acquired from the Land Processes Distributed Active Archive Center (LP DAAC) at the USGS Earth Resources Observation and Science (EROS) Center, with the standard Hierarchical Data Format (http://LPDAAC.usgs.gov). For more information on MODIS, please refer to http://modis.gsfc.nasa.gov.

2.5 SOIL MOISTURE ESTIMATION BASED ON MULTI-SENSOR DATA

Soil moisture is one of the most crucial hydrological components in the water cycle. While satellite remote sensing is the most cost-effective approach to estimate the soil moisture consistently on a large spatial scale (e.g., over the globe), the in situ soil moisture provides more accurate ground validation reference for the development of the satellite soil moisture estimation. Table 2.5 lists the sensors applied in both in situ and remote sensing soil moisture measurement from the past (AMSR-E) to the present (AMSR2 and SMOS) and (SMAP). The in situ soil moisture measurement will be elaborated in Section 2.5.1, and the SMOS and SMAP soil moisture mission will be introduced in Section 21.5.2.

2.5.1 REGIONAL-, LOCAL-, AND FIELD-SCALE STUDIES

Observation of soil water content has been accomplished worldwide in the past few decades (Meresa 2019; Shao et al. 2019; Dong 2018; Ha et al. 2018; Herman et al. 2018; Kandus et al. 2018; Dang and Kumar 2017; Gleason et al. 2017; Chert et al. 2008; Choi et al. 2008; Jackson et al. 2005; Jackson and Lettenmaier 2004; Jacobs et al. 2004; Kabela et al. 2009; Narayan et al. 2006; PIs of SMEX03

TABLE 2.5
Sensors Applied in Both In Situ and Remote Sensing Soil Moisture Measurement

	Satellite Mission/ Sensor	Sensor Type	Measurement	Spatial Resolution	Temporal Resolution
In situ (Oklahoma Mesonet)	N/A	BetaTHERM[1] Probe Scientific 229-L[2] Probe	Soil Temperature	Point-Based	15 min
Satellite Remote Sensing	AMSR-E (2002–2011)	Radiometer (6 bands: 6.9 to 89GHz)	Brightness Temperature	25 km	Daily
	AMSR2 (2011–)	Radiometer (6 bands: 6.9 to 89GHz)	Brightness Temperature	10/25 km	Daily
	SMOS (2009–)	Radiometer (L-Band 1.4Ghz)	Microwave Radiation	35–50 km	Daily
	SMAP (2014–)	Radar (L-Band 1.26GHz)	Backscattering	3km	Daily
		Radiometer (L-Band 1.4GHz)	Brightness Temperature	40km	

Note: [1] Before 2012/12/01; [2] after 2012/12/01.

TABLE 2.6
Past Successful Soil Moisture Field Campaigns

Campaign	Period	Location	Primary Data Types	Data Availability
SMEX02	June/July 2002	Iowa	Radiometer, SAR, in situ	Online
SMEX03	June/July 2003	Oklahoma, Georgia, Alabama	Radiometer, SAR, in situ	Online
SMEX04	August 2004	Arizona	Radiometer, SAR, in situ	Online
Washita 92	June 1992	Oklahoma	Radiometer, SAR, in situ	Online
SGP97	June/July 1997	Oklahoma	Radiometer, SAR, in situ	Online

2003; Service 2002; Vivoni et al. 2008; Yilmaz et al. 2008). The dimensionless variable volumetric soil moisture (VSM) that refers to the water volume in unit volume (Ulaby et al. 1986) of soil is usually used to quantitatively represent the level of water content in soil.

The Southern Great Plains (SGP) of the United States and adjacent regions include multiple arrays of in situ observations that provide estimates of soil moisture at local and regional scales. At the regional scale, an in situ observation network such as Oklahoma Mesonet, which has an average station spacing of approximately 30 km, provides high spatiotemporal soil moisture and other meteorological estimates. Table 2.6 lists some of the specific successful campaigns of the past and the ongoing experiments:

The Oklahoma Mesonet is a permanent mesoscale observing network of 120 meteorological stations across Oklahoma (Brock et al. 1995; McPherson et al. 2007). It is managed by the Oklahoma Climatological Survey in partnership with Oklahoma State University and the University of Oklahoma. Each station measures more than 20 environmental variables, including wind at 2 and 10 m, air temperature at 1.5 and 9 m, relative humidity, rainfall, pressure, solar radiation, and soil temperature and moisture at various depths (Illston et al. 2008). Mesonet data are collected and transmitted to a central point every five minutes where they are quality controlled, distributed, and archived (Shafer et al. 2000; McPherson et al. 2007). Oklahoma Mesonet data has supported a broad range of scientific research with a bibliography of over 500 peer-reviewed articles. In particular, multiple studies have used Oklahoma Mesonet data for analysis of satellite algorithms focused on soil moisture retrieval (Gu et al. 2008; Swenson et al. 2008; Pathe et al. 2009; Collow et al. 2012).

The Marena Oklahoma In Situ Sensor Testbed (MOISST, illustrated in Figure 2.7a) was installed in May 2010 as part of the calibration and validation program for the SMAP mission. The site includes more than 200 soil, vegetation, and atmospheric sensors installed over an approximately 64-ha pasture in Central Oklahoma with four main stations and multiple in situ sensors installed in profiles. Additionally, located at the site are a COsmic-ray Soil Moisture Observing System (Hornbuckle et al. 2012), global position system reflectometers, a passive distributed temperature system, an eddy correlation flux tower (installed November 2011), and a phenocam (installed May 2012). To enhance the observations from the MOISST site and to calibrate the existing sensor, field samples of soil and vegetation are routinely collected. Other in-situ measurements of soil moisture and surface atmosphere interactions can be acquired from similar instrumentations at the integrated Grassland Observing Site (iGOS) (Figure 2.7b).

Field measurements provide direct, but only point, samples of soil moisture. In contrast, remote sensing sensors on aerial or satellite platforms are capable of measuring surface VSM at the surface scale and relatively high temporal intervals but indirectly through retrieval techniques. Among a variety of bands, microwave are most frequently used for soil moisture because soil moisture affects the soil dielectric constant (permittivity) greatly at microwave frequency and thus sensitive to microwave scattering and emission.

FIGURE 2.7 Sensor arrays deployed at (a) the MOISST site near Marena, OK, and (b) the iGOS site near El Reno, OK.

2.5.2 GLOBAL SOIL MOISTURE ESTIMATION

Spatially distributed soil moisture measurements and freeze/thaw states are needed to improve our understanding of regional and global water cycles. SMOS is one of the missions that provides global observations of soil moisture over Earth's landmasses and salinity over the oceans. It is commonly known as the Water Mission and is meant to provide new insights into Earth's water cycle and climate. In addition, it aims to monitor snow and ice accumulation and to provide better weather forecasting. The mission has a low-Earth, polar, Sun-synchronous orbit at an altitude of 758 km. An important aspect of this mission is that it carries out a completely new measuring technique: the first polar-orbiting spaceborne 2D interferometric radiometer instrument called the Microwave Imaging Radiometer with Aperture Synthesis (MIRAS). This novel instrument is capable of observing both soil moisture and ocean salinity by capturing images of emitted microwave radiation around the frequency of 1.4 GHz or wavelength of 21 cm (L-band) (Kerr et al. 2001; Bayle et al. 2002; Font et al. 2004; Moran et al. 2004). The goal of this mission is to measure soil moisture with an accuracy of 4%, volumetric soil moisture at 35–50 km spatial resolution, 1–3 day revisit time, and ocean surface salinity with an accuracy of 0.5–1.5 Practical Salinity Unit (PSU) for a single observation at 200 km spatial resolution and 10–30 day temporal resolution (Delwart et al. 2008).

2.5.3 SATELLITE MISSION ON SOIL MOISTURE ESTIMATION

2.5.3.1 Soil Moisture, Active and Passive (SMAP) Mission

Soil Moisture, Active and Passive (SMAP) is one of four first-tier missions recommended by the NRC Earth Science Decadal Survey Report. SMAP provides global views of Earth's soil moisture and surface freeze/thaw state, introducing a new era in hydrologic applications and providing unprecedented capabilities to investigate the water, energy, and carbon cycles over global land surfaces. Moreover, these estimates are also helpful in understanding terrestrial ecosystems, and the processes that interlink the water, energy, and carbon cycles.

Soil moisture and freeze/thaw information provided by SMAP will lead to improved weather, flood, and drought forecasts, and predictions of agricultural productivity and climate change. This mission will contribute to the goals of the Carbon Cycle and Ecosystems, Weather, and Climate Variability and Change Earth Science focus areas as well as to hydrological science.

The SMAP mission, based on one flight system in a sun-synchronous orbit, at an altitude of 670 km with an eight-day repeat cycle. It includes synthetic aperture radar operating at L-band (frequency: 1.26 GHz; polarizations: HH, VV, HV) and an L-band radiometer (frequency: 1.41 GHz;

polarizations: H, V, U). At this altitude, the antenna scan design yields a 1000-km swath, with a 40-km radiometer resolution and 1–3-km Synthetic Aperture Radar (SAR) resolution that provides global coverage within three days at the equator and two days at boreal latitudes (> 45°N). The main goal of SMAP is to provide estimates of soil moisture in the top 5 cm of soil with an accuracy of 0.04 cm3/cm3 volumetric soil moisture, at 10-km resolution, with three-day average intervals over the global land area. These measurements will not be suitable for regions with snow and ice, mountainous topography, open water, and vegetation with total water content greater than 5 kg/m2 (Entekhabi et al. 2010).

2.5.3.2 Surface Water and Ocean Topography (SWOT) Mission

SWOT was one of the recommended missions during the 2007 National Research Council decadal survey, and was launched in 2022. It provides observations on lake and river water levels for inland water dynamics. The mission will measure water surface elevations, water surface slope, and the areal extent of lakes, wetlands, reservoirs, floodplains, and rivers globally (Alsdorf et al. 2007; Alsdorf et al. 2003). The core technology in the SWOT mission is the wide swath of Ka-band and C-band radar interferometer (KaRIN), which would achieve spatial resolution to the order of tens of meters (Alsdorf et al. 2003; http://swot.jpl.nasa.gov/).

The SWOT satellite mission is intended to indirectly estimate river flow fluctuations from remotely sensed river hydrologic features. The SWOT mission aims to offer precise river surface slope estimates; moreover, it will provide estimates on water surface elevations and inundated areas for rivers with widths greater than about 100 m (Durand et al. 2010). There are numerous inadequacies of current altimeters ranging from less frequent visits of the sensor and how the size of inland water bodies are measured, the SWOT mission compensates for these limitations (Alsdorf et al. 2003; Alsdorf et al. 2007). Preliminary case studies for this approach using synthetic data showed some promising results (Lee et al. 2010). Earth scientists and other researchers are optimistic about the SWOT mission and its application in hydrologic science and water resource management throughout the globe.

2.6 GROUND WATER ASSESSMENT FROM GRACE

The GRACE satellites have been widely used to assess drought severity and extent in different regions, including the California Central Valley, southeastern United States, and more recently to develop drought indicators for soil moisture and groundwater using data assimilation with land surface models (Famiglietti et al. 2011; Houborg et al. 2012; Scanlon et al. 2012). The large-scale nature of GRACE water storage changes can be valuable for providing a regional assessment of overall drought impacts and can also be used to estimate water requirements to overcome cumulative drought impacts. While most studies disaggregate total water storage into surface water, soil moisture, and groundwater, terrestrial water storage (TWS) is also very valuable in its own right. For instance, reservoir storage is monitored in most states and can be subtracted from GRACE TWS to estimate subsurface water storage changes. Disaggregating subsurface water storage into soil moisture and groundwater is complicated because of limited information on soil moisture and uncertainties in simulated soil moisture from land surface models, as seen from the application of GRACE during the 2011 drought in Texas. Texas has one of the most advanced groundwater-level monitoring programs of any state, making it feasible to estimate groundwater storage changes from monitoring data and comparing it with GRACE data. Soil moisture estimates from SMAP and the Oklahoma Mesonet can be compared with the simulated soil moisture from various land surface models to assess their reliability.

Correspondence between GRACE TWS and PDSI in Texas (Figure 2.8) shows that TWS can be valuable as a drought predictor (Long et al. 2013). Therefore, GRACE can provide a valuable tool in the portfolio of seasonal drought predictors for the Texas-Oklahoma region and will build on previous experience in applying this tool in Texas and throughout the High Plains.

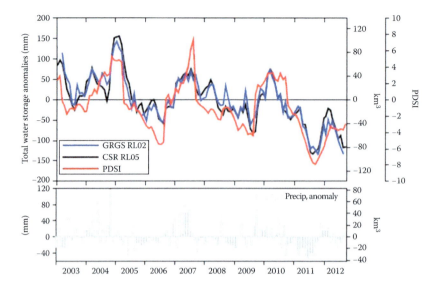

FIGURE 2.8 Relationship between GRACE Total Water Storage from two sources (GRGS and CSR) and PDSI ([Long et al. 2013]).

2.7 SUMMARY AND CONCLUSIONS

Global monitoring of geosystem processes with satellite remote sensing improved our understanding of the connections between the water, energy, and carbon cycles (Amiri et al. 2024; Saha and Pal 2024; Ali et al. 2023; Bouizrou et al. 2023; Govender et al. 2022; Kabeja et al. 2022; Duan et al. 2021; Jiao et al. 2021; Bhaga et al. 2020; Jiang and Wang 2019; Meresa 2019; Shao et al. 2019; Dong 2018; Ha et al. 2018; Herman et al. 2018; Kandus et al. 2018; Dang and Kumar 2017; Gleason et al. 2017; Roy et al. 2017; Thakur et al. 2017a; Thakur et al. 2017b; Zeng et al. 2017). Given the impact of hydrologic extremes in a changing climate, characterization of real-time precipitation, evapotranspiration, and soil moisture with higher spatiotemporal variability is crucial to monitor hydrologic processes at multiple scales. As discussed in the preceding sections, the fundamental terms in the hydrologic budget equation can now be estimated through satellite sensor. Several remote sensing–based datasets from current and future satellite missions are intended to quantify the water balance components, at different spatial and temporal scales. For instance, global precipitation can be retrieved at a high temporal resolution (three-hourly) by combining microwave and infrared-based satellite measurements (Sorooshian et al. 2000; Kummerow et al. 2001; Joyce et al. 2004; Huffman et al. 2007). Energy balance estimation and empirical models has led to moderate resolution and large-scale estimates of ET. Recent research has focused on using radar to determine soil moisture content and the depth of the groundwater table, specifically using L-band and ground-penetrating radars.

The accuracy of water budget estimates at different space and time scales is debatable and therefore an open research question. The retrieval of hydrologic variables based on single sensors might not be as reliable as expected due to the dependence on a number of ground parameters with spatial and temporal variability. The uncertainty in these estimates can be attributed to the satellite instrument type, the retrieval algorithms, and the spatial temporal resolution of these different estimates. Another possible reason is that any single satellite instrument does not measure all the water budget components simultaneously (Ferguson and Wood 2010; Sheffield et al.

2010). Robust retrieval algorithms as well as data assimilation methods that ingest multi-sensor measurements to resolve the water budget at local and regional scales are active areas of research.

This chapter focused on some of the main remote sensing datasets from satellite sensors such as GPM, SMAP NPP, GOES-R, SWOT, and other missions. Specifically, currently used three-hour GOES observations should be increased to at least one hour from GOES-R observations to better capture the diurnal variation of cloud and radiation fields. Current and future Satellites will provide high-quality, seamless, and inexpensive Earth observation. Moreover, future Earth Science missions will ensure the continuity of several key climate measurements. At both regional and global scales, the real-time operational hydrological monitoring is supported by the current and future missions.

Despite the promise of global high-resolution monitoring of the individual components of the terrestrial water budget, there remain considerable challenges in providing physically consistent and accurate estimates (Sheffield et al. 2009). The hydrologic community currently lacks both sufficient understanding from field studies and quantitative models to make reliable estimates about desired outcomes from management decisions in many hydrologic applications. There are opportunities to study and estimate the contributing variables by employing remotely sensed data and maximizing their use in hydrological studies. The main limitations for estimating local and sometimes regional-level hydrologic state variables from remote sensing techniques are the retrieval precision of land surface fluxes and the spatial heterogeneity and scaling issues. Missions such as the SWOT, GPM, and SMAP will improve our understanding of the terrestrial water cycle with a global coverage of major rivers and watershed. These datasets will supplement in situ observations for better data retrieval, combined with robust data assimilation techniques, are expected to significantly benefit water budget estimation at ungauged river basins at different spatiotemporal scales. We expect that improvements in sensor technology and novel data fusion, artificial intelligence, and machine learning techniques will lead to better estimation of hydrologic variables and water balance closure.

NOTE

1   http://science.nasa.gov/media/medialibrary/2010/07/01/Climate_Architecture_Final.pdf

REFERENCES

Adler, R. F., G. J. Huffman, A. Chang, R. Ferraro, P.-P. Xie, J. Janowiak, B. U. Rudolf et al. 2003. The version-2 Global Precipitation Climatology Project (GPCP) monthly precipitation analysis (1979—present). *Journal of Hydrometeorology* 4 (6):1147–1167.

Ali, M. H., I. Popescu, A. Jonoski, and D. P. Solomatine. 2023. Remote sensed and/or global datasets for distributed hydrological modelling: A review. *Remote Sensing* 15 (6):1642. https://doi.org/10.3390/rs15061642

Allen, R. G., M. Tasumi, A. Morse, R. Trezza, J. L. Wright, and W. Bastiaanssen et al. 2007. Satellite-based energy balance for Mapping Evapotranspiration with Internalized Calibration (METRIC)—applications. *Journal of Irrigation and Drainage Engineering* 133 (4):395–406. DOI: 10.1061(ASCE)0733-9437

Alsdorf, D., D. Lettenmaier, and C. Vörösmarty. 2003. The need for global, satellite-based observations of terrestrial surface waters. *Eos, Transactions American Geophysical Union* 84:269–276.

Alsdorf, D. E., E. Rodriguez, and D. P. Lettenmaier. 2007: Measuring surface water from space. *Reviews of Geophysics* 45:RG2002. DOI: 10.1029/2006RG000197

Amiri, A., K. Soltani, I. Ebtehaj, and H. Bonakdari. 2024. A novel machine learning tool for current and future flood susceptibility mapping by integrating remote sensing and geographic information systems. *Journal of Hydrology* 632 (2024):130936. ISSN 0022–1694, https://doi.org/10.1016/j.jhydrol.2024.130936. Available from www.sciencedirect.com/science/article/pii/S0022169424003305

Arkin, P. A., and B. N. Meisner. 1987. The relationship between large-scale convective rainfall and cold cloud over the western hemisphere during 1982–84. *Monthly Weather Review* 115:51–74.

Ba, M. B., and A. Gruber. 2001. GOES Multispectral Rainfall Algorithm (GMSRA). *Journal of Applied Meteorology and Climatology* 40:1500–1514.

Bastiaanssen, W. G. M., and N. R. Harshadeep. 2005. Managing scarce water resources in Asia: The nature of the problem and can remote sensing help, Guest Editors Allen and Bastiaanssen. *Irrigation and Drainage Systems* 19:269–284.

Bastiaanssen, W. G. M., M. Menenti, R. A. Feddes, and A. A. M. Holtslag. 1998a. The Surface Energy Balance Algorithm for Land (SEBAL): Part 2 validation. *Journal of Hydrology* 212–213:213–229.

Bastiaanssen, W. G. M., M. Menenti, R. A. Feddes, and A. A. M. Holtslag. 1998b. A remote sensing surface energy balance algorithm for land (SEBAL): 1. Formulation. *Journal of Hydrology* 212–213:198–212.

Battan, L. J. 1973. *Radar Observation of the Atmosphere*. Chicago: University of Chicago Press.

Bayle, F, J. P. Wigneron, Y. H. Kerr, P. Waldteufel, E. Anterrieu, J. C. Orlhac, A. Chanzy, O. Marloie, M. Bernardini, and S. Sobjaerg. 2002. Two-dimensional synthetic aperture images over a land surface scene. *IEEE Transactions on Geoscience and Remote Sensing* 40 (3):710–714.

Bellerby, T. J. 2004. A feature-based approach to satellite precipitation monitoring using geostationary IR imagery. *Journal of Hydrometeorology* 5:910–921.

Bhaga, T. D., T. Dube, M. D. Shekede, and C. Shoko. 2020. Impacts of climate variability and drought on surface water resources in Sub-Saharan Africa using remote sensing: A review. *Remote Sensing* 12 (24):4184. https://doi.org/10.3390/rs12244184

Bolten, J. D., V. Lakshmi, and E. G. Njoku. 2003. Soil moisture retrieval using the passive/active L-and S-band radar/radiometer. *IEEE Transactions on Geoscience and Remote Sensing* 41 (12):2792–2801.

Bouizrou, I., A. Bouadila, M. Aqnouy, and A. Gourfi. 2023. Assessment of remotely sensed precipitation products for climatic and hydrological studies in arid to semi-arid data-scarce region, central-western Morocco. *Remote Sensing Applications: Society and Environment* 30 (2023):100976, ISSN 2352–9385, https://doi.org/10.1016/j.rsase.2023.100976. Available from: www.sciencedirect.com/science/article/pii/S2352938523000587

Bringi, V., and V. Chandrasekar. 2001. *Polarimetric Doppler Weather Radar: Principles and Applications*. Cambridge, UK: Cambridge University Press.

Brock, F. V., K. C. Crawford, R. L. Elliott, G. W. Cuperus, S. J. Stadler, H. L. Johnson, and M. D. Eilts. 1995. The Oklahoma Mesonet: A technical overview. *Journal of Atmospheric and Oceanic Technology* 12 (1):5–19.

Chattopadhyay, N., and M. Hulme. 1997. Evaporation and potential evapotranspiration in India under conditions of recent and future climate change. *Agricultural and Forest Meteorology* 87 (1):55–73.

Chauhan, N. S. 1997. Soil moisture estimation under a vegetation cover: Combined active passive microwave remote sensing approach. *International Journal of Remote Sensing* 18 (5):1079–1097.

Chert, Q., Z. Li, L. Wang, Y. Shao, and T. Cheng. 2008. Soil moisture change retrieval using S-band radar data during SGP99 and SMEX02.

Choi, M., J. M. Jacobs, and D. D. Bosch. 2008. Remote sensing observatory validation of surface soil moisture using Advanced Microwave Scanning Radiometer E, Common Land Model, and ground based data: Case study in SMEX03 Little River Region, Georgia, US. *Water Resources Research* 44 (8):W08421.

Cleugh, H. A., R. Leuning, Q. Mu, and S. W. Running. 2007. Regional evaporation estimates from flux tower and MODIS satellite data. *Remote Sensing of Environment* 106:285–304.

Cohen, S., A. Ianetz, and G. Stanhill. 2002. Evaporative climate changes at Bet Dagan, Israel, 1964–1998. *Agricultural and Forest Meteorology* 111 (2):83–91.

Collow, T. W., A. Robock, J. B. Basara, and B. G. Illston. 2012. Evaluation of SMOS retrievals of soil moisture over the central United States with currently available in situ observations. *Journal of Geophysical Research: Atmospheres (1984–2012)* 117 (D9).

Dang, A. T. N. and L. Kumar. 2017. Application of remote sensing and GIS-based hydrological modelling for flood risk analysis: A case study of District 8, Ho Chi Minh city, Vietnam, Geomatics. *Natural Hazards and Risk* 8 (2):1792–1811. DOI: 10.1080/19475705.2017.1388853

Delwart, S, C. Bouzinac, P. Wursteisen, M. Berger, M. Drinkwater, M. Martin-Neira, and Y. H. Kerr. 2008. SMOS validation and the COSMOS campaigns. *IEEE Transactions on Geoscience and Remote Sensing* 46 (3):695–704.

Dong, C. 2018. Remote sensing, hydrological modeling and in situ observations in snow cover research: A review. *Journal of Hydrology* 561 (2018):573–583, ISSN 0022–1694, https://doi.org/10.1016/j.jhydrol.2018.04.027. Available from www.sciencedirect.com/science/article/pii/S0022169418302804

Duan, W., S. Maskey, P. L. B. Chaffe, P. Luo, B. He, Y. Wu, and J. Hou. 2021. Recent advancement in remote sensing technology for hydrology analysis and water resources management. *Remote Sensing* 13 (6):1097. https://doi.org/10.3390/rs13061097

Durand, M., L.-L. Fu, D. P. Lettenmaier, D. E. Alsdorf, E. Rodriguez, and D. Esteban-Fernandez. 2010. The surface water and ocean topography mission: Observing terrestrial surface water and oceanic submesoscale eddies. *Proceedings of the IEEE* 98 (5):766–779.

Entekhabi, D., E. G. Njoku, P. E. O'Neill, K. H. Kellogg, W. T. Crow, W. N. Edelstein, J. K. Entin et al. 2010. The Soil Moisture Active Passive (SMAP) mission. *Proceedings of the IEEE* 98 (5):704, 716, May. DOI: 10.1109/JPROC.2010.2043918

Eslamian, S., M. J. Khordadi, and J. Abedi-Koupai. 2011. Effects of variations in climatic parameters on evapotranspiration in the arid and semi-arid regions. *Global and Planetary Change* 78 (3–4):188–194.

Famiglietti, J., M. Lo, S. Ho et al. (2011). Satellites measure recent rates of groundwater depletion in California's Central Valley. *Geophysical Research Letters* 38:L03403.

Ferguson, C. R., and E. F. Wood. 2010. An evaluation of satellite remote sensing data products for land surface hydrology: Atmospheric infrared sounder. *Journal of Hydrometeorology* 11:1234–1262. https://doi.org/10.1175/2010JHM1217.1

Flaming, G. M. 2005. Global precipitation measurement update. 2005. Geoscience and Remote Sensing Symposium, 2005. IGARSS'05. *Proceedings. IEEE International.* 1.

Font, J, G. S. E. Lagerloef, D. M. Le Vine, A. Camps, and O. Z. Zanife. 2004. The determination of surface salinity with the European SMOS space mission. *IEEE Transactions on Geoscience and Remote Sensing* 42 (10):2196–2205.

Gleason, C. J., Y. Wada, and J. Wang. 2017. A hybrid of optical remote sensing and hydrological modeling improves water balance estimation. *Journal of Advances in Modeling Earth Systems* 10 (1):2–17.

Glenn, E. P., A. R. Huete, P. L. Nagler, and S. G. Nelson. 2008a. Relationship between remotely-sensed vegetation indices, canopy attributes and plant physiological processes: What vegetation indices can and cannot tell us about the landscape. *Sensors* 8:2136–2160.

Glenn, E. P., K. Morino, K. Didan, F. Jordan, K. C. Carroll, and P. L. Nagler et al. 2008b. Scaling sap flux measurements of grazed and ungrazed shrub communities with fine and coarse-resolution remote sensing. *Ecohydrology* 1 (4), 316–329.

Govender, T., T. Dube, and C. Shoko. 2022. Remote sensing of land use-land cover change and climate variability on hydrological processes in Sub-Saharan Africa: Key scientific strides and challenges. *Geocarto International* 37 (25):10925–10949. DOI: 10.1080/10106049.2022.2043451

Gu, Y., E. Hunt, B. Wardlow, J. B. Basara, J. F. Brown, and J. P. Verdin. 2008. Evaluation of MODIS NDVI and NDWI for vegetation drought monitoring using Oklahoma Mesonet soil moisture data. *Geophysical Research Letters* 35 (22).

Ha, L. T., W. G. M. Bastiaanssen, A. V. Griensven, A. I. J. M. Van Dijk, and G. B. Senay. 2018. Calibration of spatially distributed hydrological processes and model parameters in SWAT using remote sensing data and an auto-calibration procedure: A case study in a Vietnamese river Basin. *Water* 10 (2):212. https://doi.org/10.3390/w10020212

Herman, M. R., A. P. Nejadhashemi, M. Abouali, J. S. Hernandez-Suarez, F. Daneshvar, Z. Zhang, M. C. Anderson, A. M. Sadeghi, C. R. Hain, and A. Sharifi. 2018. Evaluating the role of evapotranspiration remote sensing data in improving hydrological modeling predictability. *Journal of Hydrology* 556 (2018):39–49, ISSN 0022–1694, https://doi.org/10.1016/j.jhydrol.2017.11.009. Available from www.sciencedirect.com/science/article/pii/S0022169417307631

Hobbins, M. T., J. A. Ramírez, and T. C. Brown. 2004. Trends in pan evaporation and actual evapotranspiration across the conterminous US: Paradoxical or complementary. *Geophysical Research Letters* 31 (13):1–5.

Hong, Y., K.-L. Hsu, S. Sorooshian, and X. Gao. 2004. Precipitation estimation from remotely sensed imagery using an artificial neural network cloud classification system. *Journal of Applied Meteorology* 43:1834–1853.

Hong, Y., K.-L. Hsu, S. Sorooshian, and X. Gao. 2005. Improved representation of diurnal variability of rainfall retrieved from the tropical rainfall measurement mission microwave imager adjusted Precipitation Estimation from Remotely Sensed Information using Artificial Neural Networks (PERSIANN) system. *Journal of Geophysical Research* 110:D06102.

Hornbuckle, B., S. Irvin, T. Franz, R. Rosolem, and C. Zweck. 2012. The potential of the COSMOS network to be a source of new soil moisture information for SMOS and SMAP. Paper read at Geoscience and Remote Sensing Symposium (IGARSS), 2012 IEEE International.

Houborg, R., M. Rodell, B. Li, et al. 2012. Drought indicators based on model-assimilated Gravity Recovery and Climate Experiment (GRACE) terrestrial water storage observations. *Water Resources Research* 48:W07525.

Hsu, K., X. Gao, S. Sorooshian, and H. V. Gupta. 1997. Precipitation estimation from remotely sensed information using artificial neural networks. *Journal of Applied Meteorology* 36:1176–1190.

Huffman, G. J., R. F. Adler, M. Morrissey, D. T. Bolvin, S. Curtis, R. Joyce, B. McGavock, J. Susskind. 2001. Global precipitation at one-degree daily resolution from multi-satellite observations. *Journal of Hydrometeorology* 2 (1):36–50.

Huffman, G. J., D. T. Bolvin, E. J. Nelkin, D. B. Wolff, R. F. Adler, G. Gu, Y. Hong, K. P. Bowman, and E. F. Stocker. 2007. The TRMM Multisatellite Precipitation Analysis (TMPA): Quasi-global, multiyear, combined-sensor precipitation estimates at fine scales. *Journal of Hydrometeorology* 8 (1):38–55.

Illston, B. G., J. B. Basara, D. K. Fischer, R. L. Elliott, C. Fiebrich, K. C. Crawford, K. Humes, and E. Hunt. 2008. Mesoscale monitoring of soil moisture across a statewide network. *Journal of Atmospheric and Oceanic Technology* 25:167–182.

Jackson, T. J., R. Bindlish, A. J. Gasiewski, B. Stankov, M. Klein, E. G. Njoku, D. Bosch, T. L. Coleman, C. A. Laymon, and P. Starks. 2005. Polarimetric scanning radiometer C-and X-band microwave observations during SMEX03. *IEEE Transactions on Geoscience and Remote Sensing* 43 (11):2418–2430.

Jackson, T. J., and D. Lettenmaier. 2004. Soil moisture experiments 2004 (SMEX04). Paper read at AGU Spring Meeting.

Jacobs, J. M., B. P. Mohanty, E. C. Hsu, and D. Miller. 2004. SMEX02: Field scale variability, time stability and similarity of soil moisture. *Remote Sensing of Environment* 92 (4):436–446.

Jiang, D., and K. Wang. 2019. The role of satellite-based remote sensing in improving simulated streamflow: A review. *Water* 11 (8):1615. https://doi.org/10.3390/w11081615

Jiao, W., L. Wang, and M. F. McCabe. 2021. Multi-sensor remote sensing for drought characterization: Current status, opportunities and a roadmap for the future. *Remote Sensing of Environment* 256 (2021):112313, ISSN 0034–4257, https://doi.org/10.1016/j.rse.2021.112313. Available from www.sciencedirect.com/science/article/pii/S0034425721000316

Joyce, R. J., J. E. Janowiak, P. A. Arkin, and P. Xie. 2004. CMORPH: A method that produces global precipitation estimates from passive microwave and infrared data at high spatial and temporal resolution. *Journal of Hydrometeorology* 5 (3):487–503.

Jung, M., M. Reichstein, P. Ciais, S. I. Seneviratne, J. Sheffield, and M. L. Goulden et al. 2010. Recent decline in the global land evapotranspiration trend due to limited moisture supply. *Nature* 467:951–954.

Justice, C., J. Townshend, E. Vermote, E. Masuoka, R. Wolfe, N. Saleous, D. Roy, and J. Morisette. 2002. An overview of MODIS Land data processing and product status. *Remote Sensing of Environment* 83 (1–2):3–15.

Kabeja, C., R. Li, D. E. Rwabuhungu Rwatangabo, and J. Duan. 2022. Monitoring land use/cover changes by using multi-temporal remote sensing for urban hydrological assessment: A case study in Beijing, China. *Remote Sensing* 14 (17):4273. https://doi.org/10.3390/rs14174273

Kabela, E. D., B. K. Hornbuckle, M. H. Cosh, M. C. Anderson, and M. L. Gleason. 2009. Dew frequency, duration, amount, and distribution in corn and soybean during SMEX05. *Agricultural and Forest Meteorology* 149 (1):11–24.

Kandus, P., P. G. Minotti, N. S. Morandeira, R. Grimson, G. G. Trilla, E. B. González, L. S. Martín, and M. P. Gayol. 2018. Remote sensing of wetlands in South America: Status and challenges. *International Journal of Remote Sensing*, 39 (4):993–1016. DOI: 10.1080/01431161.2017.1395971

Kerr, Y. H., P. Waldteufel, J. P. Wigneron, J. Martinuzzi, J. Font, and M. Berger. 2001. Soil moisture retrieval from space: The Soil Moisture and Ocean Salinity (SMOS) mission. *IEEE Transactions on Geoscience and Remote Sensing* 39 (8):1729–1735.

Khan, S. I., Y. Hong, B. Vieux, and W. Liu. 2010. Development evaluation of an actual evapotranspiration estimation algorithm using satellite remote sensing meteorological observational network in Oklahoma. *International Journal of Remote Sensing* 31 (14):3799–3819.

Kidd, C. K., D. R. Kniveton, M. C. Todd, and T. J. Bellerby. 2003: Satellite rainfall estimation using combined passive microwave and infrared algorithms. *Journal of Hydrometeorology* 4:1088–1104.

Kuligowski, R. J., 2002. A self-calibrating real-time GOES rainfall algorithm for short-term rainfall estimates. *Journal of Hydrometeorology* 3:112–130.

Kummerow, C., Y. Hong, W. S. Olson, S. Yang, R. F. Adler, J. McCollum, R. Ferraro, et al. 2001. The evolution of the Goddard Profiling Algorithm (GPROF) for rainfall estimation from passive microwave sensors. *Journal of Applied Meteorology* 40 (11):1801–1820.

Kustas, W., and Anderson, M. 2009. Advances in thermal infrared remote sensing for land surface modeling. *Agricultural and Forest Meteorology* 149:2071–2081.

Lee, H. et al. 2010. Characterization of surface water storage changes in Arctic lakes using simulated SWOT measurements. *International Journal of Remote Sensing* 31 (14):3931–3953.

Liu, W., Y. Hong, S. I. Khan, M. Huang, B. Vieux, S. Caliskan, and T. Grout. 2010. Actual evapotranspiration estimation for different land use and land cover in urban regions using Landsat 5 data. *Journal of Applied Remote Sensing* 4 (1):041873–041873–14.

Loheide, S. P. and S. M. Gorelick. 2005. A local-scale, high-resolution evapotranspiration mapping algorithm (ETMA) with hydroecological applications at riparian meadow restoration sites. *Remote Sensing of Environment* 98:182–200.

Long, D., B. R. Scanlon, L. Longuevergne, et al. 2013. GRACE satellite monitoring of large depletion in water storage in response to the 2011 drought in Texas. *Geophysical Research Letters* 40 (13):3395–3401.

McPherson, R. A., C. A. Fiebrich, K. C. Crawford, J. R. Kilby, D. L. Grimsley, J. E. Martinez, J. B. Basara, B. G. Illston, D. A. Morris, and K. A. Kloesel. 2007. Statewide monitoring of the mesoscale environment: A technical update on the Oklahoma Mesonet. *Journal of Atmospheric and Oceanic Technology* 24 (3):301–321.

Meresa, H. 2019. Modelling of river flow in ungauged catchment using remote sensing data: Application of the empirical (SCS-CN), Artificial Neural Network (ANN) and Hydrological Model (HEC-HMS). *Modeling Earth Systems and Environments* 5:257–273. https://doi.org/10.1007/s40808-018-0532-z

Monteith, J. L. 1965. Evaporation and environment. *Symposium of the Society of Experimental Biology* 19:205–224.

Moonen, A., L. Ercoli, M. Mariotti, and A. Masoni. 2002. Climate change in Italy indicated by agrometeorological indices over 122 years. *Agricultural and Forest Meteorology* 111 (1):13–27.

Moran, M. S., C. D. Peters-Lidard, J. M. Watts, and S. McElroy. 2004. Estimating soil moisture at the watershed scale with satellite-based radar and land surface models. *Canadian Journal of Remote Sensing* 30 (5):805–826.

Mu, Q., F. A. Heinsch, M. Zhao, and S. W. Running. 2007. Development of a global evapotranspiration algorithm based on MODIS and global meteorology data. *Remote Sensing of Environment* 111 (4):519–536.

Mu, Q., M. Zhao, and S. W. Running, 2011: Improvements to a MODIS global terrestrial evapotranspiration algorithm. *Remote Sensing of Environment* 115:1781–1800.

Nagler, P., J. Cleverly, D. Lampkin, E. Glenn, A. Huete, and Z. Wan. 2005. Predicting riparian evapotranspiration from MODIS vegetation indices and meteorological data. *Remote Sensing of Environment* 94:17–30.

Narayan, U., V. Lakshmi, and T. J. Jackson. 2006. High-resolution change estimation of soil moisture using L-band radiometer and radar observations made during the SMEX02 experiments. *IEEE Transactions on Geoscience and Remote Sensing* 44 (6):1545–1554.

Nishida, K., R. Nemani, J. M. Glassy, and S. W. Running. 2003. Development of an evapotranspiration index from Aqua/MODIS for monitoring surface moisture status. *IEEE Transactions on Geoscience and Remote Sensing* 41 (2).

Njoku, E. G., T. J. Jackson, V. Lakshmi, T. K. Chan, and S. V. Nghiem. 2003. Soil moisture retrieval from AMSR-E. *IEEE Transactions on Geoscience and Remote Sensing* 41 (2):215–229.

Njoku, E. G., W. J. Wilson, S. H. Yueh, S. J. Dinardo, F. K. Li, T. J. Jackson, V. Lakshmi, and J. Bolten. 2002. Observations of soil moisture using a passive and active low-frequency microwave airborne sensor during SGP99. *IEEE Transactions on Geoscience and Remote Sensing* 40 (12):2659–2673.

O'Neill, P. E., N. S. Chauhan, and T. J. Jackson, 1996. Use of active and passive microwave remote sensing for soil moisture estimation through corn. *International Journal of Remote Sensing* 17 (10):1851–1865.

Overgaard, J., D. Rosbjerg, and M. B. Butts. 2006. Land-surface modelling in hydrological perspective–a review. *Biogeosciences* 3 (2):229–241.

Peterson, T. C., V. S. Golubev, and P. Y. Groisman. 1995. Evaporation losing its strength. *Nature* 377 (6551):687–688.

Pathe, C., W. Wagner, D. Sabel, M. Doubkova, and J. B. Basara. 2009. Using ENVISAT ASAR global mode data for surface soil moisture retrieval over Oklahoma, USA. *IEEE Transactions on Geoscience and Remote Sensing* 47 (2):468–480.

Petty, G. W. 1995. The status of satellite-based rainfall estimation over land. *Remote Sensing of Environment* 51:125–137.

PIs of SMEX03. 2003. *SMEX03 Airborne Synthetic Aperture Radar (AIRSAR) Data* [cited]. Available from http://nsidc.org/data/amsr_validation/soil_moisture/smex03/index.html

Roerink, G. J., Z. Su, and M. Menenti. 2000. S-SEBI: A simple remote sensing algorithm to estimate the surface energy balance. *Physics and Chemistry of the Earth, Part B: Hydrology, Oceans and Atmosphere* 25 (2):147–157.

Rosenfeld, D., and C. W. Ulbrich. 2003. Cloud microphysical properties, processes, and rainfall estimation opportunities. *Meteorological Monographs* 30:237–237.

Roy, P. S., M. D. Behera, and S. K. Srivastav. 2017. Satellite remote sensing: Sensors, applications and techniques. *Proceedings of the National Academy of Sciences, India, Section A Physical Sciences* 87:465–472. https://doi.org/10.1007/s40010-017-0428-8

Saha, A. and S. C. Pal. 2024. Application of machine learning and emerging remote sensing techniques in hydrology: A state-of-the-art review and current research trends. *Journal of Hydrology* 632 (2024):130907, ISSN 0022-1694, https://doi.org/10.1016/j.jhydrol.2024.130907. Available from www.sciencedirect.com/science/article/pii/S0022169424003019

Scanlon, B. R., C. C. Faunt, L. Longuevergne et al. 2012. Groundwater depletion and sustainability of irrigation in the US High Plains and Central Valley. *Proceedings of the National Academy of Sciences* 109 (24):9320–9325.

Scofield, R. A., and R. J. Kuligowski. 2003. Status and outlook of operational satellite precipitation algorithms for extreme-precipitation events. *Monthly Weather Review* 18:1037–1051.

Service, A. R. 2002. *SMEX02* (03/24/2002) [cited]. Available from http://hydrolab.arsusda.gov/smex02/

Shafer, M. A., C. A. Fiebrich, D. S. Arndt, S. E. Fredrickson, and T. W. Hughes. 2000. Quality assurance procedures in the Oklahoma Mesonetwork. *Journal of Atmospheric and Oceanic Technology* 17 (4):474–494.

Shao, Z., H. Fu, D. Li, O. Altan, and T. Cheng. 2019. Remote sensing monitoring of multi-scale watersheds impermeability for urban hydrological evaluation. *Remote Sensing of Environment* 232 (2019):111338, ISSN 0034-4257, https://doi.org/10.1016/j.rse.2019.111338. Available from www.sciencedirect.com/science/article/pii/S0034425719303578

Sheffield, J., C. R. Ferguson, T. J. Troy, E. F. Wood, and M. F. McCabe. 2009. Closing the terrestrial water budget from satellite remote sensing. *Geophysical Research Letter* 26:L07403. DOI: 10.1029/2009GL037338

Sheffield, J., E. F. Wood, and F. Munoz-Arriola. 2010. Long-term regional estimates of evapotranspiration for Mexico based on downscaled ISCCP data. *Journal of Hydrometeorology* 11 (2):253–275.

Sorooshian, S., K.-L. Hsu, X. Gao, H. V. Gupta, B. Imam, and D. Braithwaite. 2000. Evaluation of PERSIANN system satellite-based estimates of tropical rainfall. *Bulletin of the American Meteorological Society* 81 (9):2035–2046.

Su, Z. 2002: The Surface Energy Balance System (SEBS) for estimation of turbulent heat fluxes. *Hydrology Earth System Sciences* 6:85–100.

Swenson, S., J. Famiglietti, J. Basara, and J. Wahr. 2008. Estimating profile soil moisture and groundwater variations using GRACE and Oklahoma Mesonet soil moisture data. *Water Resources Research* 44:W01413.

Syed, T. H., J. S. Famiglietti, M. Rodell, J. Chen, and C. R. Wilson. 2008. Analysis of terrestrial water storage changes from GRACE and GLDAS. *Water Resources Research* 44:W02433.

Tapiador, F. J., F. J. Turk, W. Petersen, A. Y. Hou, E. García-Ortega, L. A. T. Machado, C. F. Angelis, P. Salio, C. Kidd, G. J. Huffman, and M. de Castro. 2012. Global precipitation measurement: Methods, datasets and applications. *Atmospheric Research* 104–105 (0):70–97.

Tapley, B. D., S. Bettadpur, J. C. Ries, P. F. Thompson, and M. M. Watkins. 2004. GRACE measurements of mass variability in the Earth system. *Science* 305 (5683):503–505.

Thakur, P. K., B. R. Nikam, V. Garg et al. 2017a. Hydrological parameters estimation using remote sensing and GIS for Indian region: A review. *Proceedings of National Academy of Sciences, India, Section A Physical Sciences* 87:641–659. https://doi.org/10.1007/s40010-017-0440-z

Thakur, J. K., S. K. Singh, and V. S. Ekanthalu. 2017b. Integrating remote sensing, geographic information systems and global positioning system techniques with hydrological modeling. *Applied Water Science* 7:1595–1608. https://doi.org/10.1007/s13201-016-0384-5

Thies, B., T. Nauß, and J. Bendix. 2008. Precipitation process and rainfall intensity differentiation using Meteosat Second Generation Spinning Enhanced Visible and Infrared Imager data. *Journal of Geophysical Research* 113:D23206. DOI: 10.1029/2008JD010464.

Thomas, A. 2000. Spatial and temporal characteristics of potential evapotranspiration trends over China. *International Journal of Climatology* 20 (4):381–396.

Tronin, A. 2006. Remote sensing and earthquakes: A review. *Physics and Chemistry of the Earth, parts A/B/C* 31 (4):138–142.

Turk, F. J., and S. D. Miller. 2005. Toward improved characterization of remotely sensed precipitation regimes with MODIS/AMSR-E blended data techniques. *IEEE Transactions on Geoscience and Remote Sensing* 43 (5):1059–1069.

Ulaby, F. T., R. K. Moore, and A. K. Fung. 1986. *Microwave Remote Sensing: Active and Passive*. London, UK: Artech House Inc.

Vivoni, E. R., M. Gebremichael, C. J. Watts, R. Bindlish, and T. J. Jackson. 2008. Comparison of ground-based and remotely-sensed surface soil moisture estimates over complex terrain during SMEX04. *Remote Sensing of Environment* 112 (2):314–325.

Xie, P., J. E. Janowiak, P. A. Arkin, R. Adler, A. Gruber, R. Ferraro, G. J. Huffman, and S. Curtis. 2003. GPCP pentad precipitation analyses: An experimental data set based on gauge observations and satellite estimates. *Journal of Climate* 16 (2):197–2,214.

Xu, J. 2001. An analysis of the climatic changes in eastern Asia using the potential evaporation. *Journal of Japan Society of Hydrology and Water Resources* 14:151–170.

Yan, H. and S. Yang. 2007. A MODIS dual spectral rain algorithm. *Journal of Applied Meteorology and Climatology* 46:1305–1323.

Yang, H., C. Sheng, X. Xianwu, and H. Gina. 2011. Global precipitation estimation and applications. In *Multiscale Hydrologic Remote Sensing*, 371–386, Boca Raton, FL: CRC Press.

Yilmaz, M. T., E. R. Hunt, L. D. Goins, S. L. Ustin, V. C. Vanderbilt, and T. J. Jackson. 2008. Vegetation water content during SMEX04 from ground data and Landsat 5 Thematic Mapper imagery. *Remote Sensing of Environment* 112 (2):350–362.

Zeng, Z., G. Tang, Y. Hong, C. Zeng, and Y. Yang. 2017. Development of an NRCS curve number global dataset using the latest geospatial remote sensing data for worldwide hydrologic applications. *Remote Sensing Letters* 8 (6):528–536. DOI: 10.1080/2150704X.2017.1297544

LIST OF URLS

[1] ftp://ftp.cpc.ncep.noaa.gov/precip/gpi/monthly/.
[2] ftp://ftp.cpc.ncep.noaa.gov/precip/gpi/pentad/.
[3] ftp://ftp.cpc.ncep.noaa.gov/precip/gpi/daily/.
[4] http://nsidc.org/data/ae_rain.html.
[5] http://rain.atmos.colostate.edu/RAINMAP10/.
[6] berg@atmos.colostate.edu; Dr. Wesley Berg.
[7] http://pps.gsfc.nasa.gov.
[8] www.star.nesdis.noaa.gov/smcd/emb/ff/digGlobalData.php.
[9] ftp://gpcp-pspdc.gmu.edu/V6/2.5/.
[10] http://mirs.nesdis.noaa.gov; www.osdpd.noaa.gov/ml/mirs.
[11] www.ncdc.noaa.gov/oa/rsad/ssmi/gridded/index.php?name=data_access.
[12] www.star.nesdis.noaa.gov/corp/scsb/mspps/main.html; www.osdpd.noaa.gov/ml/mspps/index.html.
[13] pingping.xie@noaa.gov; Dr. Pingping Xie.

[14] www.ssmi.com.
[15] joel.susskind-1@nasa.gov; Dr. Joel Susskind.
[16] www.cpc.ncep.noaa.gov/products/janowiak/cmorph_description.html.
[17] http://sharaku.eorc.jaxa.jp/GSMaP/.
[18] http://sharaku.eorc.jaxa.jp/GSMaP_crest/.
[19] song.yang@nrlmry.navy.mil; Dr. Song Yang.
[20] ftp://pps.gsfc.nasa.gov/pub/trmmdata/3G/3G68/.
[21] ftp://trmmopen.nascom.nasa.gov/pub/merged/combinedMicro/.
[22] ftp://trmmopen.nascom.nasa.gov/pub/merged/calibratedIR/.
[23] ftp://trmmopen.nascom.nasa.gov/pub/merged/mergeIRMicro/.

3 Groundwater Targeting Using Remote Sensing

Santhosh Kumar Seelan

ACRONYMS AND DEFINITIONS

AVHRR	Advanced Very High Resolution Radiometer
DEM	Digital Elevation Model
ERS	European Remote Sensing Satellites
ERTS	Earth Resources Technology Satellites
ETM+	Enhanced Thematic Mapper Plus
FCC	False Color Composite
GIS	Geographic Information Systems
GRACE	Gravity Recovery and Climate Experiment
IR	Infrared
IRS	Indian Remote Sensing Satellites
LFC	Large Format Camera
MODIS	Moderate-Resolution Imaging Spectroradiometer
MSS	Multi-Spectral Scanner
NDVI	Normalized Difference Vegetation Index
NIR	Near-Infrared
NOAA	National Oceanic and Atmospheric Administration
RADARSAT	RADAR Satellite
SAR	Synthetic Aperture Radar
SIR	Spaceborne Imaging Radar
SPOT	Satellite Pour l'Observation de la Terre, French Earth Observing Satellites
SRTM	Shuttle Radar Topography Mission
TM	Thematic Mapper

3.1 INTRODUCTION

The availability of fresh water has determined the growth of civilizations in the past. As the world's population continues to increase, it is predicted that the availability of fresh water for human needs could be a serious limiting factor in the future. Though the estimated total volume of water on the planet is about 1.4 billion km^3, nearly 96.5% of it is held up in the oceans as salt water. An estimated 10.5 million km^3 of fresh water, which is about a third of the total fresh water available, are stored below the surface of the Earth in the form of groundwater. It is widely but unevenly distributed and is an important source for irrigation and drinking purposes. Groundwater exists whenever water infiltrates beneath the surface, the soils and rocks beneath the surface are porous and permeable enough to hold and transmit this water, and the rate of infiltration is sufficient to saturate these rocks to an appropriate degree. Groundwater is a renewable resource and therefore, if located, exploited, and managed carefully, can be sustained forever (Sadeghi-Jahani et al., 2024; Kalhor and Emaminejad, 2019; Pinto et al., 2017).

Targeting groundwater is a complex but essential task, particularly in arid and semi-arid regions, which encompass about one-fifth of the Earth's surface (EL-Bana et al., 2024; Alshehri et al., 2020;

Kim et al., 2019; Gnanachandrasamy et al., 2018; Nejad et al., 2017). The increasing population continues to expand the use of these regions for food production and living space. The shortage of water for domestic, municipal, and irrigation purposes can be eased in many regions by using groundwater stored in aquifers underlying these areas—provided we know how to locate it and use it in a sustainable way.

The excavations at Mohenjo-Daro, an archeological site in the current Sindh province of Pakistan, have revealed brick-lined, open, dug wells existing as early as 3000 BC during the Indus valley civilization. The writings of Vishnu Kautilya, who was a teacher at the ancient Takshashila University during the reign of Chandragupta Maurya, a successful ruler of the Indian sub-continent during 300 BC, indicate that groundwater was being used for irrigation purposes at that time (Regunath, 1987). Use of groundwater, a renewable resource, is an age-old practice in countries like India. The early man located groundwater by using common sense and by associating certain surface features with groundwater occurrence. The common indicators used were ant hills or termite mounds (the termites need moisture to build the mound), certain types of plants that have a root system extending down to the water table, etc. These ancient methods and rules of thumb were carefully recorded in the works of the great scholar Varahamihira—505–587 AD (Prasad, 1980).

The advent of systematic geological mapping in the 18th century and its improvements in the subsequent decades opened up the understanding of groundwater regimes in a more scientific way. The introduction of geophysical surveys in the 1930s and their refinement over the subsequent years vastly improved the targeting accuracies, as these techniques provide an insight into the third dimension viz., depth.

Aerial photographs were first used in groundwater exploration a few decades ago and are now an accepted and fundamental reconnaissance tool, as they greatly reduce the time and cost of groundwater surveys by narrowing down the target areas. Aerial photography too advanced over time with the introduction of multiband, digital cameras and sophisticated analytical tools. During the last four decades or so, remotely sensed data from space platforms have been added to the array of tools available to the groundwater scientist. With refinements in spectral and spatial resolution offered by the newer satellites, there has been an enormous increase in the capacity of the remotely sensed data to provide groundwater information and the technique has emerged as a major tool. One important factor to be emphasized here is that remote sensing does not replace the existing techniques but is a very useful additional tool (Sharma et al., 2024; Rampheri et al., 2023; Patra et al., 2018). In fact, Varahamihira's rules of thumb have withstood the test of time and are still good. But if water is to be provided quickly to the vast parched lands, a faster and cheaper procedure integrating remote sensing, geophysics, and ground-based hydrogeological surveys are a must.

The focus of this chapter is to discuss the relevant targeting keys and parameters for groundwater that can be extracted from remote sensing techniques under various geological settings. A detailed review of the literature is also made to trace the developments and status of this field.

3.2 ROLE OF REMOTE SENSING IN GROUNDWATER PROSPECTING

While groundwater is a sub-surface phenomenon, remote sensing offers information about the features on the surface of the Earth (Alshehri and Mohamed, 2023; Adams et al., 2022; Nhamo et al., 2020; Miro and Famiglietti, 2018). Therefore, what can essentially be done is to use remote sensing data to identify the surface parameters indicative of groundwater occurrence and movement. These parameters relate to geology, geomorphology, surface hydrology, soils, land use, and natural vegetation (Table 3.1). The greatest advantage of spaceborne remote sensing is that information on all the aforementioned aspects are obtainable from a single image. Features and relationships that would take a long time to put together using the much larger-scale aerial photography or surface examination can be seen in a single integrated perspective on space imagery. The conventional surveys usually start with geophysical surveys for selecting a point for drilling. But the question that invariably arises is where to conduct the geophysical survey. As geophysical surveys are manpower intensive and time consuming, one would always like to start the survey in places where the

Groundwater Targeting Using Remote Sensing 67

prospects are higher. This means reconnaissance-level information has to be generated first so that unfavorable zones can be eliminated straight away and the favorable zones can be examined more closely. This way, the geophysical and other ground surveys can be used more effectively and with better results. Hence it is ideal to use medium-resolution spaceborne data in the beginning stages of the exploration programs to extract reconnaissance-level information, followed by the more detailed interpretation of high-resolution aerial photographs or satellite imagery, and field geological and geophysical studies, before test drilling (Figure 3.1). Ideally, the investigations could be carried out in the following four simplified steps (Seelan, 1994):

1. *Medium-resolution satellite-based regional assessment:* In a regional assessment, the geographical area considered is usually very large, say, a group of districts or a state. Satellite data is interpreted for geology, geomorphology, land use, soils, drainage, and vegetation

TABLE 3.1
Overview of Groundwater Indicators Discerned from Remote Sensing

No	Major Parameters	Major Sub-sets
1	Regional geology	Lithology, trend, faults, fracture lineaments, master joints, dikes/quartz reefs
2	Regional geomorphology	Drainage pattern/drainage density, pediment zones, valley fills, ridges and valleys, cuestas and hogbacks, alluvial fans and bazadas, dunes, lineament-controlled stream channels
3	Regional hydrology	Surface water bodies, springs/seeps, floods/inundated areas/wetlands, runoff and recharge zones, groundwater discharge into streams
4	Soils	Coarse-grained soils, fine-grained soils, soil moisture
5	Hydrologic land use	Surface water irrigated cropland, urban or built-up land, forest cover
6	Natural vegetation	Vegetation indicative of soil moisture, vegetation indicative of shallow water table
7	Groundwater quality	Saline soils, salt-tolerant vegetation

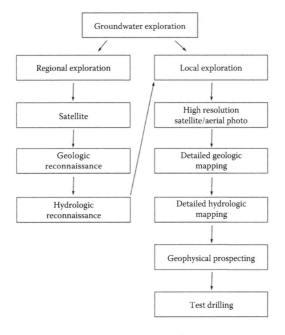

FIGURE 3.1 Role of remote sensing in groundwater exploration.

patterns and integrated to arrive at potential groundwater zones on a regional level. This is described in greater detail in subsequent sections.

2. *Aerial photo or high-resolution satellite imagery–based local assessment:* This stage includes (1) flying for aerial photography on a required scale or resolution (using manned or unmanned aircrafts) for areas identified as potential in the reconnaissance study. Where photographs are already available, fresh photography is not required; (2) interpretation of aerial photographs for detailed geology, structure, lineaments, geomorphology, soil characteristics, land use patterns, and other hydrogeological features; and (3) collection and corroboration of existing ground-based information such as the nature of aquifers, type, yield characteristics, etc. During this stage, remote sensing–derived information in sufficient detail to identify groundwater potential zones will be available.

3. *Ground-based hydrogeological and geophysical investigations:* The identified groundwater potential zones are to be subjected to verification with ground-based investigations. The potential zones demarcated should be identified on the ground for verifying their potential. Detailed hydrogeological surveys to identify the nature and type of aquifers, their yield characteristics, the depth of the aquifers and depth to the water table, piezometric gradients, etc. have to be carried out in the identified zones and adjoining areas. With this information, suitable locations for further geophysical surveys can be identified. These selected sites are to be subjected to detailed geophysical surveys to pinpoint the drilling sites. Normally for groundwater investigations, electrical methods such as resistivity surveys are observed to be very successful. To locate certain fracture/fault zones and to identify their dip direction, geophysical resistivity profiles can be carried out. All the hydrogeological and geophysical investigations in the narrow potential zones will help pinpoint the potential drilling sites where substantial yields can be expected.

4. *Exploratory drilling, testing, and well completion:* The various steps involved in the drilling of bore wells include (1) selection of a suitable drilling rig, such as direct rotary, down the hole hammer, etc., depending on the nature of rock types; (2) drilling of pilot boreholes; (3) borehole logging; (4) well construction; (5) development of the well; (6) pump testing to evaluate the transmissivity, etc.; and (7) installation of a suitable pump such as a submersible, jet, or deep well turbine pump depending on the yield.

In this procedure, a favorable zone is picked using remote sensing, investigated in detail using conventional methods, before drilling is taken up. Once the zone is proved good, most of the boreholes in the zone become successful. The unsuccessful wells are minimal. This way, there are enormous savings in the cost of ground surveys and drilling. A remote sensing–based survey therefore essentially proves a zone (rather than a point) for its groundwater potential. A remote sensing–based study cannot be independent by itself but has to be necessarily followed up by conventional surveys.

3.3 GROUNDWATER INFORMATION EXTRACTION FROM SATELLITE IMAGES

A number of studies have been carried out over the years, on the use of remote sensing in groundwater investigations (Sadeghi-Jahani et al., 2024; Rampheri et al., 2023; Adams et al., 2022; Al-Djazouli et al., 2021; Alshehri et al., 2020; Das et al., 2019; Patra et al., 2018; Hu et al., 2017). Aerial photo interpretation during the 1950s and 1960s laid the foundation for the later decades. Although photography from balloons and aircrafts started during the 18th century, it had its potential use only in military reconnaissance. After World War II the photographs found their utility in geological mapping, primarily as an offshoot of oil exploration. Early works of Mollard (1957) and Lattman and Nickelson (1958) assessed the fracture patterns revealed by aerial photographs for oil exploration. Many subsequent researchers have used aerial photographs for groundwater applications and are continuing to do so. Mollard and Patton (1961) outlined what they considered a sound approach to an integrated groundwater investigation using aerial photointerpretation inputs. They indicated the parameters, particularly in unconsolidated sedimentary terrains.

Moore (1967) studied the early satellite pictures from Skylab for mapping lineaments of hydrologic significance. With the launch of Landsat (then ERTS) and the continued availability of data from the satellite, the applications in the area of groundwater grew wider. A detailed study on deciphering groundwater from aerial photographs was made by Nefedov and Popova (1972). Their work included deciphering natural factors such as rocks, relief and surface water, determining hydrogeological conditions and also deciphering landscape components dependent on groundwater, such as soil and vegetation. Morphological indicators were also outlined. A very detailed report on the keys to detect aquifers from satellites was first brought out by Moore and Deutsch (1975). The keys to detect various types of aquifers and the procedure to extract groundwater information were outlined. The keys for interpretation also have been summarized in the *Manual of Remote Sensing*, American Society of Photogrammetry (Solomonson et al., 1983).

3.3.1 LITHOLOGY AND LINEAMENTS

The preparation of geologic maps showing lithology from reflectance differences, structures on the basis of linear and curvilinear features and the preparation of geomorphic maps are greatly improved by satellite imagery. Discrimination of rock types based on satellite imagery alone is almost impossible. However, differences in lithology are very well brought out due to their differences in reflectance. The best results are obtained when the available geological maps and data from the literature are used in conjunction with the satellite images (Sharma et al., 2024; Alshehri and Mohamed, 2023; Lee et al., 2020; Kim et al., 2019; Gemitzi et al., 2017). In any case, it must be emphasized that the satellite imagery interpretation must be considered only as a tool that can aid geological mapping. Field checking of selected areas of the interpretation is imperative. Visible, near, and middle IR and thermal images and color composites of different band combinations have been found to be very useful in differentiating rock types. Although no detailed lithological interpretations are possible due to vegetation and soil cover, differences in lithology and the boundaries of various rock outcrops can be marked. This is of direct aid in groundwater studies.

The greatest contribution of satellite images in geologic studies has been in the field of structural analysis. Structural features that are not identifiable on the ground and sometimes even on aerial photographs are easily discernable on satellite images because of their regional coverage. Major faults and fractures that are expressed as linear features on the satellite images are potential zones for groundwater. Mapping of this is very useful, especially in hard rock areas where the occurrence and movement of groundwater is mostly confined to these linear features.

Identification of lineaments is best done on visible and near IR images and color composites. They appear on the images as alignment of dark or light soil tones, elongated aligned water bodies, linear stream channels, valleys, etc. An experienced interpreter may also infer the presence of faults and fractures based on the study of topography, geomorphic features, etc. (Al-Djazouli et al., 2021; Das et al., 2019; Miro and Famiglietti, 2018; Nejad et al., 2017; Pinto et al., 2017).

Tiwari (1993) delineated lineaments using spaceborne and airborne data in the semi-arid district of Sirohi in Western Rajasthan, India, for the selection of drilling sites. All five exploratory wells drilled in the lineament zones consisted of crystalline limestone, granite, and schistose rocks, and the groundwater yields were consistently high. In order to reduce subjectivity in interpretation, Sander et al. (1997) used multiple interpreters to interpret Landsat TM and SPOT data over a project area in Ghana and concluded that 90% of the hydrologically significant lineaments were picked up by all interpreters. Koch and Mather (1997) compared the lineaments extracted from SIR-C L- and C-band SAR data and stereoscopic large format camera (LFC) over the Red Sea Hills region of Sudan and concluded that the LFC data was most useful for mapping detailed fracture patterns, while the combination of SIR-C SAR and LFC data were helpful in the location of major deep-seated fracture zones. An imagery integration approach was developed by Bruning et al. (2011) to evaluate satellite imagery for lineament analysis in a volcanic terrain in Nicaragua where human development and vegetation confound imagery interpretation. They used ATER, Landsat 7 ETM+, QuickBird, and RADARSAT-1 data in conjunction with DEM and concluded that nine of the previously mapped lineaments and 26 new ones were identified in this method. They also concluded that RADARSAT-1 products were most suitable for minimizing the anthropogenic features, but they have to be used in conjunction with optical sensor data and DEM for best overall results.

Digitally enhanced images have been found to be very useful in the demarcation of lithological boundaries and in structural mapping (Goetz et al. 1975; Abrams et al. 1977). Seelan (1980) used contrast stretching techniques, where the brightness data distribution was expanded or stretched to fill the dynamic range of the display medium, to demarcate lithological boundaries. Moore and Waltz (1983) used a five-step digital convolution procedure to extract edge and line segments and produce directionally enhanced images in order to avoid subjectivity in manual interpretation. Rana (1998) used the directional filtering technique suggested by Moore and Waltz (1983) on IRS LISS I band 4 images over the semi-arid Thar Desert region in India to identify subtle lineaments that were unnoticed otherwise. The lineaments were recognized as buried channels and zones of coarse sediments representing potential sites for the accumulation of fresh water during rains. Ahmad and Singh (2002) used data-merging techniques to fuse IRS PAN and LISS III data sets over a study area in the Indo-Gangetic plains of India to demonstrate the usefulness of such merging in mapping geomorphic features and relating to major tectonic events. Mogaji et al. (2006) used digital enhancement techniques to help extract lineaments on a Landsat 7 ETM + image. This study found that the linear contrast stretch, merging the FCC and panchromatic images, and the lineament density map extracted from these, were useful for identifying potential groundwater zones in their study area. Pal et al. (2007) fused optical and SAR data to map various lithological units over a part of the Sighbum Shear Zone and its surroundings in India. In this study, ERS-SAR data was enhanced using a Fast Fourier Transform–based filtering approach and also using Frost filtering techniques. Both the enhanced SAR images were then separately fused with histogram equalized IRS-1C LISS III image using Principal Component Analysis technique. Later, they applied the Feature-oriented Principal Components Selection technique to generate a False Color Composite from which corresponding geological maps were prepared. A semi-automatic method to infer groundwater flow paths based on the extraction of lineaments from DEMs was developed by Mallast et al. (2011) for the subterranean catchment of the Dead Sea region. They used a combined method of linear filtering and object-based classification to map the lineaments accurately. Subsequently, the lineaments were differentiated into geological and geomorphological lineaments using auxiliary information and finally evaluated in terms of hydrogeological significance. Ali et al. (2012) used contrast stretching, image ratio, image filtering, and intensity-hue saturation transformation on Landsat ETM+ data over the Abidiya area in Sudan to discriminate between mafic and ultramafic granites of ophiolitic origin.

3.3.2 Hydrogeomorphology

Geomorphic studies are indispensable in understanding the occurrence of groundwater, especially areas of late Pleistocene and recent deposits. Certain landscape elements that indicate coarse materials and near surface water table are identifiable on satellite images. Buried channels, present-day valleys filled with alluvial deposits, underfit valleys, natural levees, alluvial fans, bazadas, etc. that have their own hydrologic properties, can be mapped. The images are useful in studying the drainage pattern and drainage density, which have a bearing on the recharge conditions and permeability of the rocks.

Where geomorphology exercises a significant control over groundwater movement and occurrence, this relationship has been utilized by several authors to arrive at broad groundwater potential zones. Moore and Deutsch (1975) identified the limits of an alluvial aquifer in Vancouver, British Columbia, Canada, using false color composites of Landsat (then ERTS). They identified the limits based on the present alluvial valley of the Fraser River and an old valley south of the present river that had cultivated fields. Cultivated fields and non-cultivated fields could be separated on the imagery due to the sharp boundaries between them. Small wooded areas between the present and old river valleys had steeper slopes and were inferred to have been underlain by unconsolidated rocks. They recommended geophysical surveys and test drilling to determine the thickness, lithology, and water-bearing potential of the alluvial aquifers in the Fraser area. Seelan (1982) adopted the geomorphological approach over a study area in Central India comprising hard rocks and alluvium to arrive at potential groundwater zones. The satellite images were visually analyzed to prepare hydrogeological maps at 1:250,000 scale. This methodology was tried earlier in India through a pioneering, operational remote sensing project conducted jointly by the National Remote Sensing Agency, India and the Public Works Department of Tamil Nadu, India, during 1977–1979, aimed at delineating potential groundwater zones on a regional level (Thillaigovindarajan et al., 1985). Figures 3.2a and 3.2b illustrate this concept for a part of the study area used in this project. This approach was later adopted in several studies carried out at the National Remote Sensing Agency, India (1896a,b), including the studies under the technology mission on drinking water wherein hydrogeomorphological maps at 1:250,000 scale were generated for the entirety of India.

Gautham (1990) identified two buried channels in the area south of Allahabad, India, one west of the Tons River and the other east of it, using IRS LISS II FCC imagery. Aerial photographs were used for studies on slope, direction, position of natural levees, vegetation density, etc. It was concluded from the configuration of the buried channels that these were initially joined forming one channel that lowed from east to west, although the present master slope of the area is from west to east. The buried channels were confirmed to have good groundwater potential. Some parts of the buried channels were noticed to have been waterlogged due to seepage from the irrigation canals cutting across one of them.

Kumar and Srivastava (1991), in a study carried out in Bihar, India, tried to analyze the spatial distribution of geomorphic classes and depth-wise variation in aquifer material within the same class for determining the target horizon for further detailed investigations using remote sensing and electrical sounding. Agarwal and Mishra (1992) delineated different hydrogeomorphological units

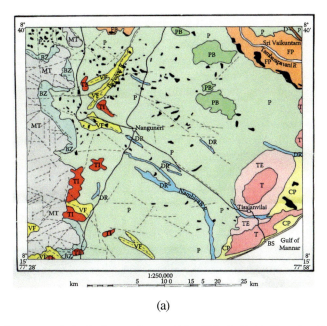

(a)

FIGURE 3.2 (a) Hydrogeomorphological map of part of Tamil Nadu, India (based on Landsat MSS imagery interpretation). (b) Legend for hydrogeomorphological map of part of Tamil Nadu, India.

Symbol	Hydrogeomorphic unit	Description	Groundwater potential
MT	Mountainous terrain	Ridge and valley complex. Lineaments prominent. Valleys controlled by lineaments.	Good recharge along valleys controlled by lineaments which can be exploited in the plains.
BZ	Bazada	Coalescence of alluvial fans. Formed along the foot hills of western ghat hills. (mountainous terrain)	Good potential towards the lower part of the fans.
TI	TOR and Inselburg complex	Peripediment zone with tors and inselburgs. Pediment fairly thick in between tors and inselburgs.	Good within the pediment zones.
PB	Pediment with thick black soil cover	Occur as patches over the pediplains. Soil cover is thick and fine grained in nature.	Poor recharge and groundwater conditions.
P	Pediplain	Fairly thick, continuous pediment cover.	Moderate within the pediment zone. Fracture lineaments offer greater potential.
VF	Valley Fill	Thick alluvial fill materials within the fracture controlled valleys.	Good groundwater potential within the alluvium.
DR	Dry river courses	Sandy, dry river beds.	Groundwater occurs as base flow immediately following the monsoon months.
T	Teri soil	Medium to coarse grained, wind blown, thick, deep red soils occuring over the pediplain.	Excellent potential.
TE	Area around teri subjected to erosion	Subjected to severe erosion by wind and water.	Excellent along the eastern and southern parts of the teri where groundwater discharges from the main teri.
D	Dune sands	Narrow strips of sand.	Does not retain water, hence not suitable for groundwater development.
FP	Flood plain	Flood plain deposits of tambaraparani river consisting of sand, silt and clay.	Good groundwater potential.
CP	Coastal plain	Sandy deposits close to the coast.	Excellent aquifers. Cautious development needed.
BS	Beach sands	Narrow sandy stretch along the coast.	Good shallow groundwater. Cautious development needed.
	Lineaments	Fracture lineaments over gneissic terrain.	Excellent potential. Detailed ground investigations needed.
	Water bodies		

(b)

FIGURE 3.2 (Continued)

in and around the immediate environs of Jhansi city in India and attempted correlation between well yields and hydrogeomorphic units prepared based on satellite data interpretation. They observed a good correlation between geomorphic units and well yields. Similarly, Shankarnarayana et al. (1996) delineated broad lithological, morphological, and structural features using Landsat TM imagery and demonstrated that the wells located on the lineament zones produced better yields. They concluded that the wells located based on lineament study yielded 14 times more than the wells located away from the lineaments. The hydrogeomorphological approach to delineate groundwater potential zones was also used by Panigrahi et al. (1996), Ravindran and Jeyaram (1997), and Rao and Reddy (1999). Dhakate et al. (2008) used ground-based vertical electrical sounding techniques to confirm the potential of geomorphological units derived from satellite imagery interpretations over a granitic terrain in Andhra Pradesh in India. They concluded that the thickness of weathered mantle is higher in areas of higher lineament density and such thicker mantle exhibits higher groundwater potential. Sethupathi et al. (2012) studied the watersheds in the northeastern Tamil Nadu region where the groundwater discharge rate was greater than the recharge rate. The study involved Landsat 7 ETM+ imagery and delineation of topography, slope, surface cover, soils, vegetation, and lineaments. They identified 194 lineaments and classified them into three categories based on length, minor, medium, and major. Areas with high lineament and high lineament intersection density had high groundwater potential along with valley fills, pediplains, and buried pediments.

Radar data has been successfully used in mapping palaeodrainage. Robinson et al. (2000) determined palaeodrainage directions in the Eastern Sahara using high-resolution, multi-wavelength, multi-polarization Spaceborne Imaging Radar (SIR-C) data and GLOBE DEM data sets. Shuttle Radar Topography Mission (SRTM) data with 90-m resolution was used to delineate mega palaeodrainage in the Eastern Sahara by Ghoneim and El-Baz (2007).

3.3.3 Soils, Land Use, and Drainage

Broad textural classes of soils are identifiable based on the visual interpretation of satellite data to a great extent. For example, fine-grained soils are generally darker than coarse-grained soils.

Groundwater Targeting Using Remote Sensing

Computer-aided analysis gives a more detailed classification but requires adequate ground truth and involves a good amount of fieldwork for accuracy. In a digital analysis, the computer is trained to identify spectral signatures of different soil types and classify them. Interpretation of soils is best done during a season when soils are well exposed and vegetation cover is minimal. Distribution of different textural classes of soils is an important factor in interpreting for groundwater occurrence and recharge. For example, fine-grained clayey soils generally tend to permit less recharge and more runoff, while coarse-grained sandy soils allow more recharge to groundwater.

An understanding of the land use pattern is very important in groundwater studies. For example, a forested area discourages runoff and encourages recharge to the groundwater, while it is the opposite in an urban, built-up area. As land use is a dynamic process, the satellite systems, because of their temporal coverage, are ideally suited for its analysis. The computer analysis of land use almost follows the same process as for soils. The accuracy and details of classification will primarily depend on the type of sensors used and the amount of ground truth available. Broad classes such as forested areas, urban areas, water bodies, cultivated fields, etc. can easily be delineated by visual interpretation itself without any enhancement aids.

Drainage density maps derived from satellite imagery can also provide vital clues to groundwater recharge conditions. Where the drainage density is higher, it is likely that recharge to groundwater is lower and where it is lower, the recharge and underground flow conditions are higher. Information on land use, soils, and drainage can form important parameters in evaluating groundwater potential of a region, as described in the later section on integrated approach.

Seelan et al. (1983) studied land utilization and landform patterns in parts of southern Uttar Pradesh State in India using Landsat data and established the control exhibited by landforms on groundwater resources situation, which in turn controls the land utilization in the region. Ghosh (1993) prepared a groundwater map of a part of the Jharia coalfield area in India using soil moisture, vegetation, and morphology as primary indicators based on aerial photo interpretation. The relationship between vegetation growth and groundwater in arid regions is an active area of research, and Xiaomei et al. (2007) extracted the vegetation information from NOAA AVHRR and NODIS NDVI and established the relationship between phreatophyte vegetation and depth to the water table in the Yinchuan Plain in China.

3.3.4 Groundwater Recharge/Discharge

Remotely sensed data has been used in the past to identify groundwater recharge/discharge zones (EL-Bana et al., 2024; Mohamed et al., 2023a; Mohamed et al., 2023b; Gaber et al., 2020; Kalhor and Emaminejad, 2019; Gemitzi et al., 2017). Thermal data has been found to be particularly useful for this purpose. Surveys made by different workers in different parts of the world have proved thermal IR sensing to be a very useful tool in locating abnormal water sources such as geysers, hot pools, hot springs, and river seepages. Airborne IR surveys over thermal areas have been carried out in Yellowstone National Park, USA (Mc Lerran, 1967); Rejkanas and Torjajoukull, Iceland (Palmason, 1970); the Tampo region of New Zealand (Hochstein and Dickinson, 1970); North Island of New Zealand (Dawson and Dickinson, 1970); and the Lake Kinneret and Dead Sea regions of Israel (Seelan, 1975). Sensing from an elevated point on the ground has also been carried out in the Lake Kinneret region (Otterman, 1971). The objective of most of these surveys was to locate thermal anomalies in regions where thermally active zones were known to occur. Thermal springs could be located successfully in most of these surveys. Palmason's surveys lead to the discovery of dozens of previously unmapped points of thermal activity.

Thermal IR images can be used to locate hot creeks fed by, say, a series of hot springs. If there are two adjacent streams, the one fed by a hot spring appears whiter. Discharges of slightly warmer waters from rivers, drainage canals, or springs into lakes could be located. Deutsch (1971) located the flow of such warmer waters into Lake Ontario with significant success. Further, the thermal IR images could be used in locating freshwater springs just off-shore of saltwater coasts and saline springs discharging into freshwater lakes. Underwater hot springs and saline springs could also be located (Otterman, 1971; Seelan, 1975). When hot saline water flows into a freshwater lake, its boundary is very sharp. This is because the saline water, due to its higher density, tends to flow under the surface without much mixing on the surface. Images taken at a particular time interval could help in outlining the flow patterns of the underwater springs. The surface thermal patterns determined from IR images should help in understanding the geologic structures that control the upflow of water and thus a better understanding of regional hydrogeology. In the Maligne Karst system of the Rocky Mountains, Alberta, Canada, the discharge point of the Karst Polje was unknown, but the thermal images indicated that the water disappearing in the Polje was reappearing as a submerged spring under the surface of Medicine Lake (Ozaray, 1975). Thermal images were also reported to have been used to locate lines of diffuse groundwater discharge along some of the deep ravines of the Lake Pukaki area, Alberta (Ozaray, 1975). Bobba et al. (1992) used Landsat imagery to delineate groundwater discharge, recharge, water table depth, and transition areas in southern Ontario, Canada. This study used Landsat 1 imagery from March 1974 and July 1974 to represent seasonal differences. A simulation of regional groundwater flow models (recharge, transition, and discharge) was developed.

Tcherepanov et al. (2005) used Landsat TM thermal imagery under different weather conditions from 1989 to 2002, combined with ground-based observations on lake temperatures to identify groundwater recharge into lakes. The study allowed for the identification of lake zones with consistently cold temperatures that were inferred to be potential groundwater zones. An integrated approach of remote sensing and GIS to identify optimal areas for artificial recharge by flooding techniques was developed by Ghayoumian et al. (2005) for the Meimeh Basin in northern Isfahan Province in Central Iran. They derived information on soil from Landsat TM imagery and integrated this information with other thematic layers, such as slope, infiltration rate, transmissivity, water table and aquifer thickness, and water quality. They concluded that 70% of the study area was suitable or very suitable for flood spreading, mostly within the central portions of the Meimeh Basin. Ghayoumian et al. (2007) also applied similar techniques at another site in Iran to identify suitable zones within alluvial fans and pediplains for artificial recharge. Wang et al. (2008) used Landsat 7 imagery to identify groundwater recharge in the Indian River and in the Rehoboth and Indian River bays in Sussex County, Delaware. They used panchromatic, NIR, and thermal bands to identify ice patterns and temperature differences in the surface water, which are indicative of groundwater discharge in the area. Khalaf and Donoghue (2012) used MODIS Level 3 MOD09Q1 and MOD11A2 products and other information to analyze the relationship between rainfall, evapotranspiration, and soil moisture and recharge rates in the West Bank region.

3.3.5 Groundwater Use for Irrigation

Areas where groundwater is being used for irrigation can be separated using satellite imagery (Mohamed et al., 2023a; Mohamed et al., 2023b; Alshehri et al., 2020; Kalhor and Emaminejad, 2019; Gemitzi et al., 2017; Hu et al., 2017). Deutsch (1974) monitored groundwater use for irrigation in the Snake River Plain lying between the northern Rocky Mountains and the Snake River in the United States. From this study, he concluded that further agricultural development was possible, as the aquifer was known to extend far beyond the presently irrigated area. Seelan (1980) mapped the groundwater-irrigated areas in the Bundelkhand granitic terrain in parts of Central India using multi-date satellite images. The groundwater use was conformed to the valley portions of the pediments, which bears the best groundwater potentials in the region. Cropped areas are discernible on the satellite pictures because of their typical high reflectance in the infrared region. In India, for example, the monsoon rainfall is restricted between the months of May and September, and the second crop grown after the monsoon is mostly irrigated. The areas irrigated by surface water were separable from the groundwater-irrigated areas because of their water impoundments, canal systems, and the distinct command areas.

Groundwater Targeting Using Remote Sensing

Groundwater-irrigated areas in regions such as in the Ogallala aquifer system in the United States, use center pivot irrigation systems that are quite easily identified using visual interpretation of satellite imagery. However, this is not so straightfoward in some sections of the Ogallala, such as in Nebraska, where the source is a combination of surface and groundwater. Most of the canal-fed fields are square in shape and thus differ from the round center-pivot groundwater-irrigated systems; however, extensive ground truth has shown that there are instances where surface water is used in center-pivot irrigation systems as well (Kurz and Seelan, 2009). In addition, there are also groundwater-fed center-pivot irrigation systems adjacent to rivers where typically surface water is the source. Thus, a differentiation between surface and groundwater-irrigated fields cannot be done based on the shape of the field alone. A Landsat image used in the study is shown in Figure 3.3.

3.3.6 Groundwater Stress

Apart from targeting groundwater sources, remote sensing techniques also find some applications related to the management of groundwater resources, particularly in the detection and monitoring of groundwater systems under stress (Mohamed et al., 2023a; Mohamed et al., 2023b; Gaber et al., 2020; Lee et al., 2020; Kim et al., 2019). However, it must be emphasized that the remotely sensed data provides information only on the objects on the surface of the Earth, while some of the systems under stress or stress-causing factors may not show up on the surface at all. Therefore, like targeting, one has to look for surface indicators of stress in order to be able to detect them on the satellite imagery or aerial photographs (Seelan, 1986).

Overextraction of groundwater lowers water tables and reduces storage. Rodell and Famiglietti (2002) described the potential of using Gravity Recovery and Climate Experiment Satellite's (GRACE) gravity calculation data on groundwater storage levels in the Ogallala aquifer. Tiwari et al. (2009) analyzed temporal changes in Earth's gravity field in northern India and its surroundings, using data obtained from GRACE, and attributed the large-scale mass loss to excessive extraction of groundwater. Combining GRACE data with hydrological models to remove natural variability, they concluded that the region lost groundwater at the rate of 54 ± 9 km^3 per year between April 2002 (the start of the GRACE mission) and June 2008. Based on this, they concluded that this is probably the largest rate of groundwater loss in any comparable-sized region on Earth and that if this trend continues, it will lead to a major water crisis in the region.

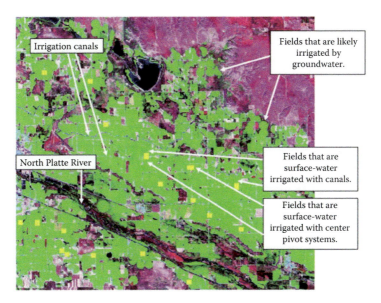

FIGURE 3.3 Surface water and groundwater-irrigated fields along the North Platte River in Nebraska.

3.3.6.1 Remote Sensing of Stress-Causing Factors

While there are several causative factors, only a few are amenable for monitoring by remote sensing, particularly from satellite altitudes (Sharma et al., 2024; Kim et al., 2019). Satellite data has been used in the past to identify areas irrigated by groundwater, as seen in an earlier section. A rapid increase in the area irrigated by groundwater over the years would indicate the tendency toward overexploitation. In the case of aerial photographs on a suitable scale, it is possible to count the number of open wells and correlate the well density with geomorphic units and pinpoint the units that have exceeded the recommended rates of exploitation (Seshubabu and Seelan, 1983).

Pollution sources, such as sewage and other solid waste disposal sites, can be identified on aerial photographs and high-resolution satellite imagery. The effects of these pollution sources and other industrial waste impoundments on groundwater is well known. In a study reported by Geraghty and Miller Inc. (1972), an aerial survey of portions of Pennsylvania State in the United States was made to determine the extent of impoundments by industry. Impoundments thus located from the air were field-checked to identify the owner, content, size, construction, and permit status of the impoundment. One such impoundment was being used by a manufacturer of batteries and cables that was dumping its untreated battery wash into a limestone quarry. The water quality of the effluent showed pH and lead levels much beyond the permissible levels and the waste disposal practices of the plant were altered in such a way as to prevent continued contamination of groundwater. The change in the groundwater controls due to unplanned mining activity in Jharia coalfields, India, was studied by Ghosh (1993). The groundwater map of the Jharia coalfields area was prepared from aerial photographs using vegetation cover as the criteria.

Agricultural activity influences the quantity of groundwater to a great extent by the use of fertilizers and due to poor management practices such as improper drainage, which results in waterlogging conditions. Although the use of fertilizer levels is not possible to gauge from remote sensing, it is possible to delineate areas under irrigation. In countries like India, fertilizer usage is maximum in irrigated areas as compared to non-irrigated agricultural areas. Waterlogged conditions have been demarcated on multispectral data from Landsat satellites (Singh, 1980), introducing the possibility of continuous monitoring for an increase or decrease in waterlogged areas. However, the data has limitations in demarcating waterlogged conditions in black soil areas (Venkataratnam, 1984). Visual and digital processing of IRS imagery have been successfully used to delineate the pre- and post-monsoon surface waterlogged areas the Gangetic plains of Bihar, India, by Chatterjee et al. (2005) and Chowdary et al. (2008).

Urbanization reduces recharge to groundwater, increases pumpage for domestic and industrial requirements, and thus can affect the quality of groundwater through the deliberate disposal of liquid and solid wastes. Therefore, the growth of urban centers can cause enormous stress on the shallow groundwater systems. A study carried out by Moore and Deutsch (1975) showed the satellite data interpretation of all population centers around the southern tip of Lake Michigan in the United States and suggested monitoring the direction and rate of urbanization, through repetitive imagery to help in determining the location and severity of future problems in aquifer management. The same study also quotes work done to locate changes in both active and newly reclaimed strip mines using satellite imagery, as the acid water drains from some of these mines have influence on the quality of groundwater. It has also been demonstrated that remote sensing can be used to study industrial discharge into lakes and rivers, which in turn have an obvious influence on groundwater. Only pollutants that add color, turbidity, or alter the temperature of the water can be detected easily by remote sensing. An investigation carried out at the National Remote Sensing Agency in India succeeded in mapping the paper mill effluent disposal pattern in the Godavari River near Rajmundry town in India using an airborne multispectral scanner data (Deekshatulu and Thiruvengadachari, 1981).

3.3.6.2 Remote Sensing of Stress Indicators

Groundwater systems, when subjected to stress, sometimes show up on the surface of the Earth through vegetation, soil, or surface water (EL-Bana et al., 2024; Alshehri and Mohamed, 2023;

Adams et al., 2022; Miro and Famiglietti, 2018). When such indications start appearing on the ground, it is possible to diagnose it using remote sensing data. Springs and seeps when characterized by luxurious growth of vegetation around them can be interpreted on satellite images. Drying of these springs will result in the vanishing of vegetation around and can be used to interpret declining water levels. Infrared photography has been found suitable for identifying springs (Robinov, 1968). Airborne thermal infrared data has been found to be useful in identifying underwater springs (Seelan, 1975; Gandino, 1983). Likewise, drying up of rivers fed by groundwater can also be an indicator of stress. Similarly, reduction in real spread of water bodies can be successfully monitored. The Landsat satellite picture of March 1983 covering the Madras (now Chennai) City and surrounding areas in India, when compared with the picture of the corresponding period in 1980 (1979–1980 being a normal rainfall year in the region) showed that the major reservoirs in the area, namely, Poondi, Arniar, Cholavaram, Red Hills, and Chembarmbakkam had reduced to 3.7%, 28.6%, 20%, 51.9%, and 21.4%, respectively, in areal spread. This was attributed to the failure of rains in 1982–1983, which resulted in poor recharge to the groundwater and extensive pumping. As these reservoirs supply water to the city, the coastal aquifers south of Madras were subjected to heavy pumping, resulting in seawater intrusion in some pockets (Seelan and Narayan, 1984).

Investigations carried out by Kruck (1976) in the Argentine Pampa area revealed that saline groundwater zones can be demarcated on the satellite picture using vegetation as an indicator. Where saline incrustations start appearing on the soils, their high reflectance levels permit easy identification on the imagery. Computer-aided interpretation helps identification of different levels of soil salinity (Ventakratnam, 1984).

3.3.7 THE INTEGRATED APPROACH

At times, a single parameter such as geomorphology would play a dominating role and should be sufficient to plan further ground exploration. But many times, it is not possible to arrive at potential groundwater zones based on one or few of these parameters. For example, a single information source such as lithology may not indicate promise, but the situation may change when viewed with corroborative evidence in structure, geomorphology, land use, and soils. Also, it is necessary to incorporate information obtained from sources other than remote sensing (Rampheri et al., 2023; Shao et al., 2020; Nejad et al., 2017); hence, the need for an integrated approach. An integrated approach where the different thematic information on geology, geomorphology, soils, land use, rainfall, drainage, etc. are studies to arrive at potential groundwater zones offers the best results. It also ensures that no relevant information is overlooked. The need for an approach integrating all parameters relating to geomorphology, soils, land use, etc. was advocated by Seelan and Thiruvengadachari (1980, 1981). A part of Tamil Nadu was studied using Landsat MSS data for geology, geomorphology, soils, and land use, which were then manually superimposed on one another via transparent overlays to extract groundwater potential zones. The geological map was used as a base, and an overlay was prepared indicating potential groundwater zones from the point of view of geology. The overlay was then transferred to the other thematic maps and the procedure was repeated. Finally, in the overlay, the regions where all or most of the themes were found favorable were marked as potential groundwater zones.

Integration of various thematic information to arrive at groundwater potential was extensively used by the National Remote Sensing Agency (1980, 1982a, 1982b, 1983, 1984) during the eighties. The use of this method was also reported by Sankar (2002) and Gopinath and Seralathan (2004).

A comparison of a hydrogeomorphic approach and the integrated approaches was made by Seelan and Thiruvengadachari (1981). It was concluded that in areas where geomorphology exercises significant control over the groundwater movement and occurrence, the extraction of geomorphic details alone can suffice, and in areas where the control is not significant, or where costs and time permit a detailed study, basic resource information with regard to geology, geomorphology, land use, and soils can be obtained and integrated to extract regional-level groundwater potential zones.

3.3.7.1 Geographic Information Systems

The manual process of integrating various thematic elements to arrive at potential groundwater zones was laborious and subjective to a great extent. With the advent of Geographic Information Systems (GIS), it became possible to overlay different thematic information digitally (Alshehri and Mohamed, 2023; Al-Djazouli et al., 2021; Das et al., 2019). The subjectivity is reduced as weightages can be assigned to various parameters. It is much faster, and a standard procedure could be used under most terrains with suitable modifications. In an early use of GIS in groundwater studies, Rundquist et al. (1991) studied the vulnerability of groundwater systems in Nebraska, USA.

In an attempt to use GIS for targeting groundwater, Seelan (1994) used remote sensing data to derive thematic information on geomorphology, land use, soils, drainage density, dikes etc., and incorporated information on rainfall and DEM (Figure 3.4) using PMAP version 2.22 GIS software to arrive at potential groundwater zones. This study was carried out in a part of Andhra Pradesh, India, over a hard rock, gneissic terrain where groundwater primarily occurs in weathered pediments and fractured valleys. The land use patterns, soils, as well as the dikes, also control the recharge, occurrence, and movement of groundwater in the region. The six digitized, thematic maps pertaining to geomorphology, land use, soils, drainage density, drainage and dikes, and rainfall were used in the GIS analysis to extract groundwater potential zones. Based on the knowledge of the various parameters and their bearing on groundwater, weightages were assigned for each of the classes in the various maps. The simplified working method for the groundwater prospects model is given in Figure 3.5. The six polygon maps with assigned weightages for each polygon were overlaid digitally to create a final groundwater potential map (Figure 3.6). The area was classified into five categories ranging from very high potential to very low potential. The study also concluded that a GIS-based integrated approach is superior to the manual approach and the geomorphic approach, as this takes into consideration the various parameters and reduces human bias in integration. However, it is important to reduce bias while assigning weightages.

With improvements in GIS technology over the years, an integrated approach to evaluating groundwater potential has become very popular. A vast number of authors have reported the use of this technique, or variations of it, in routine groundwater investigations (Sener et al., 2004; Jha et al., 2007; Kumar et al., 2007; Elewa and Qaddah, 2011; Jasmin and Mallikarjuna, 2011; Jagadesha et al., 2012; Lee et al., 2012; Magesh et al., 2012; Bagyaraj et al., 2013; Gumma and Pavelic, 2013; Khodaei and Nassery, 2013).

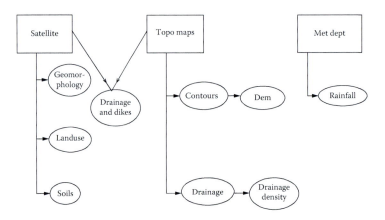

FIGURE 3.4 Sources of data and derivative maps used in GIS study.

Groundwater Targeting Using Remote Sensing

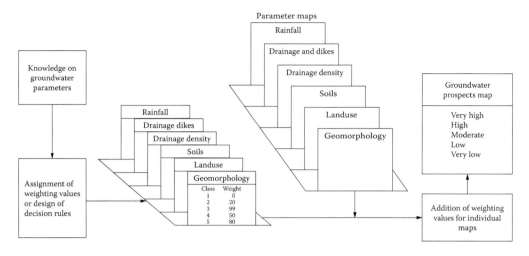

FIGURE 3.5 Simplified remote sensing and GIS-based working model method for the groundwater prospects model.

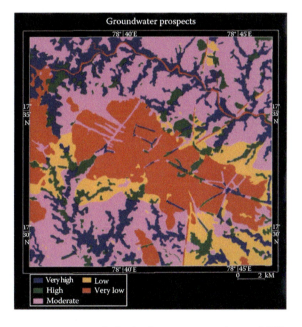

FIGURE 3.6 Groundwater prospects map derived using remote sensing and GIS.

3.4 INDICATORS AND INTERPRETATION KEYS IN UNCONSOLIDATED ROCK TERRAIN

Unconsolidated sedimentary rocks, also termed non-indurated deposits, are composed of particles of gravel, sand, silt, or clay size that are not bound or hardened by mineral cement, by pressure, or by thermal alteration of the grains. In this category of rocks are included non-indurated sediments of fluvial, aeolian, lacustrine, and marine origin. There is great variation in textural composition, depending on the provenance and conditions of transport and deposition, which is reflected in the wide range in the porosities, specific yields, and permeabilities of the rocks and in the wide differences in the yields of wells. A porosity of 20–40% is common except for fine-grained sediments like clays. The specific yield ranges from negligible in clays to about 30% in coarse-textured homogenous sediments. The hydraulic-conductivity values range from less than 1 m/day to as much as 200 m/day (Karanth, 1987).

Fluvial deposits are the materials laid down by physical processes in river channels or on flood plains. The materials are also known as "alluvial deposits." The following sections deal with fluvial materials deposited by non-glacial and glacial environments, materials that are transported by wind, followed by remote sensing parameters in these terrains.

3.4.1 ALLUVIAL AQUIFERS

The term "alluvium" is widely used to describe terrestrial sediments of recent geologic age deposited by flowing water. The sediments are composed of clastic materials of greatly varying grain size. If the particles are of a fairly uniform size, the material is said to be well sorted; if particle sizes are distributed over a wide range, the material is said to be poorly sorted. A well-sorted material contains better porosity and permeability. Aquifers in alluvial deposits are very common and constitute in many regions the only exploitable source of groundwater. They can be classified according to the environment of deposition into (1) alluvial fans and piedmont deposits, (2) valley fills, (3) alluvial plains, and (4) deltaic terrains.

3.4.1.1 Alluvial Fans and Piedmont Deposits

Alluvial fans form where a stream leaves its inclined mountain tract and enters the plain, dumping most of the sediment load because of the sudden decrease in flow velocity. The accumulation of a great mass of material forces the river, from its point of emergence from the mountains into frequent changes of course, into various directions. Thus, its sediment load is spread over a fan-shaped area. Alluvial fans, distinct near the point of emergence of the valleys from the mountain belt, tend to merge farther downstream. The resulting complex of coalescent alluvial fans is often called piedmont. The continuous belt of fans is also called bazada.

The sediments of alluvial fans are composed of particles of all sizes, from large boulders and blocks to clays with greatly differing degrees of sorting. The coarsest materials are found near the mountain border, generally mixed with finer fractions, and particle size diminishes toward the lowlands. Stratification is very imperfect in the upper ranges of the piedmont belt. Units of similar lithological composition and sorting are lens shaped in a section across the fan and string like in the direction of the river channels. It is difficult, therefore, to extend stratigraphic correlations over any appreciable distance. In addition, each period of high river flow truncates part of the previous sediments and deposits them farther downstream. In the lower part of the alluvial fans, more continuous and better-sorted layers are present and the part of the finer-grained materials increases. Thus, the thickness of the aquiferous beds in a given section is reduced, and simultaneously confined aquifers are formed. A typical section through an alluvial fan is given in Figure 3.7.

3.4.1.1.1 Exploration Parameters in Alluvial Fans and Piedmont Deposits—Humid Regions

Groundwater in alluvial fans is replenished mainly by percolation of river water, which may reach a remarkable rate over relatively short stretches, especially near the mountain border. The water may reappear in the form of springs and seepages around the toe of the fan, or it may continue its subsurface flow toward more distant downstream areas. Relatively deep drilling in the downstream part of the fan taps confined aquifers because of the interstratification of aquiferous and confining beds.

The hydrologic properties of alluvial fans depend on the physical and chemical nature of the constituent rocks. Fine and plastic components may lose their primary porosity by compaction. Calcareous rock debris can be transformed into a compact breccia or conglomerate by alternating solution and precipitation of carbonates. Where the rock material and the climate are conducive to mud flows, fans may be so rich in clay-sized particles as to be practically impervious.

Groundwater Targeting Using Remote Sensing

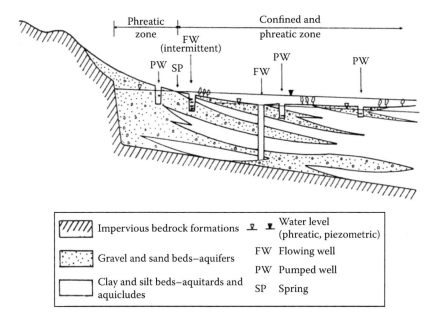

FIGURE 3.7 Typical section trough alluvial fan (after Mandel and Shiftan, 1981).

The Bhabar belt along the northern fringe of the Gangetic alluvium typifies composite fan deposits, which owe their origin to torrential streams emerging out of the Himalayan foothills. The Bhabar belt merges southward with the Tarai, the junction marked by a spring line that separates the flowing well area to the south from the recharge area toward the north.

3.4.1.1.2 Exploration Parameters in Alluvial Fans and Piedmont Deposits—Arid Regions
In arid regions, alluvial fans attain larger dimensions and are more conspicuous as landscape forms than in humid regions and also play a more important role as potential aquifers. Most of the infiltration of floodwater takes place along the braided channels of the major streams in the upper and middle reaches of the bazada belt. In the lower parts, percolation is impeded by the presence of greater amounts of clay and silt in the near surface sediments. Groundwater is phreatic below the higher parts of the piedmont belt, where the water table is often at depths that make exploitation unattractive. In the downslope direction, the water table gradually approaches the surface and finally intersects it as evidenced by the appearance of springs or seepages. In the middle and lower parts of the piedmont belt, groundwater becomes progressively confined because of the increasing number and thickness of semipermeable and impermeable layers and the simultaneous wedging out of the aquiferous and gravel beds. Some of the fine-grained beds may stem from ancient playas and can be a source of excessive groundwater salinity.

At the lower end of the piedmont belt, or below the adjoining playa or mud flat, groundwater moves upward through semi-confining layers and appears on the surface as dispersed seepages, or occasionally, as more centralized springs. Most of the emerging groundwater is used up by the vegetation or evaporates on the surface.

Best results from the well drillings are to be expected in the middle part of the piedmont belt where the water table is not too deep, a fair amount of aquiferous beds can be expected, and salinity may still be relatively low. The surface features of alluvial fans do not indicate the extent of possible aquifers.

3.4.1.2 Valley Fills
Valley fills are alluvial and colluvial deposits that lie between mountain ranges or between exposures of hard rocks. Coarse materials comprising sand, gravel, and pebbles form the bulk of the

valley fills, which are among the most productive of aquifers. Within the outcrop region of hard rocks, valley fill deposits consist of the coarsest grains, the sediments becoming finer as the distance of transport increases.

Typically, the deposits are characterized by basal gravel and pebbles, succeeded upwards by finer materials. Depending on the geologic history, sedimentation may be cyclic with repetition of the gravel-sand-clay sequence. The width of the valley fills ranges from a few tens of meters to tens of kilometers. The thickness of the valley fill varies widely depending on the configuration of the basal rocks and the land surface profile.

In hard rock terrains, the areal spread of the valley fills follows, more or less, a sinuous course of the present course of the stream, lying sometimes to one side or the other, the present ancient courses being the same at places. Usually, chances of the valley fill being thick are remote if the present-day channel deposits comprise coarse-textured materials like pebbles and boulders.

3.4.1.3 Alluvial Plains

Alluvial plains are built up by plastic material deposited by meandering or braided rivers. Of the large amount of sediments carried by streams, the coarsest and most permeable fractions are deposited along the stream channels, while the finer ones are deposited on the flood plains and backswamps. The coarsest-grained gravels and sand make up the traction load of present and ancient buried stream channels. On the inside of meander bends, coarse-grained point bar deposits are formed. Natural levees, often flanking the channels, are generally built up by fine sand and silt. During the high-water stages a slowly moving sheet of water covers the flood plain and deposits silt and clay. Coarser material may reach the flood plain when the levees are pierced during floods.

According to a schematic concept of depositional history, coarser-grained materials should prevail in the upstream part of the plain and also in older, deeper layers that were deposited during early, more vigorously erosive phases. Periodic subsidence of the depositional basin leads to the accumulation of huge thickness of alluvium, as exemplified by the Indo-Gangetic trough, which contains over 1,000 meters of alluvial sediments. The thickness of individual beds of sand, clay, etc. may range from less than 1 m to over 100 m. The percentage of coarse granular horizons gradually decreases in the downstream direction. The typical topographic forms and deposits of broad floodplains of large rivers is shown in Figure 3.8.

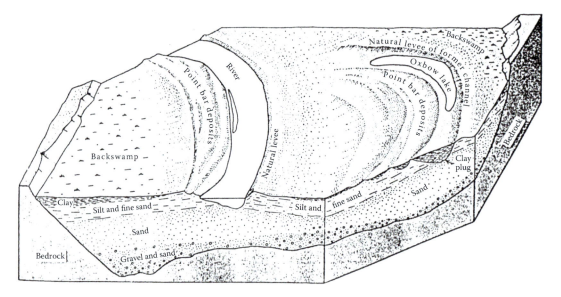

FIGURE 3.8 Topographic forms and deposits typical of broad flood plains of large rivers (after Davis and Dewiest, 1966).

3.4.1.3.1 Exploration Parameters in Alluvial Plains

Though most of the valley deposits have a simple vertical succession, from coarse sands and gravel near the bottom of the channels to silts and clays at the top, the relative thickness of the coarse and fine units depends on the type of sediments carried by the river and the geologic history of the river and the point of interest. It is practical to visualize a large alluvial plain as a complex of more or less lens-shaped elongated bodies—or continuous layers—of gravel, sand, silt, and clay, including various mixtures of these components. Figure 3.9 shows such a typical section through an alluvial plain and the successful and the unsuccessful wells depending on the sections pierced.

Braiding and meandering of streams give rise to the formation of thick and extensive granular horizons that form productive aquifers. Superimposition of backswamp deposits over meander belt deposits results in the confinement of aquifers by the fine-grained sediments like silt and clay. Cross-bedded sand, which is commonly fine or medium grained with variable content of silt and clay, is deposited on the levees and flood plains, which form good aquifers. Point bars, with coarse sands and gravel, also hold favorable hydrologic properties.

Because of great variation in the nature and thickness of aquifers and confining layers, groundwater occurs under confined, semi-confined, and unconfined conditions. Under favorable hydrogeological situations artesian-flowing conditions may be encountered in shallow wells in low-lying river terraces.

3.4.1.4 Deltaic Terrains

Deltaic deposits are formed where a river dumps much more material than the sea currents are able to sweep away. Not all rivers are capable of building deltas. Sediments consist predominantly of fine sand and silt. Gravels reach delta areas only in rare cases. Coarser sand and sandstones originate from beach sands, dunes, bars and banks, and river channel deposits. Clay and silt are deposited in tidal flats and shore lagoons as a result of the variety of depositional environments over short distances, and in time sedimentary units are lens shaped and discontinuous. Therefore, the appraisal of sub-surface conditions presents the same difficulties as on alluvial plains. A typical section through a delta is shown in Figure 3.10.

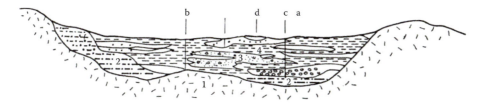

FIGURE 3.9 Section through alluvial plain (after Mandel and Shiftan, 1981).

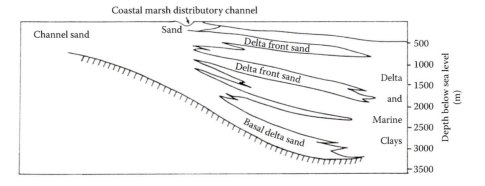

FIGURE 3.10 Hypothetical section through a delta.

One of the important features of coastal deposits is the occurrence of groundwater in interbedded alluvial and marine sands, silts, and clays deposited under beach, lagoonal, estuarine, and marine environments. Coarser deposits of gravel and pebbles are commonly found where youthful streams having rocky catchments debouch into the sea. Due to differential compaction, coastal sediments attain a seaward dip. Because of the abrupt reduction in the velocity of streamflow near the coast, the junction between coarse- and fine-grain deposits may be rather abrupt. Palaeo distributaries, particularly from youthful streams often contain coarse-grained sands.

3.4.1.4.1 Exploration Parameters in Deltaic Terrains

The distribution of freshwater aquifers is controlled by the dynamic equilibrium between hydrostatic heads in freshwater and saline water zones, influx of seawater into streams and lagoons, and relative movement of the sea in respect to the landmass.

Brackish water may be encountered even in shallow boreholes and in locations far from the sea. On the other hand, the occurrence of fresh water under certain artesian conditions, even below shallower patches of saline water, is not uncommon. It is due to the infiltration into truncated delta deposits that become confined in a downstream direction, the palaeo distributary channels are known to carry fresh groundwater in an otherwise brackish area.

3.4.2 Glacial Terrains

The retreat and advancement of glaciers due to climatic changes have given rise to a complex and often uncertain distribution of beds of sand and gravel. The coarse fractions are generally found close to the ice front or along channels of large streams. Rock debris chiefly of glacial origin is called glacial drift. Drift includes till, a highly heterogeneous and mostly non-stratified material consisting of boulder clay deposited directly from ice. The thickness of glacial deposits varies between several meters and several tens of meters, but rarely exceeds a hundred meters. They often form elongated ridges called moraines.

Glacial till was the most abundant material that was deposited on the land surface during the Pleistocene time. In the Precambrian shield region, till is generally sandy, with variable amounts of silt and little clay. Sand till forms local aquifers in some areas. In the regions of sedimentary bedrock, glacial till has considerable silt and clay and therefore low permeability.

Streams that issued from the ridge of melting ice masses picked up big loads of unconsolidated glacial sediments, dumping coarser constituents at some distance downstream. These outwash gravels occur in the form of outwash fans and outwash terraces and constitute shallow but useful aquifers. Their thickness rarely exceeds several tens of meters. Frequently the outwash gravels are intimately connected with recent river gravels. In some places the melt water was dammed up by moraines, thus forming lakes in which delicately layered, fine-grained sediments were deposited. These are aquitards or poor aquifers.

Glacial deposits also express themselves in various typical forms. Eskers are winding ridges composed of poorly sorted sand and gravel. They are the most distinctive of the various landforms composed of ice contact deposits. Eskers are the bed load deposits of former streams that occupied subglacial ice tunnels or, less commonly, streams on the ice surface. Most eskers are formed during the stagnant or near-stagnant phase of glaciation.

Kame terraces are formed by the accumulation of glacial debris along the margins of stagnant glacial ice that remains in valley areas. When the ice melts the debris is left as a terrace along the sides of the valley. The part of the terrace that was once in contact with the ice is strongly affected by collapse and slumping, thus forming irregular borders facing the centers of the valleys. Small patches of ice-contact material may be let down when the glacial ice melts. Small hills formed in this way are known as kames. Some kames also may be simply erosional remnants of larger masses of ice contact deposits. They consist of poorly sorted sand and gravel.

Kettle holes are formed by the collapse of till and ice contact sediments as isolated masses of residual ice melt. Kettle holes may be found on kame terraces, wide parts of eskers, till plains, and terminal moraines.

3.4.2.1 Exploration Parameters in Glacial Terrains

Location of areas favorable for groundwater development in glacial terrains can be very difficult. In the relatively younger deposits, much of the original topographic expression is retained and therefore is mappable. Groundwater prospection in glacial terrain is concerned with the location of outwash gravels and buried channels. Most of the forms such as moraines, eskers, and kames contain poorly sorted materials and are not favorable.

Aquifers are known to occur as extensive blanket bodies or channel deposits in the surface of buried valleys. The deposits of sand and gravel in buried valleys form aquifers that are generally many tens of kilometers long and several kilometers wide. In many cases there are no surface indications of the presence of the buried valley aquifers. The overlying till is usually some tens of meters thick or less but occasionally may be of the order of a hundred meters thick.

3.4.3 AEOLIAN DEPOSITS

Materials that are transported and deposited by wind are known as aeolian deposits. The physical process leading to the deposition of aeolian or windborne sediments include deflation whereby soil cover is stripped from the surface of rock formations and the transportation of soil particles through the air. Aeolian deposits can be divided into two types, dune sand and loess. The coarser and heavier particles are carried closest to the ground and get deposited as dunes, while the finer particles move farther up in the air and get deposited as loess several kilometers away from the source.

Sand dunes form along coasts and in inland areas such as in deserts where the rainfall is scant and surface sand is available for transportation and deposition. Aeolian sand is characterized by a lack of silt or clay fractions, by uniform texture with particles in the fine or medium grain-size range, and by well-sorted rounded grains. The sands are quite homogenous and isotropic. The sorting action of wind tends to produce deposits that are uniform on a local scale and in some cases quite uniform over large areas. Loess, on the other hand, is generally a silt-sized material, which, like dune sand, is well sorted, but unlike dune sand will vary widely in grain size. Because of small amounts of clay and calcium carbonate cement that are almost always present, loess is slightly to moderately cohesive.

3.4.3.1 Exploration Parameters in Aeolian Deposits

Aeolian sands have porosities between 30% and 45% and moderate permeability. Loess deposits have a porosity of 40–50% and low permeability. Loess is not commonly an aquifer because of its low permeability and because where its permeability is the highest it is usually in the high topographic positions where sub-surface drainage is good.

Loess deposits in arid regions are cemented to a varying degree to calcium carbonate, causing a reduction in porosity and permeability. Under favorable conditions of topographic and water table slopes extensive cemented zones may give rise to sufficient confining pressure in the lower zones to cause flowing conditions in wells. In the valleys covered in loess and bordered by hills composed of quartzite or other resistant rock types, productive boulder-gravel aquifers occur in basal sections of the valley fill.

Inter-dunal depressions form receptacles of relatively impermeable material like silt and clay washed down during periods of rainfall. The evaporation of water accumulated in the depressions precipitates calcium carbonate and gypsum in the form of impervious pans. The impervious layers may retain percolating water to give rise to perched groundwater zones occurring above the main water table. Isolated dunes amidst hard rocks may locally give rise to productive sandy aquifer. Also, aeolian action may shift stream courses, leaving behind abandoned channels that may form potential aquifers.

3.4.4 REMOTE SENSING PARAMETERS IN UNCONSOLIDATED ROCK TERRAIN

From the previous paragraphs, it is seen that in an unconsolidated rock terrain the various geomorphic units and forms have typical hydrologic properties. Therefore, the first task in the exploration of these terrains is the mapping of these unit and forms. As morphology is essentially a surface expression, and since remotely sensed data directly provides surface information, the task is accomplished through the use of such data. The basic interpretation keys are provided by the typical spectral response of these units and forms and also their tone, texture, pattern, shape, size, and association. As spectral signature alone is not conclusive enough, delineating these features via computer-aided classification is not ideal. Best results are obtained through visual interpretation of the imagery where factors such as tone, texture, pattern, shape, size, and association are taken into consideration. While the satellite imagery of medium to high spatial resolution can provide sufficient information needed in regional exploration, high-resolution satellite imagery or large-scale aerial photographs will be required in identifying objects of smaller size useful in local explorations (Adams et al., 2022; Nhamo et al., 2020; Dasho et al., 2017; Gemitzi et al., 2017).

Alluvial fan is reproduced on satellite imagery in the form of a triangle with its apex directed toward the hills and the spread-out base touching the surrounding inclined foothill plain. The tone is lighter and the vegetation is sparse. Where the fans coalesce to form bazadas, the triangular shapes are not obvious but the tone is usually lighter. Spring lines at the base marked by vegetation (red in standard false color composite) are often seen.

Stream valleys/valley fills exhibit very clear shapes or form. Where the gradient is low, meandering with large meander wavelength and with broad and only slightly incised valleys is seen. Drainage patterns imply lithology and degree of structural control. Drainage density on humid regions and drainage texture in arid regions imply grain size, compaction, and permeability (Moore, 1978). Valley fills support natural vegetation as well as irrigated agriculture. Agricultural crops in non-rainy seasons are known to show remarkable correlation with valley fills in certain hard rock terrains in India (Seelan, 1980). Underfit valleys are represented by topographically low, elongated areas with ponded drainage or with a stream meander wavelength smaller than that of the flood plain or terraces.

The natural levees are characteristically associated with stream channels where the gradients are low and seen as sharp lines close to the river channels. The tone is usually lighter, but where the deposits are fine grained, darker tones are noticed. Typical arcuate shapes characterize ox-bow lakes and other associated features, such as meander scars, point bars, etc. Elongated lakes, sinuous lakes, and aligned lakes and ponds represent old flood plains. Parallel lines of trees in alluvial plains indicate buried channels. Flood plains are distinguishable by their darker tone. River sands are white in standard false color composites. Back swamps appear dark, with little vegetation adjoining the flood plains.

The glacial and aeolian deposits are distinguishable by their unique meso relief forms and shapes. The sand dunes are elongated and arcuate and have a very light tone. The outwash gravels in the glaciated regions occur in the form of fans. Eskers form winding snake-like ridges. In the snowbound areas anomalous early melting of snow and greening of vegetation show areas of groundwater recharge.

Soil types are usually indicated by tone. Fine-grained soils commonly are darker than coarse-grained soils. Wet soils are darker than dry soils. Salt-affected soils are very light in tone, and in standard false color composites they appear as white patches. Distinctive types of native vegetation commonly show upstream extensions of drainage patterns, areas of high soil moisture, and landform outlines. Abrupt changes in land cover type or land use imply landforms that may be hydrologically significant but do not have a characteristic shape.

3.5 INDICATORS AND INTERPRETATION KEYS IN SEMI-CONSOLIDATED TO CONSOLIDATED SEDIMENTARY ROCK TERRAIN

Depending on the degree of compaction, cementation, and crystallization, consolidated, semi-consolidated, or unconsolidated sedimentary rocks grade into each other. Sedimentary rocks may occur in any stage of consolidation from incoherent granular materials like silt and sand to firmly held granular rock like siltstone and sandstone. While losing primary porosity they may develop secondary porosity due to fracturing and weathering. Important stratigraphic and structural features that have a bearing on the occurrence, movement, and availability of groundwater are stratification (in marine environments the sequence of deposition from bottom upwards is limestone, shale, and sandstone), lateral gradation, inclination, folding and faulting of strata, and unconformities.

Sedimentary rocks range from a few meters to thousands of meters in thickness and are sometimes traceable over thousands of square kilometers. When inter-bedded with clay, sandstones comprise a multi-aquifer system. If the beds are inclined, wells located along the dip direction will tap different aquifers, and high-pressure artesian conditions may occur along dip slopes. If there is more than one aquifer the pressure head may increase with the depth of the aquifer.

3.5.1 Sandstone–Shale Aquifers

Sequence of alternating sand, sandstone and clay, or clay is characteristic of many sedimentary successions. Deposition of such sequences takes place in the marine, deltaic, littoral, and arid-continental environment. The main difference between such sequence and alluvium or recent deposits is the greater age and, hence, the more advanced stage of consolidation. The sandstone and shale sequences, however, show more persistent stratigraphy when compared to recent alluvium. The primary porosity of a layer of sandstone is often strongly reduced by compaction and cementation. Zones of secondary porosity are usually found along bedding plains, joints, and fractures.

Alternating sandstone–shale formations occur under a variety of geologic conditions. On some of the continental platforms, they fill vast bowl-shaped depressions or basins and constitute large regional, often confined aquifers. In regions dominated by a normal fold pattern, conditions are also favorable for the formation of regional confined aquifers.

Sandstone is the most productive among the semi-consolidated sedimentary rocks. Shale is formed by compaction of clay sediments, and water is found in porous layers, fractures, bedding plains, and weathered zones.

3.5.2 Carbonate Rocks

Carbonate rocks in the form of limestone and dolomite consist mostly of the minerals calcite and dolomite, with very minor amounts of clay. Young carbonite rocks have high porosities but with increasing depth of burial the soft carbonate minerals are compressed and recrystallized into a denser, less porous rock mass.

Many carbonite rocks have appreciable secondary permeability as a result of fractures or openings along bedding plains. Secondary openings in carbonite rocks caused by changes in the stress conditions may be enlarged as a result of calcite or dolomite dissolution by circulating groundwater. Although some original pore space may be retained in old limestone, other forms of porosity are more important from the stand point of water production. Fractures and secondary solution openings along bedding plains probably transmit the most water.

3.5.3 Exploration Parameters in Sandstone–Shale and Carbonate Rocks

In gently folded regions and in vast bowl-shaped depositional regions a fairly complete picture of the aquiferous properties of the sandstone–shale formations can be obtained from surface observations. However, in regions with more or less horizontal strata, where outcrops are rare, the task is difficult.

Firmly cemented sandstones with low porosities will yield water to wells along fractures. The same general guiding principles apply to the location of water in these rocks as they apply to the location of groundwater in crystalline rocks of platonic origin. Most favorable areas for development of groundwater are along fault zones and within thoroughly jointed zones. Better wells will be found in broad valleys and on flat upland areas than on hill crests and valley slopes.

Dense shales, devoid of fractures are practically impervious and form confining layers or barriers. The contact zone of limestone within the underlying shale bed is usually rendered more permeable than the rest of the limestone as the presence of impervious shale limits downward circulation of groundwater, facilitating dissolution of limestone parallel to the bedding plain. However, formations composed of sandwich fashion of thin layers of carbonate rock with alternating shales offer poor prospects for groundwater. The circulation zones for groundwater within the carbonate rocks in such cases are too narrow to facilitate the dissolution of limestone.

Though individual limestone beds can be located through structural and stratigraphic studies, the yields of beds are hard to predict. The flow through large caverns resembles that of a surface stream, turbulent and cascading to different levels. Unless the development of secondary porosity is extensive, there is no regular water table. One may well strike good supplies while another one close by may be a barren one.

In areas of thick limestone or dolomite, wells located in valley bottoms are somewhat better than on valley slopes. Water storage in adjacent alluvium, together with a water table that is closer to the surface, account for some of this advantage. Wells drilled on broad uplands are also more successful than those drilled on hill slopes. Fractures and solution openings are more abundant along crests of anticlines and within synclinal troughs than they are on the flanks of the folds.

Carbonate rocks are eroded by dissolution in water containing carbon dioxide. The peculiar landscape thus formed is called karst. Mature karst morphology develops on hard, fissured carbonate rocks under humid to sub-humid climatic conditions. It is characterized by the disruption of surficial drainage patterns. River valleys are replaced by arrays of closed depressions (dolines), surface runoff disappears into sinkholes (Figure 3.11), and in extreme cases, entire rivers flow for some distance in caverns underground. Karst morphology is a strong indication of favorable aquifer zones in limestone.

3.5.4 Remote Sensing Parameters in Semi-consolidated to Consolidated Sedimentary Rock Terrain

Many sedimentary rock formations throughout the world comprise important and extensive aquifers, especially those near the surface. Groundwater exploration in sedimentary rock terrains require accurate geologic mapping as the basic data for the exploration of hydrological information such as permeability, storage characteristics, infiltration rates, specific capacity, etc. Unlike in the unconsolidated rock terrains, where geomorphic units and forms play a vital role as remote sensing parameters, here the lithology and structural information such as folds, faults, dip, and strike play a prominent role. Geomorphology too aids in understanding lithology, structure, land use and land cover patterns, and groundwater recharge conditions. Therefore, remote sensing parameters relating to lithology, structure, geomorphology, and land use are the ones to look for. Standard keys on shapes and patterns can be used while attempting to identify rock types from shapes and patterns on satellite images. However, other factors could be important locally. Previous experience of previous

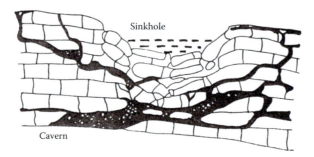

FIGURE 3.11 Sink hole in limestone (after Karanth, 1987).

knowledge of the area being interpreted is necessary for good results. Though major rock types could be delineated on satellite images, minor variations within rock types are rarely identifiable.

Landforms and topographic relief often indicate the rock types below. Hard, resistant sandstones stand out as serrated and linear ridges, softer shales and sandstones may form low-lying denudational hills. The outcrop pattern is typically a banded type for these sedimentary rocks, often outlined by vegetation in some regions.

The shape of drainage basins is a good indicator, with the drainage at times running parallel along the lithological contacts. Drainage pattern and density are important indicators of the underlying rocks. For example, sudden disappearance of drainage indicates karst topography. The other indicators could relate to relative abundance, shape, and distribution of lakes and types of native land cover. Tones are difficult to describe, as similar rock types may have different tones and vice versa depending on the cover types. However, relative tone variations within a given area can be good indicators of varying lithology.

Color composite images probably are excellent for the identification of rock types, although drainage patterns are best seen in many areas in infrared black-and-white images (band 4 in Landsat TM and in IRS). Each of the black-and-white bands has been reported by various authors to be superior to the other bands for lithological delineations; the best single band probably is determined by factors such as atmospheric conditions, time of year, and type and amount of vegetation cover. The band 7 in Landsat TM is typically meant for lithological discrimination. Rock types are thermally different from each other, and the thermal band of TM can be used for this purpose. Little geologic information is lost while making a color composite, and the ideal TM band combinations for lithological discriminations are 2, 4, 7, and 2, 6, 7.

Folded patterns and dip/strike directions are important for understanding the groundwater flow regime. Curved patterns on satellite imagery are indicative of folded beds. Cuestas and hogbacks and asymmetric ridges and valleys are indicative of folded beds. The topography is usually flat on dip slope and irregular on back slope. The vegetation distributed uniformly in dip slope is banded and parallel to ridge crest and back slope. Bazadas are formed on dip slope, and separate alluvial fans are formed on back slope. Trellis, radial, annular, and centripetal drainage patterns are indicative of folded structures. The other indicators are major deflections in drainage channels, change in meander wavelength, or changes from meandering to straight or braided patterns.

Lineaments are indicated by continuous and linear stream channels, valleys and ridges, and elongated or aligned lakes. Identical or opposite deflections in adjacent stream channels, valleys, or ridges and alignment of nearby tributaries and tributary junctions are also indicative of lineaments in sedimentary terrains. Other indicators include alignment of dark or light soil tones, elongated or aligned patterns of native vegetation, and thin strips or relatively open or dense vegetation.

Folds that may concentrate or block groundwater flows can be detected and delineated with relatively accurate results on satellite images (Alshehri and Mohamed, 2023; Kim et al., 2019; Gnanachandrasamy et al., 2018). Vegetation patterns are seen best in color composite images, but infrared images may be best for delineating drainage patterns. Vegetation patterns are prominent on green and red bands, as well as on color composite images of green, red, and infrared. Vegetation patterns can be distracting to the eye when the image is being interpreted for other features. Though vegetation has high reflectance in infrared, the vegetation patterns are least obvious in black-and-white infrared images, enabling interpretation of other features.

3.6 INDICATORS AND INTERPRETATION KEYS IN VOLCANIC TERRAIN

Volcanic rocks include basalt, rhyolite, and tuff. Commonly, they consist of several successive flows of variable thickness and lateral extent. A typical flow unit consists of a lower dense and massive horizon, passing upwards into a vesicular, amygdaloidal, or jointed horizon. The distinctive geohydrogeological feature of volcanic rocks is the significant primary porosity on the form of vesicles, lava tubes, and occasional tunnels formed due to the escape of gases. Secondary porosity is developed due to fracturing during cooling of the lavas, tectonic disturbances, and weathering.

3.6.1 Typical Profiles in Volcanic Terrain

From the viewpoint of groundwater occurrence, three principal types of volcanic terrains can be distinguished: basalt plateaus, central volcanic edifices, and mixed pyroclastic lava terrains.

Basalt plateaus are the result of repeated effusions of low viscosity that have issued from numerous fissures and sometimes contain the conspicuous mineral olivine. Preexisting relief tends to be leveled out and transformed into a flat morphology, or into a step-like one, if successive flows terminate at different distances from the erupting fissures. Clay rich soils that form by weathering between eruptions, or sands, are sometimes found between successive sheets. The original continuity of basalt plateau is often disrupted by deeply incised valleys cutting into underlying formations or by tectonic disruption, into fault blocks (Figures 3.12 and 3.13). Some volcanic plateaus are built up by ignimbrites—ashes erupted in an incandescent state and welded together by heat.

Central volcanic edifices are cone-, dome-, or shield-shaped volcanoes composed of lava flows and layers of volcanic ash and coarser materials ejected in the solid state. The surrounding lowlands are often covered with thick accumulations of volcanic ash and fine-grained pumice. Only the most copious lava flows reach large distances from the center of eruption (Figure 3.14).

Mixed lava-pyroclastic terrains of regional extent stem from prolonged periods of intensive volcanic activity. If older than the Pleistocene, the morphology is often one of maturely dissected hills

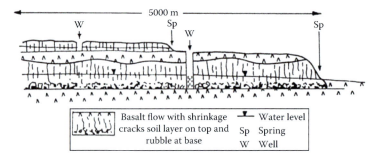

FIGURE 3.12 Groundwater in a lava plateau (after Mandel and Shiftan, 1981).

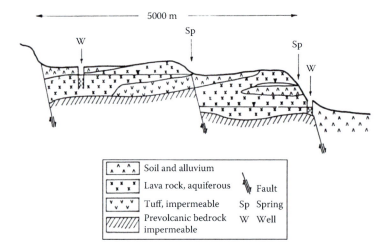

FIGURE 3.13 Groundwater in a volcanic, tilted block region (after Mandel and Shiftan, 1981).

Groundwater Targeting Using Remote Sensing

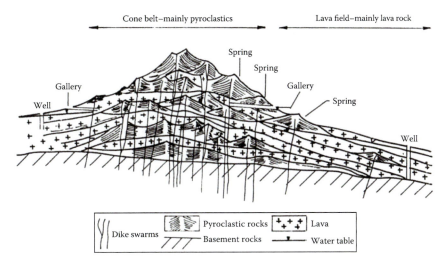

FIGURE 3.14 Groundwater in a central volcano (after Mandel and Shiftan, 1981).

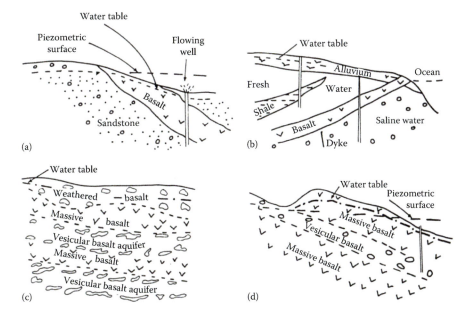

FIGURE 3.15 Basalt as an aquiclude, barrier, and aquifer (after Karanth, 1987).

that bear little resemblance to the original landscape forms. Where volcanic activity has persisted into the Pleistocene, better-preserved volcanic features are superimposed on the ancient landscape.

Depending on the period of eruption, volcanic rocks have buried preexisting rocks ranging from the Archaean to the Quaternary period. In places, they are known to rest on highly productive aquifers of sedimentary origin. Flows, nearly conformable to stratification, form effective confining layers (Figures 3.15 A & B). Alternating sequences of previous compact horizons function as a multi-aquifer system (Figure 3.15 C). If the flow dips at angles gentler than the land surface slope, artesian conditions may result with sufficient pressure to cause free flow in wells (Figure 3.15 D).

3.6.2 Groundwater Occurrence and Exploration Parameters in Volcanic Terrain

Volcanic rocks have widely varying hydrological properties, making predictions about groundwater possibilities uncertain. Some lava flows contain excellent aquifers; others are practically impermeable. The porosity of volcanic rocks varies widely, from almost negligible values for dense basalt to over 60% for pumice and other vesicular varieties (Karanth, 1987). Typically, rocks within dikes and sills will have less than 5% porosity, dense massive flow rocks will have values ranging from 1% to 10%, and vesicular volcanic rock will have porosity values ranging from 10% to 50% (Davis and De Wiest, 1966). Although porosity may be high, the permeability is largely a function of other primary and secondary structures within the rock. Many lava flows exhibit vesicular porosity caused by gas bubbles contained in the lava during eruption. But the pores thus formed are not interconnected. It appears that mainly lavas of fairly recent age (Quaternary, Late Tertiary) are aquiferous, whereas in most of the older lavas, the formation of secondary minerals and the partial disintegration into clay have clogged voids and fissures.

According to Freeze and Cherry (1979), on a large scale the permeability of the basalt is very anisotropic. The centers of lava flows are generally impervious. Buried soils that produce high permeability develop at the top of cooled lava flows. Stream deposits occur between the flows. The zones of blocky rubble generally run parallel to the flow trend. The direction of highest permeability is therefore generally parallel to the flows. The permeability is normally greatest in the direction of the steepest original dip of the flows.

Contrary to the belief that the black cotton soils formed on the basaltic plateaus are impervious and help to augment runoff without much percolation to sub-surface, there is an exceptionally good percolation down to various depth levels due to varying thickness of soil that covers the weathered zone portion of the lava flows. The percolation is enhanced by the soil profiles, which are followed by the underlying horizontally disposed lava flows with step topography on surface and sub-surface. An analysis by Dhokarikar (1991) of rainfall runoff indicates that in watersheds with good rainfall nearly 50% of the total precipitation percolates into the ground. In low rainfall areas the percolation is said to be still higher.

Acid lava rocks containing a high proportion of silica (66% or more) are generally poorer aquifers than the more frequent lavas, such as olivine basalt. Lavas formed by submarine eruptions—so-called pillow lavas—are poor aquifers, or aquifuges, since no voids or fissures are formed because of the rapid chilling and the presence of large amounts of minerals precipitated by steam. Loose pyroclastic rocks (scoria, cinders, pumice, ash) are quite permeable when fresh, but the finer-grained varieties lose much of their permeability through compaction and weathering. Mud flows that owe their origin to torrential rains on steeply inclined soft pyroclastic strata are practically impermeable. They can be discerned by morphologic characteristics and by lack of stratification.

If valleys are near volcanic eruptions, lava will flow down the valleys and bury any alluvium that may be present. Where the valley contains streams from extensive drainage systems, thick gravel may be present, which, on burial, can be important aquifers. Rivers blocked by lava will form lakes that eventually fill with silt, clay, or volcanic ash. These deposits can form important confirming beds for the underlying stream gravel.

In the basaltic plateaus, the hydrologically important porosity of basalt terrains is due to more or less vertical shrinkage cracks and to the essentially horizontal voids and rubble zones left between successive flows. Streams flowing over basalt plateaus frequently lose much water by infiltration. Large springs are formed where the contact between the basalts and an impervious substratum is exposed in incised valleys or along escarpments of the plateau. Smaller springs may issue at various elevations from perched horizons, such as ancient soils and tuffs. The presence of major fissures or faults largely determines the occurrence of the larger springs. In some basalt plateaus, permeability of the formation may be as high as in karstic limestone aquifers, and fairly evenly distributed, so that wells stand a fair chance of success where the saturated is thick enough. Under less favorable conditions well sighting may be guided by the distribution of major fissure systems that are visible on the surface or indicated by hydrologic phenomena.

In the central volcanic edifices, compacted impermeable tuff or soil layers between lava flows and sub-vertical impermeable dikes divide the sub-surface into a number of groundwater compartments, each with its own water level and its own outlet, either into an adjacent lower compartment or into a spring. Thus, the groundwater passes through a number of "steps" down the mountainside into the plain (Figure 3.14). At higher elevations, on the mountain slopes groundwater can be exploited by galleries, and this has been the traditional method in many volcanic mountains and volcanic islands. Groundwater flow is encountered when a dike is pierced. The initial flow tends to decrease with time; the galleries are therefore extended into additional groundwater compartments to maintain the supply. This practice exploits the groundwater steadily infiltrating into the galleries from above but also entails the exploitation of reserves.

On the lower parts of volcanoes at a greater distance from the center of eruptive activity, dikes are less frequent, and the aquifers pass from compartment type to the layered type common in sedimentary layered rocks. The accumulations of volcanic ash, consisting of alternating layers of larger and smaller permeability—together with lava sheets or flows—create confined sub-artesian and artesian conditions. In these parts groundwater is exploited by vertical wells. In the coastal volcanic regions, a freshwater–seawater interface problem has to be faced. In some cases, impermeable layers may isolate sections of the aquifer from contact with seawater and thus facilitate exploitation.

The best aquifers in the mixed lava-pyroclastic terrains are the most recent lavas, coarse pyroclastic (scoria, cinders), and some alluvial deposits. Fine-grained pyroclastics tend to become less permeable or impermeable through compaction. The alluvial fill of plains and valleys, consisting of gravel, sand, and clay of volcanic origin, may present important groundwater possibilities, but the alluvial deposits are as diversified and variable as those in any region.

3.6.3 Remote Sensing Parameters in Volcanic Terrain

Volcanic terrains pose many challenges to the hydrogeologist during groundwater investigations. They behave like sedimentaries where layered, like karstic where the water flows through vesicles, and as hard rocks where they are massive. Groundwater occurs in confined, semi-confined, and leaky conditions. Satellite data is extremely useful in mapping the flows, geomorphic disposition, and the fracture patterns.

In the Deccan Trap regions of India, hard and soft flows occur alternatively on the standard false color composite, the harder flows are generally lighter and devoid of vegetation. The softer flows support vegetation and are red in false color composite. The flows are mappable where steep slopes occur but are difficult in flatter terrains. Stereo viewing aids in mapping the flows.

Basalts at land surface commonly can be recognized by dark tones on black-and-white images and by dark (generally gray to bluish gray) hues on color composite images. Discharge of groundwater occurs at or near the edge of the outcrop area. The areas of discharge are indicated by patches of red color (on color composite images) due to the presence of vegetation. The dark tone of the basalt is mainly due to the black cotton soil developed on the basalts.

The dissection pattern is easily recognizable on the satellite imagery, which helps in geomorphic zonation of the different plateau. In a highly dissected plateau the stream channels are easily recognizable and are controlled by fracture systems. The soil formation is least and therefore the tone is lighter. In an undissected plateau the thickness of the weathered material is more and is darker in tone. The stream channels in the basaltic terrain are normally lineament controlled and are easily mappable on satellite imagery. Very good groundwater potential is also observed in the alluvial plains of major rivers in the basaltic region. Buried channels/old river courses are also identifiable.

Valley fills are the most important and useful information that is available in the basaltic terrains of semi-arid regions. On the FCC images it is seen as distinct red-toned patches invariably following the valleys. Presence of these red patches suggests the growth of dense, healthy, luxuriant vegetation. Survival of dense pockets of vegetation during dry periods will suggest the availability of groundwater. The confinement of such pockets mostly along the valleys will indicate the presence of thick alluvial, colluvial, and weathered material.

3.7 INDICATORS AND INTERPRETATION KEYS IN HARD ROCK TERRAIN

Locating drilling sites in hard rock terrains is considered to be one of the most difficult tasks in groundwater investigations. Extreme variations of lithology and structure coupled with highly localized water-producing zones make geological and geophysical exploration difficult. The percentage of unsuccessful wells is usually the highest in hard rock areas.

Hard rock terrains comprise a great variety of igneous and metamorphic rocks. But from a hydrological point of view they are rather homogenous in two respects. They have virtually no primary porosity compared to sandstones and other sedimentary rocks, but they have a secondary porosity due to weathering and fracturing. The storage and flow of groundwater is restricted to these zones. The general term "hard rocks" is used to describe such igneous and metamorphic rocks. The most common hard rocks are gneisses and granites.

Hard rock is a very general term used in hydrogeology for all kinds of igneous and metamorphic rocks, typical of the shield areas of the Earth. For the purpose of this chapter, the term "hard rock" excludes volcanic and carbonate rocks, as these rocks can have primary porosity and have been dealt with separately in earlier sections.

The weathering processes have considerable influence on the storage capacity of hard rocks. Mechanical disintegration, chemical solution, deposition, and the weathering effects of climate and vegetation bring about local modifications of the primary rock and its fractures. This action can imply either an increase or a decrease of the secondary porosity of the original fracture pattern of the rock. The transition zone between the weathered layer and the underlying fresh rock can function as a reasonably good aquifer, depending on the porosity of this zone.

The storage capacity of the unweathered hard rocks below the weathered zone is restricted to the interconnected system of fractures, joints, and fissures in the rock. Such openings are mainly the result of the tectonic phenomenon on the Earth's crust. Figure 3.16 shows a typical sub-surface profile in crystalline rocks.

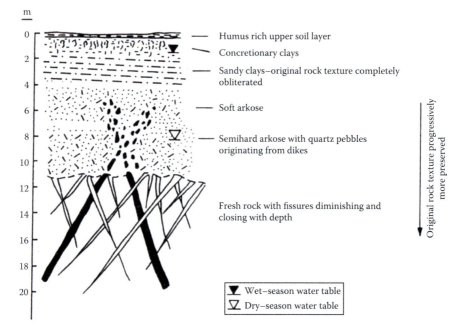

FIGURE 3.16 Profile of the sub-surface in crystalline terrains of an inter-tropical belt (after Mandel and Shiftan, 1981).

3.7.1 WEATHERED HARD ROCKS

Large regions of the continents are directly underlain by extensive batholiths of granite or by metamorphic complexes of gneiss, schist, quartzite, slate, or other meta-sediments, meta-volcanics, or associated igneous rocks. The groundwater contained in the weathered layers of these rocks is commonly tapped by thousands of wells for supply to villages, farms, and livestock. Also, groundwater discharge from the weathered layer sustains the flow of springs, and in the more humid regions the dry period baseflow of the streams.

3.7.1.1 Typical Weathered Layer Profile

The weathering process can be commonly grouped into three broad categories: physical or mechanical, chemical, and biological. The thickness, areal extent, and physical character of the weathering varies from place to place and depending on the nature of the process. Though there are variations, a typical weathered profile can be summarized as follows:

Zone(a): Sandy clays or clay sands often concretionary. Generally only a few meters thick.
Zone(b): Massive accumulation of secondary minerals (clay). Its thickness may reach up to 30 m. High porosity but low permeability.
Zone(c): Rock that is progressively altered upwards to a granular friable layer of disintegrated crystal aggregate and rock fragments. May range in thickness from a few meters to 30 m.
Zone(d): Fractured and fissured rock. May range from a few tens to several scores of meters in thickness. Low porosity but moderate permeability within the fracture system.

A typical cross-section of a weathered zone is shown in Figure 3.17.

3.7.1.2 Groundwater Potential in Weathered Zones

In general, the thicker and larger a weathered zone, the more productivities it has as an aquifer. Thin weathered zones are good aquifers only where there is good prevailing recharge, either natural or artificial. But even thin weathered zones act as excellent recharge zones for the deeper fractured systems. Many productive wells tap aquifers in Zone (c) of the weathered profile, which averages about 10–20 m in thickness. Most open wells in the Indian shield area, for example, exploit the

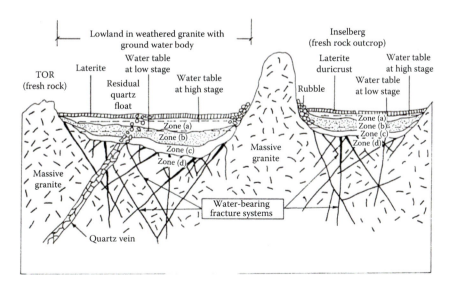

FIGURE 3.17 Idealized cross-section through a weathered hard rock terrain (after Larsson, 1984).

groundwater in the zone. Where the weathered zones are thin or absent, groundwater in the fracture systems are tapped through bore wells.

3.7.1.3 Exploration Parameters in Weathered Zones

Topography and stage of geomorphic evolution are important in understanding the development of extensive weathered zones. The weathered zones are commonly most extensive and thickest in erosional peneplains of low relief at or near base level where local relief is only a few meters and the slope of the land surface is less than 10%, for example, in the Indian peninsula. Erosional residuals such as tors and inselburgs commonly take up to about 10–15% of the gross area of these terrains in the old-age stage of the geomorphic cycle but as much as 50% or more in late mature stage. The residuals are generally devoid of weathered layers and fresh rock outcrops at the surface. The inselburgs may rise as much as 100–500 m above surrounding lowland plains, while the tors may rise only a few meters or a few tens of meters above the plains. The weathered layer profile, important from the groundwater point of view, is developed only in the lowland plains between the bare erosional residuals.

The spacing and distribution of fracture systems in the host rocks are also highly important factors in the development of the weathered layer. In granite rock terrains, for example, where the fracture systems are closely spaced, weathering agents may penetrate deeply into the lost rock to form thick weathered layers with permeable zones. On the other hand, the massive and poorly fractured granite resists weathering and forms erosional residuals, which may rise from a few meters to 100 m or more above intervening lowlands of deeply weathered rock. By this action discrete groundwater bodies are formed in the weathered layers of the lowland areas separated by unweathered uplands of fresh rock. The lowland plains are usually undulating in nature; the valleys often indicate underlying fracture systems.

3.7.2 Fractured Hard Rocks

The Precambrian shields are amongst the oldest parts of the Earth's crust. They contain hard rocks of different ages, grades of metamorphism, and structure. Many orogenic movements have affected the shields. Faulting processes have had different influences on the rocks of the shields due to differences in strength of the individual rock types. Some rock types are extremely fractured, while others are almost undisturbed, even though they belong to the same tectonic environment.

The strength of the rock or its resistance to brittle failure in its crust is a rather complicated matter. Petrographical parameters are involved, that is, grain size, grade of metamorphism, fold structures, direction of fold axis versus stress orientation, etc. These parameters play a dominant role in rock fracturing, and they are indirectly related to the occurrence of groundwater in hard rocks.

By definition, hard rocks are compact. On the other hand, the fractured pattern of the rocks creates a type of porosity that is termed fractured porosity. This means that open fractures lying below the water table levels can store water. Figure 3.18 shows typical water-bearing fracture zones in hard rock.

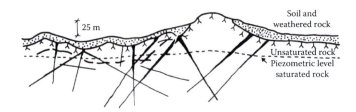

FIGURE 3.18 Typical water-bearing fracture zones in hard rock (after Larsson, 1984).

3.7.2.1 Types of Fractures

From the hydrogeological point of view, three main types of fractures can be identified in hard rock areas, viz., tensile joints, tensile fractures, and shear fractures. Tensile joints have no movement along the sides under undisturbed conditions. The most characteristic feature of these joints in metamorphic rocks is their "en echelon" lay out. That means they are not usually interconnected. At each end they are very narrow but open up in the middle, indicating their tensile origin. Tensile fractures develop parallel to the direction of compression due to parting and dilation perpendicular to compression direction. Shear fractures are a result of the differential movement of rock masses along a plane. They can range in length from many kilometers to tiny fractures a few millimeters in length. If two intersection shear fractures develop under the same stress conditions, they are called conjugate shear fractures.

3.7.2.2 Groundwater Potential in Fractured Hard Rocks

Fracturing may create significant porosity and permeability in the hard rocks and is primarily responsible for their groundwater potential. While the unfractured zones have virtually no porosity, in the fractured zones it may be as high as 30%. The direction of flow, however, is difficult to establish because of the nature of the openings and their possible relationship to fractured pattern in general.

In areas where tensile joints are predominant, the storage capacity is very low due to poor interconnection between the fractures. But in areas where tensile fractures are dominant, the storage capacity is usually high. Here the fractured systems function as large drainpipes collecting water from minor fractures belonging to the same fracture system. The storage capacity of the shear fractures is a very complex phenomenon. Heavy fracturing of hard rocks is commonly followed by intense weathering. The chemistry of the rocks plays a major role in this process. If two or more thrust faults cut each other, an axis of intersection develops, which can act as an effective drainpipe.

3.7.2.3 Exploration Parameters in Fractured Hard Rocks

The main problem for the hydrogeologist underrating groundwater exploration in hard rock areas is to find fracture pattern with maximum storage capacity. In general, shear zones and tensile fractures would offer good potential. The storage capacity also varies with different rock types. Acid-intrusive rocks such as granite, granodiorites, aplites, and pegmatites have a high storage capacity, as they are brittle rocks from a hydrogeological point of view. Fine-grained rocks are generally good aquifers. They have a characteristic type of narrowly spaced fractures.

Pegmatite intrusions are generally very brittle and therefore highly permeable. The vital point is the grain size. The more coarse grained the pegmatite, the more brittle it is. The more brittle the rock, the bigger the potential yield of groundwater. Basic intrusive rocks such as diorites and gabbros have, in general, low storage capacity. Basic rocks can be considered in field terms as tough rocks and therefore are poor aquifers.

Basic dikes constitute rather poor aquifers because of the weak interconnection between the dike and the country rock. However, the boundary zone between the dike and the country rock often contains open fractures with a high storage capacity. This characteristic usually results from thermal shrinkage at the time of cooling of the dike; consequently, open spaces develop between the dike and the country rock. The fine-grained boundary zone between the two rocks is generally more fractured than the interior part of the dike.

Dikes have another characteristic that may have local importance for groundwater. They commonly act as subterranean dams, dividing the rock into separate hydrologic units. If a mountain slope is cut by a set of dikes trending more or less parallel to the contour lines, the dikes will have a damming effect on the groundwater flow, causing the development of springs, as shown in Figure 3.19.

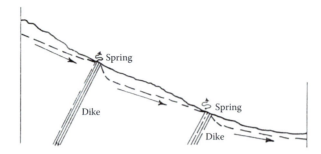

FIGURE 3.19 Damming effect of dikes at the side of a mountain (after Larsson, 1984).

3.7.3 REMOTE SENSING PARAMETERS IN HARD ROCKS

As exploration in hard rocks is a very difficult task, any additional clue on the occurrence and movement of groundwater and its recharge conditions is very valuable. Remote sensing provides valuable information on lithology, landforms, lineaments, soils, and land use, which control the occurrence movement and recharge of groundwater in hard rocks. Interpretation of lithology and landforms help in understanding and identifying potential zones in the weathered zones and near surface conditions. Interpretation and analysis of lineaments help in locating deeper aquifers in the fractured hard rocks. An understanding of soil and land use patterns is crucial in evaluating recharge conditions. Computer-aided classification is possible for soils and land use, whereas lithology, morphology, and lineaments are best done by visual interpretation. As seen in an earlier section, various digital enhancement techniques, however, are useful in enhancing objects of interest before visual interpretation. Remote sensing parameters for interpretation of various geomorphic units such as alluvial fans, valley fills, etc. that can also occur on hard rocks are described in an earlier section. The other most important landforms that are indicative of shallow groundwater conditions are inselberg/tor/pediment complexes, pediments, and buried pediments. The rock outcrops in inselberg/tor complexes are readily recognizable on the satellite imagery. The pediment zones within these complexes are characterized by lighter tones. In case of pediments buried by soil cover, the tone varies with soil cover. The black soil cover is usually represented by a darker tone, while the red soil cover is usually represented by a brighter tone (sometimes light yellow) on standard false color composites.

Where the pediment cover is thin, or where the declining water levels pose problems, the fracture zones offer the only sustainable source. Satellite remote sensing offers the best possible source for identifying these lineaments. Lineaments are all types of natural straight-line features on images. All lineaments are necessarily rock fractures, and they do not necessarily localize groundwater occurrence. On medium-resolution imagery, a few lineaments can be correlated with faults; the physical nature of most other lineaments must be investigated by indirect means by comparing lineament trends with joint trends. Commonly residual materials, soils, and vegetation cover many rock outcrops.

Since all lineaments are not necessarily productive, it is imperative to carry out a geophysical survey and test drilling to confirm the usefulness of the lineaments identified on the images. Usually, the productive lineaments follow a certain strike; therefore, it is necessary to establish the strike direction that offers the best potential.

Many fractures are vertical, and the lineaments may offer favorable locations for drilling in such cases. Most often the fractures are oblique and have a dip. In such cases, well locations should be offset to intersect the fractures below the water table but at shallow depth (nearly all fractures are progressively smaller at intersecting depths of beyond 100 m). A geophysical survey can help in interpreting the dip amount and direction.

In hard rock areas, many of the stream courses are controlled by lineaments and are being recharged by these streams. This is evidenced by the straight-line courses of these streams at these places. Though the streams may deviate after a while, the lineaments can be extended from the straight-line courses. Many of these streams are not perennial, and their beds are potentially favorable sites for test drilling. Lineaments often crisscross each other, and their meeting points are favorable points for exploration. It is not uncommon to see on the satellite images even four or five lineaments converging at a point. Lineaments are picked up more easily on the hilly regions due to their topographic expressions caused by higher erosion along the fractures, but they are of little significance on the hills where the utilization potential of the groundwater is limited. Such lineaments when they extend down to the plains are often covered by pediments and soil cover and are not visible on the images, though the utilization potential is very high on the plains. The lineaments picked up on the hills could be extended down to the plains, where further ground surveys could be carried out to confirm their potential.

Though the lineaments are easily identifiable on the images, they are often difficult to identify on the ground. Due to inherent geometric inaccuracies in the satellite images, direct transfer onto the topo maps do not help in identifying them on the ground. Very careful field investigations across the suspected lineament zones are needed before drilling starts.

Basic dikes and quartz reefs are also linear features that are easily identifiable on satellite images. They do not offer any groundwater potential themselves but act as barriers for the movement of groundwater. A careful analysis of drainage, in conjunction with dikes/quartz reefs, provides a clue as to which side of the feature offers greater potential.

Lineaments express themselves in many ways on the satellite images. It is logical that many fractures that localize the occurrence of groundwater also have an expression at land surface. A fracture that is a plane of weakness for enlargement by groundwater, therefore, may be represented by a topographic depression, a different soil tone, or a vegetation anomaly at land surface.

The lineaments appear on the satellite images as continuous and linear stream channels, valleys, and ridges. Lineaments can also be inferred from elongated or aligned lakes or native vegetation or thin strips of relatively open or dense vegetation. They are also expressed as the alignment of dark or light tone soils; identical or opposite deflections in adjacent stream channels, valleys, or ridges; and the alignment of nearby tributaries and tributary junctions.

The basic dikes appear as dark tones on images. The quartz reefs are usually brighter. The dikes and quartz reefs run for several kilometers at a stretch but are disjointed in between. They have a reasonable elevation and, where they are north-south trending, have a distinct shadow zone.

Contrast-enhanced standard false color composites are suited for linear interpretation. Black-and-white near-infrared images are also useful as vegetation patterns are least prominent and distracting on these images. Many lineaments are enhanced by low sun elevation angles.

Distribution of different textural classes of soils is an important factor in interpreting for groundwater recharge in hard rock areas. Fine-grained soils are generally darker than coarse-grained soils and offer less recharge and more runoff relatively. An understanding of land use pattern in hard rock regions is useful in evaluating a particular region for groundwater potential and also in understanding recharge conditions. For example, forested or agricultural areas allow more recharge, while a built-up area allows more run-off.

Areas that are already under irrigation by groundwater offer clues on the lineaments and landforms in hard rock areas. Groundwater-irrigated areas often correlate well with old river channels, valley fill zones, and lineaments. Groundwater-irrigated areas can be mapped fairly accurately by visual interpretation of the false color composites where the cropland areas appear pink or reddish pink and are easily separable. To differentiate between areas irrigated by ground- and surface water sources, good ground information on the existing surface irrigation schemes with their command areas is needed. The surface water bodies are also clearly seen on the images with their command areas immediately downstream. In India, for example, the ideal season for groundwater-irrigated cropland inventory would be in rabi (post-monsoon), when the crops generally subsist on irrigation. Images of February/March are ideal, as crops by then would develop canopy and be readily identifiable on the images.

3.8 CASE STUDIES AND ECONOMIC BENEFITS

The previous sections attempted to explain the keys and parameters obtainable from remote sensing for groundwater exploration in various geological terrains and the concept of using the technology in the initial stages of exploration to identify favorable groundwater zones to be followed up by conventional field methods. When remote sensing is used in the beginning stages of exploration to prove water-bearing zones, it reduces not only the cost of exploration, but produces a higher rate of success in drilling. This results in cost savings. The following sections describe two case studies where the economic benefits were calculated from the use of remote sensing to target groundwater. These studies have been reported earlier by Seelan et al. (1988) and Seelan (1996).

3.8.1 Case Study I

A remote sensing–based survey was carried out by Seelan et al. (1988) around Sullurpeta town, near the Sriharikota launch station in coastal Andhra Pradesh, India, to augment water supply to the residential quarters of the Department of Space. The town is situated about 80 kilometers north of Chennai (formerly Madras) along the east coast of India. The river Kalangi flows through the town and meets Pulicat Lake a few kilometers below the town. Pulicat is a saltwater lake with outlets/inlets into the Bay of Bengal. The lake, at sea level, extends to the east and northeast of the Sullurpeta town that has an altitude of 6 m above mean sea level. The terrain is very gently sloping toward the east and almost flat near the lake. The Kalangi has its distributary system spread over this area. Draining the western parts of Nagalapuram Hills, the Kalangi passes from the west of Sullurpeta and continues for about another 10 km before emptying itself into Pulicat Lake. The Kalangi is the only major source of water to the town, but it is dry during the summer months. Geologically the area consists of a thin layer of deltaic fluvial sediments composed of sand, silt, and clay of various sizes and grades. Below this layer, hard compact marine clay is encountered, but its thickness is unknown. Morphologically the area forms a delta of the Kalangi with distributaries draining into Pulicat Lake. The tidal limit is just a few kilometers below Sullurpeta.

A number of earlier systematic investigations have been carried out to identify possible additional sources of water supply to the town. Exploratory boreholes drilled beyond 300 m by the Central Ground Water Board (Central Ground Water Board, 1978) at a nearby place showed that the formations are predominantly marine clay and the formation waters are brine. The exploration revealed that there is no scope for the development of good-quality water at depth excepting tapping the uppermost shallow water table aquifer. Further studies around Sullurpeta by the Central Ground Water Board (1983) concluded that the medium to coarse alluvial sands along the Kalangi, a few kilometers upstream of the town, offer potential for exploitation through infiltration galleries. The colony itself suffers from brackish groundwater zones, with no perennial surface sources nearby. Accordingly, detailed cost estimates were made for different alternatives for infiltration wells along the Kalangi—collector wells, pumping mains, and pipelines—which worked out to be between $6,000 and $9,000 (costs during that period) depending on the alternative used (Civil Engineering Division, n.d.). However, before embarking on the civil works, the Civil Engineering Division of the Department of Space commissioned a study using remote sensing to look for nearby sources and cheaper options.

As the area under consideration was too small and the satellite resolutions available at that time were inadequate for local exploration, aerial photographs available for the area at a 1:10,000 scale were used in the study. The aerial photographs revealed the presence of four palaeochannels of Kalangi distributaries in the area, one of which was within 1 km from the colony site (Figure 3.20). Having obtained this very useful reconnaissance-level information from aerial photographs, detailed follow-up studies were carried out to prove the availability of fresh potable water within this zone. First, water samples were obtained from the existing shallow open wells from within and outside the palaeochannels. The analysis proved the presence of fresh water within the palaeochannels, while

Groundwater Targeting Using Remote Sensing

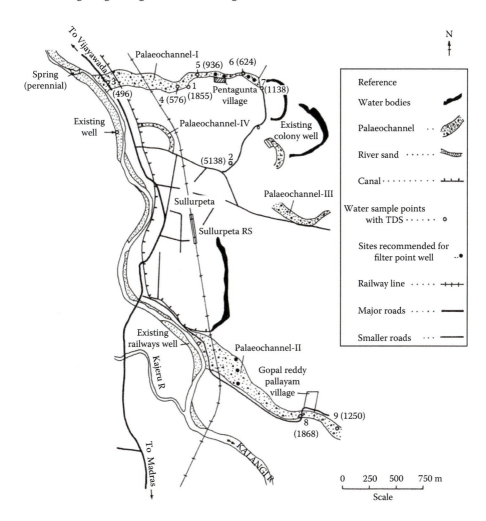

FIGURE 3.20 Map showing palaeochannels around Sullurpeta, interpreted from aerial photographs at a 1:10,000 scale.

the water was brackish outside the palaeochannels. The increasing salinity along the palaeochannels away from the main Kalangi river showed that the base flow along the palaeochannels is linked to the main Kalangi river but deteriorated away from it. As a next step, a geophysical survey was carried out across the palaeochannels. This revealed a 10-m thick sandy zone along the palaeochannels underlain by non-productive clay.

Based on these studies, exploratory shallow filter point wells were recommended for channels nearer to the colony site. A successful bore well was drilled to 10 m on the palaeochannel closest to the colony site and pre-monsoon and post-monsoon long-duration pump tests were conducted. Based on the test results, a safe discharge of 3,000 gallons per hour for 16 hours of pumping per day was recommended for the well situated on the palaeochannel to meet the housing colony's requirements. The cost of the bore well, pump house, etc. under this option worked out to $2,500.

3.8.2 Case Study II

In any natural resources planning and management, it is important that the information input is reliable and available in a timely manner to the planner/decision-maker. One of the major advantages of remote sensing is the ability to provide timely information on groundwater potential zones, which can then be used effectively to identify new water sources during situations like drought.

During a severe drought of 1986 in India, a study was commissioned by the state government of Maharashtra to interpret the satellite images covering the entire state of Maharashtra for identifying favorable indicators of groundwater on a reconnaissance level. Nearly 80% of the state of Maharashtra is covered by Deccan Trap basalts. The 18 Landsat TM scenes covering the state were visually interpreted, with limited ground truth, and groundwater potential maps were prepared at a 1:250,000 scale (Seelan et al., 1986). The fracture lineaments, valley fills, and different plateau regions on the basalts, based on the dissection pattern, were demarcated. The problem villages, which faced severe water shortages, were superimposed on the maps. The maps were provided to the state government of Maharashtra during June/July 1986. The cost of satellite images and the preparation of maps, at prevailing prices, was $1,750.

During the following year, 19,000 wells were drilled by the Maharashtra government, using the reconnaissance information provided based on Landsat TM and field surveys, out of which 14,000 wells were successful. In was reported that the increase in success rate was 5% more than the previous years when remote sensing–based inputs were not used systematically. This means more than 950 wells that otherwise might have failed were converted to successful wells. Assuming the prevailing drilling cost of $250 per well, this resulted in better utilization of funds to the tune of $11,875 in one year alone. In addition, there was the intangible benefit of being able to provide water to that many more villages during the severe drought situation, which is hard to quantify.

3.9 SUMMARY

Remote sensing provides very useful reconnaissance-level information in groundwater exploration. Remote sensing not only provides qualitative information on indicators that help in identifying groundwater potential zones, but also on areas of groundwater overexploitation (EL-Bana et al., 2024; Nainggolan et al., 2024; Sadeghi-Jahani et al., 2024; Sharma et al., 2024; Alshehri and Mohamed, 2023; Mohamed et al., 2023a; Mohamed et al., 2023b; Rampheri et al., 2023; Adams et al., 2022; Al-Djazouli et al., 2021; Alshehri et al., 2020; Gaber et al., 2020; Lee et al., 2020; Nhamo et al., 2020; Shao et al., 2020; Das et al., 2019; Kalhor and Emaminejad, 2019; Kim et al., 2019; Gnanachandrasamy et al., 2018; Miro and Famiglietti, 2018; Patra et al., 2018; Dasho et al., 2017; Gemitzi et al., 2017; Hu et al., 2017; Nejad et al., 2017; Pinto et al., 2017). The information thus obtained helps in delineating zones where further exploration could be taken up. Various parameters relating to different thematic information relevant to groundwater occurrence and movement are obtainable from satellite imagery. These, when integrated with the help of GIS techniques, can provide information on potential groundwater zones. The occurrence of groundwater, however, varies with geologic setting, and it is important to understand these variations and look for appropriate keys and parameters. It is also emphasized here that a remote sensing study, while providing valuable information, has to be followed up with ground-based surveys, as shown in the case studies, for better results.

Satellite remote sensing for groundwater applications has been practiced for the past 40 years or so, and the review showed the availability of a vast amount of literature on the topic. Much of the literature in the area of groundwater targeting has focused on developing countries, especially in the arid and semi-arid regions where the technology has been found to be most useful. Not surprisingly, India, which ranks highest in the world in groundwater abstraction rates (that is, higher than the next two nations on the list, China and the United States, combined), has the highest number of case studies reported in the literature. It is also evident from the literature review that in countries like India, where the satellite data products were made available at reasonable costs, and as a vast number of hydrogeologists were trained in the use of remote sensing for groundwater targeting, the technology could be adopted early under the nation's technology mission on drinking water, and other operational groundwater projects. With the growing world population and the increasing demand for water, remote sensing technology offers great potential, particularly in the developing regions of the world, for careful targeting, extraction, and management of groundwater resources.

ACKNOWLEDGMENTS

The author wishes to thank the Department of Space, India, for the opportunity to work on several national groundwater projects and the flexibility to experiment with new processes and procedures in extracting groundwater information from remote sensing during his tenure with the department from the mid-1970s to the mid-1990s. Special thanks are also due to Brett Sergenian, graduate student at the University of North Dakota, and the author's summer intern during 2013, for meticulously collecting and listing much of the literature cited in this chapter.

REFERENCES

Abrams, M.J., R.P. Ashby, L.C. Rowen, A.F. Goetz, and A.B. Kahale, 1977. Mapping of hydrothermal alteration in cuprite mining district, Nevada, using aircraft scanner images for the spectral region 0.46 to 2.36 μm. *Geology*, volume 5, no. 12, The Geological Society of America.

Adams, K.H., J.T. Reager, P. Rosen, D.N. Wiese, T.G. Farr, S. Rao, B.J. Haines, D.F. Argus, Z. Liu, R. Smith, J.S. Famiglietti, and M. Rodell, 2022. Remote sensing of groundwater: Current capabilities and future directions. *Water Resources Research*, volume 58, no. 10, p. e2022WR032219. https://doi.org/10.1029/2022WR032219

Agarwal, A.K., and D. Mishra, 1992. Evaluation of groundwater potential in the environs of Jhansi city Uttar Pradesh, using hydrogeomorphological assessment by satellite remote sensing technique. *Photonirvachak-Journal of the Indian Society of Remote Sensing*, volume 20, nos. 2 & 3, pp. 121–128.

Ahmad, R., and R.P. Singh, 2002. Comparison of various data fusion for surface features extraction using IRS PAN and LISS III data. *Advanced Space Research*, volume 29, no. 1, pp. 73–78.

Al-Djazouli, M.O., K. Elmorabiti, A. Rahimi, et al. 2021. Delineating of groundwater potential zones based on remote sensing, GIS and analytical hierarchical process: A case of Waddai, eastern Chad. *GeoJournal*, volume 86, pp. 1881–1894. https://doi.org/10.1007/s10708-020-10160-0

Ali, E.A., S.O. El Khidir, I.A.A. Babikir, and E.M. Abdelrahman, 2012. Landsat ETM+ digital Image processing techniques for lithological and structural lineament enhancement: Case study around Abidiya area, Sudan. *The Open Remote Sensing Journal*, volume 5, pp. 83–89.

Alshehri, F., and A. Mohamed. 2023. Analysis of groundwater storage fluctuations using GRACE and remote sensing data in Wadi As-Sirhan, Northern Saudi Arabia. *Water*, volume 15, no. 2, p. 282. https://doi.org/10.3390/w15020282

Alshehri, F., M. Sultan, S. Karki, E. Alwagdani, S. Alsefry, H. Alharbi, H. Sahour, and N. Sturchio. 2020. Mapping the distribution of shallow groundwater occurrences using remote sensing-based statistical modeling over Southwest Saudi Arabia. *Remote Sensing*, volume 12, no. 9, p. 1361. https://doi.org/10.3390/rs12091361

Bagyaraj, M., T. Ramkumar, S. Venkatramanan, and B. Gurugnanam, 2013. Application of remote sensing and GIS analysis for identifying groundwater potential zone in parts of Kodaikanal Taluk, South India. *Frontiers of Earth Science*, volume 7, no. 1, pp. 65–75.

Bobba, A.G., R.P. Bukata, and J.H. Jerome, 1992. Digitally processed satellite data as tool in detecting potential groundwater flow systems. *Journal of Hydrology*, 131, pp. 25–62.

Bruning, J.N., J.S. Gierke, and A.L. Maclean, 2011. An approach to lineament analysis for groundwater exploration in Nicaragua. *Photogrammetric Engineering & Remote Sensing*, volume 77, no. 5, pp. 509–519.

Central Ground Water Board, 1978. A report on groundwater exploration at Sriharikota Island, Nellore District, Andhra Pradesh. CGWB report, Southern Region, Hyderabad. Unpublished.

Central Ground Water Board, 1983. Hydrogeological investigations to study the feasibility of an infiltration well along Kalangi river for augmentation of water supply to ISRO facilities at Sullurpeta, Nellore District, Andhra Pradesh. CGWB report, Southern Region, Hyderabad. Unpublished.

Chatterjee, C., R. Kumar, B. Chakravorty, A.K. Lohani, and S. Kumar, 2005. Integrating remote sensing and GIS techniques with groundwater low modeling for assessment of waterlogged areas. *Water Resources Management*, volume 19, pp. 539–554.

Chowdary, V.M., R. Vinu Chandran, N. Neeti, R.V. Bothale, Y.K. Srivastava, P. Ingle, D. Ramakrishnan, D. Dutta, A. Jeyaram, J.R. Sharma, and R. Singh, 2008. Assessment of surface and sub-surface waterlogged areas in irrigation command areas of Bihar state using remote sensing and GIS. *Journal of Water Management*, volume 95, pp. 754–766.

Civil Engineering Divison, n.d. Augmentation of water supply to housing at Sullurpeta. Design Review Document, Department of Space, Government of India. Unpublished.

Das, B., S.C. Pal, S. Malik, and R. Chakrabortty. 2019. Modeling groundwater potential zones of Puruliya district, West Bengal, India using remote sensing and GIS techniques. *Geology, Ecology, and Landscapes*, 3 (3), pp. 223–237. DOI: 10.1080/24749508.2018.1555740

Dasho, O.A., E.A. Ariyibi, F.O. Akinluyi et al. 2017. Application of satellite remote sensing to groundwater potential modeling in Ejigbo area, Southwestern Nigeria. *Modeling Earth Systems Environment*, volume 3, pp. 615–633. https://doi.org/10.1007/s40808-017-0322-z

Davis, S.N., and R.J.M. De Wiest, 1966. *Hydrogeology*. John Wiley and Sons, New York, p. 463.

Dawson, G.B., and D.J. Dickinson, 1970. Heat flow studies in thermal areas of North Island of New Zealand. *Geothermics*, volume 2, part 1, pp. 466–473.

Deekshatulu, B.L., and S. Thiruvengadachari, 1981. Application of remote sensing techniques for water quality monitoring. Final project report, National Remote Sensing Agency, Hyderabad, p. 261.

Deutsch, M., 1971. Operational and experimental remote sensing in hydrology. CENTO seminar on the applications of remote sensors in the determination of natural resources, Ankara, Turkey.

Deutsch, M., 1974. Survey of remote sensing applications for groundwater exploration and management. Proceedings of the meetings of the American Association for the Advancement of Science.

Dhakate, R., V.S. Singh, B.C. Negi, S. Chandra, and V.A. Rao, 2008. Geomorphological and geophysical approach for locating favorable groundwater zones in granitic terrain, Andhra Pradesh, India. *Journal of Environmental Management*, volume 88, pp. 1373–1383.

Dhokarikar, B.G., 1991. *Groundwater resource development in basaltic rock terrain of Maharashtra*. Water Industry Publication, Pune, p. 275.

EL-Bana, E.M.M., H.M. Alogayell, M.H. Sheta, and M. Abdelfattah. 2024. An integrated remote sensing and GIS-based technique for mapping groundwater recharge zones: A case study of SW Riyadh, Central Saudi Arabia. *Hydrology*, volume 11, no. 3, p. 38. https://doi.org/10.3390/hydrology11030038

Elewa, H., and A.A. Qaddah, 2011. Groundwater potentiality mapping in the Sinai Peninsula, Egypt, using remote sensing and GIS-watershed-based modelling. *Hydrology Journal*, volume 19, pp. 613–628.

Freeze, A.R., and J.A. Cherry, 1979. *Groundwater*. Prentice-Hall, Inc., New Jersey, p. 604.

Gaber, A., A.K. Mohamed, A. ElGalladi, M. Abdelkareem, A.M. Beshr, and M. Koch. 2020. Mapping the groundwater potentiality of West Qena Area, Egypt, using integrated remote sensing and hydro-geophysical techniques. *Remote Sensing*, volume 12, no. 10, p. 1559. https://doi.org/10.3390/rs12101559

Gandino, A., 1983. Recent remote sensing technique in freshwater submarine springs monitoring: Qualitative and quantitative approach. Proceedings of the international symposium on methods and instrumentation for investigation of groundwater systems, UNESCO, Noordwijkerhout.

Gautham, A.M., 1990. Application of IRS IA data for delineating buried channels in southern part of Allahabad district of Uttar Pradesh. *Photonirvachak-Journal of Indian Society of Remote Sensing*, volume 18, no. 3, pp. 52–55.

Gemitzi, A., H. Ajami, and H. Richnow, 2017. Developing empirical monthly groundwater recharge equations based on modeling and remote sensing data—Modeling future groundwater recharge to predict potential climate change impacts. *Journal of Hydrology*, volume 546, pp. 1–13, ISSN 0022–1694, https://doi.org/10.1016/j.jhydrol.2017.01.005. Available from www.sciencedirect.com/science/article/pii/S0022169417300069

Geraghty and Miller, Inc., 1972. Groundwater contamination, an explanation of its causes and effects. A special report, Port Washington, New York.

Ghayoumian, J., B. Ghermezcheshme, S. Feiznia, and A.A. Noroozi, 2005. Integrating GIS and DSS for identification of suitable areas of artificial recharge, case study Meimeh Basin, Isfahan, Iran. *Environmental Geology*, volume 47, pp. 493–500.

Ghayoumian, J., M. Mohseni Saravi, S. Feiznia, B. Nouri, and A. Malekian, 2007. Application of GIS techniques to determine areas most suitable for groundwater recharge in a coastal aquifer in southern Iran. *Journal of Asian Earth Sciences*, volume 30, pp. 364–374.

Ghoneim, E., and F. El-Baz, 2007. The application of radar topographic data to mapping of a mega palaeodrainage in the Eastern Sahara. *Journal of Arid Environments*, volume 69, pp. 658–675.

Ghosh, R., 1993. Remote sensing for analysis of groundwater availability in an area with long unplanned mining history. *Photonirvachak-Journal of Indian Society of Remote Sensing*, volume 21, no. 3, pp. 119–126.

Gnanachandrasamy, G., Y. Zhou, and M. Bagyaraj et al. 2018. Remote sensing and GIS based groundwater potential zone mapping in Ariyalur district, Tamil Nadu. *Journal of Geological Society of India*, volume 92, pp. 484–490. https://doi.org/10.1007/s12594-018-1046-z

Goetz, A.F.H., F.C. Billingeley, A.R. Cillespic, M.J. Adams, R.L. Squires, E.M. Shoemaker, I. Lucchitta, and D.P. Elston, 1975. Application of ERTS images and image processing to regional geologic mapping in Northern Arizona: NASA Jet Propulsion Lab, California, Technical Report 32–1597.

Gopinath, G., and P. Seralathan, 2004. Identification of groundwater prospective zones using IRS ID LISS III and pump test methods. *Photonirvachak- Journal of the Indian Society of Remote Sensing*, volume 32, no. 4, pp. 329–340.

Gumma, M.K., and P. Pavelic, 2013. Mapping of groundwater potential zones across Ghana using remote sensing, geographic information systems, and spatial modeling. *Environmental Monitoring Assessment*, volume 185, pp. 3561–3579.

Hochstein, M.P., and D.J. Dickinson, 1970. Infrared sensing of thermal ground in the Tampo region, New Zealand. Proceedings of the United Nations Symposium on the Development and Utilization of Geothermal Resources, Pisa, volume 2, part 1.

Hu, K., J.L. Awange, F.E. Khandu, R.M. Goncalves, and K. Fleming, 2017. Hydrogeological characterisation of groundwater over Brazil using remotely sensed and model products. *Science of The Total Environment*, Volumes 599–600, pp. 372–386, ISSN 0048–9697, https://doi.org/10.1016/j.scitotenv.2017.04.188. Available from www.sciencedirect.com/science/article/pii/S0048969717310331

Jagadesha, D.S., D. Nagaraju, and B.C. Prabhakar, 2012. Identification of groundwater potential zones through remote sensing and GIS techniques in Muguru Add Halla in Mysore and Chamarajnagar district, Karnataka, India. *International Journal of Earth Sciences and Engineering*, volume 5, no. 5, pp. 1310–121.

Jasmin, I. and P. Mallikarjuna, 2011. Review: Satellite-based remote sensing and geographic information systems and their application in the assessment of groundwater potential, with particular reference to India. *Hydrogeology Journal*, volume 19, pp. 729–740.

Jha, M.K., A. Chowdhury, V.M. Chowdhury, and S. Peiffer, 2007. Groundwater management and development by integrated remote sensing and geographic information systems: Prospects and constraints. *Water Resources Management*, volume 21, pp. 427–467.

Kalhor, K., and Emaminejad, N. 2019. Sustainable development in cities: Studying the relationship between groundwater level and urbanization using remote sensing data. *Groundwater for Sustainable Development*, Volume 9, p. 100243, ISSN 2352–801X, https://doi.org/10.1016/j.gsd.2019.100243. Available from www.sciencedirect.com/science/article/pii/S2352801X19300712

Karanth, K.R., 1987. *Groundwater assessment development and management*. Tata McGraw Hill publishing company limited, New Delhi, p. 720.

Khalaf, A., and D. Donoghue, 2012. Estimating recharge distribution using remote sensing: A case study from the West Bank. *Journal of Hydrology*, volumes 414 and 415, pp. 354–363.

Khodaei, K., and H.R. Nassery, 2013. Groundwater exploration using remote sensing and geographic information systems in a semi-arid area (southwest of Urmieh, northwest of Iran). *Arabian Journal of Geosciences*, volume 6, pp. 1229–1240.

Kim, J.-C., H.-S. Jung, and S. Lee. 2019. Spatial mapping of the groundwater potential of the Geum river basin using ensemble models based on remote sensing images. *Remote Sensing*, volume 11, no. 19, p. 2285. https://doi.org/10.3390/rs11192285

Koch, M., and P.M. Mather, 1997. Lineament mapping for groundwater resource assessment: A comparison of digital Synthetic Aperture Radar (SAR) imagery and stereoscopic Large Format Camera (LFC) photographs in the Red Sea Hills, Sudan. *International Journal of Remote Sensing*, volume 18, no. 7, pp. 1465–1482.

Kumar, A., and S.K. Srivastava, 1991. Geomorphological units, their geohydrological characteristic and vertical electrical sounding response near Munger, Bihar. *Photonirvachak-Journal of Indian Society of Remote Sensing*, Volume 19, no. 3, pp. 205–215.

Kumar, P.K.D., G. Gopinath, and P. Serlathan, 2007. Application of remote sensing and GIS for the demarcation of groundwater potential zones in a river basin in Kerala, southwest coast of India. *International Journal of Remote Sensing*, volume 28, no. 24, pp. 5583–5601.

Kurz, B., and S.K. Seelan, 2009. Use of remote sensing to map irrigated agriculture in areas overlying the Ogallala aquifer, United States. Chapter 7, Remote Sensing of Global Croplands for Food Security, Taylor and Francis, p. 476.

Larsson, I., 1984. *Groundwater in hard rocks*. UNESCO Publication, p. 228.

Lattman, L.H., and R.P. Nickelson, 1958. Photogeologic fracture-trace mapping in Appalachian Plateau. *Bulletin of American Association of Petroleum Geologists*, volume 42, no. 9, pp. 2238–2245.

Lee, S., Y. Hyun, S. Lee, and M.-J. Lee. 2020. Groundwater potential mapping using remote sensing and GIS-based machine learning techniques. *Remote Sensing*, volume 12, no. 7, p. 1200. https://doi.org/10.3390/rs12071200

Lee, S., Y.S. Kim, and H.J. Oh, 2012. Applications of a weights-of-evidence method and GIS to regional groundwater productivity potential mapping. *Journal of Environmental Management*, volume 96, pp. 91–105.

Magesh, N.S., N. Chandrasekar, and J.P. Soundranayagam, 2012. Delineation of groundwater potential zones in Theni district, Tamil Nadu, using remote sensing, GIS, and MIF techniques. *Geosciences Frontiers*, volume 3, no. 2, pp. 189–196.

Mallast, U., R. Gloaguen, S. Geyer, T. Rodiger, and C. Siebert, 2011. Derivation of groundwater flow paths based on semi-automatic extraction of lineaments from remote sensing data. *Hydrology and Earth System Sciences*, volume 15, pp. 2665–2676.

Mandel, S., and Z.L. Shiftan, 1981. *Groundwater resources investigation and development*. Academic Press, New York, p. 269.

Mc Lerran, J.K., 1967. Infrared Sensing. *Photogrammetric Engineering*, volume XXXIII, no. 5.

Miro, M.E., and J.S. Famiglietti, 2018. Downscaling GRACE remote sensing datasets to high-resolution groundwater storage change maps of California's central valley. *Remote Sensing*, volume 10, no. 1, p. 143. https://doi.org/10.3390/rs10010143

Mogaji, K.A., O.S. Aboyeji, and G.O. Omosuyi, 2006. Mapping of lineaments for groundwater targeting in basement complex area of Ondo State using remotely sensed data. *International Journal of Water Resources and Environmental Engineering*, volume 3, no. 7, pp. 150–160.

Mohamed, A., A. Abdelrady, S.S. Alarifi, and A. Othman. 2023a. Geophysical and remote sensing assessment of Chad's groundwater resources. *Remote Sensing*, volume 15, no. 3, p. 560. https://doi.org/10.3390/rs15030560

Mohamed, A., A. Othman, W.F. Galal, and A. Abdelrady. 2023b. Integrated geophysical approach of groundwater potential in Wadi Ranyah, Saudi Arabia, using gravity, electrical resistivity, and remote-sensing techniques. *Remote Sensing*, volume 15, no. 7, p. 1808. https://doi.org/10.3390/rs15071808

Mollard, J.D., 1957. Aerial mosaics reveal fracture patterns on surface materials in Southern Saskatchewan and Manitoba. *Oil inCanada*, volume 26, pp. 1840–1864.

Mollard, J.D., and F.D. Patton, 1960. Science and Systems in groundwater investigations. Western Canada Water and Sewage Conference, proceedings, pp. 53–72; reprinted in Canadian Municipal Utilities, Canada, June 1961.

Moore, G.K., 1967. Lineaments on Skylab photographs—detection, mapping and hydrologic significance in Central Tennessee. US Geological Survey—open file report 78–196.

Moore, G.K., 1978. The role of remote sensing in groundwater exploration. Proceedings of the Indo-US workshop on remote sensing of water resources, Hyderabad.

Moore, G.K., and M. Deutsch, 1975. ERTS imagery for groundwater investigations. *Groundwater*, volume 13, no. 2, pp. 214–226.

Moore, G.K., and F.A. Waltz, 1983. Objective procedures for lineament enhancement and extraction. *Photogrammetric Engineering and Remote Sensing*, volume 49, no. 5, pp. 641–647.

Nainggolan, L., C.-F. Ni, Y. Darmawan, W.-C. Lo, I.-H. Lee, C.-P. Lin, and N.H. Hiep. 2024. Cost-effective groundwater potential mapping by integrating multiple remote sensing data and the index—overlay method. *Remote Sensing*, volume 16, no. 3, p. 502. https://doi.org/10.3390/rs16030502

National Remote Sensing Agency, 1980. Integrated remote sensing survey of natural resources of Tamil Nadu state, India. National Remote Sensing Agency. Project report, National Remote Sensing Agency, Department of Space, Government of India, Hyderabad.

National Remote Sensing Agency, 1982a. Integrated Remote Sensing Survey of Natural Resources of North Karnataka State, India. Project report, National Remote Sensing Agency, Department of Space, Government of India, Hyderabad.

National Remote Sensing Agency, 1982b. Integrated remote sensing survey of natural resources of Bundelkhand region, U.P, India. Project report, National Remote Sensing Agency, Department of Space, Government of India, Hyderabad.

National Remote Sensing Agency, 1983. Integrated Remote Sensing Survey of Natural Resources of West Coast region, India. Project report, National Remote Sensing Agency, Department of Space, Government of India, Hyderabad.

National Remote Sensing Agency, 1984. Integrated Remote Sensing Survey of Natural Resources of Upper Barak Watershed, North-East India. Project report, National Remote Sensing Agency, Department of Space, Government of India, Hyderabad.

National Remote Sensing Agency, 1986a. Report on the groundwater potential maps of drought prone districts of Maharashtra prepared based on visual interpretation of Landsat Thematic Mapper data. Project report, National Remote Sensing Agency, Department of Space, Government of India, Hyderabad.

National Remote Sensing Agency, 1986b. Report on the groundwater potential maps of drought prone districts of Karnataka prepared based on visual interpretation of Landsat Thematic Mapper data. Project report, National Remote Sensing Agency, Department of Space, Government of India, Hyderabad.

Nefedov, K.E., and T.A. Popova, 1972. *Deciphering groundwater from aerial photographs*. Translated from Russian by V.S. Kothekar. Amerind Publication Co, New Delhi, p. 191.

Nejad, S.G., F. Falah, M. Daneshfar, A. Haghizadeh, and O. Rahmati, 2017. Delineation of groundwater potential zones using remote sensing and GIS-based data-driven models. *Geocarto International*, volume 32, no. 2, pp. 167–187. DOI: 10.1080/10106049.2015.1132481

Nhamo, L., G.Y. Ebrahim, T. Mabhaudhi, S. Mpandeli, M. Magombeyi, M. Chitakira, J. Magidi, and M. Sibanda, 2020. An assessment of groundwater use in irrigated agriculture using multi-spectral remote sensing. *Physics and Chemistry of the Earth*, Parts A/B/C, volume 115, p. 102810, ISSN 1474–7065, https://doi.org/10.1016/j.pce.2019.102810. Available from www.sciencedirect.com/science/article/pii/S147470651930049X

Otterman, J., 1971. Thermal mapping of selected sites in the Lake Kinneret region, Israel. *Journal of Earth Sciences*, volume 20, no. 3.

Ozaray, G., 1975. Remote sensing in hydrogeological mapping (with special respect to India and Canada). Proceedings of the 2nd World Congress on Water Resources, New Delhi, India.

Pal, S.K., T.J. Majumdar, and A.K. Bhattacharya, 2007. ERS -2 SAR and IRS-1C LISS III data fusion: A PCA approach to improve remote sensing based geological interpretation. *ISPRS Photogrammetry and Remote Sensing*, volume 61, pp. 281–297.

Palmason, G., 1970. Aerial infrared surveys of Rejkjanes and Torjajokull thermal areas, Iceland, with a section of cost exploration surveys. Proceedings of the United Nations Symposium on the Development and Utilization of Geothermal Resources, Pisa, volume 2, part 1.

Panigrahi, B., A.K. Nayak, and S.D. Sharma, 1996. Application of remote sensing technology for groundwater potential evaluation. *Water Resources Management*, volume 9, pp. 161–173.

Patra, S., P. Mishra, and S.C. Mahapatra, 2018. Delineation of groundwater potential zone for sustainable development: A case study from Ganga Alluvial Plain covering Hooghly district of India using remote sensing, geographic information system and analytic hierarchy process. *Journal of Cleaner Production*, volume 172, pp. 2485–2502, ISSN 0959–6526, https://doi.org/10.1016/j.jclepro.2017.11.161. Available from www.sciencedirect.com/science/article/pii/S0959652617328342

Pinto, D., S. Shrestha, M.S. Babel et al. 2017. Delineation of groundwater potential zones in the Comoro watershed, Timor Leste using GIS, remote sensing and analytic hierarchy process (AHP) technique. *Applied Water Sciences*, volume 7, pp. 503–519. https://doi.org/10.1007/s13201-015-0270-6

Prasad, E.A.V., 1980. *Groundwater in Varahamihira's Brahat Samhita*. MASSLIT series, Sri Venkateswara University, Tirupati, p. 351.

Rampheri, M.B., T. Dube, F. Dondofema, and T. Dalu, 2023. Progress in the remote sensing of groundwater-dependent ecosystems in semi-arid environments. *Physics and Chemistry of the Earth*, Parts A/B/C, volume 130, p. 103359, ISSN 1474–7065, https://doi.org/10.1016/j.pce.2023.103359. Available from www.sciencedirect.com/science/article/pii/S1474706523000037

Rana, S.S., 1998. Application of directional filtering in lineament mapping for groundwater prospecting around Bhinmal—a semi-arid part of Thar desert. *Photonirvachak-Journal of the Indian Society of Remote Sensing*, volume 26, nos. 1&2, pp. 35–44.

Rao, N.S., and R.P. Reddy, 1999. Groundwater prospects in a developing satellite township of Andhra Pradesh, India, using remote sensing techniques. *Photonirvachak-Journal of the Indian Society of Remote Sensing*, volume 27, no. 4, pp. 193–202.

Ravindran, K.V., and A. Jeyaram, 1997. Groundwater prospects of Shahbad Tehsil, Baran District, Eastern Rajasthan: A remote sensing approach. *Photonirvachak-Journal of the Indian Society of Remote Sensing*, volume 25, no. 4, pp. 239–246.

Regunath, H.M., 1987. *Groundwater* (2nd edition). Wiley Eastern Limited, New Delhi, p. 563.

Robinov, C.J., 1968. The status of remote sensing in hydrology. Proceedings of the 5th symposium on remote sensing and environment, Michigan University, Ann Arbor, Michigan.

Robinson, C., F. El-Baz, M. Ozdogan, M. Ledwith, D. Blance, S. Oakley, and J. Inzana, 2000. Use of radar data to delineate palaeodrainage flow directions in the Selima sand sheet, Eastern Sahara. *Photogrammetric Engineering and Remote Sensing*, volume 66, no. 5, pp. 745–753.

Rodell, M., and J.S. Famiglietti. 2002. The potential for satellite-based monitoring of storage changes using GRACE: The High Plains aquifer, central US. *Journal of Hydrology*, 263, pp. 245–256.

Rundquist, D.C., D.A. Rodekohr, A.J. Peters, R.L. Ehrman, L. Di and G. Murrey, 1991. Statewide groundwater vulnerability assessment in Nebraska using the DRASTIC/GIS model. *Geocarto International*, volume 2, pp. 51–58.

Sadeghi-Jahani, H., H. Ketabchi, and H. Shafizadeh-Moghadam, 2024. Spatiotemporal assessment of sustainable groundwater management using process-based and remote sensing indices: A novel approach. *Science of the Total Environment*, volume 918, p. 170828, ISSN 0048–9697, https://doi.org/10.1016/j.scitotenv.2024.170828. Available from www.sciencedirect.com/science/article/pii/S0048969724009677

Sander, P., T.B. Minor, and M.M. Chesley, 1997. Ground-water exploration based on lineament analysis and reproducibility tests. *Ground Water*, volume 35, no. 5, pp. 888–895.

Sankar, K., 2002. Ebvaluation of groundwater potential zones using remote sensing data in Upper Vaigai river basin, Tamil Nadu, India. *Photonirvachak-Journal of the Indian Society of Remote Sensing*, volume 30, no. 3, pp. 119–128.

Seelan, S.K., 1975. Location of abnormal water resources by thermal infra-red technique. Diploma thesis report. Groundwater Research Center, Hebrew University of Jerusalem. Unpublished.

Seelan, S.K., 1980. Utility of image enhancement in geologic interpretation of remotely sensed data—An Indian example. Proceedings of the symposium on remote sensing in subsurface exploration. 6th annual convention of Association of Exploration Geologists, Bangalore, India.

Seelan, S.K., 1982. Landsat image derived geomorphic indicators of groundwater in parts of Central India. *Photonirvachak-Journal of Indian Society of Remote Sensing*, volume 10, no. 2, pp. 33–37.

Seelan, S.K., 1986. Detection and monitoring of groundwater systems under stress—can remote sensing help? Proceedings of the AWRS conference on groundwater systems under stress, Brisbane.

Seelan, S.K., 1994. Remote sensing applications and GIS development for groundwater investigations. Ph.D. Dissertation, Jawaharlal Nehru Technological University, Hyderabad, unpublished.

Seelan, S.K., 1996. Cost benefit aspects of remote sensing for groundwater exploration—two case studies. Book chapter, Economics of Remote Sensing, ISBN 81-86562-04-4, Manek Publications Pvt Ltd., New Delhi.

Seelan, S.K., A. Bhattacharya, and R. Venkataraman, 1988. Remote sensing for identification of fresh water zones around Sullurpeta town in Andhra Pradesh. Proceedings of the National Seminar on Groundwater Development in Coastal Tracts, Trivandrum.

Seelan, S.K., G.C. Chennaiah, and N.C. Gautam, 1983. Study of landform control over land utilization pattern in parts of southern U.P.—a remote sensing approach. *Photonirvachak-Journal of Indian Society of Remote Sensing*, volume 11, no. 1, pp. 49–53.

Seelan, S.K., V.S. Hegde, P.R. Reddy, R.S. Rao, R.K. Sood, A.K. Sharma, A.K. Gupta, G. Raj, A. Perumal, and L.R.A. Narayan, 1986. Groundwater targeting in a drought situation in Maharashtra state, India, using Landsat TM data. Proceedings of the Seventh Asian Conference on Remote Sensing, Seoul.

Seelan, S.K., and L.R.A. Narayan, 1984. Drought study around Madras city. NRSA technical report. National Remote Sensing Agency, Hyderabad, unpublished.

Seelan, S.K., and S. Thiruvengadachari, 1980. An integrated regional approach for delineation of potential groundwater zones using satellite data—an Indian case study. Abstracts of proceedings, 26th International Geological Congress, Paris.

Seelan, S.K., and S. Thiruvengadachari, 1981. Satellite sensing for extraction of groundwater resources information. Proceedings of the 15th International Symposium on Remote Sensing of the Environment, Ann Arbor, Michigan.

Sener, E., Davraz, A. and Ozcelik, M, 2004. An integration of GIS and remote sensing in groundwater investigations: A case study of Burdur, Turkey. *Hydrogeology Journal*, volume 13, pp. 826–834.

Seshubabu, K., and S.K. Seelan, 1983. Hydrogeology chapter, Godavari basin test site. Technical report, part B, Indo-FRG technical cooperation programme, National Remote Sensing Agency, Hyderabad.

Sethupathi, A.S., C. Lakshmi Narasimhan, and V. Vasanthamohan, 2012. Evaluation of hydrogeomorphological landforms and lineaments using GIS and remote sensing techniques in Bargur-Mathur sub-watersheds, Ponnaiyar River basin, India. *International Journal of Geomatics and Geosciences*, volume 3, no. 1, pp. 178–190.

Shankarnarayana, G., N. Lakhmaiah, and P.V. Prakash Goud, 1996. Hydrogeomorphological study based on remote sensing of Mulug Takuk, Warangal district, Andhra Pradesh, India. *Hydrological Sciences*, 42 (2), pp. 137–150.

Shao, Z., M.E. Huq, B. Cai, O. Altan, and Y. Li, 2020. Integrated remote sensing and GIS approach using Fuzzy-AHP to delineate and identify groundwater potential zones in semi-arid Shanxi Province, China. *Environmental Modelling & Software*, volume 134, p. 104868, ISSN 1364–8152, https://doi.org/10.1016/j.envsoft.2020.104868. Available from www.sciencedirect.com/science/article/pii/S1364815220309257

Sharma, Y., R. Ahmed, T.K. Saha, N. Bhuyan, G. Kumari, P.S. Roshani, and H. Sajjad, 2024. Assessment of groundwater potential and determination of influencing factors using remote sensing and machine learning algorithms: A study of Nainital district of Uttarakhand state, India. *Groundwater for Sustainable Development*, volume 25, p. 101094, ISSN 2352–801X, https://doi.org/10.1016/j.gsd.2024.101094. Available from www.sciencedirect.com/science/article/pii/S2352801X24000171

Singh, K.P., 1980. Application of Landsat imagery to groundwater studies in parts of Punjab and Haryana states, India. The contribution of space observations to water resources management, COSPAR. *Advances in Space Exploration*, volume 9, Pergamon Press, pp. 107–111.

Solomonson, V.V., T.J. Jackson, J.R. Lucas, G.K. Moore, A. Rango, T. Schmugge, and D. Scholz, 1983. Water resources assessment. Manual of remote sensing. *American Society of Photogrammetry*, Virginia, volume 2, pp. 1497–1570.

Tcherepanov, E.N., V.A. Zlotnik, and G.M. Henebry, 2005. Using Landsat thermal imagery and GIS for identification of groundwater discharge into shallow groundwater dominated lakes. *International Journal of Remote Sensing*, volume 26, no. 17, pp. 3649–3661.

Thillaigovindarajan, S., S.K. Seelan, M. Jayaraman, and P. Radhakrishnamoorthy, 1985. The evaluation of hydrogeological conditions in the southern part of Tamil Nadu using remote sensing techniques. *International Journal of Remote Sensing*, volume 6, nos. 3&4, pp. 447–456.

Tiwari, O.N., 1993. Lineament identification for groundwater drilling in a hard rock terrain of Sirohi District, Western Rajasthan. *Photonirvachak-Journal of the Indian Society of Remote Sensing*, volume 21, no. 1, pp. 14–19.

Tiwari, V.M., J. Wahr, and S. Swenson. 2009. Dwindling groundwater resources in Northern India, from satellite gravity observations. *Geophysical Research Letters*, volume 36, p. LI8401. DOI: 10.1029/2009GL039301

Venkataratnam, L.V., 1984. Mapping of land/soil degradation using multispectral data. Proceedings of the 8th Canadian symposium on remote sensing, Montreal.

Wang, L.T., T.E. McKenna, and T. DeLiberty, 2008. Locating groundwater discharge areas in Rehoboth and Indian River Bays and Indian River, Delaware, using Landsat 7 imagery. Report of Investigations no. 74, Delaware Geological Society, pp. 1–10.

Xiaomei, J., W. Li, Z. Youkuan, X. Zhongqi, and Y. Ying, 2007. A study of relationship between vegetation growth and groundwater in the Yinchuan Plain. *Earth Science Frontiers*, volume 14, no. 3, pp. 197–203.

Part III

Floods

4 Flood Monitoring Using the Integration of Remote Sensing and Complementary Techniques

Allan S. Arnesen, Frederico T. Genofre, Marcelo P. Curtarelli, and Matheus Z. Francisco

ACRONYMS AND DEFINITIONS

1D	One-dimensional
2D	Two-dimensional
ALOS	Advanced Land Observing Satellite
AMSR	Advanced Microwave Scanning Radiometer
ASAR	Advanced Synthetic Aperture Radar on Board ENVISAT
AVHRR	Advanced Very High Resolution Radiometer
AVNIR-2	Advanced Visible and Near Infrared Radiometer type-2
COSMO-SkyMed	Constellation of small Satellites for Mediterranean basin Observation
DEM	Digital Elevation Model
DJF	December-January-February
DMSP	Defense Meteorological Satellite Program
DS	dry smooth soil
EM	emergent macrophyte
ENVISAT	Environmental Satellite
ERS	European Remote Sensing Satellites
ETM+	Enhanced Thematic Mapper Plus
FF	Flooded Forest
FM	floating Mmcrophyte
GFDS	Global Flood Detection System
GFMS	Global Flood Monitoring System
GIS	Geographic Information Systems
GOES	Geostationary Operational Environmental Satellite
GPM	Global Precipitation Measurement
GRACE	Gravity Recovery and Climate Experiment
HMS	Hydrologic Modeling System
JERS	Japan Earth Resources Satellite
JJA	June-July-August
LiDAR	Light Detection and Ranging
MAM	March-April-May
MGB-IPH	*Modelo de Grandes Bacias—Instituto de Pesquisas Hidráulicas*

MODIS	Moderate-Resolution Imaging Spectroradiometer
NASA	National Aeronautics and Space Administration
NCEP	National Centers for Environmental Prediction
NCRS	Natural Resources Conservation Service
NFF	Non-Flooded Forest
NOAA	National Oceanic and Atmospheric Administration
PALSAR	Phased Array type L-band Synthetic Aperture Radar
POES	Polar Operational Environmental Satellites
PR	Precipitation Radar
QPE	Quantitative Precipitation Estimates
QPF	Quantitative Precipitation Forecasts
RADARSAT	RADAR Satellite
RMSE	Root Mean Square Errors
ROW	Rough Open Water
SAR	Synthetic Aperture Radar
SCS	Soil Conservation Service
SOW	Smooth Open Water
SRTM	Shuttle Radar Topography Mission
SSM/I	Special Sensor Microwave Imager
TerraSAR-X	A radar Earth observation satellite, with its phased array synthetic aperture radar
TM	Thematic Mapper
TMI	TRMM Microwave Imager
TMPA	TRMM Multi-satellite Precipitation Analysis
TRMM	Tropical Rainfall Measuring Mission
WS	Wet/Rough Soil

4.1 INTRODUCTION

Flood is defined as an overflowing by water of the normal confines of a stream or other water body, or accumulation of water by drainage over areas that are not normally submerged (Psomiadis et al., 2019; Sharma et al., 2019; WMO, 2011). It occurs when the flow capacity of the riverbed is insufficient and the water overflows, occupying adjacent areas. The frequency and magnitude of floods vary in space and time, depending on the river basin characteristics (e.g., size, land coverage, and topography) and climate conditions.

Floods are one of the greatest challenges to weather prediction because of its devastating effects in the world. Moreover, climate changes are making the flood occurrence more frequent in several countries. Using stream flow measurements and numerical simulations, Milly et al. (2002) verified, in a global approach, that the frequency of great river floods increased in the 20th century. These disasters may occur due to natural behaviors or as consequences of the urbanization process.

When long-term precipitation occurs in basins characterized by a high degree of urbanization, the flooding can be extensive, resulting in a great amount of damage and loss of life (Farhadi et al., 2022; Munawar et al., 2022; Martinis et al., 2018; Jeyaseelan, 2005). It has been shown that progressive urbanization increases the risk of inundation (Nirupama & Simonovic, 2007). Zhang et al. (2009) analyzed the impact of rapid urban expansion of an important economic zone in China, the Pearl River Delta. They pinpointed that the urbanization in this region was one of the major facts responsible for reducing the river's capability to buffer floods.

In urban areas, the increase of flood frequency at inner rivers may occur as a consequence of the suppression of minor streams and ponds, the removal of vegetation and spreading of soil sealing, increases in runoff together with the restriction of infiltration and evaporation, increases in stream sedimentation, and peak discharges (Du et al., 2012; Zhang et al., 2009).

Although flooding is a major hazard in both urban and rural areas, highly populated areas have the most severe economic impacts by this natural disaster where the value of assets at risk is greater (Neal et al., 2009). In the European Union, more than $40 billion per year is spent on flood mitigation, recovery, and compensation, and another $3 billion per year is spent on flood defense structures (van Ree et al., 2011).

The socioeconomic impacts caused by river floods increase when there is a lack of flood risk-management strategies to reduce flood damages. To minimize flood impacts over socioeconomic activities, stakeholders need reliable information to support them on evaluating flood risk, organizing aid distribution to the most severely affected areas, planning mitigation alternatives, and assessing damages. Surface and hydrological observational data are required to support these risk assessments (Tralli et al., 2005).

Remote sensing and Geographic Information Systems (GIS) are powerful tools for delineating flood zones and preparing flood hazard maps for vulnerable areas (Sadiq et al., 2023; Munawar et al., 2022; Martinis et al., 2018; Sanyal & Lu, 2004). However, one of the major limitations of remote sensing flood analyzing approaches is the scarce availability of high temporal resolution data, either from passive or active sensors.

The remote sensing instruments can be classified according to the range of the electromagnetic spectrum in which they operate. Optical or passive sensors act at the visible wavelengths of the spectrum, depending on external illumination (i.e., sunlight). The major advantage of this type of data is the ease of image visual interpretation. However, they are subject to atmospheric conditions, e.g., clouds and shadows cause losses of information.

On the other hand, active or Synthetic Aperture Radar (SAR) instruments overcome the weather condition restriction because they operate at longer wavelengths (microwave radiation) than optical data, allowing the acquisition of useful data even on cloudy days. Besides that, SAR data are independent of solar illumination conditions. Considering that flood events mostly occur on rainy days, these are fundamental characteristics of the remote sensing data used for flood extent delimitation.

Furthermore, new high temporal SAR sensors can allow flood monitoring in all phases of an extreme event. The Constellation of small Satellites for Mediterranean basin Observation (COSMO-SkyMed) mission is the best example of high temporal resolution SAR data. It is an Italian mission that offers meter-level, spatial resolution, daily acquired, X-band radar images for disaster management.

Even though the SAR sensors provide important information for flood monitoring, the integration of other data and tools is fundamental for obtaining accurate flood extent mapping. Optical and topographic data, hydrological models, and forecasting systems can also be explored to better analyze flood events.

In this chapter we review recent advances on methodological alternatives that combine remote sensing (SAR and ancillary data), GIS, and hydrological model tools for providing flood information that helps stakeholders to develop contingency plans and flood forecasting (Sadiq et al., 2023; Farhadi et al., 2022; Psomiadis et al., 2019; Schumann et al., 2018; Ban et al., 2017; Rahman & Di, 2017). After an overview of the remote sensing concepts and importance, Section 4.2.2 presents the new SAR data and algorithms used to map flood extent. Section 4.2.3 refers to the recent applications that used the integration of remote sensing–derived information and flood models. In the third sub-section (4.2.3) the main steps of flood forecasting and some applications using land cover and meteorological remote sensing data to minimize the socioeconomic impacts of floods are presented. Finally, the conclusion gathers relevant aspects highlighted throughout the chapter.

4.2 REMOTE SENSING IN FLOOD STUDIES

Remote sensing provides potential information to analyze flood events to support flood management and controlling (Farhadi et al., 2022; Domeneghetti et al., 2019; Schumann et al., 2018; Ahamed &

Bolten, 2017). One challenge of using remote sensing images for flood studies is atmospheric conditions. Optical sensors are negatively influenced by weather conditions, and when flood events occur as a consequence of rainy periods, severe cloudiness leads to completely useless data.

Active remote sensors provide data that are nearly independent of weather conditions, since the microwave radiation are about 200,000 times longer than the visible wavelengths of optical sensors. For this reason, the interaction between atmospheric micrometric particles (and water droplets) and the microwave radiation is negligible, allowing the acquisition of useful images even with complete cloud coverage over the area of interest.

This aspect by itself justifies the use of radar imagery for flood studies due to the possibility of improved temporal resolution during a specific event. However, the following section presents other considerable advantages of SAR data for flood mapping and monitoring.

4.2.1 Flood Mapping with SAR and Ancillary Data

The following sub-sections present the SAR data qualities for flood mapping, some relevant studies of different methodologies for flood monitoring, and the accuracies, errors, and uncertainties of these approaches.

4.2.1.1 SAR-Favorable Characteristics for Flood Delimitation

Besides the advantage of the nearly all-weather capability of acquiring useful data, a considerable benefit of SAR data for mapping flood extent is the interaction of the microwave radiation with flooded vegetation. Although interpreting SAR backscatter from flooded vegetation is not a straightforward task and statistical knowledge is required to obtain accurate results, SAR data allows the identification of water below canopies of forest and aquatic vegetation.

The interaction between microwave radiation and the targets is a special benefit of SAR data to map flood extent. The radiation penetration into flooded forest and macrophytes favors the identification of the total flood extent (Martinis et al., 2018; Wieland & Martinis, 2019; Arnesen et al., 2013). Three scattering mechanisms are dominant in floodplain areas: double bounce, volumetric (or canopy), and surface (or specular) (Figure 4.1).

The double bounce scattering occurs when a wave is incident on the inside corner of two flat surfaces that are joined at 90° and returns back to the radar, resulting in high backscattering values (Woodhouse, 2006). The volumetric backscattering occurs when the radiation interacts with several elements of the target being scattered in multiple directions (Henderson & Lewis, 1998). The surface scattering occurs when the radiation reaches a simple surface (open water, for instance), and in this case, the return signal depends on its roughness and SAR incidence angle.

The predominance of backscattering mechanisms depends on the SAR wavelength. For flooded forest areas, for example, the longer the wavelength, the higher the radiation penetration in the

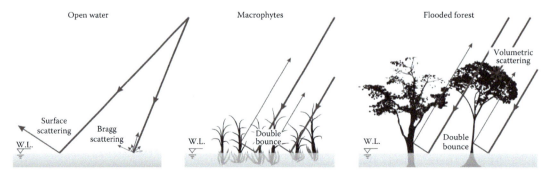

FIGURE 4.1 Resultant scattering mechanisms of the interaction between microwave radiation and usual wetland targets. Source: adapted from Arnesen (2012).

canopy will be, and consequently, high backscattering values are expected due to double bounce occurrence (Freeman & Durden, 1998). On the other hand, shorter wavelengths (X- and C-bands) have reduced penetration at the canopy, and volumetric and surface backscattering mechanisms are predominant.

Another SAR sensor parameter capable of influencing the interaction between the electromagnetic radiation and the floodable area targets is the polarization. To identify flooded forest, for instance, co-polarization configuration (HH and VV) data is preferable to cross-polarization (HV or VH) (Townsend, 2002).

Among the co-polarization data, HH data is used mose often than VV for flood mapping, especially at rural and wetland areas. Wang et al. (1995) compared the backscattering signal of a flooded forest on HH and VV for the same wavelength and incidence angle and verified that the backscattering ratio between flooded and non-flooded forest targets is higher at HH polarization than at VV polarization.

Microwave incidence angle must also be considered for mapping flood extent. Targets of interest may present different responses according to the range of SAR image incidence angle. Several studies consider that steep incidence angles can provide better results when mapping flooded forest than the shallow ones (Psomiadis et al., 2019; Tong et al., 2018; Wang & Xie, 2018; Rahman & Di, 2017; Wang et al., 1995; Lang & Kasischke, 2008).

However, the response behavior as a function of the incidence angle varies according to the target of interest. For instance, in open water areas steeper incidence angle implies lower backscattering values, while the gain of roughness significantly increases the signal. A theory that justifies the backscattering variation as a function of the incidence angle is the Bragg resonance model, usually applied to ocean surface water (Robinson, 2004). According to this model, an almost linear decrease of open water SAR backscatter occurs at the incidence angle range of 20–70°.

Using L-band ScanSAR wide swath (360 km) data from Phased Array L-band SAR (PALSAR) on board the Advanced Land Observing Satellite (ALOS) sensor, Arnesen (2012) inspected the backscatter variation along the scene range for four different targets, two flooded (forest and open water) and two non-flooded (soil and non-flooded). A convergence was verified among forest classes, once flooded forest presented a small but significant decrease of backscatter from near (19°) to far (42°) (Figure 4.2). The other two classes investigated presented a higher similarity at the PALSAR ALOS data because both open water and soil are smooth targets that favor surface scattering and signal return at L-band. This similarity is essentially high at some ranges and needs to be overcome in order to provide accurate flood extent maps.

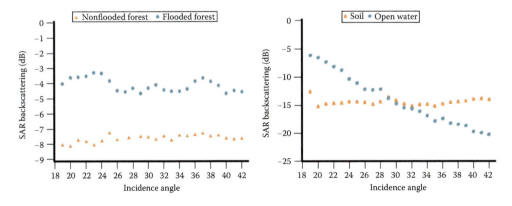

FIGURE 4.2 Samples of backscattering variation as a function of the incidence angle for the classes non-flooded forest (NFF), flooded forest (FF), open water (OW), and soil (S). Source: adapted from Arnesen (2012).

4.2.1.2 Flood Mapping Methodologies Using SAR Data

The scientific community has developed and applied several remote sensing approaches to map flood extent and/or risk using SAR data (Table 4.1). Classification accuracies can vary considerably among the SAR flood mapping approaches. According to Schumann et al. (2009), the main classification errors arise from improper image processing algorithms, variation of backscatter characteristics, unsuitable wavelengths or polarizations, unsuccessful speckle filtering, remaining geometric distortions, and inaccurate image geocoding (Schumann et al., 2018; Tong et al., 2018; Ban et al., 2017).

The digital image classification approaches can be divided into supervised and unsupervised classifications. While supervised classification includes the sample collecting step (training step), the unsupervised classification doesn't have this step and a certain algorithm segregates the image into classes according to their spectral similarities (Jensen, 2005).

SAR image studies demonstrate that both supervised and unsupervised classification approaches can have robust results when multipolarization is available (Cloude & Pottier, 1997; Frery et al., 2007). However, if only a single piece of polarization data is available, radiometric variation within a scene, resulting from acquisition effects, can implicate in lower accuracy classification results (Richards, 2009).

Thresholding radar backscatter value is frequently used in simple approaches. A binary algorithm is used to distinguish whether the area is flooded or not based on the backscatter value in decibels (dB) (Sanyal & Lu, 2004). However, as backscattering overlaps may occur among flooded and non-flooded targets, this approach can present poor results.

Another way to divide the classification approaches is by the basic processing unit of images: pixels or objects (also called regions—an aggregation of homogeneous neighbor pixels) (Jensen, 2005).

Several pixel-based studies used not only the backscattering values to discriminate the flooded or floodable areas but also, and especially, the multi-temporal criterion (Hess et al., 2003; Frappart et al., 2005; Martinez & Le Toan, 2007). Martinez and Le Toan (2007) utilized a temporal series of 21 Japan Earth Resources Satellite (JERS-1) SAR images acquired between the period of 1993 and 1997 to map the land cover types and the average duration in which each pixel is exposed to the Amazon River flood. Only the temporal changing estimator thresholds (backscattering average coefficient and temporal change) were used to segregate the classes.

However, the global accuracy of this flood mapping was limited. One of the reasons was the pixel-based classification approach that was adopted. The major limitation of pixel-based classifications of SAR data is the high radiometric variation provoked by speckle, which is the resulting effect from the interference among the coherent echoes of the individual scatters within a resolution cell (Woodhouse, 2006). Speckle can be easily explained using the complex vector representation of scattered microwaves: each individual scatter has a different location (phase) from the instrument and a different radar cross-section (signal amplitude) in such a way that the coherent sum of all the individual scattered waves is the total return signal.

This effect can be minimized on the object-based classification approaches, in which the analysis elements are image segments (or objects) composed by a group of pixels aggregated by similarity at the segmentation process.

Object-based classification methodologies have been applied to detect flooded areas in several urban studies, once high spatial resolution data are becoming available recently, such as TerraSAR-X and COSMO-SkyMed SARs (Amiri et al., 2024; Farhadi et al., 2022; Psomiadis et al., 2019; Sharma et al., 2019; Mason et al., 2010; Giustarini et al., 2013; Mason et al., 2012; Pulvirenti et al., 2011; Pierdiccca et al., 2013; Pulvirenti et al., 2013). For this kind of data, a segmentation procedure is essential because the spatial details of the images are considerably smaller than the objects' dimensions, which results in a large radiometric variance.

Taking advantage of the high temporal resolution of the COSMO-SkyMed mission, Pulvirenti et al. (2013) developed an algorithm based on an image segmentation method and a fuzzy logic

TABLE 4.1
Summary of Recent Studies Exploring Methodologies on SAR and Ancillary Data Utilization for Flood Extent Monitoring: SAR and Ancillary Data Information, Classification Aspects, Strength, Limitation, and Accuracy

SAR Data	SAR Band	Polarization	Pixel Dimensions	Flood Classification Aspects	Ancillary Data	Strength	Limitations	Accuracy	Reference
JERS-1 images	L	HH	approx. 100 m	Temporal changing thresholds: mean backscatter coefficient computed over dry and wet seasons and ratio of both seasons (change value)	Topex/Poseidon (T/P) altimetry satellite and in situ hydrograph stations	Combined use of altimetric water-level observations and inundation patterns to determine water volume variations in a wetland area	Lack of reference data for validating the classification and data acquisition of images did not coincide with the real maximum water level	90% for flooded forest and 70% for low vegetated themes	Frappart et al. (2005).
JERS-1 images	L	HH	approx. 100 m	Temporal changing thresholds: backscattering average coefficient and temporal change	Bathimetric transects of the floodplain	Temporal change to monitor the floodplain	Lack of reference data for validating the classification and data acquisition of images did not coincide with the real maximum water level	90% for all themes except from lox vegetation and flooded forest of 80%	Martinez & Le Toan al. (2005).
JERS-1 images	L	HH	approx. 100 m	Decision tree classification algorithm based on training sites of the land use types	Landsat images, DEM (digital elevation model), and ground observation	High overall accuracy	Limited accuracy for marsh (40–67%) and uncertainty to locate transition zones due to high annual variations	> 82%	Wang (2003)

TABLE 4.1 (Continued)
Summary of Recent Studies Exploring Methodologies on SAR and Ancillary Data Utilization for Flood Extent Monitoring: SAR and Ancillary Data Information, Classification Aspects, Strength, Limitation, and Accuracy

SAR Data	SAR Band	Polarization	Pixel Dimensions	Flood Classification Aspects	Ancillary Data	Strength	Limitations	Accuracy	Reference
ALOS PALSAR images (ScanSAR mode)	L	HH	approx. 100 m	Data mining tool for extracting backscatter thresholds, temporal changing thresholds, decision tree algorithm, and object-oriented classification	Landsat and MODIS images, DEM and bathymetric topographic data, and ground observation	Integrated use of SAR and ancillary data for segregation of dry and wet classes	Validation of only two dates (high and low water stages) because of lack of optical useful data	78% for low water stage, and 80% for high water stage	Arnesen et al. (2013)
ERS-1 images	C	VV	12.5 m x 12.5 m	Visual interpretations and adoption of two thresholding techinques	Landsast images, digital topographic map (1:10,000) and color aerial photographs	Temporal change was critical to monitor the floodplain	Accurate results overcoming the SAR temporal resolution constraint	96%	Brivio et al. (2002).
RADARSAT-1 images	C	HH	12.5 m x 12.5 m	Image processing with PCI and supervised classification using parallelliped	Hydrological properties of the drainage basins and Landsat images	Information extracted by remote sensing translated into hydrological model parameters for flood forecasting	Need of temporal resolution improvement	80%	Bonn & Dixon (2005)
TerraSAR-X images	C	HH	1.5 m x 1.5 m	Hybrid methodology containing backscatter thresholding, region growing, and change detection information	Very high resolution aerial photographics aquired during the flooding event	Completely unsupervised technique for flood extend mapping with satisfactory results	Requires a reference image with the same imaging properties as the SAR image	82%	Giustarini et al. (2013)

TerraSAR-X image	C	HH	3 m × 3 m	Snake algorithm applied to SAR and LiDAR data, supervised classification, water height thresholding, and seed region growing	LiDAR data, ASAR image data, and aerial photographs (0.2 m spatial resolution)	Provide flooded urban areas to enable a 2D inundation model to predict flood extent	SAR unvisible areas of the image (shadows) have lower flood mapping accuracy	76% for pixels visible to TerraSAR-X and 58% to all pixels	Mason et al. (2010)
TerraSAR-X image	C	HH	3 m × 3 m	Automatic algorithm inlcuding pre-processing, water height threshold, image segmentation, and object classification	LiDAR data, DEM, and aerial photographs	Presents operational consideration for a near-real time using high-resolution SAR data	Urban flood detection accuracy limited due to radar shadow and layover	75% for pixels visible to TerraSAR-X and 57% to all pixels	Mason et al. (2012)
COSMO-SkyMed images (spotlight mode)	X	HH	1 m × 1 m	Morphological filtering, unsupervised K-means clustering, segmentation, backscattering analysis, and object classification	DEM and Advanced Visible and Near Infrared Radiometer type-2 (AVNIR-2) data	Multi-temporal backscattering trends were used to associate image segments to different flood stages	Absence of a validation procedure for the specific flood event mapping	—	Pulvirenti et al. (2011)
COSMO-SkyMed images (stripmap and spotlight mode)	X	HH	1 m × 1 m	Image segmentation and fuzzy logic classifier	Landsat and MODIS images	Thresholding analysis of land cover types was important for flood mapping	Absence of a validation procedure for the specific flood event mapping	—	Pulvirenti et al. (2013)

classifier to map a flood that occurred in December 2009, in Tuscany, Italy. The segmentation algorithm of this study was also applied by Pulvirenti et al. (2011) and consists in an unsupervised clustering algorithm that considers multi-temporal and multi-scale features. The array of this segmentation object-based approach, fuzzy logic algorithm, and land cover information derived from ancillary data allowed the authors to generate accurate inundation maps.

A notable advantage of object-based approaches is the inclusion of objects' contextual characteristics in the structuration of a hierarchical decision tree (Popandopulo et al., 2023; DeVries et al., 2020; Wieland & Martinis, 2019; Ahamed & Bolten, 2017; Ban et al., 2017; Rahman & Di, 2017). In a decision tree, the classes are organized in a hierarchy and discriminated according to a logic sequence of decision rules that consider not only the SAR radiometric information but also other target properties such as temporal patterns, geometric information, complementary optical and topographic data, and spatial context (Benz, 2004; Silva et al., 2010).

To assess the flood extent seasonal variation in a wetland area of the lower Amazon River floodplain, Curuai Lake floodplain, Brazil, Arnesen et al. (2013) developed a hierarchical object-based classification methodology based on L-band PALSAR ALOS images and auxiliary data (optical images, water-level records, field photographs, and topographic information). The classification scheme consisted of four hierarchical levels. The first level (Level 1) is based on the annual flooding pattern of the region, segregating images into three classes: upland (non-floodable areas), floodplain (variable flooding), and permanent open water (open water surface during the lowest water level at the analyzed period). The second level (Level 2) divided floodplain into two classes assumed constant along the hydrological year: forest and non-forest. These first two levels have constant areas for all images acquired during the study period (2006–2010), while the next levels were applied for each image separately.

Therefore, at Level 3 the forest class was divided into flooded forest (FF) and non-flooded forest (NFF), while non-forest was split according to backscattering similarity into two intermediate classes: bright (including wet/rough soil—WS, emergent macrophyte—EM, and rough open water—ROW) and dark (with dry smooth soil—DS, floating Mmcrophyte—FM, and smooth open water—SOW). It is important to note that this segregation required a statistical backscattering analysis, including a data mining tool, in order to identify the main radiometric response overlapping and the need for complementary data (such as Moderate Resolution Spectroradiometer [MODIS] optical images and topographic model).

The last level of this approach (Level 4—flooding status) merged Level 3 classes according to their flood condition, the flooded class being the sum of open water (ROW and SOW), macrophytes (FM and EM), and flooded forest (FF) classes, while the non-flooded class was the represented by soil (WS and DS), non-flooded forest (NFF), and upland classes. The entire hierarchy is presented in Figure 4.3.

The classification results of Arnesen et al. (2013) had high accuracy for the period and data analyzed: 84% and 94% overall accuracies of the flood maps for low and high water stages (Figure 4.4), respectively. The authors justified the higher accuracy of the methodology for high water stages to the larger extent of open water, which was easily identified by the algorithm.

4.2.1.3 Accuracies, Errors, and Uncertainties of Flood Mapping with SAR Data

The methodological approaches for SAR mapping have inherent uncertainties due to these data characteristics. As Table 4.1 shows, the classification errors usually vary between the land use types because the polarization, incidence angle, spatial resolution, and SAR band have direct implications on the feasibility of distinguishing flooded and non-flooded classes.

While some methodologies made use of temporal variation characteristics, especially the coarse spatial resolution data (such as JERS-1 and ALOS ScanSAR mode data), with satisfactory results for flood delimitation, others have explored their fine spatial resolution to supply this task (the case of COSMO-SkyMed and TerraSAR-X applications, for example).

Flood Monitoring Using the Integration of Remote Sensing

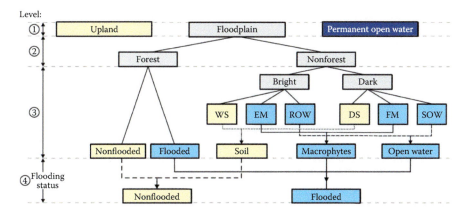

FIGURE 4.3 Classification hierarchy applied by Arnesen (2012) to map flood extent at the Curuai Lake floodplain, lower Amazon River, Brazil. The dark blue box represents the flooded class for the entire flood pulse; gray boxes are intermediate classes; beige boxes represent the non-flooded classes; and bright blue boxes represent flooded classes at each analyzed image.

Nevertheless, all of them are in one way or another subject to SAR data parameters, such as the interference of speckle noise, the occurrence of shadows and layovers, and the great similarity between dry and wet classes according to the interaction between SAR radiation and targets with similar roughness.

Therefore, the best way to overcome these inherent SAR limitations for flood mapping is integrating different data (optical, topographic, and hydrologic) with other tools are available for flood monitoring, such as GIS and hydrological models.

4.2.2 Integration of Remote Sensing, GIS, and Hydrological Models

A complementary approach that has been used to map and understand different aspects of floods in large river basins, as well as the monitoring of hydrological cycle components, is the integration of remote sensing–derived information and hydrological models (Sadiq et al., 2023; DeVries et al., 2020; Martinis et al., 2018; Wieland & Martinis, 2019; Bates, 2004; Bates et al., 2006; Matgen et al., 2007; Schumann et al., 2009; Paiva et al., 2013; Rudorff et al., 2014a, 2014b). One of the main advantages is that this approach allows not only predicting the extent and stage of the flood but also monitoring and simulating different scenarios such as storms, climate changes, different rainfall intensities, and changes in the land use and coverage. This integrated approach also allows the understanding of different environmental drivers of floods and the main mechanisms governing its dynamics.

Schumann et al. (2009) presented an extensive review of the recent progress in the integration of remote sensing–derived flood extent, water stage information, and hydrological models. According to these authors, this integration has only emerged over the last decade as a result of significant advances in SAR remote sensing techniques and high-performance computing, encouraging a boost in distributed flood modeling (Amiri et al., 2024; Schumann et al., 2018; Ban et al., 2017; Rahman & Di, 2017).

According to the literature, there are four main topics addressed in studies on integrating remote sensing–derived information (Sadiq et al., 2023; Sharma et al., 2019; Martinis et al., 2018; Tong et al., 2018; Ahamed & Bolten, 2017): (1) the retrieval and modeling of flood hydrology information from remote sensing observations, (2) the use of remote sensing data to calibrate and validate hydrological models, (3) the potential of remote sensing to understand and improve model structures, and (4) the utility of remote sensing data assimilation with models.

Despite the several advantages and the high potential to study flood dynamics, the integrated use of remote sensing–derived information and flood models requires a deep understanding of the

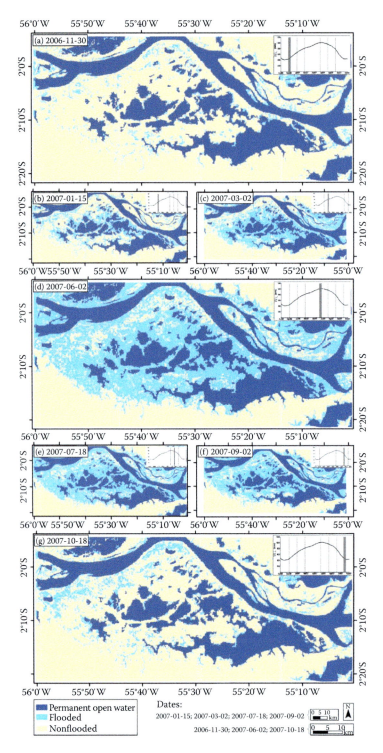

FIGURE 4.4 Flood extent variation during the 2007 flood pulse, Curuai Lake floodplain, lower Amazon River, Brazil. Graphics of the water level at each date representing the water level at Curuai gauge at the image acquisition. Source: adapted from Arnesen (2012).

Flood Monitoring Using the Integration of Remote Sensing

many factors underlying both the remote sensing and modeling parts (Schumann et al., 2009). For this reason, relatively few studies have looked at the complex interplay of these two fields. Table 4.2 shows a summary of some studies that used this approach.

Paiva et al. (2013) combined remote sensing data (SAR and optical data) and a distributed hydrological model to study the hydrodynamics and flood dynamics of the Amazon River. The authors used the MGB-IPH (*Modelo de Grandes Bacias—Instituto de Pesquisas Hidráulicas* in Portuguese) model, which is a large-scale distributed hydrological model that uses physical and conceptual equations to simulate land surface hydrological processes (Collischonn et al., 2007).

The model discretization into river reaches, catchments, hydrodynamic computational cross-sections, and parameter estimation was carried out using the digital elevation model (DEM) derived from the Shuttle Radar Topography Mission (SRTM) (Farr et al., 2007), with 15 arc-seconds

TABLE 4.2
Summary of Recent Studies That Used Integration between Remote Sensing Data and Numerical Models: Models Used, Remote Sensing Data, and Parameters and Variables Extracted from Remote Sensing Data

Model	Remote Sensing Data	Parameters and Variables Derived from Remote Sensing Data	Reference
LISFLOOD-FP	SRTM/DEM; Radarsat-1/SAR; Landsat/TM; JERS/SAR; ALOS/PALSAR	Floodplain bathymetry; roughness coefficient; flood extent	Rudorff et al. (2014a)
LISFLOOD-FP	SRTM/DEM; Radarsat-1/SAR; Landsat/TM; JERS/SAR; ALOS/PALSAR	Floodplain bathymetry; roughness coefficient; flood extent	Rudorff et al. (2014b)
MGB-IPH	TRMM/3B42 product; SRTM/DEM; ENVISAT/altimeter; SSM/I; ERS/scatterometer; AVHRR; GRACE/RL04 product	Daily precipitation; morphological parameters; water level; flood extent; terrestrial water storage	Paiva et al. (2013)
HEC-RAS	LiDAR; ENVISAT/ASAR; ERS/SAR; aerial photographs	DEM; flood extent; roughness coefficient	Schumann et al. (2007)
HEC-RAS	LiDAR; ENVISAT/ASAR	DEM; flood extent	Matgen et al. (2007)
TELEMAC-2D and Simple Finite Volume Model (SFV)	LiDAR; ENVISAT/ASAR; ESR/SAR	DEM; flood extent	Horritt et al. (2007)
LISFLOOD-FP	LiDAR/ALTM2033; ENVISAT/ASAR; ESR/SAR; Radarsat-1/SAR; ERS-2/SAR	DEM; flood extent	Bates et al. (2006)
LISFLOOD-FP	ERS-1/SAR; aerial photographs	Flood extent	Hunter et al. (2005)
LISFLOOD-FP	ERS-1/SAR	Flood extent	Hall et al. (2005)
HEC-RAS; LISFLOOD-FP and TELEMAC-2D	LiDAR; Radarsat-1/SAR; ERS-1/SAR;	DEM; flood extent	Horritt & Bates (2002)
LISFLOOD-FP and TELEMAC-2D	ERS-1/SAR	Flood extent	Horritt & Bates (2001)

resolution (approximately 500 m) and GIS-based algorithms described in Paiva et al. (2011). The simulation was validated using remote sensing–derived information (water level and flood extent).

The modeled water level obtained by Paiva et al. (2013) showed good agreement with the water level obtained using ENVISAT satellite altimetry data (Silva et al., 2010), with a Nash-Sutcliffe coefficient higher than 0.60 in 55% of the stream gauges used and correlation coefficient higher than 0.8 in 80% of the cases.

The overall inundation extent results from the MGB-IPH model obtained by Paiva et al. (2013) was similar to remote sensing estimates from Papa et al. (2010), showing that the model was able to reproduce the seasonal variation of flood extent and the north-south contrast, with flood peaks occurring in DJF (December-January-February) and MAM (March-April-May) at the Bolivian Amazon, in MAM and JJA (June-July-August) at central Amazon, and JJA in the north.

Rudorff et al. (2014a, 2014b) used a synergistic approach that combined *in situ* data, remote sensing data, and a numerical model to study the hydraulic controls and the interannual and seasonal variability of the floodplain water balance components at Curuai Lake floodplain on the lower Amazon River. These authors used the LISFLOOD-FP model (Bates & De Roo, 2000), a raster-based hydraulic model that combines one-dimensional (1D) river routing with two-dimensional (2D) overland flow.

The LISFLOOD-FP model uses different schemes to solve the river routing and overland flow equations; while the 1D river routing model used the diffusive wave scheme developed by Trigg et al. (2009), the 2D floodplain flows were solved by the inertia approximation of the St. Venant momentum equation (Bates et al., 2010). The SRTM DEM version 4.1 (Jarvis et al., 2008) was used as input data in the LISFLOOD-FP model. Before the simulations, Rudorff et al. (2014a) performed a series of corrections on the SRTM DEM and tested different approaches to combine *in situ* bathymetric data and the SRTM DEM to generate the final DEM. The authors also used floodplain roughness coefficients derived from RADARSAT and Landsat Thematic Mapper (TM) images as input in the model simulations. The simulations were validated using the water level of the Curuai gauging station and flood maps derived from SAR images produced by Hess et al. (2003) and Arnesen et al. (2013).

The results obtained by Rudorff et al. (2014a) showed that the DEM with channel adjustment yielded the best results in the simulation of the water level, showing the lowest root mean square errors (RMSE=0.17 m). The DEM corrected with channel adjustments and local hydrological flows (RMSE=0.27 m), the original SRTM DEM (RMSE=0.38 m), and the DEM without channel adjustment (RMSE=0.81 m) produced the worst results.

The simulated inundation extents also showed good agreement with SAR-derived maps for high-water conditions and lower agreement for low-water conditions. The inundation maps produced by Hess et al. (2003) based on JERS SAR images are commonly used for the assessment of accuracy in studies of inundation modeling in the central Amazon basin. However, the results obtained by Rudorff et al. (2014a) showed that these maps overestimated the actual inundation extents on the Curuai Lake floodplain at the time of the image acquisitions. Predictions of total flooded area were 36% and 20% lower in LISFLOOD-FP simulations using the rectified SRTM DEM compared to JERS mapping for high-water and low-water dates, respectively.

According to Rudorff et al. (2014a), the error in the SRTM interferometric baseline is of a long wavelength nature and is likely to span a negative elevation bias over wide regions in the Amazon basin. Forest canopies are a source of positive elevation bias. Though upstream reaches of the Amazon floodplain tend to be more predominantly covered by alluvial forest, which then becomes the main source of elevation error, it should be important to account for the interferometric baseline error along with correction for vegetation effect.

Another combined methodology was used by Weng (2001). The author evaluated the changes in runoff in the Zhujiang Delta region, the third biggest river delta in China, through the integration of GIS, remote sensing, and hydrological models and has automated the Soil Conservation Service (SCS) model.

With the support of the Landsat TM data, changes in the coverage of urban land between 1989 and 1997 was identified through classifications of land-cover techniques. Then, based on the SCS model, urban growth patterns were identified to assess the effects on surface runoff parameters. The results of this research indicated that the annual runoff depth had increased by 8.10 mm per year.

Similarly, Du et al. (2012) developed an integrated modeling system to analyze the effect of urbanization on annual runoff and flood events in the Qinhuai River basin in China. The authors used Landsat TM and Enhanced Thematic Mapper Plus (ETM+) images to represent land use changes over time. The urbanization scenarios identified were modeled in a Hydrologic Modeling System (HMS) and indicated that the values of annual runoff, daily peak flow, and flood level had increased.

As shown briefly, the integrated use of remote sensing–derived information and flood models has become established as a powerful tool, the robustness of which needs, however, to be examined further, in particular for flood forecasting (Farhadi et al., 2022; Psomiadis et al., 2019; Schumann et al., 2018; Rahman & Di, 2017). What is certainly to be gained from this development is that fundamental research questions in terms of both model evaluation and digital image processing techniques will be addressed in one way or another. Moreover, it is expected that remote sensing data and flood models that allow direct integration of such data would provide the basis for the assimilation of remotely sensed data, for instance, in operational flood-forecasting systems.

4.2.3 Flood Forecasting Using Remote Sensing–Derived Information

Floods usually affect a great amount of people, and the frequency of these extreme events is expected to increase around the world (Sadiq et al., 2023; DeVries et al., 2020; Martinis et al., 2018; Wang & Xie, 2018; Rahman & Di, 2017). Therefore, flood-forecasting systems and assessments of potential natural hazard areas are tools that decision-makers continuously desire. Flood prediction unfolds in distinct phases that are better described in the following items:

4.2.3.1 Flood Preparedness Phase: Flood Prone/Risk Zone Identification

The first step for enhancing our knowledge of a particular flood event is through flood-affected population, where results such as damage assessment, recommendations, and other important information must be documented.

Flood risk zone mapping could be presented in two different ways: a detailed mapping approach, which is required for the production of hazard assessment and updating/creating flood risk maps, or a larger-scale mapping approach that explores the general flood situation within a river basin, identifying areas under greatest risk. In both cases, the integration of remote sensing and ancillary data contributes to mapping the inundated areas, as presented in previous sections (Jeyaseelan, 2005).

4.2.3.2 Flood Prevention Phase

Although meteorological remote sensing flood monitoring can be applied from a global scale to a local storm scale, the most used is at the storm scale associated with hydrodynamic models by monitoring the intensity, movement, and propagation of the precipitation system to determine how much, when, and where the heavy precipitation is going to move during the next zero to three hours (called *nowcasting*). Meteorological satellites can detect various aspects of the hydrological cycle—precipitation (rate and accumulations), wetness transport, and surface/soil moisture (Scofield & Achutuni, 1996).

Satellite optical observations of floods have been hampered by the presence of clouds that resulted in the lack of near real-time data acquisitions. SAR sensors can achieve regular observation of the Earth's surface, even in the presence of thick cloud cover. The Advanced Very High Resolution Radiometer (AVHRR) sensor on board the National Oceanic and Atmospheric Administration (NOAA) satellites allows flood monitoring in near real-time. High-resolution infrared (10.7 μm) and visible images are the principal data sets used in this diagnosis. Soil moisture due to a heavy rainfall event or snowmelt is extremely useful information for flash flood guidance. Data from the Special Sensor Microwave Imager (SSM/I) sensor on board the Defense Meteorological Satellite Program (DMSP) can also be used in this analysis.

4.2.3.3 Flood Forecasting

An effective real-time flood forecasting system essentially requires basic structures that need to be linked in an organized manner (WMO, 2011). Parts of this structure are listed here:

1. Provision of specific forecasts relating to rainfall (both quantity and timing), for which numerical weather-precipitation models are necessary;
2. Establishment of a network of manual or automatic hydrometric stations, linked to a central control by some form of telemetry; and
3. Flood-forecasting model software, linked to the observing network and operating in real time.

Flood forecast can be issued over the areas in which remote sensing is complementary to direct precipitation and stream flow measurements, and those areas that are not instrumentally monitored (or the instruments are not working or are inaccurate). In this second category, remote sensing represents an essential tool.

Quantitative Precipitation Estimates (QPE) and Forecasts (QPF) use satellite data as one source of information to facilitate flood forecasts. New algorithms are being developed that integrate Geostationary Operational Environmental Satellite (GOES) precipitation estimates with the more physically based Polar Operational Environmental Satellites (POES) microwave estimates. An improvement in rainfall spatial distribution measurements is being achieved by integrating radar, rain gauges, and remote sensing techniques to improve real-time flood forecasting (Vicente et al., 1998).

GOES and POES weather satellites can provide climatological information on precipitation, especially for those areas not instrumentally monitored. The technique developed by Vicente et al. (1998) uses data from the infrared brightness temperature of channel 4 (10.7 μm) of GOES-8 and GOES-9 and converts from a power ratio with the rate of precipitation estimated by radar. Subsequently, the precipitation rates are adjusted to different moisture regimes, growth of convective cells, and spatial gradients of clouds. Finally, corrections for wet and dry environments are made using the average relative humidity between the surface and 500 mbar level and precipitable water at that level, obtained by the National Centers for Environmental Prediction (NCEP) Eta model.

Vicente et al. (2002) made other adjustments and the method was called Hidroestimador. The algorithm fixes the level of convective equilibrium for event hot spots, and errors due to terrain and parallax were used to automatically adjust precipitation rate. The components of direction and wind speed are extracted from the NCEP Eta model at the level of 850 mbar and the elevation and topography of the terrain are taken from geographic maps. This allows for improved accuracy of estimated rainfall in difficult areas to obtain surface data, both by conventional weather stations and weather radar.

As mentioned (Section 4.2.2), hydrological models play a major role in assessing and forecasting flood risk. These models requires several types of data as input, such as DEM data, land use, soil type, soil moisture, stream/river base flow, drainage basin size, snowpack characterization, and rainfall amount/intensity.

In 1997 the Tropical Rainfall Measuring Mission (TRMM) satellite was launched. The objective of the TRMM mission was to provide accurate global tropical precipitation estimates by using a unique combination of instruments designed purely for rainfall observation (Simpson et al., 1996; Kummerow et al., 1998).

Flood Monitoring Using the Integration of Remote Sensing

Just ten years after TRMM launched, a series of high-resolution, quasi-global, near real-time, TRMM-based precipitation estimates started to be available for the research community. The proposed Global Precipitation Measurement (GPM) mission, which would be the successor to the TRMM, envisions providing global precipitation products with temporal sampling rates ranging from three to six hours and a spatial resolution of 25–100 km^2 (Smith, 2007). A major GPM science objective is to improve prediction capabilities for floods, landslides, freshwater resources, and other hydrological applications.

The TRMM-based algorithms for flood monitoring use as input data a product named 3B42 that combines precipitation radar (PR) and TRMM Microwave Imager (TMI) adjusted with surface rain gauge measurements on a monthly basis and near real-time (3B42RT) product (Huffman et al., 2006). TRMM Multi-satellite Precipitation Analysis (TMPA) has improved algorithms in the last years (Turk & Miller, 2005; Joyce et al., 2004; Kidd et al., 2003; Sorooshian et al., 2000).

A first approach to global runoff simulation using satellite rainfall estimation was reported by Hong et al. (2007) as an approximate assessment of quasi-global runoff. The Natural Resources Conservation Service (NCRS) runoff curve number method incorporated rainfall data and other remote sensing products in a rainfall-runoff approach.

Kugler & De Groeve (2007) demonstrated in their report the verification and validation of the model that supports the Global Flood Detection System (GFDS). GFDS provides a systematic monitoring of ongoing flood events around the world and is updated every day. The model can also be used to reconstruct historical flood events that were not recorded in a systematic way.

Another satellite-based tool that has been running and receiving improvements during the last few years is called Global Flood Monitoring System (GFMS—http://flood.umd.edu). The validation and analysis based on the recent flood events over the Upper Mississippi Valley from the GFMS real-time system demonstrated that the real-time GFMS had a satisfactory response in flood detection, evolution, and magnitude calculation according to observed daily stream flow data (Wu et al., 2014). Some of the monitoring tools of flooding are shown at Table 4.3.

4.3 CONCLUSIONS

This chapter provides a comprehensive assessment of and approach to flood monitoring using multi-sensor remote sensing along with modeling and other GIS data (Amiri et al., 2024; Colacicco et al., 2024; Popandopulo et al., 2023; Sadiq et al., 2023; Farhadi et al., 2022; Munawar et al., 2022; DeVries et al., 2020; Domeneghetti et al., 2019; Psomiadis et al., 2019; Sharma et al., 2019;

TABLE 4.3

Global Flood Monitoring Tools Using Input Data from the Tropical Rainfall Measurement Mission—NASA/TRMM

Flood Monitoring Tool	Input Data	Spatial Coverage and Resolution	Reference
NASA—Tropical Rainfall Measuring Mission—TRMM NRT	TRMM/TMPA-RT	50°S-50°N 12 km	Huffman et al. (2006)
MODIS NRT	Terra and Aqua/MODIS	50°S-50°N 12 km	Turk & Miller (2005)
Dartmouth Flood Observatory—DFO	TRMM/TMI and Aqua/AMSR-RT	Global 250 m and 10 km	Brakenridge & Anderson (2005)
Global Flood Detection System—GFDS	TRMM/TMI and Aqua/AMRS-E TRRM/TMI	Global 10 km	Kugler & De Groeve (2007)
Global Flood Monitorating System—GFMS	TRMM/TMPA-RT MERRA	Global 10 km	Wu et al. (2014)

Martinis et al., 2018; Notti et al., 2018; Schumann et al., 2018; Tong et al., 2018; Wang & Xie, 2018; Wieland & Martinis, 2019; Ahamed & Bolten, 2017; Ban et al., 2017; Rahman & Di, 2017). The major limitation of remote sensing data for flood monitoring is the low frequency of data acquisition during a specific natural disaster. Cloud coverage at rainy events makes optical remote sensing data nearly impossible to be used in order to extract flood extent. SAR data is almost independent of weather conditions, which may consequently provide more frequent cloud-free images and have other potential advantages for flood monitoring as the interaction of microwave radiation with flooded targets.

Moreover, recently new remote sensing satellites are providing daily SAR data that allows a certain flood event to be mapped during all stages, such as the COSMO-SkyMed mission. All the SAR sensor characteristics are reflected in the backscatter, and studies that explore these properties can achieve high flood mapping accuracy. Nevertheless, overlaps between classes may occur at SAR images, and ancillary data (such as optical images) represent an effective alternative to minimize classification confusion.

There are several digital image processing methodologies available for extracting flood areas from remote sensing data (Amiri et al., 2024; Colacicco et al., 2024; Popandopulo et al., 2023; Sadiq et al., 2023; Farhadi et al., 2022; Munawar et al., 2022; DeVries et al., 2020; Domeneghetti et al., 2019; Psomiadis et al., 2019; Sharma et al., 2019; Martinis et al., 2018; Notti et al., 2018; Schumann et al., 2018; Tong et al., 2018; Wang & Xie, 2018; Wieland & Martinis, 2019; Ahamed & Bolten, 2017; Ban et al., 2017; Rahman & Di, 2017). However, this chapter showed that the most successful of them are techniques that integrate different sources of data and tools. The integration of GIS, hydrological models, and remote sensing is advantageous for flood mapping approaches in urban, rural, and wetland areas since the input data is reliable and accurate (gauging and topographic data, for instance).

As this chapter has presented, the integration of remote sensing, GIS, and hydrological modeling tools can be part of a major flood-forecasting system. Prediction models of potential flood extent can help stakeholders to develop contingency plans, facilitate a more effective response, and minimize the socioeconomic impacts of these natural disasters.

REFERENCES

Ahamed, A., & Bolten, J.D. 2017. A MODIS-based automated flood monitoring system for Southeast Asia. *International Journal of Applied Earth Observation and Geoinformation*, 61, 104–117, ISSN 1569-8432, doi: 10.1016/j.jag.2017.05.006. (www.sciencedirect.com/science/article/pii/S0303243417301095)

Amiri, A., Soltani, K., Ebtehaj, I., & Bonakdari, H. 2024. A novel machine learning tool for current and future flood susceptibility mapping by integrating remote sensing and geographic information systems. *Journal of Hydrology*, 632, 130936, ISSN 0022-1694, doi: 10.1016/j.jhydrol.2024.130936. (www.sciencedirect.com/science/article/pii/S0022169424003305)

Arnesen, A.S. 2012. *Monitoramento da área inundada na Planície de Inundação do Lago Grande de Guruai (PA) por meio de imagens ScanSAR/ALOS e dados auxiliares*. Master Science Thesis, Instituto Nacional de Pesquisas Espaciais, São José dos Campos, Brazil.

Arnesen, A.S., Silva, T.S.F., Hess, L.L., Novo, E.M.L.M., Rudorff, C.M., Chapman, B.D., & McDonald, K.C. 2013. Monitoring flood extent in the lower Amazon River floodplain using ALOS/PALSAR ScanSAR images. *Remote Sensing of Environment*, 130, 51–61, doi: 10.1016/j.rse.2012.10.035.

Ban, H.-J., Kwon, Y.-J., Shin, H., Ryu, H.-S., & Hong, S. 2017. Flood monitoring using satellite-based RGB composite imagery and refractive index retrieval in visible and near-infrared bands. *Remote Sensing*, 9, no. 4: 313, doi: 10.3390/rs9040313.

Bates, P.D. 2004. Remote sensing and flood inundation modelling. *Hydrological Processes*, 18, 2593–2597, doi: 10.1002/hyp.5649.

Bates, P.D., & De Roo, A.P.J. 2000. A simple raster-based model for flood inundation simulation. *Journal of Hydrology*, 236, 54–77, doi: 10.1016/S0022-1694(00)00278-X.

Bates, P.D., Horritt, M.S., & Fewtrell, T.J. 2010. A simple inertial formulation of the shallow water equations for efficient two-dimensional flood inundation modelling. *Journal of Hydrology*, 387, 33–45, doi: 10.1016/j.jhydrol.2010.03.027.

Bates, P.D., Wilson, M.D., Horritt, M.S., Mason, D.C., Holden, N., & Currie, A. 2006. Reach scale floodplain inundation dynamics observed using airborne synthetic aperture radar imagery: Data analysis and modelling. *Journal of Hydrology*, 328, 306–318, doi: 10.1016/j.jhydrol.2005.12.028.

Benz, U. 2004. Multi-resolution, object-oriented fuzzy analysis of remote sensing data for GIS-ready information. *ISPRS Journal of Photogrammetry and Remote Sensing*, 58, no. 3–4: 239–258, doi: 10.1016/j.isprsjprs.2003.10.002.

Bonn, F., & Dixon, R. 2005. Monitoring flood extent and forecasting excess runoff risk with RADARSAT-1 data. *Natural Hazards*, 35, 377–393, doi: 10.1007/s11069-004-1798-1.

Brakenridge, G.B., & Anderson, E. 2005. MODIS-based flood detection, mapping, and measurement: The potential for operational hydrological applications. *Transboundary Floods, Proc. of NATO Advanced Research Workshop*, Baile Felix &Oradea, Romania, May 4–8.

Brivio, P.A., Colombo, R., Maggi, M., & Tomasoni, R. 2002. Integration of remote sensing data and GIS for accurate mapping of flooded areas. *International Journal of Remote Sensing*, 23, 429–441, doi: 10.1080/01431160010014729.

Cloude, S.R., & Pottier, E. 1997. An entropy based classification scheme for land applications of polarimetric SAR. *IEEE Transactions on Geoscience and Remote Sensing*, 35, no. 1: 68–78, doi: 10.1109/36.551935.

Colacicco, R., Refice, A., Nutricato, R., Bovenga, F., Caporusso, G., D'Addabbo, A., La Salandra, M., Lovergine, F.P., Nitti, D.O., & Capolongo, D. 2024. High-resolution flood monitoring based on advanced statistical modeling of Sentinel-1 multi-temporal stacks. *Remote Sensing*, 16, no. 2: 294, doi: 10.3390/rs16020294.

Collischonn, W., Allasia, D.G., Silva, B.C., & Tucci, C.E.M. 2007. The MGB-IPH model for large-scale rainfall-runoff modeling. *Hydrology Science Journal*, 52, 878–895, doi: 10.1623/hysj.52.5.878.

DeVries, B., Huang, C., Armston, J., Huang, W., Jones, J.W., & Lang, M.W. 2020. Rapid and robust monitoring of flood events using Sentinel-1 and Landsat data on the Google Earth Engine. *Remote Sensing of Environment*, 240, 111664, ISSN 0034-4257, doi: 10.1016/j.rse.2020.111664. (www.sciencedirect.com/science/article/pii/S003442572030033X)

Domeneghetti, A., Schumann, G.J.-P., & Tarpanelli, A. 2019. Preface: Remote sensing for flood mapping and monitoring of flood dynamics. *Remote Sensing*, 11, no. 8: 943, doi: 10.3390/rs11080943.

Du, J., Qian, L., Rui, H., Zuo, T., Zheng, D., Xu, Y., & Xu, C.-Y. 2012. Assessing the effects of urbanization on annual runoff and flood events using an integrated hydrological modeling system for Qinhuai River basin, China. *Journal of Hydrology*, 464–465, 127–139, doi: 10.1016/j.jhydrol.2012.06.057.

Farhadi, H., Esmaeily, A., & Najafzadeh, M. 2022. Flood monitoring by integration of remote sensing technique and multi-criteria decision making method. *Computers & Geosciences*, 160, 105045, ISSN 0098-3004, doi: 10.1016/j.cageo.2022.105045. (www.sciencedirect.com/science/article/pii/S0098300422000127)

Farr, T.G., Rosen, P.A., Caro, E., Crippen, R., Duren, R., Hensley, S., Kobrick, M., Paller, M., Rodriguez, E., Roth, L., Seal, D., Shaffer, S., Shimada, J., Umland, J., Werner, M., Oskin, M., Burbank, M., & Alsdorf, D. 2007. The shuttle radar topography mission. *Reviews of Geophysics*, 45, RG2004, doi: 10.1029/2005RG000183.

Frappart, F., Seyler, F., Martinez, J.-M., León, J.G., & Cazenave, A. 2005. Floodplain water storage in the Negro River basin estimated from microwave remote sensing of inundation area and water levels. *Remote Sensing of Environment*, 99, 387–399, doi: 10.1016/j.rse.2005.08.016.

Freeman, A., Member, S., & Durden, S.L. 1998. A three-component scattering model for polarimetric SAR data. *Jet Propulsion*, 36, 963–973, doi: 10.1109/36.673687.

Frery, A.C., Correia, A.H., & Freitas, C.C. 2007. Classifying multifrequency fully polarimetric imagery with multiple sources of statistical evidence and contextual information. *IEEE Transactions on Geoscience and Remote Sensing*, 45, no. 10: 3098–3109, doi: 10.1109/TGRS.2007.903828.

Giustarini, L., Hostache, R., Matgen, P., Schumann, G.J.-P., Bates, P.D., & Mason, D.C. 2013. A change detection approach to flood mapping in urban areas using TerraSAR-X. *IEEE Transactions on Geoscience and Remote Sensing*, 51, 2417–2430, doi: 10.1109/TGRS.2012.2210901.

Hall, J.W., Tarantola, S., Bates, P.D., & Horritt, M.S. 2005. Distributed sensitivity analysis of flood inundation model calibration. *Journal of Hydraulic Engineering*, 131, 117–126, doi: 10.1061/(ASCE)0733-9429(2005)131:2(117).

Hapuarachchi, H.A.P., & Wang, Q.J. 2008. A review of methods and systems available for flash flood forecasting. *CSIRO: Water for a Healthy Country National Research Flagship*. Water for a Healthy Country Flagship Report Series, ISSN 1835-095X.

Henderson, F., & Lewis, A. 1998. *Manual of Remote Sensing: Principles and Applications of Imaging Radar*, 3rd ed. New York: Wiley, p. 896.

Hess, L.L., Melack, J.M., Novo, E.M.L.M., Barbosa, C.C.F., & Gastil, M. 2003. Dual-season mapping of wetland inundation and vegetation for the central Amazon basin. *Remote Sensing of Environment*, 87, 404–428, doi: 10.1016/j.rse.2003.04.001.

Hong, Y., Adler, R.F., Hossain, F., Curtis, S., & Huffman, G.J. 2007. A first approach to global runoff simulation using satellite rainfall estimation. *Water Resources Research*, 43, doi: 10.1029/2006WR005739.

Horritt, M.S., & Bates, P.D. 2001. Predicting floodplain inundation: Raster-based modelling versus the finite-element approach. *Hydrological Process*, 15, 825–842, doi: 10.1002/hyp.188.

Horritt, M.S., & Bates, P.D. 2002. Evaluation of 1D and 2D numerical models for predicting river flood inundation. *Journal of Hydrology*, 268, 87–99, doi: 10.1016/S0022-1694(02)00121-X.

Horritt, M.S., Di Baldassarre, G., Bates, P.D., & Brath, A. 2007. Comparing the performance of a 2-D finite element and a 2-D finite volume model of floodplain inundation using airborne SAR imagery. *Hydrological Process*, 21, 2745–2759, doi: 10.1002/hyp.6486.

Huffman, G.J., Adler, R.F., Bolvin, D.T., Gu, G., Nelkin, E.J., Bowman, K.P., Stocker, E.F., & Wolff, D.B. 2006. The TRMM multi-satellite precipitation analysis: Quasi-global, multi-year, combined-sensor precipitation estimates at fine scale. *Journal of Hydrometeorology*, 8, 38–55, doi: 10.1175/JHM560.1.

Hunter, N.M., Bates, P.D., Horritt, M.S., de Roo, P.J., & Werner, M.G.F. 2005. Utility of different data types for calibrating flood inundation models within a GLUE framework. *Hydrology and Earth System Sciences*, 9, 412–430, doi: 10.5194/hess-9-412-2005.

Jarvis, A., Reuter, H.I., Nelson, A., & Guevara, E. 2008. *Hole-Filled Seamless SRTM Data V4*. International Center for Tropical Agriculture, Cali, Colombia, Available at: http://srtm.csi.cgiar.org.

Jensen, J.R. 2005. *Introductory Digital Image Processing: A Remote Sensing Perspective*, 3rd ed. Englewood Cliffs: Prentice Hall, p. 526.

Jeyaseelan, A.T. 2005. Droughts and floods assessment and monitoring using remote sensing and GIS. *Proceedings of the Satellite Remote Sensing and GIS Applications in Agriculture Meteorology*, pp. 291–313.

Joyce, R.J., Janowiak, J.E., Arkin, P.A., & Xie, P. 2004. CMORPH: A method that produces global precipitation estimates from passive microwave and infrared data at high spatial and temporal resolution. *Journal of Hydrometeorology*, 5, 487–503, doi: 10.1175/1525-7541(2004)005.

Kidd, C.K., Kniveton, D.R., Todd, M.C., & Bellerby, T.J. 2003. Satellite rainfall estimation using combined passive microwave and infrared algorithms. *Journal of Hydrometeorology*, 4, 1088–1104, doi: 10.1175/1525-7541(2003)004.

Kugler, Z., & De Groeve, T. 2007. The Global Flood Detection System. *Joint Research Centre, Institute for the Protection and Security of the Citizen, European Commission*. EUR 23303 EN, p. 44.

Kummerow, C., Barnes, W., Kozu, T., Shiue, J., & Simpson, J. 1998. The Tropical Rainfall Measuring Mission TRMM sensor package. *Journal of Atmospheric and Oceanic Technology*, 15, 809–817, doi: 10.1175/1520-0426(1998)015.

Lang, M.W., & Kasischke, E.S. 2008. Using C-band synthetic aperture radar data to monitor forested wetland hydrology in Maryland's Coastal Plain, USA. *Wetlands*, 46, 535–546, doi: 10.1109/TGRS.2007.909950.

Martinez, J.M., & Toan, T.L. 2007. Mapping of flood dynamics and spatial distribution of vegetation in the Amazon floodplain using multitemporal SAR data. *Remote Sensing of Environment*, 108, 209–223, doi: 10.1016/j.rse.2006.11.012.

Martinis, S., Plank, S., & Ćwik, K. 2018. The use of Sentinel-1 time-series data to improve flood monitoring in arid areas. *Remote Sensing*, 10, no. 4: 583, doi: 10.3390/rs10040583.

Mason, D.C., Davenport, I.J., Neal, J.C., Schumann, G.J.-P., & Bates, P.D. 2012. Near real-time flood detection in urban and rural areas using high-resolution synthetic aperture radar images. *IEEE Transactions on Geoscience and Remote Sensing*, 50, 3041–3052, doi: 10.1109/TGRS.2011.2178030.

Mason, D.C., Speck, R., Devereux, B., Schumann, G.J., Member, A., Neal, J.C., & Bates, P.D. 2010. Flood detection in urban areas using TerraSAR-X. *IEEE Transactions on Geoscience and Remote Sensing*, 48, 882–894, doi: 10.1109/TGRS.2009.2029236.

Matgen, P., Schumann, G., Henry, G.-B., Hoffmann, L., & Pfister, L. 2007. Integration of SAR-derived river inundation areas, high-precision topographic data and a river flow model toward near real-time flood management. *International Journal of Applied Earth Observation and Geoinformation*, 9, 247–263, doi: 10.1016/j.jag.2006.03.003.

Milly, P.C.D., Wetherald, R.T., Dunne, K.A., & Delworth, T.L. 2002. Increasing risk of great floods in a changing climate. *Nature*, 415, 514–517, doi: 10.1038/415514a.

Munawar, H.S., Hammad, A.W.A., & Waller, S.T. 2022. Remote sensing methods for flood prediction: A review. *Sensors*, 22, no. 3: 960, doi: 10.3390/s22030960.

Neal, J.C., Bates, P.D., Fewtrell, T.J., Hunter, N.M., Wilson, M.D., & Horritt, M.S. 2009. Distributed whole city water level measurements from the Carlisle 2005 urban flood event and comparison with hydraulic model simulations. *Journal of Hydrology*, 368, 42–55, doi: 10.1016/j.jhydrol.2009.01.026.

Nirupama, N., & Simonovic, S.P. 2007. Increase of flood risk due to urbanisation: A Canadian example. *Natural Hazards*, 40, 25–41, doi: 10.1007/s11069-006-0003-0.

Notti, D., Giordan, D., Caló, F., Pepe, A., Zucca, F., & Galve, J.P. 2018. Potential and limitations of open satellite data for flood mapping. *Remote Sensing*, 10, no. 11: 1673, doi: 10.3390/rs10111673.

Paiva, R.C.D., Buarque, D.C., Collischonn, W., Bonnet, M.-P., Frappart, F., Calmant, S., & Mendes, C.A.B. 2013. Large-scale hydrologic and hydrodynamic modeling of the Amazon River basin. *Water Resource Research*, 49, 1226–1243, doi: 10.1002/wrcr.20067.

Paiva, R.C.D., Collischonn, W., & Tucci, C.E.M. 2011. Large scale hydrologic and hydrodynamic modeling using limited data and a GIS based approach. *Journal of Hydrology*, 406, 170–181, doi: 10.1016/j.jhydrol.2011.06.007.

Papa, F., Prigent, C., Aires, F., Jimenez, C., Rossow, W.B., & Matthews, E. 2010. Interannual variability of surface water extent at the global scale, 1993–2004. *Journal of Geophysical Research*, 115, D12111, doi: 10.1029/2009JD012674.

Popandopulo, G., Illarionova, S., Shadrin, D., Evteeva, K., Sotiriadi, N., & Burnaev, E. 2023. Flood extent and volume estimation using remote sensing data. *Remote Sensing*, 15, no. 18: 4463, doi: 10.3390/rs15184463.

Psomiadis, E., Soulis, K.X., Zoka, M., & Dercas, N. 2019. Synergistic approach of remote sensing and GIS techniques for flash-flood monitoring and damage assessment in Thessaly Plain Area, Greece. *Water*, 11, no. 3: 448, doi: 10.3390/w11030448.

Pulvirenti, L., Chini, M., Pierdicca, N., Guerriero, L., & Ferrazzoli, P. 2011. Flood monitoring using multitemporal COSMO-SkyMed data: Image segmentation and signature interpretation. *Remote Sensing of Environment*, 115, 990–1002, doi: 10.1016/j.rse.2010.12.002.

Pulvirenti, L., Pierdicca, N., Chini, M., Member, S., & Guerriero, L. 2013. Monitoring flood evolution in vegetated areas using COSMO-SkyMed Data : The Tuscany 2009 case study. *IEEE Journal of Selected Topics in Applied Earth Observations and Remote Sensing*, 6, 1807–1816, doi: 10.1109/JSTARS.2012.2219509.

Rahman, M.S., & Di, L. 2017. The state of the art of spaceborne remote sensing in flood management. *Natural Hazards*, 85, 1223–1248, doi: 10.1007/s11069-016-2601-9.

Richards, J.A. 2009. *Remote Sensing with Imaging Radar*, 3rd ed. Berlin: Springer-Verlag, p. 361.

Robinson, I.S. 2004. *Measuring the Oceans from Space: The Principles and Methods of Satellite Oceanography*, Chichester: Springer-Praxis.

Rudorff, C.M., Melack, J.M., & Bates, P.D. 2014a. Flooding dynamics on the lower Amazon floodplain: 1. Hydraulic controls on water elevation, inundation extent, and river-floodplain discharge. *Water Resource Research*, 50, 1–16, doi: 10.1002/2013WR014091.

Rudorff, C.M., Melack, J.M., & Bates, P.D. 2014b. Flooding dynamics on the lower Amazon floodplain: 2. Seasonal and interannual hydrological variability. *Water Resource Research*, 50, 1–15, doi: 10.1002/2013WR014714.

Sadiq, R., Imran, M., & Ofli, F. 2023. Remote sensing for flood mapping and monitoring. In: Singh, A. (ed.) *International Handbook of Disaster Research*. Singapore: Springer, doi: 10.1007/978-981-16-8800-3_178-1.

Sanyal, J., & Lu, X.X. 2004. Application of remote sensing in flood management with special reference to monsoon Asia: A Review. *Natural Hazards*, 33, 283–301, doi: 10.1023/B:NHAZ.0000037035.65105.95.

Schumann, G., Bates, P.D., Horritt, M.S., Matgen, P., & Pappenberger, F. 2009. Progress in integration of remote sensing—derived flood extent and stage data and hydraulic models. *Reviews of Geophysics*, 47, RG4001, doi: 10.1029/2008RG000274.

Schumann, G., Matgen, P., Hoffmann, L., Hostache, R., Pappenberger, F., & Pfister, L. 2007. Deriving distributed roughness values from satellite radar data for flood inundation modelling. *Journal of Hydrology*, 344, 96–111, doi: 10.1016/j.jhydrol.2007.06.024.

Schumann, G.J.-P., Brakenridge, G.R., Kettner, A.J., Kashif, R., & Niebuhr, E. 2018. Assisting flood disaster response with earth observation data and products: A critical assessment. *Remote Sensing*, 10, no. 8: 1230, doi: 10.3390/rs10081230.

Scofield, R.A., & Achutuni, R. 1996. The satellite forecasting funnel approach for predicting flash floods. *Remote Sensing Reviews*, 14, 251–282, doi: 10.1080/02757259609532320.

Sharma, T.P.G., Zhang, J., Koju, U.A., Zhang, S., Bai, Y., & Suwal, M.K. 2019. Review of flood disaster studies in Nepal: A remote sensing perspective. *International Journal of Disaster Risk Reduction*, 34, 18–27, ISSN 2212-4209, doi: 10.1016/j.ijdrr.2018.11.022. (www.sciencedirect.com/science/article/pii/S221242091831029X)

Silva, T.S.F., Costa, M.P.F., & Melack, J.M. 2010. Assessment of two biomass estimation methods for aquatic vegetation growing on the Amazon floodplain. *Aquatic Botany*, 92, 161–167, doi: 10.1016/j.aquabot.2009.10.015.

Simpson, J., Kummerow, C., Tao, W.-K., & Adler, R.F. 1996. On the Tropical Rainfall Measuring Mission (TRMM). *Meteorology Atmospheric Physics*, 60, 19–36, doi: 10.1007/BF01029783.

Smith, E. 2007. The international Global Precipitation Measurement (GPM) program and mission: An overview. In: Levizzani, V., & Turk, F.J. (ed.) *Measuring Precipitation from Space: URAINSAT and the Future*. Berlin: Springer-Verlag, pp. 611–653.

Sorooshian, S., Hsu, K.-L., Gao, X., Gupta, H., Imam, B., & Braithwaite, D. 2000. Evaluation of PERSIANN system satellite-based estimates of tropical rainfall. *Bulletin of the American Meteorological Society*, 81, 2035–2046, doi: 10.1175/1520-0477(2000)081.

Tong, X., Luo, X., Liu, S., Xie, H., Chao, W., Liu, S., Liu, S., Makhinov, A.N., Makhinova, A.F., & Jiang, Y. 2018. An approach for flood monitoring by the combined use of Landsat 8 optical imagery and COSMO-SkyMed radar imagery. *ISPRS Journal of Photogrammetry and Remote Sensing*, 136, 144–153, ISSN 0924-2716, doi: 10.1016/j.isprsjprs.2017.11.006. (www.sciencedirect.com/science/article/pii/S0924271616304142)

Townsend, P.A. 2002. Estimating forest structure in wetlands using multitemporal SAR. *Remote Sensing of Environment*, 79, 288–304, doi: 10.1016/S0034-4257(01)00280-2.

Tralli, D.M., Blom, R.G., Zlotnicki, V., Donnellan, A., & Evans, D.L. 2005. Satellite remote sensing of earthquake, volcano, flood, landslide and coastal inundation hazards. *ISPRS Journal of Photogrammetry and Remote Sensing*, 59, 185–198, doi: 10.1016/j.isprsjprs.2005.02.002.

Trigg, M.A., Wilson, M.D., Bates, P.D., Horritt, M.S., Alsdorf, D.E., Forsberg, B.R., & Vega, M.C. 2009. Amazon flood wave hydraulics. *Journal of Hydrology*, 374, 92–105, doi: 10.1016/j.jhydrol.2009.06.004.

Turk, F.J., & Miller, S.D. 2005. Toward improved characterization of remotely sensed precipitation regimes with MODIS/AMSR-E blended data techniques. *IEEE Transactions on Geoscience and Remote Sensing*, 43, 1059–1069, doi: 10.1109/TGRS.2004.841627.

Van Ree, C.C.D.F., Van, M.A., Heilemann, K., Morris, M.W., Royet, P., & Zevenbergen, C. 2011. FloodProBE: Technologies for improved safety of the built environment in relation to flood events. *Environmental Science and Policy*, 14, 874–883, doi: 10.1016/j.envsci.2011.03.010.

Vicente, G.A., Davenport, J.C., & Scofield, R.A. 2002. The role of orographic and parallax corrections on real time high resolution satellite estimation. *International Journal of Remote Sensing*, 23, 221–230, doi: 10.1080/01431160010006935.

Vicente, G.A., Scofield, R.A., & Menzel, W.P. 1998. The operational GOES infrared rainfall estimation technique. *Bulletin of American Meteorology Society*, 79, 1883–1898, doi: 10.1175/1520-0477(1998)079.

Wang, X., & Xie, H. 2018. A review on applications of remote sensing and Geographic Information Systems (GIS) in water resources and flood risk management. *Water*, 10, no. 5: 608, doi: 10.3390/w10050608.

Wang, Y. 2003. Seasonal change in the extent of inundation on floodplains detected by JERS-1 Synthetic Aperture Radar data. *International Journal of Remote Sensing*, 25, 2497–2508, doi: 10.1080/01431160310001619562

Wang, Y., Hess, L.L., Filoso, S., & Melack, J.M. 1995. Understanding the radar backscattering from flooded and nonflooded Amazonian forests: Results from canopy backscatter modeling. *Remote Sensing of Environment*, 54, 324–332, doi: 10.1016/0034-4257(95)00140-9.

Weng, Q. 2001. Modeling urban growth effects on surface runoff with the integration of remote sensing and GIS. *Environmental Management*, 28, 737–748, doi: 10.1007/s002670010258.

Wieland, M., & Martinis, S. 2019. A modular processing chain for automated flood monitoring from multispectral satellite data. *Remote Sensing*, 11, no. 19: 2330, doi: 10.3390/rs11192330.

Woodhouse, I.H. 2006. *Introduction to Microwave Remote Sensing*. Boca Raton: CRC, p. 370.

World Meteorological Organization (WMO). 2011. *Manual on Flood Forecasting and Warning*, WMO—Nº 1072.

Wu, H., Adler, R.F., Tian, Y., Huffman, G.J., Li, H., & Wang, J. 2014. Real-time global flood estimation using satellite-based precipitation and a coupled land surface and routing model. *Water Resources Research*, 50, 2693–2717, doi: 10.1002/2013WR014710.

Zhang, S., Na, X., Kong, B., Wang, Z., Jiang, H., Yu, H., & Dale, P. 2009. Identifying wetland change in China's Sanjiang Plain using remote sensing. *Wetlands*, 29, 302–313, doi: 10.1672/08-04.1.

5 Flood Studies Using Synthetic Aperture Radar Data

Sandro Martinis, Claudia Kuenzer, and André Twele

ACRONYMS AND DEFINITIONS

ACMs	Active Contour Models
ALOS	Advanced Land Observing Satellite
ANN	Artificial Neural Networks
ASAR	Advanced Synthetic Aperture Radar on Board ENVISAT
ASTER	Advanced Spaceborne Thermal Emission and Reflection Radiometer
CNES	National Centre of Space Research
COSMO-SkyMed	Constellation of small Satellites for Mediterranean basin Observation
CSK	COSMO-SkyMed constellation
DEMs	Digital Elevation Models
DLR	German Aerospace Center
EEC	Enhanced Ellipsoid Corrected
ENVISAT	Environmental Satellite
EO	Earth Observation
EOC	Earth Observation Center
ERS	European Remote Sensing Satellites
ESA	European Space Agency
FNEA	Fractal Net Evolution Approach
GEC	Ground Ellipsoid Corrected
GIM	Geocoded Incidence Angle Mask
GIS	Geographic Information Systems
GUI	Graphical User Interface
HAND	Height Above Nearest Drainage Index
JERS	Japanese Earth Resources Satellite
LAI	Leaf Area Index
LiDAR	Light Detection and Ranging
ML	Machine Learning
MMU	Minimum Mapping Unit
MRF	Markov random field
OGC	Open Geospatial Consortium
PALSAR	Phased Array type L-band Synthetic Aperture Radar
RADARSAT	RADAR Satellite
RISAT	Radar Imaging Satellite
SAR	Synthetic Aperture Radar
SEASAT	First satellite designed for remote sensing of the Earth's oceans with synthetic aperture radar (SAR)
SIR	Spaceborne Imaging Radar
SOMs	Self-Organizing Maps
SRTM	Shuttle Radar Topography Mission

SWBD	SRTM Water Body Mask
TerraSAR-X	A radar Earth observation satellite, with its phased array synthetic aperture radar
TFS	TerraSAR-X Flood Service
VM	Virtual Machine
WAM	Water Indication Mask
WPS	Web Processing Service
WSM	Wide Swath Mode

5.1 INTRODUCTION

The demand for crisis information on natural disasters, humanitarian emergency situations, and civil security issues has substantially increased worldwide in recent years. The use of Earth observation (EO) data in disaster management is essential to provide large-scale crisis information. Especially in the preparedness, emergency, and reconstruction stage of the disaster cycle remote sensing data have been proven to be indispensable for various national and international initiatives. For this reason, the European Space Agency (ESA) and the National Centre of Space Research (CNES) of France initiated the International Charter "Space and Major Disasters" in 1999 (www.disasterscharter.org). This consortium of space agencies and satellite data providers aims at providing a unified system of rapid satellite data acquisition and delivery in case of major natural or man-made disasters. A rising awareness of satellite-based crisis information has led to an increase in requests to corresponding value adders to support civil-protection and relief organizations with disaster-related mapping and analysis. Examples for value adders are: Sertit (Service Régional de Traitement d'Image et de Télédétection), ZKI (Center for Satellite Based Crisis Information), ITHACA (Information Technology for Humanitarian Assistance, Cooperation and Action), e-Geos, and UNOSAT (United Nations Institute for Training and Research Operational Satellite Applications Programme).

Flood is not only one of the most widespread natural disasters, which regularly causes large numbers of casualties with rising economic loss, extensive homelessness, and disaster-induced disease, but it is also the most frequent disaster type. According to statistics of the International Charter "Space and Major Disaster" ~52.3% of the total number of activations between the years 1999 and 2013 are related to floods.

But not only in the context of disaster situations does knowledge on flood extent play a crucial role. Many places on Earth are subject to water-level fluctuations. Especially in tropical and subtropical regimes, large areas of natural and inhabited sphere are frequently inundated due to monsoon-driven rainy seasons. Extreme changes in inundation extent, inundation depth, as well as intra-annual inundation frequency define phenolgical patterns, animal migration routes, and last but not least human living space, including future development plans for the expansion of infrastructure. Some selected examples for regions with extreme variations in inundation are the large inland lakes of China (e.g., Poyang Lake, Dongting Lake), which frequently double in their extent during flood seasons; the Tonle Sap Lake—the largest freshwater lake in Southeast Asia—in Cambodia, with annual variations strongly related to the flood pulse of the Mekong; the Okavango inland delta in Botswana, Africa, which—when inundated via annual flood pulse and rainfall—is a major biodiversity hot spot on the African continent; or the Mekong Delta in Vietnam, where the annual rainy season brings with it "the beautiful flood" that people are used to and which enables them to fish and to irrigate their paddy rice fields and orchards. Thus, not every situation of flood or inundation is automatically disastrous or a natural hazard. As EO delivers a neutral representation of what is happening on the ground, the categorization into an unwanted or even catastrophic flood, an annual river pulse-related flood that people are used to, or simply a short-term inundation event has to be undertaken by an image analyst.

What all floods and inundation situations have in common is that they often cover large regions that are difficult to access from the ground. Spaceborne remote sensing data are a well-suited information source to obtain a synoptic view about large-scale flood situations and their spatio-temporal evolution in a time- and cost-efficient manner. This is especially valid in regions where hydrological information is difficult to obtain due to inaccessibility or sparse distribution of gauging stations.

Optical satellite sensors have successfully been used in the past to detect flood areas (e.g., Surampudi and Kumar, 2023, Aziz et al., 2020, Rahman and Thakur, 2018, Blasco et al., 1992, Smith, 1997, Wang et al., 2002, Peinado et al., 2003, Van der Sande et al., 2003, Ahtonen et al., 2004, Brakenridge and Anderson, 2005, Ottinger et al., 2013). A detailed review can be found in Marcus and Fonstad (2008). If available, optical data are the preferred information source for flood and inundation mapping due to their straightforward interpretability and rich information content. However, as flood events often occur during long-lasting periods of precipitation and persistent cloud cover, a systematic monitoring by optical imaging instruments is usually impossible. This fact drastically decreases the regular utilization of spaceborne optical sensors in an operational rapid mapping context. It is further a particular obstacle in small- to medium-sized watersheds where inundations often recede before meteorological conditions improve (Hitouri et al., 2024, Mason et al., 2021, Schumann et al., 2007).

The use of the microwave region (1 mm–1 m) of the electromagnetic spectrum offers some clear advantages compared to sensors operating in the visible, infrared, or thermal range. Being an active monostatic instrument, and therefore providing its own source of illumination in the microwave range, synthetic aperture radar (SAR) is characterized by nearly all-weather/day-night imaging capabilities, as the emitted radar signal is able to penetrate clouds and the imaging process is independent from solar radiation (Halder and Bose, 2024, Mason et al., 2021, Natsuaki and Hiroto, 2020, Shen et al., 2019a, Ramanuja, 2018). Thus, principally all acquired data sets can be used for flood detection. SAR sensors in recent years underwent a striking improvement in spatial and temporal resolution and are therefore an ideal choice for near real-time assessments in emergency situations. An overview for spatial resolution categories of remote sensing sensors is given in Table 5.1.

SAR-derived flood extent maps can be an important information source for effective flood disaster management by supporting humanitarian relief organizations and decision-makers (Voigt et al., 2007). Furthermore, such maps provide valuable distributed calibration and validation data for hydraulic models of river flow processes (e.g., Surampudi and Kumar, 2023, Mason et al., 2018, Natsuaki and Hiroto, 2020, Ramanuja, 2018, Bates et al., 1997, Horritt, 2000, Aronica et al., 2002, Hunter et al., 2005, Horritt, 2006, Pappenberger et al., 2007, Schumann et al., 2009, Hostache et al., 2009, Matgen et al., 2010) and support the derivation of spatially accurate hazard maps in terms of flood prevention activities, insurance risk management, and spatial planning (e.g., De Moel et al., 2009).

Over the last decades, spaceborne SAR systems (Figure 5.1) have increasingly been used for flood extent mapping. While past and current SAR satellite and space shuttle radar missions with spatial

TABLE 5.1
Categories for Spatial Resolution of Remote Sensing Sensors (European Commission 2011)

Category	Acronym	Resolution [m]
Very high resolution 1	VHR1	≤ 1
Very high resolution 2	VHR2	> 1–≤ 4
High resolution 1	HR1	> 4–≤ 10
High resolution 2	HR2	> 10–≤ 30
Medium resolution 1	MR1	> 30–≤ 100
Medium resolution 2	MR2	> 100–≤ 300
Low resolution	LR	> 300

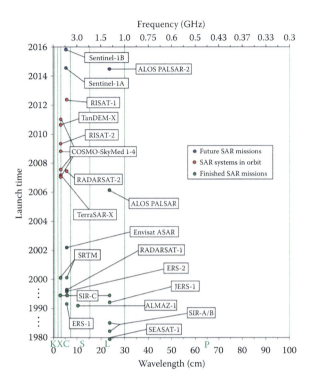

FIGURE 5.1 Launch of past and future civil spaceborne SAR missions between the years 1978 and 2016 in dependence of the system's wavelength. Corresponding frequencies and commonly used spectral bands are also illustrated (Source: modified based on Martinis, 2010).

resolutions of the categories HR2 to MR1 (Table 5.1) have a proven track record for large-scale flood and inundation mapping in the X- (SIR-C/X-SAR, SRTM), C- (ERS-1/2 AMI, Envisat ASAR, RADARSAT-1, RISAT-1, SIR-C/X-SAR), and L-band (SEASAT-1, JERS-1, ALOS PALSAR, SIR-A/B/C/X-SAR) domain, their capability for deriving flood parameters in complex and small-scale scenarios is limited.

Since 2007 the successful launch of the European platforms TerraSAR-X/TanDEM-X and the COSMO-SkyMed constellation (CSK), consisting of four satellites, marks a new generation of civil X-band SAR systems suitable for flood mapping purposes (Berezowski et al., 2020, Quang et al., 2019, Landuyt et al., 2019). These satellites provide data up to the 0.24 m spatial resolution (TerraSAR-X Staring Spotlight mode), permitting an operational derivation of detailed hydrological parameters from space during rapid mapping activities. The potential of these data has been demonstrated by several studies to support flood emergency situations (e.g., Aziz et al., 2020, Berezowski et al., 2020, Gebremichael et al., 2020, Shen et al., 2019a, Shen et al., 2019b, Giustarini et al., 2012, Kuenzer et al., 2013a, Kuenzer et al., 2013b, Martinis et al., 2009, 2011, 2013, 2014, Martinis and Twele, 2010, Mason et al., 2010, Matgen et al., 2011, Pulvirenti et al., 2011, 2012, Pierdicca et al., 2013, Schumann et al., 2010).

The upcoming European Space Agency's satellite mission Sentinel-1, a constellation of two polar orbiting C-band SAR sensors, will enable a systematic large-scale flood monitoring with a spatial resolution of 5 x 20 m in the standard interferometric wide swath mode and a high temporal resolution of six days. Sentinel-1 is designed to operate in a pre-programmed, conflict-free mode, which ensures a consistent long-term data archive for flood mapping purposes (Torres et al., 2012). An overview of the key characteristics of current, past, and planned future spaceborne SAR systems is given in Table 5.2.

TABLE 5.2

Key Characteristics of Current, Past, and Planned Future Civil Spaceborne SAR Missions (based on Lillesand et al., 2004, eoPortal Directory, 2014)

SAR System	Band	Polarization	Look Angle [°]	Swath [km]	Resolution [m]
ALMAZ-1	S	HH	20–70	350	10–30
ALOS PALSAR-1	L	HH, VV, HV, VH	10–51	40–350	6.25–100
ALOS PALSAR-2	L	HH, VV, HV, VH	8–70	25–350	1–100
COSMO-SkyMED 1–4	X	HH, VV, HV, VH	20–59.5	10–200	1–100
Envisat ASAR	C	HH, VV, HV, VH	14–45	58–405	30–1000
ERS-1/2 AMI	C	VV	23	100	30
JERS-1	C	HH	35	75	18
RADARSAT-1	C	HH	10–60	45–500	8–100
RADARSAT-2	C	HH, VV, HV, VH	10–60	10–500	3–100
RISAT-1	C	HH, VV, HV, VH	12–55	10–225	1–50
RISAT-2	X	HH, VV, HV, VH	20–45	10–50	1–8
SEASAT-1	L	HH	20–26	100	25
Sentinel-1A/B	C	HH, VV, HV, VH	20–45	20–400	5–100
SIR-A	L	HH	47–53	40	40
SIR-B	L	HH	15–60	10–60	15–45
SIR-C	X, C, L	HH, VV, HV, VH	15–60	15–90	15–45
SRTM	X, C	HH, VV	45–52	50–250	30, 90
TanDEM-X	X	HH, VV, HV, VH	15–60	5–200	0.24–40
TerraSAR-X	X	HH, VV, HV, VH	15–60	5–200	0.24–40

This chapter aims to give an overview of the strengths and limitations of spaceborne SAR to monitor flood and inundation events. First, the physical basics of the interaction of the radar signal with water surfaces under different conditions, as well as the difficulties that may arise in detecting water using SAR data are described. An overview about the state of the art concerning SAR-based water detection is given. Finally, three operational algorithms from recent literature—all developed by the Earth Observation Center (EOC) of the German Aerospace Center (DLR)—are presented, which are targeted at different application domains: an object-based algorithm optimized for semi-automatic water detection during flood rapid mapping activities by an active image interpreter (Rapid Mapping of Flooding—RaMaFlood), an automatic and open source distributable software tool enabling continuous water monitoring from TerraSAR-X and Envisat ASAR on a local to global scale (Water Mask Processor—WaMaPro), and a fully automatic processing chain for flood mapping based on TerraSAR-X data on a global scale (TerraSAR-X Flood Processor). These different tools all have their own advantages and limitations and will be presented in detail. Furthermore, current trends and open research gaps in SAR-based flood mapping will be elucidated and discussed (Mason et al., 2021, Natsuaki and Hiroto, 2020, Lin et al., 2019).

5.2 INTERACTION OF SAR SIGNAL AND WATER BODIES

This section describes the physical basics of the interaction of the SAR signal with water bodies under various environmental conditions (smooth and rough open water surfaces, partially submerged vegetation areas, urban areas) and addresses the difficulties and limitations of using SAR in water mapping and flood detection (Hitouri et al., 2024, Aziz et al., 2020, Nemni et al., 2020, Zeng et al., 2020).

The interaction between actively emitted microwaves and water bodies depends both on environmental factors and acquisition parameters of the satellite system. In principal, the detectability of water in SAR imagery is controlled by the contrast between water areas and the surrounding land, which is highly influenced by surface roughness characteristics and the system-specific parameters wavelength λ, incidence angle θ_i, and polarization. The following reflection and scattering types can be observed and are schematically illustrated in Figure 5.2: specular reflection, corner reflection, diffuse surface scattering, diffuse volume scattering, and Bragg scattering. These effects occur when the radar signal interacts with smooth and rough open water surfaces, partially submerged flooded vegetation, or flooded urban settlements. Factors that lead to misclassifications of the flood extent are summarized in Table 5.3 and partially visualized in Figure 5.3. The preferred radar system parameters for monitoring floods in dependence of different environmental parameters are listed in Table 5.4.

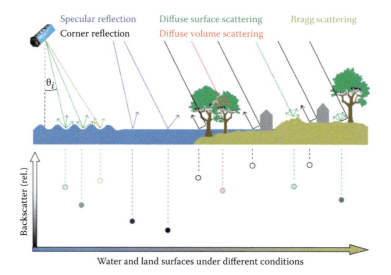

FIGURE 5.2 Scattering mechanisms of water and land surfaces under different environmental conditions as well as specular and diffuse components of surface-scattered radiation as a function of SAR incidence angle and surface roughness (Source: Martinis, 2010).

TABLE 5.3
Factors Leading to Misclassification of Flooding in SAR Data as Well as Their Occurrence and Impact on the Flood Classification Result (Feature Range: High +++; Medium ++; Low +).

Overestimation of Flooding		Underestimation of Flooding	
Factor	Occurrence/Impact	Factor	Occurrence/Impact
Shadowing effects behind vertical objects (e.g., vegetation, topography, anthropogenic structures)	+++	Volume scattering of partially submerged vegetation and water surfaces completely covered by vegetation	+++
Smooth natural surface features (e.g., sand dunes, salt and clay pans, bare ground)	+++	Double bounce scattering of partially submerged vegetation	++
Smooth anthropogenic features (e.g., streets, airstrips)	++	Anthropogenic features on the water surface (e.g., ships, debris)	+
Heavy rain cells	+	Roughening of the water surface by wind, heavy rain, or high flow velocity	+
		Layover effects on vertical objects (e.g., topography, urban structures, vegetation)	+

Flood Studies Using Synthetic Aperture Radar Data 141

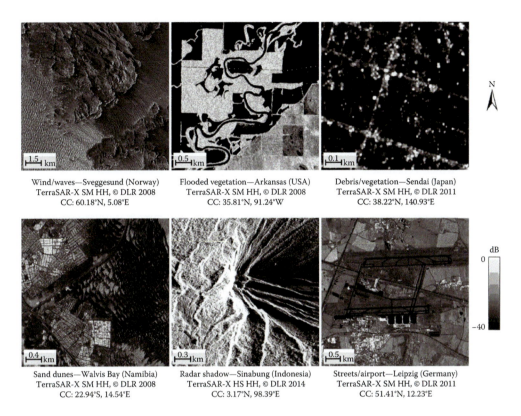

FIGURE 5.3 Examples of SAR scenes [dB] with potential features leading to under- (upper row) and over-estimations (lower row) in water and flood mapping.

TABLE 5.4
Preferred Radar System Parameters for Observing Inundations in Dependence of Environmental Conditions

Flood Type	Band	Incidence Angle	Polarization	Reason
Smooth water	X	shallow	HH	Higher contrast between non-water/water surfaces with decreasing wavelength
Rough water	L	shallow	HH	Lower sensitivity of the SAR signal to roughened water surfaces with increasing wavelength
Flooded dense vegetation	L	moderate/ steep	HH/VV	Increased probability of double bounce effects; enhanced contrast between non-flooded/flooded vegetation
Flooded short/ sparse vegetation	X/C	moderate/ steep	HH/VV	Increased probability of double bounce effects; enhanced contrast between non-flooded/flooded vegetation
Urban areas	X/C	steep	HH	Reduced shadowing effects; enhanced contrast between non-water/water areas

5.2.1 SMOOTH OPEN WATER

Surface roughness, which determines the angular distribution of surface scattering, is the main environmental factor affecting backscattering of the SAR signal, whereas the dielectric properties of a target control the penetration depth and therefore the intensity of the returned energy. Generally, an increasing moisture content of objects results in enhanced reflectivity and stronger surface scattering (Ulaby et al., 1982).

The ideal case in detecting the extent of a water body is that the water surface is smoother than the surrounding land with respect to wavelength and incident angle of the transmitted pulse. A water surface that is smooth in relation to the surrounding dry land surface with respect to wavelength and incident angle of the SAR radiation facilitates the extraction of the water extent (Mason et al., 2021, Mason et al., 2018, Nemni et al., 2020). An open water body may be simplistically modeled as a perfectly smooth boundary separating two semi-infinite media of high dielectric constant (Ulaby et al., 1982), which acts as a specular reflector directing the incident microwave signal away from a side-looking SAR sensor (e.g., Surampudi and Kumar, 2023, Gebremichael et al., 2020, Natsuaki and Hiroto, 2020, Smith, 1997, Horritt et al., 2003, Mason et al., 2007). Due to a very low signal return, open smooth water surfaces appear relatively dark in the data. These regions contrast to surrounding non-water areas of enhanced roughness, which are characterized by increased diffuse surface scattering. According to the Rayleigh criterion, the contrast between water and other land cover classes rises with increasing local incidence angle θ_{loc} (e.g., Ulaby et al., 1982):

$$\sigma_{rms} = \frac{\lambda}{8cos\theta_{loc}} \quad (1)$$

where σ_{rms} is the root mean square height of the surface variations.

With decreasing system wavelength the sensitivity of a smooth water surface to diffuse scattering increases. However, as the number of possible objects on the land that might appear smooth and have a similar backscatter to water is reduced at longer wavelengths, a higher contrast ratio between water and the land areas occurs at higher system frequencies (Drake and Shuchman, 1974). Consequently, water monitoring using X-band SAR appears to be more suitable than using longer wavelengths, e.g., in the C- and L-band domain. Imaging at low incidence angles also means an increase in the occurrence of radar shadowing effects (Lewis, 1998). Radar shadow mainly occurs behind steep vertical features or slopes when the radar beam is no longer able to illuminate the ground surface (Figure 5.3). The increasing incidence angle from near to far range is accompanied by a higher frequency of shadow effects. In high-resolution SAR data shadowing occurs also behind single vertical objects such as anthropogenic structures (e.g., buildings) and natural features (e.g., vegetation). These areas of low radar backscatter are easily mixed up with smooth open water areas and lead to an overestimation of the water extent. These errors also occur at look-alike areas of low surface roughness, such as sand dunes (Figure 5.3), bare ground, agricultural crop land, airport runways (Figure 5.3), and streets (Figure 5.3). A further phenomenon primarily occurring in X-band data are cloud shadows, which are generated by an attenuation of the traversing signal due to hydrometeors in rain cells (Danklmayer et al., 2009).

In contrast, imaging with steep incidence angles increases the probability of radar layover, which can be observed if the incidence angle is smaller than the slope of the object facing the sensor. This phenomenon is related to information loss, which is particularly obstructive in identifying narrow water bodies bordered by high trees, banks (Henderson, 1987), as well as between buildings. Anthropogenic features on the water surface such as ships and debris (Figure 5.3) generally lead to underestimations of the water extent.

Also, the type of polarization plays an important role in detecting open water bodies, which describes the restriction of electromagnetic waves to a single plane perpendicular to the direction of propagation of the SAR signal. Polarimetric SAR systems transmit either in a horizontal (H) or vertical (V) plane, which also can also be received horizontally or vertically. Thus, there can be two possibilities of like-polarization (HH, VV) and cross-polarization (HV, VH). Generally, HH polarization provides the best discrimination between water and non-water terrain (e.g., Ahtonen et al., 2004, Henry et al., 2006, Schumann et al., 2007). This is caused by a low scattering of the horizontal component of the signal from the smooth open water surface. An increase in surface roughness reduces the ability to discriminate between water and land, comparably more using VV

Flood Studies Using Synthetic Aperture Radar Data

than using HH polarization. Over smooth water surfaces, like-polarization offers enhanced class separability in comparison to cross-polarization (e.g., Martinis, 2010). Several studies showed the superiority of cross-polarization HV (Horritt et al., 2003, Henry et al., 2006) and VH (Schumann et al., 2007) over like-polarization VV in terms of a roughened water surface, given the fact that VV polarized electromagnetic waves are more sensitive to ripples and waves.

5.2.2 Rough Open Water

The influence of wind and heavy rain leads to the appearance of small perturbations (ripples) (Lewis, 1998) in the scale of millimeters to centimeters and longer waves with wavelength in the order of meters and kilometers on water surfaces. An increasing roughness causes a higher backscatter of the SAR signal with similar intensities to the surrounding non-water regions (Figure 5.2). Therefore, the existence of roughened water surfaces potentially causes an underestimation of the water extent in SAR-based water mapping, which depends on the size of the roughened water surface within a SAR scene (Halder and Bose, 2024, Mason et al., 2018, Nemni et al., 2020).

Bragg scattering occurs for slightly rough water surfaces with tiny capillary waves and short gravity waves at incidence angles > 30° (Ulaby et al., 1982). It is a special case of scattering and can be expressed by the Bragg equation, which describes the relationship between the wavelengths of periodically spaced surface patterns. If the scatterer positions are oriented in such a way that they have geometric structures aligned with the phase fronts of the illumination and if they are spaced periodically in range, the backscattering strongly increases through constructive inference at certain incidence angles (Raney, 1998).

The larger a water body, the more sensible it is for the formation of waves. Narrow rivers enclosed by riparian vegetation are rarely affected by wind, whereas on oceans roughness structures appear more frequently. Generally, the visible roughness structure of roughened inland water bodies does not show regular wave patterns. Regular patterns are mainly observable on ocean surfaces (Figure 5.3), whereas irregular patterns occur more frequently on the surface of large inland water bodies such as lakes or large inundated areas—e.g., rice paddy land in Southeast Asia or other monsoon-influenced regions.

5.2.3 Partially Submerged Vegetation

In addition to the roughening of a water surface, partially submerged vegetation may cause a backscatter increase over water bodies (Figure 5.3). Microwaves have the capability to penetrate into media. Therefore, in comparison to optical sensors SAR offers the unique opportunity to detect—to a certain extent—inundation beneath vegetation (Hitouri et al., 2024, Berezowski et al., 2020, Zeng et al., 2020). This is enabled by multiple-bounce effects: the penetrated radar pulse is backscattered from the horizontal water surface and lower sections of the vegetation (trunks and branches). This results in an increased signal return (e.g., Richards et al., 1987, Townsend, 2001, Hong et al., 2010) in comparison to non-flood conditions (Figure 5.2) as diffuse scattering on the dry ground reduces the corner reflection effect.

However, the signal return from partially submerged vegetation is very complex and strongly depends on system parameters (e.g., wavelength, incidence angle, and polarization) and environmental parameters (canopy type, structure, and density). Theoretical scattering models can be used to describe the interaction of these parameters (e.g., Sundaram et al., 2023, Berezowski et al., 2020, Gebremichael et al., 2020, Lin et al., 2019, Ormsby et al., 1985, Richards et al., 1987, Wang et al., 1995, Kasischke and Bourgeau-Chavez, 1997).

According to Kasischke and Bourgeau-Chavez (1997) and Townsend (2002), the backscatter coefficient $\sigma_{0,w}$ of wetlands dominated by woody vegetation such as shrubs and trees can be described by

$$\sigma_{0,w} = \sigma_{0,c} + \tau_c^2 \tau_t^2 \left(\sigma_{0,s} + \sigma_{0,t} + \sigma_{0,d} + \sigma_{0,m} \right) \tag{2}$$

where $\sigma_{0,c}$ is the backscatter coefficient of the vegetation canopy, τ_c the transmission coefficient of the vegetation canopy, τ_t the attenuation of the SAR signal by the tree trunks, $\sigma_{0,s}$ the backscatter from the ground surface, $\sigma_{0,t}$ the direct backscatter from the tree trunks, $\sigma_{0,d}$ the double-bounce scattering between the trunks and the water surface, and $\sigma_{0,m}$ the backscatter from multi-path scattering between the ground surface and the canopy (Figure 5.4).

With very dense covering vegetation canopies (e.g., a dense mangrove forest canopy above inundated areas) water surface might also remain undetected if the SAR pulse does not reach the water surface and is caught in volume scattering within the canopy.

Next to water surfaces covered by forest canopies, a very common and special case is the backscatter characteristics of paddy rice—a land use class extensively distributed on Earth (Gebremichael et al., 2020, Landuyt et al., 2019, Kuenzer and Knauer, 2013). Here, vegetation emerges sub-aquatically (very low backscatter of a typical water surface) and then reaches the water surface so that rice plant components influence the return pulse (increase in backscatter). During the ripening phase backscatter then decreases again due to plant geometry (Kuenzer and Knauer, 2013).

Floating aquatic plants such as water hyacinths are a major problem in water mapping as the plant structure increases the radar signal in comparison to open water areas. In cases where the radar backscatter over water hyacinths is higher than the surrounding non-water surfacesclassification accuracies can be significantly increased (e.g., Martinis and Twele, 2010).

Generally, the capability of microwaves to penetrate into vegetation canopy increases with the system's wavelength. L-band SAR sensors have proven to be very effective at detecting flooding in forests (e.g., Mason et al., 2021, Mason et al., 2018, Nemni et al., 2020, Shen et al., 2019a, Shen et al., 2019b, Ormsby et al., 1985, Richards et al., 1987, Hess et al., 1990, Hess and Melack, 1994, Hess et al., 1995, Townsend and Walsh, 1998, Hess et al., 2003). In these wavelengths, the double-bounce trunk-ground signal interactions generate bright signatures in the data (Richards et al., 1987). In C-band and especially X-band, canopy attenuation, volume, and surface scattering from the top layer of the forest canopy is usually higher (Richards et al., 1987). This is related to a decreased backscatter ratio between forests with dry and flooded conditions.

Some studies state that also C-band SAR data can be used to map inundation beneath selected floodplain forest canopies (Townsend and Walsh, 1998, Townsend, 2001, 2002, Lang et al., 2008). A decrease in leaf area index (LAI) increases the transmissivity of the crown layer (Townsend, 2001, 2002) and therefore increases the amount of microwave energy reaching the forest floor (Lang et al., 2008). Therefore, higher classification accuracies can generally be derived during leaf-off conditions (Townsend, 2001, 2002). For example, Townsend (2001) achieved differences in classification accuracy of the class flood, which is increased more than 17% between leaf-on and leaf-off Radarsat-1 data over forests in the lower Roanoke River floodplain in eastern North Carolina.

Increasing backscatter using C-band SAR over floating aquatic macrophytes and emergent shrubs in floodplain lakes is reported by Alsdorf et al. (2000). Usually, high X-band double-bounce

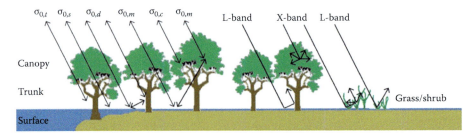

FIGURE 5.4 Conceptual illustration of the major sources of backscatter from partially inundated vegetation and effect of flooded vegetation on X- and L-band SAR (Source: Martinis, 2010, based on Kasischke et al., 1997, Lang et al., 2008, Ormsby et al., 1985).

returns from flooded foliated forests only occur at the edges as the SAR signal is not able to penetrate the vegetation canopy (Henderson, 1995).

Double bouncing may also occur with decreasing wavelength at shorter or sparser vegetation with thin branches and small diameter trunks. Horritt et al. (2003) observed backscatter increases over marshlands in C-band rather than in L-band. The reason for this lies in the capability of the C-band signal to penetrate the sparse canopy and to interact with the water surface and lower parts of the vegetation. This causes an enhanced signal return. Sparse vegetation may be very transparent for the SAR signal at L-band (Figure 5.4). In that case no interaction of the water body and the vegetation occurs. Results reported by Ormsby et al. (1985) and Ramsey (1995) indicate enhanced backscattering in marshland environments in C-band and even in X-band, respectively. Voormansik et al. (2013) used TerraSAR-X data to successfully map flooding in deciduous and coniferous forests of the temperate zone in Estonia during the no-foliage season. In this study an average backscatter increase of 3.2 dB could be stated over mixed forests. The difference in average backscatter during non-flood and flood situations offered values of 6.2 dB and 4.0 dB over deciduous and coniferous forests, respectively. Increased double bounce mechanisms could also be stated over flood surfaces covered by olive groves and deciduous forests in Italy using Cosmo-SkyMed data (Pulvirenti et al., 2012) and covered by grassland and foliated shrubs in Caprivi, Namibia, using TerraSAR-X data (Martinis and Twele, 2010). Therefore, even in X-band there lies a certain potential to map flooding beneath vegetation.

Other properties that need to be considered for flood mapping beneath vegetation are incidence angle and polarization (Halder and Bose, 2024, Hitouri et al., 2024, Aziz et al., 2020, Zeng et al., 2020, Ramanuja, 2018). Several studies have shown that steep incidence angles are better suited for flood mapping in forests than shallow ones (e.g., Richards et al., 1987, Hess et al., 1990, Wang and Imhoff, 1993, Wang et al., 1995, Bourgeau-Chavez et al., 2001, Lang et al., 2008). This generalization can be attributed to a shorter path length of the SAR signal through the canopy, which increases the transmissivity in the crown layer. Thus, more microwave energy is available for ground-trunk interactions. In contrast, shallower incidence angle signals interact more strongly with the canopy, resulting in increased volume scattering (Hess et al., 1990, Lang et al., 2008). For example, Lang et al. (2008) stated a decrease of 2.45 dB between incidence angles of 23.5° and 47° and of 0.62 dB between incidence angles of 23.5° and 43.5° in Radarsat-1 data over flooded forests at the Roanoke River, North Carolina.

Radar systems with multiple polarizations provide more information on inundated vegetation areas than single-polarized SAR (Mason et al., 2021, Natsuaki and Hiroto, 2020, Lin et al., 2019, Hess and Melack, 2003, Horritt et al., 2003, Gebhardt et al., 2012, Kuenzer and Knauer, 2013). Many studies employing multi-polarized data indicate advantages of like-polarization (HH or VV) for separating flooded and non-flooded forests (e.g., Evans et al., 1986, Wu and Sader, 1987, Martinis, 2010). Backscattering is generally very low for cross-polarization (HV or VH), as depolarization does not occur for ideal corner reflectors (Leckie, 1998). According to Wang et al. (1995) and Townsend (2002), the backscatter ratio between flooded and non-flooded forest is higher at HH polarization than at VV polarization.

5.2.4 FLOODING IN URBAN AREAS

In urban settlements the detectability of flooding is strongly reduced in comparison to rural areas. This is a common problem in SAR-based water detection, which is widely discussed in the literature (e.g., Surampudi and Kumar, 2023, Gebremichael et al., 2020, Landuyt et al., 2019, Lin et al., 2019, Rahman and Thakur 2018, Ramanuja, 2018, Giacomelli et al., 1995, Martinis, 2010, Mason et al., 2010, Giustarini et al., 2012, Kuenzer et al., 2013a). Dihedral and trihedral reflection from anthropogenic structures such as buildings, as well as the presence of metal surfaces, leads to enhanced backscatter and strong contributions from side-lobes, which are nearly identical for non-flooded and flooded situations. Also, illumination phenomena which are related to the side-looking geometry of imaging radar systems restrict the ability to detect urban flooding. Frequently, water surfaces might not be visible due to shadowing and layover effects of urban structures. The likelihood to detect flooding in urban areas generally increases with decreasing incidence angle, increasing spatial resolution of the SAR sensor and increasing distance between anthropogenic structures in range direction. However, as non-flooded roads and other smooth man-made structures commonly also appear dark due to specular reflection, these targets can hardly be separated from smooth water surfaces.

5.3 STATE OF THE ART IN SAR-BASED WATER DETECTION

Various approaches have been proposed in the literature for extracting the inundation extent from SAR data (Mason et al., 2021, Berezowski et al., 2020, Zeng et al., 2020, Shen et al., 2019a, Shen et al., 2019b). This section provides an overview of the state of the art regarding SAR-based water mapping algorithms. Using uni-temporal data, these techniques are only able to detect water surfaces. The comparison of the derived water extent with auxiliary reference water masks of normal water-level conditions or with pre-event remote sensing data sets allows the separation between flooded and permanently standing water areas. The same holds true for the analyses of long intra-annual or multi-year time series, which then allows for the extraction of permanently inundated water bodies.

Classification accuracies of water/flood surfaces vary considerably within the literature and rarely exceed 90%. Besides the method and type of SAR data used, the quality of the classification result depends on the complexity of the flood situation covered by the SAR data. In Table 5.5 the strengths and limitations of the image processing approaches commonly used in water/flood

TABLE 5.5
Strengths and Limitations of Methods Commonly Applied for SAR-Based Water/Flood Extent Mapping

Flood/Water Detection Method	Subtype	Strengths	Limitations
Visual interpretation	-	Straightforward applicability, high quality	Very subjective, time-demanding, requires an experienced image interpreter
Thresholding	Manual trial-and-error thresholding; automatic thresholding	Very fast, high potential of automation, moderate complexity, basis for other methods (e.g., change detection, integration of contextual and auxiliary information)	May fail in cases of low contrast between water/non-water surfaces
Change detection	Post-classification comparison; analysis of feature maps; repeat-pass interferometry	Reduction of water look-alikes; improved detection of flooded vegetation areas, separation between flooded and permanent water areas	Availability of reference data of non-flood conditions, possibly high complexity
Contextual classification	Texture based; object based; ACMs; Markov models (spatial/hierarchical)	Improvement of classification accuracy by considering relationships between pixels	Possibly high complexity and high computational demands

detection are summarized. Some of these approaches can be combined (e.g., automatic thresholding of SAR difference images). Further improvements can be achieved by integrating auxiliary data into the classification process (Section 5.3.5).

Flood mapping using SAR data is generally carried out either by visual interpretation or by digital image analysis.

5.3.1 Visual Interpretation

Visual interpretation and manual digitalization of the land/water boundary gives a reasonably accurate assessment of the water extent (Sanyal and Lu, 2003). As a major disadvantage, this approach is very time-consuming, especially in large-scale flood situations. The results are further subjective and hard to reproduce since they can vary according to the experience of the image interpreter.

5.3.2 Thresholding

Digital image processing techniques classify each pixel of the SAR raster data into water and non-water areas using spectral properties as well as contextual and auxiliary information (Mason et al., 2021, Mason et al., 2018, Nemni et al., 2020, Rahman and Thakur, 2018, Ramanuja, 2018).

Thresholding is the most popular image processing techniques in water detection. Commonly, a threshold value is used to label all image elements of the SAR data as water or non-water. Due to its simplicity, this method is computationally very fast and therefore suitable for rapid mapping purposes. The quality of thresholding procedures for detecting water using SAR imagery depends on the contrast between water and non-water areas. In low to moderate roughness conditions of open water surfaces, usually most of the water extent can be extracted when employing this technique. All image elements of the SAR amplitude or intensity data with gray values lower than a defined threshold are assigned to the water class. The thresholding process may fail if increased surface roughness conditions reduce the contrast between water and non-water areas. The occurrence of double bouncing with partially flooded vegetation areas demands the definition of an additional threshold value, which labels all pixels above a certain backscatter value to an additional class, "flooded vegetation."

An adequate threshold can be determined in a supervised manner using visual inspection of the global scene histogram or manual trial-and-error approaches (e.g., Halder and Bose, 2024, Hitouri et al., 2024, Sundaram et al., 2023, Berezowski et al., 2020, Quang et al., 2019, Ramanuja, 2018, Townsend and Walsh, 1998, Townsend, 2001, Brivio et al., 2002, Henry et al., 2006, Matgen et al., 2007, Lang et al., 2008, Gstaiger et al., 2012). Several studies also state that empirically defined default threshold values can be successfully used for detecting smooth water surfaces in case areas of stable environmental conditions are repeatedly being monitored or data are acquired with the same SAR system parameters. In this context Kuenzer et al. (2013a, 2013b) and Gstaiger et al. (2012) performed water extent mapping based on Envisat ASAR Wide Swath Mode (WSM) and TerraSAR-X data in the Mekong Delta. Wendleder et al. (2012) conducted global water body detection based on amplitude and bi-static coherence information of TanDEM-X VV-polarized StripMap data. The combined approach based on amplitude and bi-static coherence information results in a significantly higher accuracy (98.7%) than the watermask solely based on amplitude information (92.5%) for a test site at the river Elbe, in Germany. As the bi-static coherence is only generated in the course of the TanDEM-X mission for the generation of a global DEM in a pre-defined acquisition plan, this information is commonly not available for flood disaster mapping.

Only a few approaches for the automatic extraction of water-related threshold values based on image statistics can be found in the literature (Martinis et al., 2009, Schumann et al., 2010, Matgen et al., 2011, Pulvirenti et al., 2012). However, automatic threshold extraction approaches should be favored for near real-time applications and systematic satellite-based flood monitoring systems. Schumann et al. (2010) compute a threshold value of −8.5 dB from the global gray-level histogram of ASAR wide swath mode data using Otsu's method (Otsu, 1979), which derives a criterion measure to evaluate the between-class variance of water and non-water areas. Matgen et al. (2011) perform thresholding by modeling the flood class using a non-linear fitting algorithm under the gamma distribution assumption. Martinis et al. (2009) present an automatic tile-based thresholding approach that solves the flood detection problem in large-size TerraSAR-X amplitude data even with small *a priori* class probabilities by applying the KI thresholding approach (Kittler and Illingworth, 1986) on selected image tiles, which are likely to represent a bimodal distribution of the classes to be separated. This method is enhanced in robustness and adapted to SAR data radiometrically calibrated to sigma naught in Martinis et al. (2014) and adapted to calibrated Cosmo-SkyMed data by Pulvirenti et al. (2012).

5.3.3 Change Detection

Change detection based on multi-temporal image analysis is an effective tool for delineating varying inundation extent and inundation frequency. In the context of flood mapping, change detection is usually performed by comparing pre-flood reference data with in-flood imagery using post-classification comparison (e.g., Herrera-Cruz and Koudogbo, 2009) as well as by analyzing feature maps such as difference (e.g., Halder and Bose, 2024, Mason et al., 2018, Nemni et al., 2020, Zeng et al., 2020, McMillan et al., 2006, Matgen et al., 2011, Giustarini et al., 2012), normalized difference (Nico et al., 2000, Martinis et al., 2011), ratio (e.g., Rémi and Hervé, 2007), and log ratio data (e.g., Bazi et al., 2005). Change detection is ideally performed using data acquired with the same sensor and with similar system parameters. Change detection may help to enhance the flood mapping result obtained from the analysis of a single flood image by reducing overestimations of inundated areas related to water look-alike areas. Also it allows separating areas permanently covered by water from temporally flooded terrain. Amplitude as well as coherence change detection are applied in the SAR domain. In the amplitude approach, regions are labeled as flooded where the backscatter has considerably decreased in cases of smooth water surfaces or increased in cases of double-bouncing vegetation areas from pre- to post-disaster data. Phase information derived from SAR interferometry also has the potential to be used for flood mapping. Various studies (e.g., Marinelli et al., 1997, Dellepiane et al., 2000) state that repeat-pass SAR interferometry can be employed to identify water as regions of low interferometric phase correlation, which can be separated from dry land of higher coherence. Single-pass interferometry can be used to eliminate temporal decorrelation effects and therefore to enhance the quality of water masks (Wendleder et al., 2012, Warth and Martinis, 2013).

5.3.4 Contextual Classification

Some approaches integrate spatial-contextual information from a local neighborhood within the flood-detection workflow: Ahtonen et al. (2004) present an automatic surface water procedure, which integrates local textural features into the labeling scheme. This method uses a machine learning (ML) classifier trained by unsupervised thresholding of log-mean data. The classification process is performed on a 3D feature space composed of logarithmically transformed occurrence measures within a kernel of size 5 x 5.

A supervised flood mapping algorithm SAR using Kohonen's self-organizing maps (SOMs) (Kohonen, 1995) based on artificial neural networks (ANN) is proposed by Kussul (2008). For considering spatial connections between neighboring pixels, the network is trained in an unsupervised manner using backscatter values from sliding windows in Envisat ASAR, ERS-2, and Radarsat-1 data.

In the past, several methodologies based on region growing have been used in waterline detection. Commonly, seeded regions using semi-automatic or automatic algorithms are dilated according to their statistical properties until stopping criteria are reached (Mason et al., 2021, Gebremichael et al., 2020, Zeng et al., 2020, Rahman and Thakur, 2018, Ramanuja, 2018, Malnes et al., 2002, Matgen et al., 2011, Martinis et al., 2014, Mason et al., 2010, 2012).

In recent years, statistical active contour models (ACMs), so-called snake algorithms, gained in popularity for delineating land/water boundaries in single-polarized SAR data. These sophisticated region growing procedures make use of a dynamic curvilinear contour to iteratively search through the 2D image space until they settle upon object boundaries, driven by an energy function that is attracted to edge points. ACMs have proven useful for converting unconnected or noisy image edges into smooth continuous vector boundaries. Therefore, these algorithms are suitable for segmenting speckle-affected SAR imagery. Based on the study of Horritt (1999), a semi-automatic ACM (Psnake NT) was developed by Horritt (1999). This tool identifies inundated areas as pixels of homogeneous speckle statistics accounting for the gamma-distribution intensity of SAR data. This method was widely applied for river flood delineation in rural areas using SAR data in the HR2 and MR1 resolution domain (e.g., Mason et al., 2018, Nemni et al., 2020, Shen et al., 2019a, De Roo et al., 1999, Horritt et al., 2001, Ahtonen et al., 2004, Matgen et al., 2007, Schumann et al., 2009). Further, it is successfully applied for computing polygonal approximations of rough seawater surfaces (Horritt, 2000). Mason et al. (2007) modified the algorithm in such a way that the snake was conditioned both on SAR backscatter values and on LiDAR digital elevation models (DEMs). Using 3D rather than 2D curvatures, the waterline became smoothly varying in elevation. One disadvantage of Psnake NT is that this algorithm is dependent on significant user input. Several initializations of the contour line by manually set seed vectors are necessary to obtain satisfying results. Further, as Psnake NT belongs to the groups of parametric ACMs that have a rigid topography, additional seeds are necessary to delineate isolated flood regions (Mason et al., 2010). This, however, is critical in high-resolution SAR data of the categories VHR1-HR1, where, in contrast to data of coarser resolution (HR2-LR), the inundation area is commonly separated in multiple isolated flood regions by, e.g., vegetation areas or man-made objects, which prohibit the expansion of the snake. In this case, geometric snake models (e.g., Malladi et al., 1995), which permit topology changes due to flexible level sets to simultaneously detect several water objects, seem to be more suitable. Within this context, a semi-automatic flood detection algorithm based on region-based level sets is proposed by Silveira and Heleno (2009).

A Bayesian segmentation technique to separate land and sea regions in X-band SAR data is proposed by Ferreira and Bioucas-Dias (2008). The class conditional densities are estimated by a finite mixture of Gamma distributions whose parameters are estimated from manually selected training samples. The *a priori* probability of the labels is modeled by a Markov random field (MRF), which promotes local continuity of the classification result given a spatial neighborhood system. The maximum *a posteriori* estimation is performed by using graph cuts (Kolmogorov and Zabih, 2004).

Several studies present object-based classifications for flood mapping purposes (e.g., Sundaram et al., 2023, Berezowski et al., 2020, Quang et al., 2019, Landuyt et al., 2019, Hess et al., 2003, Martinis et al., 2009, 2011, Martinis and Twele, 2010, Herrera-Cruz and Koudogbo, 2009, Mason et al., 2012, Pulvirenti et al., 2012). They are based on the concept that important information necessary for image analysis is not always represented in single pixels but in homogeneous image segments and their mutual relations (Baatz and Schäpe, 1999). Within this context, a hybrid multi-contextual Markov model for unsupervised near real-time flood detection in X-band SAR data has been developed. The Markov model is initialized by an automatic tile-based thresholding procedure (Martinis et al., 2009). Scale-dependent (Martinis et al., 2011) and optional spatio-temporal contextual information (Martinis and Twele, 2010) is integrated into the segment-based classification process by combining causal with noncausal Markov image modeling related to hierarchical directed and planar un-directed irregular graphs, respectively.

5.3.5 INTEGRATION OF AUXILIARY DATA

The integration of auxiliary data sets can significantly support the flood mapping process. Some studies make use of digital topographic information to improve classification results by detecting flooding beneath vegetation or by removing look-alike areas according to simplified hydrological assumptions (e.g., Hitouri et al., 2024, Mason et al., 2018, Nemni et al., 2020, Lin et al., 2019, Wang et al., 2002, Horritt et al., 2003, Mason et al., 2007, Mason et al., 2010, Martinis et al., 2009). Other hydrologically relevant layers, such as the height above nearest drainage index (HAND) (Rennó et al., 2008), can be used to filter out regions where the probability of flood occurrence is low (Westerhoff et al., 2013).

Within the last years fuzzy logic techniques (Zadeh, 1965) have increasingly been used in flood monitoring to combine ambiguous information sources by accounting for their uncertainties as opposed to only relying on crisp data sets. Martinis and Twele (2010) apply fuzzy theory for quantifying the uncertainty in the labeling of each image element in flood possibility masks. The proposed method combines marginal posterior entropy-based confidence maps with spatio-temporal relationships of potentially submerged vegetation to smooth open water areas. A pixel- and object-based fuzzy logic approach for inundation mapping based on Pierdicca et al. (2008) is described in Pulvirenti et al. (2011, 2013). It integrates theoretical SAR scattering models, simplified hydrologic assumptions, and local context in the form of intensity, topographical, and land cover information. Based on the availability of pre-flood scenes the semi-automatic algorithm is able to detect both open water areas and submerged vegetation areas. A fully automatic TerraSAR-X-based flood mapping service, which uses a fuzzy logic–based algorithm combining SAR backscatter information with digital elevation and slope information as well as the size of water bodies for the refinement of the initial thresholding result, is proposed by Martinis et al. (2014). The fuzzy logic–based post-classification process results in an improvement of the flood mask by mainly reducing water look-alike areas in mountainous regions related to radar shadowing. In a TerraSAR-X ScanSAR test data set of Albania/Montenegro this approach enhanced the user accuracy of the class flood from ~20.9% to ~82.4% and the producer accuracy from ~51.7% to ~83.7%.

5.4 CASE STUDIES

In this chapter the use of medium- and high-resolution SAR (HR1-MR1) data for flood monitoring is demonstrated. In three test scenarios three different and recently published operational flood/water detection algorithms are presented in relation to the desired application domain. The first test case describes the application of a semi-automatic, object-based water detection algorithm (RaMaFlood) on multi-temporal TerraSAR-X data in the Caprivi region of Namibia to map both open flood water surface and flooded vegetation areas (Section 5.4.1). This method commonly is used in the context of flood-related rapid mapping activities by an active image interpreter. In the second test case, a time-efficient automatic algorithm for continuous water monitoring (WaMaPro) is presented (Section 5.4.2). The algorithm is applied on a time series of Envisat ASAR WSM and TerraSAR-X data sets acquired with similar acquisition parameters, respectively, in the Mekong Delta in Vietnam. The third case study describes the use of a fully automatic and globally applicable TerraSAR-X flood service (TFS), which is designed to deliver robust results for a broad range of biomes and acquisition conditions (Section 5.4.3). The processing chain is exemplarily applied on TerraSAR-X StripMap (SM) and ScanSAR (SC) scenes acquired during flood situations in Nepal 2008, Germany 2011, and Albania/Montenegro 2013. The main characteristics of these methods are qualitatively compared in Table 5.6.

5.4.1 Semi-automatic Object-Based Flood Detection (RaMaFlood)

In this chapter an object-based water detection algorithm, RaMaFlood, is described (Martinis et al., 2009, 2011). It can be run both automatically and semi-automatically on pre-processed SAR data within the the eCognition Developer software. However, as a graphical user interface (GUI) has been implemented for this method, which allows an image interpreter to control and modify the parameters of each single processing step in a WYSIWYG manner, it is especially

Flood Studies Using Synthetic Aperture Radar Data

TABLE 5.6
Comparison of the State of the Art Flood and Water Detection Algorithms RaMaFlood, WaMaPro, and TFS (Ranking: Very High ++; High +; Medium 0; Low -; Very Low—)

	Semi-automatic Object-Based Water Detection (RaMaFlood)	Automatic Pixel-Based Water Detection (WaMaPro)	Fully Automatic Pixel-Based Flood Detection (TFS)
Accuracy	++	+	+
Look-alike elimination	++	−	+
Automation	o	+	++
Processing time	+	++	++
Multi-sensor capability	++	++	+
Flood possibility mask	−−	−−	++
Open source	−−	++	−−

useful when applied in a semi-automatic way. Using this method, detailed information about the extent of open water bodies can be derived. Also, partially submerged vegetation areas can be extracted semi-automatically by an active image interpreter. This is hardly feasible in a completely automatic way due to the complexity of double-bounce scattering mechanisms. By intersecting the derived water masks with auxiliary reference water masks of normal water-level conditions the separation between flooded and permanently standing water areas is accomplished. The derived crisis information is integrated into map products generated by GIS experts and disseminated to end users. During numerous flood-related rapid mapping activities of DLR's Center for Satellite Based Crisis Information (ZKI), this method has proven its effectiveness for SAR data acquired by sensors in the X-Band (TerraSAR-X/TanDEM-X, COSMO-SkyMed), C-band (Envisat ASAR, Radarsar-1/2), and L-Band (ALOS PALSASR) domain.

5.4.1.1 Methodology

RaMaFlood is experimentally applied to a multi-temporal data set of four TerraSAR-X StripMap scenes covering the evolution of a flood situation over a period of three and a half months in the Zambezi floodplain. The study area is mainly situated in the Caprivi Strip in northeastern Namibia, which is surrounded by Zambia in the north and Botswana in the south (Figure 5.5). The Caprivi Strip is regularly affected by flooding related to heavy seasonal rainfalls.

The SAR data are all acquired in HH polarization with the same beam mode (minimum incidence angle: ~29.4°; maximum incidence angle: ~32.4°) in ascending orbit. The pixel spacing is 2.75 m (see Table 5.7).

Before applying the segmentation and classification step, the data are radiometrically calibrated to sigma naught. Even if this step is optional, it may have the advantage of removing topographic effects within the SAR data. For a reduction of the SAR data–inherent speckle effect, a Gamma-MAP filter (Lopes et al., 1990) of window size 3 × 3 is used. The filtering step also minimizes the statistical overlap between class distributions and, therefore, causes improved class separability.

Within the sequence of the four acquisitions, the evolution of a large-scale flood situation is visible. In comparison to the surrounding dry land, open water surfaces appear dark due to specular reflection of the incident radar signal. In contrast, flooded vegetation causes very distinct and bright signatures. Indeed, X-band SAR has a strongly reduced ability to detect inundation beneath dense vegetation such as forest due to increased canopy attenuation and volume scattering in comparison to the longer C- and L-band signals (e.g., Richards et al., 1987). In this study area, however, the

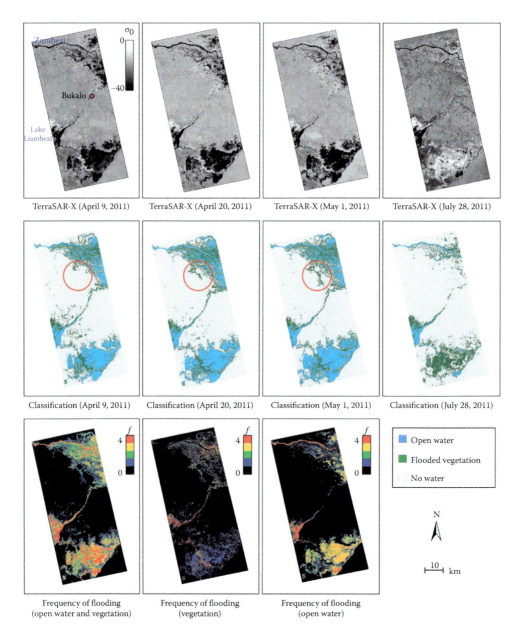

FIGURE 5.5 Time series of TerraSAR-X data (top row) and respective inundation maps (middle row) in Caprivi, Namibia. Flood frequency (*f*) maps (bottom row): total flooding, flooded vegetation areas, and open water surfaces. Coordinates: UL: 17.450°S, 24.245°E; UR: 17.390°S, 24.539°E; LL: 18.173°S, 24.402°E; LR: 18.111°S, 24.702°E.

emergent vegetation is mainly composed of grassland, foliated shrubs, and aquatic plants, whose structure causes a multiple-bounce effect, which leads to an interaction of the penetrated radar pulse with the water surface and lower sections of the vegetation. This phenomenon causes a high signal return.

TABLE 5.7
Acquisition Parameters of Multi-Temporal TerraSAR-X SM Scenes Covering the Caprivi Strip in Northeastern Namibia

Acquisition Time	Beam Mode	Polarization	Pixel Spacing (m)
2011–04–09, 16:42 UTC	SM_007R	HH	2.75
2011–04–20, 16:42 UTC	SM_007R	HH	2.75
2011–05–01, 16:42 UTC	SM_007R	HH	2.75
2011–07–28, 16:42 UTC	SM_007R	HH	2.75

The workflow of the semi-automatic, object-based flood detection approach consists of the following main processing steps (Martinis et al., 2009, 2011):

- Image segmentation
- Thresholding
- Post-classification refinement

The first processing step of the proposed method is the segmentation of the SAR data. Segmentation is the basic step in low-level image processing (Lucchese and Mitra, 1998), in which an image is subdivided into disjoint regions that are uniform with respect to homogeneity criteria such as spectral or textural parameters (Haralick and Sharipo, 1985). In comparison to pixel-based applications, this offers the advantage that besides spectral-related information, also object parameters such as contextual information, texture, and object geometry can be used for improving classification accuracy. Object-based image analysis has constantly gained importance in EO applications during the last decade (Baatz and Schäpe, 1999, Blaschke and Strobl, 2001). This is particularly related to the strongly increased spatial resolution of remote sensing data, which demands image analysis techniques that are specifically adapted to the increased intraclass and decreased interclass variability of images (Bruzzone and Carlin, 2006). In particular, for data of the recently launched high-resolution SAR sensors (TerraSAR-X, COSMO-SkyMed, Radarsat-2), the use of the per-parcel methods appears promising. These data are, in comparison to SAR imagery, of coarser resolution, characterized by higher variances in backscattering properties of different land cover classes due to the reduced mixed-pixel phenomenon and the SAR-intrinsic speckle effect. Therefore, semantic image information is less represented in a single pixel but rather in homogeneous image objects and their mutual relations (Baatz and Schäpe, 1999). The decomposition of the images can be accomplished by several segmentation techniques described in the literature (Haralick and Sharipo, 1985, Zhang, 1996, Carleer et al., 2005). This study is based on the Fractal Net Evolution Approach (FNEA) (Baatz and Schäpe, 1999), which enables a multi-scale representation of the data using a bottom-up region merging approach. The multi-level representation of an image Y can be represented as a connected graph Ψ_L with L levels composed of a set of nodes S. An irregular graph with three levels is generated according to the method proposed in Martinis et al. (2011) to integrate the advantages of small-, medium-, and large-scale objects into the classification process (Martinis et al., 2009). In order to prevent over- and under-segmentation of the data, the graph is automatically generated by modeling the segmentation parameters to decompose the image at each level by a mean number of objects intended by the user. This is accomplished by the procedure described in Martinis et al. (2011): First, several subsets of the SAR image are automatically selected to describe the heterogeneity of the backscatter of the SAR data. Then, a pre-segmentation of the subsets is performed by the FNEA approach. The homogeneity parameter is estimated, which leads to a decomposition of the entire image with average object sizes of the segments at each level which come close to those intended by the user. This is accomplished by generating a database, which contains models describing the relationship between homogeneity parameter and object size according to data of different SAR sensor types and image contents. Finally, this model is selected for creating the whole graph that best fits the pre-segmentation result at each segmentation level. The irregular graph is built with a relative object size of ~50% between adjacent graph levels.

The classification of smooth open water areas is initialized by labeling all image elements with a backscatter value lower than a defined threshold to the class "water." Thresholding algorithms only extract adequate threshold values if the histogram is not uni-modal. Therefore, the capability of approaches to detect an adequate threshold in the histogram depends on the *a priori* probability of the classes to be separated. If, e.g., the spatial extent of the water bodies in large SAR scenes is low, the class distributions cannot be modeled sufficiently. In this study the threshold value is automatically derived using a tile-based thresholding procedure proposed by Martinis et al. (2009, 2014) that solves the flood detection problem in even large-size SAR data with small *a priori* class probabilities. The thresholding approach consists of the following processing steps: image tiling, tile selection, and sub-histogram-based thresholding of a small number of tiles of the entire SAR image.

First, based on the SAR scene, a bi-level quadtree structure is generated. The SAR data are divided into N quadratic non-overlapping sub-images of user-defined size c^2 on level S^+. Each parent object is represented by four quadratic child objects of size $(c/2)^2$ on level S^-. The variable c is empirically defined to 400 pixels. A limited number of tiles are selected out of N according to the probability of the tiles to contain a bi-modal mixture distribution of the classes "open water" and "non-water." This selection step is based on statistical hierarchical relations between parent and child objects in a bi-level quadtree structure. The parametric Kittler and Illingworth minimum error thresholding approach (Kittler and Illingworth, 1986) is used to derive local threshold values using a cost function that is based on statistical parameterization of the sub-histograms of all selected tiles as bi-modal Gaussian mixture distributions. One global threshold position is derived by computing the arithmetic mean of the local thresholds. This is used for initially separating open water surfaces and non-water areas in the SAR data.

For post-classification refinement multi-scale image information is combined with thresholding (Martinis et al., 2009). The initial threshold is first applied to the coarsest segmentation level S_1. Objects on this level contain some variations in the spectral properties of the pixels. Most of the inundation is identified by this step; however, fine tuning is subsequently reached, progressively enforcing the spectral homogeneity constraints of non-flood objects in the neighborhood around open water and flooded vegetation objects.

The preliminary extracted water bodies on S_1 are used as seeds for dilating the water regions based on medium-scale objects on S_2. This is repeated accordingly by performing region growing of small-scale objects on S_3 adjacent to seeds defined by the thresholding process on the S_1 and S_2. Only image elements located in the neighborhood of flood areas identified at level S_{l+1} are scanned. Thus, the risk of detecting flood look-alikes distant from initially labeled flood objects is reduced.

Finally digital elevation models can be integrated into the classification process to improve the quality of the flood masks in a hydrological plausible way (Martinis et al., 2009).

5.4.1.2 Results

By applying the RaMaFlood tool to the TerraSAR-X test data set of Namibia the finest level S_1 is partitioned into an intended mean object size of ~250 m². Accordingly, the mean object size increases to ~500 m² and ~1,000 m² at S_2 and S_3, respectively. Open water areas are derived by automatically extracting the threshold value using a tile-based thresholding procedure. The threshold values are defined by $\tau_1 = -16.2$ dB (2011–04–09), $\tau_1 = -19.4$ dB (2011–04–20), $\tau_1 = -19.0$ dB (2011–05–01), and $\tau_1 = -18.9$ dB (2011–07–28) (Table 5.8). Flooded vegetation areas appear more heterogeneous than

Flood Studies Using Synthetic Aperture Radar Data

TABLE 5.8
Threshold Values Used for Separating Open Water and Non-water Areas (τ_1) as Well as Non-water and Partially Flooded Vegetation Areas (τ_2)

Acquisition Time	τ_1	τ_2
2011–04–09, 16:42 UTC	−16.2 dB	−8.5 dB
2011–04–20, 16:42 UTC	−19.4 dB	−8.5 dB
2011–05–01, 16:42 UTC	−19.0 dB	−9.0 dB
2011–07–28, 16:42 UTC	−18.9 dB	−9.0 dB

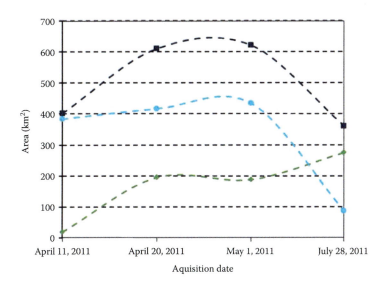

FIGURE 5.6 The evolution of the total flood area, the open water areas, and flooded vegetation areas in a time series of four TerraSAR-X StripMap data acquired over Caprivi, Namibia.

open water areas due to the occurrence of different vegetation types and various vegetation densities, which alter the intensity of the double bounce effect. Therefore, better results are derived by setting the threshold between the class conditional densities of non-water areas and flooded vegetation using empirically derived values: $\tau_2 = -8.5$ dB (2011–04–09), $\tau_2 = -8.5$ dB (2011–04–20), $\tau_2 = -9.0$ dB (2011–05–01), and $\tau_2 = -9.0$ dB (2011–07–28) (Table 5.8). The threshold values are used for an initial separation of open water and flooded vegetation areas from non-water areas using coarse-scale image information.

These regions are grown incorporating information of finer object scales into the classification process. The final post-classification step contains the removal of flood objects with an area less than 300 m² and the closing of non-water areas with an area less than 1,000 m² and 2,500 m² completely surrounded by open water areas and flooded vegetation areas, respectively.

The flood extent increases from 2011–04–09 until 2011–05–01 and reaches its lowest extent on 2011–07–28 (see Figures 5.5 and 5.6). The frequency of the flooding for open water surfaces and flooded vegetation areas of the time series is shown in the lower row of Figure 5.5.

The main flood extent consists of open water areas on the data of 2011–04–09 until 2011–05–01. Flooded vegetation areas mainly exist at the borders of lakes. Some smaller rivers and tributaries are completely covered by vegetation. The filling of dry riverbeds covered by vegetation is clearly apparent in some regions where the backscatter increases due to the transfer from vegetation-related volume scattering to double bounce scattering (see, e.g., the area surrounded by the red circle in Figure 5.5).

The reverse behavior can be observed with decreasing water levels between 2011–05–01 and 2011–07–28. The occurrence of vegetation-induced double bounce effects is considerable within the scenes. Its proportion of the total flood extent increases from ~5% on 2011–04–09 to ~30% on 2011–04–20/2011–05–01. If double bounce effects would not be considered in the classification process the flood extent would be considerably underestimated. This would especially be the case for the data acquired on 2011-07-28, where the percentage of the flooded vegetation increases to ~75% of the total flood extent. At this time the water level decreases. Vegetation completely flooded on 2011–04–09 and on 04–20/05–01 gradually emerges through the water surface on 2011–07–28. The backscatter behavior is therefore reversed, changing from specular reflecting surfaces to double-bouncing vegetation.

5.4.2 Automatic Pixel-Based Water Detection (WaMaPro)

In the following section the tool WaMaPro (Huth et al., 2009, Gstaiger et al., 2012, Kuenzer et al., 2013a, 2013b) is introduced, which is a pixel- and threshold-based, open-source software tool, able to handle TerraSAR-X and Envisat ASAR data of different spatial resolution to derive water masks in a fast and efficient manner. Most of the case study material presented here has been elaborated on in detail in Kuenzer et al. (2013a, 2013b).

5.4.2.1 Methodology

Study area for the WaMaPro tool is the Mekong Delta in the south of Vietnam—one of the world's largest river deltas, covering 39,000 km², located between 8°30′ and 11°30′N and 104°30′ and 106°50′E. The Mekong is a single-peak pulsing river, with an annual discharge of 475 km³, and it defines the flood pulse pattern and sediment delivery to the delta. Flood pulse of the Mekong as well as precipitation in the Mekong Delta itself are characterized by accentuated dry and rainy seasons, defined by the southwest Indian and northwest Pacific monsoon. Whereas the rainy season lasts from early June to December, the dry season lasts from December to May. During the rainy season large parts of the delta are frequently flooded. Despite these floods and the fact that most of the delta area is located well below 3 m a.s.l. the local inhabitants term the annual flood waters "the beautiful flood." Flooding sets the base for their livelihood: the Mekong Delta is the agricultural base for Vietnam, often termed the country's "rice bowl." The frequent natural floods—paired with fertile soils and warm climate—enable up to three rice harvests per year. Furthermore, fruit tree orchards, and aquaculture activities are prominent in the delta. Annual flood waters bring with them nutrient-rich sediments, enable local inhabitants to fish and irrigate, and improve the navigability of 55,000 km of man-made canals. Of course, the floods also have some negative impacts in the region, such as inundation on the ground floors of houses, which are not elevated, and also fatalities have occurred in the past. However, the region is—in the first place—not a region facing "disastrous," "hazardous" floods, but rather experiencing an annually returning, natural phenomenon, which is also welcomed. Nevertheless, a good understanding of the flood dynamics in the region are crucial—especially in an environment of rapid socioeconomic development, including a strong increase in urban and settled space, expanding infrastructure networks and industry, and increasing mobility. Planners of the region need to know which areas are frequently flooded and for how long, how floods proceed, and which areas are rarely or never flooded, and thus pose a safe ground for, e.g., construction and development.

The flood dynamics of the delta were investigated using a five-year time series of 60 Envisat ASAR WSM data sets at 150 m pixel spacing, as well as multi-temporal TerraSAR-X data in ScanSAR (five scenes) and StripMap mode (four scenes) for selected regions within the delta, at 8.25 and 2.5 m, respectively. An overview of this data can be found in Table 5.9.

Before applying WaMaPro, when using Envisat ASAR WSM data a fully automatic preprocessing code triggers the geocorrection of the SAR data as well as an incidence angle correction in the ESA's NEST software. Data is corrected to the normalized radar backscatter coefficient (sigma naught) to an incidence angle of 30°. TerraSAR-X data—due to its excellent geolocation accuracy and limited swath width—does not need any pre-treatment. WaMaPro is automatically triggered if data availability is indicated via an email of the data provider.

The automatic processing of water masks proceeds as already depicted in Huth et al. (2009), Gstaiger et al. (2012), and Kuenzer et al. (2013a, 2013b). As WaMaPro has been developed in the context of several EO-based research projects with partners in developing and emerging countries, the main requirement for the algorithm was an intuitive simplicity, speedy performance, the ability to process SAR data of different sensors, and especially the independence of licensed software or complex processing infrastructures. Furthermore, it was found to be convenient to embed WaMaPro in a Web Processing Service (WPS). WaMaPro was first coded in Matlab, then re-coded in C++, and is currently developed toward an open-source plugin for Q-GIS. It is based on a simple threshold method, which allows for the separation of land and water pixels. As laid out in Gstaiger et al. (2012), to firstly reduce the typical SAR inherent speckle noise, the first step of the algorithm is to apply a standard convolution median filter with a kernel size of 5 × 5 pixels, resulting in a filtered and speckle-reduced image P1. After this pre-processing two empirically chosen thresholds divide water from non-water pixels (processed image P5). Here, the first threshold, T1, which has a lower value than the final water threshold, defines confident water areas, leading to P2. The second threshold, T2, which has a higher value than the land threshold, classifies confident land areas, leading to image product P3. Then buffer zones of two pixels, which are only generated via dilatation, are applied to P", which results in product P4. The buffers define the transition zone from

---

TABLE 5.9
Data Analyzed in This Test Case Study (Source: Kuenzer et al., 2013a)

SAR Data Type		Acquired Date
Envisat ASAR WSM data	2007	2007–06–14, 2007–07–03, 2007–07–10, 2007–07–19, 2007–08–07, 2007–08–14, 2007–08–23, 2007–09–11, 2007–09–18, 2007–10–16, 2007–10–23, 2007–11–01, 2007–11–20, 2007–11–27, 2007–12–06
	2008	2008–06–01, 2008–06–17, 2008–06–24, 2008–07–03, 2008–07–22, 2008–08–14, 2008–08–23, 2008–08–26, 2008–09–11, 2008–09–30, 2008–10–07, 2008–10–16, 2008–11–04, 2008–11–11, 2008–11–20, 2008–11–23, 2008–12–16, 2008–12–25
	2009	2009–06–02, 2009–06–18, 2009–07–04, 2009–08–27, 2009–10–01, 2009–12–10, 2009–12–13
	2010	2010–01–14, 2010–01–17, 2010–02–18, 2010–03–25, 2010–04–29, 2010–05–02, 2010–06–03, 2010–07–08, 2010–08–12, 2010–08–15, 2010–09–16, 2010–10–08, 2010–10–21
	2011	2011–01–03, 2011–01–14, 2011–02–02, 2011–03–04, 2011–03–15, 2011–04–03, 2011–06–21
TerraSAR-X ScanSAR data	2008	2008–06–18, 2008–08–23, 2008–09–25, 2008–10–28, 2008–11–30
TerraSAR-X StripMap data	2008	2008–08–01, 2008–09–03, 2008–10–06, 2008–11–08

water to land, also represented by mixed pixels. The second threshold now enables the inclusion of the water pixels within this zone in the initial binary water mask. The temporary results P3 and P4 are now compared, and if coincidence occurs the value (water or land, 0 or 1) is written to P5. Otherwise, the value from P2 is written to P5 (P4 & P3 || P2). In this way overestimated water pixels are excluded (Kuenzer et al., 2013b, p. 694). Isolated pixels are removed via morphological image closing (P6) (Figure 5.7). The elimination of so-called islands and lakes according to a pre-defined maximum size (T3, T4) is mainly of relevance for high-resolution SAR data (e.g., TerraSAR-X) but does not affect Envisat ASAR WSM-derived results. The results reached with WaMaPro have been compared with other water mask or flood mask derivation algorithms and could reach or partially also exceed the accuracies of other methods tested, as elaborated in Gstaiger et al. (2012). WaMaPro enables automatic processing of the water masks right at data delivery announced via email by the data provider, and the chain automatically performs all processing steps from the retrieval of the data to product generation—a process that also has been encapsulated in a WPS. Two exemplary outputs of the water mask processing are presented in Figure 5.8.

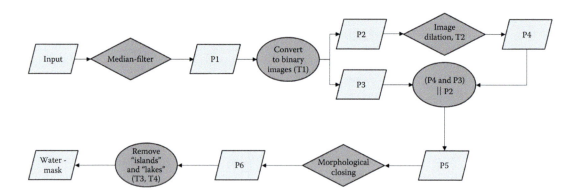

FIGURE 5.7 Processing chain of the histogram-based approach for water mask derivation (modified from Gstaiger et al., 2012, Kuenzer et al., 2013b).

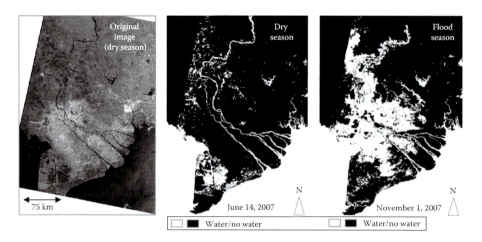

FIGURE 5.8 Envisat ASAR WSM amplitude image of 2007–06–14 (left), and water mask derived from this data set (center). The differing extent of water between the start of the rainy season (center) and the flood peak around the end of the rainy season (water masks derived from a data set in December, right) is obvious. Extent: UL: 12°N, 104°15′E, LR: 8°30′N, 106°50′E (Source: Kuenzer et al., 2013b).

5.4.2.2 Results

The joint visualization of individual water masks from—for example—one time series of rainy season data acquired by Envisat ASAR WSM data yields results like the one shown in Figure 5.9, presenting the progression of water extent in the Mekong Delta for the year 2007.

Furthermore, the complete Envisat ASAR WSM-derived water mask time series of 60 data sets, covering the years 2007–2011 has been analyzed. Such a unique, long-term data set at a resolution of 150 m enables the visualization of flood-frequencies during the time span and is a good source of information for the extraction of areas, which are unusually often flooded, or which are hardly ever flooded. In Figure 5.9 we can see that regions that are always flooded are, of course, the arms of the Mekong River, entering the South China Sea. However, also some regions at the southwestern tip of the Mekong Delta (province of Ca Mao) are very frequently flooded.

At the same time, the northern part of the delta is also very often flooded, which reflects the flood pulse rhythm typical for the Mekong Delta. Floods protrude into the delta via overland flow and overbank flow mainly coming from the north. So the flood in the delta moves from the north to

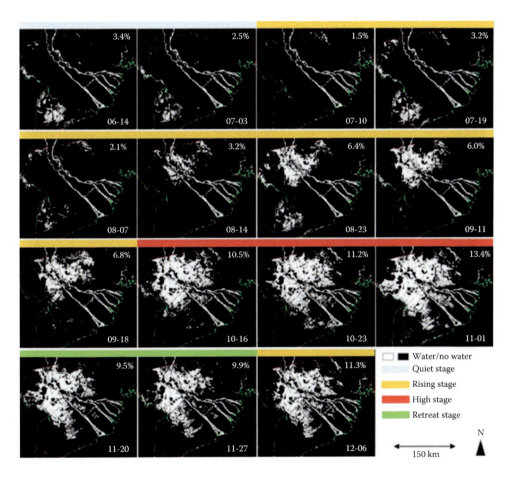

FIGURE 5.9 Flood progression in the Mekong Delta over the course of the year 2007 as derived from 15 amplitude Envisat ASAR WSM data sets using WaMaPro (Source: Kuenzer et al., 2013b). Black areas indicate regions that are not flooded. White areas indicate inundated/flooded areas. In the lower right corner the month and day of the acquisition is depicted. In the upper right corner of each water mask is what percent of the Mekong Delta, covering 39,000 km², is flooded. Flooded areas usually extend during the rainy season (rising stage) between June and September, flood extent reaches its peak between October and November, and then the annual flood waters recede again.

the south, and radially away from the Mekong's main stem (see also Figure 5.10). The coastal areas, as well as regions in the southern center of the delta, are least often flooded—these are regions of fruit orchards, slightly elevated land, or coastal forest. The high-resolution snaps from Google Earth depict regions that are never flooded: such as elevated hills and small mountains (Figure 5.10a), strongly diked land, such as presented in (Figure 5.10b), (a national park), dike-protected research farms (Figure 5.10c,d), and orchard areas on slightly elevated ground (Figure 5.10f). Only for the very dense mangrove areas in the Mekong Delta it is not possible to say if the ground below the very dense canopies is flooded or not as the penetrability of Envisat ASAR is limited.

Figure 5.11 depicts the results of a scale-related comparison between water mask derivation from four observations undertaken more or less during the same time in the rainy season of 2008 with multi-temporal Envisat ASAR WSM data (Figure 5.11a), TerraSAR-X ScanSAR data (Figure 5.11b), and TerraSAR-X StripMap data (11c). The three subsets depict a region in Can Tho City, located in the center of the Mekong Delta. Their extent is about 2 x 3 km. Little white boxes within the black background are digitized houses, which are usually located alongside canals. Different shades of blue indicate how often a pixel has been detected as flooded during the four observations. It can be observed that the better the spatial resolution the better and more precise the mapping result of the flood map. While in Envisat ASAR-derived water masks the narrow rivers and canals in the region cannot be extracted, TerraSAR-X StripMap data in particular allow for the extraction of the smallest

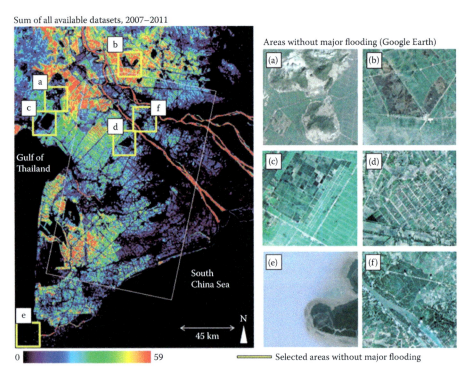

FIGURE 5.10 Inundation in the Mekong Delta from 2007 to 2011 derived from all available Envisat ASAR WSM data sets, enabling the visualization of spatial patterns of flood frequency (Source: Kuenzer et al., 2013b). Sixty observations were available, and 60 water masks could be derived. Dark blue areas are rarely flooded, while reddish tones depict areas that are always water covered (e.g., the Mekong River branches). It is obvious that the northern and southwestern parts of the Mekong Delta are most frequently flooded. In the northern part of the delta triple season rice crop is grown. Rarely flooded are the fruit orchard regions in the center and east of the delta, as well as well diked areas. The little subsets on the right side of the figure depict regions that are rarely flooded, such as elevated hills (a), well-diked regions (b, c, d), fruit orchards at higher elevation (f), or regions where flood water cannot be detected—such as under dense mangrove canopies (e).

Flood Studies Using Synthetic Aperture Radar Data 161

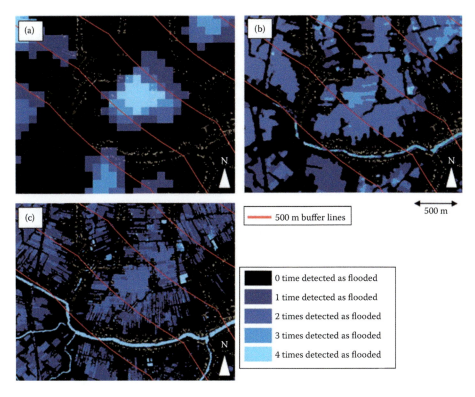

FIGURE 5.11 Capability of Envisat ASAR WSM (a), TerraSAR-X ScanSAR (b), and TerraSAR-X StripMap (c) for urban flood mapping (Source: modified based on Kuenzer et al., 2013a).

canals. In the Envisat-derived water masks it appears like several houses are flooded, whereas it can be clearly noted in the two TerraSAR-X-based results that the areas are not really flooded (problem of mixed pixels in coarser-resolution SAR data). For any kind of flood impact assessment it is thus of utmost importance to evaluate the results with respect to the observation scale chosen.

Thresholds in WaMaPro are defined empirically. This is not as elegant as a threshold automatically derived from the image histogram itself. However, WaMaPro has been applied to study sites not only in the Mekong Delta, but also to large areas in West Africa, China's Dongting Lake region, the Chinese Yellow River Delta, as well as to study sites in Russia (Huth et al., 2014). We can see here that an empirical threshold selected for TerraSAR-X processing can work in a "one size fits all" manner for these regions (T1: 60 DN, T2: 90 DN). Thus, WaMaPro could theoretically be used for a global processing based on TerraSAR-X data. For Envisat ASAR data a "one size fits all" threshold can only be transferred from one region to another if the data is properly corrected geometrically, as well as for incidence angle effects. This challenge also applies for upcoming Sentinel-1 data. Depending on incidence angle and mode (spatial resolution will vary between 20 x 40 m and 5 x 2 m) swath widths between 400 and 80 km can occur.

Generally, WaMaPro can meet difficulties in regions with heavy terrain. Here, the integration of a DEM into the processing chain (such as presented for the TerraSAR-X flood service in Section 5.4.3) seems recommendable. However, for the study of typical wetland dynamics or large overland flooding (usually not located in very rugged terrain, but in flat tundras, savannas, river deltas, coastal or lakeshore and river floodplain regions) WaMaPro has also so far worked sufficiently without a DEM. Problems that arise in these regions are mainly related to volume and double bouncing of partially submerged vegetation areas, which reduce the detectability of the water extent.

An advantage of WaMaPro is its very easy handling. The software is encapsulated in a virtual machine (VM) and can be handed on to interested users, which can apply WaMaPro as a ready-to-use Q-GIS plugin (Huth and Kuenzer, 2014). The software is therefore independent of any software package inflicting license costs. WaMaPro runs exceptionally fast: the processing of an Envisat ASAR WSM scene at 150 m resolution, and a frame size of about 200 x 200 km to a binary water mask is performed in well below one minute per scene (30–40 seconds) on an Intel 8-core CPU with 2.4 GHz and 32GB of RAM. The processing of a water mask from a TerraSAR-X ScanSAR scene at 8.25 m pixel spacing with a frame size of 100 x 150 km or a TerraSAR-X StripMap scene with 2.5 m pixel spacing and a frame extent of about 30 x 60 km takes about 4–5 minutes per scene.

5.4.3 Fully Automatic Pixel-Based Flood Detection (TerraSAR-X Flood Service)

This section presents a fully automated processing chain for near real-time pixel-based flood detection at a global level. Compared to semi-automatic flood mapping approaches (Section 5.4.1) commonly applied in the rapid mapping community, automatic processing chains enable the reduction of the critical time span from the delivery of satellite data after flood events to the provision of satellite-derived crisis information (i.e., flood extent) to emergency management and decision-makers. The thematic results of automatic processing chains can be directly ingested in web mapping applications to visualize and intersect the derived flood information with a number of other relevant geodata (e.g., DEMs, reference water levels, gauge data, hydrological features, and critical infrastructure).

When considering the broad range of different biomes and acquisition scenarios related to a global flood mapping approach, a major concern in the design and implementation of the algorithm is to reach a high level of robustness. This is a particular challenge when the algorithm is part of an automatic processing chain since a user-based classification refinement using different post-processing options (see Section 5.4.1) is no longer possible. Hence, the methodology needs to be as universally applicable as possible, delivering satisfying results independent of varying environmental conditions and acquisition parameters (e.g., beam mode, incidence angle). For this purpose, a classification methodology based on the previous work of Martinis et al. (2009) was substantially refined and extended for purposes of robustness and transferability (Martinis et al., 2014). The methodology is described in Section 5.4.3.1. The robustness and accuracy of the approach is then tested for several study sites located in different biomes (Section 5.4.3.2).

5.4.3.1 Methodology

The workflow described in this chapter has been tested for a comprehensive set of TerraSAR-X scenes acquired during flood situations all over the world with different sensor configurations. Three TerraSAR-X scenes out of this data set are analyzed in detail, and the classification result is discussed in Section 5.4.3.2. The scenes correspond to flood events in Nepal 2008, Germany 2011, and Albania/Montenegro 2013. Table 5.10 lists the main acquisition parameters.

While the automatic processing chain has been designed for enhanced ellipsoid corrected (EEC) and ground ellipsoid corrected (GEC) TerraSAR-X amplitude imagery of different acquisition

---

TABLE 5.10

Acquisition Parameters of TerraSAR-X Scenes Selected from the Test Dataset

Location	Acquisition Time	Beam Mode	Polarization	Pixel Spacing (m)
Nepal	2008–09–05, 12:09 UTC	SC_006R	HH	8.25
Germany	2011–01–17, 16:52 UTC	SM_008R	HH	2.75
Albania/Montenegro	2013–03–20, 16:32 UTC	SC_008R	HH	8.25

Flood Studies Using Synthetic Aperture Radar Data

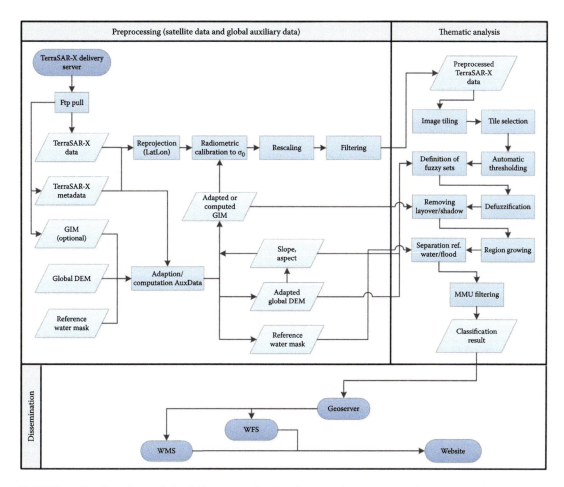

FIGURE 5.12 Workflow of the fully automatic TerraSAR-X Flood Service (Source: modified based on Martinis et al., 2015).

modes (SpotLight, StripMap, ScanSAR, Wide ScanSAR), it can principally be extended to other SAR satellite systems (Mason et al., 2018, Nemni et al., 2020, Lin et al., 2019, Rahman and Thakur, 2018). The TerraSAR-X amplitude data is commonly delivered via an ftp server. In order to ensure immediate data processing, the data download is triggered automatically through a Python script after the reception of a delivery email. When the download to the local file system has been completed the data is extracted and the corresponding file structure is searched for all relevant files, i.e., the SAR data, the metadata file, and, optionally, the so-called Geocoded Incidence Angle Mask (GIM). The GIM can be ordered as an optional add-on layer together with the EEC product and provides information on the local incidence angle for each image element of the geocoded SAR scene and on the presence of shadow and layover areas. In case no GIM was ordered jointly with the TerraSAR-X scene, this auxiliary layer is generated automatically during the following preprocessing steps:

In order to ensure the same coordinate systems for all data products on a global level, the delivered TerraSAR-X images are first reprojected to geographical coordinates (lat/lon, WGS84). This target system is also used for all global auxiliary data products. Besides TerraSAR-X satellite imagery, two types of auxiliary data are included in the process: reference water masks and digital elevation models (DEM). Since the SAR-based methodology detects all open water areas irrespective of their context (permanent water bodies, seasonally inundated areas, extreme and disastrous flood events), reference water masks are needed for separating between permanent water bodies and flooded areas. For this purpose, the SRTM water body mask (SWBD) (SWBD 2005) with a horizontal resolution of 30 m is extracted and resampled to fit each respective SAR scene. While not at the same spatial resolution as high-resolution TerraSAR-X data, there is currently no alternative to SWBD with respect to spatial resolution and near global coverage. Since the SWBD is primarily based on data recorded during the space shuttle flight in February 2000 it needs to be considered that due to seasonal effects, the derived water body extents can differ considerably from an assumed normal water level. In the future, the inclusion of new global reference water datasets such as the TanDEM-X Water Indication Mask (WAM) (Wendleder et al., 2012) with a horizontal resolution of 12 m might be considered instead of the SWBD. For a refinement of the flood mask, the ASTER Global Digital Elevation Model Version 2 (GDEM V2) (Wendleder et al., 2012) with a pixel size of one arc second is employed. The same DEM is also used for the optional computation of a GIM in the pre-processing step. For the thematic analysis, the slope information in degrees for each pixel (x, y), i.e., the local steepness of the terrain, is computed. Further, the local incidence angle θ_{loc} for each pixel (x, y), i.e., the GIM, is used for the radiometric calibration of a SAR scene to sigma naught σ_0 [dB].

The pre-processing of the TerraSAR-X amplitude data includes a radiometric calibration of the data to normalized radar cross-section σ_0. This is performed to account for incidence angle–linked SAR backscatter variations in range direction and for the reduction of topographic effects that can both negatively influence the automatic threshold derivation. σ_0 is rescaled to a value range of [0,400] in order to derive positive values during all following processing steps. For the reduction of the SAR-inherent speckle effect, a median filter of kernel size 3 x 3 is finally applied on the rescaled pixel values.

For the unsupervised initialization of the flood processor, a parametric tile-based thresholding procedure is applied (Martinis et al., 2009). This approach was originally developed to automatically detect the inundation extent in SAR amplitude data with even small class *a priori* probabilities in a time-efficient manner. For a detailed description of the algorithm, the reader is directed to Section 5.4.1.1. In the following section, the main emphasis is laid on the description of the enhancements required to meet the robustness and transferability demands of a fully automated processing chain operational for the global scale.

From the automatic thresholding procedure, one global threshold τ_g is obtained by computing the arithmetic mean of the locally derived thresholds. The standard deviation σ_τ of the local thresholds can be used as an indicator for a successful thresholding. If σ_τ exceeds an empirically derived critical threshold τ_σ (e.g., 5.0 dB) a (sub-)histogram merging strategy is applied by computing τ_g directly from a merged histogram, which is a combination of the distributions of the individual tiles. If the tile selection or the derivation of a reasonable threshold value fails (in this study the maximum possible threshold is set to −10 dB), it can be assumed that (1) either no water areas exist in the covered region, (2) the water extent is very small, or (3) water bodies do not appear as dark backscatter regions due to, e.g., wind-induced roughening of the water surface or protruding vegetation leading to volume or double bounce scattering of the radar signal. In this case the threshold is approximated by the following equation, which expresses the linear relationship between the global threshold value τ_g [dB] separating water and non-water areas and the scene center incidence angle θ_c:

$$\tau_g = -0.1002 * \theta_c - 12.08 \qquad (3)$$

This regression describes the backscatter decrease over calm water areas with increasing incidence angle and was derived empirically by analyzing a test data set of 150 TerraSAR-X HH scenes of flood events acquired in different acquisition modes and incidence angles.

After the initial classification result is derived by the application of the global threshold, a fuzzy logic–based algorithm is employed for post-classification purposes. Fuzzy logic (Zadeh, 1965) is a valuable tool for combining ambiguous information sources by accounting for their uncertainties as opposed to only relying on crisp data sets. Over the last years fuzzy logic techniques have increasingly been used for the improvement of flood monitoring algorithms. Martinis and Twele (2010) apply fuzzy theory for the quantification of the uncertainty in the labeling of each image element in flood possibility masks. The algorithm combines marginal posterior entropy-based confidence maps with spatio-temporal relationships of potentially flooded double bouncing vegetation to open water areas. A pixel- and object-based fuzzy-logic approach for flood detection based on Pierdicca et al. (2008) is described in Pulvirenti et al. (2011) and Pulvirenti et al. (2013), respectively, integrating theoretical electromagnetic scattering models, simplified hydrologic assumptions, and contextual information. For the purposes of the underlying study a fuzzy set of four elements is built consisting of: SAR backscatter (σ_0), digital elevation (h) and slope (s) information, as well as the extent (a) of water bodies. The elements of the fuzzy set are defined by standard S and Z membership functions (Pal and Rosenfeld, 1988), which express the degree of an element's membership m_f to the class water. The degree of membership is defined by real numbers within the interval [0, 1], where 0 denotes minimum and 1 maximum class membership. The membership degree depends on the position of the crossover point x_c (i.e., the half width of the fuzzy curve), which is defined by the fuzzy thresholds x_1 and x_2.

The fuzzy threshold values for each element are either determined according to statistical computations or are set empirically. Incorrectly labeled water regions are commonly caused by classifying objects with a low surface roughness and therefore low backscatter characteristics similar to calm water surfaces, such as roads, "smooth" agricultural crop land, or radar shadow. Digital elevation models (GDEM V2) are integrated into the post-classification step to improve the classification accuracy through simple hydrological assumptions, i.e., by reducing the membership degree of an image element in dependence of the height above the main water area by applying the standard Z membership function. The open water surface is derived by applying the global threshold τ_g to image Y. Subsequently, the separation between flooded regions and standing water areas is performed using a reference water mask.

The fuzzy thresholds of the elevation information are defined as:

$$x_{1[h]} = \mu_{h(water)} \tag{4}$$

and

$$x_{2[h]} = \mu_{h(water)} + f_\sigma * \sigma_{h(water)}, \tag{5}$$

where $\mu_{h(water)}$ and $\sigma_{h(water)}$ are the mean and standard deviation of the elevation of all initially derived water objects. Using this fuzzy set, the number of look-alike areas in regions significantly higher in elevation than the mean water areas is reduced, e.g., in mountainous terrain. The factor f_σ is defined by:

$$f_\sigma = \sigma_{h(water)} + 3.5 \tag{6}$$

This function was integrated to reduce the influence of the DEM in areas of low topography. The minimum value of f_σ is defined by 0.5.

The standard Z function is used for describing the membership degree to open water areas according to the radar backscatter. Full membership is assigned to image elements with a backscatter lower than the fuzzy threshold

$$x_{1[\sigma_0]} = \mu_{\sigma_0}(\tau_g), \tag{7}$$

where $\mu_{\sigma_0}(\tau_g)$ is the mean backscatter of the initial flood classification result by applying τ_g to Y. No membership degree [0] is assigned to pixels greater than

$$x_{2[\sigma_0]} = \tau_g \tag{8}$$

Topographic slope information derived from globally available digital elevation data is integrated as a third element in the fuzzy system by using the Z membership function with parameters $x_{1[sl]} = 0°$, $x_{2[sl]} = 15°$. Using this auxiliary information layer, water look-alikes in steep terrain are removed.

The S membership function is applied to the size a of the water bodies to reduce the number of dispersed small areas of low backscatter, which are commonly related to water look-alike areas. No membership degree is further assigned to elements with a size lower than $x_{1[a]} = 250 m²$, maximum grade to elements with a size greater than $x_{2[a]} = 1,000 m²$.

In order to combine all fuzzy elements into one composite fuzzy set, the average of the membership degrees is computed for each pixel. Subsequently, the flood mask is created through a threshold defuzzification step, which transforms each image element with a membership degree > 0.6 into a crisp value, i.e., a discrete thematic class.

In order to integrate image elements at the boundary of flood water surfaces and non-flooded regions and to increase the spatial homogeneity of the detected flood plain a region growing step is performed. The preliminary extracted water bodies of the defuzzified classification result are used as seeds for dilating the water regions. The water areas are iteratively enlarged until a tolerance criterion is reached. Only image elements located in the neighborhood of the flood areas are scanned to avoid the detection of water look-alikes distant from initially labeled water surfaces. The region growing tolerance criterion is defined by a relaxed fuzzy threshold of > 0.45. Therefore, the region growing step is controlled by both the SAR backscatter information and auxiliary data (topographic slope, elevation, and size of water bodies).

To eliminate open water look-alikes in areas affected by radar layover and radar shadow, the GIM is integrated into the classification process. Using a minimum mapping unit (MMU) of 30 pixels, small isolated flood objects with a pixel count less than this threshold are removed from the water mask. Small land objects (i.e., islands) that are fully enclosed by water are reclassified to water based on the same MMU value. The classification result is subsequently matched to a global reference water mask (SWBD) to differentiate between flooded areas and standing water bodies.

For dissemination of the results, the final flood mask, the fuzzy mask, which can be used for the quantification of the uncertainty in the labeling of each image element, and satellite footprints are stored in a database and visualized through a web-based user interface. The process chain is based on a framework of WPS standard-compliant to the Open Geospatial Consortium (OGC).

5.4.3.2 Results

Due to specular reflection of the incident radar signal, open water areas visually appear dark in the test scenes (see Table 5.10 upper row and Figure 5.13) and can be discerned from land surfaces, where diffuse scattering is predominant. Water look-alike areas comprise features with a low surface roughness (e.g., roads or airport runways) and areas of radar shadow. The scene histograms are hence marked by a bi-modal distribution of the water and non-water classes. The histogram of each tile selected for threshold computation is modeled by statistical parameterization of local bi-modal class-conditional density functions, and reliable thresholds are derived using minimum error thresholding. Since the standard deviations τ_σ of locally derived thresholds are significantly lower than the critical value of 5.0 dB for each scene, they are combined to global thresholds by computing their arithmetic mean. Accordingly, global threshold values of −16.7 dB (Nepal), −15.9 dB (Germany), and −19.2 dB (Albania/Montenegro) are subsequently employed for initializing the post-classification process.

The fuzzy logic–based post-classification refinement is used for combining SAR backscatter information with the fuzzy elements terrain height, slope, and water body size (see Section 5.4.3.1). The fuzzy maps shown in Figure 5.13 indicate the membership degree [0, 1] of the composite fuzzy set for each pixel. The test data acquired for the Albania/Montenegro 2013 floods show

Flood Studies Using Synthetic Aperture Radar Data 167

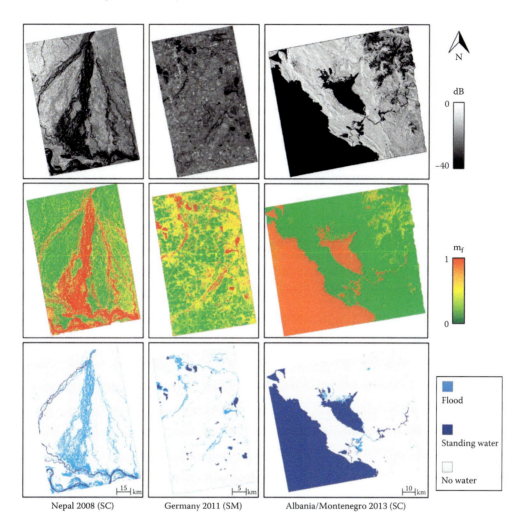

FIGURE 5.13 Radiometrically calibrated TerraSAR-X data (top row), fuzzy maps (middle row), and final classification results (bottom row) for three test areas (Source: modified based on Martinis et al., 2015).

mountainous terrain with steep slopes in the northeastern parts of the scene. The areas facing away from the sensor are thus marked by a low signal return. Using a classification approach purely based on SAR backscatter, such areas would potentially be labeled as flooded. However, as depicted on the fuzzy map, the steep slope and high altitude of these areas led to a reduction in the total membership degree of these pixels. In the threshold defuzzification step, which transforms each image element with a membership degree > 0.6 into a crisp value, i.e., a discrete label, most of these water look-alikes are successfully excluded from the final flood mask. In contrast to the Albania/Montenegro 2013 data set, the scene recorded during the Germany 2011 floods (middle row) is characterized by much lower differences in altitude. Here, SAR backscatter information has a major influence on the final membership degree of the composite fuzzy set. Typical water look-alikes such as airport runways and roads in flat areas without much height difference in comparison to initially found flood pixels can thus erroneously get labeled as flooded, even after a fuzzy logic–based post-classification refinement.

Classification problems can also occur due to artifacts in DEM data. In our examples, ASTER GDEM V2 data occasionally contained spikes in regions of normal water bodies. Accordingly, the high slope values surrounding these spikes can lower the membership degree of the composite fuzzy set below 0.6, which would lead to a misclassification of these pixels to the non-water class.

The algorithm generally performs very well in rural areas. In urban areas the detectability of water regions is reduced due to enhanced signal returns, with backscatter values frequently higher than extracted threshold. Further, floods in urban areas might simply not be visible to the SAR instrument when they are situated within areas of radar shadow cast by buildings (Mason et al., 2010). The low signal return from these areas can then lead to a potential confusion with flooded areas that exhibit comparable backscatter levels. Although the presence of shadow and layover in urban areas can be simulated using high-resolution digital surface models such as LiDAR data (Soergel et al., 2003), this type of data is expensive and not available on a global scale.

5.5 CONCLUSION

SAR-derived water and flood extent maps are an important information source for an effective flood disaster management (Halder and Bose, 2024, Hitouri et al., 2024, Sundaram et al., 2023, Surampudi and Kumar, 2023, Mason et al., 2021, Mason et al., 2018, Aziz et al., 2020, Berezowski et al., 2020, Gebremichael et al., 2020, Martinis et al., 2022; Natsuaki and Hiroto, 2020, Nemni et al., 2020, Zeng et al., 2020, Quang et al., 2019, Landuyt et al., 2019, Lin et al., 2019, Shen et al., 2019a, Shen et al., 2019b, Rahman and Thakur, 2018, Ramanuja, 2018). Various approaches have been proposed in the literature for extracting the inundation extent from SAR data. In this study three SAR-based operational water and flood detection algorithms of the recent literature are described, which are developed for different application domains.

Semi-automatic object-based algorithms can be used to derive very detailed flood maps with high classification accuracy during rapid mapping activities. For example, the overall accuracy of the RaMaFlood algorithm is stated to be 95.44% (producer accuracy: 82,01%; user accuracy: 98.65%) for a TerraSAR-X StripMap scene covering a flood situation in Tewkesbury, UK, in 2007 (Martinis et al., 2009). Commonly, commercial and non-commercial value-adding entities provide flood maps extracted from remote sensing data in map products upon request by, e.g., relief organizations or political decision-makers several hours after satellite data delivery. The disadvantage is that in most cases a certain amount of user interaction is needed for SAR data pre-processing, the collection and adaptation of auxiliary data useful for classification refinement, the thematic analysis as well as the preparation and dissemination of the crisis information to end users.

Compared to semi-automatic water/flood mapping approaches automatic processing chains significantly reduce the time delay between satellite data delivery and product dissemination.

In this context an automatic and open source distributable processing chain (WaMaPro) enabling continuous water monitoring from TerraSAR-X and Envisat ASAR on a local to global level using empirically pre-defined thresholds is presented. This software has been successfully used for detecting smooth water surfaces in areas of stable environmental conditions that are repeatedly being monitored with data acquired with similar SAR system parameters.

Further, a global fully automatic TerraSAR-X-based flood mapping approach is proposed. The processing chain includes the download and pre-processing of TerraSAR-X data, computation and adaption of global auxiliary data, unsupervised class initialization, post-classification refinement, and dissemination of the flood masks via a web-based user interface. The computational effort of the whole processing chain is less than one hour. The methodology is universally applicable, delivering satisfying results independent from acquisition parameters, and requires no user input. These promising qualitative results have been further confirmed by means of a quantitative accuracy assessment performed for two study sites in Germany and Thailand, achieving encouraging overall accuracies of ~91.6% and ~87.5%, respectively.

Within the last decade, fully automatic algorithms and processing chains have been developed for unsupervised mapping of open water surfaces using various SAR sensors. Further improvements in classification accuracy could be reached by integrating upcoming high-resolution auxiliary data such as digital elevation models, reference water masks, and land cover information. Future work should focus on improving and automating the algorithms in complex situations such as flooded urban areas and partially submerged vegetation areas. By adapting the proposed methods to data of the upcoming ESA's Sentinel-1 C-Band SAR mission, which is characterized by a systematic acquisition strategy and therefore enables a continuous monitoring of water surfaces, some of the presented limitations of SAR-based water mapping could be reduced by automatically exploiting the growing archive during the lifetime of the mission.

REFERENCES

Ahtonen, P., M. Euro, M. Hallikainen, S. Solbø, B. Johansen, and I. Solheim. 2004. SAR and optical based algorithms for estimation of water bodies. Technical Report, FloodMan Project, Helsinki University of Technology, Helsinki, Finland.

Alsdorf, D. E., J. M. Melack, T. Dunne, L. A. K. Mertes, L. L. Hess, and L. C. Smith. 2000. Interferometric radar measurements of water level changes on the Amazon flood plain. *Nature* 404:174–177.

Aronica, G., P. D. Bates, and M. S. Horritt. 2002. Assessing the uncertainty in distributed model predictions using observed binary pattern information within GLUE. *Hydrol Process* 16:2001–2016.

Aziz, M. A., M. Moniruzzaman, A. Tripathi, et al. 2020. Delineating flood zones upon employing synthetic aperture data for the 2020 flood in Bangladesh. *Earth Syst Environ* 6:733–743 (2022). https://doi.org/10.1007/s41748-022-00295-0

Baatz, M., and A. Schäpe. 1999. Object-oriented and multi-scale image analysis in semantic networks. Proceedings of the 2nd International Symposium on Operationalization of Remote Sensing, Enschede, Netherlands.

Bates, P. D., M. S. Horritt, C. N. Smith, and D. C. Mason. 1997. Integrating remote sensing observations of flood hydrology and hydraulic modelling. *Hydrol Process* 11:1777–1795.

Bazi, Y., L. Bruzzone, and F. Melgani. 2005. An unsupervised approach based on the generalized Gaussian model to automatic change detection in multitemporal SAR images. *IEEE T Geosci Remote* 43:874–887.

Benz, U., P. Hofmann, G. Wilhauck, I. Lingenfelder, and M. Heynen. 2004. Multi-resolution, object-oriented fuzzy analysis of remote sensing data for GIS-ready information. *ISPRS J Photogramm* 58:239–258.

Berezowski, T., T. Bieliński, and J. Osowicki, 2020. Flooding extent mapping for synthetic aperture radar time series using river gauge observations. *IEEE J Sel Top Appl Earth Obs Remote Sens* 13:2626–2638. https://doi.org/10.1109/JSTARS.2020.2995888

Blaschke, T., and J. Strobl. 2001. What's wrong with pixels? Some recent developments interfacing remote sensing and GIS. *GIS—Zeitschrift für Geoinformationssysteme* 6:12–17.

Blasco, F., M. F. Bellan, and M. U. Chaudury. 1992. Estimating the extent of floods in Bangladesh using SPOT data. *Remote Sens Environ* 39:167–178.

Bourgeau-Chavez, L. L., E. S. Kasischke, S. M. Brunzell, J. P. Mudd, K. B. Smith, and A. L. Frick. 2001. Analysis of space-borne SAR data for wetland mapping in Virginia riparian ecosystems. *Int J Remote Sens* 22:3665–3687.

Brakenridge, G. R., and E. Anderson. 2005. MODIS-based flood detection, mapping, and measurement: The potential for operational hydrological applications. Proceedings of the NATO Advanced Research Workshop, Baile Felix—Oradea, Romania.

Brivio, P. A., R. Colombo, M. Maggi, and R. Tomasoni. 2002. Integration of remote sensing data and GIS for accurate mapping of flooded areas. *Int J Remote Sens* 23:429–441.

Bruzzone, L., and L. Carlin. 2006. A multilevel context-based system for classification of very high spatial resolution images. *IEEE T Geosci Remote* 44:2587–2600.

Carleer, A. P., O. Debeir, and E. Wolff. 2005. Assessment of very high spatial resolution satellite image segmentations. *Photogrammetric Eng Rem S* 71:1285–1294.

Carroll, M., J. Townshend, C. DiMiceli, P. Noojipady, and R. Sohlberg. 2009. A new global raster water mask at 250 meter resolution. *Int J of Digital Earth* 2:291–308.

Danklmayer, A., B. J. Döring, M. Schwerdt, and M. Chandra. 2009. Assessment of atmospheric propagation effects in SAR images. *IEEE T Geosci Remote* 47:3507–3518.

Dellepiane, S., S. Monni, G. Bo, and C. Buck. 2000. SAR images and interferometric coherence for flood monitoring. Proceedings of the IEEE Geoscience and Remote Sensing Symposium, Honolulu, Hawaii, USA, 2608–2610.

De Moel, H., J. Van Alphen, and J. Aerts. 2009. Flood maps in Europe, methods, availability and use. *Nat Hazard Earth Sys* 9:289–301.

De Roo, A., J. Van der Knijff, M. S. Horritt, G. Schmuck, and S. De Jong. 1999. Assessing flood damages of the 1997 Oder flood and the 1995 Meuse flood. Proceedings of the 2nd International Symposium on Operationalisation of Remote Sensing, Enschede, The Netherlands.

Drake, B., and R. A. Shuchman. 1974. Feasibility of using multiplexed SLAR imagery for water resources management and mapping vegetation communities. Proceedings of the 9th International Symposium on Remote Sensing of Environment, Environmental Research Institute of Michigan, Ann Arbor, USA, 714–724.

eoPortal Directory. 2014. https://directory.eoportal.org/

Europen Commission. 2011. GMES data access specifications of the earth observation needs over the period 2011–2013. Brussel, Belgium.

Evans, D. C., T. G. Farr, J. P. Forf, T. W. Thompson, and C. L. Werner. 1986. Multipolarization radar images for geologic mapping and vegetation discrimination. *IEEE T Geosci Remote* 24:246–257.

Ferreira, J. P. G., and J. M. Bioucas-Dias. 2008. Bayesian land and sea segmentation of SAR imagery. Proceedings of the 3rd TerraSAR-X Science Team Meeting, Oberpfaffenhofen, Germany.

Gebhardt, S., J. Huth, N. Lam Dao, A. Roth, and C. Kuenzer. 2012. A comparison of TerraSAR-X quadpol backscattering with RapidEye multispectral vegetation indices over rice fields in the Mekong Delta, Vietnam. *Int J Remote Sens* 33:7644–7661.

Gebremichael, E., A. L. Molthan, J. R. Bell, Lori A. Schultz, and C. Hain. 2020. Flood hazard and risk assessment of extreme weather events using synthetic aperture radar and auxiliary data: A case study. *Remote Sensing* 12, no. 21:3588. https://doi.org/10.3390/rs12213588

Giacomelli, A., M. Mancini, and R. Rosso. 1995. Assessment of flooded areas from ERS-1 PRI data: An application to the 1994 flood in Northern Italy. *Phys Chem Earth* 20:469–474.

Giustarini, L., R. Hostache, P. Matgen, G. Schumann, P. D. Bates, and D. C. Mason. 2012. A change detection approach to flood mapping in urban areas using TerraSAR-X. *IEEE T Geosci Remote* 51:2417–2430.

Gstaiger, V., J. Huth, S. Gebhardt, T. Wehrmann, and C. Kunezer. 2012. Multi-sensoral and automated derivation of inundated areas using TerraSAR-X and ENVISAT ASAR data. *Int J Remote Sens* 33: 7291–7304.

Halder, S., and S. Bose. 2024. Sustainable flood hazard mapping with GLOF: A google earth engine approach. *Nat Hazards Res*, ISSN 2666–5921. https://doi.org/10.1016/j.nhres.2024.01.002. (www.sciencedirect.com/science/article/pii/S2666592124000027).

Haralick R. M., and L. G. Sharipo. 1985. Image segmentation techniques. *Comput Vis Graph Image Process* 29:100–132.

Henderson, F. M. 1987. Consistency of open surface detection with L-band SEASAT SAR imagery and confusion with other hydrologic features. Proceedings of the 13th Annual Conference of Remote Sensing Society (Advances in Digital Image Processing), University of Nottingham, UK, 69–78.

Henderson, F. M. 1995. Environmental factors and the detection of open surface water using X-band radar imagery. *Int J Remote Sens* 16:2423–2437.

Henry, J. B., P. Chastanet, K. Fellah, and Y. L. Desnos. 2006. ENVISAT multi-polarised ASAR data for flood mapping. *Int J Remote Sens* 27:1921–1929.

Herrera-Cruz, V., and F. Koudogbo. 2009. TerraSAR-X Rapid mapping for flood events. Proceedings of the International Society for Photogrammetry and Remote Sensing (Earth Imaging for Geospatial Information), Hannover, Germany, 170–175.

Hess, L. L., and J. M. Melack. 1994. Mapping wetland hydrology and vegetation with synthetic aperture radar. *Int J Ecology Environ Sci* 20:197–205.

Hess, L. L., and J. M. Melack. 2003. Remote sensing of vegetation and flooding on Magela Creek floodplain (Northern Territory, Australia) with the SIR-C synthetic aperture radar. *Hydrobiologia* 500:65–82.

Hess, L. L., J. M. Melack, and D. S. Simonett. 1990. Radar detection of flooding beneath the forest canopy: A review. *Int J Remote Sens* 11:1313–1325.

Hess, L. L., J. M. Melack, E. M. Novo, C. C. Barbosa, and M. Gastil. 2003. Dual season mapping of wetland inundation and vegetation for the central Amazon basin. *Remote Sens Environ* 87:404–428.

Hess, L. L., J. M. Melack, S. Filoso, and Y. Wang. 1995. Delineation of inundated area and vegetation along the Amazon floodplain with the SIR-C synthetic aperture radar. *IEEET Geosci Remote* 33:896–904.

Hitouri, S., M. Mohajane, M. Lahsaini, S.K. Ajim Ali, T. A. Setargie, G. Tripathi, P. D'Antonio, S. K. Singh, and A. Varasano. 2024. Flood susceptibility mapping using SAR data and machine learning algorithms in a small watershed in Northwestern Morocco. *Remote Sens* 16, no. 5: 858. https://doi.org/10.3390/rs16050858

Hong, S.-H., S. Wdowinski, and S.-W. Kim. 2010. Evaluation of TerraSAR-X observations for wetland InSAR application. *IEEE T Geosci Remote* 48:864–873.

Horritt, M. 1999. A statistical active contour model for SAR image segmentation. *Image Vision Computing*:213–224.

Horritt, M. S. 2000. Calibration of a two-dimensional finite element flood flow model using satellite radar imagery. *Water Resour Res* 36:3279–3291.

Horritt, M. S. 2006. A methodology for the validation of uncertain flood inundation models. <cite lang="cy">*J Hydrol* 326:153–165.

Horritt, M. S., D. C. Mason, and A. J. Luckman. 2001. Flood boundary delineation from synthetic aperture radar imagery using a statistical active contour model. *Int J Remote Sens* 22:2489–2507.

Horritt, M. S., D. C. Mason, D. M. Cobby, I. J. Davenport, and P. Bates. 2003. Waterline mapping in flooded vegetation from airborne SAR imagery. *Remote Sens Environ* 85:271–281.

Hostache, R., P. Matgen, G. Schumann, C. Puech, L. Hoffmann, and L. Pfister. 2009. Water level estimation and reduction of hydraulic model calibration uncertainties using satellite SAR images of floods. *IEEE T Geosci Remote* 47:431–441.

Hunter, N. M., P. D. Bates, M. S. Horritt, P. J. De Roo, and M. Werner. 2005. Utility of different data types for flood inundation models within a GLUE framework. *Hydrol Earth Syst Sc* 9:412–430.

Huth, J., and C. Kuenzer. 2014. WaMaPro handbook. Version 1.0, Unpublished software manual, p. 24.

Huth, J., M. Ahrens, C. Kuenzer. 2014. WaMaPro: The water mask processing hand book for WaMaPro version 2.2.0, Status April, p. 38.

Huth, J., S. Gebhardt, T. Wehrmann, I. Schettler, C. Kuenzer, M. Schmidt, and S. Dech. 2009. Automated inundation monitoring using TerraSAR-X multi-temporal imagery. In: Proc. of the European Geosciences Union General Assembly, 19–24 April 2009, Vienna, Austria.

Ivins, J., and J. Porill. 1995. Active region models for segmenting textures and colours. *Image Vision Comput* 13:431–438.

Kasischke, E. S., J. M. Melack, and M. C. Dobson. 1997. The use of imaging radars for ecological applications: A review. *Remote Sens Environ* 59:141–156.

Kasischke, E. S., and L. L. Bourgeau-Chavez. 1997. Monitoring south Florida wetlands using ERS-1 SAR imagery. *Photogramm Eng and Rem S* 33:281–291.

Kittler, J., J. Illingworth. 1986. Minimum error thresholding. *Pattern Recogn* 19:41–47.

Kohonen, T. 1995. Self-organizing maps. Third edition, Springer-Verlag, Heidelberg, Germany.

Kolmogorov, V., and R. Zabih. 2004. What energy functions can be minimized via graph cuts? *IEEE T Pattern Anal* 26:147–159.

Kuenzer, C., H. Guo, I. Schlegel, V. Q. Tuan, X. Li, and S. Dech. 2013a. Varying scale and capability of Envisat ASAR-WSM, TerraSAR-X Scansar and TerraSAR-X Stripmap data to assess urban flood situations: A case study of the Mekong delta in Can Tho province. *Remote Sens* 5:5122–5142.

Kuenzer, C., G. Huadong, J. Huth, P. Leinenkugel, L. Xinwu, and S. Dech. 2013b. Flood mapping and flood dynamics of the Mekong Delta: ENVISAT-ASAR-WSM based time series analyses. *Remote Sens* 5:687–715.

Kuenzer, C., and K. Knauer. 2013. Remote sensing of rice crop areas: A review. *Int J Remote Sens* 34:2101–2139.

Kussul, N. 2008. Intelligent computations for flood monitoring. 14th International Conference of Knowledge-Dialogue-Solution, Varna, Bulgaria, 48–54.

Landuyt, L., A. V. Wesemael, G. J. P. Schumann, R. Hostache, N. E. C. Verhoest, and F. M. B. Van Coillie, 2019. Flood mapping based on synthetic aperture radar: An assessment of established approaches. *IEEE Trans Geosci Remote Sens* 57, no. 2:722–739, February. https://doi.org/10.1109/TGRS.2018.2860054.

Lang, M. W., P. A. Townsend, and E. S. Kasischke. 2008. Influence of incidence angle on detecting flooded forests using C-HH synthetic aperture radar data. *Remote Sens Environ* 112:3898–3907.

Leckie, D. G. 1998. Forestry applications using imaging radar. In Henderson, F. M., and A. J. Lewis, (eds.): Manual of remote sensing: Principles and applications of imaging radar. Third edition, John Wiley and Sons, New York, USA.

Lewis, A. J. 1998. Geomorphic and hydrologic applications of active microwave remote sensing. In Henderson, F. M., and A. J. Lewis, (eds.): Manual of remote sensing: Principles and applications of imaging radar. Third edition, John Wiley and Sons, New York, USA.

Lillesand, T. M., R. W. Kiefer, and J. W. Chipman. 2004: Remote sensing and image interpretation. Fifth edition, John Wiley and Sons, New York, USA.

Lin, Y. N., S.-H. Yun, A. Bhardwaj, and E. M. Hill. 2019. Urban flood detection with sentinel-1 multi-temporal Synthetic Aperture Radar (SAR) observations in a bayesian framework: A case study for hurricane matthew. *Remote Sens* 11, no. 15:1778. https://doi.org/10.3390/rs11151778

Lopes, A., E. Nezry, R. Touzi, and H. Laur. 1990. Maximum a posteriori speckle filtering and first order texture models in SAR images. Proceedings of IEEE International Geoscience and Remote Sensing Symposium (IGARSS 1990), College Park, MD, 3:2409–2412.

Lucchese, L., and S. K. Mitra. 1998. An algorithm for unsupervised color image segmentation. Proceedings IEEE 2nd Workshop Multimedia Signal Process, Redondo Beach, CA, 33–38.

Malladi, R., J. Sethian, and B. Vemuri. 1995. Shape modeling with front propagation: A level set approach. *IEEE T Pattern Anal* 17:158–175.

Malnes, E., T. Guneriussen, and K. A. Høgda. 2002. Mapping of flood-area by RADARSAT in Vannsjø, Norway. Proceedings of the 29th International Symposium on Remote Sensing of the Environment, Buenos Aires, Argentina.

Marcus, W. A., and M. A. Fonstad. 2008. Optical remote mapping of rivers at sub-meter resolution and watershed extents. Earth Surf. Processes Landforms, 33:4–24.

Marinelli, L., R. Michel, A. Beaudoin, and J. Astier. 1997. Flood mapping using ERS tandem coherence image: A case study in south France. Proceedings of the 3rd ERS Symposium, Florence, Italy, 531–536.

Martinis, S. 2010. Automatic near real-time flood detection in high resolution X-band synthetic aperture radar satellite data using context-based classification on irregular graphs. PhD Thesis, Ludwig-Maximilians-University Munich, Munich, Germany.

Martinis, S., and A. Twele. 2010. A hierarchical spatio-temporal Markov model for improved flood mapping using multi-temporal X-band SAR data. *Remote Sens* 2:2240–2258.

Martinis, S., A. Twele, and S. Voigt. 2009. Towards operational near real-time flood detection using a split-based automatic thresholding procedure on high resolution TerraSAR-X data. Nat Hazard Earth Syst 9:303–314.

Martinis, S., Groth, S., Wieland, M., Knopp, L., Rättich, M. 2022. Towards a global seasonal and permanent reference water product from Sentinel-1/2 data for improved flood mapping. *Remote Sensing of Environment*, 278.

Martinis, S., A. Twele, C. Strobl, J. Kersten, and E. Stein. 2013. A multi-scale flood monitoring system based on fully automatic MODIS and TerraSAR-X processing chains. *Remote Sens* 5:5598–5619.

Martinis, S., A. Twele, and S. Voigt. 2011. Unsupervised extraction of flood-induced backscatter changes in SAR data using Markov image modeling on irregular graphs. *IEEE T Geosci Remote* 49:251–263.

Martinis, S., J. Kersten, and A. Twele. 2015. A fully automated TerraSAR-X based flood service. *ISPRS J Photogramm*, Volume 104, 203–212.

Mason, D. C., J. Bevington, S. L. Dance, B. Revilla-Romero, R. Smith, S. Vetra-Carvalho, and H. L. Cloke. 2021. Improving urban flood mapping by merging synthetic aperture radar-derived flood footprints with flood hazard maps. *Water* 13, no. 11:1577. https://doi.org/10.3390/w13111577

Mason, D. C., I. J. Davenport, I. J. C. Neal, G. J.-P. Schumann, and P. D. Bates. 2012a. Near real-time flood detection in urban and rural areas using high-resolution Synthetic Aperture Radar images. *IEEE T Geosci Remote* 50:3041–3052.

Mason, D. C., M. S. Horritt, J. T. Dall'Amico, T. R. Scott, and P. D. Bates. 2007. Improving river flood extent delineation from synthetic aperture radar using airborne laser altimetry. *IEEE T Geosci Remote* 45:3932–3943.

Mason, D. C., R. Speck, B. Devereux, G. J.-P. Schumann, J. C. Neal, and P. D. Bates. 2010. Flood detection in urban areas using TerraSAR-X. *IEEE T Geosci Remote* 48:882–894.

Mason, D.C., S.L. Dance, S. Vetra-Carvalho, and H. L.Cloke, 2018. Robust algorithm for detecting floodwater in urban areas using synthetic aperture radar images. *J Appl Remote Sens* 12, no. 4: 045011, 5 November. https://doi.org/10.1117/1.JRS.12.045011

Matgen, P., G. Schumann, J. B. Henry, L. Hoffmann, and L. Pfister. 2007. Integration of SAR derived river inundation areas, high precision topographic data and a river flow model toward near real-time flood management. *Int J Appl Earth Obs* 9:247–263.

Matgen, P., M. Montanari, R. Hostache, L. Pfister, L. Hoffmann, D. Plaza, V. R. N. Pauwels, G. J. M. De Lannoy, R. De Keyser, and H. H. G. Savenije. 2010. Towards the sequential assimilation of SAR-derived water stages into hydraulic models using the particle filter: Proof of concept. *Hydrol Earth Syst Sc* 14:1773–1785.

Matgen, P., R. Hostache, G. Schumann, L. Pfister, L. Hoffman, and H. H. G. Svanije. 2011. Towards an automated SAR based flood monitoring system: Lessons learned from two case studies. *Phys Chem Earth* 36:241–252.

McMillan, A., J. G. Morley, B. J. Adams, and S. Chesworth. 2006. Identifying optimal SAR imagery specifications for urban flood monitoring: A hurricane Katrina case study. 4th International Workshop on Remote Sensing for Disaster Response, Magdalene College, Cambridge.

Natsuaki, R., and H. Nagai. 2020. Synthetic aperture radar flood detection under multiple modes and multiple orbit conditions: A case study in Japan on Typhoon Hagibis, 2019. *Remote Sens* 12, no. 6:903. https://doi.org/10.3390/rs12060903

Nemni, E., J. Bullock, S. Belabbes, and L. Bromley. 2020. Fully convolutional neural network for rapid flood segmentation in synthetic aperture radar imagery. *Remote Sens* 12, no. 16:2532. https://doi.org/10.3390/rs12162532

Nico, G., M. Pappalepore, G. Pasquariello, S. Refice, and S. Samarelli. 2000. Comparison of SAR amplitude vs. coherence flood detection methods: A GIS application. *Int J Remote Sens* 21:1619–1631.

Ormsby, J. P., B. J. Blanchard, and A. J. Blanchard. 1985. Detection of lowland flooding using active microwave systems. *Photogramm Eng Rem S* 51:317–328.

Otsu, N. 1979. A threshold selection method from gray-level histograms. *IEEE Sys Man Cybern* 9:62–66.

Ottinger, M., C. Kuenzer, G. Liu, S. Wang, and S. Dech. 2013: Monitoring land cover dynamics in the Yellow River Delta from 1995 to 2010 based on Landsat 5 TM. *Appl Geogr* 44:53–68.

Pal, S. K., and A. Rosenfeld. 1988. Image enhancement and thresholding by optimization of fuzzy compactness. *Pattern Recogn Letters* 7: 77–86.

Pappenberger, F., K. Frodsham, K. Beven, R. Romanowicz, and P. Matgen. 2007. Fuzzy set approach to calibrating distributed flood inundation models using remote sensing observations. *Hydrol Earth Syst Sc* 11:739–752.

Peinado, O, C. Kuenzer, S. Voigt, P. Reinartz, and H. Mehl. 2003. Fernerkundung und GIS im Katastrophenmangement—die Elbe Flut 2003. In: Strobl, J., T. Blaschke, and G. Griesebner, (eds.): Angewandte geographische informationsverarbeitung XV. Beitraege zum AGIT-Symposium Salzburg 2003. Heidelberg, Wichmann, 342–348.

Pierdicca, N., L. Pulvirenti, M. Chini, L. Guerriero, and L. Candela. 2013. Observing floods from space: Experience gained from COSMO-SkyMed observations. *Acta Astronaut* 84:122–133.

Pierdicca, N., M. Chini, L. Pulvirenti, and F. Macina. 2008. Integrating physical and topographic Information into a fuzzy scheme to map flooded area by SAR. *Sensors* 8: 4151–4164.

Pulvirenti, L., M. Chini, F. S. Marzano, N. Pierdicca, S. Mori, L. Guerriero, G. Boni, and L. Candela. 2012. Detection of floods and heavy rain using Cosmo-SkyMed data: The event in Northwestern Italy of November 2011. Proceedings of IEEE International Geoscience and Remote Sensing Symposium (IGARSS 2012), Munich, Germany, 22–27 July, 3026–3029.

Pulvirenti, L., N. Pierdicca, M. Chini, and L. Guerriero. 2011. An algorithm for operational flood mapping from Synthetic Aperture Radar (SAR) data using fuzzy logic. *Nat Hazard Earth Sys* 11:529–540.

Pulvirenti, L., Pierdicca, N., Chini, M., and Guerriero, L. 2013. Monitoring flood evolution in vegetated areas using COSMO-SkyMed data: The Tuscany 2009 case study. *IEEE J Sel Top Appl* 99:1–10.

Quang, N. H., V. A. Tuan, N. T. P. Hao, L. T. T. Hang, N. M. Hung, V. L. Anh, L. T. M. Phuong, and R. Carrie. 2019. Synthetic aperture radar and optical remote sensing image fusion for flood monitoring in the Vietnam lower Mekong basin: A prototype application for the Vietnam Open Data Cube, *Eur J Remote Sens* 52:1, 599–612. https://doi.org/10.1080/22797254.2019.1698319

Rahman, M. R., and P. K. Thakur 2018. Detecting, mapping and analysing of flood water propagation using synthetic aperture radar (SAR) satellite data and GIS: A case study from the Kendrapara District of Orissa State of India. Egypt J Remote Sens Space Sci 21, Supplement 1:S37–S41, ISSN 1110–9823. https://doi.org/10.1016/j.ejrs.2017.10.002. (www.sciencedirect.com/science/article/pii/S1110982317301126).

Ramanuja, M. 2018. Review of synthetic aperture radar frequency, polarization, and incidence angle data for mapping the inundated regions. *J Appl Remote Sens* 12, no. 2:021501, 15 May. https://doi.org/10.1117/1.JRS.12.021501

Ramsey, E. W. 1995. Monitoring flooding in coastal wetlands by using radar imagery and ground-based measurements. *Int J Remote Sens* 16:2495–2502.

Raney, R. K. 1998. Radar fundamentals: Technical perspective. In Henderson, F. M., and A. J. Lewis, (eds.): Manual of remote sensing: Principles and applications of imaging radar. Third edition, John Wiley and Sons, New York, USA.

Rémi, A., and Y. Hervé. 2007. Change detection analysis dedicated to flood monitoring using ENVISAT Wide Swath mode data. Proceedings of the ENVISAT Symposium 2007, Montreux, Switzerland, SP-636.

Rennó, C. D., A. D. Nobre, L. A. Cuartas, J. V. Soares, M. G. Hodnett, J. Tomasella, and M. J. Waterloo. 2008. HAND, a new terrain descriptor using SRTM-DEM: Mapping terra-firme rainforest environments in Amazonia. *Remote Sens Environ* 112:3469–3481.

Richards, J. A., P. W. Woodgate, and A. K. Skidmore. 1987. An explanation of enhanced radar backscattering from flooded forests. *Int J Remote Sens* 8:1093–1100.

Sanyal, J., and X. X. Lu. 2003. Application of remote sensing in flood management with special reference to Monsoon Asia: A review. *Nat Hazards* 33:283–301.

Schumann, G., G. D. di Baldassarre, D. Alsdorf, and P. D. Bates. 2010. Near real-time flood wave approximation on large rivers from space: Application to the River Po, Italy. *Water Resour Res* 46:1–8.

Schumann, G., G. D. di Baldassarre, and P. D. Paul. 2009. The utility of spaceborne radar to render flood inundation maps based on multialgorithm ensembles. *IEEE T Geosci Remote* 47:2801–2807.

Schumann, G., R. Hostache, C. Puech, L. Hoffmann, P. Matgen, F. Pappenberger, and L. Pfister. 2007. High-resolution 3-D flood information from radar imagery for flood hazard management. *IEEE T Geosci Remote* 45:1715–1725.

Shen, X., D. Wang, K. Mao, E. Anagnostou, and Y. Hong. 2019b. Inundation extent mapping by synthetic aperture radar: A review. *Remote Sens.* 11, no. 7:879. https://doi.org/10.3390/rs11070879

Shen, X., E. N. Anagnostou, G. H. Allen, G. R. Brakenridge, and A. J. Kettner. 2019a. Near-real-time non-obstructed flood inundation mapping using synthetic aperture radar *Remote Sens Envir* 221:302–315, ISSN 0034–4257. https://doi.org/10.1016/j.rse.2018.11.008. (www.sciencedirect.com/science/article/pii/S0034425718305169).

Silveira, M., and S. Heleno. 2009. Separation between water and land in SAR images using region-based level sets. *IEEE Geosci Remote S* 6:471–475.

Smith, L. C. 1997. Satellite remote sensing of river inundation area, stage, and discharge: A review. *Hydrol Process* 11:1427–1439.

Soergel, U., U. Thoennessen, and U. Stilla. 2003. Visibility analysis of man-made objects in SAR images. Proceedings of the 2nd GRSS/ISPRS Joint Workshop on Data Fusion and Remote Sensing Over Urban Areas, 120–124.

Sundaram, S., S. Devaraj, and K. Yarrakula. 2023. Mapping and assessing spatial extent of floods from multi-temporal synthetic aperture radar images: A case study over Adyar watershed, India. *Environ Sci Pollut Res* 30:63006–63021. https://doi.org/10.1007/s11356-023-26467-7

Surampudi, S., and V. Kumar. 2023. Flood depth estimation in agricultural lands from L and C-band synthetic aperture radar images and digital elevation model. *IEEEAccess* 11:3241–3256. https://doi.org/10.1109/ACCESS.2023.3234742

Torres, R., P. Snoeij, D. Geudtner, D. Bibby, M. Davidson, E. Attema, P. Potin, B. Rommen, N. Floury, et al. 2012. GMES Sentinel-1 mission. *Remote Sens Environ* 120:9–24.

Townsend, P. A. 2001. Mapping seasonal flooding in forested wetlands using multi-temporal RADARSAT SAR. *Photogramm Eng and Rem S* 67:857–864.

Townsend, P. A. 2002. Relationships between forest structure and the detection of flood inundation in forest wetlands using C-band SAR. *Int J Remote Sens* 23:332–460.

Townsend, P. A., and S. J. Walsh. 1998. Modeling floodplain inundation using an integrated GIS with radar and optical remote sensing. *Geomorphology* 21:295–312.

Ulaby, F. T., R. K. Moore, and A. K. Fung. 1982. Microwave remote sensing: Active and passive: Vol. II—Radar remote sensing and surface scattering and emission theory. Addison-Wesley Publishing Company, Advanced Book Program, Reading, Massachusetts, USA.

Van der Sande, C. J., S. M. de Jong, and A. P. J. De Roo. 2003. A segmentation and classification approach of IKONOS-2 imagery for land cover mapping to assist flood risk and flood damage assessment. *Int J Appl Earth Obs* 4:217–229.

Voigt, S., T. Kemper, T. Riedlinger, R. Kiefl, K. Scholte, and H. Mehl. 2007. Satellite image analysis for disaster and crisis-management support. *IEEE T Geosci Remote* 45:1520–1528.

Voormansik, K., J. Praks, O. Antropov, J. Jagomägi, and K. Zalite. 2013. Flood mapping with TerraSAR-X in forested regions in Estonia. *IEEE J Sel Top Appl* x:x–x.

Wang, Y., J. D. Colby, and K. A. Mulcahy. 2002. An efficient method for mapping flood extent in a coastal flood plain using Landsat TM and DEM data. *Int J Remote Sens* 23:3681–3696.

Wang, Y., L. L. Hess, S. Filoso, and J. M. Melack. 1995. Understanding the radar backscattering from flooded and non-flooded Amazonian forests: Results from canopy backscatter modeling. *Remote Sens Environ* 54:324–332.

Wang, Y., and M. L. Imhoff. 1993. Simulated and observed L-HH radar backscatter from tropical mangrove forests. *Int J Remote Sens* 14:2819–2828.

Warth, G., and S. Martinis. 2013. Improved flood detection by using bistatical coherence data of the TanDEM-X mission. Proceedings of 4. TanDEM-X Science Team Meeting. 4. TanDEM-X Science Team Meeting, Oberpfaffenhofen, Germany.

Wendleder, A., B. Wessel, A. Roth, M. Breunig, K. Martin, and S. Wagenbrenner. 2012. TanDEM-X water indication mask: Generation and first evaluation results. *IEEE J Sel Top Appl Earth Obs Remote Sens* 6:171–179.

Westerhoff, R. S., M. P. H. Kleuskens, H. C. Winsemius, H. J. Huizinga, G. R. Brakenridge, and C. Bishop. 2013. Automated global water mapping based on wide-swath orbital synthetic-aperture radar. *Hydrol Earth Syst Sc* 17:651–663.

Wu, S. T., and S. A. Sader. 1987. Multipolarization SAR data for surface feature delineation and forest vegetation characterization. *IEEE T Geosci Remote* 25:67–76.

Zadeh, L. A. 1965. Fuzzy sets. *Informa Control* 8:338–353.

Zeng, Z., Y. Gan, A. J. Kettner, Q. Yang, C. Zeng, G. R. Brakenridge, and Y. Hong. 2020. Towards high resolution flood monitoring: An integrated methodology using passive microwave brightness temperatures and sentinel synthetic aperture radar imagery. *Journal of Hydrology* 582:124377, ISSN 0022–1694. https://doi.org/10.1016/j.jhydrol.2019.124377. (www.sciencedirect.com/science/article/pii/S0022169419311126).

Zhang, Y. J. 1996. A survey on evaluation methods for image segmentation. *Pattern Recogn* 29:1335–1346.

6 Remote Sensing of Mangrove Forests

Le Wang, Jing Miao, and Ying Lu

ACRONYMS AND DEFINITIONS

LiDAR	Light Detection and Ranging
SIF	Solar-Induced Chlorophyll Fluorescence
UAV	Unmanned Aerial Vehicle
LUE	Light Use Efficiency
APAR	Absorbed Photosynthetically Active Radiation
SRG	Seeded Region Growing
MCWS	Marker-Controlled Watershed Segmentation
CHM	Canopy Height Model
ITDD	Individual Tree Detection and Delineation
LAI	Leaf Area Index
NDVI	Normalized Difference Vegetation Index
TLS	Terrestrial Laser Scans
VHOI	Vegetation Horizontal Occlusion Index
PRI	Photochemical Reflectance Index
GPP	Gross Primary Productivity
IKONOS	A commercial earth observation satellite typically collecting sub-meter to 5 m data
SRTM	Shuttle Radar Topography Mission
SLA	Specific Leaf Area
ITCs	Individual Tree Crowns
TH	Tree Height
CD	Crown Diameter
RTM	Radiative Transfer Models
VHOI	Vegetation Horizontal Occlusion Index
EC	Eddy Covariance
MRI	Mangrove Recognition Index
SMRI	Submerged Mangrove Recognition Index
LCLU	Land Cover and Land Use
DBH	Diameter at Breast Height

6.1 INTRODUCTION

Mangrove forests are among the most productive ecosystems in the coastal zone given the fact they are among the very few salt-tolerant plant communities inhabiting tropical and subtropical intertidal areas. As of 2021, the total area covered by mangrove forests worldwide was 15 million hectares, accounting for 1% of global tropical forests, which are distributed in 123 countries and territories (Maurya et al., 2021). Mangrove ecosystems have demonstrated indispensable ecological and economic benefits, such as the sequestration of carbon dioxide from the atmosphere, the protection of coastal inhabitants from storm surges, and the provision of habitat for a wide range of invaluable

Remote Sensing of Mangrove Forests

marine organisms (Bouillon et al., 2008; Wang et al., 2019). Unfortunately, mangrove forests have undergone significant disturbance in many areas around the world as a result of both anthropogenic and natural drivers. It is estimated that approximately 50% of the world's mangrove forests have disappeared over the past 50 years alongside ongoing deforestation taking place at 1–2% per year globally (Alongi, 2012). The disappearance of mangrove forests triggers a ripple effect in coastal ecosystems, impacting biodiversity and disrupting their intricate equilibrium. Furthermore, the decline of mangrove forests constitutes a multifaceted menace to human well-being, encompassing heightened vulnerability to natural disasters, economic distress, food insecurity, and exacerbation of climate change impacts. Therefore, there is a significant demand for the protection and restoration of mangrove forests.

Remote sensing of mangrove forests, encompassing a continuous spectrum of spatial scales ranging from leaf to individual tree to forest and finally to mangrove ecosystem, aim to retrieve vital functional indicators at each scale (Figure 6.1). To be more specific, the mangrove leaf is the smallest unit for research, and its characteristics, such as salt tolerance, unique shape, leathery texture, and adaptation for oxygen exchange largely determine the characteristics of the mangrove species (Noor et al., 2015; Parida & Jha, 2010). At the individual tree scale, the delineation of mangrove tree height and canopy structure can respond to stressors and disturbances that can differ significantly among various species and even among individuals within the same species (Lassalle et al., 2022). Furthermore, individual trees in the same ecological setting are more likely to interact in ecological as well as biochemical processes so as to form a mangrove forest. We can use the findings from the mangrove canopy to obtain valuable information for climate change mitigation and sustainable coastal management (Friess et al., 2022; Husain et al., 2020). Furthermore, the tidal patterns, soil composition, and the habitats that collectively constitute the entire mangrove ecosystem are of significant importance (Sasekumar, 1974). Observing mangroves at different spatial scales can lead our understanding in mangroves' interaction with its surrounding socioeconomic environment, as well as with global climate change.

However, traditional filed-based observations are often time-consuming and labor-intensive. The results of such studies depend heavily on the representativeness and accuracy of the samples

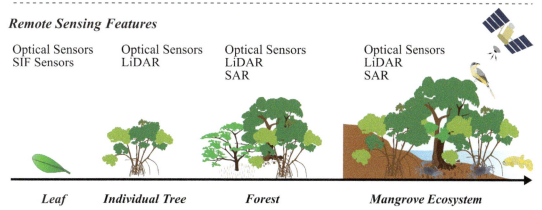

FIGURE 6.1 Remote sensing technologies uncover key functional indicators across different mangrove scales, from leaf to individual tree to forests, and to the entire mangrove ecosystem.

collected in the field. Fortunately, remote sensing technologies can conduct the necessary observations with a cost-effective means. Remote sensing observations at different scales have revealed various characteristics, such as spectral reflectance and LiDAR characteristics. These spectral and structural features from remote sensing data proved to be successful in identifying the extent of mangrove forests, interspecific classification, extraction of mangrove individuals, assessment of mangrove health, and description of biochemical processes (Wang et al., 2019). More importantly, remote sensing has helped to enhance the spatial scale and temporal scale of research, which have become indispensable and established tools for studying mangroves from leaf to individual tree, to canopy, to ecosystem scales.

In the following sections, we will delve into the methodology for extracting valuable features of mangrove forests at four progressive scales with various types of remote sensing data. Moreover, after effective processing and analysis, we will ultimately derive the essential functional indicators we mentioned earlier.

6.2 LEAF LEVEL

Mangrove leaves have adapted to the challenging and dynamic conditions of the intertidal zone along the coastlines of tropical and subtropical regions, where they are subjected to inundation by seawater, fluctuating water levels, and high levels of solar radiation (Mitra, 2013; Shank et al., 2010). In consequence, the characteristics discerned through remote sensing data exhibit notable disparities when compared to those of terrestrial forests and divergent land use. These disparities are frequently manifested in six key leaf attributes, i.e., specific leaf area (SLA), representing the leaf's surface area per unit of dry weight; photosynthetic capacity; the quantified nitrogen content within the leaf; the quantified phosphorus content within the leaf; the rate of dark respiration exhibited by the leaf; and the temporal duration characterizing leaf lifespan (longevity) (Saenger & West, 2016). Based on these six traits, remote sensing was utilized to collect spectral and even ecological process descriptive information of the leaves, which accurately revealed the different mangrove species, the health status of the mangrove forests, and the ecological processes of the mangrove forests.

Mangrove leaf with different species is important to provide a more detailed inventory of mangrove biodiversity to support more complicated management of coastal ecosystems. However, it is difficult to identify different mangrove species from multispectral remote sensing imagery acquired at medium spatial resolution due to limitations in the range and bandwidth of bands of the imagery. In contrast, hyperspectral remote sensing can provide more accurate and rich spectral information for species mapping, often yielding more favorable species classification results compared to multispectral imagery (Figure 6.2a). Furthermore, it is noteworthy that mangrove at leaf scale, when assessed across varying states of health, display discernible variations in hyperspectral information (Figure 6.2b). It is evident that the utilization of reflectance ratio indices, derived from hyperspectral bands, facilitates the extraction of pertinent spectral information. For instance, the spectral reflectance ratios R605/R760, R695/R760, and R710/R760 exhibit the capacity to discriminate between stressed and non-stressed red mangrove leaves. Conversely, in the case of white mangroves, the ratio R695/R420 emerges as the sole indicator proficient in detecting stress presence. Remarkably, all four ratios exhibit sensitivity to stress conditions in black mangroves. Moreover, Jiang et al. (2021) found that first derivative of hyperspectral provided good estimation accuracy in leaf traits of mangroves. To be more specific, the red edge position with the maximum first derivative reflectance value in the spectral range of 680–760 nm was 702.7, 701.3, 701.3, and 702.7 nm under the severity of 0–5%, 5–15%, 15–25%, and over 25%, respectively (Figure 6.3).

Observation of mangroves at leaf scale has shown new possibilities along with the advancement of remote sensors. For example, solar-induced chlorophyll fluorescence (SIF) remote sensing has emerged as a promising method for directly replacing photosynthetic activity in mangrove ecosystems. SIF, as a surrogate of photosynthetic processes, is an effective tool for assessing

Remote Sensing of Mangrove Forests

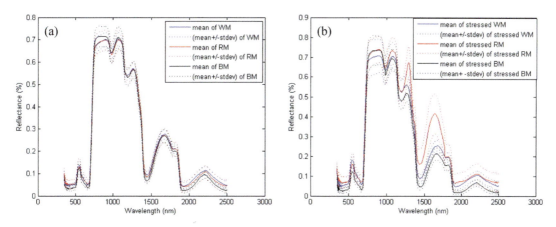

FIGURE 6.2 Hyperspectral reflectance (400–2400 nm) of three dominant mangrove species in Panama: white mangrove, red mangrove, and black mangrove (a) and their reflectance under stressed conditions (b) (from Wang and Sousa, 2009).

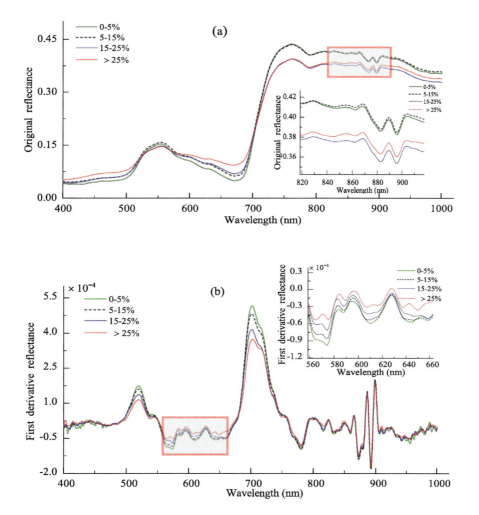

FIGURE 6.3 The mean original reflectance (a) and first derivative reflectance (b) spectra (400–1000 nm) derived from SOC hyperspectral images under the pest and disease severity of 0–5%, 5–15%, 15–25%, and > 25% (from Jiang et al., 2021).

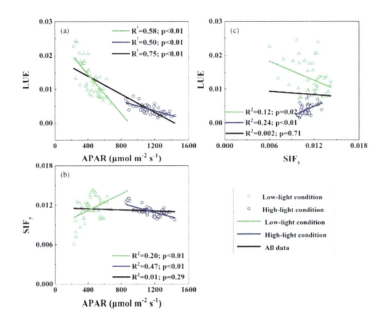

FIGURE 6.4 The correlation between SIF$_y$ from UAV observation, LUE, and APAR under different light conditions (from Wu et al., 2022).

photosynthesis in mangrove environments. Wu et al. (2022), for the first time, collected SIF$_y$ observations from ground and unmanned aerial vehicle (UAV) platform and provided an exploratory attempt to downscale SIF from canopy level to leaf level. Moreover, this research assessed the relationship between SIF$_y$, LUE, and absorbed photosynthetically active radiation (APAR) under different light conditions. It was found that the relationship between mangrove photosynthesis and SIF varied depending on the light conditions, with a negative correlation in low light and a positive correlation in high light (Figure 6.4). Similarly, the use of remote sensing data to reveal subtle biochemical processes in mangrove leaves may be the way forward for future research.

6.3 INDIVIDUAL TREE LEVEL

Conservation of mangrove forests depends on a comprehensive understanding of their complex structure, expanding to the level of individual trees (Yin & Wang, 2019). Ecological models centered on the dynamics of individual trees reveal valuable insights into how mangroves respond to factors such as inter-tree competition and changes in nutrient levels and anthropogenic impacts (Grueters et al., 2014; Weber et al., 2008; Zhang & DeAngelis, 2020). These models are valuable tools for guiding effective mangrove management practices. Studying the life cycle of individual mangrove trees is indispensable for documenting their growth patterns, phenological behavior, and mortality (Berger et al., 2008; Chamberlain et al., 2021). However, the challenges due to the dense canopy and muddy understory impede the measurements of individual mangroves.

Lassalle et al. (2022) utilized WorldView imagery combined with deep learning–based to enhance individual tree crowns (ITCs) and employed a marker-controlled watershed segmentation algorithm for their delineation. The MT-EDv3 neural network was employed for computing the normalized Euclidean distance between the crown pixels and the tree's apex. Following this, a Gaussian Lapp filter was applied to the resultant image to augment the delineation of the crown boundary prior to segmentation. This study utilized information enhancement and extraction methods to exploit the high spatial resolution spectra, which efficiently enabled the delineation of mangrove ITCs.

However, the efficiency of traditional optical remote sensing technologies in the measurement of mangrove ITCs is largely limited due to these structural characteristics. It requires the development of new methods and the use of innovative remote sensing data for more accurate delineation.

To this end, UAV-LiDAR provides valid observation information for the extraction of individual mangrove trees. Based on this method, tree height (TH) and crown diameter (CD) play a pivotal role in characterizing these fundamental attributes. Smaller bias and standard deviation can be achieved by UAV-LiDAR, which has a higher point cloud density, in tasks such as measuring the TH and CD of individual trees. Yin and Wang (2019) collected the point cloud of mangrove canopy by UAV-LiDAR, as shown in Figure 6.5. Their research proposed that the spatial resolution of individual tree detection and delineation (ITDD) should be less than 1/4 of the canopy diameter in mangroves, which is also applicable to other forest models. Finally, comparing the performance of two ITDD algorithms: seeded region growing (SRG) algorithm and marker-controlled watershed segmentation (MCWS) algorithm, MCWS was chosen to be superimposed on a canopy height model (CHM) with a resolution of 0.25 m (1/4 of the canopy diameter in the research area) to depict the individual mangrove forest.

However, there is still a small amount of work on the depiction of mangrove ITCs based on different types of remote sensing imagery. Key challenges in individual mangrove studies include the absence of ground-based LiDAR data, difficulties in distinguishing mangroves from other vegetation types, and the limited variation in canopy heights among neighboring mangroves. To enhance research in this field, potential avenues for improvement include integrating multiple remote sensing data features, such as ground-based LiDAR and high-resolution imagery, and enhancing the precision of extraction algorithms.

6.4 FOREST LEVEL (CANOPY)

Canopy structure pertains to the spatial arrangement of aboveground plant organs within a plant community (Campbell & Norman, 1989). In the context of mangrove forests, this structure is

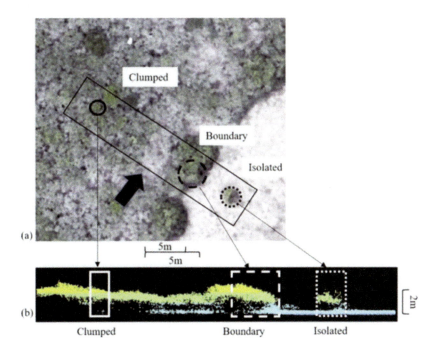

FIGURE 6.5 Point cloud from UAV-LiDAR on clumped, boundary, and isolated mangrove trees (from Yin and Wang, 2019).

notably characterized by the intricate interplay of branches and leaves in the crowns of mangrove trees (Nagelkerken et al., 2008). These branches often weave together to create a densely woven network. Additionally, the canopy features an abundance of dark green leaves that have evolved to withstand the harsh conditions of the mangrove environment, which include salinity, periodic tidal inundation, and elevated temperatures (Li et al., 2014; Noor et al., 2015; Parida & Jha, 2010). Importantly, the canopy generates a distinctive microclimate within the forest, marked by slightly cooler temperatures, increased humidity levels, and reduced wind speeds compared to the lower layers of the ecosystem (D'odorico et al., 2013; Devaney et al., 2017; Medellu & Berhimpon, 2012). The microclimate condition can significantly alternate the growth and survival of various species inhabiting the canopy. Consequently, delving into research within the mangrove canopy can offer valuable insights into the ecology, biodiversity, and ecosystem services that mangrove forests contribute to the natural world (Malik et al., 2015). Understanding canopy structure enables researchers to gain valuable insights into the functioning and resilience of these ecosystems, which is vital for conservation, land management, and our understanding of Earth's natural processes.

The utilization of remote sensing techniques for monitoring mangrove canopies offers an effective means to assess the overall health and productivity of regional mangrove forests. Traditional methods for leaf area index (LAI) measurement involve both direct contact and indirect transmittance techniques, which can be particularly challenging in the intricate and densely vegetated understory of mangrove ecosystems. Therefore, it becomes imperative to leverage a diverse array of available sensors, while also employing spectral and structural features to derive mangrove LAI. Tian et al. (2017) mapped mangrove LAI using UAV imagery, offering a comparative analysis against WorldView-2 imagery. Their findings revealed that the average Normalized Difference Vegetation Index (NDVI) yielded the highest accuracy for WorldView-2 (R^2 = 0.778), whereas scaled NDVI achieved optimal accuracy for UAV data (R^2 = 0.817). Furthermore, Zhao et al. (2023) demonstrated that hyperspectral data from the Zhuhai-1 satellite, coupled with radiative transfer models (RTM) and machine learning technology (XGBoost), significantly enhanced the precision of mangrove LAI inversion (R^2 = 0.86). Additionally, comprehensive insights into the three-dimensional structural attributes of mangrove forests, obtained through terrestrial laser scans (TLS), prove instrumental in overcoming challenges related to clumped canopy structures. Guo et al. (2018) introduced a novel metric known as the vegetation horizontal occlusion index (VHOI), which enhances the assessment of TLS data quality. By combining UAV imagery with TLS data, this approach facilitates a more accurate measurement of the LAI of mangrove ecosystems. In summary, the integration of various remote sensing technologies, spectral analyses, and structural assessments offers a comprehensive toolkit for evaluating the health, productivity, and structural intricacies of mangrove canopies, thereby advancing our understanding of these critical coastal ecosystems.

Studies quantifying the biochemical parameters as well as dynamic processes of mangrove canopies have benefited from the integration of remote sensing at different spatial scales, the varying data types, and even the in situ collection of biochemical sampling data. Heenkenda et al. (2015) mapped LAI of mangroves based on high spatial resolution bands of WorldView-2 satellite imagery, and their spectral transformations based the digital cover photography method. The spatial distribution of canopy chlorophyll variation was then estimated from correlations between LAI and canopy chlorophyll content. In addition to the structural characteristics, there are other studies that focus on the carbon dynamics within mangrove canopies. Zhu et al. (2019) conducted a year-long experiment that involved, uninterrupted collection of high-resolution spectral data, half-hourly measurements of eddy covariance (EC) carbon fluxes, and the monitoring of numerous environmental parameters (Figure 6.6). Their study explored connections between mangrove photosynthetic activity (Photochemical Reflectance Index, PRI) and carbon dynamics, examining how various environmental factors influence these relationships over different time scales (daily and seasonal). Based on the same EC data, Zhu et al. (2021) explored SIF-GPP relationship in subtropical mangrove forests and its environmental controls with one-year time-series measurements. It revealed that the

Remote Sensing of Mangrove Forests

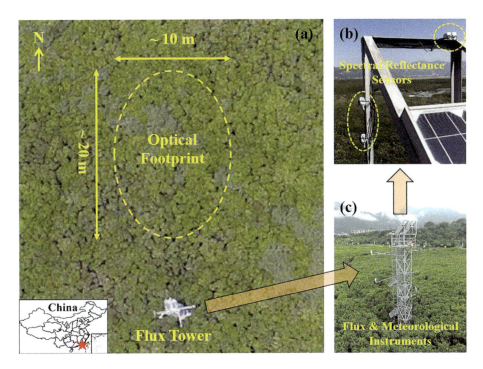

FIGURE 6.6 Aerial view of mangrove forests near flux tower overlaid with approximate footprint of spectral reflectance sensors deployed on the top of the flux tower, and flux and deployed meteorological instruments (from Zhu et al., 2019).

daytime correlation between PRI and carbon dynamics fluctuated in response to meteorological shifts. Furthermore, it indicated that PRI measures linked to physiological processes outperformed conventional vegetation indices based on structural characteristics in accurately monitoring mangrove photosynthetic parameters.

Studies on mangrove canopy are more established than those at other scales. However, they are still lagging behind those studies of terrestrial forests. This lag is primarily due to frequent cloud cover and intermittent tidal inundation in the coastal area (Wang et al., 2019; Xia et al., 2020). The accumulation of long time series remote sensing imagery presents the opportunity to address these challenges. Recent studies utilized the long-term remote sensing imageries to extract and characterize the mangrove canopy at varying tidal levels, which achieved impressive performances (Gao & Zhou, 2023; Xia et al., 2020; Zhang et al., 2017). However, existing studies have only addressed the boundary extraction of mangrove forests under complex tidal and cloudy conditions while the characterization of biochemical indicators as well as energy exchange-based remote sensing technologies has not yet been adequately addressed. Integration of multiple remote sensing data, assisted by the accumulation of ground data, may offer new possibilities for solving the aforementioned problems.

6.5 ECOSYSTEM LEVEL

Mangrove ecosystems include both the biotic communities living in mangrove forests and the abiotic components that exist in mangrove areas. The complex structure and unique location of mangrove forests contributes to their richness in nutrients and biodiversity at the mangrove ecosystem level (Alongi, 2009; Macintosh & Ashton, 2002). These characteristics further equip mangrove ecosystems with a remarkable ability to protect coastal regions from the erosive impacts of tides and the destructive forces of hurricanes (Kazemi et al., 2021; Spalding et al., 2014). Additionally, tides in mangrove ecosystems facilitate the nutrient exchange between mangrove forests and the adjacent communities (Alongi, 2009). For a more comprehensive understanding of mangrove ecosystems, the subsequent sections provide insights from four pivotal perspectives: tidal-driven nutrient exchange, biodiversity, carbon conservation, and coastal area protection. Within each section, the applications of remote sensing are summarized.

6.5.1 Tidal-Driven Nutrient Exchange

Tides are the foundation for the flourishing of both mangrove forests and the adjacent ecosystems (Alongi, 2009). By regularly moving in and out, tides facilitate the exchange of water and sediments within mangrove ecosystems. For mangrove ecosystems, it is a major factor influencing the availability of non-organic nutrients (Reef et al., 2010). Due to the sealing effects made by tides and sedimentation, the decomposition rate in mangrove forests is very low, resulting in limited production of inorganic nutrients by the ecosystems themselves (Alongi, 2009; Ola & Lovelock, 2021; Reef et al., 2010). Therefore, mangrove ecosystems depend heavily on external sources of non-organic nutrients, especially nitrates and phosphates, rather than generating them internally. On the other hand, it means that mangrove ecosystems are rich in organic matters from leaf litters and animal detritus (Ola & Lovelock, 2021). Tides helps transport these organic matters from mangrove ecosystems to the adjacent ecosystems, making mangrove ecosystems significant organic nutrient pools for other creatures adjacent to the mangrove ecosystems. Therefore, it is of great significance to trace the situation of tides in mangrove forests.

Remote sensing, with its spatially and temporally continuous observations, provide a significant opportunity to monitor the degrees and heights of tides. Several indices have been designed for this purpose, such as the mangrove recognition index (MRI) (Zhang & Tian, 2013) and the submerged mangrove recognition index (SMRI) (Xia et al., 2021). These indices are proven to be effective in identifying mangrove forests that have been merged by tides (Figure 6.7). Particularly, the SMRI was investigated when the tides were low, medium, and high in mangrove forests. An exponential relationship between SMRI and tide level has been validated. Therefore, with these indices, the tides in mangrove forests can be effectively monitored.

Furthermore, while directly monitoring nutrient exchange within mangrove ecosystems is challenging with remote sensing, factors influencing the degree of nutrient exchange can be discerned. It is estimated that the extent of the nutrient exchange between a mangrove ecosystem and the adjacent ocean ecosystem is influenced by a variety of factors, such as net forest primary production, tidal range, ratio of mangrove to watershed area, total mangrove area, frequency of storms, amount of rainfall, and volume of water exchange (Twilley, 1988). The majority of these factors can be detected using remote sensing. For instance, remote sensing has long been employed to estimate the biomass of mangrove forests by building relationships between field measurement biomass data with satellite imagery data (Pham et al., 2019; Wang et al., 2019). In addition, the availability of numerous remote sensing–derived mangrove maps, such as the Global Mangrove Watch (Bunting et al., 2018), provides comprehensive information of mangrove extents worldwide. Land cover and land use (LCLU) products, such as the National Land Cover Database based on Landsat imagery (Homer et al., 2012) and the WorldCover database based on Sentinel imagery (Zanaga et al., 2022), enable the detection of land categories within mangrove regions, including water bodies and mangrove vegetation (Figure 6.8). Furthermore, high-resolution remote sensing imageries, such as of the planet, empowers researchers to extract a wealth of detailed information. This includes accurate mapping of mangrove forest densities, the distribution of watershed and mangrove areas, and other pertinent details. Consequently, remote sensing stands as a potential and effective tool for unraveling the effects of tides within mangrove ecosystems.

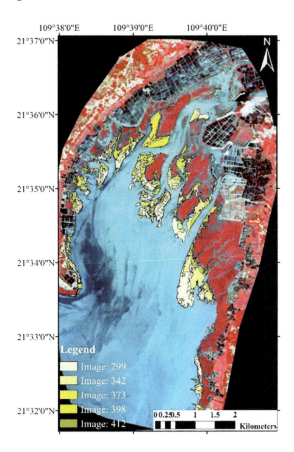

FIGURE 6.7 Area of submerged mangrove forests at various tide heights distinguished by SMRI (from Xia et al., 2021).

6.5.2 BIODIVERSITY

Due to the rich food webs and stable environment, mangrove forests are excellent sites for obtaining food and providing shelters. They support a wide range of organisms, from bacteria to big predators, within both the forests and adjacent areas (Das et al., 2022; Macintosh & Ashton, 2002). The strong and tangled root systems in mangrove forests create several small niches that serve as perfect shelters for juveniles of countless organisms, such as invertebrates, algae, crabs, and fish (Alongi, 2009). Organic matters in mangrove soil ensure a stable food supply for these inhabitants. In regions like South Florida, about 75% of the game fish and 90% of the commercial species rely on mangrove forests at least once during their lifespan (Wilkie & Fortuna, 2003). The waste produced by these organisms, in turn, contributes to the enrichment of organic matters within the mangrove forests. Furthermore, with shallow waters, exposed mudflats, dense and sturdy tree branches, and abundant food sources, mangrove forests provide ideal habitats for birds, reptiles, and amphibians (Luther, 2014; Mohd-Azlan et al., 2015). These creatures often prey upon fish, crabs, and smaller organisms residing within the mangrove ecosystem. Additionally, a variety of mammals visit mangrove forests in search of food, such as tigers, foxes, deer, and rabbits. Snakes, turtles, and crocodiles are also commonly found in mangrove forests (FIELD et al., 1998; Macintosh & Ashton, 2002). Collectively, these diverse inhabitants highlight the uniqueness and significance of mangrove forest biodiversity.

However, it is almost impossible to evaluate the biodiversity of mangrove forests using remote sensing alone. This is because remote sensing primarily relies on detecting and analyzing electromagnetic

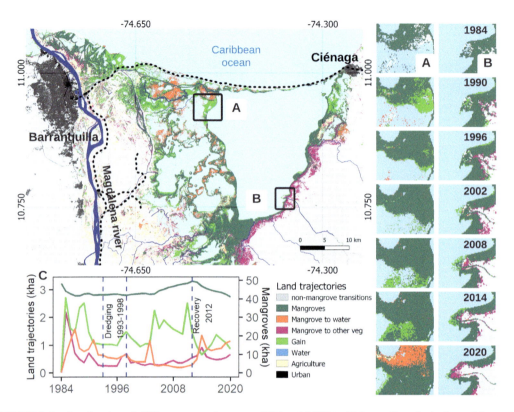

FIGURE 6.8 Landsat-based difference map between 1984 and 2020 in Ciénaga Grande de Santa Marta. (a) Image subsets represented mangrove gain with mangrove to open water detected in 2020. (b) Image subsets show mangrove to other vegetation inland and gain since 2008. (c) The time-series plot shows the land trajectories of change and dredging periods between 1993 and 1998 and mangrove recovery from 2004 to its maximum in 2012 (from Murillo-Sandoval et al., 2022).

radiation emitted or reflected from the Earth's surface, making it challenging to detect the traces of animals directly. Thus, existing studies of the biodiversity in mangrove forests mainly focus on the classification of mangrove species. Wang et al. (2004) was the first study that successfully mapped mangrove species. The IKONOS 1-m panchromatic and 4-m multispectral images are utilized to classify mangrove species along the Caribbean coast of Panama. They found that as mangrove species are clustered as narrow strips and small patches, high spatial resolution images and object-based image analysis are essential.

Additionally, the utilization of remote sensing LiDAR data is promising for future biodiversity studies of mangrove forests, since LiDAR is capable of measuring the structural complexity of forests, a widely acknowledged and robust indicator of biodiversity. Additionally, the structural complexity of a forest is also listed as one of the Essential Biodiversity Variables (Pereira et al., 2013). The laser pulses emitted by LiDAR systems can penetrate the dense canopy of mangrove trees, thereby offering detailed, three-dimensional information of mangrove structures. However, it is crucial to emphasize that research in this domain remains relatively limited. A notable example comes from the work of Yin and Wang (2019). They successfully utilized the UAV LiDAR to delineate the distribution of individual trees in a mangrove forest in China. They found that a high point density is required for effective and accurate detection of mangrove structures. This is attributed to the fact that mangrove trees are highly clumped and overlapped. Nevertheless, it is evident that further comprehensive studies are still needed to explore the application of LiDAR data in this domain.

6.5.3 CARBON CONSERVATION

Mangrove forests play a pivotal role in the conservation of carbon, making them invaluable ecosystems for climate change mitigation and carbon sequestration. Growing in tidal environments where nitrogen and phosphorus contents are extremely low, plants in mangrove forests develop their own nutrient-conserving processes. Most plants in mangrove forests are evergreen with low growth rate, which means a smaller nutrient investment for new leaves and lower nutrient loss rates than deciduous plants (Alongi, 2020a; Holmer & Olsen, 2002). They increase the efficiency of metabolic processes, reabsorb nutrients before the leaves fall and reuse old roots (Alongi, 2009). Also, after the leaves fall, through the process of leaf litter decomposition, some nutrients are immobilized back into the mangrove forests. Moreover, due to the tides and sealing muds, oxygen is deficient, resulting in a very low decomposition rate in mangrove forests, which means that mangroves are rich in organic matter from leaf litter and animal detritus (Alongi, 2020b). Thus, mangrove forests are very good at sequestering organic carbon within the ecosystem. It is estimated that the carbon density of the mangrove ecosystem is four times that of other tropical forests despite the fact that they only account for 0.4–7% of the total global carbon sink (Alongi, 2020a; Spalding & Leal, 2021).

Remote sensing has been used for estimating mangrove carbon stock for decades. These studies can be divided into two main categories. The first category aims at developing allometric equations to estimate the relationship between carbon stock and biophysical parameters derived from remote sensing data. The engaged parameters typically include diameter at breast height (DBH), tree height, crown cover, and so on. A representative study was conducted by Fatoyinbo et al. (2008). They successfully estimated mangrove biomass carbon stocks in a mangrove forest in Mozambique, utilizing tree height data derived from the Shuttle Radar Topography Mission (SRTM). Alternatively, the second category of carbon stock study in mangrove forests involves the development of regression models from vegetation indices. A pioneering study by Wicaksono et al. (2011). This was the first study that focused on the carbon stock mapping of mangrove ecosystems. Both aboveground carbon and belowground carbon were estimated in their study site. In addition, the effectiveness of various vegetation indices was compared. Finally, the combination of linear regression and global environment monitoring index was considered as the most effective. To summarize, remote sensing is indispensable for the carbon stock estimation of mangrove forests.

Additionally, the photosynthetic capacity of mangrove forests is a pivotal component in determining carbon sink efficiency (Alongi, 2009). It can be evaluated by biological traits, such as leaf chlorophyll content, leaf nitrogen content, light use efficiency, and fluorescence. Traditionally, these traits are directly collected in the fields. Considerable time and labor are required for studies even at local scales, making large-scale studies almost impossible. Thanks to remote sensing technologies, we have found that these biological traits have a significant relationship with the structure of mangrove forests and the environment in mangrove ecosystems that can be detected through reliable satellite measurements. Forest structure indices, such as LAI, biomass, and tree height have been used to evaluate the carbon sink of terrestrial forests since 1981. For instance, Chen et al. (2019) successfully evaluated the global carbon sink of terrestrial forests through the status of LAI, CO_2 fertilization, nitrogen deposition, and climate information derived from various remote sensing images and stationary data. Their methods have a high potential to be applied in mangrove forests. We believe, with satellite meteorological data such as temperature and precipitation, we will be able to simulate the changes of these biological traits in mangrove forests.

6.5.4 COASTAL AREA PROTECTION

With rich species and complex root systems, mangrove forests are effective physical barriers safeguarding the coastal areas from the storm surges, tropical cyclones, high winds, tidal bores, seawater seepage, and intrusion (Del Valle et al., 2020; Sun & Carson, 2020). They slow down water velocities and deposit suspended sediments in waves. The destructive force of storms and tsunamis is decreased when they approach the upland. The rate at which water passes to upland is declined. So, mangrove forests are considered buffer zones between the coast and ocean that stabilize the coasts and filter the water.

The use of remote sensing for costal area protection has encountered some challenges. Although a few efforts have been made, we consider two potential aspects: the evaluation of hurricane strength and the evaluation of property loss. The strength of waves and storms are key factors representing the influence of a hurricane. However, evaluating or quantifying these factors are difficult. Traditionally, the data came from field station observations, whereas the wind energy can be recorded consistently with remote sensing. For instance, Blended Sea Winds offers the wind speed every six hours at 0.25° resolution. In addition, the process and extent of storm surges are influenced by coastal topography, which can be derived from SRTM or LiDAR images. Thus, we believe that, equipped with tide gauge data, a spatial inundation model at different water levels can be built with remote sensing images. Thus, remote sensing can boost the evaluation of hurricane strength. Additionally, the loss of coastal properties after a hurricane is a significant indicator for the role of mangrove forests in coastal protection. Although we cannot measure the casualties with remote sensing, the loss of buildings, forests, or other land covers can all be detected using change detection with optical images or radar images. In addition, the nighttime light remote sensing was proven to be effective in estimating the change of economic activities caused by hurricanes (Del Valle et al., 2020). Therefore, with the availability of advanced images, the role of mangrove forests in coastline protection is promising for future investigation.

6.6 CONCLUSION

Our chapter serves as an introductory exploration of the diverse applications of remote sensing techniques in mangroves across four different spatial scales: leaf, individual tree, forest, and ecosystem. At the leaf level, hyperspectral features stand out as the most useful approach. Hyperspectral efficiently differentiates several mangrove species and successfully detects the external stress conditions without direct contact or any disruption to the delicate mangrove environment. For the individual tree scale, LiDAR takes the spotlight as the most efficient method, given its capacity to penetrate the mangrove canopy. Thus, it offers an efficient means of delineating the boundaries between individual trees within interconnected mangrove forests. At the forest level, observations primarily revolve around long time-series data and large spatial coverage imagery. This expansion in both spatial and temporal dimensions empowers researchers to overcome the challenges posed by the often-unpredictable coastal weather where mangroves thrive. Finally, when focusing on the entire mangrove ecosystem, a complex interplay of marine, terrestrial, urban, and various other organisms is revealed. Here, integration of different types of remote sensing images becomes indispensable for comprehensive ecosystem analysis. In summary, remote sensing emerges as an essential tool for the study of mangroves at a full spectrum of spatial scales, offering invaluable insights and data for understanding these vital coastal ecosystems.

REFERENCES

Alongi, D. M. (2009). *The Energetics of Mangrove Forests*. Springer Science & Business Media.
Alongi, D. M. (2012). Carbon sequestration in mangrove forests. *Carbon Management*, *3*(3), 313–322.
Alongi, D. M. (2020a). Global significance of mangrove blue carbon in climate change mitigation. *Sci*, *2*(3), 67.
Alongi, D. M. (2020b). Nitrogen cycling and mass balance in the world's mangrove forests. *Nitrogen*, *1*(2), 167–189.
Berger, U., Rivera-Monroy, V. H., Doyle, T. W., Dahdouh-Guebas, F., Duke, N. C., Fontalvo-Herazo, M. L., Hildenbrandt, H., Koedam, N., Mehlig, U., & Piou, C. (2008). Advances and limitations of individual-based models to analyze and predict dynamics of mangrove forests: A review. *Aquatic Botany*, *89*(2), 260–274.

Bouillon, S., Borges, A. V., Castañeda-Moya, E., Diele, K., Dittmar, T., Duke, N. C., ... & Twilley, R. R. (2008). Mangrove production and carbon sinks: A revision of global budget estimates. *Global Biogeochemical Cycles, 22*(2).

Bunting, P., Rosenqvist, A., Lucas, R. M., Rebelo, L.-M., Hilarides, L., Thomas, N., Hardy, A., Itoh, T., Shimada, M., & Finlayson, C. M. (2018). The global mangrove watch—a new 2010 global baseline of mangrove extent. *Remote Sensing, 10*(10), 1669.

Campbell, G. S., & Norman, J. M. (1989). The description and measurement of plant canopy structure. In B. Marshall, G. Russell, & P. G. Jarvis (Eds.), *Plant Canopies: Their Growth, Form and Function* (pp. 1–20). Cambridge University Press.

Chamberlain, D. A., Phinn, S. R., & Possingham, H. P. (2021). Mangrove forest cover and phenology with Landsat dense time series in central Queensland, Australia. *Remote Sensing, 13*(15), 3032.

Chen, J. M., Ju, W., Ciais, P., Viovy, N., Liu, R., Liu, Y., & Lu, X. (2019). Vegetation structural change since 1981 significantly enhanced the terrestrial carbon sink. *Nature Communications, 10*(1), 4259.

Das, S. C., Thammineni, P., & Ashton, E. C. (2022). Mangroves: A unique ecosystem and its significance. In *Mangroves: Biodiversity, Livelihoods and Conservation* (pp. 3–11). Springer.

Del Valle, A., Eriksson, M., Ishizawa, O. A., & Miranda, J. J. (2020). Mangroves protect coastal economic activity from hurricanes. *Proceedings of the National Academy of Sciences, 117*(1), 265–270.

Devaney, J. L., Lehmann, M., Feller, I. C., & Parker, J. D. (2017). *Mangrove Microclimates Alter Seedling Dynamics at the Range Edge*. Wiley Online Library.

D'odorico, P., He, Y., Collins, S., De Wekker, S. F., Engel, V., & Fuentes, J. D. (2013). Vegetation—microclimate feedbacks in woodland—grassland ecotones. *Global Ecology and Biogeography, 22*(4), 364–379.

Fatoyinbo, T. E., Simard, M., Washington-Allen, R. A., & Shugart, H. H. (2008). Landscape-scale extent, height, biomass, and carbon estimation of Mozambique's mangrove forests with Landsat ETM+ and Shuttle Radar Topography Mission elevation data. *Journal of Geophysical Research:Biogeosciences, 113*(G2).

Field, C., Osborn, J., Hoffman, L., Polsenberg, J., Ackerly, D., Berry, J., Björkman, O., Held, A., Matson, P., & Mooney, H. (1998). Mangrove biodiversity and ecosystem function. *Global Ecology & Biogeography Letters, 7*(1), 3–14.

Friess, D. A., Adame, M. F., Adams, J. B., & Lovelock, C. E. (2022). Mangrove forests under climate change in a 2 C world. *Wiley Interdisciplinary Reviews:Climate Change, 13*(4), e792.

Gao, E., & Zhou, G. (2023). Spatio-temporal changes of mangrove-covered tidal flats over 35 years using satellite remote sensing imageries: A case study of Beibu Gulf, China. *Remote Sensing, 15*(7), 1928.

Grueters, U., Seltmann, T., Schmidt, H., Horn, H., Pranchai, A., Vovides, A., Peters, R., Vogt, J., Dahdouh-Guebas, F., & Berger, U. (2014). The mangrove forest dynamics model mesoFON. *Ecological Modelling, 291*, 28–41.

Guo, X., Wang, L., Tian, J., Yin, D., Shi, C., & Nie, S. (2018). Vegetation horizontal occlusion index (VHOI) from TLS and UAV image to better measure mangrove LAI. *Remote Sensing, 10*(11), 1739.

Heenkenda, M. K., Joyce, K. E., Maier, S. W., & de Bruin, S. (2015). Quantifying mangrove chlorophyll from high spatial resolution imagery. *ISPRS Journal of Photogrammetry and Remote Sensing, 108*, 234–244.

Holmer, M., & Olsen, A. B. (2002). Role of decomposition of mangrove and seagrass detritus in sediment carbon and nitrogen cycling in a tropical mangrove forest. *Marine Ecology Progress Series, 230*, 87–101.

Homer, C. H., Fry, J. A., & Barnes, C. A. (2012). The national land cover database. *USGeological Survey Fact Sheet, 3020*(4), 1–4.

Husain, P., Al Idrus, A., & Ihsan, M. S. (2020). The ecosystem services of mangroves for sustainable coastal area and marine fauna in Lombok, Indonesia: A review. <cite lang="id">*Jurnal Inovasi Pendidikan dan Sains, 1*(1), 1–7.

Jiang, X., Zhen, J., Miao, J., Zhao, D., Wang, J., & Jia, S. (2021). Assessing mangrove leaf traits under different pest and disease severity with hyperspectral imaging spectroscopy. *Ecological Indicators, 129*, 107901.

Kazemi, A., Castillo, L., & Curet, O. M. (2021). Mangrove roots model suggest an optimal porosity to prevent erosion. *ScientificReports, 11*(1), 9969.

Lassalle, G., Ferreira, M. P., La Rosa, L. E. C., & de Souza Filho, C. R. (2022). Deep learning-based individual tree crown delineation in mangrove forests using very-high-resolution satellite imagery. *ISPRS Journal of Photogrammetry and Remote Sensing, 189*, 220–235.

Li, Q., Lu, W., Chen, H., Luo, Y., & Lin, G. (2014). Differential responses of net ecosystem exchange of carbon dioxide to light and temperature between spring and neap tides in subtropical mangrove forests. *The Scientific World Journal, 2014*.

Luther, D. (2014). The invasion of terrestrial fauna into marine habitat: Birds in mangroves. *Invasion Biology and Ecological Theory:Insights from a Continent in Transformation*, *15*(10), 103-117.

Macintosh, D. J., & Ashton, E. C. (2002). A review of mangrove biodiversity conservation and management. Centre for Tropical Ecosystems Research, University of Aarhus, Denmark.

Malik, A., Fensholt, R., & Mertz, O. (2015). Mangrove exploitation effects on biodiversity and ecosystem services. *Biodiversity and Conservation*, *24*, 3543–3557.

Maurya, K., Mahajan, S., & Chaube, N. (2021). Remote sensing techniques: Mapping and monitoring of mangrove ecosystem—a review. *Complex &Intelligent Systems*, *7*, 2797–2818.

Medellu, C. S., & Berhimpon, S. (2012). The influence of opening on the gradient and air temperature edge effects in mangrove forests. *International Journal of Basic & Applied Sciences IJBAS-IJENS*, *12*, 53–57.

Mitra, A. (2013). *Sensitivity of Mangrove Ecosystem to Changing Climate* (Vol. 62). Springer.

Mohd-Azlan, J., Noske, R. A., & Lawes, M. J. (2015). The role of habitat heterogeneity in structuring mangrove bird assemblages. *Diversity*, *7*(2), 118–136.

Murillo-Sandoval, P. J., Fatoyinbo, L., & Simard, M. (2022). Mangroves cover change trajectories 1984–2020: The gradual decrease of mangroves in Colombia. *Frontiers in Marine Science*, *9*, 892946.

Nagelkerken, I., Blaber, S., Bouillon, S., Green, P., Haywood, M., Kirton, L., Meynecke, J.-O., Pawlik, J., Penrose, H., & Sasekumar, A. (2008). The habitat function of mangroves for terrestrial and marine fauna: A review. *Aquatic Botany*, *89*(2), 155–185.

Noor, T., Batool, N., Mazhar, R., & Ilyas, N. (2015). Effects of siltation, temperature and salinity on mangrove plants. *European Academic Research*, *2*(11), 14172–14179.

Ola, A., & Lovelock, C. E. (2021). Decomposition of mangrove roots depends on the bulk density they grew in. *Plant and Soil*, *460*, 177–187.

Parida, A. K., & Jha, B. (2010). Salt tolerance mechanisms in mangroves: A review. *Trees*, *24*(2), 199–217.

Pereira, H. M., Ferrier, S., Walters, M., Geller, G. N., Jongman, R. H., Scholes, R. J., Bruford, M. W., Brummitt, N., Butchart, S. H., & Cardoso, A. (2013). Essential biodiversity variables. *Science*, *339*(6117), 277–278.

Pham, T. D., Yokoya, N., Bui, D. T., Yoshino, K., & Friess, D. A. (2019). Remote sensing approaches for monitoring mangrove species, structure, and biomass: Opportunities and challenges. *Remote Sensing*, *11*(3), 230.

Reef, R., Feller, I. C., & Lovelock, C. E. (2010). Nutrition of mangroves. *Tree Physiology*, *30*(9), 1148–1160.

Saenger, P., & West, P. W. (2016). Determinants of some leaf characteristics of Australian mangroves. *Botanical Journal of the Linnean Society*, *180*(4), 530–541.

Sasekumar, A. (1974). Distribution of macrofauna on a Malayan mangrove shore. *The Journal of Animal Ecology*, *43*(1), 51–69.

Shank, G. C., Zepp, R. G., Vähätalo, A., Lee, R., & Bartels, E. (2010). Photobleaching kinetics of chromophoric dissolved organic matter derived from mangrove leaf litter and floating Sargassum colonies. *Marine Chemistry*, *119*(1–4), 162–171.

Spalding, M. D., & Leal, M. (2021). The state of the world's mangroves 2021. *Global Mangrove Alliance*, 79.

Spalding, M. D., Mcivor, A., Tonneijck, F., Tol, S., & Eijk, P. V. (2014). *Mangroves for Coastal Defence*. Wetlands International and the Nature Conservancy.

Sun, F., & Carson, R. T. (2020). Coastal wetlands reduce property damage during tropical cyclones. *Proceedings of the National Academy of Sciences*, *117*(11), 5719–5725.

Tian, J., Wang, L., Li, X., Gong, H., Shi, C., Zhong, R., & Liu, X. (2017). Comparison of UAV and WorldView-2 imagery for mapping leaf area index of mangrove forest. *International Journal of Applied Earth Observation and Geoinformation*, *61*, 22–31.

Twilley, R. R. (1988). Coupling of mangroves to the productivity of estuarine and coastal waters. Coastal-Offshore Ecosystem Interactions: Proceedings of a Symposium Sponsored by SCOR, UNESCO, San Francisco Society, California Sea Grant Program, and US Dept. of Interior, Mineral Management Service held at San Francisco State University, Tiburon, CA, April 7–22, 1986.

Wang, L., Jia, M., Yin, D., & Tian, J. (2019). A review of remote sensing for mangrove forests: 1956–2018. *Remote Sensing of Environment*, *231*, 111223.

Wang, L., & Sousa, W. P. (2009). Distinguishing mangrove species with laboratory measurements of hyperspectral leaf reflectance. *International Journal of Remote Sensing*, *30*(5), 1267–1281.

Wang, L., Sousa, W. P., Gong, P., & Biging, G. S. (2004). Comparison of IKONOS and QuickBird images for mapping mangrove species on the Caribbean coast of Panama. *Remote Sensing of Environment*, *91*(3–4), 432–440.

Weber, P., Bugmann, H., Fonti, P., & Rigling, A. (2008). Using a retrospective dynamic competition index to reconstruct forest succession. *Forest Ecology and Management, 254*(1), 96–106.

Wicaksono, P., Danoedoro, P., Hartono, H., Nehren, U., & Ribbe, L. (2011). Preliminary work of mangrove ecosystem carbon stock mapping in small island using remote sensing: Above and below ground carbon stock mapping on medium resolution satellite image. Remote Sensing for Agriculture, Ecosystems, and Hydrology XIII, Prague, Czech Republic.

Wilkie, M. L., & Fortuna, S. (2003). Status and trends in mangrove area extent worldwide. *Forest Resources Assessment Programme*. Working Paper (FAO). Rome, Italy.

Wu, L., Wang, L., Shi, C., & Yin, D. (2022). Detecting mangrove photosynthesis with solar-induced chlorophyll fluorescence. *International Journal of Remote Sensing, 43*(3), 1037–1053.

Xia, Q., Jia, M., He, T., Xing, X., & Zhu, L. (2021). Effect of tide level on submerged mangrove recognition index using multi-temporal remotely-sensed data. *Ecological Indicators, 131*, 108169.

Xia, Q., Qin, C.-Z., Li, H., Huang, C., Su, F.-Z., & Jia, M.-M. (2020). Evaluation of submerged mangrove recognition index using multi-tidal remote sensing data. *Ecological Indicators, 113*, 106196.

Yin, D., & Wang, L. (2019). Individual mangrove tree measurement using UAV-based LiDAR data: Possibilities and challenges. *Remote Sensing of Environment, 223*, 34–49.

Zanaga, D., Van De Kerchove, R., Daems, D., De Keersmaecker, W., Brockmann, C., Kirches, G., Wevers, J., Cartus, O., Santoro, M., & Fritz, S. (2022). ESA WorldCover 10 m 2021 v200. https://doi.org/10.5281/zenodo.5571936

Zhang, B., & DeAngelis, D. L. (2020). An overview of agent-based models in plant biology and ecology. *Annals of Botany, 126*(4), 539–557.

Zhang, X., & Tian, Q. (2013). A mangrove recognition index for remote sensing of mangrove forest from space. *Current Science, 105*(8), 1149.

Zhang, X., Treitz, P. M., Chen, D., Quan, C., Shi, L., & Li, X. (2017). Mapping mangrove forests using multi-tidal remotely-sensed data and a decision-tree-based procedure. *International Journal of Applied Earth Observation and Geoinformation, 62*, 201–214.

Zhao, D., Zhen, J., Zhang, Y., Miao, J., Shen, Z., Jiang, X., Wang, J., Jiang, J., Tang, Y., & Wu, G. (2023). Mapping mangrove leaf area index (LAI) by combining remote sensing images with PROSAIL-D and XGBoost methods. *Remote Sensing in Ecology and Conservation, 9*(3), 370–389.

Zhu, X., Hou, Y., Zhang, Y., Lu, X., Liu, Z., & Weng, Q. (2021). Potential of sun-induced chlorophyll fluorescence for indicating mangrove canopy photosynthesis. *Journal of Geophysical Research:Biogeosciences, 126*(4), e2020JG006159.

Zhu, X., Song, L., Weng, Q., & Huang, G. (2019). Linking in situ photochemical reflectance index measurements with mangrove carbon dynamics in a subtropical coastal wetland. *Journal of Geophysical Research:Biogeosciences, 124*(6), 1714–1730.

Part IV

Wetlands

7 Remote Sensing of Mangrove Wetlands

Chandra Giri

ACRONYMS AND DEFINITIONS

ASTER	Advanced Spaceborne Thermal Emission and Reflection Radiometer
DEM	Digital Elevation Model
DOC	Dissolved Organic Carbon
EROS	Earth Resources Observation and Science
ETM+	Enhanced Thematic Mapper Plus
FAO	Food and Agriculture Organization
GLS	Global Land Survey
IKONOS	A commercial earth observation satellite typically collecting sub-meter to 5 m data
IRS	Indian Remote Sensing Satellites
LiDAR	Light Detection and Ranging
MODIS	Moderate-Resolution Imaging Spectroradiometer
MRV	Monitoring, Reporting, and Verification
NDVI	Normalized Difference Vegetation Index
NGA	National Geospatial-Intelligence Agency
OLI	Operational Land Imager
PALSAR	Phased Array type L-band Synthetic Aperture Radar
RADARSAT	RADAR Satellite
REDD	Reduced Emissions from Deforestation and Degradation
RSLR	Relative Sea Level Rise
SPOT	Satellite Pour l'Observation de la Terre, French Earth Observing Satellites
SRTM	Shuttle Radar Topography Mission
TOA	Top-of-Atmosphere
USGS	United States Geological Survey

7.1 INTRODUCTION

Mangrove wetlands are distributed in the intertidal region between the sea and the land in the tropical and subtropical regions of the world, spanning mostly between 35°N and 35°S latitude (Hemati et al., 2024; Li et al., 2024; Abdelmajeed et al., 2023; Demarquet et al., 2023; Du et al., 2023; Fluet-Chouinard et al., 2023; Yu et al., 2023; Zhang et al., 2023; Faruque et al., 2022; Li et al., 2022; Thamaga et al., 2022; Dang et al., 2021; Lu and Wang, 2021; Maurya et al., 2021; Chamberlain et al., 2020; Gxokwe et al., 2020; Liao et al., 2019; Pham et al., 2019a, 2019b; Wang et al., 2019; Bunting et al., 2018; Mahdavi et al., 2018; Hu et al., 2018; Oostdijk et al., 2018; Valderrama-Landeros et al., 2018; Guo et al., 2017; Giri et al., 2011b). They grow in harsh environmental settings such as high salinity, high temperature, extreme tides, high sedimentation, and muddy anaerobic soils (Kathiresan and Bingham, 2001). The total area of mangroves in the year 2000 was 137,760 km² spread across 118 countries and territories (Giri et al., 2011b). At least 35% of the area of mangrove forests—wooded, tropical wetlands—was deforested from

1980 to 2000 (Valiela et al., 2001). Much of what remains in these deforested areas is in degraded condition (Duke et al., 2007). Coastal habitats around the world are under heavy population and development pressures and are subjected to frequent storms and other natural disturbances.

The continued decline of mangrove forests is caused by conversion to agriculture, aquaculture, tourism, urban development, and over-exploitation (Zhang et al., 2023; Liao et al., 2019; Pham et al., 2019a, 2019b; Wang et al., 2019; Bunting et al., 2018; Mahdavi et al., 2018; Hu et al., 2018; Oostdijk et al., 2018; Valderrama-Landeros et al., 2018; Guo et al., 2017; Alongi, 2014; Giri et al., 2008a). The remaining mangrove forests have been declining at a faster rate than inland tropical forests and coral reefs (Caldeira, 2012). Relative sea level rise could be the greatest threat to mangroves in the future (Gilman et al., 2008). Predictions suggest that 30–40% of coastal wetlands (IPCC, 2007) and functionality of mangrove forests could be lost in the next 100 years if the present rate of loss continues. As a result, important ecosystem goods and services (e.g., natural barrier, carbon sequestration, biodiversity) provided by mangrove forests will be diminished or lost (Duke et al., 2007).

Mangrove forests are among the most productive and biologically important ecosystems in the world because they provide important and unique ecosystem goods and services to human society and coastal and marine systems (Zhang et al., 2023; Oostdijk et al., 2018). The forests help stabilize shorelines and reduce the devastating impacts of natural disasters such as tsunamis and hurricanes. They also provide breeding and nursing grounds for marine and pelagic species, and food, medicine, fuel, and building materials for local communities. Mangroves, including associated soils, can sequester more carbon than other tropical ecosystems (Figure 7.1). Covering only 0.1% of the Earth's continental surface, mangrove forests account for 11% of the total input of terrestrial carbon into the ocean (Jennerjahn and Ittekkot, 2002) and 10% of the terrestrial dissolved organic carbon (DOC) exported to the ocean (Dittmar et al., 2006). Rapid disappearance and degradation of mangroves could have negative consequences on the transfer of materials into the marine systems and influence the atmospheric composition and climate.

Despite the importance of mangrove forests, our understanding of their present status and distributions is inadequate. Previous estimates of the total area of global mangroves range from 110,000 to 240,000 km^2 (Zhang et al., 2023; Wilkie and Fortune, 2003). The Food and Agriculture Organization (FAO) of the United Nations estimate was based on a compilation of disparate and incompatible geospatial and statistical data sources and did not provide spatial information with sufficient detail (Zhang et al., 2023; Wilkie and Fortune, 2003). Global estimates also have been computed using published literature (Alongi, 2002); however, these

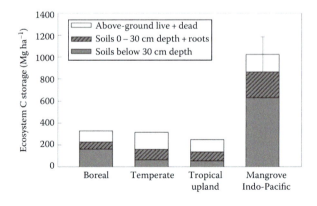

FIGURE 7.1 Carbon stock potential (per hectare) of mangroves compared to other ecosystems (Donato et al., 2011a). Although mangroves occupy only ~0.1% of the Earth's surface, they sequester a disproportionately high percentage of above- and belowground carbon.

Remote Sensing of Mangrove Wetlands

estimates were inconsistent across space and time. Local studies to map mangroves are abundant; however, they do not cover large areal extents, often limited to very small study areas that use different remote sensing or non–remote sensing data sources, classification approaches, and classification systems.

Thus, regular mapping of mangroves using consistent data sources and methodologies was needed, especially over large areas and worldwide (Fluet-Chouinard et al., 2023; Yu et al., 2023; Zhang et al., 2023; Faruque et al., 2022). To fulfill this need, the US Geological Survey Earth Resources Observation and Science (USGS EROS) Center mangrove research team has been mapping and monitoring global mangroves since 2007. This chapter presents a summary of major activities conducted by the team in the context of how remote sensing can be used for mangrove mapping and monitoring from local to global scales.

7.2 DEFINING MANGROVES FOR MAPPING AND MONITORING

A mangrove is a tree, shrub, palm, or ground fern that grows in the coastal intertidal zone in the tropical and subtropical regions of the world. Mangrove species are classified into two categories: true mangroves and mangrove associates. Tomlinson (1986) defined true mangroves with the following features: (1) occurring only in a mangrove environment and not extending into terrestrial communities, (2) presence of morphological specialization (aerial roots, vivipary), (3) possessing a physiological mechanism for salt exclusion and/or salt excretion, and (4) taxonomic isolation from terrestrial relatives. This definition has been accepted widely. True mangroves generally possess a distinct signature in optical remote sensing data (Figure 7.2).

7.3 REMOTE SENSING DATA

A large variety of remote sensing data is being used to obtain the information needed for effective management of mangrove forests. Initially, aerial photographs were used to obtain information about mangrove forests, but now a wide variety of remotely sensed data such as Landsat/SPOT/IRS/CBERES, IKONOS, QuickBird, GeoEye, Radar (PALSAR, Radarsat, SRTM)*, hyperspectral data, MODIS, and aerial videography are available for use. In recent years, unmanned aerial vehicle imagery has been used to collect data (Wan et al., 2014). A brief summary of recent applications of remote sensing of mangrove wetlands is presented in Table 7.1.

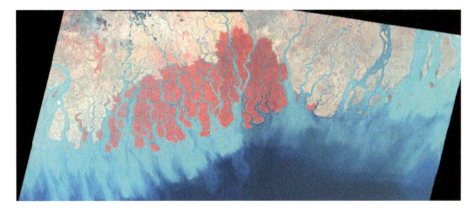

FIGURE 7.2 Mangroves (dark red) in the Sundarbans (Bangladesh and India) in a Landsat false color composite (band combinations 4R, 3G, 2B).

TABLE 7.1
Non-exhaustive List of Examples Illustrating How Remote Sensing Can Be Used for Remote Sensing of Mangrove Wetlands

Data	Applications	References
Aerial photographs,	Mapping (extent and distribution), monitoring (deforestation and regrowth), forest degradation, species discrimination, forest density	(Heenkenda et al., 2014; Jeanson et al., 2014)
Landsat/SPOT/IRS/CBERES	Mapping (extent and distribution), monitoring (deforestation and regrowth), forest degradation, species discrimination, forest density, disturbance (storms), habitat fragmentation, ecosystem health, mangrove coastal retreat monitoring	(Giri et al., 2011b; Gebhardt et al., 2012; Kalubarme, 2014; Santos et al., 2014)
IKONOS, QuickBird, GeoEye	Mapping (extent and distribution), monitoring (deforestation and regrowth), forest degradation, species discrimination, forest density, disturbance (storms)	(Leempoel et al., 2013; Giri et al., 2011a; Heenkenda et al., 2014)
Radar (PALSAR, Radarsat, SRTM),	Tree height/biomass, disturbance (storms), mapping (extent and distribution)	(Hamdan et al., 2014; Lucas et al., 2007)
Hyperspectral data	Mapping (extent and distribution), monitoring (deforestation and regrowth), forest degradation, species discrimination, forest density	(Mitchell and Lucas, 2001)
MODIS	Mapping (extent and distribution), monitoring (deforestationand& regrowth), forest degradation	(Rahman et al., 2013)
Aerial video imagery	Species discrimination	(Everitt et al., 1991)

7.4 SCALE ISSUES

Moderate-resolution satellite data such as Landsat contain enough detail to capture mangrove forest distribution and dynamics. However, very small patches (< 900 m²) of mangrove forests, including newly colonized individual mangrove stands, small island mangroves, or relatively small, fragmented, and linear patches of mangrove stands, cannot be mapped using Landsat. High-resolution satellite data (e.g., IKONOS, QuickBird) (Figure 7.3), aerial photographs, or subpixel classification are needed to assess and monitor these small areas. It should be noted, however, that these small areas, while important for many applications, will not make a substantial difference in the global total (Wilkie and Fortune, 2003).

Mangrove mapping and monitoring at 1–2 m resolution will be needed in the future to create precise information on individual land cover features such as species composition and alliances, ecosystem shifts, water courses, infrastructure, and developed areas. High-resolution monitoring, deemed impossible until recently, is becoming a reality due to the increasing availability of high-resolution (< 5 m) multispectral satellite data, improvements in image processing algorithms, and advancements in computing resources.

7.5 METHODS OF MAPPING MANGROVES USING REMOTE SENSING

Both visual and digital image classification using both supervised and unsupervised classification approaches are being used for mapping and monitoring mangrove wetlands from local to global scales (Zhang et al., 2023; Faruque et al., 2022; Chamberlain et al., 2020; Bunting et al., 2018; Guo et al., 2017). Visual interpretation/on-screen digitizing, pixel-based classification, and object-based classification are some of the common approaches. All of these techniques are being used for interpreting aerial photography, aerial videography, aerial digital imagery, high-resolution imageries

Remote Sensing of Mangrove Wetlands 199

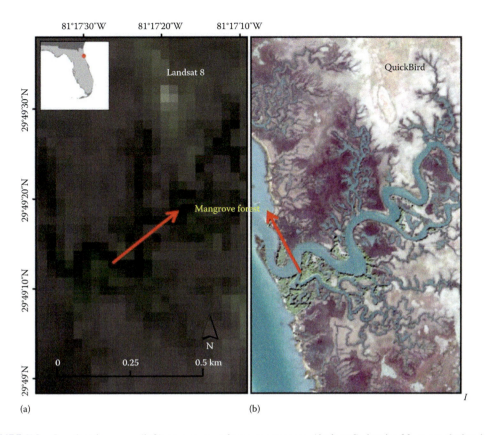

FIGURE 7.3 Landsat imagery (left) can capture large mangroves (dark red), but its 30-m resolution is too coarse to accurately detect and classify the small, fragmented, and linear patterns of Florida's mangroves. QuickBird imagery (right), with a spatial resolution of 2.44 m, is better suited for mapping and monitoring smaller, fragmented, and linear mangrove stands.

(QuickBird, IKONOS), medium-resolution imageries (ASTER, SPOT, Landsat, IRS, CBERS), hyperspectral, and radar data. The pixel-based classification approach is most commonly used for interpreting (Kuenzer et al., 2011). Collection and use of *in situ* data plays a critical role in the classification of remote sensing imageries.

There are numerous image classification and change detection techniques documented in the literature (Civco et al., 2004; Singh, 1989). For change analysis, Civco et al. (2004) compared four techniques: traditional post-classification, cross-correlation analysis, neural networks, and object-oriented classification. They concluded that there are advantages and disadvantages of each method, suggesting there is no single best method. Myint et al. (2014) evaluated and compared the effectiveness of different band combinations and classifiers (unsupervised, supervised, object-oriented nearest neighbor, and object-oriented decision rule) for quantifying mangrove forest changes using multi-temporal Landsat data. The study was conducted in three tropical areas in Asia: the Sundarbans (Bangladesh and India), the Irrawaddy Delta (Myanmar), and Trang (Thailand). A ranking system of 36 change maps produced by evaluating and comparing the effectiveness and efficiency of 13 change detection approaches was used.

Myint et al. (2014) found that the performance of the change detection approaches varied within and among study areas. The overall accuracy increased with decreasing environmental complexity. Which change detection approach performed best or worst depended on the criterion considered (e.g., overall, producer or user accuracy, or the mean of all accuracies). However, several clear patterns emerged when aggregating the results for all study areas and considering, overall, producer and user accuracies simultaneously.

First, the choice of band combination had a greater impact on change detection accuracies than the choice of classifier (e.g., supervised, unsupervised, or object-oriented nearest neighbor). Second, a combination of bands 2, 5, and 7 was the most effective at detecting mangrove loss and persistence, followed by a combination of principal component bands 1, 2, and 3. It was determined that more bands do not necessarily mean better results; in fact, the inclusion of additional bands led to more signature confusion and thus lowered change detection accuracy. Third, discriminant analysis can be effective at identifying the optimal bands for differentiating between mangrove loss and other categories. Finally, alternative approaches, such as a decision-rule object-oriented approach based on principal component band 3, are useful for mapping mangrove loss (Myint et al., 2014).

The automated decision-rule approach as well as a composite of bands 2, 5, and 7 with the unsupervised method and the same composite with the object-oriented nearest neighbor classifier are the most effective approaches to monitoring mangrove deforestation. Principal component bands 1, 2, and 3 generated from a composite of all bands of start and end dates with the object-oriented nearest neighbor classifier and the same composite band with the unsupervised method can be expected to be similarly effective in detecting mangrove loss.

7.6 GLOBAL MANGROVE MAPPING

Even with the availability of more than 40 years of moderate resolution satellite data (i.e., Landsat) and with a distinct signature of mangrove forests in the visible and near-infrared portion of the electromagnetic spectrum, mapping of mangrove forests at this resolution at the global scale was never attempted. Cost and computing facilities have been the primary limitations to using Landsat data for global studies. However, with the availability of free Global Land Survey (GLS), advent of making the Landsat archive freely available, and improvements in computing facilities, global mapping at the Landsat scale became possible. Global data on the extent and conditions of mangrove forests (Zhang et al., 2023) could provide critical information needed for policymaking and resource management. Giri et al. (2011b) used state-of-the-science remote sensing to prepare a wall-to-wall map of the mangrove forests of the world at 30-m resolution (Figure 7.4).

This database is the first most comprehensive, globally consistent, and highest resolution (30 m) global mangrove database (Fluet-Chouinard et al., 2023; Yu et al., 2023; Zhang et al., 2023; Chamberlain et al., 2020; Wang et al., 2019; Guo et al., 2017). Results from the Giri et al. (2011b)

FIGURE 7.4 Mangrove forest distributions of the world for circa 2000 based on Landsat 30-m data. The study established that the total area of mangroves (shown in green) is 137,760 km^2 spread across 118 countries and territories in the tropical and subtropical regions of the world. Approximately 75% of mangroves worldwide are located in just 15 countries.

study showed that the remaining mangrove forest cover of the world is less than previously thought. This new estimate is ~12.3% smaller than the most recent estimate by the Food and Agriculture Organization of the United Nations (FAO). The present extent is 137,760 km² spread across 118 countries and territories in the tropical and subtropical regions of the world. The database includes mangrove vegetation, excluding water bodies and barren lands necessary for global carbon accounting. The largest extent of mangroves is found in Asia (42%), followed by Africa (20%), North and Central Americas (15%), Oceania (12%), and South America (11%). Approximately 75% of mangroves are concentrated in just 15 countries worldwide (Table 7.2).

Mangrove area decreases with an increase in latitude, except between 20° N and 25° N latitude (Figure 7.5), which is where the Sundarbans are located (at the confluence of the Ganges, Brahmaputra, and Meghna rivers in Bangladesh and India on the Bay of Bengal). The Sundarbans are the largest tract of mangrove forests in the world (~10,000 km²). The study confirms earlier findings that biogeographic distribution of mangroves is generally confined to the tropical and subtropical regions of the world but can be found as far as 32°20' N in Bermuda and 38°45' S in Australia (Giri et al., 2011b). The largest percentage of mangrove is found in +5° to –5° degrees latitude (Figure 7.5).

7.7 MANGROVE MONITORING

Time-series remote sensing data can be used to regularly monitor mangrove forests. Giri et al. (2008b) monitored the regional (e.g., tsunami-impacted countries of Indonesia, Malaysia, Thailand, Myanmar, Bangladesh, India, and Sri Lanka) mangrove forest cover dynamics and identified their rates and causes of change from 1975 to 2005 using Landsat data. Results were derived using multitemporal satellite data and field observations. The repetitive coverage of satellite data provides an up-to-date and consistent overview of the extent, distribution, and dynamics of mangrove forests with better spatial and thematic details than the existing coarse-resolution and field inventory data. The analysis addressed the following research questions:

- How much mangrove forest remains?
- Where are these mangrove forests located?

TABLE 7.2
Fifteen Most Mangrove-Rich Countries and Their Cumulative Percentages

SN	Country	Area (square meters)	Cumulative %	Region
1	Indonesia	3112989.48	22.60	Asia
2	Australia	977975.46	29.70	Oceania
3	Brazil	962683	36.68	South America
4	Mexico	741917.00	42.07	North and Central Americas
5	Nigeria	653669.10	46.82	Africa
6	Malaysia	505386.00	50.48	Asia
7	Myanmar	494584.00	54.07	Asia
8	Papua New Guinea	480121.00	57.56	Oceania
9	Bangladesh	436570.00	60.73	Asia
10	Cuba	421538.00	63.79	North and Central Americas
11	India	368276.00	66.46	Asia
12	Guinea Bissau	338652.09	68.92	Africa
13	Mozambique	318851.10	71.23	Africa
14	Madagascar	278078.13	73.25	Africa
15	Philippines	263137.41	75.16	Asia

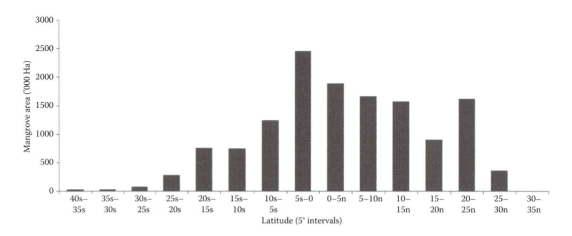

FIGURE 7.5 Latitudinal distribution of mangrove forests of the world.

- What is the spatial and temporal rate of change?
- What are the main reasons for the change?
- What are potential areas for rehabilitation/regeneration?

In this research, Landsat satellite data were pre-processed and classified for 1975, 1990, 2000, and 2005. Post-classification change analysis was performed subtracting the classification maps, 1975–1990s, 1975–2000s, 1975–2005, 1990s–2000s, 1990s–2005, and 2000–2005. The change areas were interpreted visually with the help of secondary data and ground data information to identify the factors responsible for the change. Once the mangrove/non-mangrove areas were calculated for each period, the annual rate of change for the region and for each country was calculated. The time-series analysis revealed a net loss of 12% of mangrove forests in the region from 1975 to 2005 (Giri et al., 2008b). Major hot-spot areas were also identified (Figure 7.6).

The rate of deforestation was not uniform in both spatial and temporal domains. The annual rate of deforestation fromg 1975 to 2005 was highest (~1%) in Myanmar compared to Thailand (0.73%), Indonesia (0.33%), Malaysia (0.2%), and Sri Lanka (0.08%). Mangrove areas in India and Bangladesh remained unchanged or increased slightly. Giri et al. (2008b) identified the major deforestation fronts in the Ayeyarwady Delta (Figure 7.7), Rakhine, and Tanintharyi in Myanmar; Swettenham and Bagan in Malaysia; Belawan, Pangkalanbrandan, and Langsa in Indonesia; and southern Krabi and Ranong in Thailand. Major reforestation and afforestation areas are located on the southeastern coast of Bangladesh, and Pichavaram, Devi Mouth, and Godavari in India.

Similarly, Giri et al. (2007), Giri and Muhlhausen (2008), Long et al. (2013), and Giri et al. (2014) performed change studies in the Sundarbans (Bangladesh and India), Madagascar, the Philippines, and South Asia, respectively. In the Sundarbans, multi-temporal satellite data from the 1970s, 1990s, and 2000s were used to monitor the deforestation and degradation of mangrove forests. The spatiotemporal analysis showed that despite having the highest population density in the world in its immediate periphery, the areal extent of the mangrove forest of the Sundarbans has not changed substantially in the last ~25 years. However, the forest is constantly changing due to erosion, aggradation, deforestation, and mangrove rehabilitation programs. The net forest area increased by 1.4% from the 1970s to 1990 and decreased by 2.5% from 1990 to 2000. The change is insignificant in the context of classification errors and the dynamic nature of mangrove forests. The Sundarbans is an excellent example of the co-existence of humans with terrestrial and aquatic plant and animal life. The strong commitment of governments under various protection measures such

Remote Sensing of Mangrove Wetlands 203

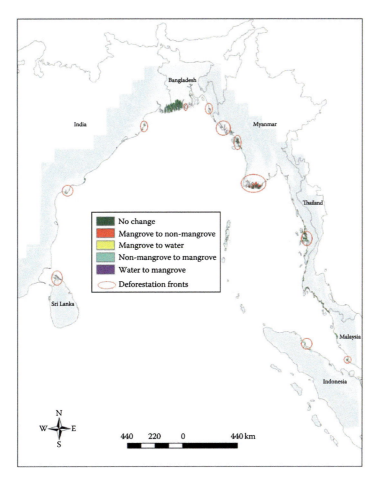

FIGURE 7.6 Major mangrove forest cover change areas from 1975 to 2005 in the tsunami-impacted regions of South and Southeast Asia.

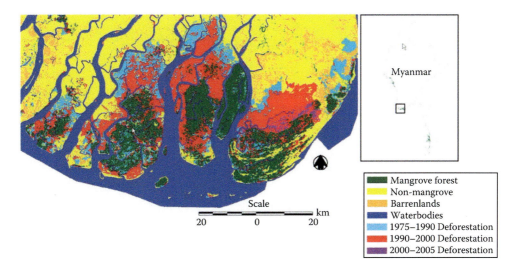

FIGURE 7.7 Spatial distribution of mangrove deforestation in the Ayeyarwady Delta, Myanmar, from 1975 to 1990 (cyan), 1990 to 2000 (red), and 2000 to 2005 (purple).

as forest reserves, wildlife sanctuaries, national parks, and international designations is believed to be responsible for keeping this forest relatively intact (at least in terms of area). Nevertheless, the forest is under threat from natural and anthropogenic forces that could lead to forest degradation, primarily due to top-dying and overexploitation of forest resources.

Time-series analysis revealed that the mangrove forests of Madagascar are declining, albeit at a much slower rate (~1.5% per year) than the global average. The forests are declining due to logging, overexploitation, clear-cutting, degradation, and conversion to other land uses. In this research, Giri and Muhlhausen (2008) interpreted time-series Landsat data from 1975, 1990, 2000, and 2005 using a hybrid supervised and unsupervised classification approach. Landsat data were geometrically corrected to an accuracy of ±0.5 pixels, an accuracy necessary for change analysis. The results showed that Madagascar lost 7% of mangrove forests from 1975 to 2005, to a present extent of ~27.97 km^2. Deforestation rates and causes varied both spatially and temporally. The forests increased by 5.6% (212 km^2) from 1975 to 1990, decreased by 14.3% (455 km^2) from 1990 to 2000, and decreased by 2.6% (73 km^2) from 2000 to 2005. Major changes occurred in Bombekota Bay, Mahajamba Bay, the coast of Ambanja, the Tsiribihina River, and Cape St. Vincent. The main factors responsible for mangrove deforestation include conversion to agriculture (35%), logging (16%), conversion to aquaculture (3%), and urban development (1%).

In the Philippines, several national mangrove estimates existed; however, information was unavailable at sufficient spatial and thematic detail for change analysis. Long et al. (2013) prepared a historical and contemporary mangrove distribution database of the Philippines for 1990 and 2010 at nominal 30-m spatial resolution using Landsat data. Image classification was performed using a supervised decision tree classification approach. Additionally, decadal land cover change maps from 1990 to 2010 were prepared to depict changes in mangrove area using a post-classification technique. Total mangrove area decreased 10.5% from 1990 to 2010. A comparison of estimates produced from this study with selected historical mangrove area estimates revealed that total mangrove area in the Philippines decreased by approximately half (51.8%) from 1918 to 2010.

Mangrove forests in South Asia occur along the tidal sea edge of Bangladesh, India, Pakistan, and Sri Lanka. These forests provide important ecosystem goods and services to the region's dense coastal populations and support important functions of the biosphere. Mangroves are under threat from both natural and anthropogenic stressors; however, the current status and dynamics of the region's mangroves is poorly understood. Giri et al. (2014) mapped the current extent of mangrove forests in South Asia and identified mangrove forest cover change (gain and loss) from 2000 to 2012 using Landsat data. Three case studies were also conducted in the Indus Delta (Pakistan), Goa (India), and the Sundarbans (Bangladesh and India) to identify rates, patterns, and causes of change in greater spatial and thematic detail than a regional assessment of mangrove forests.

Giri et al. (2014) found that the areal extent of mangrove forests in South Asia is approximately 11,874.76 km^2, representing ~7% of the global total. Approximately 921.35 km^2 of mangroves were deforested and 804.61 km^2 were reforested, with a net loss of 116.73 km^2 from 2000 to 2012. In all three case studies, mangrove areas have remained unchanged or have increased slightly; however, the turnover was greater than the net change. Both natural and anthropogenic factors are responsible for the change.

Giri et al. (2014) found that although the major causes of forest cover change are similar throughout the region, specific factors are dominant in specific areas. The major causes of deforestation in South Asia include (1) conversion to other land use (e.g., conversion to agriculture, shrimp farms, development, human settlement), (2) over-harvesting (e.g., grazing, browsing and lopping, fishing), (3) pollution, (4) decline in freshwater availability, (5) flooding, (6) reduction of silt deposition, (7) coastal erosion, and (8) disturbances from tropical cyclones and tsunamis. The forests are changing for distinct reasons in some locations, including sea salt extraction in the Indus Delta in Pakistan, over-harvesting of fruits in the Sundarbans, and garbage disposal in Mumbai, India. Conversely, mangrove areas are increasing in some regions because of aggradation, plantation efforts, and natural regrowth. The protection of existing mangrove areas is facilitating regrowth. The region's diverse socioeconomic and environmental conditions highlight complex patterns of mangrove distribution and change. Results from this study provide important insight to the conservation and management of the important and threatened South Asian mangrove ecosystem.

7.8 SPECIES DISCRIMINATION

Accurate and reliable information on the spatial distribution of mangrove species is needed for a wide variety of applications, including the sustainable management of mangrove forests, conservation and reserve planning, ecological and biogeographical studies, and invasive species management (Du et al., 2023; Fluet-Chouinard et al., 2023; Dang et al., 2021; Wang et al., 2019; Bunting et al., 2018; Mahdavi et al., 2018; Hu et al., 2018). Remotely sensed data have been used for such purposes with mixed results. Myint et al. (2008) employed an object-oriented approach with the use of a lacunarity technique to identify different mangrove species and their surrounding land use and land cover classes in southern Thailand using Landsat data. Results from the study showed that the mangrove zonation could be mapped using Landsat (Figure 7.8). It was also found that the object-oriented approach with lacunarity-transformed bands is more accurate (overall accuracy 94.2%; kappa coefficient = 0.91) than traditional per-pixel classifiers (overall accuracy 62.8%; kappa coefficient = 0.57). Besides multispectral images, hyperspectral data have been used to discriminate mangrove species (Chakravortty et al., 2014).

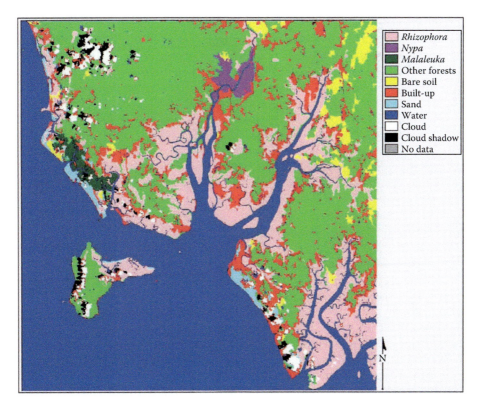

FIGURE 7.8 Species zonation map generated by the object-oriented approach using lacunarity-transformed bands in southern Thailand.

7.9 IMPACT/DAMAGE ASSESSMENT FROM NATURAL DISASTERS

Information regarding the present condition, historical status, and dynamics of mangrove forests was needed to study the impacts of the Gulf of Mexico oil spill of 2010. Such information was unavailable for Louisiana at sufficient spatial and thematic detail. Giri et al. (2011a) prepared mangrove forest distribution maps of Louisiana (before and after the oil spill) at 1-m and 30-m spatial resolution using aerial photographs and Landsat data, respectively. Image classification was performed using a decision-tree classification approach. Maps were prepared of mangrove forest cover change pairs for 1983, 1984, and every two years from 1984 to 2010 depicting "ecosystem shifts" (e.g., expansion, retraction, disappearance).

Direct damage to mangroves from the oil spill was minimal, but long-term impacts need to be monitored. This new spatiotemporal information can be used to assess long-term impacts of the oil spill on mangroves. The study also proposed an operational methodology based on remote sensing (Landsat, Advanced Spaceborne Thermal Emission and Reflection Radiometer [ASTER], hyperspectral, light detection and ranging [LiDAR], aerial photographs, and field inventory data) to monitor the existing and emerging mangrove areas and their disturbance and regrowth patterns. Several parameters, such as spatial distribution, ecosystem shifts, species composition, and tree height/biomass, can be measured to assess the impact of the oil spill and mangrove recovery and restoration. Future research priorities will be to quantify the impacts and recovery of mangroves considering multiple stressors and perturbations, including the oil spill, winter freeze, sea level rise, land subsidence, and land use/land cover change for the entire US Gulf Coast.

Similarly, Landsat imagery is being used to map mangrove damage caused by Typhoon Haiyan on November 8, 2013, in the Philippines (Long et al. manuscript under development). The Normalized Difference Vegetation Index (NDVI) was used as a standardized measure. NDVI is one of the most widely used vegetation indices (Tucker, 1979) to measure and monitor plant growth, vegetation cover, and biomass production. NDVI values range from 1.0 to −1.0 with dense vegetated areas (e.g., closed canopy tropical forest) generally yielding high NDVI values (0.6–0.8), sparsely vegetated areas (e.g., open shrub and grasslands) yielding moderate values (0.2–0.3), and non-vegetated (e.g., rock, sand, and snow) yielding low NDVI values (0.1 and below). Numerous studies have employed repeated measures of NDVI to monitor mangrove vegetation response from varying disturbances (Giri et al., 2011a), but few studies have applied this approach to monitor mangrove disturbance from typhoons (Wang, 2012) and still fewer at a 30-m spatial resolution.

Typhoon Haiyan transected five Landsat path/row footprints in the Philippines. Landsat 7 and Landsat 8 (ETM + and OLI) imagery was used for the study. Image pre-processing for all imagery included converting from digital numbers to top-of-atmosphere (TOA) reflectance, stacking, masking for atmospheric contamination, and NDVI transformation. Mangrove areas were mapped prior to Haiyan for the year 2013 using a supervised decision tree classification approach applied to the Landsat imagery captured before November 8, 2013 (Long et al., 2013). Independent variables used for mangrove classification included Landsat TOA imagery captured before November 8, 2013, a 30-m Shuttle Radar Topography Mission (SRTM) Digital Elevation Model (DEM), a slope index derived from the 30-m DEM, NDVI index transformed from Landsat TOA imagery, and a 2010 mangrove land cover map. Next, the 2013 mangrove land cover map was applied to mask values from all NDVI indices (i.e., before and after) for mangrove areas. The NDVI images before and after the storm were differenced to create NDVI change maps for all five Landsat footprints. These NDVI change maps illustrate where substantial changes occurred in NDVI following Typhoon Haiyan. Larger decreases in NDVI values indicate greater damage in mangrove canopy. Last, mangrove NDVI change maps were quality checked with field photos and field notes. Field data collection was independent from mapping and useful for quality checking mangrove damage maps.

Differing spatial patterns of mangrove damage resulted from variations in storm surge intensity, bathymetry and topography, wind speed and direction, and type and condition of the existing mangrove vegetation (Figure 7.9). Overall, a general pattern of mangrove damage with the greatest decreases in NDVI values, indicating greater mangrove disturbance, occurring in proximity nearest

Remote Sensing of Mangrove Wetlands

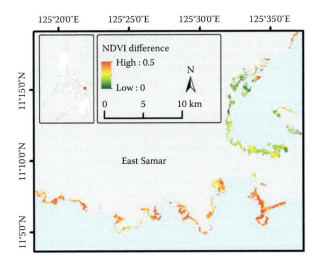

FIGURE 7.9 NDVI difference before and after Typhoon Haiyan in the Philippines.

Typhoon Haiyan's eye transect, especially north of the eye-wall path, and decreasing with greater distance north and south from the eye path. Mangrove damage was also generally highest in the eastern Philippines and generally lessened westward corresponding with decreasing storm intensity from east to west. Average NDVI mangrove values prior to Haiyan were approximately 0.8, indicating vigorous vegetation growth. In this analysis, mangrove damage was considered significant if NDVI values reduced by more than 0.5.

The total mangrove area affected (i.e., experiencing some decrease in NDVI) by Haiyan was estimated to be 214.45 km^2, about 9% of the Philippines' total mangrove area. Not all mangrove areas were included in the analysis because of insufficient data resulting from cloud cover in a few areas. Of the total affected mangrove area, only 6.53 km^2 of mangroves experienced a decrease in NDVI of 0.5 or greater, indicating substantial damage to mangrove health. Areas of substantial damage are considerably smaller than the mangrove areas experiencing minor damage.

7.10 HIGH-RESOLUTION MANGROVE MAPPING

As mentioned in the previous section, Landsat-based monitoring can capture the dynamic nature of mangrove forests (Hemati et al., 2024; Li et al., 2022; Thamaga et al., 2022; Bunting et al., 2018; Mahdavi et al., 2018; Hu et al., 2018; Oostdijk et al., 2018; Valderrama-Landeros et al., 2018). However, higher resolution (< 5-m spatial resolution) remotely sensed data are required to capture newly colonized individual stands or relatively small patches of mangrove stands that cannot be captured with 30-m Landsat imagery. At this resolution, it is possible to create precise information on individual land cover features such as species composition and alliances, ecosystem shifts, water courses, infrastructure, and developed areas.

High-resolution monitoring, deemed impossible until recently, is becoming a reality due to increasing availability of high-resolution (< 5-m) multispectral satellite data, improvement in image processing algorithms, and advancement in computing resources. Processing "big geo data" under this emerging paradigm is a challenge because of the data volume, software constraints, and computing needs. It is difficult, time consuming, and expensive to store, manage, visualize, and analyze these datasets using conventional image processing tools and computing resources. The United States Geological Survey (USGS) mangrove research team proposes to develop a cloud computing environment to visualize and analyze high-resolution satellite data for land change monitoring focusing on mangrove mapping and monitoring of Florida. This new methodology will develop the next generation of tools to analyze and visualize "big geo data" and improve our scientific understanding of mangrove forest cover change needed for decision-making. Through the National Geospatial-Intelligence Agency (NGA), USGS, and commercial vendors, the research team has access to all available (unclassified) high-resolution satellite data (e.g., WorldView-2, GeoEye, IKONOS) from 2000 onward.

The USGS research team proposes a web-based mapping and monitoring platform and application to pre-process, analyze, and visualize the data. This application environment, based mainly on open source tools, will be deployed in a cloud-based system for improved performance, effectively unlimited storage capacity, reduced software costs, scalability, elasticity, device independence, and increased data reliability. Using this platform, the team plans to map and monitor mangrove distribution and change (area, density, and species composition), and species expansion or squeeze due to climate change. The team also plans to quantify the impact and recovery from natural disasters such as hurricanes. High-resolution data are suitable to capture newly colonized individual mangrove stands, small island mangroves, or relatively small, fragmented, and linear patches of mangrove stands that cannot be captured with 30-m Landsat imagery (Figure 7.8).

The prototype system will improve the availability and usability of high-resolution data, algorithms, analysis tools, and scientific results through a centralized environment that fosters knowledge sharing, collaboration, innovation, and direct access to computing resources. Additional data such as Landsat and LiDAR can be added for similar analysis. The information generated from this study will improve our scientific understanding of distribution and dynamics of mangroves in Florida. The innovative tools and methods developed from this study will pave the way for future land change monitoring at high spatial resolution from local to national scales.

7.11 CONCLUSIONS AND RECOMMENDATIONS

This chapter demonstrates how remote sensing data (Landsat and high-resolution satellite data) can be used for mangrove forest cover mapping and monitoring (Hemati et al., 2024; Li et al., 2024; Abdelmajeed et al., 2023; Demarquet et al., 2023; Du et al., 2023, Fluet-Chouinard et al., 2023; Yu et al., 2023; Zhang et al., 2023; Faruque et al., 2022; Li et al., 2022; Thamaga et al., 2022; Dang et al., 2021; Lu and Wang, 2021; Maurya et al., 2021; Chamberlain et al., 2020; Gxokwe et al., 2020; Liao et al., 2019; Pham et al., 2019a, 2019b; Wang et al., 2019; Bunting et al., 2018; Mahdavi et al., 2018; Hu et al., 2018; Oostdijk et al., 2018; Valderrama-Landeros et al., 2018; Guo et al., 2017). These studies show that global mangroves can be mapped and monitored using Landsat; however, for specific use in localized areas, high-resolution satellite data or aerial photographs are needed. High-resolution data are useful to detect early colonization, small patches of mangrove forests, and changes. With the availability of high-resolution satellite data, advancements in computer technology, and improvement in classification algorithms, it will be possible to map and monitor global mangroves using high-resolution satellite data in the future. Remote sensing can help answer the following five major research topics:

1. Observed and predicted changes in relative sea level rise (RSLR), temperature, precipitation, atmospheric CO_2 concentration, hydrology, more frequent and destructive tropical storms, land use change in neighboring ecosystems, and human response to climate change are major components of environmental change that threaten the distribution and health of mangrove ecosystems (Alongi, 2008; Gilman et al., 2008; Saintilan et al., 2014). Monitoring mangrove forests using remote sensing could serve as an indicator of global climate change. For example, poleward migration beyond their historical range could indicate a changing climate.
2. Our scientific understanding of the impact of climate change on mangrove ecosystems and adaptation options is limited. RSLR is expected to be the major climate change

component affecting mangrove ecosystems globally (Krauss et al., 2013; McKee et al., 2007). Mangroves may respond to RSLR by migrating upslope or maintaining their surface elevation through vertical accretion or sedimentation. However, several case studies around the world show that mangrove systems do not have the ability to keep pace with predicted RSLR rates (McKee et al., 2007), resulting in contraction or squeeze in extent at the shoreline from erosion and/or submergence, suffocation, and eventual death. RSLR could also impact the species composition of mangrove stands and related floral and faunal biodiversity. In contrast, "mangroves could be able to keep pace with sea-level rise in some places" (McIvor et al., 2013). Therefore, global impact will likely be site-specific, driven by the rate and degree of RSLR and other biophysical factors. Remote sensing observations could be used to quantify these changes spatially and temporally.
3. Mangrove forests are among the most carbon-rich habitats on the planet (Alongi, 2014; Donato et al., 2011b) compared to other terrestrial forests, owing to the ecosystems' high productivity, rich organic input in long-term soil carbon storage, and low emissions of methane in the saltwater environment. These inter-tidal ecosystems are strong carbon sinks. Therefore, the forests could contribute to climate change mitigation efforts through carbon financing mechanisms such as Reduced Emissions from Deforestation and Degradation (REDD+) and other coastal marine conservation and mitigation initiatives (a.k.a. "Blue Carbon"). Remote sensing could be used to quantify carbon stock and carbon stock change in support of Monitoring, Reporting, and Verification (MRV) of REDD+.
4. It has been suggested that mangrove ecosystems have expanded poleward in the last three decades primarily due to an increase in temperature or decrease in the frequency of winter freeze (Cavanaugh et al., 2013; Saintilan et al., 2014). However, other reports contradict these findings, stating that the observation period is inadequate (Giri et al., 2011a). For example, Cavanaugh et al. (2014) used 1984–2011 Landsat data coupled with climate data and concluded that mangrove extent was expanding poleward along the northeast coast of Florida beyond its northern historical range. Giri and Long (2014) and Giri et al. (2011a) conducted similar studies but included one additional year of Landsat data (i.e., 1983–2011) and concluded that although mangroves expanded from 1984 to 2010, the area has not reached the 1983 extent. The dramatic decrease in mangrove extent was caused by a severe winter freeze in December 1983 that reduced mangrove extent in the northern boundary of Florida and Louisiana by approximately 90%. Historic Landsat data dating back to 1972 and high-resolution satellite data could be used to answer whether mangrove is expanding poleward or not.
5. The Indian Ocean tsunami of December 2004, Hurricane Katrina along the US Gulf Coast in 2005, Cyclone Nargis in Myanmar in 2008, and similar natural disasters in the past decade have highlighted the importance of the mangrove ecosystem as a "bio-shield" or "natural barrier" in protecting vulnerable coastal communities from natural disasters. Several studies after the 2004 tsunami concluded that mangrove ecosystems provided protection to life and property (Barbier, 2006; Dahdouh-Guebas et al., 2005; Danielsen et al., 2005; Kathiresan and Rajendran, 2005; UNEP, 2005). Subsequent publications argued against this claim and concluded that mangrove ecosystems did not provide protection during the tsunami (Kerr and Baird, 2007; Vermaat and Thampanya, 2006). Studies conducted in India showed that villages located behind mangrove forests suffered less damage than those directly exposed to the coast (Kathiresan and Rajendran, 2005). In addition, Kathiresan (2012) suggested that the destructive power of the storm surge was exacerbated during Cyclone Nargis by the recent loss of mangroves in Myanmar, although no primary evidence to support these statements was presented. Some researchers who are skeptical about the ability of mangroves to protect against tsunamis have noted that mangroves might be more capable of protecting against tropical storm surges. Storm surges differ from tsunamis in that they have shorter

wavelengths and relatively more of their energy is near the water surface. The current consensus is seemingly that mangroves provide protection to a certain extent, but proving it scientifically is a challenge. Increased frequency and intensity of extreme events could have profound impacts on coastal development and human safety in the future. Remote sensing data can provide unbiased and transparent information to help address some of these issues.

ACKNOWLEDGMENTS

I would like to thank the two anonymous reviewers who helped improve the manuscript. Any use of trade, firm, or product names is for descriptive purposes only and does not imply endorsement by the US government.

REFERENCES

Abdelmajeed, A.Y.A., Albert-Saiz, M., Rastogi, A., Juszczak, R., 2023. Cloud-based remote sensing for wetland monitoring—A review. Remote Sensing 15, no. 6, 1660. https://doi.org/10.3390/rs15061660

Alongi, D.M., 2002. Present state and future of the world's mangrove forests. Environmental Conservation 29, 331–349.

Alongi, D.M., 2008. Mangrove forests: Resilience, protection from tsunamis, and responses to global climate change. Estuarine, Coastal and Shelf Science 76, 1–13.

Alongi, D.M., 2014. Carbon cycling and storage in mangrove forests. Annual Review of Marine Science 6, 195–219.

Barbier, E.B., 2006. Natural barriers to natural disasters: Replanting mangroves after the tsunami. Frontiers in Ecology and the Environment 4, 124–131.

Bunting, P., Rosenqvist, A., Lucas, R.M., Rebelo, L-M., Hilarides, L., Thomas, N., Hardy, A., Itoh, T., Shimada, M., Finlayson, C.M., 2018. The global mangrove watch—A new 2010 global baseline of mangrove extent Remote Sensing 10, no. 10, 1669. https://doi.org/10.3390/rs10101669

Caldeira, K., 2012. Avoiding mangrove destruction by avoiding carbon dioxide emissions. Proceedings of the National Academy of Sciences 109, 14287–14288.

Cavanaugh, K.C., Kellner, J.R., Forde, A.J., Gruner, D.S., Parker, J.D., Rodriguez, W., Feller, I.C., 2013. Poleward expansion of mangroves is a threshold response to decreased frequency of extreme cold events. Proceedings of the National Academy of Sciences, 201315800.

Cavanaugh, K.C., Kellner, J.R., Forde, A.J., Gruner, D.S., Parker, J.D., Rodriguez, W., Feller, I.C., 2014. Reply to Giri and Long: Freeze-mediated expansion of mangroves does not depend on whether expansion is emergence or reemergence. Proceedings of the National Academy of Sciences, 201401809.

Chakravortty, S., Shah, E., Chowdhury, A.S., 2014. Application of Spectral Unmixing Algorithm on Hyperspectral Data for Mangrove Species Classification, Applied Algorithms. Springer, pp. 223–236.

Chamberlain, D., Phinn, S., Possingham, H., 2020. Remote sensing of mangroves and estuarine communities in central Queensland, Australia Remote Sensing 12, no. 1: 197. https://doi.org/10.3390/rs12010197

Civco, D., Hurd, J.D., Wilson, E.H., Song, M., Zhang, Z., 2004. A comparison of land use and land cover change detection methods. In ASPRS-ACSM Annual Conference 21, 18–33.

Dahdouh-Guebas, F., Jayatissa, L.P., Di Nitto, D., Bosire, J.O., Lo Seen, D., Koedam, N., 2005. How effective were mangroves as a defence against the recent tsunami? Current Biology 15, R443–R447.

Dang, A.T.N., Kumar, L., Reid, M., Nguyen, H., 2021. Remote sensing approach for monitoring coastal wetland in the Mekong Delta, Vietnam: Change trends and their driving forces. Remote Sens 13, 3359. https://doi.org/10.3390/rs13173359

Danielsen, F., Sørensen, M.K., Olwig, M.F., Selvam, V., Parish, F., Burgess, N.D., Hiraishi, T., Karunagaran, V.M., Rasmussen, M.S., Hansen, L.B., 2005. The Asian tsunami: A protective role for coastal vegetation. Science (Washington) 310, 643.

Demarquet, Q., Rapinel, S., Dufour, S., Hubert-Moy, L., 2023. Long-term wetland monitoring using the landsat archive: A review. Remote Sensing 15, no. 3, 820. https://doi.org/10.3390/rs15030820

Dittmar, T., Hertkorn, N., Kattner, G., Lara, R.J., 2006. Mangroves, a major source of dissolved organic carbon to the oceans. Global Biogeochemical Cycles 20, no. 1.

Donato, D.C., Kauffman, J.B., Murdiyarso, D., Kurnianto, S., Stidham, M., Kanninen, M., 2011a. Mangroves among the most carbon-rich forests in the tropics. Nature Geosciences 4, no. 5, 293–297.

Donato, D.C., Kauffman, J.B., Murdiyarso, D., Kurnianto, S., Stidham, M., Kanninen, M., 2011b. Mangroves among the most carbon-rich forests in the tropics. Nature Geoscience 4, 293–297.

Du, C., Khan, S., Ke, Y., Zhou, D., 2023. Assessment of spatiotemporal dynamics of mangrove in five typical mangrove reserve Wetlands in Asia, Africa and Oceania. Diversity 15, no. 2, 148. https://doi.org/10.3390/d15020148

Duke, N.C., Meynecke, J.-O., Dittmann, S., Ellison, A.M., Anger, K., Berger, U., Cannicci, S., Diele, K., Ewel, K.C., Field, C.D., 2007. A world without mangroves? Science 317, 41–42.

Everitt, J., Escobar, D., Judd, F., 1991. Evaluation of airborne video imagery for distinguishing black mangrove (Avicennia germinans) on the lower Texas Gulf Coast. Journal of Coastal Resources 7 no. 4, 1169–1173.

Faruque, M.J., Vekerdy, Z., Hasan, M.Y., Islam, K.Z., Young, B., Ahmed, M.T., Monir, M.U., Shovon, S.M., Kakon, J.F., Kundu, P., 2022. Monitoring of land use and land cover changes by using remote sensing and GIS techniques at human-induced mangrove forests areas in Bangladesh. Remote Sensing Applications: Society and Environment 25, 100699, ISSN 2352-9385. https://doi.org/10.1016/j.rsase.2022.100699. (www.sciencedirect.com/science/article/pii/S2352938522000076)

Fluet-Chouinard, E., Stocker, B.D., Zhang, Z. et al., 2023. Extensive global wetland loss over the past three centuries. Nature 614, 281–286. https://doi.org/10.1038/s41586-022-05572-6

Gebhardt, S., Nguyen, L.D., Kuenzer, C., 2012. Mangrove ecosystems in the mekong delta—Overcoming uncertainties in inventory mapping using satellite remote sensing data. The Mekong Delta System. Springer, pp. 315–330.

Gilman, E.L., Ellison, J., Duke, N.C., Field, C., 2008. Threats to mangroves from climate change and adaptation options: A review. Aquatic Botany 89, 237–250.

Giri, C., Long, J., Abbas, S., Murali, R.M., Qamer, F.M., Pengra, B., Thau, D., 2014. Distribution and dynamics of mangrove forests of South Asia. Journal of Environmental Management 148, 101–111.

Giri, C., Long, J., Tieszen, L., 2011a. Mapping and monitoring Louisiana's mangroves in the aftermath of the 2010 Gulf of Mexico oil spill. Journal of Coastal Resources 27, 1059–1064.

Giri, C., Muhlhausen, J., 2008. Mangrove forest distributions and dynamics in Madagascar (1975–2005). Sensors-Basel 8, 2104–2117.

Giri, C., Ochieng, E., Tieszen, L., Zhu, Z., Singh, A., Loveland, T., Masek, J., Duke, N., 2011b. Status and distribution of mangrove forests of the world using earth observation satellite data. Global Ecology and Biogeography 20, no. 1, 154–159.

Giri, C., Pengra, B., Zhu, Z.L., Singh, A., Tieszen, L.L., 2007. Monitoring mangrove forest dynamics of the Sundarbans in Bangladesh and India using multi-temporal satellite data from 1973 to 2000. Estuarine Coastal and Shelf Science 73, 91–100.

Giri, C., Zhu, Z., Tieszen, L., Singh, A., Gillette, S., Kelmelis, J., 2008a. Mangrove forest distributions and dynamics (1975–2005) of the tsunami-affected region of Asia†. Journal of Biogeography 35, 519–528.

Giri, C., Zhu, Z., Tieszen, L.L., Singh, A., Gillette, S., Kelmelis, J.A., 2008b. Mangrove forest distributions and dynamics (1975–2005) of the tsunami-affected region of Asia. Journal of Biogeography 35, 519–528.

Giri, C.P., Long, J., 2014. Mangrove reemergence in the northernmost range limit of eastern Florida. Proceedings of the National Academy of Sciences, 201400687.

Gxokwe, S., Dube, T., Mazvimavi, D., 2020. Multispectral remote sensing of wetlands in semi-arid and arid areas: A review on applications, challenges and possible future research directions. Remote Sensing 12, no. 24, 4190. https://doi.org/10.3390/rs12244190

Guo, M., Li, J., Sheng, C., Xu, J., Wu, L., 2017. A review of wetland remote sensing. Sensors 17, no. 4, 777. https://doi.org/10.3390/s17040777

Hamdan, O., Khali Aziz, H., Mohd Hasmadi, I., 2014. L-band ALOS PALSAR for biomass estimation of Matang Mangroves, Malaysia. Remote Sensing of Environment 155, 69–78.

Heenkenda, M.K., Joyce, K.E., Maier, S.W., Bartolo, R., 2014. Mangrove species identification: Comparing world view-2 with aerial photographs. Remote Sensing 6, 6064–6088.

Hemati, M., Mahdianpari, M., Shiri, H. Mohammadimanesh, F., 2024. Integrating SAR and optical data for aboveground biomass estimation of coastal wetlands using machine learning: Multi-scale approach. Remote Sensing 16, no. 5, 831. https://doi.org/10.3390/rs16050831

Hu, L., Li, W. Xu, B. 2018. The role of remote sensing on studying mangrove forest extent change. International Journal of Remote Sensing 39, no. 19, 6440–6462. https://doi.org/10.1080/01431161.2018.1455239

IPCC, 2007. Climate Change 2007: The Physical Science Basis, Contribution of Working Group I to the Fourth Assessment Report of the Intergovernmental Panel on Climate Change, in: al., S.S.e. (Ed.). Cambridge University Press.

Jeanson, M., Anthony, E.J., Dolique, F., Cremades, C., 2014. Mangrove Evolution in Mayotte Island, Indian Ocean: A 60-Year Synopsis Based on Aerial Photographs. Wetlands 34, 459–468.

Jennerjahn, T.C., Ittekkot, V., 2002. Relevance of mangroves for the production and deposition of organic matter along tropical continental margins. Naturwissenschaften 89, 23–30.

Kalubarme, M.H., 2014. Mapping and monitoring of mangroves in the coastal districts of Gujarat state using remote sensing and geo-informatics. Asian Journal of Geoinformatics 14.

Kathiresan, K., 2012. Importance of mangrove ecosystem. International Journal of Marine Science 2.

Kathiresan, K., Bingham, B.L., 2001. Biology of mangroves and mangrove ecosystems. Advances in Marine Biology 40, 81–251.

Kathiresan, K., Rajendran, N., 2005. Coastal mangrove forests mitigated tsunami. Estuarine. Coastal and Shelf Science 65, 601–606.

Kerr, A.M., Baird, A.H., 2007. Natural barriers to natural disasters. BioScience 57, 102–103.

Krauss, K.W., McKee, K.L., Lovelock, C.E., Cahoon, D.R., Saintilan, N., Reef, R., Chen, L., 2013. How mangrove forests adjust to rising sea level. New Phytologist, n/a-n/a.

Kuenzer, C., Bluemel, A., Gebhardt, S., Quoc, T.V., Dech, S., 2011. Remote sensing of mangrove ecosystems: A review. Remote Sensing 3, 878–928.

Leempoel, K., Bourgeois, C., Zhang, J., Wang, J., Chen, M., Satyaranayana, B., Bogaert, J., Dahdouh-Guebas, F., 2013. Spatial heterogeneity in mangroves assessed by GeoEye-1 satellite data: A case-study in Zhanjiang Mangrove National Nature Reserve (ZMNNR), China. Biogeosciences Discussions 10, 2591–2615.

Li, A., Song, K., Chen, S., Mu, Y., Xu, Z., Zeng, Q. 2022. Mapping African wetlands for 2020 using multiple spectral, geo-ecological features and Google Earth Engine. ISPRS Journal of Photogrammetry and Remote Sensing 193, 252–268, ISSN 0924-2716. https://doi.org/10.1016/j.isprsjprs.2022.09.009. (www.sciencedirect.com/science/article/pii/S0924271622002532)

Li, S., Zhu, Z., Deng, W., Zhu, Q., Xu, Z., Peng, B., Guo, F., Zhang, Y., Yang, Z. 2024. Estimation of aboveground biomass of different vegetation types in mangrove forests based on UAV remote sensing. Sustainable Horizons 11, 100100, ISSN 2772-7378. https://doi.org/10.1016/j.horiz.2024.100100. (www.sciencedirect.com/science/article/pii/S2772737824000129)

Liao, J., Zhen, J., Zhang, L., Metternicht, G., 2019. Understanding dynamics of mangrove forest on protected areas of Hainan Island, China: 30 Years of evidence from remote sensing. Sustainability 11, no. 19, 5356. https://doi.org/10.3390/su11195356

Long, J., Napton, D., Giri, C., Graesser, J., 2013. A mapping and monitoring assessment of the Philippines' mangrove forests from 1990 to 2010. Journal of Coastal Research 30, 260–271.

Lu, Y., Wang, L., 2021. How to automate timely large-scale mangrove mapping with remote sensing. Remote Sensing of Environment 264, 112584, ISSN 0034-4257. https://doi.org/10.1016/j.rse.2021.112584. (www.sciencedirect.com/science/article/pii/S0034425721003047)

Lucas, R.M., Mitchell, A.L., Rosenqvist, A., Proisy, C., Melius, A., Ticehurst, C., 2007. The potential of L-band SAR for quantifying mangrove characteristics and change: Case studies from the tropics. Aquatic Conservation: Marine and Freshwater Ecosystems 17, 245–264.

Mahdavi, S., Salehi, B., Granger, J., Amani, M., Brisco, B., Huang, W., 2018. Remote sensing for wetland classification: A comprehensive review. GIScience & Remote Sensing, 55, no. 5, 623–658. https://doi.org/10.1080/15481603.2017.1419602

Maurya, K., Mahajan, S., Chaube, N., 2021. Remote sensing techniques: Mapping and monitoring of mangrove ecosystem—A review. Complex Intell Syst 7, 2797–2818. https://doi.org/10.1007/s40747-021-00457-z

McIvor, A., Spencer, T., Möller, I., Spalding, M., 2013. The response of mangrove soil surface elevation to sea level rise. The Nature Conservancy and Wetlands International, Natural Coastal Protection Series: Report 3.

McKee, K.L., Cahoon, D.R., Feller, I.C., 2007. Caribbean mangroves adjust to rising sea level through biotic controls on change in soil elevation. Global Ecology and Biogeography 16, 545–556.

Mitchell, A., Lucas, R., 2001. Integration of aerial photography, hyperspectral and SAR data for mangrove characterization. Geoscience and Remote Sensing Symposium, 2001. IGARSS'01. IEEE 2001 International. IEEE, pp. 2193–2195.

Myint, S.W., Franklin, J., Buenemann, M., Kim, W.K., Giri, C.P., 2014. Examining change detection approaches for tropical mangrove monitoring. Photogrammetric Engineering & Remote Sensing 80, 983–993.

Myint, S.W., Giri, C.P., Le, W., Zhu, Z.L., Gillette, S.C., 2008. Identifying mangrove species and their surrounding land use and land cover classes using an object-oriented approach with a lacunarity spatial measure. Giscience and Remote Sensing 45, 188–208.

Oostdijk, M., Santos, M.J., Whigham, D., Verhoeven, J., Silvestri, S. 2018. Assessing rehabilitation of managed mangrove ecosystems using high resolution remote sensing. Estuarine, Coastal and Shelf Science 211, 238–247, ISSN 0272-7714. https://doi.org/10.1016/j.ecss.2018.06.020. (www.sciencedirect.com/science/article/pii/S0272771417300434)

Pham, T.D., Xia, J., Thang Ha, N., Bui, D.T., Le, N.N., Tekeuchi, W., 2019a. A review of remote sensing approaches for monitoring blue carbon ecosystems: Mangroves, seagrassesand salt marshes uring 2010–2018. Sensors 19, no. 8, 1933. https://doi.org/10.3390/s19081933

Pham, T.D., Yokoya, N., Tien Bui, D., Yoshino, K., Friess, D.A., 2019b. Remote sensing approaches for monitoring mangrove species, structure, and biomass: Opportunities and challenges. Remote Sensing 11, no. 3, 230. https://doi.org/10.3390/rs11030230

Rahman, A.F., Dragoni, D., Didan, K., Barreto-Munoz, A., Hutabarat, J.A., 2013. Detecting large scale conversion of mangroves to aquaculture with change point and mixed-pixel analyses of high-fidelity MODIS data. Remote Sensing of Environment 130, 96–107.

Saintilan, N., Wilson, N.C., Rogers, K., Rajkaran, A., Krauss, K.W., 2014. Mangrove expansion and salt marsh decline at mangrove poleward limits. Global Change Biology 20, 147–157.

Santos, L.C.M., Matos, H.R., Schaeffer-Novelli, Y., Cunha-Lignon, M., Bitencourt, M.D., Koedam, N., Dahdouh-Guebas, F., 2014. Anthropogenic activities on mangrove areas (São Francisco River Estuary, Brazil Northeast): A GIS-based analysis of CBERS and SPOT images to aid in local management. Ocean Coast Manage 89, 39–50.

Singh, A., 1989. Review article digital change detection techniques using remotely-sensed data. International Journal of Remote Sensing 10, 989–1003.

Thamaga, K.H., Dube, T., Shoko, C., 2022. Advances in satellite remote sensing of the wetland ecosystems in Sub-Saharan Africa. Geocarto International 37, no. 20, 5891–5913. htpps://doi.org/10.1080/10106049.2021.1926552

Tomlinson, P., 1986. The Botany of Mangroves: Cambridge Tropical Biology Series. Cambridge University Press.

Tucker, C.J., 1979. Red and photographic infrared linear combinations for monitoring vegetation. Remote Sensing of Environment 8, 127–150.

UNEP, 2005. After the Tsunami: Rapid Environmental Assessment Report—February 22, 2005. Nairobi, Kenya. https://wedocs.unep.org/20.500.11822/8372

Valderrama-Landeros, L., Flores-de-Santiago, F., Kovacs, J.M., Flores-Verdugo, F., 2018. An assessment of commonly employed satellite-based remote sensors for mapping mangrove species in Mexico using an NDVI-based classification scheme. Environmental Monitoring and Assessment 190, 1–13.

Valiela, I., Bowen, J.L., York, J.K., 2001. Mangrove forests: One of the world's threatened major tropical environments at least 35% of the area of mangrove forests has been lost in the past two decades, losses that exceed those for tropical rain forests and coral reefs, two other well-known threatened environments. Bioscience 51, 807–815.

Vermaat, J.E., Thampanya, U., 2006. Mangroves mitigate tsunami damage: A further response. Estuarine, Coastal and Shelf Science 69, 1–3.

Wan, H., Wang, Q., Jiang, D., Fu, J., Yang, Y., Liu, X., 2014. Monitoring the invasion of spartina alterniflora using very high resolution unmanned aerial vehicle imagery in Beihai, Guangxi (China). The Scientific World Journal 2014, 638296.

Wang, L., Jia, M., Yin, D., Tian, J. 2019. A review of remote sensing for mangrove forests: 1956–2018. Remote Sensing of Environment 231, 111223, ISSN 0034-4257. https://doi.org/10.1016/j.rse.2019.111223. (www.sciencedirect.com/science/article/pii/S0034425719302421)

Wang, Y., 2012. Detecting vegetation recovery patterns after hurricanes in South Florida using NDVI time series (Doctoral dissertation, University of Miami).

Wilkie, M.L., Fortuna, S., 2003. Status and trends in mangrove area extent worldwide. Forest Resources Assessment Programme. Working Paper (FAO) 63.

Yu, C., Liu, B., Deng, S., Li, Z., Liu, W., Ye, D., Hu, J., Peng, X., 2023. Using medium-resolution remote sensing satellite images to evaluate recent changes and future done development trends of Mangrove Forests on Hainan Island, China. Forests 14, no. 11, 2217. https://doi.org/10.3390/f14112217

Zhang, X., Liu, L., Zhao, T., Chen, X., Lin, S., Wang, J., Mi, J., Liu, W. 2023. GWL_FCS30: A global 30 m wetland map with a fine classification system using multi-sourced and time-series remote sensing imagery in 2020. Earth Syst Sci Data, 15, 265–293. https://doi.org/10.5194/essd-15-265-2023, 2023

8 Wetland Mapping Methods and Techniques Using Multi-Sensor, Multi-Resolution Remote Sensing

Successes and Challenges

Deepak R. Mishra, X. Yan, Shuvan Ghosh, Christine Hladik, Jessica L. O'Connell, and Hyun Jung ("J.") Cho

ACRONYMS AND DEFINITIONS

AGB	Aboveground Biomass
AI	Artificial Intelligence
ALOS	Advanced Land Observing Satellite
ANN	Artificial Neural Network
AOPs	Apparent Optical Properties
ASD	Analytical Spectral Devices
AVHRR	Advanced Very High Resolution Radiometer
CA	California
CART	Classification and Regression Trees
CBERS	China–Brazil Earth Resources Satellite
CCC	Canopy Chlorophyll Content
CDR	Climate Data Record
Chl	Chlorophyll
CNN	Convolutional Neural Network
CO	Colorado
COSMO-SkyMed	Constellation of Small Satellites for Mediterranean Basin Observation
CRD	Coastal Resources Division
CSP	Carbon Sequestration Potential
CVA	Change Vector Analysis
DEM	Digital Elevation Model
DL	Deep Learning
DNR	Department Of Natural Resources
EPA	Environmental Protection Agency
ERS	European Remote Sensing Satellites
ESA	European Space Agency
EV	Explained Variance
EVI2	Enhanced Vegetation Index 2
FOV	Field-of-View

FL	Florida
FROM-GLC	Finer Resolution Observation and Monitoring of Global Land Cover
GA	Georgia
GBM	Green Biomass
GEE	Google Earth Engine
GIMMS	Global Inventory Modeling and Mapping Studies
GLAI	Green Leaf Area Index
GLC	Global Land Cover
GLWD	Global Lakes and Wetlands Database
GMW	Global Mangrove Watch
GPP	Gross Primary Productivity
GPWD	Global Potential Wetland Distribution Dataset
GRanD	Global Reservoir and Dam Database
HH	Horizontal Transmit and Horizontal Receive Polarization
IFOV	Instantaneous Field of View
IL	Illinois
IOPs	Inherent Optical Properties
ISODATA	Iterative Self-Organizing Data Analysis Techniques
JERS	Japanese Earth Resources Satellite
JRC	Joint Research Center
KNN	K-Nearest Neighbor
LAI	Leaf Area Index
LCC	Leaf Chlorophyll Content
LEDAPS	Landsat Ecosystem Disturbance Adaptive Processing System
LiDAR	Light Detection and Ranging
LSWI	Land Surface Water Index
LUE	Light Use Efficiency
MD	Minimum Distance
MERIS	Medium-Resolution Imaging Spectrometer
MGIC	Murray Global Intertidal Change
ML	Many Machine Learning
MNDWI	Modified Normalized Difference Water Index
MODIS	Moderate-Resolution Imaging Spectroradiometer
MSS	Multi-Spectral Scanner
N	Nitrogen
NASA	National Aeronautics and Space Administration
NDVI	Normalized Difference Vegetation Index
NDWI	Normalized Difference Water Index
NE	Nebraska
NH	New Hampshire
NIR	Near-Infrared
NN	Neural Network
NOAA	National Oceanic and Atmospheric Administration
nRMSEP	Normalized RMSEP
OBIA	Object-Based Image Analysis
OLS	Ordinary Least Squares
P	Phosphorus
PA	Pennsylvania
PALSAR	Phased Array type L-band Synthetic Aperture Radar
PDSI	Palmer Drought Severity Index
PIFs	Pseudo-Invariant Features

PLS	Partial Least Squares
PTWI	Rainfall Topographic Moisture Index
RADARSAT	RADAR Satellite
RF	Random Forest
RMSE	Root Mean Square Error
RMSEP	Root Mean Square Error of Prediction
RNN	Recurrent Neural Networks
RRN	Relative Radiometric Normalization
SAR	Synthetic Aperture Radar
SAVI	Soil Adjusted Vegetation Index
SIR	Shuttle Imaging Radar
SPOT	Satellite Pour l'Observation de la Terre, French Earth Observing Satellites
SRF	Spectral Response Function
SVM	Support Vector Machine
SWIR	Shortwave Infrared
TM	Thematic Mapper
TOC	Top of Canopy
UAVs	Unmanned Aerial Vehicles
UCL	University of Leuven
US	United States
USA	United States of America
VARI	Visible Atmospherically Resistant Index
VCI	Vegetation Condition Index
VF	Vegetation Fraction
VHP	Vegetation Health Products
Vis	Vegetation Indices
WDRVI	Wide Dynamic Range Vegetation Index
WI	Wisconsin

8.1 OVERVIEW

Wetlands occur as ecotones between terrestrial and aquatic systems, and they are sites where nutrient concentrations change as water flows between two ecosystems and are thus essential buffers between uplands and open waters (Naiman et al., 1990). These ecosystems vary in location and size within landscapes, hydrology, presence/absence and plant functional types, productivity, and biogeochemical characteristics, bringing great diversity to wetlands (Keddy, 2010). Because of this, the definition of wetlands has been variable, with different organizations and studies using different definitions. According to the US Environmental Protection Agency (EPA), wetlands are areas where water covers the soil or is present either at or near the surface of the soil all year or for varying periods during the year, including during the growing season. The Ramsar Convention defined wetlands as areas of marsh, fen, peat land, or water, whether natural or artificial, permanent or temporary, with water that is static or flowing, fresh, brackish, or salty, including areas of marine water, the depth of which at low tide does not exceed 6 m (Ramsar Information Bureau 1998). Another common wetland definition is the one by Cowardin et al. (1979), subsequently adopted by the US Fish and Wildlife Service. According to this definition, wetlands share three common distinguishing features: the presence of water within the plant root zone, unique hydric soil conditions, and the presence of hydrophytes during some portion of the hydroperiod (Cowardin et al., 1979). Overall, water, soil, and vegetation are three main elements of wetlands, while wetlands vary widely because of regional and local differences in topography, climate, hydrology, and other factors, including human disturbance. These factors make wetland ecosystem transitional, spatial heterogeneity, and temporal dynamics and determine that wetland is more difficult to distinguish from other land cover/use types (Yan & Niu, 2021).

Wetlands are distributed in all climate zones from the tropics to the tundra and on every continent except Antarctica (Hu et al., 2017). Wetland types vary from freshwater (inland or nontidal) to saltwater flood (coastal or tidal) regimes. In inland zones, they can be unvegetated barrens, moss-dominated acidic bogs and fens, and inland herbaceous wetlands, including wet meadows, ephemeral ponds, prairie potholes, and semi-arid playas. In coastal zones, wetland types include fresh, brackish, and saltwater herbaceous marshes. Woody wetlands include freshwater bottomland hardwood swamps and saltwater mangroves (Keddy, 2010; Mitsch & Gosselink, 2007).

Remote sensing techniques offer timely, up-to-date, and relatively accurate information for sustainable and effective management of wetland ecosystems (Adam et al., 2010). It can help inventory and monitor wetland distribution, productivity, and ecological status (Mayer & Lopez, 2011; Ozesmi & Bauer, 2002). As an important part of wetland research, wetland remote sensing has developed rapidly with the advancement of computer technology and artificial intelligence (AI) and has attracted attention from researchers and policymakers.

This chapter presents a review of the wetland remote sensing techniques and the challenges of wetland remote sensing. Following this, three case studies are presented outlining a detailed methodology for new remote sensing approaches. These new approaches promise to solve ubiquitous problems in mapping wetland species and understanding ecosystem dynamics with wetland habitats.

8.2 EVOLUTION OF WETLAND REMOTE SENSING

The history of remote sensing within wetlands mirrors that of terrestrial environments. Its development history unfolds from several dimensions: data, methods, and application scope.

8.2.1 DATA

Remote sensing instruments measure and record the reflection and the emission of electromagnetic radiation from the target area (Abd El-Ghany et al., 2020). Sensors can be divided into passive and active sensors, whereas platforms range from Earth observation satellites, planes, and helicopters to unmanned aerial vehicles (UAVs) with fixed wings and rotaries (Lechner et al., 2020). Recently, the advent of the Google Earth Engine (GEE, https://earthengine.google.com [accessed on February 18, 2024]) has provided a platform to assist geoscientists interested in the geo–big data analysis (Mahdianpari et al., 2020). Earth Engine's public data archive includes more than 40 years of historical imagery and scientific datasets, updated and expanded daily, which have been used in many geoscience fields.

8.2.1.1 Active Remote Sensors

Active sensors send out a pulse of energy and detect the changes in the return signal. The radar instrument uses the antenna to emit regular pulses of microwave energy of a known wavelength where waves generated electronically vibrate in a predetermined orientation. Microwave energy is sensitive to soil moisture variations and inundation, and is least attenuated by wetland canopies (Klemas et al., 2014). The common types of active sensors are laser altimeter, light detection and ranging (LiDAR), radar, ranging instrument, scatterometer, and sounder. Some studies have mapped the wetlands using synthetic aperture radar (SAR) and airborne LiDAR (Millard & Richardson, 2013). SAR data are more responsive to differences in water content, size/roughness, and relatively broad-scale structural differences than optical sensors (Kaplan & Avdan, 2018). Therefore, SAR has been shown to be excellent for surface water monitoring, largely for open water (Canisius et al., 2019), and it also has shown its potential for monitoring water under vegetation canopies and flooded areas, with its penetration and double bounce capabilities (HESS et al., 1990). Investigations in monitoring coastal and inland wetlands that exploited C- and L-band data from Shuttle Imaging Radar missions (SIR-A, SIR-B, and SIR-C), ERS-1 and ERS-2 (Fiaschi & Wdowinski, 2020), JERS (Hess et al., 2015) and RADARSAT-1 (Kwoun & Lu, 2009), RADARSAT-2 (LaRocque et al., 2020), and Sentinel-1 (Niculescu et al., 2020) have also contributed significantly to wetland research. L-band radar data generally distinguishes between flooded and non-flooded forests and between forest and marsh vegetation compared to optical Landsat TM sensor, which could not detect water under the canopy. The TerraSAR-X and COSMO-SkyMed constellations operate in X-band, enabling them to deliver complementary information for monitoring wetlands (Adeli et al., 2020). Active remote sensors such as radar backscatter provide different information than optical sensors and make them highly important, complementing data from optical satellites for land cover classification, as well as for wetland mapping and monitoring, because radar systems can acquire images at any time of day and night, almost under any weather conditions, and they can pass through gases and clouds and can even penetrate solids to a limited degree (depending on the density) (Dronova et al., 2021). Radar imaging applications in forested wetlands have found that longer wavelengths had greater penetration in forest canopies. Radar imagery has been used in wetland locations ranging from tropics to boreal regions. However, the radar signal is often reduced in wetlands dominated by lower-biomass herbaceous vegetation when a layer of water is present primarily due to specular reflectance (Kasischke et al., 1997).

8.2.1.2 Passive Remote Sensors

The development of wetland remote sensing relies on passive remote sensors primarily due to the high cost of active sensors. They vary in cost of acquisition, spatial coverage, and spatial, spectral, and temporal resolution, and they often cannot simultaneously have high temporal/spatial and spectral resolution. High spatial resolution optical data (< 10 m), such as QuickBird and IKONOS data, have been used to produce accurate classifications of wetland vegetation communities and water resources for small regions (Rapinel et al., 2015; Sawaya et al., 2003). Optical data with moderate spatial resolution (10–100 m), such as Landsat, Sentinel, and SPOT (from French "Satellite pour l'Observation de la Terre") series, have been widely used for mapping wetland systems in sample regions (Kaplan & Avdan, 2017; Niculescu et al., 2016). Among them, Landsat sensor data have been the traditional choice of wetland mapping, owing to the wide geographic coverage, temporal depth of the archive, and the availability of the data at no charge to users (Gallant, 2015). Many studies conducted long-term wetland monitoring and change mapping research based on the Landsat series data (Niu et al., 2012). Regional, continental, and global changes in wetland land cover and water resources have been monitored using optical data with coarse spatial resolution (>100 m), such as the National Oceanic and Atmospheric Administration Advanced Very High Resolution Radiometer (NOAA AVHRR) (Moreau et al., 2003), and Moderate Resolution Imaging Spectroradiometer (MODIS) (Bansal et al., 2017; Chen et al., 2018).

However, satellite imagery also has limitations compared to aerial photography. Because of the spatial resolution of most satellite imagery (> 10 m), it is difficult to identify small or long, narrow wetlands (Ozesmi & Bauer, 2002). Hyperspectral sensors on both airborne and spaceborne platforms have significantly improved wetland remote sensing by providing narrowband spectra. High spectral resolution can aid in distinguishing certain biophysical and biochemical data. Both hyperspectral and multi-spectral data can estimate multiple vegetation parameters and identify broad-scale patterns in wetland emergent properties.

Aerial photography has an advantage in spatial resolution and data acquisition time; however, wetland studies use these data typically in narrow coastal areas or along rivers because of the generally small cover areas, the limited acquisition cost, and the availability (Guo et al., 2017). In most cases, aerial photographs are combined with other satellite images to study wetlands on a regional or national scale. The suitability of each sensor for wetland remote sensing applications requires matching sensor strengths and limitations to study goals.

8.2.1.3 Wetland-Related Datasets

There are some global-level wetland-related datasets that could help the wetland research, including the land cover and land use dataset, such as the GlobeLand30 and FROM-GLC, water, wetland, or mangroves mapping. The following are the details about some of them.

GlobeLand30 is a global land cover dataset (Chen et al., 2017) that comes from China's National High-Tech Research and Development Program (863 Program) "Global Land Cover Remote Sensing Mapping and Key Technology Research" project (www.globallandcover.com/). The acquisition period is 2010, and the spatial resolution is 30 m. The GlobeLand30 dataset is the world's first 30-m resolution global land cover dataset, contains rich and detailed information on the spatial distribution of global land cover, and can better describe most human land use activities and the resulting landscape patterns. The dataset contains 10 main types of land cover: cultivated land, forest, grassland, shrubland, wetland, water body, tundra, artificial surfaces, bare land, and permanent snow and ice.

FROM-GLC is a global land cover dataset (http://data.ess.tsinghua.edu.cn/) (Yu et al., 2013) that comes from the Tsinghua University team, acquired in 2015, with a spatial resolution of 30 m. Wetland-related types include water bodies and wetlands (marshland, mudflat, marshland, leaf-off). The team updated the dataset and increased the spatial resolution to 10 m.

GlobCover is a global land cover dataset (http://due.esrin.esa.int/page_globcover.php) (Arino et al., 2012) that was jointly completed by ESA and the University of Leuven (UCL). The purpose was to produce a bi-monthly and annual MERIS fine-resolution global land cover map in 2009. This dataset was completed in 2010 using the 2009 MERIS image, with a spatial resolution of 300 m. There are 22 categories in the dataset, of which categories related to wetlands include water bodies; freshwater submerged broad-leaved forests; saltwater submerged semi-deciduous or evergreen broad-leaved forests; and freshwater, brackish water, or saltwater regularly submerged vegetation (grassland, shrubs, and woody vegetation).

GLC_FCS30_2020 is a global land cover dataset (https://zenodo.org/record/4280923#.YowCEaiZOUk) (Zhang et al., 2021) that was completed by the Institute of Aerospace Information Innovation, Chinese Academy of Sciences in 2020, with a spatial resolution of 30 m. This dataset is based on the 2015 Global 30 m Land Cover Product Fine Classification System (GLC\UFCS30–2015), combined with the 2019–2020 time series land satellite surface reflectance data; Sentinel-1 SAR data, DEM terrain elevation data, global thematic auxiliary datasets, and prior knowledge datasets are generated. The dataset divides the world into 30 land cover types. Among them, wetland-related types include water bodies and wetlands.

The Joint Research Centre Global Surface Water Survey and Mapping map (from now on referred to as JRC global water body) (Pekel et al., 2016) is a global surface water distribution dataset that contains a surface water range and time distribution map from 1984 to 2019. The dataset is based on the 4,185,439 Landsat 5, 7, and 8 satellite images acquired during the period from March 16, 1984, to December 31, 2019, and the expert system was used to classify each pixel as water/non-water, and the results were sorted into monthly history and two periods (1984–1999, 2000–2019) data for change detection. The dataset can be accessed on the Google Earth Engine platform (Joint Research Centre Global Surface Water Mapping Layers, v1.2).

The Global Reservoir and Dam Database (GRanD) (Lehner et al., 2011) is a global reservoir dataset, information for which 11 participating institutions collected from many sources (https://globaldamwatch.org/data/). The dataset is managed by McGill University and provides a reliable and geographically clear database for the scientific community. GRanD v1.3 contains records of 7320 reservoirs and related dams, and the global reservoir storage capacity is 6863.5 km^3.

Global Mangrove Watch (GMW) is a global mangrove distribution dataset (Bunting et al., 2018). It is part of the 2011 JAXA Kyoto & Carbon Initiative. It was completed by multiple universities in cooperation with international organizations and was also a pilot project of the Ramsar Global Wetland Observing System. GMW used ALOS-PALSAR and Landsat (optical) data to generate the 2010 global mangrove baseline map based on JERS-1 SAR, ALOS-PALSAR, and ALOS-2 PALSAR-2 to obtain six periods from 1996 to 2016 baseline changes.

The Global Lakes and Wetlands Database (GLWD) (Lehner & Döll, 2004) is a global lakes and wetlands database jointly developed and created by the World Wide Fund for Nature, the Environmental Systems Research Center, and the University of Kassel in Germany using existing maps, data, and information. The spatial resolution is 1 km, and the acquisition time is 2004. There are many detailed wetlands and water types, such as lakes, reservoirs, rivers, flooded swamps, forest swamps, coastal swamps, saline-alkali land, peatlands, temporary wetlands/lakes, etc. The GLWD has been proven in numerous written studies to represent a comprehensive database of global lakes larger than 1 km^2 and can well represent the largest global wetland range.

Murray Global Intertidal Change (from here on referred to as MGIC) is a global intertidal change dataset (Murray et al., 2019) that contains 707,528 Landsat images along the global coastline from 60°N to 60°S. The global intertidal ecosystem map is generated by classification; each pixel is divided into tidal flats, permanent water, or other reference data. The spatial resolution is 30 m, and the temporal resolution is 11 periods (1984–1986; 1987–1989; 1990–1992; 1993–1995; 1996–1998; 1999–2001; 2002–2004; 2005–2007; 2008–2010; 2011–2013; 2014–2016).

The Global Potential Wetland Distribution Dataset (GPWD) (Hu et al., 2017) is a global potential wetland distribution dataset that has a spatial resolution of 1 km. GPWD simulated the potential distribution of global wetlands, including inland wetlands and coastal wetlands, such as mangrove swamps, and excluded marine wetlands and other coral reefs and seaweeds by using the new Rainfall Topographic Moisture Index (PTWI) and global remote sensing training samples.

8.2.2 Wetland Remote Sensing Application Scopes and Related Techniques

There is a diverse range of applications for wetland remote sensing due to the mixture characteristics of wetlands, including the wetland mapping/classification and change detection, remote sensing of wetland vegetation, water body inundation remote sensing, estimating biophysical and biochemical parameters of wetland vegetation, soil characteristics, and so on.

8.2.2.1 Wetland Classification and Change Detection

Despite many advances in remote sensing technology, wetland classification is still challenging from the remote sensing perspective (Corcoran et al., 2011; Landmann et al., 2010). A significant reason for this difficulty is that although each wetland class has several distinctive characteristics, they share some ecological similarities with each other and other non-wetland classes (Henderson & Lewis, 2008; Mahdavi et al., 2018). Therefore, different wetland types exhibit similar spectral and/or backscattering information in remote sensing imagery. At the same time, wetlands have strong spatial heterogeneity and temporal dynamics, such as seasonality. Therefore, remote sensing tools have significantly contributed to wetland mapping and monitoring in various aspects, including classification (Mohammadimanesh et al., 2018) and change detection (Brisco et al., 2017).

As one of the important elements, classification features play a big role in the whole process of classification. Besides the spectral features, such as the bands of optical and SAR images, researchers also developed some indices based on the object reflection properties to improve classification accuracy, such as the Normalized Difference Vegetation Index (NDVI) (Rouse et al., 1974), the Normalized Difference Water Index (NDWI) (B. Gao, 1996), the Modified Normalized Difference Water Index (MNDWI) (Xu, 2006), and the Land Surface Water Index (LSWI) (Xiao et al., 2004). First (i.e., range, mean, standard deviation/variance, and entropy) and second (i.e., angular second moment, contrast, correlation, entropy, homogeneity, dissimilarity) order texture features have also been commonly used in remote sensing studies; this sort of contextual identification can improve the classification accuracy of high-resolution images (Szantoi et al., 2013). Geometric objects are also used in wetlands, although wetlands rarely exhibit regular or consistent shapes and sizes (Dronova, 2015). The most commonly used contextual variables for identifying wetland classes are distance, proximity, adjacency, and relative border to specific classes, such as water bodies (Dronova, 2015). This is useful, especially when there is a need to integrate the environmental variables (e.g., slope, soil type, and rainfall) in the mapping process (Adam et al., 2010).

Incorporating the spatial feature within spectral information may contribute to differentiating complex land cover because some of these classes may have very similar spectral characteristics (Zhao & Du, 2016). Several approaches have been proposed to evaluate the efficiency of integrating spectral-spatial features for classification, including kernel methods, Bayesian models, Markov random field, and conditional random field (Rezaee et al., 2018). However, it is hard to identify a common standard for setting the proper parameters and suitable features for the different research objects and areas. Inspired by the high efficiency of the human brain in object recognition, high-level spatial features produced by hierarchical learning have attracted substantial interest in several applications, such as object recognition, scene labeling, and document analysis (DiCarlo et al., 2012).

On the other hand, the classification method is also a decisive factor for accurate wetland classification. Generally, wetland classification methodology using remotely sensed data is categorized into pixel-based or object-based classifications, the latter of which is also referred to as object-based image analysis (OBIA) (Amani et al., 2017). In wetlands, OBIA context information can aid classification, especially when high spatial heterogeneity, soil moisture, and inundation confound spectral analysis by other methods (Wright & Gallant, 2007). Techniques used with passive remote sensors include unsupervised, supervised, semi-supervised, self-supervised learning, and reinforcement learning classification. Besides revolutionizing the processing and analysis of open-source earth observation data, the GEE cloud platform introduced automatic training sample migration possibilities and supports many machine learning (ML) methods, such as k-nearest neighbor (KNN), support vector machine (SVM), minimum distance (MD), classification and regression trees (CART), and random forest (RF) (Gorelick et al., 2017), which greatly facilitates researchers. The ML and deep learning (DL) use of algorithms with remote sensing data has improved the accuracy of wetland classification and change detection (Ahmed et al., 2021). The supervised classifiers use labeled training data as input to define models for predicting the class label of the test data (Cord & Cunningham, 2008), and the examples listed previously belong to supervised algorithms. The popular unsupervised algorithms include the Iterative Self-Organizing Data Analysis Techniques (ISODATA) Field (Memarsadeghi et al., 2007) and the k-means clustering (Lloyd, 1982). Whether pixels or objects are used for image classification, the extracted information (spectral and/or textural) can be subjected to different classifiers. The advanced neural network (NN) has recently been a valuable tool for machines to learn dynamic non-linear associations (Lary et al., 2016). Various upgraded NNs for standard land-cover classification were proposed, such as convolutional neural network (CNN) (Günen, 2022), artificial neural network (ANN) (Zun-You et al., 2015), recurrent neural networks (RNN) (Sharma et al., 2018), region-based CNN (Ma et al., 2022), U-Net (Garg et al., 2019), and Mask R-CNN (He et al., 2017).

Remote sensing has been used to successfully monitor vegetation colonization in restored tidal wetlands through short-term change detection. In addition, change detection is possible using short- and long-term time-series imagery, allowing forward-looking trajectories of change, as well as a historical understanding of system processes and change stimuli (Tuxen et al., 2008). Change vector analysis (CVA), similar to pixel vector modulus and cross-correlation analysis, is a change detection technique that can determine the direction and magnitude of changes in a multidimensional spectral space (Baker et al., 2007). The application of current remote sensing techniques on historical imageries not only helps quantify ecological changes in wetlands over time, but such analysis can help to link environmental changes to anthropogenic drivers (Ballanti et al., 2017). Many studies focused on extracting wetland characteristics, which were used to model spatial-temporal changes historically using multitemporal Landsat images (Chouari, 2021). It is worth mentioning that the Landsat series satellites are widely used in wetland change detection research due to their moderate resolution and long-term data acquisition capabilities.

8.2.2.2 Remote Sensing Mapping of Wetland Vegetation and Biomass

Applications of wetland vegetation remote sensing include monitoring invasive wetland vegetation, mapping and monitoring vegetation species distribution, understanding the seasonal phenological dynamics of vegetation, and so on. They are critical technical tasks in the sustainable management of wetlands (Adam et al., 2010).

There is ongoing interest in developing remote sensing methods for mapping and monitoring the spatial distribution and biomass of mangrove (Aslan et al., 2016). Remote sensing of mangroves was not limited to mapping their extents but also many other complex topics, such as biophysical parameter inversion and ecosystem process characterization (Wang et al., 2019, pp. 1956–2018). In addition, remotely sensed proxies of wetland phenology are sensitive to spatial heterogeneity. While the heterogeneity and spatial resolution affect phenological agreement among sensors, the study showed that the higher spatial resolution may increase error and complexity in phenological fitting and satellite-based phenology may elucidate links among canopy structure and function (Dronova et al., 2021). Many researchers have used time-series images (spectral bands or vegetation indices) to identify vegetation interference by extracting time-series features, significantly improving efficiency and monitoring accuracy (De Marzo et al., 2021). Recently, the combined approach of remote sensing and hydrodynamic or earth system models was also applied to reveal the relationship between the spatial patterns of vegetation communities and other factors (Zhiqiang et al., 2016). Remote sensing was used to track the wetland function degradation and function loss process, which plays a crucial role in ecological balance and global change (Shen et al., 2019). Remotely sensed data have become the primary source for biomass estimation since the spectral signatures or vegetation indices are correlated with biomass estimation. The techniques include a bottom-up algorithm (valley-following and directional texture), a top-down algorithm (multiscale edge segments, threshold-based spatial clustering, a double-aspect method, and a vision expert system), and template matching (Lu, 2006).

A different approach for estimating vegetation patterns is multivariate analysis, such as Partial Least Squares (PLS) regression. While vegetation indices use either data mining or expert knowledge to select appropriate bands for indexing vegetation parameters, PLS regression uses Eigen vector-based techniques to reduce full-spectrum data predictor components that maximize vegetation parameter prediction (Mevik & Wehrens, 2007) and factor out non-target cover type influence from estimates (Chen et al., 2009). The use of PLS regression in wetland studies is illustrated in Study 3.

8.2.2.3 Surface Water Mapping Based on Remote Sensing

Surface water is the main element of wetlands and the most essential resource and environmental factor in maintaining human survival and ecosystem stability; therefore, timely and accurate information on dynamic surface water is urgently needed (Li et al., 2020). In recent years, many lakes and rivers experienced expansion and shrinkage processes over both short- and long-term scales, resulting in significant hydrological, ecological, and economic problems. The remote sensing technique successfully captured the temporal and spatial change processes of inundation in different zones (Wu & Liu, 2015). The advantages of using remotely sensed data in surface water are almost near real-time surface water extent, the extensive spatial coverage of the data, the effectiveness and robustness of the surface water mapping methods, and the relatively low cost for mapping in a large aerial extent (Munasinghe et al., 2018).

Remote sensing of inland waters is based on the optical properties of water constituents. These properties can be divided into two categories: (1) properties that depend on the medium and the directional structure of the ambient light field, known as apparent optical properties (AOPs), and (2) those that depend only on the medium and are independent of the ambient light field, known as inherent optical properties (IOPs) (Mishra et al., 2017). Using three million Landsat satellite images, Pekel et al. (2016) quantified changes in global surface water from 1984 to 2015 at 30-m resolution and recorded the months and years when water was present, where occurrence changed, and what form changes took in terms of seasonality and persistence. This study demonstrates the potential of optical imaging for surface water mapping. The techniques for big data exploration and information extraction were exploited, namely, expert systems (Lu & Weng, 2007), visual analytics (Keim et al., 2008), and evidential reasoning (Yang & Xu, 2002).

Surface water detection is realistically feasible with SAR imagery. Horizontal smooth surfaces reflect nearly all incident radiation away and decreasing the amount of returned radiation, represented by dark tonality on radar images, which can be clearly delineated from the non-inundated areas due to the surface roughness of water and land (Rahman & Thakur, 2018). The magnitude of the deteriorating effects in a SAR water image is a function of wavelength, incidence angle, and polarization (Schumann & Moller, 2015). Although they found that all polarization modes can be employed for surface water mapping, there is better land–water surface discrimination in the HH polarization (Schumann & Moller, 2015). Many SAR image-processing techniques exist to successfully derive water area or extent, including simple visual interpretation (Smiley & Hambati, 2020), image histogram thresholding (Chini et al., 2017), automatic classification algorithms (Yang & Cervone, 2019), image texture algorithms (Dasgupta et al., 2018), multi-temporal change detection methods (Clement et al., 2018), complex auto-logistic regression (Bee et al., 2008; Jiang et al., 2015), and principal component analysis (Nandi et al., 2016; See & Porio, 2015).

8.3 THE CHALLENGES AND POTENTIAL SOLUTIONS OF WETLAND REMOTE SENSING

8.3.1 Proximal Sensing of Wetlands

Remote sensing within wetlands has particular challenges. For example, wetlands are patchy, and individual patches can be smaller than some sensors' spatial resolution. Wetlands can have high species diversity, particularly freshwater wetlands (PHINN, 1998). Water inundation also influences spectral reflectance, reducing near-infrared spectra and shifting the red-edge position, altering NDVI and red-edge-type indices (Kearney et al., 2009; Turpie, 2013). As such, wetland-specific indices are needed to account for the wetland's optical properties. A method for developing such indices is described in this section.

Remote sensing of terrestrial vegetation using medium-resolution satellite data has been successful because of the plant pigments, chlorophyll *a* and *b*, that absorb energy in the blue (centered at 450 nm) and the red (centered at 670 nm) wavelengths, as well as the leaf internal spongy mesophyll structure that is responsible for the high reflectance in the near-infrared (NIR) region (700–1300 nm) (Lillesand et al., 2008). For terrestrial vegetation, spectral indices can be formulated as simple ratios, normalized ratios, three-band combinations, or other transformations, including the Simple Vegetation Index and NDVI, and they have been successfully used to correlate with diverse plant characteristics at a broad span of scales from individual leaf areas to global vegetation dynamics (Graetz, 1990; Perroy et al., 2018; Sagan et al., 2020; Xue & Su, 2017). These simple multispectral indices have also been used on wetland and aquatic vegetation, especially floating and emergent aquatics (Cho et al., 2008; Kiage & Walker, 2009; Peñuelas et al., 1993; Shekede et al., 2008). They can be divided into three broad categories of use: structural, biochemical, and physiological (Perroy et al., 2018). For example, NDVI calculated with Landsat MSS and TM data has been used to map the types and spatial extent of invasion by freshwater emergent and floating weeds (Shekede et al., 2008). However, surface water film or background water can affect the utility of these indices for monitoring wetland vegetation. Cho et al. (2008) found that the slight submergence of water hyacinth (*Eichhornia crassipes*) reduced the mean NIR reflectance from 76% to 6.3% during the close-range spectroscopy study in experimental outdoor tanks (Cho et al., 2008). In order to improve the classification accuracy of wetland plants, conventionally used vegetation indices have been modified to include wetness factors in addition to plant greenness factors, i.e., Vegetation Water Index (Wang et al., 2012). Moisture levels in the background mud were included in the mapping of salt marsh using vegetation ratios in order to help discriminate areas with dead marsh plants from areas with sparse live marsh plants (Ramsey & Rangoonwala, 2006). In order to re-parameterize an existing vegetation index for wetland vegetation, proximal spectroscopy studies are needed to gain further understanding of wetland reflectance patterns and the effect of water background on the NIR reflectance. Next is an example of the top-of-canopy reflectance data acquisition from tidal salt marsh vegetation using an *in situ* hyperspectral spectroradiometer (Mishra et al., 2012).

A dual-fiber system, with two inter-calibrated Ocean Optics USB4000 hyperspectral radiometers (OceanOptic Inc., Dunedin, FL, USA), mounted on a sturdy frame was used to acquire the top of canopy (TOC) spectral percent reflectance (%R) data in the range of 200–1100 nm, with a sampling interval of 0.3 nm (Rundquist et al., 2004). The first radiometer with a field-of-view (FOV) of 25° pointed downwards to acquire upwelling radiance (L; Wm^{2}sr^{-1}), while the second radiometer, equipped with a cosine corrector, pointed upward to acquire downwelling irradiance (E; Wm–2) simultaneously. Based on the FOV and the height of the frame (5 m), the spatial resolution (IFOV) of the sensor was calculated to be 1.83 m (Figure 8.1), using the following relationship:

$$d = 2\{h \times \left(tan\frac{\alpha}{2}\right) \tag{1}$$

where, d = diameter of the IFOV, h = height of the sensor from the target, and α = FOV of the sensor. The radiometers were inter-calibrated by comparing incident irradiance to the measured upwelling radiance of a 99% white Spectralon reflectance panel (Labsphere, Inc., North Sutton, NH, USA), made of barium sulfate. In case of changing sky conditions, the sensor was recalibrated at regular intervals (Figure 8.1). Noise removal in the raw hyperspectral data was accomplished by smoothing using a moving window average of 7 nm, after which the smoothed data was further interpolated at 1 nm intervals. Four scans of radiance and irradiance acquired per study plot were converted to four percent reflectance (%R) readings. Mean %R was estimated from the four individual scans to obtain composite spectra of the study plot.

FIGURE 8.1 Spectral reflectance data acquisition in salt marsh environments—a: *in situ* spectral reflectance acquisition using Ocean Optics sensor; b and c: sensor's IFOV, d: sensor calibration using 99% Spectralon reflectance panel.

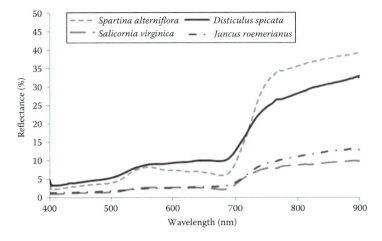

FIGURE 8.2 Sample hyperspectral spectral reflectance of four different salt marsh species as acquired by the Ocean Optics sensor.

To examine inundation effects on NIR reflectance, the %R of four salt marsh species was acquired based on the aforementioned experimental setup (Figure 8.2). The overall trend in %R showed a less prominent red-edge in all species as compared to terrestrial vegetation, which might be attributed to soil moisture and standing water background. Moisture/water and vegetation have contrasting spectral responses in the near infra-red (NIR) region of the spectrum; while moisture/water shows more absorption and less/no scattering, vegetation scatters considerably, owing to foliar content and cellular structure. The differences in the NIR spectral response at the species level were very much evident and understandable, due to substantial difference in the foliar structure and canopy architecture. In addition to %R spectra such as these, to calibrate and validate vegetation indices–based models using proximal or satellite data, plot-level data on wetland biophysical properties are required.

8.3.2 Multi-Sensor Joint Observation

The high temporal and spatial change rate of wetlands has a strong demand for high time-frequency and spatial resolution remote sensing data, but the existing single remote sensing satellite data source often makes it hard to meet the needs of wetland observation limited to weather conditions and other factors. Synergistic use of multi-source remote sensing data to increase the availability of high-spatial-resolution time-series remote sensing data to achieve high time-frequency wetland classification has become one of the most urgent problems to be solved. Although the recognition ability of different sensors for wetlands is not consistent, the ability of different wavelengths to distinguish between wetlands, water bodies, and non-wetlands is determined in the spectral space, which provides feasibility for the unification of sensors. For example, as the most widely used multispectral sensors, combining Landsat, MODIS, Sentinel-2, and Sentinel-1 can provide better data complementation for wetland classification. Therefore, studying the coordination of sensors has a great application value in wetland classification.

Diverse satellite images enhance the information for observing the Earth's surface and bring additional uncertainties in the applications using multi-sensor images, such as change detection, multitemporal analysis, and image fusion. To address this challenge, the researcher developed a multi-rule-based relative radiometric normalization (RRN) method for multi-sensor satellite images, which is a widely used method for enhancing the radiometric consistency among multitemporal satellite images and involves the identification of spectral- and spatial-invariant pseudo-invariant features (PIFs) and a PLS regression–based RRN model using neighboring target pixels around PIFs (Xu et al., 2023).

The spectral response function (SRF), also known as the bandpass function, slit function, instrument transfer function, or instrument line shape, is the detector element's response to monochromatic incident light through a slit (Dirksen et al., 2006; Lee et al., 2017, p. 2). Thus, the measured spectra can be represented by the convolution of the instrument SRF and the incoming spectra at the original resolution (Kang et al., 2022). Furthermore, the SRF includes the overall performance of the instrument mirror reflectivity, entrance slit width, and detector responsivities (Hewison et al., 2013). It characterizes the sensitivity of each spectral band and has been recognized as one of the most important sources of uncertainty for using multi-sensor data. Therefore, an accurate understanding of the SRF plays a vital role in the calibration process and multi-sensor joint observation. While some investigations of SRF effects for cross-calibration focus on comparing sensors at the reflectance band level, many include spectral indices that provide information on plant biophysical parameters (Cundill et al., 2015). Gonsamo et al. compared the SRF differences among 21 Earth observation satellite sensors and their cross-sensor corrections for red, near-infrared (NIR), and shortwave infrared (SWIR) reflectance and normalized difference vegetation index (NDVI), and the training dataset to derive the SRF cross-sensor correction coefficients was generated from state-of-the-art radiative transfer models (Gonsamo & Chen, 2013). The results indicate that reflectance and NDVI from different satellite sensors cannot be regarded as directly equivalent. Variations in processing strategies, non-spectral differences, and algorithm preferences among sensor systems and data streams hinder cross-sensor spectra and NDVI comparability and continuity. Similarly,

Cundill et al. (2015) compared 48 spectral indices for cultivated grasslands using simulated data of ten very high spatial resolution sensors and found that the index values calculated from the sensor data with differing spectral response functions are not directly comparable. To reduce band setting differences between sensors, it's necessary to consider the effects of SRF.

NEW METHODS IN WETLAND REMOTE SENSING: CASE STUDIES

Three case studies have been incorporated in this chapter as examples of new methods in wetland remote sensing. These studies cover a broad geographic area and wetland diversity, from freshwater wetlands to tidal salt marshes. The first study explored the response of fresh wetland vegetation NDVI to climate change in China from 1981 to 2015 based on Global Inventory Modeling and Mapping Studies (GIMMS) Normalized Difference Vegetation Index product (NDVI3g), National Oceanic and Atmospheric Association (NOAA) Vegetation Health Products (VHP), and climate data. The second study was conducted on the salt marsh habitats of the entire Georgia coast. It discusses a MODIS-based mapping protocol to map the biophysical characteristics of salt marsh vegetation accurately. The third study was conducted in a low-diversity freshwater marsh environment to estimate and map belowground biomass and foliar nitrogen (N) using satellite data and hybrid modeling. These methods represent some of the recent developments in wetland remote sensing techniques. This chapter aims to provide an efficient and non-destructive mapping protocol for emergent wetlands. The techniques can be used for mapping the spatial distribution of wetland species and their biophysical properties, delineating critical hot spots of wetland stress, assessing the success of previous marsh restoration projects, identifying areas of degradation as candidates for future restoration actions, and evaluating the overall productivity trend of wetlands at any geographic location.

STUDY 1: RESPONSE OF WETLAND NDVI TO CLIMATE CHANGE

1. INTRODUCTION

Wetland vegetation is the totality of all plant communities in a wetland area, and it plays an important role in environmental function, water storage, regulation of flood flows, and maintenance of the regional water balance (Yan et al., 2022; Zhu et al., 2011). It is also an excellent indicator of early signs of any physical or chemical degradation in wetland environments (Dennison et al., 1993). In the past few decades, climate change has had various effects on all continental ecosystems (Gao et al., 2016; Rustad, 2008), one of which is that it has a profound impact on the growth and function of vegetation (Peng et al., 2013). With the frequent occurrence of extreme phenomena such as El Niño and La Niña in recent years, understanding the response of wetland vegetation to climate change is important for predicting these changes (Shen et al., 2019). Remote sensing techniques offer timely, up-to-date, and relatively accurate information for sustainable and effective management of wetland vegetation (Adam et al., 2010). Specifically, NDVI-based time series are fundamental to the remote sensing of vegetation phenology and to extract numerical observations related to vegetation dynamics (Martínez & Gilabert, 2009). The NDVI relies on the absorption of red radiation by chlorophyll and other leaf pigments, and the strong scattering of near-infrared radiation by foliage. Research shows seasonal variations of NDVI are closely related to vegetation phenology, such as green-up, peak, and offset of development (Beck et al., 2006). In particular, analysis of contemporary vegetation dynamics using NDVI is especially effective for large transboundary geographical regions of the Northern Hemisphere with variable terrain, diverse vegetation, and multiple types of land use (Chu et al., 2019). However, monitoring wetland vegetation is more difficult than monitoring non-wetland vegetation. The main reason is that the reflectance spectra of wetland vegetation canopies are often very similar and are combined with reflectance spectra of the underlying soil, hydrologic regime, and atmospheric vapor (Noda et al., 2021). Therefore, in addition to NDVI, other indexes are needed to assist in monitoring the growth status of wetland vegetation, such as the Vegetation Condition Index (VCI). Some previous studies have analyzed the correlation between vegetation and climate variables (precipitation and temperature) under different locational and regional conditions (Ichii et al., 2002, pp. 1982–1990; Schultz & Halpert, 1993). However, the timeliness and resolution of the data used for these studies were low, the method was relatively simple and single, the influence of El Niño and La Niña was not considered, and there were few details about China. At present, some scholars have explored the response of vegetation in different regions to climate change across China. The influence of climate change on vegetation varied significantly with different climate zones. For example, zones with rising temperatures promoted the growth of vegetation and included the wet area of the central subtropical zone, as well as the wet, semi-wet, semi-arid, and arid areas of the plateau temperature zone (Bing & Lin, 2017). The zones in which temperature rise inhibits vegetation were the semi-wet area and the arid area of the warm temperature zone (A et al., 2016). In addition, zones with increasing precipitation promoted vegetation development and included the semi-arid and arid areas of the medium temperature zone and the plateau temperature zone, and the semi-arid area of the plateau sub-frigid zone (Du et al., 2015). As a unique vegetation type, wetland vegetation keeps its roots moist all year round, leading to its inconsistent response to climate change with other vegetation types. In the context of global climate change, some scholars suggested wetland management and restoration (Erwin, 2009). Some previous studies have focused on the response of specific wetland vegetation types (seagrasses, tidal marsh plants, mangroves, and so on) to climate change (Short et al., 2016). However, few studies analyzed the response of overall wetland vegetation to climate change at the national scale over time. Therefore, how does climate change impact wetland vegetation? The objectives of this study are to (1) analyze the trend change of wetland vegetation in China from 1981 to 2015 and (2) explore the responses of wetland vegetation to climate change.

2. Methods

2.1 Indices and Data Sources

This study selected NDVI as the indicator to reflect the change in wetland vegetation. NDVI is the most commonly used vegetation index for characterizing vegetation growth and coverage by remote sensing methods (Vrieling et al., 2014). The relationship between NDVI and global climate change has been widely confirmed both regionally and globally (Ji & Peters, 2003; Kim et al., 2012; Prasad et al., 2007). The value of vegetation NDVI usually ranges from 0 to 1. The higher the NDVI value, the better the growth and coverage of vegetation. The NDVI data used in this study was derived from GIMMS NDVI3g (NDVI 3rd generation), which was developed using data from Advanced Very High Resolution Radiometer (AVHRR), with a temporal resolution of 15 days and spatial resolution of 1/12°. It has been calibrated for cloud testing and sensor shifting, and it eliminates the effects of many factors, such as the solar zenith angle (Piao et al., 2011). The GIMMS NDVI3g dataset had a conservative measurement uncertainty on the order of ±0.005. This low measurement uncertainty makes the detection of nonstationary seasonal and inter-annual climate, making use of the GIMMS NDVI3g data possible. Data from AVHRR covers the time series from 1981 to 2015 and supplies the only long-term updated large-scale dataset of vegetation greenness (Pinzon & Tucker, 2014). VCI was expressed as an NDVI anomaly relative to a 25-year climatology estimate based on bio-physical and ecosystem laws (law-of-minimum, law-of-tolerance, and carrying capacity). VCI is a proxy for moisture conditions, and it can detect drought and measure its impacts on vegetation (Kogan & Sullivan, 1993). Therefore, VCI could reflect the environmental conditions of wetlands. The value of VCI ranges from 0 to 100. Generally, if VCI is less than 40, it indicates that the wetland condition is poor; if VCI is greater than 40 and less than 70, it means that the wetland condition is moderate; and if VCI is greater than 70, it shows that the wetland condition is good (Kogan & Sullivan, 1993). VCI is a derivative of NOAA VHP data, which is a long-term sequence dataset newly developed by the US Oceanic and Atmospheric Administration. The data comes from two polar-orbiting satellites, NOAA-9 and NOAA-11, and have minimized cloud pollution and atmospheric effects in the weekly synthesis. Its temporal resolution is seven days, and its spatial resolution is 4 km. The climate indicators this study chose were temperature and precipitation. The near-surface temperature and ground precipitation rate can reflect the temperature and precipitation of the growing environment of wetland vegetation. China's regional ground meteorological elements dataset was downloaded from the Cold and Arid Science Data Center and was used to analyze climate change, and it is a set of near-surface meteorological and environmental element reanalysis datasets (Yang et al., 2016). Its temporal resolution is three hours, and its spatial resolution is 0.1°. Among the seven data types it contains, this study selected near-surface temperature and ground precipitation rate. In addition, this study used national wetland classification data to extract wetland vegetation samples. This dataset was developed based on Landsat and CBERS-02B imagery in 1978, 1990, 2000, and 2008 across China through manual visual interpretation and a large quantity of field verifications (Niu et al., 2012). They were synthesized from a spatial resolution of 30 m to 1 km. This study extracted marshes and swamps in the national wetland classification data for selecting samples. Finally, this study used the latest China climate zoning dataset as the climate zoning data, which used the daily meteorological observation data of 609 meteorological stations from 1971 to 2000. It divided China into 12 temperature zones, 24 dry-wet zones, and 56 climate zones.

2.2 Selection of Wetland Vegetation Samples

Considering the hydrodynamics of wetlands, to obtain stable wetland vegetation samples, this study used the national wetland maps to identify wetland patches that overlapped from each of the datasets. These wetland patches included coastal and inland marshes and swamps with an area of more than 9 km², which appeared in each of the four national wetland maps from 1978 to 2008 (Niu et al., 2012). In a 3 × 3 grid, the middle patch is considered the most stable. Therefore, considering the spatial and temporal stability of the wetland, this study selected the wetland patches with an area of more than 9 km², which can ensure the stability of the intermediate patch. Then, this study extracted the geometric central points of these patches. The geolocations of those points were used to extract the NDVI value of wetland vegetation from the GIMMS NDVI3g data. To ensure that the extracted points belong to the wetland vegetation type, this study checked them one by one based on the temporal trajectory changes in NDVI value over the past 35 years (Figure 8.3) and removed the samples that did not belong to the vegetation (they may have been flooded for a period of time). In this way, a total of 117 stable wetland vegetation samples were obtained nationwide (Figure 8.4).

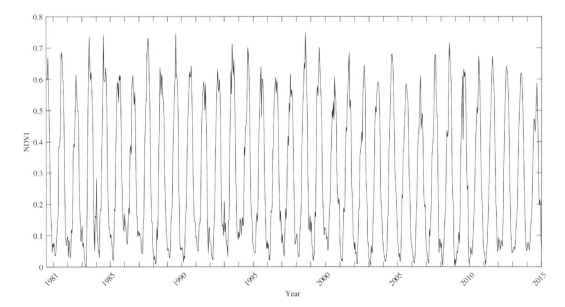

FIGURE 8.3 The temporal trajectory variations of NDVI of stable wetland vegetation sample from 1981 to 2015.

Wetland Mapping Methods and Techniques 229

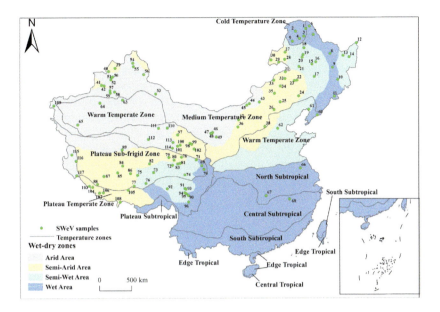

FIGURE 8.4 The distribution of stable wetland vegetation samples in wet-dry zones and temperature zones.

2.3 Ordinary Least Squares Linear Regression

VCI and climate factors can be considered background information for wetland vegetation samples. In addition to focusing on its original value, this study pays more attention to its average overall change trend (upwards or downwards). Ordinary least squares linear regression is a common and sufficient technique for estimating it according to our research goal, so this study calculates the average change rate of indicators (VCI, near-surface temperature, and ground precipitation) in the growing season by the ordinary least squares linear regression method (OLS):

$$Y(t) = \alpha_0 + \beta_0 + \varepsilon \qquad (2)$$

where α_0 and β_0 are the fitting intercept and slope, respectively. The months selected for the growing season are May to September. This study uses the two-tailed T-test to identify the trend significance level of the regression equation. If the obtained P-value is no more than 0.01, the trend is extremely significant. The slope of the indicator trend line can by calculated by:

$$SLOPE = \frac{n \times \sum_{i=1}^{n} i \times Aveseasonal_i - \left(\sum_{i=1}^{n} i\right)\left(\sum_{i=1}^{n} iAveseasonal_i\right)}{n \times \sum_{i=1}^{n} i^2 - \left(\sum_{i=1}^{n} i\right)^2} \qquad (3)$$

where n is the number of years studied, i is the serial number of the year, and $Aveseasonal_i$ is the average of the indicators for the i-year growing season. If *SLOPE* is positive, the indicator shows an increasing trend in n years; otherwise, it shows a decreasing trend. If *SLOPE* is equal to zero, the indicator doesn't show significant trend change during the study period.

2.4 Wavelet Transform Method

NDVI is the main research object in this study, and to obtain more detailed information beyond the overall trend of changes, such as the turning points, this study applied a wavelet transform in the NDVI analysis. The wavelet transform is a signal analysis method that can automatically adapt to time-frequency analysis requirements (Kulkarni, 2011). It is widely used in time-series feature extraction (Kędra & Wiejaczka, 2018; Kumar & Foufoula-Georgiou, 1997). Meanwhile, wavelet transform was compared with other traditional methods (trend regression method, moving average method, and the Mann–Kendall test method). The conclusion was that the theory of extracting trend items from a time series based on wavelet transform is better. Wavelet transform can decompose the time series into a high-frequency part and a low-frequency part. High frequency can be used to extract the random item and the periodic item of the time series, while the remaining low-frequency part can be used to extract trend items of the time series (Yan et al., 2022). In this study, wavelet transform is mainly used to extract trend items of the NDVI time series.

The wavelet transform is to shift (b) the wavelet basis function ψ(t) and then the inner product with the time series $x(t)$ at different scales (a):

$$WT_X(a,b) = a^{-\frac{1}{2}} \int_{\infty}^{-\infty} X(t) \Psi^*\left(\frac{t-b}{a}\right) dt, \ a > 0 \tag{4}$$

When applying wavelet transform to extract features of a time series, the wavelet basis function and decomposition level should be determined. The wavelet basis function is related to whether it has regularity, orthogonality, symmetry, vanishing moment, and tight support. The decomposition level is determined based on the complexity of the signal. The more complex the signal is, the more decomposition layers should be selected. The selected wavelet basis function is the db8 wavelet with lower compactness and a higher vanishing moment. The time series of the study is the half-month wetland vegetation NDVI in 35 years, so the signal is relatively complex. At this time, the decomposition level is selected as eight layers. The low-frequency information a8 is used to obtain the inter-annual variation trend of stable wetland vegetation NDVI.

2.5 Correlation Analysis

Correlation analysis studies whether there is a certain dependency relationship between phenomena and then explores the relevant direction and extent. Its calculation formula is:

$$R_{xy} = \frac{\sum_{i=1}^{n}(x_i - \bar{x})(y_i - \bar{y})}{\sqrt{\sum_{i=1}^{n}(x_i - \bar{x})^2}\sqrt{\sum_{i=1}^{n}(y_i - \bar{y})^2}} \tag{5}$$

where R_{xy} is the correlation coefficient of the variables x and y, x_i is the value of NDVI of i-year growing season, $\bar{x}$ is the average of NDVI for the growing season of all study years, y_i is the value of climate factors of i-year growing season, and $\bar{y}$ is the average of climate factors for the growing season of all study years. Wavelet transform analysis was used to determine whether a turning point exists. For wetland vegetation samples with turning points in NDVI changes, this study calculated the partial correlation coefficients between NDVI and climate variables (near-surface temperature and ground precipitation rate) before and after the turning points to analyze whether temperature or precipitation contributed to NDVI changes. When studying the correlation between NDVI and temperature, the precipitation rate was set as the control variable. Moreover, this study controlled temperature when studying the correlation between NDVI and precipitation rate. The formula for calculating the partial correlation coefficient is:

$$R_{xy.z} = \frac{R_{xy} - R_{xz}R_{yz}}{\sqrt{(1 - R_{xz}^2)(1 - R_{yz}^2)}} \tag{6}$$

where $R_{xy.z}$ is the partial correlation coefficient of x and y under the control factor z.

Wetland Mapping Methods and Techniques

3. RESULTS AND DISCUSSION

3.1 Spatio-Temporal Trend Change of NDVI in Stable Wetland Vegetation Samples

The growth conditions and luxuriance of wetland vegetation in different temperature zones and wet-dry zones are diverse, leading to significantly different wetland vegetation NDVI. The highest NDVI average value being above 0.75 only occurs in the wet area of the cold temperature zone. Among all the wet-dry zones, the average value of wetland vegetation NDVI in dry areas (arid and aemi-arid) was generally lower than the humid areas (wet and semi-wet) (Figure 8.5a). Among all the temperature zones, the average value of wetland vegetation NDVI in the cold temperature, north subtropical, and central subtropical zones were generally higher than the other zones (Figure 8.5b). The corresponding wet-dry zones of the three temperature zones in China were all wet areas. Through statistics (Table 8.1), this study found that 82% of stable wetland vegetation samples with NDVI ≥ 0.6 were located in the wet and semi-wet areas. Meanwhile, all stable wetland vegetation samples with NDVI ≤ 0.3 were located in the semi-arid and arid areas. Therefore, this study concluded that wetland vegetation NDVI was significantly related to the degree of wetness. Wetland vegetation has an overall trend in long-term sequence, and there are differences in different regions. In this study, trend information of stable wetland vegetation samples for 35 years were extracted from the trend items of the wavelet transform. A turning point is the year in which the trend changes. Mathematically, a turning point is a point at which the derivative changes sign, which can be obtained by calculating its derivative. For example, the year was a turning point if the trend changed from an upward trend to a downward trend after that year. Firstly, the amplitude of NDVI was calculated to judge whether it had a normal fluctuation. It was obtained by dividing the fluctuation value and the multi-year average value of NDVI. In the analysis of ecosystem disturbances with time-series changes, 10% was generally used as the reference threshold of the natural rhythm of the ecosystem (Pettorelli et al., 2005). When the range of amplitude was within ±10%, it was considered a normal fluctuation; otherwise, NDVI had obvious trend changes from 1981 to 2015. This study found that the NDVI changes of seven stable wetland vegetation samples were within the normal fluctuation range, and their locations were not concentrated. The NDVI of the remaining 110 stable wetland vegetation samples showed a significant trend change. Among the 110 stable wetland vegetation samples that had a significant variation trend, 83 of them had a turning point in the 35-year NDVI trend change. Most of the years in which the turning point was located were 2003 and 2004 (Figure 8.6a). Meanwhile, NDVI of 80 stable wetland vegetation points first increased and then decreased after the turning point (Figure 8.6b). For the rest of the stable wetland vegetation samples, there were no turning points in NDVI change. Among them, the NDVI of 19

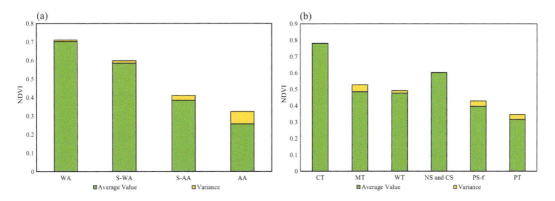

FIGURE 8.5 The average value and variance of NDVI of wetland vegetation samples in wet-dry zones (a) and temperature zones (b). (WA is wet area; S-WA is semi-wet area; S-AA is semi-arid area; AA is arid area. CT is cold temperature zone; MT is medium temperature zone; WT is warm temperature zone; NS is north subtropical; CS is central subtropical; PS-f is plateau sub-frigid zone; PT is plateau temperature zone).

TABLE 8.1
The Percentage of Samples with Different NDVI Averages in Each Wet-Dry Zone (Unit: %)

	NDVI ≥ 0.6	0.3 < NDVI < 0.6	NDVI ≤ 0.3
WA	44	4	0
S-WA	38	24	0
S-AA	12	57	41
AA	6	15	59

Note: WA is wet area; S-WA is semi-wet area; S-AA is semi-arid area; AA is arid area.

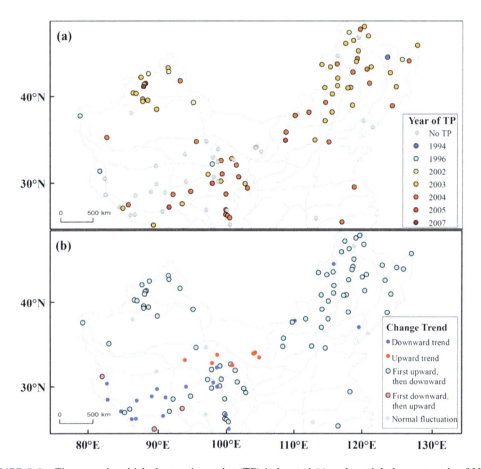

FIGURE 8.6 The years in which the turning point (TP) is located (a) and spatial change trends of NDVI (b) of 117 stable wetland vegetation samples.

stable wetland vegetation samples showed a downward trend. Overall, about 80% of the wetland vegetation NDVI showed a downward trend after 2004, and about 10% of the wetland vegetation NDVI showed an upward trend after 2003.

3.2 Spatio-Temporal Trend Change of VCI in Stable Wetland Vegetation Samples

VCI can reflect the environmental status of wetland vegetation. If VCI is less than 40, it indicates that the wetland condition is poor (Kogan & Sullivan, 1993). This study defines the type in which

Wetland Mapping Methods and Techniques   233

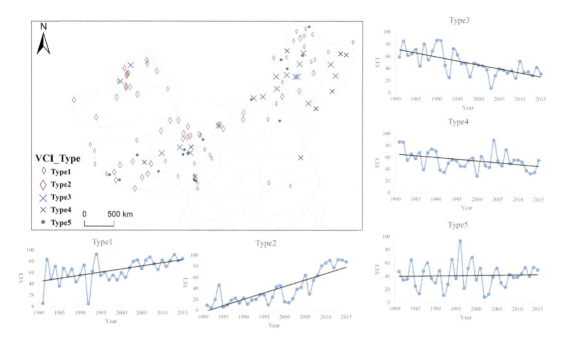

FIGURE 8.7 Spatial change trends of VCI of 117 stable wetland vegetation samples.

the value of VCI increases from 40 upwards over time as Type 1 (Figure 8.7). When VCI increases from less than 40 to more than 40, this study defines them as Type 2, which means that the environment of wetland vegetation is significantly better. When VCI shows a downward trend in the long time series, and it decreases from more than 40 to less than 40, this study defines them as Type 3, which means that the environment of wetland vegetation is significantly worse. Type 4 happens when the value of VCI is still more than 40 but shows a decreasing trend. This study defines Type 5 as being when the value of VCI is always around 40. When the trend of VCI is Type 1, it means that the environment of wetland vegetation is better, but its impact on wetland vegetation is less than Type 2. Similarly, when the trend of VCI is Type 4, it means that the environment of wetland vegetation is worse, but its impact on wetland vegetation is less than Type 3. Type 5 shows that the trend of VCI and the environment of wetland vegetation have not changed significantly. Among the 117 samples, 40 of them had a VCI trend change of Type 2, which had a significantly better wetland environment. The wetland environment of 18 samples with VCI trend change Type 3 indicated that the environments were significantly worse. There was no significant trend (Type 5) in VCI of 16 samples. Generally, stable wetland vegetation samples whose VCI trend changed Type 2 and Type 1 accounted for 63% of the total; correspondingly, the wetland vegetation environments were in a gradually improving state. Meanwhile, stable wetland vegetation samples whose VCI trend change were Type 3 and Type 4 accounted for 23% of the total, corresponding with wetland vegetation environments in a worse state.

3.3 Climate Trend Change Corresponding to Stable Wetland Vegetation Samples

This study used the near-surface temperature of China's regional ground meteorological elements dataset to analyze the temperature changes in the regions corresponding to the stable wetland vegetation samples. The SLOPE of near-surface temperature characterized the trend change of the temperature of the samples. Results showed that the near-surface temperature of most sample points (110/117) showed an upward trend (Figure 8.8a), meaning an increasing temperature across those sample regions during the past 35 years. By contrast, the near-surface temperatures of the other seven samples showed a downward trend. This study used the ground precipitation rate of China's

234 Remote Sensing Handbook, Volume V

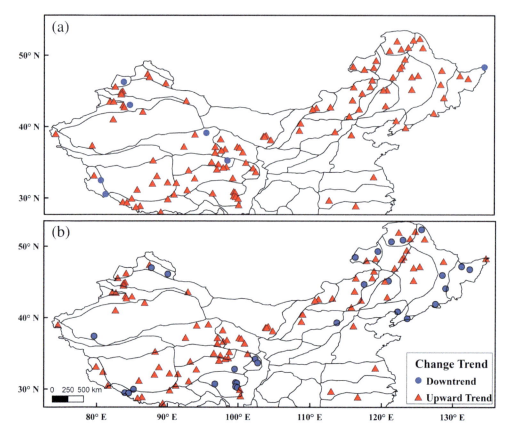

FIGURE 8.8 Spatio-temporal change trends of near-surface temperature (a) and ground precipitation rate (b)

regional ground meteorological elements dataset to analyze the precipitation changes of the stable wetland vegetation samples. The results showed that the precipitation rate of most sample points (88/117) showed an upward trend in the past 35 years (Figure 8.8b), meaning the general increasing trend of precipitation across China. By contrast, the downward trend of the precipitation rate occurred across the other 29 sample points.

3.4 Responses of Wetland Vegetation NDVI to Climate Variables

This study calculated the partial correlation coefficient between NDVI and temperature and precipitation in the growing season. After the wavelet transform, the NDVI trend change and turning point of every wetland vegetation sample were obtained. For those samples with a turning point in NDVI change, this study calculates their correlations both before and after the turning point, respectively. The percentage of samples in each climate zone with different correlation relations is shown in Table 8.2. The response of wetland vegetation NDVI to climate variables was obviously different among the various dry-wet climate zones. Wetland vegetation NDVI always negatively correlated with precipitation in wet climate zones (wet and semi-wet areas), and in the samples with turning points, the degree of negative correlation in the wet area was stronger than in the semi-wet area. However, wetland vegetation NDVI positively correlated with precipitation in arid climate zones (semi-arid and arid areas). In contrast, wetland vegetation NDVI negatively correlated with temperature only appeared in semi-arid area samples without turning points. When samples have NDVI turning points, the wetland vegetation NDVI and temperature are always positively correlated. The response of wetland vegetation NDVI to climate changes is also different in various

TABLE 8.2
The Percentage of Samples with Partial Correlation between Stable Wetland Vegetation NDVI and Climate Indicators in Each Climate Zone (Unit: %)

	\multicolumn{8}{c}{TP Exists}	\multicolumn{4}{c}{TP Doesn't Exist}
	\multicolumn{4}{c}{Before TP}	\multicolumn{4}{c}{After TP}				
	$PCorr_{NDVI\&T}$		$PCorr_{NDVI\&P}$		$PCorr_{NDVI\&T}$		$PCorr_{NDVI\&P}$		$PCorr_{NDVI\&T}$		$PCorr_{NDVI\&P}$	
	+	−	+	−	+	−	+	−	+	−	+	−
WA	71	29	21	79	57	43	14	86				
S-WA	50	50	45	55	75	25	35	65	67	33	0	100
S-AA	57	43	77	23	57	43	63	37	33	67	67	33
AA	68	32	79	21	68	32	63	37	89	11	67	33
CT	75	25	25	75	75	25	0	100				
MT	50	50	67	33	60	40	48	52	67	33	67	33
WT	67	33	100	0	33	67	100	0				
PT	46	54	46	54	62	38	54	46	77	23	69	31
PS-f	92	8	62	38	85	15	54	46	14	86	43	57

Note: "PCorr_NDVI&T" is the partial correlation between NDVI and temperature; "PCorr_NDVI&P" is the partial correlation between NDVI and precipitation. TP is the turning point; WA is the wet area; S-WA is the semi-wet area; S-AA is the semi-arid area; AA is the arid area; CT is the cold temperature zone; MT is the medium temperature zone; WT is the warm temperate zone; PS-f is the plateau sub-frigid zone; PT is the plateau temperate zone. "+" represents positive correlation; "−" represents negative correlation.

climate temperature zones. When there was a turning point in temperature-limited zones (e.g., cold temperature zone and plateau sub-frigid zone), wetland vegetation NDVI positively correlated with temperature (Table 8.2), and a more significant correlation was observed in the plateau sub-frigid zone. However, no obvious correlation existed in the medium temperature zone. The trend change of wetland NDVI before and after the turning point in the warm temperature zone and the plateau temperature zone was opposite, but the correlation was not very high, indicating the complexity of wetland vegetation change from climate change. When samples didn't have a turning point, NDVI significantly positively correlated with temperature in the medium temperature and plateau temperature zones and significantly negatively correlated with temperature in the plateau sub-frigid zone. In contrast, the response of wetland vegetation NDVI to precipitation showed no obvious relationship among various temperature zones. Wetland vegetation with high NDVI was negatively correlated with precipitation, while wetland vegetation with low NDVI was positively correlated with precipitation (Tables 8.1 and 8.2). This was because the wetland vegetation with high NDVI was mainly distributed in the area with high humidity. At the same time, the NDVI value of wetland vegetation was relatively low in areas with low humidity. Therefore, the increase in precipitation had an inhibitory effect on wetland vegetation with high NDVI, while it had a positive effect on wetland vegetation with low NDVI.

3.5 The Impact of Climate Change on Wetland Vegetation

El Niño and La Niña are effective indicators of global climate change. El Niño refers to the phenomenon of large-scale rising of water temperature that occurs over several years along the coast of Peru and on the eastern equatorial Pacific Ocean. La Niña refers to the abnormal cooling of the water temperature on the ocean surface (Zheng et al., 2015). The high-frequency part of the time series obtained by wavelet transform corresponds to the NDVI time series period item, and the low-frequency part corresponds to the trend item. In the study, wavelet transform can also be used to remove the short period (seasonal change in the year) in the wetland vegetation NDVI time series change and get the medium- and long-term periodic change characteristics (inter-annual change characteristics), which have good corresponding relationships with El Niño and La Niña (Figure 8.9).

When the periodicity of wetland vegetation NDVI is corresponding to the seven El Niño years from 1981 to 2015 (a. 1982–1983; b. 1986–1987; c. 1991–1994; d. 1997–1998; e. 2002–2007; f. 2009–2010; g. 2014–2015), we find that the El Niño year is corresponding to the part of NDVI that shows a downward trend. Out of the 117 wetland samples, 69 showed three or more corresponding time intervals with the decrease in NDVI and an El Niño year. We have calculated the corresponding situation of seven El Niño year intervals and decreasing NDVI trends and found that there are 106 wetland vegetation samples with decreasing NDVI corresponding to 2014–2015 El Niño and 62 corresponding to 2009–2010 El Niño. The corresponding quantity between El Niño year intervals a, b, c, d, and e and wetland vegetation samples with declining NDVI are 38, 30, 16, 31, and 12, respectively.

When the periodicity of wetland vegetation NDVI is compared to the six La Niña years from 1981 to 2015 (a. 1984–1985; b. 1988–1989; c. 1995–1996; d. 1999–2000; e. 2007–2009; f. 2011–2012), we find that the La Niña year is corresponding to the part of NDVI that shows an upward trend. Among the 117 wetland samples, the NDVI of 69 samples corresponded with at least three La Niña year intervals. We have calculated the corresponding situation of six La Niña year intervals and rising NDVI trends and found that there are 64 wetland vegetation samples with rising NDVI trends corresponding to the 1984–1985 La Niña, 52 corresponding to the 1988–1989 La Niña, and 73 corresponding to the 1999–2000 and 2011–2012 La Niña. The corresponding quantity between La Niña year interval c and e and wetland vegetation samples with rising NDVI are 28 and 17, respectively.

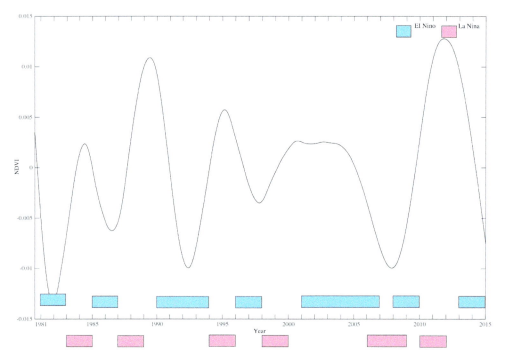

FIGURE 8.9 One example of inter-annual variation trend of stable wetland vegetation NDVI corresponds to El Niño and La Niña years.

El Niño and La Niña have opposite effects on climate. El Niño will disturb the consistent climate characteristics of the region; that is, the precipitation in rainy seasons (or regions) will measurably decrease, and the temperature in low-temperature seasons (or regions) will increase abnormally. La Niña, on the other hand, will significantly enhance the climate characteristics of the region; that is, the humid land will be wetter, and the arid land will be drier.

3.6 The Complexity of Wetland Vegetation Response to Climate Change

When regional precipitation cannot meet the requirements of wetland vegetation growth, the increase of precipitation plays a positive role. The arid area of the medium temperature zone, which is located in the western inner Mongolia Plateau, has a very dry climate. It is inland, so humid airflow that comes from the ocean is difficult to reach. The same situation occurs in the plateau temperature zone's semi-arid and arid areas. The increase of wetland vegetation NDVI in these areas is likely to be affected by the increase in precipitation.

In dry climates, temperature rise will have a negative impact on wetland vegetation growth. In semi-arid areas, which are located in Hulunbuir Plain, west Liaohe Plain, and the eastern inner Mongolia Plateau, the increase of precipitation is the promoting factor of vegetation growth. However, the increase in temperature will greatly enhance the transpiration of wetland vegetation and destroy the water balance. Therefore, in these areas, the negative impact of temperature increase is greater than the positive impact of increased precipitation and is a possible reason for the downward trend of wetland vegetation NDVI. In the semi-arid area, which is located in an alpine valley in southern Tibet, precipitation increased slightly over the long-term sequence. Because of the limited climate characteristics, such limited precipitation and large amounts of evaporation, the temperature increase plays a leading role in the decline of wetland vegetation coverage.

In humid climates, the increase in precipitation will inhibit the growth of wetland vegetation. Excessive moisture can cause damage to vegetation because it causes hypoxia and inhibits aerobic respiration of vegetation. In semi-wet areas of Songliao Plain, Sanjiang Plain, and its southern mountainous areas, the humidity is relatively high and the decline of wetland vegetation NDVI is likely to be affected by the increase of precipitation.

Some wetland vegetation changes are due to the influence of local climate and geomorphological environmental characteristics. Batang County, Ganzi Tibetan Autonomous Prefecture, Sichuan Province, which is located in the semi-wet area of the plateau temperature zone. The topography here is complex and the vertical difference of climate is obvious under the joint influence of plateau airflow and monsoons. In this region, the temperature rises rapidly in spring, and summer there is hot and dry. The maximum temperature in summer exceeds 35°C. After fall, the hot and cold air alternates and the microclimate frequently occurs. The wetland vegetation NDVI in this area is significantly negatively correlated with temperature. With the increase of temperature, NDVI shows a downward trend in the long-term sequence. Because the growth season studied in this paper is from May to September, mainly including the end of spring, the whole summer and the beginning of autumn, it is inferred that the temperature increase exacerbates the hot and dry climate characteristics, and the transpiration increases, resulting in a decline in wetland vegetation NDVI.

On the national scale, the main controlling factor of vegetation change is the comprehensive regulation of natural factors such as temperature and precipitation. However, for wetland vegetation located at a landscape scale, such as the Zoige Plateau, human activities are important factors that cannot be ignored. In the wet areas located in the Lesser Khingan Mountains and Changbai Mountains, large-scale production activities such as peat extraction, drainage, damming, and mining have led to a rapid decline in wetland areas in recent years (Yangwu et al., 2009). A large amount of swamps were transformed into pastures in the Zoige Plateau, so human activities may be one of the key incentives for wetland vegetation degradation in this region. In that situation, the change of wetland vegetation NDVI is related to the superposition of climate fluctuations and human activities.

Existing studies have shown that higher temperatures in the middle and high latitudes of the Northern Hemisphere will increase vegetation growth activity (Nemani et al., 2003). In areas where precipitation can meet the requirement of wetland vegetation growth, the increase in temperature will promote the growth activity of vegetation, which may be the reason why 63% of wetland vegetation is in a better environment with the increase in temperature. However, in areas with dry climates and insufficient precipitation, vegetation growth activity is mainly limited by water factors. The increase in temperature will lead to a decrease in soil moisture, which will lead to a worse environment for wetland vegetation.

STUDY 2: BIOPHYSICAL PARAMETER MAPPING USING MULTISPECTRAL SATELLITE DATA

1. Introduction

Biophysical variables such as canopy chlorophyll (Chl) content (CCC), green leaf area index (LAI) (a ratio of green foliage area vs. ground area), green vegetation fraction (VF) (percent of green canopy cover), and aboveground green biomass (GBM) are necessary parameters to provide quantitative insight into the health and physiological status of vegetation. Further, these parameters can serve as direct proxies of primary productivity and nitrogen content. Monitoring these characteristics through remotely sensed data can help infer the overall health and productivity of these valuable natural resources on a larger scale so that effective management strategies can be implemented to high priority areas.

Chl is one of the most important foliar biochemicals, and the content within a vegetation canopy is related closely to both vegetation productivity and absorbed photosynthetically active radiation. The importance of studying Chl in vegetation has been recognized for decades (Danks, 1983). Changes in Chl are related to photosynthetic capacity (thus, productivity), developmental stage, and canopy stresses (e.g., Ustin et al., 1998). It has been suggested that Chl may appear to be the one community property most directly related to the prediction of productivity (Whittaker & Likens, 1973).

The synoptic view provided by airborne and spaceborne sensors has the potential for estimating Chl on a regional and global basis. Variations in leaf Chl produce large differences in leaf reflectance and transmittance spectra; however, canopy reflectance is also strongly affected by other factors (e.g., canopy architecture, canopy Chl distribution, LAI, soil background) that mask and confound changes in canopy reflectance caused by leaf Chl, thus making retrieval of total Chl in a vegetative canopy both complicated and challenging. Several remote sensing techniques using reflectance in the red and near-infrared (NIR) spectral regions have been proposed to estimate Chl in leaves and canopies. Tucker (1977) reported that red reflectance is very sensitive to low amounts of both biomass and Chl and might be the best spectral region to estimate low amounts of Chl. However, saturation of red reflectance at intermediate to high levels of Chl (Buschmann & Nagel, 1993; Kanemasu, 1974; Myneni et al., 1997; Tucker, 1977) limits the applicability of the techniques that use red and NIR bands for Chl assessment, primarily due to choices of band location and width (Baret & Guyot, 1991; Gitelson et al., 1996b; Jenkins et al., 2002; SELLERS, 1985). Further, the mathematical formulation of the NDVI limits this vegetation index in CHL studies because the normalization procedure makes the NDVI insensitive to variation in R_{red} when $R_{NIR} \gg R_{red}$ (Gitelson, 2004). These limitations prevent accurate estimation of Chl in moderate-to-high vegetation densities.

Other key variables required for estimating coast salt marsh status are green LAI, VF, and GBM. Remote estimations of these characteristics can be performed using transformations of spectral reflectance, called vegetation indices (VIs) (Rouse et al., 1974). A physically based algorithm for estimating LAI from NDVI observations has been developed (Myneni et al., 1997). However, the relationship between NDVI and LAI is essentially non-linear and suffers a rapid decrease of sensitivity at moderate-to-high densities of photosynthetic green biomass (Gitelson, 2004; Gitelson, Viña, et al., 2003; Myneni et al., 1997). Alternative methods have been proposed that yield more linear relationships between remotely sensed data and VF, LAI, and GBM (Chen & Cihlar, 1996; Gao et al., 2000; Gitelson, Verma, et al., 2003; Gitelson, Viña, et al., 2003).

Monitoring these biophysical parameters not only helps in assessing overall wetland dynamics but also facilitates prioritization of restoration in areas that require immediate restoration and conservation measures. A robust biophysical mapping protocol is also critical in assessing the success/failure of previous restoration efforts (Friess et al., 2012; Hinkle & Mitsch, 2005). The MODIS instrument holds considerable potential for advancing our capabilities to estimate and monitor the biophysical characteristics of salt marshes across large geographic areas. MODIS provides a near-daily global coverage of moderate resolution data in the red and NIR spectral regions (250 m) and in the green range (500 m) that are well calibrated, atmospherically corrected, and have relatively high geolocational accuracy (Wolfe et al., 2002). However, the utility of MODIS 250-m and 500-m data for quantitatively estimating biophysical characteristics of wetland/marshland has not been investigated and used to date. In Study 2, a method for re-parameterization of existing vegetation indices is presented. This method uses field data and MODIS surface reflectance data to map the biophysical characteristics of salt marsh habitats along the Georgia coast.

2. METHODS

2.1 Field Data Collection of Wetland Biophysical Properties

2.1.1 Leaf Chlorophyll Content (LCC)

The Minolta 502 SPAD Chlorophyll meter (Spectrum Technologies Inc., East Plainfield, IL, USA) was used to measure the *in situ* leaf level of Chl (Figure 8.10). A total of 20 stratified random SPAD readings were acquired from each sampling location across varying Chl levels inside the IFOV of the sensor. The 20 readings were averaged and converted to absolute Chl values (mg/m^2) by using coefficients derived from a calibration experiment conducted in the beginning of the field season. The calibration procedure involved laboratory-based analytical extraction of Chl from a few leaf samples and the development of a linear statistical relationship between the analytical Chl, and the corresponding SPAD readings (Gitelson et al., 2005).

2.1.2 Green Biomass (GBM)

Aboveground green biomass data were collected by destructive sampling from a 0.09 m^2 (1 ft^2) subplot within each study plot using a PVC frame and clippers (Figure 8.10). Biomass samples were sorted to separate the live biomass from the dead, and oven dried at 65°C overnight (~24 h) to get rid of moisture. The dry weight was recorded using a standard measuring balance. Precautions were taken to avoid moisture absorption by the dried green biomass during dry weight measurement. The dry green biomass weights (g/ft^2) were then rescaled to g/m^2.

2.1.3 Vegetation Fraction (VF)

Percent green vegetation fraction (VF) was estimated from a circular crop of vertical digital photographs of the study plots acquired by OLYMPUS E—400 digital SLR camera (Olympus America Inc., Centre Valley, PA, USA). The camera was installed on the frame along with a laser pointer next to the hyperspectral radiometer. The laser pointer marked the center of the digital photograph and the IFOV. The digital photograph was cropped to match the IFOV of the hyperspectral radiometer, and the VF was estimated by the ratio of the number of green pixels to the total number of pixels in each photograph (Figure 8.10).

2.1.4 Leaf Area Index (LAI)

LAI Plant Canopy Analyzer 2000 (LICOR Biosciences Inc., Lincoln, NE, USA) (Gitelson, 2004) and AccuPAR LP-80 Ceptometer (Decagon Devices Inc., Pullman, WI, USA) (Delalieux et al., 2008; Kovacs et al., 2009) were used to estimate leaf area index (LAI) (Figure 8.10). The median of four LAI readings taken in each study plot was used as the LAI of the study plot. Each LAI measurement involved one above-canopy and four below-canopy readings. This study estimated green LAI (GLAI) as the product of LAI and vegetation fraction (VF).

FIGURE 8.10 Biophysical data collection—a and b: biomass collection from study plot; c and d: leaf Chl content measurement using SPAD 502 Chl meter; e and f: vegetation fraction measured from the IFOV of the sensor; g and h: leaf area index measurement using LICOR LAI Plant Canopy Analyzer 2000 and AccuPAR LP-80 Ceptometer.

Further, canopy-level Chl (mg/m^2) was calculated as the product of LAI and LLC based on Gitelson et al. (2005).

The biophysical parameters were acquired from roughly 200 study plots during multiple field trips spanning over 2010–2011 for MODIS-based model calibration and validation.

2.2 Satellite Data

Multi-temporal eight-day Level 1B atmospherically corrected surface reflectance composites for the Georgia coast were acquired from the National Aeronautics and Space Administration (NASA) (http://modis-land.gsfc.nasa.gov) for the growing seasons (April–October) from 2000 through 2013. Both 250 m and 500 m scenes from the MODIS sensor were downloaded and mosaicked. Salt marsh subsets were prepared using digital boundaries acquired from the National Wetlands Inventory (www.fws.gov/wetlands/index.html).

2.3 Model Calibration and Validation

Following the initial pre-processing of satellite data, *in situ* sampling locations were used to extract pixel values from MODIS images. MODIS scenes were chosen based on the proximity of the dates between the image acquisition and field data collection. Pre-existing and well-established vegetation indices were derived from the extracted pixel values for model calibration for each biophysical parameter (Table 8.3). For pixels containing multiple sampling locations, the average value of the individual biophysical parameter was calculated and used in model calibration. During this process, roughly 69 sampling plots acquired in 2010 were reduced to 10–15 MODIS 250 m and 7–10 MODIS 500 m pixels. Further, an independent field dataset was acquired during the field campaigns in 2011 containing another 121 sampling plots, subsequently reduced to 10–12 MODIS 250 m pixels and ~10 MODIS 500 m pixels, and were used for model validation (Table 8.4). Performance uncertainties were analyzed based on the percent RMSE (Table 8.4).

2.4 Monthly Composite Products and Phenology Extraction

Following successful calibration and validation, eight-day time series composites were generated in ERDAS Imagine 2011 (Leica Geosystems, Heerbrugg, Canton St. Gallen, Switzerland) for GLAI, VF, Chl, and GBM for the growing seasons (March–November) from 2000 through 2013, using the best-fit models. Composites were generated for both 250-m and 500-m resolution images. For each growing season month, four composites per parameter per resolution were created; therefore, over nine months, 144 composites were created. As such, for all four parameters, composites generated for 14 growing seasons amounted to almost 4032 composites (combining both 250 m and 500 m composites for coastal Georgia). These composites were used for qualitative assessments of salt

---

TABLE 8.3
List of Satellite Image–Derived Vegetation Indices Used for Calibration and Validation of Models to Estimate Marsh Biophysical Parameters. * R_{red} Was Used Instead of $R_{red\text{-}edge}$ as Described in Mishra et al. (2012)

Vegetation Index	Formula	Reference
Normalized Difference Vegetation Index (NDVI)	$(R_{NIR} - R_{Red})/(R_{NIR} + R_{Red})$	Rouse et al. (1974)
Enhanced Vegetation Index 2 (EVI2)	$\{2.5 \times (R_{NIR} - R_{Red})/(R_{NIR} + 2.4 \times R_{Red} + 1)\}$	Huete et al. (2002)
Chlorophyll Index Red (CI$_{red}$)*	$(R_{NIR} - R_{Red})/R_{Red}$	Gitelson et al. (2006)
Wide Dynamic Range Vegetation Index (WDRVI)	$(\alpha \times R_{NIR} - R_{Red})/(\alpha \times R_{NIR} + R_{Red})$	Gitelson (2004)
Soil Adjusted Vegetation Index (SAVI)	$(R_{NIR} - R_{Red}) \times (1 + L)/(R_{NIR} + R_{Red} + L)$	Huete (1988)
Chlorophyll Index Green (CI$_{green}$)	$(R_{NIR} - R_{Green})/R_{Green}$	Gitelson et al. (2006)
Visible Atmospherically Resistant Index (VARI)	$(R_{Green} - R_{Red})/(R_{Green} + R_{Red})$	Gitelson et al. (2002)

TABLE 8.4
Coefficients of Determination (R²) and Percent Root Mean Square Error (%RMSE) Values for MODIS 250 m– and 500 m–Derived Best-Fit Models

		GLAI	VF (%)	GBM (g/m²)	CHL (mg/m²)	Best Fit Model
250 m	R²	0.821	0.845	0.85	0.837	WDRVI ($\alpha = 0.1$)
	RMSE (%)	14.967	22.055	19.616	25.817	
500 m	R²	0.91	0.98	0.938	0.864	VARI
	RMSE (%)	32.474	24.344	17.34	21.998	

marsh habitats pre– and post–significant natural and anthropogenic events such as hurricanes and droughts, and to evaluate the performance of the best-fit models chosen for the mapping of biophysical parameters. In addition, phenology charts for site-specific salt marsh patches were derived from these time-series composites, using the spatial analysis module in ArcGIS 10.0 desktop for the entire 14-year growing season dataset. The phenology charts were examined for effects of discrete natural and anthropogenic events on the health of marsh patches, as well as for long-term trends in the biophysical values over the past 14 years.

3. RESULTS AND DISCUSSION

After an extensive testing of numerous VIs on MODIS data using aforementioned methods for calibration and validation, the Wide Dynamic Range Vegetation Index (WDRVI, $\alpha = 0.1$) (Gitelson, 2004) was selected for estimating the biophysical parameters (GLAI, CHL, VF, GBM) for the 250 m dataset, whereas the Visible Atmospheric Resistant Index (VARI) (Gitelson et al., 1996a) was selected for predicting the biophysical parameters for the 500 m dataset (Table 8.4). The performance of several red-NIR based VIs on 250 m data were fairly close to each other; however, WDRVI was selected for map composite preparation because it provided the best combination of R² and %RMSE for all biophysical parameters (results not shown). In the case of the 500 m data, VARI outperformed all other VIs by a large margin (results not shown). No significant pattern of over- or underestimation was observed in the model residuals (results not shown). Time-series composites were generated for each biophysical parameter and for both resolutions separately, using the best-fit models for the growing seasons from 2000 through 2013 (Figure 8.11). The time-series composites provided a qualitative estimation of the physiological health and productive capacity of the salt marshes, both at the site-specific and landscape level. These composites elucidate seasonal variation in biophysical parameters, and help restoration managers and conservationists identify critical patches of marsh stress (Figure 8.11).

Phenology charts for the last decade were extracted for selected wetland patch locations using a 3 × 3 or higher window size, depending on the extent of marsh patch (i.e., the patch size of marsh undergoing marsh dieback). The monthly averages of biophysical parameters were extracted to construct a 14-year phenology chart for wetland patches (Figure 8.12). For example, Figure 8.12 shows the 14-year phenology of a marsh patch along the Jerico River, which has been repeatedly affected by dieback events since 2000 (J. Mackinnon, CRD, GA DNR, personal communication). A similar analysis was performed on dieback-affected patches throughout coastal Georgia. This analysis was restricted to patches covering an area of at least 4 ha, ensuring study patches were detectable using the MODIS 250 m data. The yellow boxes indicate years when dieback was reported for these sites by Coastal Resources Division (CRD), Georgia. MODIS 250 m data isolated the dieback signal in all four biophysical parameters for this site (Figure 8.12). This study also noticed a decreasing trend in marsh productivity at the site, indicating that it is experiencing increased stress over the last 15 years.

Wetland Mapping Methods and Techniques 243

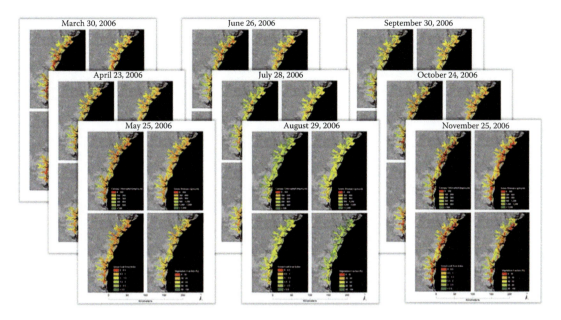

FIGURE 8.11 Examples of weekly map composites comprising the four biophysical parameters produced using MODIS 250 m data–based models for the 2006 growing season.

Since marsh diebacks have been linked to persistent drought in Georgia, this study correlated the trends and fluctuations in biophysical parameters with monthly Palmer Drought Severity Index (PDSI) estimates retrieved for each study area (Figure 8.12). As a measure of dryness based on precipitation and temperature, PDSI shows drought in terms of negative numbers and wet conditions in terms of positive numbers (Heim, 2002). Correlations among PDSI and mean biophysical values were analyzed using Pearson's r, and linear regression was conducted using Ordinary Least Squares Regression during the peak of the growing season (July, August, and September) (results not shown). The phenological charts demonstrated that PDSI and biophysical variables co-varied together over the growing seasons of the last 14 years; low PDSI for the years 2002–2003, 2007–2008, and 2011–2012 (major years of drought) coincided with low biophysical values as predicted by the model. Further, for the wet growing seasons (2004–2006 and 2009–2010), high PDSI values concurred with high levels of marsh productivity. Linear regression results suggested that PDSI has significant positive effects on each biophysical parameter during the peak growing season (July, August, and September) (results not shown). As PDSI decreased (increased drought conditions), there was a decreasing trend in all four biophysical parameters, suggesting increased stress in salt marshes. The time-series composites and phenological charts provide the end user with qualitative and quantitative information on the wetland's health and physiological status.

STUDY 3: BELOWGROUND BIOMASS AND FOLIAR NITROGEN (N) MAPPING USING HYBRID MODELING

1. Introduction

Remote sensing models of aboveground biomass are commonly derived from optical data (Lu & Weng, 2007; Smith et al., 2002); however, monitoring belowground biomass to understand whole-plant productivity remains challenging. Belowground biomass estimates are needed because roots and rhizomes are the dominant source of soil organic carbon in wetlands (Moore, 1987; Rasse et al., 2005). In addition, wetland belowground biomass contributes to soil stability and accretion

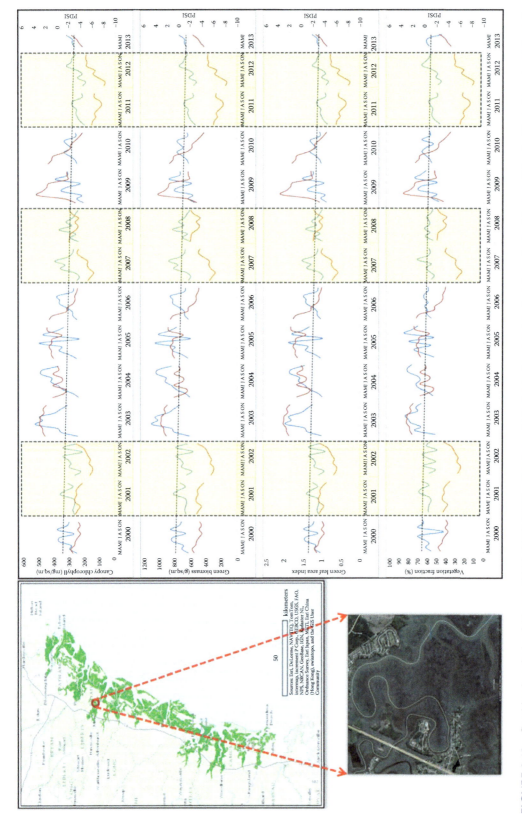

FIGURE 8.12 Combined analysis of 14-year (2000–2013) PDSI estimates and biophysical parameter phenology derived from a dieback-affected marsh patch on the Jerico River, showing the overall spatio-temporal growth trend of the patch. Years with observed marsh dieback events at the location are highlighted in yellow.

by adding organic matter volume (Miller et al., 2008; Nyman et al., 1993). Belowground biomass may consequently help maintain or build wetland elevation to prevent subsidence below sea level (Nyman et al., 2006). Therefore, tools for predicting and promoting wetland plant belowground biomass can support wetland management for resilience and provision of ecosystem benefits.

Measuring belowground biomass across sites is difficult because of wide fluctuations in aboveground vs. belowground growth within and among species (Mokany et al., 2006). Thus, aboveground biomass often is a poor predictor of belowground conditions. To reliably estimate belowground production, this study needs models that describe plant allocation among above- and belowground vegetation. Two such growth models exist. The first has been called balanced growth (Shipley & Meziane, 2002) or optimal partitioning (Kobe et al., 2010; Mccarthy & Enquist, 2007) and suggests that plants allocate growth toward growth limiting resources. In this view, new growth is directed belowground when water and nutrients are limiting and aboveground otherwise. Conversely, the isometric allocation growth model (Mccarthy & Enquist, 2007) suggests above- and belowground growth scales together as a result of physiological constraints. Consequently, nutrient addition stimulates total net plant production, fertilizing above- and belowground biomass equivalently. For either balanced or isometric growth, leaf N concentration, a measure of plant nutrient uptake, can inform biomass estimates. In the balanced growth case, additional nutrients, such as N and P, increase shoot growth without equally increasing belowground biomass. Therefore, leaf N scales with root:shoot ratios and can be used to estimate belowground biomass (O'Connell et al., 2014). Alternatively, plants responding isometrically to nutrient addition have stimulated below- and aboveground biomass and constant root:shoot ratios with respect to N. Reliably modeling wetland fates therefore requires studies that resolve relationships among plant nutrient uptake and biomass allocation across a range of environmental conditions.

Remote sensing may assist with whole-plant biomass estimation by providing site-wide spectral reflectance estimates of plant nutrient uptake and aboveground biomass. For example, foliar N may be a good indicator of long-term plant available nutrients (Boyer et al., 2001; Cohen & Fong, 2005) and can be estimated with remote sensing approaches (Curran, 1989a; Townsend et al., 2003). Several analytical techniques relate spectral data to canopy N and plant biomass, such as band-depth analysis (Kokaly & Clark, 1999), spectral matching techniques (Gao & Goetzt, 1995), and partial least squares (PLS) regression (Byrd et al., 2014; Smith et al., 2002). In particular, PLS regression, a full-spectrum multivariate method, is ideal for maximizing explained variance among plant responses and spectral reflectance signals (Geladi & Kowalski, 1986; Mevik & Cederkvist, 2004). PLS uses an Eigen vector-based approach to reduce many correlated predictors, such as spectral data, to a few independent components. These components maximize predictive value and minimize background effects introduced by non-target cover types such as water and dead vegetation (Chen et al., 2009). Such background effects commonly reduce the predictive value of spectral reflectance signals in emergent wetlands, as discussed earlier.

In an early study, O'Connell et al. (2014) explored the predictive capacity of spectral reflectance estimates of foliar N to model belowground biomass. They compared the whole-plant productivity of *Schoenoplectus acutus*, a common freshwater macrophyte, across a simple experimental N-addition and control treatment. Results suggested *S. acutus* exhibited balanced growth, reducing belowground biomass in fertilized plots (O'Connell et al., 2014). Additionally, spectral reflectance differed among high and low N plants. Particular differences were at 550 nm (peak greenness), 1250 nm (corresponding to leaf water content), 1750 nm, and 2100 nm (both corresponding to spectral N absorption features) (Curran, 1989b; Thenkabail et al., 2002, 2012). Thus, spectral reflectance estimates of N-driven biomass allocation might correlate with belowground biomass within *S. acutus* marshes.

This study presents a method for identifying within-site variability in belowground biomass and root:shoot ratios in a low-diversity coastal freshwater marsh. The effectiveness of multispectral remote sensing (Landsat 7) for estimating aboveground biomass and % foliar N is evaluated and compared to optimal remote sensing data, hyperspectral field spectroradiometer data. Ultimately, a hybrid modeling approach is used to linearly relate reflectance-based PLS regression estimates of foliar N and aboveground biomass to belowground biomass and root:shoot ratios. Initially several satellite platforms and species-specific vs. cross-species models were compared (O'Connell et al., 2015). Here only the best-fit satellite reflectance models are presented: species-specific models based on Landsat 7 satellite reflectance data.

2. METHODS

2.1 Study Sites

Three impounded freshwater marsh units in the Sacramento–San Joaquin Delta, CA, USA, were sampled: Twitchell Island east unit (38.1073°, −121.6483°), Twitchell Island west unit (38.1069°, −121.6449°), and Mayberry Slough southeast unit on Sherman Island (38.0490°, −121.7660°). All units historically were part of extensive freshwater perennial peat marshes of the Sacramento-San Joaquin Delta. Emergent vegetation in all three units was mostly *Schoenoplectus acutus* and *Typha domingensis, latifolia, angustifolia*, and *Typha* spp. hybrids.

2.2 Experimental Design

Above- and belowground emergent wetland plant biomass and % foliar N were estimated using field methods on a semi-monthly basis during one summer growing season. These data were used to build relationships with spectral reflectance. Aboveground biomass was estimated within 1 m^2 plots using validated species-specific allometric equations (Byrd et al., 2014). For belowground biomass, ingrowth root cores were used to estimate relative variation in belowground biomass within 30 cm of the wetland surface, where the majority of root production occurs (McKee et al., 2007; McKee, 2011). At foliar N plots, leaf samples were collected for estimation of % foliar N. Sample locations were from permanent plots along four transects per wetland. These transects were at least 30 m long, spanning the width of Landsat 7 pixels. Landsat 7 pixels were used as the sample unit, and a single 1 m^2 estimate of vegetation parameters (above- and belowground biomass, and foliar N) was calculated by averaging all values within each pixel for each vegetation survey.

2.3 Spectral Reflectance Data Collection

To build spectral models, *in situ* hyperspectral reflectance was collected using an ASD field spectroradiometer (FieldSpec Pro FR, Analytical Spectral Devices, Inc., Boulder, CO, USA). Field spectroradiometer collection was conducted for plots within 2 m of a boardwalk or wetland edge, allowing access with a spectroradiometer. Spectral readings were sampled every 1.4 nm over 350–1000 nm and 2 nm over 1000–2500 nm using a 25° field of view fore-optics at nadir 1 m above the vegetation canopy. For plots inaccessible by portable spectroradiometer, spectral measurements from satellite platforms were relied on. These *in situ* hyperspectral data were used to build empirical models of % foliar N and aboveground biomass. Field spectrometer-based hyperspectral models were compared with the satellite-based multispectral models to calculate differences in model explanatory power. Landsat 7 CDR surface reflectance images (available via earthexplorer.usgs.gov) were acquired, which are atmospherically corrected surface reflectance data products processed by USGS using the LEDAPS algorithm (Masek et al., 2012).

2.4 Spectral Reflectance Model Development

To develop models predicting belowground biomass and root:shoot ratio, first two spectral reflectance model sets were generated to estimate % foliar N and aboveground biomass (model sets 1–2, Table 8.5). Spectral reflectance-derived predicted outcomes from models 1 and 2 were used to develop linear regressions for estimating belowground biomass and root:shoot ratio, i.e., hybrid reflectance models. These hybrid models for predicting belowground biomass included a balanced growth model (model set 3, Table 8.5), an isometric model (model set 4, Table 8.5) and a fusion balanced and isometric growth model (model set 5, Table 8.5). Isometric growth was identified as equivalency in scale for above- and belowground biomass, where root:shoot ratios are roughly constant. For balanced growth, % foliar N should be related to root:shoot ratio. A similar set of models were built for root:shoot ratio (model sets 6–7, Table 8.5).

TABLE 8.5
Spectral Reflectance and Hybrid Spectral Reflectance Model Set Development. Hybrid Models 3 and 6 Represent Tests of Balanced Growth Models, 4 Isometric Model Tests, and 5 and 7 Fusion Balanced and Isometric Growth Models

Model Set	Model
Spectral models	
1	Predicted % foliar N ~ spectral reflectance
2	Predicted aboveground biomass ~ spectral reflectance
Hybrid spectral models	
3	Field belowground biomass ~ predicted N (model set 1)
4	Field belowground biomass ~ predicted aboveground biomass (model set 2)
5	Field belowground biomass ~ predicted aboveground biomass (model set 2) + predicted N (model set 1)
6	Field root:shoot ratio ~ predicted N (model set1)
7	Field root:shoot ratio ~ predicted belowground biomass (model set 3) / predicted aboveground biomass (model set 2)

The PLS package in R (Mevik & Wehrens, 2007) was used to build PLS regression models that used full-spectrum reflectance as predictors for estimating % foliar N and aboveground biomass (each separately). This study minimized model over-fitting by selecting components corresponding to the first local minima for root mean square error of prediction (RMSEP, e.g., averaged difference in predicted vs. measured values). RMSEP was estimated using leave-one-out cross-validation (Mevik & Wehrens, 2007). This study used a random subset as training data (70% of samples) and compared R^2 or explained variance and RMSEP with a testing dataset (30% of samples). This study also calculated normalized RMSEP (nRMSEP = RMSEP/[response maximum − response minimum]). For hyperspectral models, loadings plots for the first three PLS regression components (i.e., those explaining the majority of variance) were examined to determine wavelengths most associated with foliar N and aboveground biomass. These loadings were used to guide the selection of an appropriate multispectral sensor. Landsat 7 ultimately was chosen because its measurement capabilities best corresponded to key prediction spectra in PLS regression loadings plots. Landsat prediction accuracy (RMSE and nRMSE) also was greater than for other sensors tested (Hyperion and World View-2) (data not presented). To compare the performance of Landsat 7 under a variety of conditions, models were run with all sites included and then again with the shallowest and most spatially heterogeneous site excluded (Twitchell Island West Unit). The purpose was to generate recommendations for conditions where freely available, but spatially and spectrally coarse Landsat 7 might be utilized for productivity and N monitoring.

3. RESULTS AND DISCUSSION

3.1 Estimating Foliar N and Aboveground Biomass with Hyper- and Multispectral Reflectance

Spectral reflectance of *Typha* spp. and *S. acutus* species were plotted and had subtle differences in visible (500–600 nm), near infrared (800–1000 nm), and shortwave infrared (>1500 nm) wavelengths (Figure 8.13). PLS regression of *in situ* hyperspectral reflectance correlated well with % foliar N (Table 8.6). Models for *Typha* spp. explained the greatest variance in % foliar N (56%). Loadings plots from these analyses suggested bands in the visible spectra (500–700 nm), in the near infrared spectra (1100–1300 nm), and in the shortwave infrared spectra (> 1700 nm) were important for associating % foliar N with reflectance. Landsat 7 data collection matched some PLS regression loadings peaks for % foliar N, particularly bands 5 (1550–1750 nm) and 7 (2080–2350 nm) (Figure 8.14). However, other

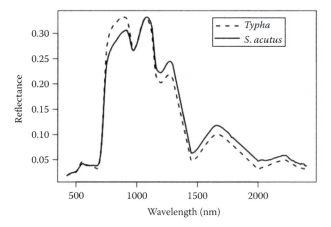

FIGURE 8.13 Loadings values of % foliar N from PLS regression of hyperspectral data for A) *Typha* spp. and B) *S. acutus*. The mid-point of bands measured by Landsat 7 are superimposed and labeled with band number (b1–7). Measured vs. predicted values for training and testing datasets are presented in panels C and E for *Typha* and in panels D and F for *S. acutus*. EV is explained variance and RMSEP is root mean squared error of prediction.

TABLE 8.6

PLS Regression Models of % Foliar N and Aboveground Biomass (AG) (g) from Spectroradiometer-Based Hyperspectral and Satellite-Based Multispectral Reflectance. Model Set Numbers Refer to Table 8.5. C Was the Component Number, EV (% Explained Variance), RMSEP (Root Mean Square Error of Prediction), and nRMSEP (Normalized RMSEP). Training and Testing Data Were 70% and 30% of Samples

Model			Training C	EV	RMSEP	nRMSEP	Testing RMSEP	nRMSEP
Field spectroradiometer hyperspectral reflectance models								
1a.	% Foliar N *Typha* sp.	= hyperspectral reflectance	7	56	0.56	0.18	0.66	0.22
1b.	% Foliar N *S. acutus*	= hyperspectral reflectance	5	46	0.53	0.26	0.43	0.26
Landsat 7 multispectral reflectance models								
1a.	% Foliar N *Typha* sp.	= Landsat reflectance	3	27.0	0.67	0.32	0.77	0.47
1b.	% Foliar N *S. acutus*	= Landsat reflectance	3	43.6	0.57	0.28	0.47	0.35
2a.	AG *Typha* sp.	= Landsat reflectance	3	59	448.2	0.24	309.3	0.25
2b.	AG *S. acutus*	= Landsat reflectance	2	67	155.5	0.20	262.7	0.34

spectra with high loadings were missed, particularly 1250 nm (Figure 8.14). Multispectral Landsat 7 satellite reflectance was related to % foliar N, but for *Typha* spp., explained variance was reduced 25% from hyperspectral spectroradiometer reflectance data (Table 8.6; Figure 8.15). Conversely, *S. acutus* Landsat 7 foliar N models were roughly equivalent to hyperspectral models (explained variance 43–46% each) (Table 8.6, Figure 8.15). Landsat 7 PLS regression also estimated aboveground biomass (explained variance > 60%, Table 8.6).

Wetland Mapping Methods and Techniques 249

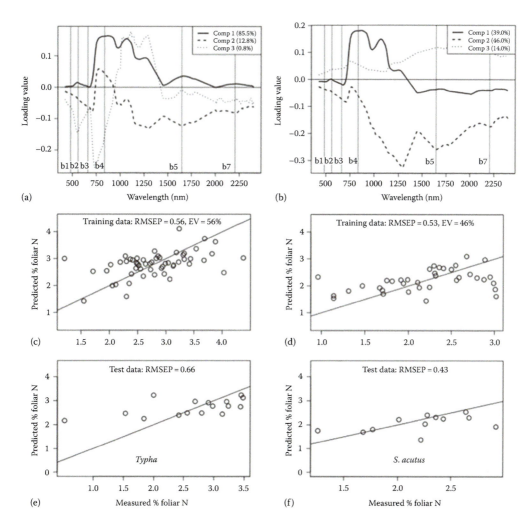

FIGURE 8.14 Measured vs. predicted % foliar N developed from PLS regression of Landsat 7 for training and testing data are plotted in panels A and C for *Typha* spp. and B and D for *S.acutus*.

3.2 Predicting Belowground Biomass Using Satellite-Derived Multispectral Hybrid Models

A linear relationship was observed for reflectance-derived estimates of % foliar N and belowground biomass (Table 8.7; Figure 8.15), with the strongest relationships observed for *Typha* spp. (29% RMSEP 56.6 g, nRMSEP 17%). Excluding the Twitchell west unit, where the greatest interspersion of cover types within pixels was observed, increased explanatory power for *S. acutus* models (34%, RMSEP 184.7 g, nRMSEP 23%). Spectral estimators of aboveground biomass had no detectable relationship to field-measured belowground biomass (P ranged from 0.23 to 0.48 for all models, Table 8.7). Further, combined liner models of % foliar N and aboveground biomass spectral estimators did not improve model fit over models relying only on % foliar N (e.g., model set 5 from Table 8.5).

3.3 Predicting Root: Shoot Ratio Using Satellite-Derived Multispectral Hybrid Models

Hybrid modeling of root:shoot ratio had greater correspondence to measured values than modeling of belowground biomass (Table 8.7; Figure 8.16). Spectrally estimated % foliar N had a significant linear relationship with field-measured root:shoot ratio, but best-fit models combined spectral

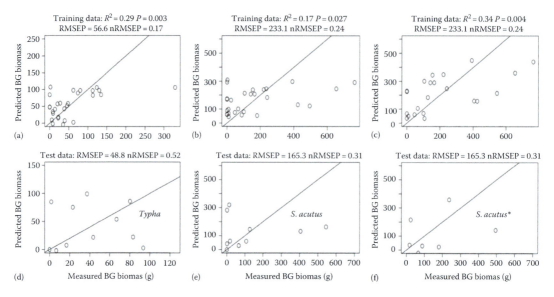

FIGURE 8.15 Measured vs. predicted belowground biomass (g) developed from hybrid model set 3 from training and testing data are plotted in panels A and D for *Typha* spp., B and E for *S. acutus*, and C and F for *S. acutus*, excluding the west unit.

estimates of aboveground biomass with spectral estimates of belowground biomass (predicted values from model set 3 derived from spectral estimates of % foliar N). As with previous *S. acutus* models, excluding the Twitchell west unit improved linear fit (explained variance with west unit 32%, without west unit 80.4%; Table 8.7; Figure 8.16).

3.4 Overall Performance

Remote-sensing measures were useful for biomass estimation. In our study, field-measured (data not presented) and spectrally estimated % foliar N both were associated with belowground biomass and biomass allocation in *S. acutus* and *Typha* spp., while measures of aboveground biomass alone were poor predictors. For root:shoot ratio, aboveground biomass helped parameterize models and improved model fit. Field spectroradiometer-based reflectance measurements explained greater than 45% of the variation in foliar N concentration. Landsat 7 bands overlapped with some hyperspectral wavelengths associated with % foliar N. However, predictive power was lost by transitioning from hyperspectral, high spatial resolution field reflectance data to multispectral, moderate-resolution satellite data. Nonetheless, spectral estimators derived from Landsat 7 had relationships with foliar N, suggesting this public data source might be utilized for coarse canopy N estimation. Further, estimates of % foliar N from Landsat 7 bands had significant relationships with belowground biomass.

Biomass allocation ratios were estimated with greater precision than belowground biomass, benefiting from spectral estimates of aboveground biomass and % foliar N together. Emergent wetland macrophytes in these sites appeared to conform to a balanced growth model, as root:shoot ratios were variable and negatively related to environmental nutrients (O'Connell et al., 2015). Plants may increase the allocation of photosyntate to belowground biomass when nutrients are limited, both to increase absorptive area to maximize nutrient capture and to increase rhizomatic carbohydrate storage as a buffer against future scarcity (Kobe et al., 2010). While plants in our study exhibited balanced growth, the model development process described here also would detect isometric growth and might assist with whole-plant biomass estimates.

TABLE 8.7
Hybrid Models, Where Predictions from Models in Table 8.6 Are Regressors of Belowground Biomass (BG) (g) or Root:Shoot Ratio (RS). Source Model Is Indicated in Parentheses. Aboveground Biomass (AG) (g), EV (% Explained Variance), RMSEP (Root Mean Square Error of Prediction, and nRMSEP (Normalized RMSEP) Are Presented for Significant Models. Training and Testing Data Were 70% and 30% of Samples. * West Unit Was Not Included.

Hybrid Reflectance–Linear Regression Models			Training F	df	P	EV	RMSEP	nRMSEP	Testing RMSEP	nRMSEP
3a.	BG *Typha* sp.	= Predicted % foliar N (1b)	10.7	1,26	0.003	29.1	56.6	0.17	48.8	0.52
3b.	BG *S. acutus*	= Predicted % foliar N (1c)	5.5	1,26	0.027	17.5	184.7	0.24	310	0.32
3b.	BG *S. acutus**	= Predicted % foliar N (1c)	10.5	1,20	0.004	34.4	176.5	0.23	309.3	0.32
4a.	BG *Typha* sp.	= Predicted AG (2b)	0.73	1,26	0.400	2.7	-	-	-	-
4b.	BG *S. acutus*	= Predicted AG (2c)	0.5	1,26	0.480	1.9	-	-	-	-
4b.	BG *S. acutus**	= Predicted AG (2c)	1.5	1,22	0.230	6.4	-	-	-	-
6a.	RS *Typha* sp.	= Predicted % foliar N (1b)	9.6	1,23	0.005	29.5	0.23	0.21	0.09	0.28
6b.	RS *S. acutus*	= Predicted % foliar N (1c)	2.3	1,23	0.145	9	1.7	0.23	0.37	0.22
6b.	RS *S. acutus**	= Predicted % foliar N (1c)	5.7	1,16	0.029	26.4	1.8	0.23	0.36	0.21
7a.	RS *Typha* sp.	= Predicted BG (3b) / Predicted AG (2b)	28.9	1,19	<0.001	60.3	0.15	0.14	0.26	0.31
7b.	RS *S. acutus*	= Predicted BG (3c) / Predicted AG (2c)	9.9	1,21	0.005	32.1	0.72	0.24	2.27	0.30
7b.	RS *S. acutus**	= Predicted BG (3c) / Predicted AG (2c)	65.7	1,16	<0.001	80.4	0.89	0.12	0.68	0.22

3.5 Correlations among Foliar N, Aboveground Biomass and Spectral Reflectance

Remote-sensing studies often estimate leaf N concentration (Tian et al., 2011; Stroppiana et al., 2012; Abdel-Rahman et al., 2013). In our study, where full-spectrum field spectrometer reflectance data were available, estimation of foliar N also was possible. Reflectance data over 500–700 nm, 1100–1300 nm, and greater than 1700 nm were particularly informative for both *Typha* and *S. acutus*. Reflectance over 500–700 nm encompasses peak greenness and the red edge. Reflectance at 550 nm, peak greenness, is associated with chlorophyll (Thenkabail et al., 2002; Townsend et al., 2003), which contains 6.5% N by weight: 30–50% of leaf N is within ribulose-1,5-biphosphate carboxylase-oxygenase (rubisco) inside chloroplasts (Kokaly et al., 2009). Chlorophyll and foliar N are thus

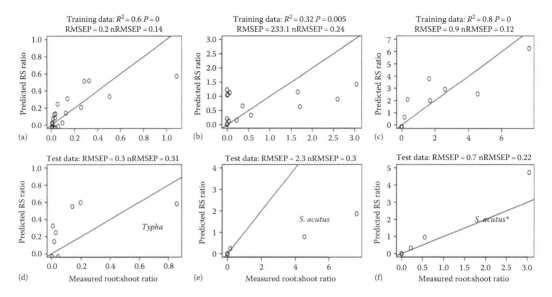

FIGURE 8.16 Measured vs predicted root:shoot ratio developed from hybrid model set 7 from training and testing data are plotted in panels A and D for *Typha* spp., B and E for *S. acutus*, and C and F for *S. acutus*, excluding the west unit.

associated with photosynthesis and plant productivity. Red edge reflectance also has been associated with plant stress and N accumulation (Thenkabail et al., 2012). For *S. acutus*, 1250 nm was the highest loading value estimating % foliar N from hyperspectral reflectance. This wavelength also had the greatest correspondence to % foliar N in earlier N-addition experimental investigations of *S. acutus* (O'Connell et al., 2014), suggesting consistency of spectral reflectance response to increased foliar N. Reflectance at 1250 nm has been related to leaf water content (Thenkabail et al., 2002, 2012) and is also associated with plant turgor pressure, thermoregulation, and vigor (Kokaly et al., 2009).

Reducing hyperspectral to multispectral data resulted in some precision loss. Landsat 7 bands corresponded to some high loadings values from hyperspectral analysis, particularly within shortwave infrared regions. Others demonstrated % foliar N was associated with shortwave infrared absorption at 1730, 2180, and 2240 nm (Curran, 1989). In particular, increased absorption around 2050–2170 was associated with amide bond vibration within proteins (Kokaly et al., 2009). Landsat 7 bands 5 (1550–1750 nm) and 7 (2080–2350 nm) (landsat.usgs.gov; Figure 8.12) capture these important N spectra, suggesting Landsat 7 may be useful for regional N monitoring, particularly for detecting hotspots where foliar N differences may be very high. Landsat 8 (launched February 2013, landsat.usgs.gov) measures similar spectra and may prove useful in future studies.

3.6 Predicting Belowground Biomass and Root:Shoot Ratio Using Multispectral Hybrid Models

Percent foliar N predicted species-specific belowground biomass and also had strong relationships with root:shoot ratios. In combination with an earlier experimental study (O'Connell et al., 2014), this methodology presents the first use of spectral data to estimate belowground biomass and biomass allocation indirectly. This effort links several relationships commonly observed in literature: remote sensing can be related to aboveground biomass (Smith et al., 2002; Kokaly et al., 2009), and foliar N concentration (Ryu et al., 2009; Stroppiana et al., 2009, 2012; Mokhele and Ahmed, 2010; Tian et al., 2011; Abdel-Rahman et al., 2013), foliar N is linked to environmental N (Bedford et al., 1999; Sartoris et al., 1999; Kao et al., 2003; Larkin et al., 2012), and aboveground biomass and environmental N have strong relationships with root:shoot ratios (Darby and Turner, 2008a; Turner et al., 2009; Deegan et al., 2012; Morris et al., 2013).

Within the west unit, *S. acutus* was poorly predicted. The west unit was shallowest (< 20 cm vs. > 50 cm elsewhere) and dominated by *Typha* spp., which are competitively dominant over *S. acutus* in shallower water. The west unit also had greater interspersion of dead thatch with live emergent vegetation. For *S. acutus*, thatch accumulation over lower portions of vertically oriented stems can mask biomass reflectance. In contrast, *Typha* spp. have broad, horizontally oriented leaves. Horizontal broad canopies cover thatch accumulation at plant bases, reducing the influence of thatch in spectral signals. PLS regression using multispectral data has less capacity than hyperspectral analyses to minimize background effects from non-target cover types. Areas with high heterogeneity probably should be measured by hyperspectral sensors, such as AVIRIS. AVIRIS measures 400–2500 nm at 4–20 m spatial resolution, depending on the platform (aviris.jpl.nasa.gov). However, Landsat is promising for measuring foliar N and whole-plant biomass hot and cool spots, particularly for pixels that have a high cover of target species.

CONCLUSIONS FROM THE CASE STUDIES AND FUTURE RESEARCH

STUDY 1

As an important part of wetlands, wetland vegetation and its environmental conditions are experiencing unprecedented changes with climate change. The use of remote sensing technology can help identify its vegetation status, thereby providing a basis for wetland protection and management. This study aimed to characterize the trend change of wetland vegetation in a time series. NDVI of wetland vegetation in China showed a significant trend change from 1981 to 2015 overall. Among them, 75% of NDVI trend change has reached a turning point. About 80% of the wetland vegetation NDVI showed a downward trend after 2004, and about 10% of the wetland vegetation NDVI showed an upward trend after 2003. Of the wetland vegetation environment, 63% was in a gradually improving state, while 23% was in a worse state. Wetland vegetation NDVI was negatively correlated with precipitation in humid areas, while it was positively correlated with precipitation in dry areas. In temperature-limited zones, wetland vegetation NDVI is positively correlated with temperature. Extreme weather events such as El Niño and La Niña may affect wetland vegetation NDVI.

STUDY 2

The biophysical mapping methodology developed through this study can be used for identifying critical hot spots of wetland degradation due to factors other than natural disasters (e.g., developmental activities, localized drought, and urban pollutant runoff). These products have the potential to facilitate the prioritization of restoration efforts through the identification of areas in need of immediate attention and also help in comparing the health and growth of wetland patches/habitats pre- and post-implementation of restoration activities. Therefore, biophysical mapping will provide government regulators and restoration managers with important information to develop restoration and restoration plans. Time-series composites can be further used to detect progressive change in the wetland habitats both inter- and intra-seasonally. Further, long-term phenology plots provide documentation of the progressive improvement or decline in wetland health following natural or human-induced disasters. In addition, MODIS-derived, long-term, time-series biophysical products can be used to develop light use efficiency (LUE) models in order to map the carbon sequestration potential of the salt marsh habitats. Future research will include the use of MODIS-derived biophysical products to map the Gross Primary Productivity (GPP) or Carbon Sequestration Potential (CSP) of salt marsh habitats.

STUDY 3

Partial least squares regression of hyperspectral and multispectral data is promising for estimating % foliar N in wetlands and also for indirectly estimating variation in belowground biomass, particularly where large differences exist as hot spots. Species-specific investigations are necessary to parameterize new modeling efforts at other sites. Currently, measures of belowground biomass are difficult to obtain. Publically available Landsat data provide global and long-term estimates of landscape change (Wulder et al., 2012). When properly supported by field and experimental studies, a hybrid modeling approach may reduce time and cost of broad-scale studies of belowground biomass and root:shoot ratios. Future studies should focus on Landsat 8 and on hyperspectral sensors, where available.

When working with multispectral data, our approach may have the greatest utility in sites with fairly homogeneous cover. Examples include salt and brackish marshes, typically dominated by one to a few species. Such area also are vulnerable to sea level rise, and there is an urgent need for understanding whole plant productivity patterns. For example, the approach outlined might help resolve controversy concerning the influence of eutrophication within coastal marshes. Some suggest elevated environmental nutrients results in loss of belowground biomass and accelerated vegetated marsh loss (Darby & Turner, 2008b, 2008c; Deegan et al., 2012; Turner et al., 2009). Others argue eutrophication fertilizes belowground production (Morris et al., 2013). Hybrid reflectance modeling may help estimate correlations among eutrophication, plant uptake as foliar N, and resulting productivity responses, helping to resolve these issues and reduce uncertainty in managing wetlands for resiliency and provisioning of ecosystem benefits.

REFERENCES

A, D., Zhao, W., Qu, X., Jing, R., & Xiong, K. (2016). Spatio-temporal variation of vegetation coverage and its response to climate change in North China plain in the last 33 years. *International Journal of Applied Earth Observation and Geoinformation*, 53, 103–117. https://doi.org/10.1016/j.jag.2016.08.008

Abd El-Ghany, N. M., Abd El-Aziz, S. E., & Marei, S. S. (2020). A review: Application of remote sensing as a promising strategy for insect pests and diseases management. *Environmental Science and Pollution Research*, 27(27), 33503–33515. https://doi.org/10.1007/s11356-020-09517-2

Abdel-Rahman, E. M., Ahmed, F. B., & Ismail, R. (2013). Random forest regression and spectral band selection for estimating sugarcane leaf nitrogen concentration using EO-1 Hyperion hyperspectral data. *International Journal of Remote Sensing*, 34(2), 712–728. https://doi.org/10.1080/01431161.2012.713142

Adam, E., Mutanga, O., & Rugege, D. (2010). Multispectral and hyperspectral remote sensing for identification and mapping of wetland vegetation: A review. *Wetlands Ecology and Management*, 18(3), 281–296. https://doi.org/10.1007/s11273-009-9169-z

Adeli, S., Salehi, B., Mahdianpari, M., Quackenbush, L. J., Brisco, B., Tamiminia, H., & Shaw, S. (2020). Wetland monitoring using SAR data: A meta-analysis and comprehensive review. *Remote Sensing*, 12(14), Article 14. https://doi.org/10.3390/rs12142190

Ahmed, K. R., Akter, S., Marandi, A., & Schüth, C. (2021). A simple and robust wetland classification approach by using optical indices, unsupervised and supervised machine learning algorithms. *Remote Sensing Applications:Society and Environment*, 23, 100569. https://doi.org/10.1016/j.rsase.2021.100569

Amani, M., Salehi, B., Mahdavi, S., Granger, J. E., Brisco, B., & Hanson, A. (2017). Wetland classification using multi-source and multi-temporal optical remote sensing data in Newfoundland and Labrador, Canada. *Canadian Journal of Remote Sensing*, 43(4), 360–373. https://doi.org/10.1080/07038992.2017.1346468

Arino, O., Ramos Perez, J. J., Kalogirou, V., Bontemps, S., Defourny, P., & Van Bogaert, E. (2012). Global land cover map for 2009 (GlobCover 2009) [dataset]. In <cite lang="fr">© *European Space Agency (ESA) & Université catholique de Louvain (UCL)*. PANGAEA. https://doi.org/10.1594/PANGAEA.787668

Aslan, A., Rahman, A. F., Warren, M. W., & Robeson, S. M. (2016). Mapping spatial distribution and biomass of coastal wetland vegetation in Indonesian Papua by combining active and passive remotely sensed data. *Remote Sensing of Environment*, 183, 65–81. https://doi.org/10.1016/j.rse.2016.04.026

Baker, C., Lawrence, R. L., Montagne, C., & Patten, D. (2007). Change detection of wetland ecosystems using Landsat imagery and change vector analysis. *Wetlands*, 27(3), 610–619. https://doi.org/10.1672/0277-5212(2007)27[610:CDOWEU]2.0.CO;2

Ballanti, L., Byrd, K. B., Woo, I., & Ellings, C. (2017). Remote sensing for wetland mapping and historical change detection at the Nisqually River Delta. *Sustainability*, 9(11), Article 11. https://doi.org/10.3390/su9111919

Bansal, S., Katyal, D., & Garg, J. K. (2017). A novel strategy for wetland area extraction using multispectral MODIS data. *Remote Sensing of Environment*, *200*, 183–205. https://doi.org/10.1016/j.rse.2017.07.034

Baret, F., & Guyot, G. (1991). Potentials and limits of vegetation indices for LAI and APAR assessment. *Remote Sensing of Environment*, *35*(2), 161–173. https://doi.org/10.1016/0034-4257(91)90009-U

Beck, P. S. A., Atzberger, C., Høgda, K. A., Johansen, B., & Skidmore, A. K. (2006). Improved monitoring of vegetation dynamics at very high latitudes: A new method using MODIS NDVI. *Remote Sensing of Environment*, *100*(3), 321–334. https://doi.org/10.1016/j.rse.2005.10.021

Bedford, B. L., Walbridge, M. R., & Aldous, A. (1999). Patterns in nutrient availability and plant diversity of temperate North American wetlands. *Ecology*, *80*(7), 2151–2169. https://doi.org/10.1890/0012-9658(1999)080[2151:PINAAP]2.0.CO;2

Bee, M., Benedetti, R., & Espa, G. (2008). Spatial models for flood risk assessment. *Environmetrics*, *19*(7), 725–741. https://doi.org/10.1002/env.932

Bing, G., & Lin, J. (2017). *Driving Mechanism of Vegetation Coverage Change in the Yarlung Zangbo River Basin under the Stress of Global Warming*. www.semanticscholar.org/paper/Driving-Mechanism-of-Vegetation-Coverage-Change-in-Bing-Lin/2a070e626d4885ecb00fdd7750cf25485b908604

Boyer, K. E., Fong, P., Vance, R. R., & Ambrose, R. F. (2001). Salicornia virginica in a southern California salt marsh: Seasonal patterns and a nutrient-enrichment experiment. *Wetlands*, *21*(3), 315–326. https://doi.org/10.1672/0277-5212(2001)021[0315:SVIASC]2.0.CO;2

Brisco, B., Ahern, F., Murnaghan, K., White, L., Canisus, F., & Lancaster, P. (2017). Seasonal change in wetland coherence as an aid to wetland monitoring. *Remote Sensing*, *9*(2), Article 2. https://doi.org/10.3390/rs9020158

Bunting, P., Rosenqvist, A., Lucas, R. M., Rebelo, L.-M., Hilarides, L., Thomas, N., Hardy, A., Itoh, T., Shimada, M., & Finlayson, C. M. (2018). The global mangrove watch—A new 2010 global baseline of mangrove extent. *Remote Sensing*, *10*(10), Article 10. https://doi.org/10.3390/rs10101669

Buschmann, C., & Nagel, E. (1993). In vivo spectroscopy and internal optics of leaves as basis for remote sensing of vegetation. *International Journal of Remote Sensing*, *14*(4), 711–722. https://doi.org/10.1080/01431169308904370

Byrd, K. B., O'Connell, J. L., Di Tommaso, S., & Kelly, M. (2014). Evaluation of sensor types and environmental controls on mapping biomass of coastal marsh emergent vegetation. *Remote Sensing of Environment*, *149*, 166–180. https://doi.org/10.1016/j.rse.2014.04.003

Canisius, F., Brisco, B., Murnaghan, K., Van Der Kooij, M., & Keizer, E. (2019). SAR backscatter and In SAR coherence for monitoring wetland extent, flood pulse and vegetation: A study of the Amazon Lowland. *Remote Sensing*, *11*(6), Article 6. https://doi.org/10.3390/rs11060720

Chen, B., Chen, L., Huang, B., Michishita, R., & Xu, B. (2018). Dynamic monitoring of the Poyang Lake wetland by integrating Landsat and MODIS observations. *ISPRS Journal of Photogrammetry and Remote Sensing*, *139*, 75–87. https://doi.org/10.1016/j.isprsjprs.2018.02.021

Chen, J., Cao, X., Peng, S., & Ren, H. (2017). Analysis and applications of GlobeLand30: A review. *ISPRS International Journal of Geo-Information*, *6*(8), Article 8. https://doi.org/10.3390/ijgi6080230

Chen, J., Gu, S., Shen, M., Tang, Y., & Matsushita, B. (2009). Estimating aboveground biomass of grassland having a high canopy cover: An exploratory analysis of in situ hyperspectral data. *International Journal of Remote Sensing*, *30*(24), 6497–6517. https://doi.org/10.1080/01431160902882496

Chen, J. M., & Cihlar, J. (1996). Retrieving leaf area index of boreal conifer forests using Landsat TM images. *Remote Sensing of Environment*, *55*(2), 153–162. https://doi.org/10.1016/0034-4257(95)00195-6

Chini, M., Hostache, R., Giustarini, L., & Matgen, P. (2017). A hierarchical split-based approach for parametric thresholding of SAR images: Flood inundation as a test case. *IEEE Transactions on Geoscience and Remote Sensing*, *55*(12), 6975–6988. https://doi.org/10.1109/TGRS.2017.2737664

Cho, H. J., Kirui, P., & Natarajan, H. (2008). Test of multi-spectral vegetation index for floating and canopy-forming submerged vegetation. *International Journal of Environmental Research and Public Health*, *5*(5), 477–483. https://doi.org/10.3390/ijerph5050477

Chouari, W. (2021). Wetland land cover change detection using multitemporal Landsat data: A case study of the Al-Asfar wetland, Kingdom of Saudi Arabia. *Arabian Journal of Geosciences*, *14*(6), 523. https://doi.org/10.1007/s12517-021-06815-y

Chu, H., Venevsky, S., Wu, C., & Wang, M. (2019). NDVI-based vegetation dynamics and its response to climate changes at Amur-Heilongjiang River Basin from 1982 to 2015. *Science of the Total Environment*, *650*, 2051–2062. https://doi.org/10.1016/j.scitotenv.2018.09.115

Clement, M. A., Kilsby, C. G., & Moore, P. (2018). Multi-temporal synthetic aperture radar flood mapping using change detection. *Journal of Flood Risk Management*, *11*(2), 152–168. https://doi.org/10.1111/jfr3.12303

Cohen, R. A., & Fong, P. (2005). Experimental evidence supports the use of δ15N content of the opportunistic green macroalga enteromorpha intestinalis (chlorophyta) to determine nitrogen sources to estuaries1. *Journal of Phycology*, *41*(2), 287–293. https://doi.org/10.1111/j.1529-8817.2005.04022.x

Corcoran, J., Knight, J., Brisco, B., Kaya, S., Cull, A., & Murnaghan, K. (2011). The integration of optical, topographic, and radar data for wetland mapping in northern Minnesota. *Canadian Journal of Remote Sensing*, *37*(5), 564–582. https://doi.org/10.5589/m11-067

Cord, M., & Cunningham, P. (2008). *Machine Learning Techniques for Multimedia: Case Studies on Organization and Retrieval*. Springer, Berlin. https://doi.org/10.1007/978-3-540-75171-7.

Cowardin, L. M., Carter, V., Golet, F. C., & LaRoe, E. T. (1979). Classification of wetlands and deepwater habitats of the United States. In *FWS/OBS (79/31)*. U.S. Fish and Wildlife Service. https://pubs.usgs.gov/publication/2000109

Cundill, S. L., Van der Werff, H. M. A., & Van der Meijde, M. (2015). Adjusting spectral indices for spectral response function differences of very high spatial resolution sensors simulated from field spectra. *Sensors*, *15*(3), Article 3. https://doi.org/10.3390/s150306221

Curran, P. J. (1989). Remote sensing of foliar chemistry. *Remote Sensing of Environment*, *30*(3), 271–278. https://doi.org/10.1016/0034-4257(89)90069-2

Danks, S. M. (1983). *Photosynthetic Systems: Structure,Function, and Assembly*. https://cir.nii.ac.jp/crid/1130282271983997312

Darby, F. A., & Turner, R. E. (2008a). Below- and aboveground biomass of Spartina alterniflora: Response to nutrient addition in a Louisiana salt marsh. *Estuaries and Coasts*, *31*(2), 326–334.

Darby, F. A., & Turner, R. E. (2008b). Below- and aboveground biomass of spartina alterniflora: Response to nutrient addition in a Louisiana Salt Marsh. *Estuaries and Coasts*, *31*(2), 326–334. https://doi.org/10.1007/s12237-008-9037-8

Darby, F. A., & Turner, R. E. (2008c). Effects of eutrophication on salt marsh root and rhizome biomass accumulation. *Marine Ecology Progress Series*, *363*, 63–70.

Dasgupta, A., Grimaldi, S., Ramsankaran, R. A. A. J., Pauwels, V. R. N., & Walker, J. P. (2018). Towards operational SAR-based flood mapping using neuro-fuzzy texture-based approaches. *Remote Sensing of Environment*, *215*, 313–329. https://doi.org/10.1016/j.rse.2018.06.019

Deegan, L. A., Johnson, D. S., Warren, R. S., Peterson, B. J., Fleeger, J. W., Fagherazzi, S., & Wollheim, W. M. (2012). Coastal eutrophication as a driver of salt marsh loss. *Nature*, *490*(7420), 388–392. https://doi.org/10.1038/nature11533

Delalieux, S., Somers, B., Hereijgers, S., Verstraeten, W. W., Keulemans, W., & Coppin, P. (2008). A near-infrared narrow-waveband ratio to determine Leaf Area Index in orchards. *Remote Sensing of Environment*, *112*(10), 3762–3772. https://doi.org/10.1016/j.rse.2008.05.003

De Marzo, T., Pflugmacher, D., Baumann, M., Lambin, E. F., Gasparri, I., & Kuemmerle, T. (2021). Characterizing forest disturbances across the Argentine Dry Chaco based on Landsat time series. *International Journal of Applied Earth Observation and Geoinformation*, *98*, 102310. https://doi.org/10.1016/j.jag.2021.102310

Dennison, W. C., Orth, R. J., Moore, K. A., Stevenson, J. C., Carter, V., Kollar, S., Bergstrom, P. W., & Batiuk, R. A. (1993). Assessing water quality with submersed aquatic vegetation: Habitat requirements as barometers of chesapeake bay health. *BioScience*, *43*(2), 86–94. https://doi.org/10.2307/1311969

DiCarlo, J. J., Zoccolan, D., & Rust, N. C. (2012). How does the brain solve visual object recognition? *Neuron*, *73*(3), 415–434. https://doi.org/10.1016/j.neuron.2012.01.010

Dirksen, R., Dobber, M., Voors, R., & Levelt, P. (2006). Prelaunch characterization of the ozone monitoring instrument transfer function in the spectral domain. *Applied Optics*, *45*(17), 3972–3981. https://doi.org/10.1364/AO.45.003972

Dronova, I. (2015). Object-based image analysis in wetland research: A review. *Remote Sensing*, *7*(5), Article 5. https://doi.org/10.3390/rs70506380

Dronova, I., Taddeo, S., Hemes, K. S., Knox, S. H., Valach, A., Oikawa, P. Y., Kasak, K., & Baldocchi, D. D. (2021). Remotely sensed phenological heterogeneity of restored wetlands: Linking vegetation structure and function. *Agricultural and Forest Meteorology*, *296*, 108215. https://doi.org/10.1016/j.agrformet.2020.108215

Du, J., Jiaerheng, A., Zhao, C., Fang, G., Yin, J., Xiang, B., Yuan, X., & Fang, S. (2015). Dynamic changes in vegetation NDVI from 1982 to 2012 and its responses to climate change and human activities in Xinjiang, China. *Ying Yong Sheng Tai Xue Bao = The Journal of Applied Ecology*, *26*(12), 3567–3578.

Erwin, K. L. (2009). Wetlands and global climate change: The role of wetland restoration in a changing world. *Wetlands Ecology and Management*, *17*(1), 71–84. https://doi.org/10.1007/s11273-008-9119-1

Fiaschi, S., & Wdowinski, S. (2020). Local land subsidence in Miami Beach (FL) and Norfolk (VA) and its contribution to flooding hazard in coastal communities along the U.S. Atlantic coast. *Ocean & Coastal Management*, *187*, 105078. https://doi.org/10.1016/j.ocecoaman.2019.105078

Friess, D. A., Spencer, T., Smith, G. M., Möller, I., Brooks, S. M., & Thomson, A. G. (2012). Remote sensing of geomorphological and ecological change in response to saltmarsh managed realignment, The Wash, UK. *International Journal of Applied Earth Observation and Geoinformation*, *18*, 57–68. https://doi.org/10.1016/j.jag.2012.01.016

Gallant, A. L. (2015). The challenges of remote monitoring of wetlands. *Remote Sensing*, *7*(8), Article 8. https://doi.org/10.3390/rs70810938

Gao, B.-C. (1996). NDWI—A normalized difference water index for remote sensing of vegetation liquid water from space. *Remote Sensing of Environment*, *58*(3), 257–266. https://doi.org/10.1016/S0034-4257(96)00067-3

Gao, B.-C., & Goetzt, A. F. H. (1995). Retrieval of equivalent water thickness and information related to biochemical components of vegetation canopies from AVIRIS data. *Remote Sensing of Environment*, *52*(3), 155–162. https://doi.org/10.1016/0034-4257(95)00039-4

Gao, Q., Guo, Y., Xu, H., Ganjurjav, H., Li, Y., Wan, Y., Qin, X., Ma, X., & Liu, S. (2016). Climate change and its impacts on vegetation distribution and net primary productivity of the alpine ecosystem in the Qinghai-Tibetan Plateau. *Science of the Total Environment*, *554–555*, 34–41. https://doi.org/10.1016/j.scitotenv.2016.02.131

Gao, X., Huete, A. R., Ni, W., & Miura, T. (2000). Optical—Biophysical relationships of vegetation spectra without background contamination. *Remote Sensing of Environment*, *74*(3), 609–620. https://doi.org/10.1016/S0034-4257(00)00150-4

Garg, L., Shukla, P., Singh, S., Bajpai, V., & Yadav, U. (2019). Land use land cover classification from satellite imagery using mUnet: A modified unet architecture. *Proceedings of the 14th International Joint Conference on Computer Vision,Imaging and Computer Graphics Theory and Applications*, 359–365. https://doi.org/10.5220/0007370603590365

Geladi, P., & Kowalski, B. R. (1986). Partial least-squares regression: A tutorial. <cite lang="la">*Analytica Chimica Acta*, *185*, 1–17. https://doi.org/10.1016/0003-2670(86)80028-9

Gitelson, A. A. (2004). Wide dynamic range vegetation index for remote quantification of biophysical characteristics of vegetation. *Journal of Plant Physiology*, *161*(2), 165–173. https://doi.org/10.1078/0176-1617-01176

Gitelson, A. A., Kaufman, Y. J., & Merzlyak, M. N. (1996a). Use of a green channel in remote sensing of global vegetation from EOS-MODIS. *Remote Sensing of Environment*, *58*(3), 289–298. https://doi.org/10.1016/S0034-4257(96)00072-7

Gitelson, A. A., Kaufman, Y. J., Stark, R., & Rundquist, D. (2002). Novel algorithms for remote estimation of vegetation fraction. *Remote Sensing of Environment*, *80*(1), 76–87. https://doi.org/10.1016/S0034-4257(01)00289-9

Gitelson, A. A., Keydan, G. P., & Merzlyak, M. N. (2006). Three-band model for noninvasive estimation of chlorophyll, carotenoids, and anthocyanin contents in higher plant leaves. *Geophysical Research Letters*, *33*(11). https://doi.org/10.1029/2006GL026457

Gitelson, A. A., Merzlyak, M. N., & Lichtenthaler, H. K. (1996b). Detection of red edge position and chlorophyll content by reflectance measurements near 700 nm. *Journal of Plant Physiology*, *148*(3), 501–508. https://doi.org/10.1016/S0176-1617(96)80285-9

Gitelson, A. A., Verma, S. B., Viña, A., Rundquist, D. C., Keydan, G., Leavitt, B., Arkebauer, T. J., Burba, G. G., & Suyker, A. E. (2003). Novel technique for remote estimation of CO_2 flux in maize. *Geophysical Research Letters*, *30*(9). https://doi.org/10.1029/2002GL016543

Gitelson, A. A., Viña, A., Arkebauer, T. J., Rundquist, D. C., Keydan, G., & Leavitt, B. (2003). Remote estimation of leaf area index and green leaf biomass in maize canopies. *Geophysical Research Letters*, *30*(5). https://doi.org/10.1029/2002GL016450

Gitelson, A. A., Viña, A., Ciganda, V., Rundquist, D. C., & Arkebauer, T. J. (2005). Remote estimation of canopy chlorophyll content in crops. *Geophysical Research Letters*, *32*(8). https://doi.org/10.1029/2005GL022688

Gonsamo, A., & Chen, J. M. (2013). Spectral response function comparability among 21 satellite sensors for vegetation monitoring. *IEEE Transactions on Geoscience and Remote Sensing*, *51*(3), 1319–1335. https://doi.org/10.1109/TGRS.2012.2198828

Gorelick, N., Hancher, M., Dixon, M., Ilyushchenko, S., Thau, D., & Moore, R. (2017). Google earth engine: Planetary-scale geospatial analysis for everyone. *Remote Sensing of Environment*, *202*, 18–27. https://doi.org/10.1016/j.rse.2017.06.031

Graetz, R. D. (1990). Remote sensing of terrestrial ecosystem structure: An ecologist's pragmatic view. In R. J. Hobbs & H. A. Mooney (Eds.), *Remote Sensing of Biosphere Functioning* (pp. 5–30). Springer. https://doi.org/10.1007/978-1-4612-3302-2_2

Günen, M. A. (2022). Performance comparison of deep learning and machine learning methods in determining wetland water areas using EuroSAT dataset. *Environmental Science and Pollution Research*, *29*(14), 21092–21106. https://doi.org/10.1007/s11356-021-17177-z

Guo, M., Li, J., Sheng, C., Xu, J., & Wu, L. (2017). A review of wetland remote sensing. *Sensors*, *17*(4), Article 4. https://doi.org/10.3390/s17040777

He, K., Gkioxari, G., Dollar, P., & Girshick, R. (2017). *Mask R-CNN*, 2961–2969. https://openaccess.thecvf.com/content_iccv_2017/html/He_Mask_R-CNN_ICCV_2017_paper.html

Heim, R. R. (2002). A review of twentieth-century drought indices used in the United States. *Bulletin of the American Meteorological Society*, *83*(8), 1149–1166. https://doi.org/10.1175/1520-0477-83.8.1149

Henderson, F. M., & Lewis, A. J. (2008). Radar detection of wetland ecosystems: A review. *International Journal of Remote Sensing*, *29*(20), 5809–5835. https://doi.org/10.1080/01431160801958405

Hess, L. L., Melack, J. M., Affonso, A. G., Barbosa, C., Gastil-Buhl, M., & Novo, E. M. L. M. (2015). Wetlands of the Lowland Amazon Basin: Extent, vegetative cover, and dual-season inundated area as mapped with JERS-1 synthetic aperture radar. *Wetlands*, *35*(4), 745–756. https://doi.org/10.1007/s13157-015-0666-y

Hess, L. L., Melack, J. M., & Simonett, D. S. (1990). Radar detection of flooding beneath the forest canopy: A review. *International Journal of Remote Sensing*, *11*(7), 1313–1325. https://doi.org/10.1080/01431169008955095

Hewison, T. J., Wu, X., Yu, F., Tahara, Y., Hu, X., Kim, D., & Koenig, M. (2013). GSICS inter-calibration of infrared channels of geostationary imagers using Metop/IASI. *IEEE Transactions on Geoscience and Remote Sensing*, *51*(3), 1160–1170. https://doi.org/10.1109/TGRS.2013.2238544

Hinkle, R. L., & Mitsch, W. J. (2005). Salt marsh vegetation recovery at salt hay farm wetland restoration sites on Delaware Bay. *Ecological Engineering*, *25*(3), 240–251. https://doi.org/10.1016/j.ecoleng.2005.04.011

Hu, S., Niu, Z., & Chen, Y. (2017). Global wetland datasets: A review. *Wetlands*, *37*(5), 807–817. https://doi.org/10.1007/s13157-017-0927-z

Huete, A., Didan, K., Miura, T., Rodriguez, E. P., Gao, X., & Ferreira, L. G. (2002). Overview of the radiometric and biophysical performance of the MODIS vegetation indices. *Remote Sensing of Environment*, *83*(1), 195–213. https://doi.org/10.1016/S0034-4257(02)00096-2

Huete, A. R. (1988). A soil-adjusted vegetation index (SAVI). *Remote Sensing of Environment*, *25*(3), 295–309. https://doi.org/10.1016/0034-4257(88)90106-X

Ichii, K., Kawabata, A., & Yamaguchi, Y. (2002). Global correlation analysis for NDVI and climatic variables and NDVI trends: 1982–1990. *International Journal of Remote Sensing*, *23*(18), 3873–3878. https://doi.org/10.1080/01431160110119416

Jenkins, J. P., Braswell, B. H., Frolking, S. E., & Aber, J. D. (2002). Detecting and predicting spatial and interannual patterns of temperate forest springtime phenology in the eastern U.S. *Geophysical Research Letters*, *29*(24), 54-1–54-4. https://doi.org/10.1029/2001GL014008

Ji, L., & Peters, A. J. (2003). Assessing vegetation response to drought in the northern great plains using vegetation and drought indices. *Remote Sensing of Environment*, *87*(1), 85–98. https://doi.org/10.1016/S0034-4257(03)00174-3

Jiang, W., Chen, Z., Lei, X., Jia, K., & Wu, Y. (2015). Simulating urban land use change by incorporating an autologistic regression model into a CLUE-S model. *Journal of Geographical Sciences*, *25*(7), 836–850. https://doi.org/10.1007/s11442-015-1205-8

Kanemasu, E. T. (1974). Seasonal canopy reflectance patterns of wheat, sorghum, and soybean. *Remote Sensing of Environment*, *3*(1), 43–47. https://doi.org/10.1016/0034-4257(74)90037-6

Kang, M., Ahn, M.-H., Ko, D. H., Kim, J., Nicks, D., Eo, M., Lee, Y., Moon, K.-J., & Lee, D.-W. (2022). Characteristics of the spectral response function of geostationary environment monitoring spectrometer analyzed by ground and in-orbit measurements. *IEEE Transactions on Geoscience and Remote Sensing*, *60*, 1–16. https://doi.org/10.1109/TGRS.2021.3091677

Kao, J. T., Titus, J. E., & Zhu, W.-X. (2003). Differential nitrogen and phosphorus retention by five wetland plant species. *Wetlands*, *23*(4), 979–987. https://doi.org/10.1672/0277-5212(2003)023[0979:DNAPRB]2.0.CO;2

Kaplan, G., & Avdan, U. (2017). Mapping and monitoring wetlands using sentinel-2 satellite imagery. *ISPRS Annals of the Photogrammetry, Remote Sensing and Spatial Information Sciences*, *IV-4-W4*, 271–277. https://doi.org/10.5194/isprs-annals-IV-4-W4-271-2017

Kaplan, G., & Avdan, U. (2018). Monthly analysis of wetlands dynamics using remote sensing data. *ISPRS International Journal of Geo-Information*, *7*(10), Article 10. https://doi.org/10.3390/ijgi7100411

Kasischke, E., Bourgeau-Chavez, L., Smith, K. B., Romanowicz, E., & Richardson, C. (1997). *Monitoring Hydropatterns in South Florida Ecosystems Using ERS SAR Data*. www.semanticscholar.org/paper/Monitoring-hydropatterns-in-south-florida-using-ERS-Kasischke-Bourgeau-Chavez/b1a3b4a38c22fd0f5e64266216c369a6fd4b3ee9

Kearney, M. S., Stutzer, D., Turpie, K., & Stevenson, J. C. (2009). The effects of tidal inundation on the reflectance characteristics of coastal marsh vegetation. *Journal of Coastal Research*, *2009*(256), 1177–1186. https://doi.org/10.2112/08-1080.1

Keddy, P. A. (2010). *Wetland Ecology: Principles and Conservation*. Cambridge: Cambridge University Press.

Kędra, M., & Wiejaczka, Ł. (2018). Climatic and dam-induced impacts on river water temperature: Assessment and management implications. *Science of the Total Environment*, *626*, 1474–1483. https://doi.org/10.1016/j.scitotenv.2017.10.044

Keim, D. A., Mansmann, F., Schneidewind, J., Thomas, J., & Ziegler, H. (2008). Visual analytics: Scope and challenges. In S. J. Simoff, M. H. Böhlen, & A. Mazeika (Eds.), *Visual Data Mining: Theory, Techniques and Tools for Visual Analytics* (pp. 76–90). Springer. https://doi.org/10.1007/978-3-540-71080-6_6

Kiage, L. M., & Walker, N. D. (2009). Using NDVI from MODIS to monitor duckweed bloom in Lake Maracaibo, Venezuela. *Water Resources Management*, *23*(6), 1125–1135. https://doi.org/10.1007/s11269-008-9318-9

Kim, Y., Kimball, J. S., Zhang, K., & McDonald, K. C. (2012). Satellite detection of increasing Northern Hemisphere non-frozen seasons from 1979 to 2008: Implications for regional vegetation growth. *Remote Sensing of Environment*, *121*, 472–487. https://doi.org/10.1016/j.rse.2012.02.014

Klemas, V., Finkl, C. W., & Kabbara, N. (2014). Remote sensing of soil moisture: An overview in relation to coastal soils. *Journal of Coastal Research*, *30*(4), 685–696. https://doi.org/10.2112/JCOASTRES-D-13-00072.1

Kobe, R. K., Iyer, M., & Walters, M. B. (2010). Optimal partitioning theory revisited: Nonstructural carbohydrates dominate root mass responses to nitrogen. *Ecology*, *91*(1), 166–179. https://doi.org/10.1890/09-0027.1

Kogan, F., & Sullivan, J. (1993). Development of global drought-watch system using NOAA/AVHRR data. *Advances in Space Research*, *13*(5), 219–222. https://doi.org/10.1016/0273-1177(93)90548-P

Kokaly, R. F., Asner, G. P., Ollinger, S. V., Martin, M. E., & Wessman, C. A. (2009). Characterizing canopy biochemistry from imaging spectroscopy and its application to ecosystem studies. *Remote Sensing of Environment*, *113*, Supplement 1, S78—S91. https://doi.org/10.1016/j.rse.2008.10.018

Kokaly, R. F., & Clark, R. N. (1999). Spectroscopic determination of leaf biochemistry using band-depth analysis of absorption features and stepwise multiple linear regression. *Remote Sensing of Environment*, *67*(3), 267–287. https://doi.org/10.1016/S0034-4257(98)00084-4

Kovacs, J. M., King, J. M. L., Flores de Santiago, F., & Flores-Verdugo, F. (2009). Evaluating the condition of a mangrove forest of the Mexican Pacific based on an estimated leaf area index mapping approach. *Environmental Monitoring and Assessment*, *157*(1), 137–149. https://doi.org/10.1007/s10661-008-0523-z

Kulkarni, J. S. (2011). Wavelet transform applications. *2011 3rd International Conference on Electronics Computer Technology*, *1*, 11–17. https://doi.org/10.1109/ICECTECH.2011.5941550

Kumar, P., & Foufoula-Georgiou, E. (1997). Wavelet analysis for geophysical applications. *Reviews of Geophysics*, *35*(4), 385–412. https://doi.org/10.1029/97RG00427

Kwoun, O., & Lu, Z. (2009). Multi-temporal RADARSAT-1 and ERS backscattering signatures of coastal wetlands in Southeastern Louisiana. *Photogrammetric Engineering &Remote Sensing*, *75*(5), 607–617. https://doi.org/10.14358/PERS.75.5.607

Landmann, T., Schramm, M., Colditz, R. R., Dietz, A., & Dech, S. (2010). Wide area wetland mapping in semi-arid Africa using 250-meter MODIS metrics and topographic variables. *Remote Sensing*, *2*(7), Article 7. https://doi.org/10.3390/rs2071751

Larkin, D. J., Lishawa, S. C., & Tuchman, N. C. (2012). Appropriation of nitrogen by the invasive cattail Typha × glauca. *Aquatic Botany*, *100*(0), 62–66. https://doi.org/10.1016/j.aquabot.2012.03.001

LaRocque, A., Leblon, B., Woodward, R., & Bourgeau-Chavez, L. (2020). Wetland mapping in new brunswick, Canada with landsat5-TM, alos-palsar, and radarsat-2 imagery. *ISPRS Annals of the Photogrammetry, Remote Sensing and Spatial Information Sciences, V-3–2020*, 301–308. https://doi.org/10.5194/isprs-annals-V-3-2020-301-2020

Lary, D. J., Alavi, A. H., Gandomi, A. H., & Walker, A. L. (2016). Machine learning in geosciences and remote sensing. *Geoscience Frontiers*, *7*(1), 3–10. https://doi.org/10.1016/j.gsf.2015.07.003

Lechner, A. M., Foody, G. M., & Boyd, D. S. (2020). Applications in remote sensing to forest ecology and management. *One Earth*, *2*(5), 405–412. https://doi.org/10.1016/j.oneear.2020.05.001

Lee, R. A. M., O'Dell, C. W., Wunch, D., Roehl, C. M., Osterman, G. B., Blavier, J.-F., Rosenberg, R., Chapsky, L., Frankenberg, C., Hunyadi-Lay, S. L., Fisher, B. M., Rider, D. M., Crisp, D., & Pollock, R. (2017). Preflight spectral calibration of the orbiting carbon observatory 2. *IEEE Transactions on Geoscience and Remote Sensing*, *55*(5), 2499–2508. https://doi.org/10.1109/TGRS.2016.2645614

Lehner, B., & Döll, P. (2004). Development and validation of a global database of lakes, reservoirs and wetlands. *Journal of Hydrology*, *296*(1), 1–22. https://doi.org/10.1016/j.jhydrol.2004.03.028

Lehner, B., Liermann, C. R., Revenga, C., Vörösmarty, C., Fekete, B., Crouzet, P., Döll, P., Endejan, M., Frenken, K., Magome, J., Nilsson, C., Robertson, J. C., Rödel, R., Sindorf, N., & Wisser, D. (2011). High-resolution mapping of the world's reservoirs and dams for sustainable river-flow management. *Frontiers in Ecology and the Environment*, *9*(9), 494–502. https://doi.org/10.1890/100125

Li, Y., Niu, Z., Xu, Z., & Yan, X. (2020). Construction of high spatial-temporal water body dataset in China based on sentinel-1 archives and GEE. *Remote Sensing*, *12*(15), Article 15. https://doi.org/10.3390/rs12152413

Lillesand, T. M., Kiefer, R. W., & Chipman, J. W. (2008). *Remote Sensing and Image Interpretation* (6th ed). Hoboken, NJ: John Wiley & Sons. http://catdir.loc.gov/catdir/enhancements/fy0814/2007028033-t.html

Lloyd, S. (1982). Least squares quantization in PCM. *IEEE Transactions on Information Theory*, *28*(2), 129–137. https://doi.org/10.1109/TIT.1982.1056489

Lu, D. (2006). The potential and challenge of remote sensing-based biomass estimation. *International Journal of Remote Sensing*, *27*(7), 1297–1328. https://doi.org/10.1080/01431160500486732

Lu, D., & Weng, Q. (2007). A survey of image classification methods and techniques for improving classification performance. *International Journal of Remote Sensing*, *28*(5), 823–870. https://doi.org/10.1080/01431160600746456

Ma, Y., Zhou, Z., She, X., Zhou, L., Ren, T., Liu, S., & Lu, J. (2022). Identifying dike-pond system using an improved cascade R-CNN model and high-resolution satellite images. *Remote Sensing*, *14*(3), Article 3. https://doi.org/10.3390/rs14030717

Mahdavi, S., Salehi, B., Granger, J., Amani, M., Brisco, B., & Huang, W. (2018). Remote sensing for wetland classification: A comprehensive review. *GIScience &Remote Sensing*, *55*(5), 623–658. https://doi.org/10.1080/15481603.2017.1419602

Mahdianpari, M., Jafarzadeh, H., Granger, J. E., Mohammadimanesh, F., Brisco, B., Salehi, B., Homayouni, S., & Weng, Q. (2020). A large-scale change monitoring of wetlands using time series Landsat imagery on Google Earth Engine: A case study in Newfoundland. *GIScience &Remote Sensing*, *57*(8), 1102–1124. https://doi.org/10.1080/15481603.2020.1846948

Martínez, B., & Gilabert, M. A. (2009). Vegetation dynamics from NDVI time series analysis using the wavelet transform. *Remote Sensing of Environment*, *113*(9), 1823–1842. https://doi.org/10.1016/j.rse.2009.04.016

Masek, J. G., Vermote, E. F., Saleous, N., Wolfe, R., Hall, F. G., Huemmrich, F., Gao, F., Kutler, J., & Lim, T. K. (2012). LEDAPS landsat calibration, reflectance, atmospheric correction preprocessing code. In *Oak RidgeNational Laboratory Distributed Active ; Archive Center, Oak Ridge, Tennessee, U.S.A.* http://dx.doi.org/10.3334/ORNLDAAC/1080

Mayer, A. L., & Lopez, R. D. (2011). Use of remote sensing to support forest and wetlands policies in the USA. *Remote Sensing*, *3*(6), Article 6. https://doi.org/10.3390/rs3061211

Mccarthy, M. C., & Enquist, B. J. (2007). Consistency between an allometric approach and optimal partitioning theory in global patterns of plant biomass allocation. *Functional Ecology*, *21*(4), 713–720. https://doi.org/10.1111/j.1365-2435.2007.01276.x

McKee, K. L. (2011). Biophysical controls on accretion and elevation change in Caribbean mangrove ecosystems. *Estuarine, Coastal and Shelf Science*, *91*(4), 475–483. https://doi.org/10.1016/j.ecss.2010.05.001

McKee, K. L., Cahoon, D. R., & Feller, I. C. (2007). Caribbean mangroves adjust to rising sea level through biotic controls on change in soil elevation. *Global Ecology and Biogeography*, *16*(5), 545–556. https://doi.org/10.1111/j.1466-8238.2007.00317.x

Memarsadeghi, N., Mount, D. M., Netanyahu, N. S., & Le Moigne, J. (2007). A fast implementation of the isodata clustering algorithm. *International Journal of Computational Geometry & Applications*, *17*(1), 71–103. https://doi.org/10.1142/S0218195907002252

Mevik, B.-H., & Cederkvist, H. R. (2004). Mean Squared Error of Prediction (MSEP) estimates for Principal Component Regression (PCR) and Partial Least Squares Regression (PLSR). *Journal of Chemometrics*, *18*(9), 422–429. https://doi.org/10.1002/cem.887

Mevik, B.-H., & Wehrens, R. (2007). The pls package: Principal component and partial least squares regression in R. *Journal of Statistical Software*, *18*, 1–23. https://doi.org/10.18637/jss.v018.i02

Millard, K., & Richardson, M. (2013). Wetland mapping with LiDAR derivatives, SAR polarimetric decompositions, and LiDAR—SAR fusion using a random forest classifier. *Canadian Journal of Remote Sensing*, *39*(4), 290–307. https://doi.org/10.5589/m13-038

Miller, R. L., Fram, M., Fujii, R., & Wheeler, G. (2008). Subsidence reversal in a re-established wetland in the Sacramento-San Joaquin Delta, California, USA. *San Francisco Estuary and Watershed Science*, *6*(3). https://doi.org/10.15447/sfews.2008v6iss3art1

Mishra, D. R., Cho, H. J., Ghosh, S., Fox, A., Downs, C., Merani, P. B. T., Kirui, P., Jackson, N., & Mishra, S. (2012). Post-spill state of the marsh: Remote estimation of the ecological impact of the Gulf of Mexico oil spill on Louisiana Salt Marshes. *Remote Sensing of Environment*, *118*, 176–185. https://doi.org/10.1016/j.rse.2011.11.007Srefj

Mishra, D. R., Ogashawara, I., & Gitelson, A. A. (2017). *Bio-Optical Modeling and Remote Sensing of Inland Waters*. Elsevier, Amsterdam.

Mitsch, W. J., & Gosselink, J. G. (2007). *Wetlands*. Hoboken, NJ: John Wiley & Sons.

Mohammadimanesh, F., Salehi, B., Mahdianpari, M., Brisco, B., & Motagh, M. (2018). Multi-temporal, multi-frequency, and multi-polarization coherence and SAR backscatter analysis of wetlands. *ISPRS Journal of Photogrammetry and Remote Sensing*, *142*, 78–93. https://doi.org/10.1016/j.isprsjprs.2018.05.009

Mokany, K., Raison, R. J., & Prokushkin, A. S. (2006). Critical analysis of root: Shoot ratios in terrestrial biomes. *Global Change Biology*, *12*(1), 84–96. https://doi.org/10.1111/j.1365-2486.2005.001043.x

Mokhele, T. A., & Ahmed, F. B. (2010). Estimation of leaf nitrogen and silicon using hyperspectral remote sensing. *Journal of Applied Remote Sensing*, *4*. https://doi.org/10.1117/1.3525241

Moore, P. D. (1987). Ecological and hydrological aspects of peat formation. *Geological Society, London, Special Publications*, *32*(1), 7–15. https://doi.org/10.1144/GSL.SP.1987.032.01.02

Moreau, S., Bosseno, R., Gu, X. F., & Baret, F. (2003). Assessing the biomass dynamics of Andean bofedal and totora high-protein wetland grasses from NOAA/AVHRR. *Remote Sensing of Environment*, *85*(4), 516–529. https://doi.org/10.1016/S0034-4257(03)00053-1

Morris, J. T., Shaffer, G. P., & Nyman, J. A. (2013). Brinson review: Perspectives on the influence of nutrients on the sustainability of coastal wetlands. *Wetlands*, *33*(6), 975–988. https://doi.org/10.1007/s13157-013-0480-3

Munasinghe, D., Cohen, S., Huang, Y.-F., Tsang, Y.-P., Zhang, J., & Fang, Z. (2018). Intercomparison of satellite remote sensing-based flood inundation mapping techniques. *JAWRA Journal of the American Water Resources Association*, *54*(4), 834–846. https://doi.org/10.1111/1752-1688.12626

Murray, N. J., Phinn, S. R., DeWitt, M., Ferrari, R., Johnston, R., Lyons, M. B., Clinton, N., Thau, D., & Fuller, R. A. (2019). The global distribution and trajectory of tidal flats. *Nature*, *565*(7738), 222–225. https://doi.org/10.1038/s41586-018-0805-8

Myneni, R. B., Ramakrishna, R., Nemani, R., & Running, S. W. (1997). Estimation of global leaf area index and absorbed par using radiative transfer models. *IEEE Transactions on Geoscience and Remote Sensing*, *35*(6), 1380–1393. https://doi.org/10.1109/36.649788

Naiman, R. J., Décamps, H. (Henri), & Unesco. (1990). *The Ecology and Management of Aquatic-Terrestrial Ecotones /: Edited by R.J. Naiman and H. Décamps*. UNESCO. https://digitallibrary.un.org/record/194620

Nandi, A., Mandal, A., Wilson, M., & Smith, D. (2016). Flood hazard mapping in Jamaica using principal component analysis and logistic regression. *Environmental Earth Sciences*, *75*(6), 465. https://doi.org/10.1007/s12665-016-5323-0

Nemani, R. R., Keeling, C. D., Hashimoto, H., Jolly, W. M., Piper, S. C., Tucker, C. J., Myneni, R. B., & Running, S. W. (2003). Climate-driven increases in global terrestrial net primary production from 1982 to 1999. *Science*, *300*(5625), 1560–1563. https://doi.org/10.1126/science.1082750

Niculescu, S., Boissonnat, J.-B., Lardeux, C., Roberts, D., Hanganu, J., Billey, A., Constantinescu, A., & Doroftei, M. (2020). Synergy of high-resolution radar and optical images satellite for identification and mapping of wetland macrophytes on the danube delta. *Remote Sensing*, *12*(14), Article 14. https://doi.org/10.3390/rs12142188

Niculescu, S., Lardeux, C., Grigoras, I., Hanganu, J., & David, L. (2016). Synergy between LiDAR, RADARSAT-2, and spot-5 images for the detection and mapping of wetland vegetation in the danube delta. *IEEE Journal of Selected Topics in Applied Earth Observations and Remote Sensing*, *9*(8), 3651–3666. https://doi.org/10.1109/JSTARS.2016.2545242

Niu, Z., Zhang, H., Wang, X., Yao, W., Zhou, D., Zhao, K., Zhao, H., Li, N., Huang, H., Li, C., Yang, J., Liu, C., Liu, S., Wang, L., Li, Z., Yang, Z., Qiao, F., Zheng, Y., Chen, Y., . . . Gong, P. (2012). Mapping wetland changes in China between 1978 and 2008. *Chinese Science Bulletin*, *57*(22), 2813–2823. https://doi.org/10.1007/s11434-012-5093-3

Noda, H. M., Muraoka, H., & Nasahara, K. N. (2021). Phenology of leaf optical properties and their relationship to mesophyll development in cool-temperate deciduous broad-leaf trees. *Agricultural and Forest Meteorology*, *297*, 108236. https://doi.org/10.1016/j.agrformet.2020.108236

Nyman, J. A., DeLaune, R. D., Roberts, H. H., & Patrick, W. H. (1993). Relationship between vegetation and soil formation in a rapidly submerging coastal marsh. *Marine Ecology Progress Series*, *96*(3), 269–279.

Nyman, J. A., Walters, R. J., Delaune, R. D., & Patrick, W. H. (2006). Marsh vertical accretion via vegetative growth. *Estuarine, Coastal and Shelf Science*, *69*(3), 370–380. https://doi.org/10.1016/j.ecss.2006.05.041

O'Connell, J. L., Byrd, K. B., & Kelly, M. (2014). Remotely-sensed indicators of N-related biomass allocation in schoenoplectus acutus. *PLoS One*, *9*(3), e90870. https://doi.org/10.1371/journal.pone.0090870

O'Connell, J. L., Byrd, K. B., & Kelly, M. (2015). A hybrid model for mapping relative differences in belowground biomass and root: Shoot ratios using spectral reflectance, Foliar N and plant biophysical data within coastal marsh. *Remote Sensing*, *7*(12), Article 12. https://doi.org/10.3390/rs71215837

Ozesmi, S. L., & Bauer, M. E. (2002). Satellite remote sensing of wetlands. *Wetlands Ecology and Management*, *10*(5), 381–402. https://doi.org/10.1023/A:1020908432489

Pekel, J.-F., Cottam, A., Gorelick, N., & Belward, A. S. (2016). High-resolution mapping of global surface water and its long-term changes. *Nature*, *540*(7633), Article 7633. https://doi.org/10.1038/nature20584

Peng, S., Piao, S., Ciais, P., Myneni, R. B., Chen, A., Chevallier, F., Dolman, A. J., Janssens, I. A., Peñuelas, J., Zhang, G., Vicca, S., Wan, S., Wang, S., & Zeng, H. (2013). Asymmetric effects of daytime and night-time warming on Northern Hemisphere vegetation. *Nature*, *501*(7465), Article 7465. https://doi.org/10.1038/nature12434

Peñuelas, J., Gamon, J. A., Griffin, K. L., & Field, C. B. (1993). Assessing community type, plant biomass, pigment composition, and photosynthetic efficiency of aquatic vegetation from spectral reflectance. *Remote Sensing of Environment*, *46*(2), 110–118. https://doi.org/10.1016/0034-4257(93)90088-F

Perroy, D. A. R., Roth, K. L., Wetherley, E. B., Meerdink, S. K., & Ryan, L. (2018). Hyperspectral vegetation indices. In *Hyperspectral Indices and Image Classifications for Agriculture and Vegetation* (2nd ed.). CRC Press, Boca Raton, FL.

Pettorelli, N., Vik, J. O., Mysterud, A., Gaillard, J.-M., Tucker, C. J., & Stenseth, N. Chr. (2005). Using the satellite-derived NDVI to assess ecological responses to environmental change. *Trends in Ecology & Evolution*, *20*(9), 503–510. https://doi.org/10.1016/j.tree.2005.05.011

Phinn, S. R. (1998). A framework for selecting appropriate remotely sensed data dimensions for environmental monitoring and management. *International Journal of Remote Sensing*, *19*(17), 3457–3463. https://doi.org/10.1080/014311698214136

Piao, S., Cui, M., Chen, A., Wang, X., Ciais, P., Liu, J., & Tang, Y. (2011). Altitude and temperature dependence of change in the spring vegetation green-up date from 1982 to 2006 in the Qinghai-Xizang Plateau. *Agricultural and Forest Meteorology*, *151*(12), 1599–1608. https://doi.org/10.1016/j.agrformet.2011.06.016

Pinzon, J. E., & Tucker, C. J. (2014). A non-stationary 1981–2012 AVHRR NDVI3g time series. *Remote Sensing*, *6*(8), Article 8. https://doi.org/10.3390/rs6086929

Prasad, A. K., Sarkar, S., Singh, R. P., & Kafatos, M. (2007). Inter-annual variability of vegetation cover and rainfall over india. *Advances in Space Research, 39*(1), 79–87. https://doi.org/10.1016/j.asr.2006.02.026

Rahman, M. R., & Thakur, P. K. (2018). Detecting, mapping and analysing of flood water propagation using Synthetic Aperture Radar (SAR) satellite data and GIS: A case study from the Kendrapara District of Orissa State of India. *The Egyptian Journal of Remote Sensing and Space Science, 21*, S37–S41. https://doi.org/10.1016/j.ejrs.2017.10.002

Ramsey III, E.W., & Rangoonwala, A. (2006). Canopy reflectance related to marsh dieback onset and progression in Coastal Louisiana. *Photogrammetric Engineering and Remote Sensing, 72*(6), 641–652. https://doi.org/10.14358/PERS.72.6.641

Rapinel, S., Hubert-Moy, L., & Clément, B. (2015). Combined use of LiDAR data and multispectral earth observation imagery for wetland habitat mapping. *International Journal of Applied Earth Observation and Geoinformation, 37*, 56–64. https://doi.org/10.1016/j.jag.2014.09.002

Rasse, D. P., Rumpel, C., & Dignac, M.-F. (2005). Is soil carbon mostly root carbon? Mechanisms for a specific stabilisation. *Plant and Soil, 269*(1), 341–356. https://doi.org/10.1007/s11104-004-0907-y

Rezaee, M., Mahdianpari, M., Zhang, Y., & Salehi, B. (2018). Deep convolutional neural network for complex wetland classification using optical remote sensing imagery. *IEEE Journal of Selected Topics in Applied Earth Observations and Remote Sensing, 11*(9), 3030–3039. https://doi.org/10.1109/JSTARS.2018.2846178

Rouse, J. W., Haas, R. H., Schell, J. A., & Deering, D. W. (1974, January 1). *Monitoring Vegetation Systems in the Great Plains with ERTS*. https://ntrs.nasa.gov/citations/19740022614

Rundquist, D., Perk, R., Leavitt, B., Keydan, G., & Gitelson, A. (2004). Collecting spectral data over cropland vegetation using machine-positioning versus hand-positioning of the sensor. *Computers and Electronics in Agriculture, 43*(2), 173–178. https://doi.org/10.1016/j.compag.2003.11.002

Rustad, L. E. (2008). The response of terrestrial ecosystems to global climate change: Towards an integrated approach. *Science of the Total Environment, 404*(2), 222–235. https://doi.org/10.1016/j.scitotenv.2008.04.050

Ryu, C., Suguri, M., & Umeda, M. (2009). Model for predicting the nitrogen content of rice at panicle initiation stage using data from airborne hyperspectral remote sensing. *Biosystems Engineering, 104*(4), 465–475. https://doi.org/10.1016/j.biosystemseng.2009.09.002

Sagan, V., Peterson, K. T., Maimaitijiang, M., Sidike, P., Sloan, J., Greeling, B. A., Maalouf, S., & Adams, C. (2020). Monitoring inland water quality using remote sensing: Potential and limitations of spectral indices, bio-optical simulations, machine learning, and cloud computing. *Earth-Science Reviews, 205*, 103187. https://doi.org/10.1016/j.earscirev.2020.103187

Sartoris, J. J., Thullen, J. S., Barber, L. B., & Salas, D. E. (1999). Investigation of nitrogen transformations in a southern California constructed wastewater treatment wetland. *Ecological Engineering, 14*(1–2), 49–65. https://doi.org/10.1016/S0925-8574(99)00019-1

Sawaya, K. E., Olmanson, L. G., Heinert, N. J., Brezonik, P. L., & Bauer, M. E. (2003). Extending satellite remote sensing to local scales: Land and water resource monitoring using high-resolution imagery. *Remote Sensing of Environment, 88*(1), 144–156. https://doi.org/10.1016/j.rse.2003.04.006

Schultz, P. A., & Halpert, M. S. (1993). Global correlation of temperature, NDVI and precipitation. *Advances in Space Research, 13*(5), 277–280. https://doi.org/10.1016/0273-1177(93)90559-T

Schumann, G. J.-P., & Moller, D. K. (2015). Microwave remote sensing of flood inundation. *Physics and Chemistry of the Earth, Parts A/B/C, 83–84*, 84–95. https://doi.org/10.1016/j.pce.2015.05.002

See, J. C. G., & Porio, E. E. (2015). Assessing social vulnerability to flooding in metro manila using principal component analysis. *Philippine Sociological Review, 63*, 53–80.

Sellers, P. J. (1985). Canopy reflectance, photosynthesis and transpiration. *International Journal of Remote Sensing, 6*(8), 1335–1372. https://doi.org/10.1080/01431168508948283

Sharma, A., Liu, X., & Yang, X. (2018). Land cover classification from multi-temporal, multi-spectral remotely sensed imagery using patch-based recurrent neural networks. *Neural Networks, 105*, 346–355. https://doi.org/10.1016/j.neunet.2018.05.019

Shekede, M. D., Kusangaya, S., & Schmidt, K. (2008). Spatio-temporal variations of aquatic weeds abundance and coverage in Lake Chivero, Zimbabwe. *Physics and Chemistry of the Earth, Parts A/B/C, 33*(8), 714–721. https://doi.org/10.1016/j.pce.2008.06.052

Shen, G., Yang, X., Jin, Y., Xu, B., & Zhou, Q. (2019). Remote sensing and evaluation of the wetland ecological degradation process of the Zoige Plateau Wetland in China. *Ecological Indicators, 104*, 48–58. https://doi.org/10.1016/j.ecolind.2019.04.063

Shen, X., Liu, B., Xue, Z., Jiang, M., Lu, X., & Zhang, Q. (2019). Spatiotemporal variation in vegetation spring phenology and its response to climate change in freshwater marshes of Northeast China. *Science of the Total Environment, 666*, 1169–1177. https://doi.org/10.1016/j.scitotenv.2019.02.265

Shipley, B., & Meziane, D. (2002). The balanced-growth hypothesis and the allometry of leaf and root biomass allocation. *Functional Ecology, 16*(3), 326–331. https://doi.org/10.1046/j.1365-2435.2002.00626.x

Short, F. T., Kosten, S., Morgan, P. A., Malone, S. L., & Moore, G. E. (2016). Impacts of climate change on submerged and emergent wetland plants. *Aquatic Botany*. https://doi.org/10.1016/j.aquabot.2016.06.006

Smiley, S. L., & Hambati, H. (2020). Using photograph interpretation to understand perceptions of floods in Dar es Salaam, Tanzania. *Papers in Applied Geography, 6*(2), 159–173. https://doi.org/10.1080/23754931.2020.1755887

Smith, M.-L., Ollinger, S. V., Martin, M. E., Aber, J. D., Hallett, R. A., & Goodale, C. L. (2002). Direct estimation of aboveground forest productivity through hyperspectral remote sensing of canopy nitrogen. *Ecological Applications, 12*(5), 1286–1302. https://doi.org/10.1890/1051-0761(2002)012[1286:DEOAFP]2.0.CO;2

Stroppiana, D., Boschetti, M., Brivio, P. A., & Bacchi, S. (2009). Plant nitrogen concentration in paddy rice from field canopy hyperspectral radiometry. *Field Crops Research, 111*, 119–129.

Stroppiana, D., Fava, F., Baschetti, M., & Brivio, P. A. (2012). Estimation of nitrogen content in crops and pastures using hyperspectral vegetation indices. In P. S. Thenkabail, J. G. Lyon, & A. Huete (Eds.), *Hyperspectral Remote Sensing of Vegetation* (pp. 245–262). CRC Press. www.crcnetbase.com/doi/pdf/10.1201/b11222-16

Szantoi, Z., Escobedo, F., Abd-Elrahman, A., Smith, S., & Pearlstine, L. (2013). Analyzing fine-scale wetland composition using high resolution imagery and texture features. *International Journal of Applied Earth Observation and Geoinformation, 23*, 204–212. https://doi.org/10.1016/j.jag.2013.01.003

Thenkabail, P. S., Lyon, J. G., & Huete, A. (2012). Advances in hyperspectral remote sensing of vegetation and agricultural croplands. In P. S. Thenkabail, J. G. Lyon, & A. Huete (Eds.), *Hyperspectral Remote Sensing of Vegetation* (pp. 28–29). Taylor and Francis Group, Boca Raton, FL.

Thenkabail, P. S., Smith, R. B., & De-Pauw, E. (2002). Evaluation of narrowband and broadband vegetation indices for determining optimal hyperspectral wavebands for agricultural crop characteristics. *Photogrammetric Engineering and Remote Sensing, 68*(6), 607–621.

Tian, Y. C., Yao, X., Yang, J., Cao, W. X., Hannaway, D. B., & Zhu, Y. (2011). Assessing newly developed and published vegetation indices for estimating rice leaf nitrogen concentration with ground- and space-based hyperspectral reflectance. *Field Crops Research, 120*(2), 299–310. https://doi.org/10.1016/j.fcr.2010.11.002

Townsend, P. A., Foster, J. R., Chastain, R. A., & Currie, W. S. (2003). Application of imaging spectroscopy to mapping canopy nitrogen in the forests of the central Appalachian Mountains using Hyperion and AVIRIS. *Geoscience and Remote Sensing, IEEE Transactions On, 41*(6), 1347–1354. https://doi.org/10.1109/tgrs.2003.813205

Tucker, C. J. (1977). Asymptotic nature of grass canopy spectral reflectance. *Applied Optics, 16*(5), 1151–1156. https://doi.org/10.1364/AO.16.001151

Turner, R. E., Howes, B. L., Teal, J. M., Milan, C. S., Swenson, E. M., & Toner, D. D. G. (2009). Salt marshes and eutrophication: An unsustainable outcome. *Limnology and Oceanography, 54*(5), 1634–1642. https://doi.org/10.4319/lo.2009.54.5.1634

Turpie, K. R. (2013). Explaining the spectral red-edge features of inundated marsh vegetation. *Journal of Coastal Research, 29*(5), 1111–1117. https://doi.org/10.2112/JCOASTRES-D-12-00209.1

Tuxen, K. A., Schile, L. M., Kelly, M., & Siegel, S. W. (2008). Vegetation colonization in a restoring tidal marsh: A remote sensing approach. *Restoration Ecology, 16*(2), 313–323. https://doi.org/10.1111/j.1526-100X.2007.00313.x

Ustin, S. L., Roberts, D. A., Pinzón, J., Jacquemoud, S., Gardner, M., Scheer, G., Castañeda, C. M., & Palacios-Orueta, A. (1998). Estimating canopy water content of chaparral shrubs using optical methods. *Remote Sensing of Environment, 65*(3), 280–291. https://doi.org/10.1016/S0034-4257(98)00038-8

Vrieling, A., Meroni, M., Shee, A., Mude, A. G., Woodard, J., de Bie, C. A. J. M. (Kees), & Rembold, F. (2014). Historical extension of operational NDVI products for livestock insurance in Kenya. *International Journal of Applied Earth Observation and Geoinformation, 28*, 238–251. https://doi.org/10.1016/j.jag.2013.12.010

Wang, L., Dronova, I., Gong, P., Yang, W., Li, Y., & Liu, Q. (2012). A new time series vegetation—Water index of phenological—Hydrological trait across species and functional types for Poyang Lake wetland ecosystem. *Remote Sensing of Environment, 125*, 49–63. https://doi.org/10.1016/j.rse.2012.07.003

Wang, L., Jia, M., Yin, D., & Tian, J. (2019). A review of remote sensing for mangrove forests: 1956–2018. *Remote Sensing of Environment, 231*, 111223. https://doi.org/10.1016/j.rse.2019.111223

Whittaker, R. H., & Likens, G. E. (1973). Primary production: The biosphere and man. *Human Ecology, 1*(4), 357–369. https://doi.org/10.1007/BF01536732

Wolfe, R. E., Nishihama, M., Fleig, A. J., Kuyper, J. A., Roy, D. P., Storey, J. C., & Patt, F. S. (2002). Achieving sub-pixel geolocation accuracy in support of MODIS land science. *Remote Sensing of Environment, 83*(1), 31–49. https://doi.org/10.1016/S0034-4257(02)00085-8

Wright, C., & Gallant, A. (2007). Improved wetland remote sensing in Yellowstone National Park using classification trees to combine TM imagery and ancillary environmental data. *Remote Sensing of Environment, 107*(4), 582–605. https://doi.org/10.1016/j.rse.2006.10.019

Wu, G., & Liu, Y. (2015). Capturing variations in inundation with satellite remote sensing in a morphologically complex, large lake. *Journal of Hydrology, 523*, 14–23. https://doi.org/10.1016/j.jhydrol.2015.01.048

Wulder, M. A., Masek, J. G., Cohen, W. B., Loveland, T. R., & Woodcock, C. E. (2012). Opening the archive: How free data has enabled the science and monitoring promise of Landsat. *Remote Sensing of Environment, 122*, 2–10. https://doi.org/10.1016/j.rse.2012.01.010

Xiao, X., Hollinger, D., Aber, J., Goltz, M., Davidson, E. A., Zhang, Q., & Moore, B. (2004). Satellite-based modeling of gross primary production in an evergreen needleleaf forest. *Remote Sensing of Environment, 89*(4), 519–534. https://doi.org/10.1016/j.rse.2003.11.008

Xu, H. (2006). Modification of Normalised Difference Water Index (NDWI) to enhance open water features in remotely sensed imagery. *International Journal of Remote Sensing, 27*(14), 3025–3033. https://doi.org/10.1080/01431160600589179

Xu, H., Zhou, Y., Wei, Y., Guo, H., & Li, X. (2023). A multirule-based relative radiometric normalization for multisensor satellite images. *IEEE Geoscience and Remote Sensing Letters, 20*, 1–5. https://doi.org/10.1109/LGRS.2023.3298505

Xue, J., & Su, B. (2017). Significant remote sensing vegetation indices: A review of developments and applications. *Journal of Sensors, 2017*, e1353691. https://doi.org/10.1155/2017/1353691

Yan, X., & Niu, Z. (2021). Reliability evaluation and migration of wetland samples. *IEEE Journal of Selected Topics in Applied Earth Observations and Remote Sensing, 14*, 8089–8099. https://doi.org/10.1109/JSTARS.2021.3102866

Yan, X., Wang, R., & Niu, Z. (2022). Response of China's wetland NDVI to climate changes. *Wetlands, 42*(6), 55. https://doi.org/10.1007/s13157-022-01568-0

Yang, J.-B., & Xu, D.-L. (2002). On the evidential reasoning algorithm for multiple attribute decision analysis under uncertainty. *IEEE Transactions on Systems, Man, and Cybernetics—Part A: Systems and Humans, 32*(3), 289–304. https://doi.org/10.1109/TSMCA.2002.802746

Yang, L., & Cervone, G. (2019). Analysis of remote sensing imagery for disaster assessment using deep learning: A case study of flooding event. *Soft Computing, 23*(24), 13393–13408. https://doi.org/10.1007/s00500-019-03878-8

Yang, X., Liu, S., Yang, T., Xu, X., Kang, C., Tang, J., Wei, H., Ghebrezgabher, M. G., & Li, Z. (2016). Spatial-temporal dynamics of desert vegetation and its responses to climatic variations over the last three decades: A case study of Hexi region in Northwest China. *Journal of Arid Land, 8*(4), 556–568. https://doi.org/10.1007/s40333-016-0046-3

Yangwu, Z., Tianli, Z., Tijiu, C. A. I., & Yane, M. A. (2009). Restoration technique of the degraded ecosystem of swamp in Xiaoxing'an Mountains. *Forest Resources Wanagement, 0*(5), 73.

Yu, L., Wang, J., Clinton, N., Xin, Q., Zhong, L., Chen, Y., & Gong, P. (2013). FROM-GC: 30 m global cropland extent derived through multisource data integration. *International Journal of Digital Earth, 6*(6), 521–533. https://doi.org/10.1080/17538947.2013.822574

Zhang, X., Liu, L., Chen, X., Gao, Y., Xie, S., & Mi, J. (2021). GLC_FCS30: Global land-cover product with fine classification system at 30 m using time-series Landsat imagery. *Earth System Science Data, 13*(6), 2753–2776. https://doi.org/10.5194/essd-13-2753-2021

Zhao, W., & Du, S. (2016). Learning multiscale and deep representations for classifying remotely sensed imagery. *ISPRS Journal of Photogrammetry and Remote Sensing, 113*, 155–165. https://doi.org/10.1016/j.isprsjprs.2016.01.004

Zheng, F., Feng, L., & Zhu, J. (2015). An incursion of off-equatorial subsurface cold water and its role in triggering the "double dip" La Niña event of 2011. *Advances in Atmospheric Sciences*, *32*(6), 731–742. https://doi.org/10.1007/s00376-014-4080-9

Zhiqiang, T., Qi, Z., Mengfan, L., Yunliang, L., Xiuli, X., & Jiahu, J. (2016). A study of the relationship between wetland vegetation communities and water regimes using a combined remote sensing and hydraulic modeling approach. *Hydrology Research*, *47*(S1), 278–292. https://doi.org/10.2166/nh.2016.216

Zhu, W.-L., Cui, L.-H., Ouyang, Y., Long, C.-F., & Tang, X.-D. (2011). Kinetic adsorption of ammonium nitrogen by substrate materials for constructed wetlands. *Pedosphere*, *21*(4), 454–463. https://doi.org/10.1016/S1002-0160(11)60147-1

Zun-You, K., Ru, A., & Xiang-Juan, L. (2015). ANN based high spatial resolution remote sensing wetland classification. 2015 14thInternational Symposium on Distributed Computing and Applications for Business Engineering and Science (DCABES), 180–183. https://doi.org/10.1109/DCABES.2015.52

9 Inland Valley Wetland Cultivation and Preservation for Africa's Green and Blue Revolution Using Multi-Sensor Remote Sensing

Murali Krishna Gumma, Pardhasaradhi Teluguntla, Pranay Panjala, Birhanu Zemadim Birhanu, and Pavan Kumar Bellam

ACRONYMS AND DEFINITIONS

AGRA	Alliance for a Green Revolution in Africa
ALI	Advanced Land Imager
ALOS	Advanced Land Observing Satellite
ASTER	Advanced Spaceborne Thermal Emission and Reflection Radiometer
AVHRR	Advanced Very High Resolution Radiometer
CARD	Coalition for African Rice Development
CGIAR	Consultative Group on International Agricultural Research
CSI	Consortium of Spatial Information
DEMs	Digital Elevation Models
DSS	Decision Support System
EO	Earth Observation
ERS	European Remote Sensing Satellites
ET	Evapotranspiration
ETM+	Enhanced Thematic Mapper Plus
FAO	Food and Agricultural Organization
FCCs	False Color Composites
GEE	Google Earth Engine
GIS	Geographic Information Systems
GR	Green Ratio
IITA	International Institute for Tropical Agriculture
IKONOS	A commercial earth observation satellite typically collecting sub-meter to 5 m data
IRS	Indian Remote Sensing Satellites
IV	Inland Valley
JERS	Japanese Earth Resources Satellite
LGP	Length of Growing Period
LULC	Land Use/Land Cover
MIR	Mid-Infrared Ratio
MODIS	Moderate-Resolution Imaging Spectroradiometer

MSS	Multi-Spectral Scanner
NASA	National Aeronautics and Space Administration
NDVI	Normalized Difference Vegetation Index
NDWI	Normalized Difference Water Index
NGOs	nongovernmental organizations
NIR	Near-Infrared
PALSAR	Phased Array type L-band Synthetic Aperture Radar
RGB	Red, Green, Blue
RVI	Ratio Vegetation Index
SAR	Synthetic Aperture Radar
SPOT	Satellite Pour l'Observation de la Terre, French Earth Observing Satellites
SRTM	Shuttle Radar Topography Mission
SWIR	Shortwave infrared
TIR	Thermal Infrared
TM	Thematic Mapper
TWI	Tasseled-cap Wetness Index
USGS	United States Geological Survey
VNIR	visible and near-infrared
WCA	West and Central Africa

9.1 INTRODUCTION

Africa is the second largest continent by land area after Asia with a total area of 30.22 million km^2 (including the adjacent islands). It has great rivers such as the River Nile, which is the longest in the world and flows a distance of 6,650 kilometers, and the River Congo, which is the deepest in the world, as well as the second largest in the world in terms of discharge or flow. Yet, Africa also has vast stretches of arid, semi-arid, and desert lands with little or no water. Further, Africa's population is projected to increase by four times by the year 2100, reaching about 4 billion, up from the current population of just over 1 billion (Grinin and Korotayev, 2023). Food insecurity, a nexus of poverty, diseases, and malnutrition, is already highest in Africa (Heidhues *et al.*, 2004; Adeyeye *et al.*, 2023; Khan and Ali, 2023), and the challenge of meeting the food security needs of the fastest-growing continent in the 21st century is daunting. Many solutions are being considered to ensure food security in Africa. These ideas include measures such as center pivot irrigation systems, by increasing irrigation in a continent that currently has just about 2% of the global irrigated areas (Thenkabail *et al.*, 2009a; Thenkabail *et al.*, 2010; Darko *et al.*, 2020; Chen *et al.*, 2023). Other ideas involve improving crop productivity (kg/m^2) and increasing crop water productivity (kg/m^3). However, an overwhelming proportion of Africa's agriculture occurs in areas where productivity increases are hard to attain, particularly in rice-based farming systems, on uplands that have poor soil fertility and water availability (Scholes, 1990; Abegunde *et al.*, 2019; Tschora and Cherubini, 2020; Rodenburg and Saito, 2022). Consequently, interest in developing sustainable agriculture in Africa's lowland wetlands, considered by some as the "new frontier" in agriculture, has swiftly increased in recent years (Alemayehu *et al.*, 2022, Djagba *et al.*, 2018). The lowland wetland systems include the big wetland systems that are prominent and widely recognized (Figure 9.1) as well as the less prominent but more widespread inland valley (IV) wetlands (Figures 9.2–9.8) that are found along the first to highest-order river systems.

Africa's big wetland ecosystems (Figure 9.1; MAW, 2014) are estimated to cover more than 131 million hectares (4.33% of total geographic area of the Continent) that vary in type from saline coastal lagoons in West Africa to fresh and brackish water lakes in East Africa. They deliver a wide range of ecosystem services that contribute to human well-being, such as nutrition, water supply and purification, climate and flood regulation, coastal protection, feeding and nesting sites, recreational opportunities, and increasingly, tourism (ESA, 2014). In contrast, the inland valley wetland

Inland Valley Wetland Cultivation and Preservation

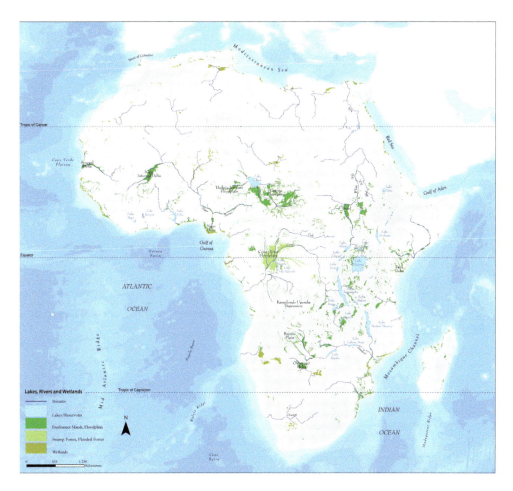

FIGURE 9.1 African wetlands (MAW, 2014). These are: "Areas of marsh, fen, peatland or water, whether natural or artificial, permanent or temporary, with water that is static or flowing" (RAMSAR, 2004). But these do not include inland valley wetlands.

systems (Figures 9.2–9.5) occupy roughly 6–20% of various agroecosystems with higher percentage areas in the wetter agroecosystems and the lower percentage areas in the drier agroecosystems (Thenkabail *et al.*, 2000b; Kotzé *et al.*, 2016; Debray *et al.*, 2019). Wetlands, with their abundant supply of fresh water, generally fertile soils, and high productivity, therefore, play a central role in the economy of all river basins and coastal zones. They provide fish, water for agriculture, household uses, as well as transport. Additionally, many distant communities as well as entire cities and regions benefit from wetlands.

In this chapter, we will provide a focused study of the wetlands of West and Central Africa (WCA) and demonstrate their richness and importance in ensuring Africa's food security. Throughout WCA, there is increasing pressure for agricultural development due to population growth and efforts to increase food security. The inland valley wetlands have high potential for growing agricultural crops due to: (1) easy access to the river water, (2) significantly longer duration of adequate soil moisture to grow crops when compared with adjoining uplands, and (3) rich soils (depth and fertility) (FAO, 2005; WARDA, 2006; Tiner, 2009). However, 90% of current agriculture in WCA is concentrated in uplands, characterized by poor soils and limited water resources. In spite of such huge advantages over uplands, inland valley wetlands in WCA are highly under-utilized, mainly as a result of: (1) waterborne diseases such as malaria, bilharzias, trypanosomiasis (sleeping sickness), onchocerciasis (river blindness), and dracontiasis (guinea worm), and (2) difficulty in accessing them from roads-settlements-markets (WARDA, 2003; Lafferty, 2009; Wingard et al., 2017). But these difficulties can be overcome with advances in satellite remote sensing (Topp et al., 2020), modern health care (Hetzel et al., 2007; Thamaga et al., 2022), and qualitative assessment of planning practices in transport infrastructure (Woodhouse, 2009; Antonson et al., 2021).

Considering this background, there is a growing consensus that expediting WCA's green revolution (more crop per unit area) and blue revolution (more crop per unit of water) hinges on prioritizing its soil-water rich and largely under-utilized inland valley wetlands, comprising approximately 80% of WCA's total wetlands, with the remainder being river flood plains (12%) and coastal wetlands (8%) (Lyon et al., 2001; Mitsch and Gosselink, 2007; Thenkabail et al., 2009b; Paul et al., 2020). The WCA has yet to experience a green revolution, which is crucially needed for the food security and economic progress of these countries, particularly for its subsistence farmers, who represent the vast majority of WCA's population of 350 million. The green revolution technologies developed in Asia in terms of improved agronomic, genetic traits, and better water management can be adopted with minor modifications to WCA's own green revolution (Otsuka and Muraoka, 2017). The importance of inland valley wetlands is particularly high for rice cultivation, as it is becoming a major staple in WCA. Records show a rapid increase of rice consumption in West Africa from 1 million tons in 1964 to 8.6 million tons in 2004 (WARDA, 2003; FAO, 2005; Zenna et al., 2017; Akpoti et al., 2021). IV wetlands have higher crop yields than the equivalent upland areas (Rodenburg et al., 2014; Owino et al., 2020). For example, potential yields of rice in IVs were estimated at 2.5–4.0 ton/ha compared to 1.5–2.0 ton/ha on uplands (WARDA, 2006). Also, an important link in achieving food security is plant population improvement via recurrent selection (Pereira de Castro et al., 2023).

Balancing the need to allocate more land for agriculture by converting land from other uses or natural cover are ecological regarding the environmental impacts of land cover changes, such as wetland development (and the surrounding catchments) and the profound social and economic repercussions for people reliant on these natural resources and ecosystem functions. IV wetlands play an important role in bio-geochemical cycling, flood control, and recharging of aquifers. They are considered to be one of the richest and most productive biomes, serving as cradles of biological diversity that support diversified flora and fauna (RAMSAR, 2004; Couvreur et al., 2021). Anthropogenic pressures on coastal wetlands not only affects the potential breeding sites for waterfowl but also impacts their role as significant carbon sinks in soils and plants (Lal et al., 2002; Mitsch and Gosselink, 2007; Newton et al., 2020).

Clearly, it is essential to incorporate wetlands explicitly within a natural resource management framework. There is a need only to develop technologies adapted to farmers' economic needs to facilitate Africa's long-awaited Green Revolution and support its Blue Revolution, but also to sustain the integrity of the globally valuable WCA ecosystems. At present, the basis for making decisions relating to sustainable utilization of wetland resources is weak (Gliessman, 2007; Musasa and Marambanyika, 2020). Given the fact that the characteristics of wetlands are known to vary dramatically within and across agroecosystems (Andriesse et al., 1994; Buchanan et al., 2020; Krause et al., 2021), it is useful to map, characterize, and model different wetland systems using remote sensing–derived environmental variables, archive field data, and artificial intelligence (Gumma et al., 2009a; Gumma et al., 2011b; Slagter et al., 2020; Rapinel et al., 2023). This will provide impetus and enable the development of appropriate technologies for maximizing food production along with transportation (food security) with minimum ecological and environmental disturbance. A pre-requisite for sustainable management of IV wetlands is greater understanding of the interaction between climate, soil, topography, water, biophysical, health, and socioeconomic factors that influence both wetland utilization and the impacts that result, including societal benefits.

Given the previous discussion, the three goals presented in this chapter are:

First, identify, delineate, map, classify, and characterize wetlands of the entire WCA region using data fusion involving satellite multi-sensor data (e.g., Landsat ETM+, JERS SAR, ALOS PALSAR, MODIS, IKONOS/QuickBird; see Table 9.1 and Table 9.2), secondary data (SRTM, FAO soils,

TABLE 9.1
Wetland Delineation, Mapping, and Characterization Using Sensor Data Fusion. Characteristics of Data are Listed (Sources: Ahmed et al., 2023; Radočaj et al., 2020; Thenkabail et al., 2010).

Sensor	Spatial (meters)	Spectral (#)	Radiometric (bit)	Band range (µm)	Band widths (µm)	Irradiance (W m^{-2}sr^{-1}µm^{-1})	Data Points (#per hectares)	Frequency of Revisit (days) Data period
				A. Moderate resolution				
1. MODIS Terra\Aqua	250, 500	2/7	12	0.62–0.67	0.05	1528.2	0.16, 0.04	8-day reflectance
				0.84–0.876	0.036	974.3	0.16, 0.04	2000–present
				0.459–0.479	0.02	2053		(wall to wall, Figure 9.1)
				0.545–0.565	0.02	1719.8		
				1.23–1.25	0.02	447.4		
				1.63–1.65	0.02	227.4		
				2.11–2.16	0.02	86.7		
				B. High resolution optical				
2. Sentinel—2A/B	10	13	12	0.43–0.45	0.02	–		5-day (both)
				0.45–0.52	0.07			
				0.54–0.57	0.03			
				0.64–0.67	0.03			
				0.69–0.71	0.03			
				0.73–0.74	0.01			
				0.77–0.79	0.02			
				0.78–0.88	0.1			
				0.85–0.87	0.02			
				0.93–0.95	0.02			
				1.35–1.38	0.03			
				1.56–1.65	0.09			
				2.11–2.28	0.17			
2. Landsat- TM\ETM+	30	7	8	0.45–0.52	0.07	1970	11.1	16
				0.52–0.60	0.80	1843		GLS2005

(Continued)

TABLE 9.1 (Continued)
Wetland Delineation, Mapping, and Characterization Using Sensor Data Fusion. Characteristics of Data are Listed (Sources: Ahmed et al., 2023; Radočaj et al., 2020; Thenkabail et al., 2010).

Sensor	Spatial (meters)	Spectral (#)	Radiometric (bit)	Band range (μm)	Band widths (μm)	Irradiance (W m^{-2}sr^{-1}μm^{-1})	Data Points (#per hectares)	Frequency of Revisit (days) Data period
				0.63–0.69	0.60	1555		(wall to wall, Figure 9.1)
				0.76–0.90	0.14	1047		
				1.55–1.74	0.19	227.1		
				10.4–12.5	2.10	0		
				2.08–2.35	0.25	80.53		
C. Radar
3. Sentinel—1A/B	20	C-band	12	3.75 to 7.5 cm	5.405 GHz	-		6 days (both) consolidated 1996
3a. JERS/SAR	100, 500	L-band	8	23.5 cm	L band	-	1, 0.04	Two periods (wall to wall, Figure 9.1)
and/or								
3b. ALOS PALSAR	9–157	L-band	8	23.5 cm	14–28 Mhz	-	123, 0.4	2006–present For benchmark areas see Figure 9.2
D. Very high resolution optical
WorldView-1–4	Pan- 31 cm MS- 1.2–1.8 m	12		0.450–0.800	0.35			min - 1 day
				0.400–450	0.05			
				0.450–0.510	0.06			
				0.510–0.580	0.07			
				0.585–0.625	0.04			
				0.630–0.690	0.06			
				0.705–0.745	0.04			
				0.770–0.895	0.12			
				0.860–1.040	0.18			

Inland Valley Wetland Cultivation and Preservation

Pleiades-1A/B	Pan—50 cm MS—3 m	5	1.195–1.225 1.550–1.590 1.640–1.680 0.45–0.80 0.43–0.55 0.49–0.61 0.60–0.72 0.75–0.95	0.1 0.04 0.04 0.35 0.12 0.12 0.12 0.2			min. 1 day
KOMPSAT	Pan—40–55 cm MS—2.2 m	5	0.450–0.900 0.450–0.520 0.520–0.600 0.630–0.690 0.760–0.900	0.45 0.07 0.08 0.06 0.14			min. 1 day
Planet	3–5 m	4	0.455–0.515 0.500–0.590 0.590–0.670 0.780–0.860	0.06 0.09 0.08 0.08			min. 1 day
4a. IKONOS	1–4	4	11		0.71 0.89 0.66	1930.9 1854.8 1156.5	10000, 625	5 For benchmark areas see Figure 9.2
			0.445–0.516 0.506–0.595 0.632–0.698	0.96	1156.9			
and/or			0.757–0.853					
4b. QUICKBIRD	0.61–2.44	4	11		0.07 0.08 0.06	1381.79 1924.59 1843.08	14872, 625	5 For benchmark areas see Figure 9.2
			0.45–0.52 0.52–0.60 0.63–0.69					
			0.76–0.89		0.13	1574.77		

TABLE 9.2
Satellite Sensor Data That Can Be Used in Wetland Studies (Adopted from Thenkabail et al., 2010)

Sensor	Spatial (meters)	Spectral (#)	Radiometric (bit)	Band range (μm)	Band widths (μm)	Irradiance (W m^{-2}sr^{-1}μm^{-1})	Data Points (#per hectares)	Frequency of Revisit (days)
A. Coarse resolution sensors
1. AVHRR	1000	4	11	0.58–0.68	0.10	1390	0.01	daily
				0.725–1.1	10.95–11.65	1410		
				3.55–3.93	0.38	1510		
				10.30–10.95	0.65	0		
				10.95–11.65	0.7	0		
2. MODIS	250, 500, 1000	36/7	12	0.62–0.67	0.05	1528.2	0.16, 0.04, 0.01	daily
				0.84–0.876	0.036	974.3	0.16, 0.04, 0.01	
				0.459–0.479	0.02	2053		
				0.545–0.565	0.02	1719.8		
				1.23–1.25	0.02	447.4		
				1.63–1.65	0.02	227.4		
				2.11–2.16	0.05	86.7		
B. Multi-spectral sensors
3. Landsat-1, 2, 3 MSS	56X79	4	6	0.5–0.6	0.1	1970	2.26	16
				0.6–0.7	0.1	1843		
				0.7–0.8	0.1	1555		
				0.8–1.1	0.3	1047		
4. Landsat-4, 5 TM	30	7	8	0.45–0.52	0.07	1970	11.1	16
				0.52–0.60	0.80	1843		
				0.63–0.69	0.60	1555		
				0.76–0.90	0.14	1047		
				1.55–1.74	0.19	227.1		
				10.4–12.5	2.10	0		
				2.08–2.35	0.25	80.53		

5. Landsat-7 ETM+	30	8	8	0.45–0.52	0.65	1970	44.4, 11.1	16
				0.52–0.60	0.80	1843		
				0.63–0.69	0.60	1555		
				0.50–0.75	0.150	1047		
				0.75–0.90	0.200	227.1		
				10.0–12.5	2.5	0		
				1.75–1.55	0.2	1368		
				0.52–0.90(p)	0.38	1352.71		
5b. Landsat-8	30	11	8	0.433–0.453	0.02	1970	44.4, 11.1	16
				0.45–0.515	0.065	1843		
				0.53–0.60	0.07	1555		
				0.63–0.68	0.05	1047		
				0.845–0.885	0.04	227.1		
				1.56–1.66	0.1	0		
				2.10–2.30	0.2	1368		
				0.50–0.68	0.18	1352.71		
				1.360–1.390	0.03	1368		
				10.6–11.2	0.6	1352.71		
				11.5–12.5	1.0	1368		
6. ASTER	15, 30, 90	15	8	0.52–0.63	0.11	1846.9	44.4, 11.1, 1.23	16
				0.63–0.69	0.06	1546.0		
				0.76–0.86	0.1	1117.6		
				0.76–0.86	0.1	1117.6		
				1.60–1.70	0.1	232.5		
				2.145–2.185	0.04	80.32		
				2.185–2.225	0.04	74.96		
				2.235–2.285	0.05	69.20		
				2.295–2.365	0.07	59.82		
				2.360–2.430	0.07	57.32		
			12	8.125–8.475	0.35	0		
				8.475–8.825	0.35	0		
				8.925–9.275	0.35	0		
				10.25–10.95	0.7	0		

(Continued)

TABLE 9.2 (Continued)
Satellite Sensor Data That Can Be Used in Wetland Studies (Adopted from Thenkabail et al., 2010)

Sensor	Spatial (meters)	Spectral (#)	Radiometric (bit)	Band range (μm)	Band widths (μm)	Irradiance (W m^{-2}sr^{-1}μm^{-1})	Data Points (#per hectares)	Frequency of Revisit (days)
7. ALI	30	10	12	10.95–11.65	0.7	0	11.1	16
				.048–0.69 (p)	0.64	1747.8600		
				0.433–0.453	0.20	1849.5		
				0.450–0.515	0.65	1985.0714		
				0.425–0.605	0.80	1732.1765		
				0.633–0.690	0.57	1485.2308		
				0.775–0.805	0.30	1134.2857		
				0.845–0.890	0.45	948.36364		
				1.200–1.300	1.00	439.61905		
				1.550–1.750	2.00	223.39024		
				2.080–2.350	2.70	78.072727		
8. SPOT-1	2.5–20	15	16	0.50–0.59	0.09	1858	1600, 25	3–5
-2				0.61–0.68	0.07	1575		
-3				0.79–0.89	0.1	1047		
-4				1.5–1.75	0.25	234		
				0.51–0.73 (p)	0.22	1773		
9. IRS-1C	23.5	15	8	0.52–0.59	0.07	1851.1	18.1	16
				0.62–0.68	0.06	1583.8		
				0.77–0.86	0.09	1102.5		
				1.55–1.70	0.15	240.4		
				0.5–0.75 (P)	0.25	1627.1		
10. IRS-1	23.5	15	8	0.52–0.59	0.07	1852.1	18.1	16
				0.62–0.68	0.06	1577.38		
				0.77–0.86	0.09	1096.7		
				1.55–1.70	0.15	240.4		
				0.5–0.75 (P)	0.25	1603.9		
11. IRS-P6-AWiFS	56	4	10	0.52–0.59	0.07	1857.7	3.19	16

Inland Valley Wetland Cultivation and Preservation 277

Sensor		Bands	Wavelength (μm)		Values	Resolution
12. CBERS -2 -3B	20 m pan, 20 m MS / 20 m pan, 20 m MS		0.62–0.68	0.06	1556.4	
			0.77–0.86	0.09	1082.4	
-3	5 m pan,	11	1.55–1.70	0.15	239.84	
-4	20 m MS		0.51–0.73	0.22	1934.03	25, 25
			0.45–0.52	0.07	1787.10	
			0.52–0.59	0.07	1587.97	400, 25
			0.63–0.69	0.06	1069.21	
			0.77–0.89	0.12	1664.3	

C. Hyper-spectral sensor

Sensor		Bands	Wavelength (μm)		Values	Resolution
1. Hyperion	30	196	196 effective Calibrated bands VNIR (band 8 to 57 427.55 to 925.85 nm SWIR (band 79 to 224) 932.72 to 2395.53 nm	10 nm wide (approx.) for all 196 bands	See data in Neckel and Labs (1984). Plot it and obtain values for Hyperion bands	11.1 16

D. Hyper-spatial sensor

Sensor		Bands	Wavelength (μm)		Values	Resolution
1. WorldView-2	0.46–1.84	8	0.4–0.45	0.05	1758.2229	10000, 625 3.7
			0.45–0.51	0.06	1974.2416	
			0.51–0.58	0.07	1856.4104	
			0.585–0.625	0.035	1738.4791	
			0.63–0.69	0.06	1559.4555	
			0.705–0.745	0.04	1342.0695	
			0.770–0.895	0.125	1069.7302	
			0.860–0.900	0.4	861.2866	
		PAN	0.860–0.900	0.4	1580.814	
2. IKONOS	1–4	4	0.445–0.516	0.71	1930.9	10000, 625 5
			0.506–0.595	0.89	1854.8	
			0.632–0.698	0.66	1156.5	
			0.757–0.853	0.96	1156.9	
3. QUICKBIRD	0.61–2.44	4	0.45–0.52	0.07	1381.79	14872, 625 5

(Continued)

TABLE 9.2 (Continued)
Satellite Sensor Data That Can Be Used in Wetland Studies (Adopted from Thenkabail et al., 2010)

Sensor	Spatial (meters)	Spectral (#)	Radiometric (bit)	Band range (μm)	Band widths (μm)	Irradiance (W m^{-2}sr^{-1}μm^{-1})	Data Points (#per hectares)	Frequency of Revisit (days)
4. RESOURCESAT	5.8	3	10	0.52–0.60	0.08	1924.59	33.64	24
				0.63–0.69	0.06	1843.08		
				0.76–0.89	0.13	1574.77		
5. RAPID EYE-A - E	6.5	5	12	0.52–0.59	0.07	1853.6	236.7	1–2
				0.62–0.68	0.06	1581.6		
				0.77–0.86	0.09	1114.3		
				0.44–0.51	0.07	1979.33		
				0.52–0.59	0.07	1752.33		
				0.63–0.68	0.05	1499.18		
				0.69–0.73	0.04	1343.67		
				0.77–0.89	0.12	1039.88		
6. WORLDVIEW	0.55	1	11	0.45–0.51	0.06	1996.77	40000	1.7–5.9
7. FORMOSAT-2	2–8	5	11	0.45–0.52	0.07	1974.93	2500, 156.25	daily
				0.52–0.60	0.08	1743.12		
				0.63–0.69	0.06	1485.23		
				0.76–0.90	0.14	1041.28		
				0.45–0.90(p)	0.45	1450		
8. KOMPSAT-2	1–4	5	10	0.5–0.9	0.4	1379.46	10000, 625	3–28
				0.45–0.52	0.07	1974.93		
				0.52–0.6	0.08	1743.12		
				0.63–0.59	0.04	1485.23		
				0.76–0.90	0.14	1041.28		

Notes:

Of the 242 bands, 196 are unique and calibrated. These are: (A) band 8 (427.55 nm) to band 57 (925.85 nm) that are acquired by visible and near-infrared (VNIR) sensor; and (B) band 79 (932.72 nm) to band 224 (2395.53 nm) that are acquired by short-wave infrared (SWIR) sensor.

First band is panchromatic, rest multi-spectral.

precipitation), and in situ data (Fujii *et al.*, 2010; Sjöström *et al.*, 2023). IV wetlands are often too small to be depicted on most maps, leading wetland surveys to focus primarily on localized areas (Gilmore *et al.*, 2008; Wdowinski *et al.*, 2008; Mahdavi *et al.*, 2018; Suárez *et al.*, 2023) and limit themselves to large flood plains, swamps, and water bodies with or without irrigated areas. However, recent studies (Thenkabail and Nolte, 2000; Lan and Zhang, 2006; Becker *et al.*, 2007; Islam *et al.*, 2008; Gumma *et al.*, 2022) have identified the potential of satellite remote sensing data and techniques for mapping different types of wetlands. None, however, have done so over very large areas such as nations, continents, and the world. Therefore, we propose utilizing multi-data fusion to effectively identify, map, classify, and characterize IV wetlands at high resolution (nominal 30 m) across the entire WCA rapidly and accurately, employing automated and semi-automated methods.

Second, develop a decision support system (DSS) through spatial modeling to perform land suitability analysis in order to determine which of the IV wetland areas are best suited for: (1) agricultural development or (2) preservation. The goal is to balance food security and economic development with environmental conservation. Since the need is to maximize crop yields sustainably with minimal ecological and environmental impacts for the IV wetland ecosystems, we need to take into consideration climatic, soil, topographic, water, biophysical, health, and socioeconomic factors and potential societal benefits from the IV wetland ecosystem and use them in decision-support systems. Stakeholders (e.g., Coalition for African Rice Development CARD/Alliance for a Green Revolution in Africa [AGRA] network, Consultative Group on International Agricultural Research [CGIAR] network, International Institute for Tropical Agriculture [IITA]) will be involved in assigning weights to various spatial data layers used in the models of the DSS and hence will represent the collective knowledge of experts.

Third, provide access to data and products through USGS/NASA as well as stakeholder (e.g., CARD/AGRA network, CGIAR Consortium of Spatial Information [CSI] network, IITA) through public domain web/data portals. This will assist stakeholders in providing farmers and policymakers with scientifically sound information, enabling them to identify the most suitable sites for promoting sustainable farming systems. The products will include: (1) IV wetland maps, (2) wetland characteristics (e.g., phenology, land cover), (3) DSS, and (4) model outputs showing IV wetlands that are most suitable for (1) development as agricultural land and (2) conservation of biological diversity (outputs of goal 2).

Carbon budget of wetlands: Wetlands globally store approximately 771 billion tons of carbon, which accounts for roughly 20% of all the carbon on Earth (Pelley, 2008; Tiner, 2009; Suárez *et al.*, 2023). This amount is comparable to the current carbon content in the atmosphere. However, they also release methane, a greenhouse gas (Pelley, 2008; Bao *et al.*, 2023) that is 22 times more potent than CO_2, on a per-unit-mass basis, in absorbing long-wave radiation on a 100-year time horizon (Zhuang *et al.*, 2009; Xia *et al.*, 2024). Nearly 60% of the planet's wetlands have been destroyed in the past 100 years, mostly for agriculture (Fluet-Chouinard *et al.*, 2023, Hu *et al.*, 2017).

In Africa, since most wetlands (Figure 9.1) are still intact, there is immense pressure to develop them to ensure African food security. Indeed, many consider wetlands as the best hope for Africa's Green and Blue Revolutions (Otsuka and Muraoka, 2017, WARDA, 2006) and a far better option for food security than the alternative of building large dams that will result in greater destruction of pristine rainforests (FAO, 2005). Given the previous discussions, WCA represents an unparalleled opportunity to guide agricultural expansion while being mindful of critical conservation goals and curtail the need for future remediation.

9.2 DEFINITIONS AND STUDY AREAS

9.2.1 Definition Used for Mapping Wetlands

(1) "Areas of marsh, fen, peatland or water, whether natural or artificial, permanent or temporary, with water that is static or flowing" (RAMSAR, 2004), and (2) "Seasonally or permanently waterlogged, including lakes, rivers, estuaries, and freshwater marshes; an area of low-lying land submerged or inundated periodically" (Gallant *et al.*, 2014). Mapping various types of wetlands, including irrigated agriculture, freshwater bodies, salt pans, lagoons, mangroves, riparian vegetation, permanent

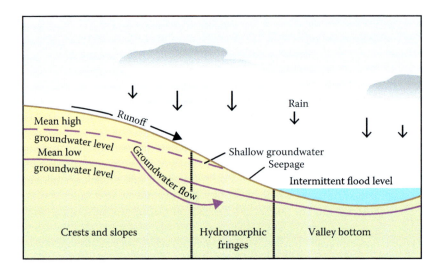

FIGURE 9.2 Depiction of wetlands (Source: WARDA, 2006).

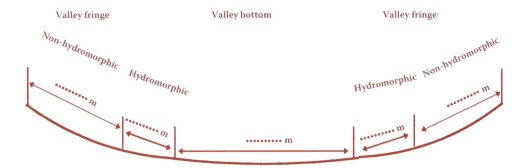

FIGURE 9.3 Inland valley wetlands consist of valley bottoms, hydromorphic valley fringes, and non-hydromorphic valley fringes.

marshes, water bodies with or without aquatic plants, and seasonal wetlands will support wetland conservation, carbon sequestration, and preserving flora and fauna. Clearly demarcated IV wetlands that occur overwhelmingly on 1–4th order streams and roughly constitute about 80% of all wetlands in WCA (Andriesse et al., 1994; Gumma et al., 2009a) will allow for determining which wetlands need to be preserved for maintaining flora and fauna as well as to sequester carbon, and which can be developed for agriculture. Hydromorphism is considered a permanent or temporary state of water saturation in the soil associated with conditions of reduction (Figure 9.3). This condition easily occurs in the soil each time when water stagnates and is not replenished. This is, for instance, the case in clayey soils with a slow internal drainage (Aguilar et al., 2003; Cheng et al., 2020; Hersi et al., 2023). Typical inland valleys (IVs) at field level are illustrated in Figure 9.4 as found in different agroecological zones (AEZs). These IVs are clearly seen in various satellite imagery and are distinct from uplands and hence can be delineating (Figure 9.5) using methods and approaches described in this chapter. We will describe delineation and characterizations of IVs in study areas spread across distinct AEZs of West and Central Africa (WCAs) (Figure 9.6).

9.3 REMOTE SENSING DATA FOR IV WETLAND CHARACTERIZATION

With the availability of multiple sensors offering varying spatial and temporal resolutions, and easy access to the scientific community, it is now scientists who are harnessing such data for diverse applications. The critical ecosystems services and agro-economic services provided by wetlands underscores their importance for conservation and restoration efforts. In this context, the identification and characterization of IV wetlands emerge as a priority to sustain food production for a growing population where cultivable land is becoming scarce and water resources are in high demand. For instance, studies by Thenkabail and Nolte (1995a, 1995b, 1996), Thenkabail *et al.* (2000b), and Gupta *et al.* (2020) have used various sensors and advanced techniques to map and characterize IV ecosystems in West Africa. Gumma et al. (2009b) have modeled different layers of information derived from satellite imagery to identify suitable areas for rice cultivation in the IV wetlands of Ghana. The effective use of remotely sensed data for such ecosystems also relies on understanding the bio-physical characteristics and extent of the IV wetlands. Morphometric characteristics of the river basin, including drainage network, density influenced by lithology and soils, are crucial factors in selecting appropriate remotely sensed imagery. Spatial resolution significantly impacts the mapping, characterization, and modeling of IV wetlands. The resolution of the sensor determines the level of LULC classification achievable. Moreover, elevation data, such as Digital Elevation Models (DEMs), are essential for delineating stream orders in various-sized IV wetlands. While bands sensitive to water absorption, such as MIR and FIR, are useful for mapping such wetlands, specialized sensors like ASTER (VNIR, SWIR, and TIR subsystems) have been developed for modeling wetland habitat vulnerability (Rebelo *et al.*, 2009; Khatun *et al.*, 2021).

9.4 STUDY AREA AND ECOREGIONAL APPROACH

The 24 WCA nations provide an ideal location for mapping IV wetlands, studied at a nominal resolution of 30 m for the entire area (Figure 9.7, Table 9.3). Results are reported on an ecoregional basis across WCA using the climate length of the growing period (LGP) method, FAO/UNESCO soils, and elevation (Figure 9.7) (Fischer *et al.*, 2021). The 18 large ecoregions, each spanning 10 million ha or more (Figure 9.7), cover > 90% of WCA's geographic area and are identified and mapped based on the definitions provided in Section 9.2.2 and Figure 9.4. Next, IV wetlands are categorized and characterized using time-series MODIS Terra/Aqua data (Figure 9.5), as well as other temporal and spatial measures, including texture derivatives from very high resolution imagery (e.g., IKONOS, QuickBird, GeoEye; McCarty *et al.*, 2021), along with environmental variables derived from topography, soils, and other existing datasets. Information on habitat mapping of the species of flora and fauna that are identified for conservation is also generated. Finally, spatial models are developed to determine IV wetlands most suited for cultivation and conservation. For example, IV wetlands that form an isolated patch may be best to preserve, especially if they are part of a wildlife migration corridor, whereas wetlands near a population center, close to transportation, and with less-developed over-story vegetation may be best to cultivate.

> **Field-plot data:** We adopted multiple strategies to collect field-plot data. First, we used a large and rich collection (1023 points) of field-plot data on IV wetlands spread across WCA (see distribution and source of these points in Figure 9.6). For each point, we have data on: (1) type of wetland (e.g., hydromorphic, non-hydromorphic), (2) wetland order (e.g., first, second), (3) wetland bottom width, (4) land use type (e.g., natural or cultivated), (5) moisture level, (6) land cover percentages (e.g., trees, shrubs, grasses, waterbody, cultivated), and (7) digital photos. Second, through collaboration with CARD/AGRA, CGIAR/CSI, and other African networks of national and international institutes that are actively involved in Africa's wetland issues. These data will include IV wetland point data as well as spatial data on socioeconomics and numerous other datasets (e.g., Figure 9.6). Third, such data will be sourced from previous work in West Africa (Gumma *et al.*, 2009a; Fujii *et al.*, 2010; Krishna *et al.*, 2010). Fourth, very high resolution data (e.g., WorldView-2) are used as ground truth.

TABLE 9.3
Parameters Describing the Level I Agroecological and Soil Zones

Level I AESZa	Agroecological Zone According to IITA's Definition	LGPb (days)	Major FAO Soil Groupingc	Aread (million ha)
1	Northern Guinea savanna	151–180	Luvisols	25.2
2	Southern Guinea savanna	181–210	Luvisols	18.4
3	Southern Guinea savanna	181–210	Acrisols	12.4
4	Southern Guinea savanna	181–210	Ferralsols	11.9
5	Southern Guinea savanna	181–210	Lithosols	10.7
6	Derived savanna	211–270	Ferralsols	47.2
7	Derived savanna	211–270	Luvisols	24.9
8	Derived savanna	211–270	Nitosols	14.2
9	Derived savanna	211–270	Arenosols	14.0
10	Derived savanna	211–270	Acrisols	11.7
11	Derived savanna	211–270	Lithosols	10.8
12	Humid forest	> 270	Ferralsols	150.1
13	Humid forest	> 270	Nitosols	27.2
14	Humid forest	> 270	Gleysols	19.2
15	Humid forest	> 270	Arenosols	18.9
16	Humid forest	> 270	Acrisols	18.0
17	Mid-altitude savanna*e*		Ferralsols	45.4
18	Mid-altitude savanna*f*		Nitosols	12.3

Notes:

[a] AESZ: level I agroecological and soil zones.
[b] LGP: length of growing period.
[c] Names refer to the soil classification scheme of FAO/UNESCO (1974).
[d] The area figures are for WCA and were determined using the AREA procedure of IDRISI (Eastman, 1992).
[e] Area distribution of LGP (Fischer et al., 2021) in AEZ 17 is: 151–180 days 11%, 181–210 days 9%, 211–270 days 59%, > 270 days 21%.
[f] Area distribution of LGP in AEZ 18 is: 151–180 days 2%, 181–210 days 5%, 211–270 days 53%, > 270 days 40%.

9.5 METHODS OF RAPID AND ACCURATE IV WETLAND MAPPING OF WCA

9.5.1 EXISTING METHODS OF WETLAND MAPPING

There are several studies that discuss methods of wetland mapping using remote sensing images (Thenkabail et al., 2000a; Ozesmi and Bauer, 2002; Wright and Gallant, 2007; Gumma et al., 2009a; Jones et al., 2009; Gxokwe et al., 2020; Mirmazloumi et al., 2021; Sahour et al., 2021; Yang et al., 2021; Zhang et al., 2022; Huang et al., 2023). High levels of accuracy in delineating and mapping wetlands are feasible when multi-temporal, multi-spectral, and very high spatial resolution imagery are used (e.g., (Gilmore et al., 2008; Simioni et al., 2020; Hasanlou and Seydi, 2021; Onojeghuo et al., 2021). Ramsey III et al. (1998) found that an integrated ERS SAR-optical (TM and CIR) improved the accuracy of wetland classes by up to 20%. The SAR data are sensitive to soil moisture and are quite ideal for delineating lowlands (with high moisture) and uplands (with lower moisture) (Wagner et al., 2007; Adeli et al., 2020; Singh et al., 2021; Zhang et al., 2024). Others (Thenkabail and Nolte, 2000; Islam et al., 2008; Adeli et al., 2020) demonstrated the ability to attain high levels of accuracy in delineating and mapping wetlands using SAR data. These data include (Table 9.4): (1) Global Land Survey 2005 (GLS2005) Landsat 30m, (2) Japanese Earth

FIGURE 9.4   Inland valley wetland illustration. The photos show valley bottoms (Gumma *et al.*, 2009b).

Resources Satellite Synthetic Aperture Radar (JERS SAR) 100m, (3) MODIS 250–500m, (4) Space Shuttle Topographic Mission (SRTM) 90m, and (5) secondary datasets (e.g., soils).

9.5.2 Automated Methods of Wetland Delineation and Mapping

Automated methods of wetland delineation involve (Table 9.4; Islam *et al.*, 2008; O'Neil *et al.*, 2020; López-Tapia *et al.*, 2021): (1) algorithms to rapidly delineate wetland streams using SRTM DEM data, (2) thresholds of SRTM-derived slopes, (3) thresholds of spectral indices and wavebands, and (4) automated classification techniques. First, wetlands are topographical lowlands, and hence the DEM data offer a significant opportunity to delineate lowlands from uplands. Automated methods involving the SRTM-derived wetland boundaries have four known limitations (Islam *et al.*, 2008): (1) generating non-existent or spurious wetlands, (2) providing non-smooth alignment, (3) resulting in spatial dislocation of streams, and (4) absence of stream width. Second, the SRTM DEM data are used to derive local slope maps in degrees using the slope function of ArcInfo Workstation GIS. A threshold (Table 9.2) of degree slope provides areas of wetlands or low-lying areas and non-wetlands. Third, the wetlands in the images can be highlighted by enhancing hyperspectral images (Bhatnagar *et al.*, 2020; Mao *et al.*, 2020; Xie *et al.*, 2021). The thresholds of indices and wavebands will automatically delineate wetlands from non-wetlands using time-series Earth observation data and machine learning algorithm (MLA) (Schowengerdt, 2007; Goyal *et al.*, 2022). Numerous researchers have also attempted to separate wetlands using automated classification techniques on various remotely sensed data (Fuller *et al.*, 2006; Lan and Zhang, 2006; Mahdianpari *et al.*, 2020) without initially identifying and separating wetland areas based on their location in the toposequence. However, as Ozesmi and Bauer (2002) point out, this approach leads to difficulties in coastal wetland categorization due to spectral confusion (Lan and Zhang, 2006; Xing *et al.*, 2023). This is because automated classification techniques are applied to entire image areas, encompassing wetlands and other land units that often possess significantly similar spectral properties. Classification accuracies improve when multi-temporal data are used in conjunction with ancillary

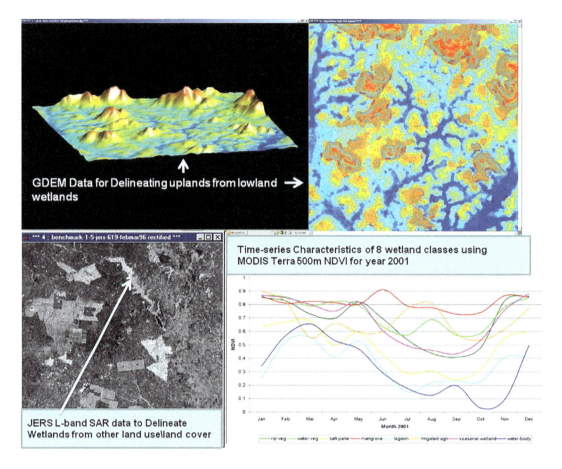

FIGURE 9.5 Delineating uplands from lowlands using various satellite imagery: (a) IKONOS 4 m DEM data shown in 3D (top left); (b) IKONOS 4-m DEM (top right); (c) JRTS SAR data (bottom left); and (d) MODIS temporal NDVI signatures of wetland classes. Inland valleys (IV) are seen in blue in the top two images. In the bottom left (JERS SAR), very high backscatter are areas of oil palm plantations. High backscatter (see arrow pointer) shows IV wetlands (Thenkabail et al., 2004).

data such as soils and topography (Ozesmi and Bauer, 2002) in the GIS modeling framework (Sader et al., 1995; Lyon et al., 2001; Fuller et al., 2006). Automated methods are rapid but need to be supplemented with semi-automated methods to increase accuracies and decrease errors of omissions and commissions.

9.5.3 Semi-Automated Methods of IV Wetland Delineation and Mapping

The semi-automated methods: (1) check any omissions or commissions of IV wetlands derived using automated methods and (2) apply appropriate corrections to improve the mapping accuracies. The semi-automated methods involve (Thenkabail et al., 2000a; Gage et al., 2020; Siervo et al., 2023): (1) image enhancement techniques involving ratio indices and applying simple thresholds were investigated for delineating wetlands automatically (Table 9.4; Lyon and McCarthy, 1995; Thenkabail and Nolte, 2000; Krause et al., 2023); (2) enhanced displays in red, green, blue (RGB) false color composites (FCCs) in different combinations of the ETM+ bands were also able to highlight wetland boundaries. The RGB FCCs that best highlight wetlands from other areas (Thenkabail et al., 2000a) were: (i) ETM+4/ETM+7, ETM+4/ETM+3, ETM+4/ETM+2; (ii) ETM+4, ETM+3, ETM+5; (iii) ETM+7, ETM+4, ETM+2; and (iv) ETM+3,

Inland Valley Wetland Cultivation and Preservation

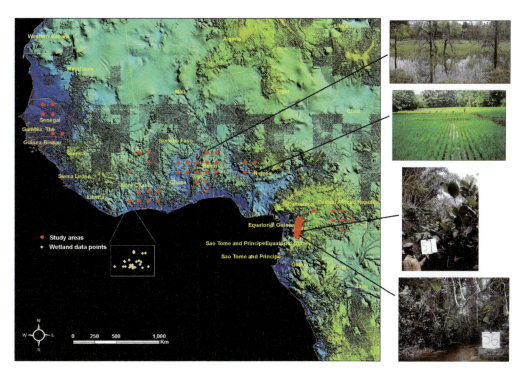

FIGURE 9.6 Inland valley wetland study areas across WCA. Note: background image is GTOPO30 1-km DEM data. Red dots are study areas. Sand color shows ground data points. Photos on right show typical IV wetlands.

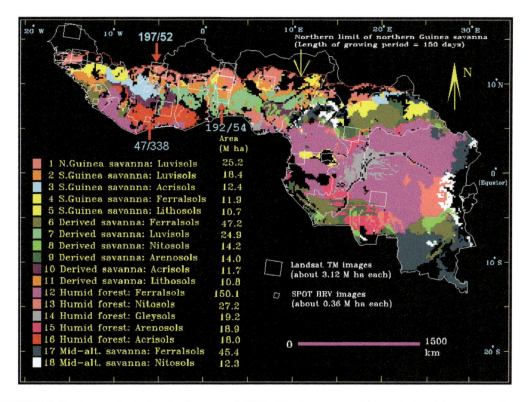

FIGURE 9.7 Agroecological and soil zones of WCA. The datasets used in producing this map are shown in Table 9.3 and consist of: (a) International Institute of Tropical Agriculture's (IITA's) agroecological zones defined by the length of growing period (LGP), and (b) FAO soils.

TABLE 9.4
Automated Methods to Separate Wetlands, Including Inland Valley Wetlands, from Non-wetlands (Gumma et al., 2009, Islam et al., 2008, Thenkabail and Nolte, 1995b)

Index or Parameter	Definition	Range −1.0 to 1.0 Dimensionless or 0 to 100%	Threshold Values That Best Delineated Wetlands
a. Slope derived from SRTM DEM	This is the percentage slope derived using spatial analyst tools available in ArcGIS	0 to 100	< 0.5 %
b. Normalized Difference Vegetation Index (NDVI) (Rouse et al., 1974)	$NDVI = \dfrac{\rho_4 - \rho_3}{\rho_4 + \rho_3}$ where ρ_3 and ρ_4 are the reflectance values derived from bands 3 (Red) and 4 (NIR) of Landsat ETM+ data, respectively.	−1.0 to +1.0	−0.25 to 0.10
c. Tasseled-cap Wetness Index (TWI) (Crist and Cicone, 1984)	TWI = ([**B1**] * 0.1509 + [**B2**] * 0.1973 + [**B3**] * 0.3279 + [**B4**] * 0.3406 + [**B5**] * −0.7112 + [**B7**] * −0.4572) Where, B1 to B7 are the DN values of the respective bands of Landsat ETM+ data. This index represents the overall degree of wetness over the area as reflected by the image data.	0 to 100	0 to 30
d. Normalized Difference Water Index (NDWI) (McFeeters, 1996)	$NDWI = \dfrac{\rho_2 - \rho_4}{\rho_2 + \rho_4}$ where ρ_2 and ρ_5 are the reflectance values derived from bands 2 (Green) and 4 (NIR) of Landsat ETM+ data, respectively.	−1.0 to +1.0	−0.15 to 0
e. Mid-Infrared Ratio (MIR) (Coppin and Bauer, 1994)	$MIR = \dfrac{Band 4}{Band 5}$ where, bands 4 and 5 are NIR and Mid-Infrared bands of Landsat ETM + data, respectively.	0 to 4	> 0.25
f. Ratio Vegetation Index (RVI) (Tucker, 1979)	$RVI = \dfrac{Band 4}{Band 3}$ where, bands 4 and 3 are NIR and Red bands of Landsat ETM + data, respectively.	0 to 6	< 0.6
g. Green Ratio (GR)	$GR = \dfrac{Band 4}{Band 2}$ where, bands 4 and 2 are NIR and Green bands of Landsat ETM + data, respectively.	0 to 4	0.5 to 0.8
h. Ratio of indices (this study)	RoI = B4/B7 * B4/B3 * B4/B2	0–240	12.5–20
i. Reflectance of SWIR 1 band (this study)	Band 5 where, band 5 is the shortwave Infrared band 1 of Landsat ETM + data.	0 to 47	< 1

ETM+2, ETM+1; and (3) once the images are enhanced (Section 9.4.1) and displayed (Section 9.4.2), they are subjected to object-oriented image analysis using eCognition software (Lemons and Karlin, 2020) to delineate wetlands and non-wetlands (Bock *et al.*, 2005; Granger *et al.*, 2021; Peng *et al.*, 2023), and then the results are compared, with the IV wetland maps derived using automated methods. Studies (Islam *et al.*, 2008; Berhanu *et al.*, 2023) have established that accuracies between 88–97% are attainable using ETM+ and SRTM data and the automated and semi-automated methods.

9.6 CHARACTERIZATION AND CLASSIFICATION OF IV WETLANDS

The IV wetland areas are highlighted using various types of remote sensing and ancillary data (e.g., Figure 9.5). Any of the images with 30 m spatial resolution or better (see Tables 9.1 and 9.2) can be used to delineate, characterize, and map IV wetlands based on methods and approaches described in the previous section and sub-section (Section 9.4 and its sub-section; also Table 9.4).

CASE STUDIES OF A LOCATION IN COTE D'VOIRE AND ENTIRE GHANA

The IV wetland maps using SPOT HRV 20 m resolution imagery are illustrated in Figure 9.8 for a location in Gagnoa, Cote d'Ivoire (see Figure 9.7 for the location). The land use and land cover (LULC) classes of the entire study area (that includes IV lowlands as well as uplands) are shown in Figure 9.9. The IV lowlands of this area is in Figure 9.8. Subsequently, these wetlands are classified using optimized layered classification, as detailed in Wright and Gallant (2007), Taddeo and Dronova (2018), and Mu *et al.* (2020), using a standard classification scheme such as the USGS Anderson (Table 9.5) (Jin *et al.*, 2019; Anderson *et al.*, 1976).

The land-use categories derived from the imagery in this study include uplands, valley fringes, valley bottoms, and others. A comparison is made with an equivalent level 1 class of the USGS classification system (Jin *et al.*, 2019, Anderson *et al.*, 1976), which aligns well with the focus on agriculture within the IVs of the present study. The classification system employed also reflects the toposequence, elucidating the type of land use/cover in the IVs. For instance, the class "significant farmland" in the uplands corresponds to agricultural land in the USGS system, while in the valley fringes, it can be agricultural land or range land due to slope conditions. In the valley bottoms, these areas are classified as wetlands in the USGS system, potentially suitable for rice cultivation. Comparisons with a standard classification system aid in understanding and relating different systems at various levels and across scales. Examination of statistics in Table 9.6 reveals the distribution of land use/land cover (LULC) in the study area. Despite the uplands occupying approximately 40% of the total area, the valley fringes and valley bottoms together amount to 58%, suggesting potential rice cultivation areas. The resulting outcome for the Gagnoa, Cote d'Ivoire, study area is illustrated in Figure 9.9, with its legend in the same figure and accompanying class bispectral plots in Figure 9.10. Figure 9.11 illustrates the approach of utilizing tassel cap bispectral plots of land use classes to define and differentiate various distinct classes. Table 9.7 further provides the percentage of land cover types in each of the 16 land use classes.

Other classifications not presented here may involve rule-based wetland mapping using fused MODIS, Landsat, secondary data such as GDEM (e.g., Figure 9.5), as well as wetland change probability mapping (Wdowinski *et al.*, 2008; Sibanda and Ahmed, 2021). Additionally, incorporating geostatistical evaluation of fine-scale spatial structure (e.g., Wallace and Marsh, 2005; Koike *et al.*, 2022) can stratify wetlands based on overall canopy characteristics. Clustering algorithms, such as canonical correlation, can then group the wetlands into similar types based on various suites of environmental variables and their derivatives. The goal is to quantify different characteristics of the wetlands to enable comparison for suitability based on a set of criteria. For example, if two wetlands differ in total canopy cover but are otherwise similar, it might be preferable to develop the wetland with less canopy since the cost for land clearing would be lower. However, it is essential to note that

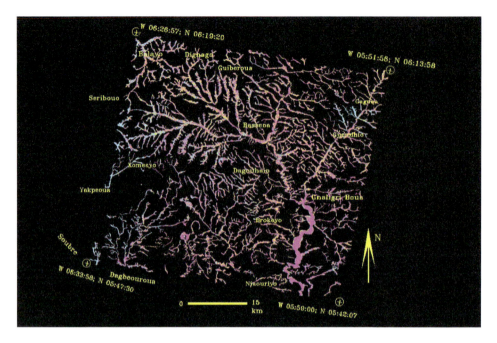

FIGURE 9.8 Delineated inland valley wetlands using SPOT HRV data based on semi-automated methods (see Section 9.4.3) described in this chapter (Thenkabail and Nolte, 1995b).

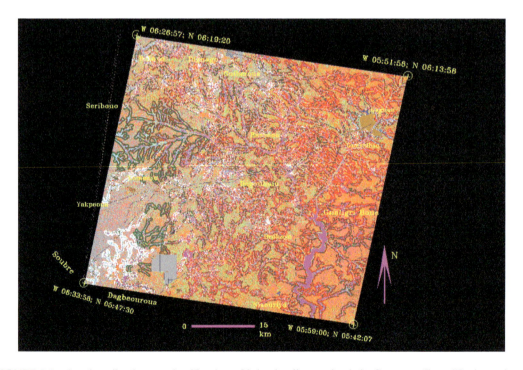

FIGURE 9.9 Land use/land cover classification of inland valley wetlands in Gaganoa, Cote d'Ivoire, using SPOT HRV images and semi-automated methods (Thenkabail and Nolte, 1995b).

TABLE 9.5
Comparison of the Land Use/Land Cover Classification System Used in This Study with the US Geological Survey (USGS) Classification System (Anderson et al., 1976)

Classification System Used in This Study	Classification of USGS Level I	Classification of USGS Level II
	Upland	
1 significant farmlands	2 agricultural land	21 cropland and pasture
2 scattered farmlands	2 agricultural land, or 3 rangeland	
3 insignificant farmlands	3 rangeland	32 herbaceous rangeland
		33 mixed rangeland
4 wetland/marshland		
5 dense forest		
6 very dense forest		
	Valley fringe	
7 significant farmlands	2 agricultural land, or	
	3 rangeland, or	33 mixed range land
	4 forest land	43 mixed forest land
8 scattered farmlands	3 rangeland, or	33 mixed rangeland
	2 agricultural land, or	
	4 forest land	43 mixed forest land
9 insignificant farmlands	4 forest land, or	43 mixed forest land
	2 agricultural land, or	
	3 rangeland	33 mixed rangeland
	Valley bottom	
10 significant farmlands	6 wetland	
11 scattered farmlands	6 wetland	
12 insignificant farmlands	6 wetland	61 forested land
	Others	
13 water	5 water	
14 built-up area/settlements	1 urban or built-up land	
15 roads	1 urban or built-up land	14 transportation communication and utilities
16 barren land or desert land	7 barren land	

while canopy cover may influence the type of action taken on a wetland, the biological diversity within that wetland must also be considered before any action is pursued.

The concept of IV delineation and mapping was applied for an entire country of Ghana (Figure 9.12) as described in detail in Gumma et al., 2009, 2011b). The delineation of IVs of entire country of Ghana (Figure 9.12) was performed using Landsat ETM+ 30 m data (Figure 9.13) using the methods and approaches described in section 9.5 and 9.6 and their sub-sections

TABLE 9.6
Land Use Distribution in the Gagnoa, Cote d'IvVoire, Study Area[a,b] (Thenkabail and Nolte, 1995b)

			Full Study Area		
No.	Land Use Category	Color	Area (ha)	Study Area (Percent of Total)	Mean NDVI
	Uplands		157,601	40.1	
1	Significant farmlands	gray	22,589	5.8	0.29
2	Scattered farmlands	seafoam	31,992	8.1	0.34
3	Savanna vegetation[c]	violet	0	0	–
4	Wetlands/marshland	mocha	7,024	1.8	0.25
5	Dense vegetation	rose	54,619	13.9	0.34
6	Very dense vegetation	red-orange	41,377	10.5	0.39
	Valley fringes		158,606	40.3	
7	Significant farmlands	white	26,299	6.7	0.31
8	Scattered farmlands	pine-green	39,376	10.0	0.32
9	Insignificant farmlands[d]	red	92,931	23.6	0.38
	Valley bottom		70,638	18.0	
10	Significant farmlands	cyan	11,490	2.9	0.29
11	Scattered farmlands	yellow	19,058	4.9	0.33
12	Insignificant farmlands[e]	magenta	40,090	10.2	0.35
	Others		**6,268**	**1.6**	
13	Water	blue	358	0.1	–0.07
14	Built-up area/settlements	tan	2,703	0.7	0.11
15	Roads	navy	2,194	0.5	0.09
16	Barren land or desert lands	sand	1,013	0.3	0.13

Notes:

[a] The study area falls entirely into agroecological zone 16 of the level I map (Figure 9.1 and Table 9.1).

[c] Class 3 occurs only in Guinea savanna zones.

[d] Spectral characteristic of vegetation in class 9 is similar to that of classes 5, 6, and 12; the difference is mainly in the toposequence position.

[e] Mainly riparian vegetation; spectral characteristics of vegetation similar to classes 5, 6, and 9; the difference is mainly in the toposequence position.

FIGURE 9.10 Land use/land cover class legend for Figure 9.15 (Thenkabail and Nolte, 1995b).

Inland Valley Wetland Cultivation and Preservation 291

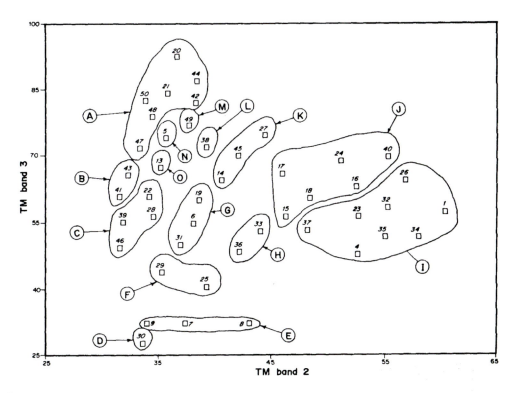

FIGURE 9.11 Land use/land cover classes depicted in Figures 9.8 and 9.9 (Thenkabail and Nolte, 1995b). The X-axis represents mean land cover class reflectance in Landsat Thematic Mapper (TM) band #2, and the Y-axis represents mean land cover class reflectance in Landsat Thematic Mapper ™ band #3. The "A," "B," etc. circled are the class crouping. For example, classes 47, 48, 42, 44, 21, 20, and 50 are grouped under class "A," which subsequently gets renumbered into one of the land use/land cover (LULC) classes, as in Figures 9.8–9.10.

TABLE 9.7
Percentage Distribution of Land Cover Types in the 16 Land Use Classes for SPOT HRV K:44, J:338 Covering the Region of Gagnoa, Côte d'Ivoire (Thenkabail and Nolte, 1995b) (See the Area in Figure 9.8)

Code of Land Use Classes[a]	\multicolumn{10}{c}{Code of Land Cover Types}
	1	2	3	4	5	6	7	8	9	10
1		4	14	12	58	12	0			0
2		20	30	25	10	15	0			0
3[b]										
4		21	31	27	4	7	1			9
5		48	25	0	0	0	0			27
6		83	17	0	0	0	0			0
7		10	19	4	57	6	3			1
8		19	39	6	13	6	2			15
9		30	55	5	2	1	1			6
10		7	6	6	60	21	0			0
11		17	6	5	17	0	0			0
12		32	52	11	5	0	0			0
13	100									

(Continued)

TABLE 9.7 (Continued)
Percentage Distribution of Land Cover Types in the 16 Land Use Classes for SPOT HRV K:44, J:338 Covering the Region of Gagnoa, Côte d'Ivoire (Thenkabail and Nolte, 1995b) (See the Area in Figure 9.8)

Code of Land Use Classes[a]	Code of Land Cover Types									
	1	2	3	4	5	6	7	8	9	10
14								100		
15									100	
16							100			

Notes:
[a] See land use class names in Table 9.5.
[b] Class 3 (savanna vegetation) does not exist in this study area.

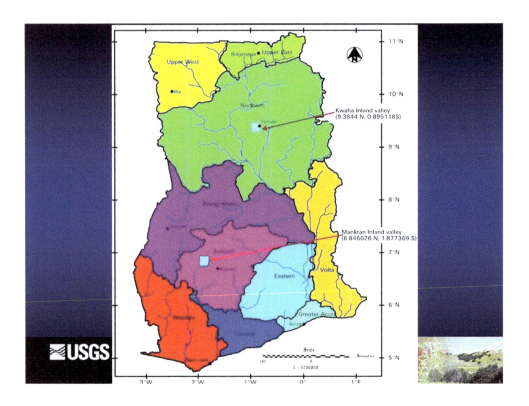

FIGURE 9.12 Map of the Country of Ghana. IV wetlands were mapped for the entire country of Ghana using Landsat ETM+ images. For the Mankran and Kwaha study areas, greater details of IV wetlands were studied using QuickBird imagery (Gumma et al., 2009).

of this book chapter. The results showed that there was 2,714946 hectares of IV wetlands in Ghana (Figure 9.13), which is 11.5% of the total geographic area (23,853,300 hectares) of the country of Ghana (Figure 9.12). For a small sub-area IV wetlands were delineated using Landsat 30m ETM+ data (Figure 9.14a) as well as IKONOS 5 m data (Figure 9.14b). Interestingly, both

Inland Valley Wetland Cultivation and Preservation 293

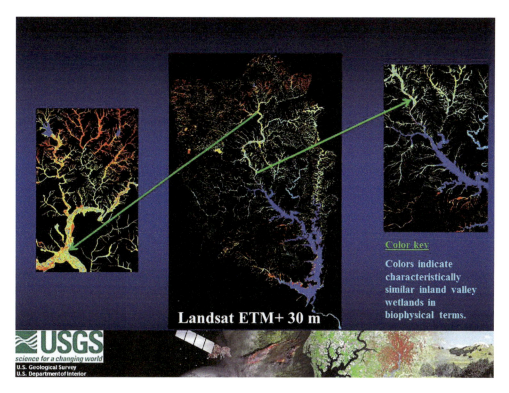

FIGURE 9.13 IV wetlands were delineated using Landsat ETM+ imagery based on semi-automated methods described in this chapter. Results showed that 11.4% (2,714,946 ha) of the total geographic area (23,853,300 ha) of Ghana was IV wetlands. Only 5% (130,000 hectares) of IV wetlands is currently cultivated (Gumma *et al.*, 2009).

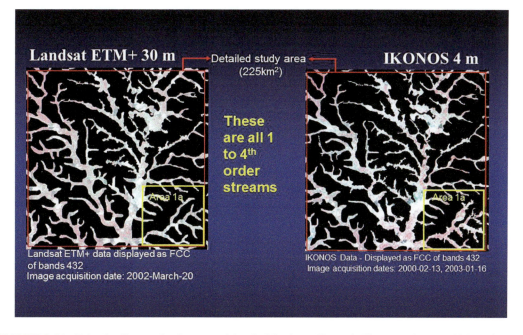

FIGURE 9.14 Inland valley wetlands mapped for the Mankran, Kumasi, Ghana study area (225 Km^2). The left image is derived from Landsat ETM+ 30 m and the right image using IKONOS 4 m. Total area of IV wetlands was determined as 27.72% using Landsat ETM+ and 28.50% using IKONOS (Gumma *et al.*, 2009).

data showed about the same percentage of the IVs in the study area (Figure 9.14). This clearly demonstrates that accurate IV delineation is equally possible using Landsat 30m ETM+ data as well as IKONOS 5 m data.

9.7 CLOUD COMPUTING TO DELINEATE IV LOWLANDS

We have explored the possibilities of delineating IV wetlands using the Google Earth Engine (GEE) cloud-computing platform using the Shuttle Radar Topography Mission (SRTM) digital elevation data (DEM) data. Initially, the existing SRTM product containing elevation data was converted into slope values using, e.g., Terrain slope. Typically, lowlands are distinguished by slopes measuring less than 3%.

The classification of slope categories is as follows:

—Slopes less than 1% are classified as highly suitable.
—Slopes between 1% and 2% are classified as moderate.
—Slopes between 2% and 3% are classified as low.
—Slopes greater than 3% are considered not suitable for crop cultivation.

This led to GEE cloud computing generated IV wetlands (Figure 9.15) using SRTM DEM Data. The sample code can be found here (use your Gmail account to log in for access):

https://code.earthengine.google.com/5a076a803f7fff34a5e1a5c6998647f1

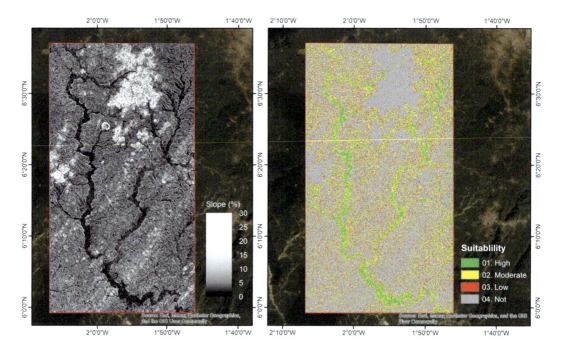

FIGURE 9.15 IV suitability delineation using GEE. The figure illustrates the delineation of the sample area. The first part of the illustration represents the slope, while the second part highlights the valley areas deemed suitable for crop cultivation based on the specified slope categories.

Inland Valley Wetland Cultivation and Preservation

9.8 SPATIAL DATA WEIGHTS MODELS FOR IDENTIFYING AREAS FOR AGRICULTURE VS. CONSERVATION

The goal of the spatial modeling (e.g., Figures 9.16 and 9.17) is to identify IV wetland areas most suitable for (1) cultivation and (2) conservation using spatial data layers (Figure 9.16) and their relative weights (Table 9.8). For example, as a result of our extensive knowledge of the wetlands in WCA

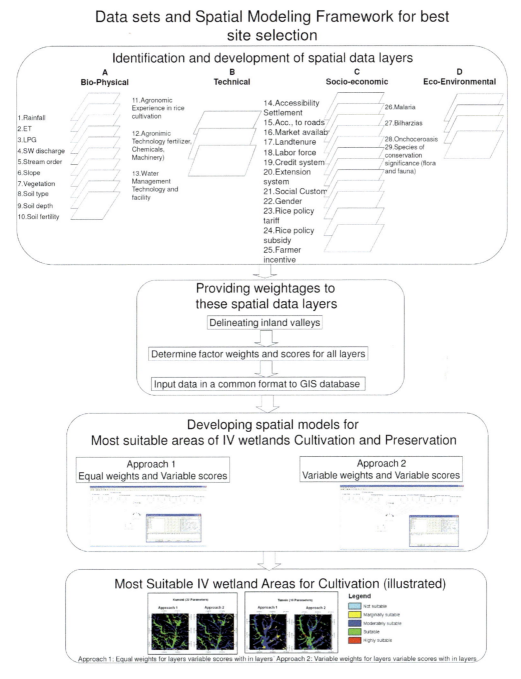

FIGURE 9.16 Spatial model steps involved in selecting the most suitable areas for rice cultivation in IV wetlands (Gumma *et al*., 2009).

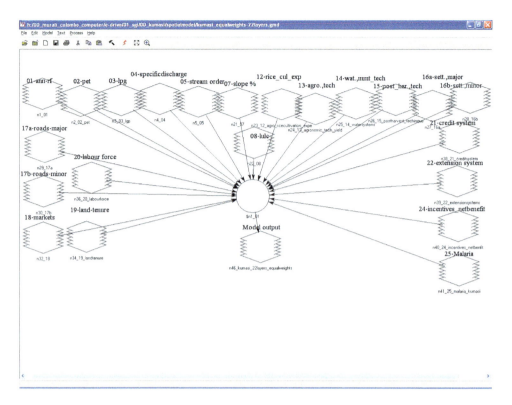

FIGURE 9.17 Illustration of a typical spatial model built in ERDAS (Gumma *et al.*, 2009).

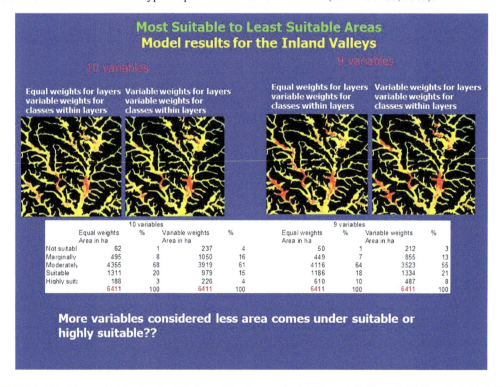

FIGURE 9.18 Most suitable sites for IV rice cultivation are in (a) Kumasi (left), and (b) Tamale (right). For each location the results and statistics are provided considering nine variables and two approaches: (1) equal weight for layer, variable weight for classes within the layer; and (b) variable weight for layer, variable weight for classes within layer.

TABLE 9.8
Process of Assigning Weightage to Spatial Data Layers and Classes within Each Spatial Data Layer Based on Expert Opinion from Stakeholders Documented during a Three-Day Workshop Held in Accra, Ghana. Illustrated for IV Wetlands of Ghana (Gumma et al., 2009). The Variable Number and Names Follow This Protocol. Biophysical Parameters Are Rainfall, ET, LGP, and so on. Similarly, Technical Factors Are Water Quality, Agronomic Technology, and so On

A. Bio-Physical Variables		Weight	Scores
1	**Rainfall**	1	
	< 700		1
	700–1000		2
	1000–1300		3
	1300–1600		4
	> 1600		5
2	ET	1	
	< 700		1
	700–1000		2
	1000–1300		3
	1300–1600		4
	> 1600		5
3	LGP	1	
	90–120		1
	120–150		1
	150–180		2
	180–210		3
	210–240		4
	240–270		5
	> 270		5
4	water resources: surface water unit discharge	2	
	very high		5
	high		4
	moderate		3
	low		2
	very low		1
5	water resources: sream order	5	
	1		1
	2		2
	3		3
	4		4
	5		5
6	Slope	1	
	< 0.5		5
	0.5–1		5
	1–1.5		4
	1.5–2		3
	2.0–3.0		2
	3.0–5		1
	> 5		1
7	Vegetation	4	
	dense forest natural vegetation		1

(Continued)

TABLE 9.8 (*Continued*)
Process of Assigning Weightage to Spatial Data Layers and Classes within Each Spatial Data Layer Based on Expert Opinion from Stakeholders Documented during a Three-Day Workshop Held in Accra, Ghana. Illustrated for IV Wetlands of Ghana (Gumma et al., 2009). The Variable Number and Names Follow This Protocol. Biophysical Parameters Are Rainfall, ET, LGP, and so on. Similarly, Technical Factors Are Water Quality, Agronomic Technology, and so On

		Weight	Scores
	fragmented natural vegetation		2
	moderate natural vegetation		3
	sparse natural vegetation		4
	fallow lands and farmlands		5
8	soil type	3	
	Type 1		5
	Type 2		4
	Type 3		3
	Type 4		2
	Type 5		1
9	soil depth	4	
	< 10		1
	10–20		2
	20–30		3
	30–40		4
	> 40		5
10	soil fertility	3	
	Type 1		5
	Type 2		4
	Type 3		3
	Type 4		2
	Type 5		1
B. Technical Factors	**Water Quality**		
11	agronomic experience in rice cultivation	2	
	< 2 yrs experience		1
	2–5 years experience		2
	5–10 years experience		3
	10–15 years experience		4
	> 15 years experience		5
12	agronomic technology (fertilizer, chemicals, machinary)	3	
	very high tech		5
	high tech		4
	moderate		3
	low tech		2
	very low tech		1
13	water management technology and facility	3	
	major irrigation canal systems		5
	minor canal systems		4
	pump and lift irrigation		3
	dug well and manual		2
	rainfed		1
C. Socioeconomic Factors	**Post-harvest**		
14a	accessibility settlements: major (> 500 people)	5	
	< 500m		5

TABLE 9.8 (*Continued*)
Process of Assigning Weightage to Spatial Data Layers and Classes within Each Spatial Data Layer Based on Expert Opinion from Stakeholders Documented during a Three-Day Workshop Held in Accra, Ghana. Illustrated for IV Wetlands of Ghana (Gumma et al., 2009). The Variable Number and Names Follow This Protocol. Biophysical Parameters Are Rainfall, ET, LGP, and so on. Similarly, Technical Factors Are Water Quality, Agronomic Technology, and so On

		Weight	Scores
	500m–1000m		4
	1000–2000		3
	2000–4000		2
	> 4000		1
14b	accessibility settlements: minor (< 500 people)	3	
	< 500m		3
	500m–1000m		2
	1000–2000		1
	2000–4000		1
	> 4000		1
15a	accessibility roads: Major	3	
	< 500m		5
	500m–1000m		4
	1000–2000		3
	2000–4000		2
	> 4000		1
15b	accessibility roads: Minor	1	
	< 500m		3
	500m–1000m		2
	1000–2000		1
	2000–4000		1
	> 4000		1
16a	Market: major (> 50000 people): define by size of settlement	3	
	< 500m		5
	500m–1000m		4
	1000–2000		3
	2000–4000		2
	> 4000		1
16b	Market: moderate (10,000–50,000 people) define by size of settlement	2	
	< 500m		4
	500m–1000m		3
	1000–2000		2
	2000–4000		1
	> 4000		1
16c	Market: minor (2000–10000 people) define by size of settlement	1	
	< 500m		3
	500m–1000m		2
	1000–2000		1
	2000–4000		1
	> 4000		1

(*Continued*)

TABLE 9.8 (*Continued*)
Process of Assigning Weightage to Spatial Data Layers and Classes within Each Spatial Data Layer Based on Expert Opinion from Stakeholders Documented during a Three-Day Workshop Held in Accra, Ghana. Illustrated for IV Wetlands of Ghana (Gumma et al., 2009). The Variable Number and Names Follow This Protocol. Biophysical Parameters Are Rainfall, ET, LGP, and so on. Similarly, Technical Factors Are Water Quality, Agronomic Technology, and so On

		Weight	Scores
17	land tenure	3	
	ownership individual		5
	ownership community\family		4
	lease < 80 GHC per ha		3
	Lease 80–100		2
	Lease > 100		1
18	labor force	3	
	labor force enough		5
	labor force ok		3
	labor force shortage		2
	labor force extremely short		1
19	credit systems	3	
	credit fully available		5
	credit available		4
	credit difficult		3
	credit very difficult		2
	credit not available		1
20	extension system	1	
	available		5
	inadequate		3
	not available		1
21	social customs	1	
22	gender	3	
	female gender obstacle		1
	female gender not obstacle		3
	male gender obstacle		1
	male gender not obstacle		3
23	rice policy tariff	3	
	no tariff		1
	tariff 10%		2
	tariff 10–20%		3
	tariff 21–30%		4
	tariff > 30 %		5
24	rice policy subsidy	4	
	no subsidy		1
	low subsidy		2
	moderate subsidy		3
	high subsidy		5
25	farmers incentive		
D. Eco-environmental Factors			
26	malaria	2	
	Very high incidence		1

TABLE 9.8 (*Continued*)
Process of Assigning Weightage to Spatial Data Layers and Classes within Each Spatial Data Layer Based on Expert Opinion from Stakeholders Documented during a Three-Day Workshop Held in Accra, Ghana. Illustrated for IV Wetlands of Ghana (Gumma et al., 2009). The Variable Number and Names Follow This Protocol. Biophysical Parameters Are Rainfall, ET, LGP, and so on. Similarly, Technical Factors Are Water Quality, Agronomic Technology, and so On

		Weight	Scores
	high incidence		2
	moderate incidence		3
	low incidence		4
	negligible incidence		5
27	bilhazias	1	
	Very high incidence		1
	high incidence		2
	moderate incidence		3
	low incidence		4
	negligible incidence		5
28	onchocercasis	3	
	Very high incidence		1
	high incidence		2
	moderate incidence		3
	low incidence		4
	negligible incidence		5
29	species of conservation significance flora and fauna		
	critically endangered		1
	endangered species		2
	vulnerable		3
	not endangered		5

(see Fujii *et al.*, 2010; Kominoski *et al.*, 2021) a total of 29 biophysical, technical, socioeconomic, and eco-environmental factors (e.g., Table 9.8, Figure 9.17) are deemed significant. In this project, weights were assigned to these spatial data layers (e.g., Table 9.8) using expert knowledge gathered from stakeholder networks (e.g., CARD/AGRA, CGIAR CSI, IITA) (Gumma *et al.*, 2009). Socioeconomic factors encompass aspects such as accessibility of settlements, road networks, markets, land tenure, labor force, credit systems, extension systems, social customs, gender, rice policy tariff, rice policy subsidy, and farmers' incentives. The models, implemented algebraically (e.g., coded in ERDAS modeler [Nelson and Khorram, 2018]; Figure 9.17), yield outputs determining suitability for cultivation and/or conservation.

Two sets of data and four scenarios were analyzed to identify suitable areas in the IV wetlands (Gumma *et al.*, 2009). A dataset comprising ten variables, with equal weights assigned to layers and varying weights for classes within the layers, as well as a dataset comprising nine variables under similar scenarios, generated outputs displaying relatively lower areas classified as "suitable." However, the nine-variable dataset produced higher areas classified as suitable (Figure 9.18). For example, if two wetlands differ solely in their proximity to transportation and markets, it may be preferable to develop the wetland closer to the markets. Similarly, if two wetlands are comparable but one serves as an isolated habitat crucial for migratory wildlife, prioritizing conservation efforts for that wetland may be warranted.

9.9 ACCURACIES, ERRORS, AND UNCERTAINTIES

The thematic accuracy of the wetland maps is assessed through both error matrix analysis and regression analysis. A number of statistical considerations, including appropriate sampling scheme, sample size, and sample unit, are taken into account (Congalton and Green, 2008). Error matrices, comprising overall accuracy, producer accuracy and user accuracy (Congalton, 2009), are reported. The study used 1023 wetland data points that were already available (e.g., Figure 9.7), as well as data obtained through our African network partners during the project.

9.10 CONCLUSIONS

The chapter provides a comprehensive overview of mapping inland valley (IV) wetlands in West and Central Africa (WCA) using remote sensing and GIS. Wetlands are at the center of the development *versus* preservation debate in Africa. Africa's food security, especially given that its population is projected to be four times (reaching about 4 billion) by the year 2100 relative to its present population of just over 1 billion, calls for an urgent need to utilize IV wetlands for agriculture. At the same time, preserving the unique flora and fauna and the carbon sequestered in the wetlands is of utmost importance.

First, the chapter provides a roadmap for consistent IV wetland characterization and mapping at various spatial resolutions using a multitude of remote sensing data. For this, the chapter uses WCA nations as case studies. Second, the chapter demonstrates wetland land use/land cover classification and the study of their time-series phenological characteristics (Gumma *et al.*, 2011a; Gumma *et al.*, 2014). Third, the remote sensing–derived products, along with secondary data (e.g., length of growing period, soils, slope, elevation, temperature, agroecological zones), as well as a number of other data such as the biophysical data and socioeconomic data, were assigned weights by experts for their importance, harmonized, standardized, and built into a decision support spatial model that pinpointed IV wetland areas that are (1) best suited for cultivation and (2) prioritized for conservation.

The chapter shows approaches and methods of utilizing Earth observation (EO) for the purposes of: (1) understanding IV wetlands as land units for Africa's Green and Blue Revolutions and (2) balancing inevitable developmental activities with environmental/ecological solutions that inform which areas to preserve. The outputs and outcomes of such a study is expected to benefit: (1) farmers to make decisions on where to focus their IV wetland agriculture based on pinpointed areas most suitable for cultivation; (2) national governments to make decisions on promoting IV wetland cultivation and conservation; (3) financial institutions (e.g., African Development Bank) to make educated decisions on where to invest to accelerate Africa's Green and Blue Revolutions; (4) researchers and NGOs (nongovernmental organizations) working in Africa.

ACKNOWLEDGMENTS

Any use of trade, firm, or product names is for descriptive purposes only and does not imply endorsement by the US government.

REFERENCES

Abegunde, V.O., Sibanda, M., Obi, A., 2019. The dynamics of climate change adaptation in sub-Saharan Africa: A review of climate-smart agriculture among small-scale farmers. Climate 7, 132.

Adeli, S., Salehi, B., Mahdianpari, M., Quackenbush, L.J., Brisco, B., Tamiminia, H., Shaw, S., 2020. Wetland monitoring using SAR data: A meta-analysis and comprehensive review. Remote Sensing 12, 2190.

Adeyeye, S.A.O., Ashaolu, T.J., Bolaji, O.T., Abegunde, T.A., Omoyajowo, A.O., 2023. Africa and the nexus of poverty, malnutrition and diseases. Critical Reviews in Food Science and Nutrition 63, 641–656.

Aguilar, J., Fernandez, J., Dorronsoro, C., Stoops, G., Dorronsoro, B., 2003. Hydromorphy in soils. http://edafologia.ugr.es/hidro/index.htm.

Ahmed, M.R., Ghaderpour, E., Gupta, A., Dewan, A., Hassan, Q.K., 2023. Opportunities and challenges of spaceborne sensors in delineating land surface temperature trends: A review. IEEE Sensors Journal 23(7), 6460–6472, 1 April, 2023. https://doi.org/10.1109/JSEN.2023.3246842.

Akpoti, K., Dossou-Yovo, E.R., Zwart, S.J., Kiepe, P., 2021. The potential for expansion of irrigated rice under alternate wetting and drying in Burkina Faso. Agricultural Water Management 247, 106758.

Alemayehu, T., Assogba, G.M., Gabbert, S., Giller, K.E., Hammond, J., Arouna, A., Dossou-Yovo, E.R., van de Ven, G.W.J., 2022. Farming systems, food security and farmers' awareness of ecosystem services in inland valleys: A study from Côte d'Ivoire and Ghana. Front. Sustain. Food Syst. 6, 892818. https://doi.org/10.3389/fsufs.2022.892818.

Anderson, J.R., Hardy, E.E., Roach, J.T., Witmer, R.E., 1976. A Land Use and Land Cover Classification System for Use with Remote Sensor Data. Volume 964. US Government Printing Office, Washington, DC, USA.

Andriesse, W., Fresco, L., Van Duivenbooden, N., Windmeijer, P., 1994. Multi-scale characterization of inland valley agro-ecosystems in West Africa. NJAS Wageningen Journal of Life Sciences 42, 159–179.

Antonson, H., Buckland, P., Nyqvist, R., 2021. A society ill-equipped to deal with the effects of climate change on cultural heritage and landscape: A qualitative assessment of planning practices in transport infrastructure. Climatic Change 166, 18.

Bao, T., Jia, G., Xu, X., 2023. Weakening greenhouse gas sink of pristine wetlands under warming. Nature Climate Change 13, 462–469.

Becker, B.L., Lusch, D.P., Qi, J., 2007. A classification-based assessment of the optimal spectral and spatial resolutions for Great Lakes coastal wetland imagery. Remote Sensing of Environment 108, 111–120.

Berhanu, M., Suryabhagavan, K.V., Korme, T., 2023. Wetland mapping and evaluating the impacts on hydrology, using geospatial techniques: A case of Geba Watershed, Southwest Ethiopia. Geology, Ecology, and Landscapes 7, 293–310.

Bhatnagar, S., Gill, L., Regan, S., Naughton, O., Johnston, P., Waldren, S., Ghosh, B., 2020. Mapping vegetation communities inside wetlands using Sentinel-2 imagery in Ireland. International Journal of Applied Earth Observation and Geoinformation 88, 102083.

Bock, M., Xofis, P., Mitchley, J., Rossner, G., Wissen, M., 2005. Object-oriented methods for habitat mapping at multiple scales—case studies from Northern Germany and Wye Downs, UK. Journal for Nature Conservation 13, 75–89.

Buchanan, S.W., Baskerville, M., Oelbermann, M., Gordon, A.M., Thevathasan, N.V., Isaac, M.E., 2020. Plant diversity and agroecosystem function in riparian agroforests: Providing ecosystem services and land-use transition. Sustainability 12, 568.

Chen, F., Zhao, H., Roberts, D., Van de Voorde, T., Batelaan, O., Fan, T., Xu, W., 2023. Mapping center pivot irrigation systems in global arid regions using instance segmentation and analyzing their spatial relationship with freshwater resources. Remote Sensing of Environment 297, 113760.

Cheng, G., Zhu, H.-H., Wen, Y.-N., Shi, B., Gao, L., 2020. Experimental investigation of consolidation properties of nano-bentonite mixed clayey soil. Sustainability 12, 459.

Congalton, R.G., 2009. Accuracy and error analysis of global and local maps: Lessons learned and future considerations. Remote Sensing of Global Croplands for Food Security, 441.

Congalton, R.G., Green, K., 2008. Assessing the Accuracy of Remotely Sensed Data: Principles and Practices (2nd ed.). CRC Press, Boca Raton, FL. https://doi.org/10.1201/9781420055139.

Coppin, P.R., Bauer, M.E., 1994. Processing of multitemporal Landsat TM imagery to optimize extraction of forest cover change features. IEEE Transactions on Geoscience and Remote Sensing 32(4), 918–927.

Couvreur, T.L., Dauby, G., Blach-Overgaard, A., Deblauwe, V., Dessein, S., Droissart, V., Hardy, O.J., Harris, D.J., Janssens, S.B., Ley, A.C., 2021. Tectonics, climate and the diversification of the tropical African terrestrial flora and fauna. Biological Reviews 96, 16–51.

Crist, E.P., Cicone, R.C., 1984. A physically-based transformation of Thematic Mapper data---The TM Tasseled Cap. IEEE Transactions on Geoscience and Remote Sensing GE-22(3), 256–263.

Darko, R.O., Liu, J., Yuan, S., Sam-Amoah, L.K., Yan, H., 2020. Irrigated agriculture for food self-sufficiency in the sub-Saharan African region. International Journal of Agricultural and Biological Engineering 13, 1–12.

Debray, V., Wezel, A., Lambert-Derkimba, A., Roesch, K., Lieblein, G., Francis, C.A., 2019. Agroecological practices for climate change adaptation in semiarid and subhumid Africa. Agroecology and Sustainable Food Systems 43, 429–456.

Djagba, J.F., Sintondji, L.O., Kouyaté, A.M., Baggie, I., Agbahungba, G., Hamadoun, A., Zwart, S.J., 2018. Predictors determining the potential of inland valleys for rice production development in West Africa. Applied Geography 96, 86–97, ISSN 0143–6228, https://doi.org/10.1016/j.apgeog.2018.05.003. www.sciencedirect.com/science/article/pii/S0143622817308640.

Eastman, J.R., 1992. IDRISI: Version 4.0, March 1992. Clark University, Graduate School of Geography.

ESA, 2014. Europian Space Agency. www.esa.int/Our_Activities/Observing_the_Earth/World_Wetlands_Day_focuses_on_agriculture. (accessed in 06 May 2014).

FAO, 2005. Limpopo Basin Profile Report. www.arc.agric.za/limpopo/profile.htm. (accessed in 01 May 2014).

Fischer, G., Nachtergaele, F.O., van Velthuizen, H.T., Chiozza, F., Franceschini, G., Henry, M., Muchoney, D., Tramberend, S., 2021. Global Agro-Ecological Zones (GAEZ v4) Model Documentation. FAO & IIASA, p. 303.

Fluet-Chouinard, E., Stocker, B.D., Zhang, Z. et al., 2023. Extensive global wetland loss over the past three centuries. Nature 614, 281–286. https://doi.org/10.1038/s41586-022-05572-6.

Fujii, H., Gumma, M., Thenkabail, P., Namara, R., 2010, August. Suitability evaluation for lowland rice in inland valleys in West Africa. Journal Transactions of the Japanese Society of Irrigation, Drainage and Rural Engineering 78(4), 47–55, August, 2010.

Fuller, L.M., Morgan, T.R., Aichele, S.S., 2006. Wetland delineation with IKONOS high-resolution satellite imagery, Fort Custer Training Center, Battle Creek, Michigan, 2005. (accessed in 03 March 2006). Scientific Investigations Report. United States Geological Survey, p. 20.

Gage, E., Cooper, D.J., Lichvar, R., 2020. Comparison of USACE three-factor wetland delineations to national wetland inventory maps. Wetlands 40, 1097–1105.

Gallant, A.L., Kaya, S.G., White, L., Brisco, B., Roth, M.F., Sadinski, W., Rover, J., 2014. Detecting emergence, growth, and senescence of wetland vegetation with polarimetric synthetic aperture radar (SAR) data. Water 6(3), 694–722.

Gilmore, M.S., Wilson, E.H., Barrett, N., Civco, D.L., Prisloe, S., Hurd, J.D., Chadwick, C., 2008. Integrating multi-temporal spectral and structural information to map wetland vegetation in a lower Connecticut River tidal marsh. Remote Sensing of Environment 112, 4048–4060.

Gliessman, S.R., 2007. Agroecology: The Ecology of Sustainable Food Systems. CRC Press, Boca Raton, FL.

Goyal, A., Upreti, M., Chowdary, V., Jha, C., 2022. Delineation and monitoring of wetlands using time series earth observation data and machine learning algorithm: A case study in Upper Ganga river stretch. In: Geospatial Technologies for Resources Planning and Management. Springer International Publishing, Cham, pp. 123–139.

Granger, J.E., Mahdianpari, M., Puestow, T., Warren, S., Mohammadimanesh, F., Salehi, B., Brisco, B., 2021. Object-based random forest wetland mapping in Conne River, Newfoundland, Canada. Journal of Applied Remote Sensing 15, 038506–038506.

Grinin, L., Korotayev, A., 2023. Africa: The continent of the future. Challenges and opportunities. In: Sadovnichy, V., Akaev, A., Ilyin, I., Malkov, S., Grinin, L., Korotayev, A. (eds) Reconsidering the Limits to Growth. World-Systems Evolution and Global Futures. Springer, Cham. https://doi.org/10.1007/978-3-031-34999-7_13.

Gumma, M., Thenkabail, P.S., Fujii, H., Namara, N., 2009. Spatial models for selecting the most suitable areas of rice cultivation in the inland valley Wetlands of Ghana using remote sensing and geographic information systems. Journal of Applied Remote Sensing 3(1), 033537, 1 June, 2009. https://doi.org/10.1117/1.3182847.

Gumma, M., Thenkabail, P.S., Fujii, H., Namara, R., 2009a. Spatial models for selecting the most suitable areas of rice cultivation in the inland valley Wetlands of Ghana using remote sensing and geographic information systems. Journal of Applied Remote Sensing 3, 033537–033537–033521.

Gumma, M.K., Amede, T., Getnet, M., Pinjarla, B., Panjala, P., Legesse, G., Tilahun, G., Van den Akker, E., Berdel, W., Keller, C., 2022. Assessing potential locations for flood-based farming using satellite imagery: A case study of afar region, ethiopia. Renewable Agriculture and Food Systems 37, S28–S42.

Gumma, M.K., Nelson, A., Thenkabail, P.S., Singh, A.N., 2011a. Mapping rice areas of South Asia using MODIS multitemporal data. Journal of Applied Remote Sensing 5, 053547.

Gumma, M.K., Thenkabail, P.S., Fujii, H., Namara, R., 2009b. Spatial models for selecting the most suitable areas of rice cultivation in the inland valley Wetlands of Ghana using remote sensing and geographic information systems. Journal of Applied Remote Sensing 3, 033537.

Gumma, M.K., Thenkabail, P.S., Hideto, F., Nelson, A., Dheeravath, V., Busia, D., Rala, A., 2011b. Mapping irrigated areas of Ghana using fusion of 30 m and 250 m resolution remote-sensing data. Remote Sensing 3, 816–835.

Gumma, M.K., Thenkabail, P.S., Maunahan, A., Islam, S., Nelson, A., 2014. Mapping seasonal rice cropland extent and area in the high cropping intensity environment of Bangladesh using MODIS 500m data for the year 2010. ISPRS Journal of Photogrammetry and Remote Sensing 91, 98–113.

Gupta, G., Khan, J., Upadhyay, A.K., Singh, N.K., 2020. Wetland as a sustainable reservoir of ecosystem services: Prospects of threat and conservation. Restoration of Wetland Ecosystem: A Trajectory Towards a Sustainable Environment, 31–43.Gxokwe, S., Dube, T., Mazvimavi, D., 2020. Multispectral remote sensing of wetlands in semi-arid and arid areas: A review on applications, challenges and possible future research directions. Remote Sensing 12, 4190.

Hasanlou, M., Seydi, S.T., 2021. Use of multispectral and hyperspectral satellite imagery for monitoring waterbodies and wetlands. In: Southern Iraq's Marshes: Their Environment and Conservation. Springer International Publishing, Cham, pp. 155–181.

Heidhues, F., Atsain, A., Nyangito, H., Padilla, M., Ghersi, G., Vallée, L., 2004. Development strategies and food and nutrition security in Africa: An assessment. 2020 Vision Discussion Papers Papers 38, International Food Policy Research Institute (IFPRI).

Hersi, N.A., Mulungu, D.M., Nobert, J., 2023. Groundwater recharge estimation under changing climate and land use scenarios in a data-scarce Bahi (Manyoni) catchment in internal drainage basin (IDB), Tanzania using soil and water assessment tool (SWAT). Groundwater for Sustainable Development 22, 100957.

Hetzel, M.W., Iteba, N., Makemba, A., Mshana, C., Lengeler, C., Obrist, B., Schulze, A., Nathan, R., Dillip, A., Alba, S., 2007. Understanding and improving access to prompt and effective malaria treatment and care in rural Tanzania: The ACCESS Programme. Malaria Journal 6, 83.

Hu, S., Niu, Z., Chen, Y., Li, L., Zhang, H., 2017. Global wetlands: Potential distribution, wetland loss, and status. Science of the Total Environment 586, 319–327, ISSN 0048–9697, https://doi.org/10.1016/j.scitotenv.2017.02.001. www.sciencedirect.com/science/article/pii/S0048969717302425.

Huang, Y., Peng, J., Chen, N., Sun, W., Du, Q., Ren, K., Huang, K., 2023. Cross-scene wetland mapping on hyperspectral remote sensing images using adversarial domain adaptation network. ISPRS Journal of Photogrammetry and Remote Sensing 203, 37–54.

Islam, M.A., Thenkabail, P., Kulawardhana, R., Alankara, R., Gunasinghe, S., Edussriya, C., Gunawardana, A., 2008. Semi-automated methods for mapping wetlands using Landsat ETM+ and SRTM data. International Journal of Remote Sensing 29, 7077–7106.

Jin, S., Homer, C., Yang, L., Danielson, P., Dewitz, J., Li, C., Zhu, Z., Xian, G., Howard, D., 2019. Overall methodology design for the United States national land cover database 2016 products. Remote Sensing 11, 2971. https://doi.org/10.3390/rs11242971.

Jones, K., Lanthier, Y., van der Voet, P., van Valkengoed, E., Taylor, D., Fernández-Prieto, D., 2009. Monitoring and assessment of wetlands using earth observation: The GlobWetland project. Journal of Environmental Management 90, 2154–2169.

Khan, Z., Ali, A., 2023. Global food insecurity and its association with malnutrition. Emerging Challenges in Agriculture and Food Science 2.

Khatun, R., Talukdar, S., Pal, S., Saha, T.K., Mahato, S., Debanshi, S., Mandal, I., 2021. Integrating remote sensing with swarm intelligence and artificial intelligence for modelling wetland habitat vulnerability in pursuance of damming. Ecological Informatics 64, 101349.

Koike, K., Kiriyama, T., Lu, L., Kubo, T., Heriawan, M.N., Yamada, R., 2022. Incorporation of geological constraints and semivariogram scaling law into geostatistical modeling of metal contents in hydrothermal deposits for improved accuracy. Journal of Geochemical Exploration 233, 106901.

Kominoski, J.S., Pachón, J., Brock, J.T., McVoy, C., Malone, S.L., 2021. Understanding drivers of aquatic ecosystem metabolism in freshwater subtropical ridge and slough wetlands. Ecosphere 12, e03849.

Kotzé, E., Loke, P., Akhosi-Setaka, M., Du Preez, C., 2016. Land use change affecting soil humic substances in three semi-arid agro-ecosystems in South Africa. Agriculture, Ecosystems & Environment 216, 194–202.

Krause, J.R., Oczkowski, A.J., Watson, E.B., 2023. Improved mapping of coastal salt marsh habitat change at Barnegat Bay (NJ, USA) using object-based image analysis of high-resolution aerial imagery. Remote Sensing Applications: Society and Environment 29, 100910.

Krause, S., Beach, T.P., Luzzadder-Beach, S., Cook, D., Bozarth, S.R., Valdez Jr, F., Guderjan, T.H., 2021. Tropical wetland persistence through the Anthropocene: Multiproxy reconstruction of environmental change in a Maya agroecosystem. Anthropocene 34, 100284.

Krishna, G.M., Prasad, T.S., Bubacar, B., 2010. Delineating shallow ground water irrigated areas in the Atankwidi Watershed (Northern Ghana, Burkina Faso) using Quickbird 0.61–2.44 meter data. African Journal of Environmental Science and Technology 4, 455–464.

Lafferty, K.D., 2009. The ecology of climate change and infectious diseases. Ecology 90, 888–900.

Lal, R., Hansen, D.O., Uphoff, N., Lal, R., 2002. Food Security and Environmental Quality in the Developing World (1st ed.). CRC Press, Boca Raton, FL. https://doi.org/10.1201/9781420032215.

Lan, Z., Zhang, D., 2006. Study on optimization-based layered classification for separation of wetlands. International Journal of Remote Sensing 27, 1511–1520.

Lemons, R., Karlin, A.L., 2020. GIS Tips and Tricks- working with big data in eCognition. Photogrammetric Engineering and Remote Sensing 86(5), 269–270(2). https://doi.org/10.14358/PERS.86.5.269.

López-Tapia, S., Ruiz, P., Smith, M., Matthews, J., Zercher, B., Sydorenko, L., Varia, N., Jin, Y., Wang, M., Dunn, J.B., 2021. Machine learning with high-resolution aerial imagery and data fusion to improve and automate the detection of wetlands. International Journal of Applied Earth Observation and Geoinformation 105, 102581.

Lyon, J., Lopez, R.D., Lyon, J.G., Lyon, L.K., Lopez, D.K., 2001. Wetland Landscape Characterization: GIS, Remote Sensing and Image Analysis (1st ed.). CRC Press, Boca Raton, FL, p. 135. https://doi.org/10.1201/9781420022681

Lyon, J., McCarthy, J., 1995. Wetland and Environmental Applications of GIS. CRC/Lewis Publishers, Boca Raton, FL, p. 368.

Mahdavi, S., Salehi, B., Granger, J., Amani, M., Brisco, B., Huang, W., 2018. Remote sensing for wetland classification: A comprehensive review. GIScience & Remote Sensing 55, 623–658.

Mahdianpari, M., Granger, J.E., Mohammadimanesh, F., Salehi, B., Brisco, B., Homayouni, S., Gill, E., Huberty, B., Lang, M., 2020. Meta-analysis of wetland classification using remote sensing: A systematic review of a 40-year trend in North America. Remote Sensing 12, 1882.

Mao, D., Wang, Z., Du, B., Li, L., Tian, Y., Jia, M., Zeng, Y., Song, K., Jiang, M., Wang, Y., 2020. National wetland mapping in China: A new product resulting from object-based and hierarchical classification of Landsat 8 OLI images. ISPRS Journal of Photogrammetry and Remote Sensing 164, 11–25.

MAW, 2014. Map of African Wetlands, African Wetlands. http://africa.wetlands.org/Africanwetlands/. (accessed in 06 May 2014).

McCarty, W., Patrick, O., Damon, M.R., Hall, A.A., 2021. Science utilizing data from spire global as part of the NASA commercial smallsat data acquisition program. In: <cite lang="en">2021 IEEE International Geoscience and Remote Sensing Symposium IGARSS.</cite> Brussels, Belgium, pp. 604–607. https://doi.org/10.1109/IGARSS47720.2021.9554232.

McFeeters, S.K., 1996. The use of the Normalized Difference Water Index (NDWI) in the delineation of open water features. International Journal of Remote Sensing 17(7), 1425–1432.

Mirmazloumi, S.M., Moghimi, A., Ranjgar, B., Mohseni, F., Ghorbanian, A., Ahmadi, S.A., Amani, M., Brisco, B., 2021. Status and trends of wetland studies in Canada using remote sensing technology with a focus on wetland classification: A bibliographic analysis. Remote Sensing 13, 4025.

Mitsch, W.J., Gosselink, J.G., 2007. Wetlands. John Wiley and Sons., New York, NY, p. 582.

Mu, S., Li, B., Yao, J., Yang, G., Wan, R., Xu, X., 2020. Monitoring the spatio-temporal dynamics of the wetland vegetation in Poyang Lake by Landsat and MODIS observations. Science of the Total Environment 725, 138096.

Musasa, T., Marambanyika, T., 2020. Threats to sustainable utilization of wetland resources in ZIMBABWE: A review. Wetlands Ecology and Management 28, 681–696.

Nelson, S.A.C., Khorram, S., 2018. Image Processing and Data Analysis with ERDAS IMAGINE® (1st ed.). CRC Press, Boca Raton, FL. https://doi.org/10.1201/b21969.

Newton, A., Icely, J., Cristina, S., Perillo, G.M., Turner, R.E., Ashan, D., Cragg, S., Luo, Y., Tu, C., Li, Y., 2020. Anthropogenic, direct pressures on coastal wetlands. Frontiers in Ecology and Evolution 8, 144.

O'Neil, G.L., Goodall, J.L., Behl, M., Saby, L., 2020. Deep learning using physically-informed input data for wetland identification. Environmental Modelling & Software 126, 104665.

Onojeghuo, A.O., Onojeghuo, A.R., Cotton, M., Potter, J., Jones, B., 2021. Wetland mapping with multi-temporal sentinel-1 &-2 imagery (2017–2020) and LiDAR data in the grassland natural region of alberta. GIScience & Remote Sensing 58, 999–1021.

Otsuka, K., Muraoka, R., 2017. A green revolution for sub-Saharan Africa: Past failures and future prospects. Journal of African Economies 26, i73–i98. https://doi.org/10.1093/jae/ejx010.

Owino, C.N., Kitaka, N., Kipkemboi, J., Ondiek, R.A., 2020. Assessment of greenhouse gases emission in smallholder rice paddies converted from Anyiko Wetland, Kenya. Frontiers in Environmental Science 8, 80.

Ozesmi, S.L., Bauer, M.E., 2002. Satellite remote sensing of wetlands. Wetlands Ecology and Management 10, 381–402.

Paul, A., Tikwe, K., Nakwe, S., 2020. Cassava and vegetable farming on wet land among farmers in Ibaji Local Government Area, Kogi State, Nigeria. Journal of Agricultural Extension 25, 48–61.

Pelley, J., 2008. Can wetland restoration cool the planet? Environmental Science & Technology 42, 8994–8994.

Peng, K., Jiang, W., Hou, P., Wu, Z., Ling, Z., Wang, X., Niu, Z., Mao, D., 2023. Continental-scale wetland mapping: A novel algorithm for detailed wetland types classification based on time series Sentinel-1/2 images. Ecological Indicators 148, 110113.

Pereira de Castro, A., Breseghello, F., Furtini, I.V., Utumi, M.M., Pereira, J.A., Cao, T.-V., Bartholomé, J., 2023. Population improvement via recurrent selection drives genetic gain in upland rice breeding. Heredity 131, 201–210.

Radočaj, D.; Obhođaš, J.; Jurišić, M.; Gašparović, M., 2020. Global open data remote sensing satellite missions for land monitoring and conservation: A review. Land 9, 402. https://doi.org/10.3390/land9110402.

RAMSAR, 2004. Ramsar Convention. www.ramsar.org. (accessed in 01 May 2014).

Ramsey III, E.W., Nelson, G.A., Sapkota, S.K., 1998. Classifying coastal resources by integrating optical and radar imagery and color infrared photography. Mangroves and Salt Marshes 2, 109–119.

Rapinel, S., Panhelleux, L., Gayet, G., Vanacker, R., Lemercier, B., Laroche, B., Chambaud, F., Guelmami, A., Hubert-Moy, L., 2023. National wetland mapping using remote-sensing-derived environmental variables, archive field data, and artificial intelligence. Heliyon 9.

Rebelo, L.-M., Finlayson, C.M., Nagabhatla, N., 2009. Remote sensing and GIS for wetland inventory, mapping and change analysis. Journal of Environmental Management 90, 2144–2153.

Rodenburg, J., Saito, K., 2022. Towards sustainable productivity enhancement of rice-based farming systems in sub-Saharan Africa. Field Crops Research 287, 108670.

Rodenburg, J., Zwart, S.J., Kiepe, P., Narteh, L.T., Dogbe, W., Wopereis, M.C., 2014. Sustainable rice production in African inland valleys: Seizing regional potentials through local approaches. Agricultural Systems 123, 1–11.

Sader, S.A., Ahl, D., Liou, W.-S., 1995. Accuracy of Landsat-TM and GIS rule-based methods for forest wetland classification in Maine. Remote Sensing of Environment 53, 133–144.

Sahour, H., Kemink, K.M., O'Connell, J., 2021. Integrating SAR and optical remote sensing for conservation-targeted wetlands mapping. Remote Sensing 14, 159.

Scholes, R., 1990. The influence of soil fertility on the ecology of southern African dry savannas. Journal of Biogeography, 415–419.

Schowengerdt, R., 2007. Remote Sensing: Models and Models for Image Processing. Elsevier, New York.

Sibanda, S., Ahmed, F., 2021. Modelling historic and future land use/land cover changes and their impact on wetland area in Shashe sub-catchment, Zimbabwe. Modeling Earth Systems and Environment 7, 57–70.

Siervo, V., Pescatore, E., Giano, S.I., 2023. Geomorphic analysis and semi-automated landforms extraction in different natural landscapes. Environmental Earth Sciences 82, 128.

Simioni, J.P., Guasselli, L.A., de Oliveira, G.G., Ruiz, L.F., de Oliveira, G., 2020. A comparison of data mining techniques and multi-sensor analysis for inland marshes delineation. Wetlands Ecology and Management 28, 577–594.

Singh, G., Das, N.N., Panda, R.K., Mohanty, B.P., Entekhabi, D., Bhattacharya, B.K., 2021. Soil moisture retrieval using SMAP L-band radiometer and RISAT-1 C-band SAR data in the paddy dominated tropical region of India. IEEE Journal of Selected Topics in Applied Earth Observations and Remote Sensing 14, 10644–10664.

Sjöström, J.K., Cortizas, A.M., Nylund, A., Hardman, A., Kaal, J., Smittenberg, R.H., Risberg, J., Schillereff, D., Norström, E., 2023. Complex evolution of Holocene hydroclimate, fire and vegetation revealed by molecular, minerogenic and biogenic proxies, Marais Geluk wetland, eastern Free State, South Africa. Quaternary Science Reviews 314, 108216.

Slagter, B., Tsendbazar, N.-E., Vollrath, A., Reiche, J., 2020. Mapping wetland characteristics using temporally dense Sentinel-1 and Sentinel-2 data: A case study in the St. Lucia wetlands, South Africa. International Journal of Applied Earth Observation and Geoinformation 86, 102009.

Suárez, F., Sarabia, A., Sanzana, P., Latorre, C., Muñoz, J.F., 2023. The Quebrada Negra wetland study: An approach to understand plant diversity, hydrology, and hydrogeology of high-Andean wetlands. Wiley Interdisciplinary Reviews: Water, e1683.

Taddeo, S., Dronova, I., 2018. Indicators of vegetation development in restored wetlands. Ecological Indicators 94, 454–467.

Thamaga, K.H., Dube, T., Shoko, C., 2022. Advances in satellite remote sensing of the wetland ecosystems in sub-Saharan Africa. Geocarto International 37, 5891–5913.

Thenkabail, P., Biradar, C., Noojipady, P., Dheeravath, V., Li, Y., Velpuri, M., Gumma, M., Reddy, G., Turral, H., Cai, X., Vithanage, J., Schull, M., Dutta, R., 2009a. Global irrigated area map (GIAM) for the end of the last millennium derived from remote sensing. International Journal of Remote Sensing 30(14), 3679–3733.

Thenkabail, P., Hanjra, M., Dheeravath, V., Gumma, M., 2010. A holistic view of global croplands and their water use for ensuring global food security in the 21st century through advanced remote sensing and non-remote sensing approaches. Remote Sensing 2, 211–261.

Thenkabail, P., Lyon, J.G., Turral, H., Biradar, C., 2009b. Remote Sensing of Global Croplands for Food Security (1st ed.). CRC Press, Boca Raton, FL. https://doi.org/10.1201/9781420090109

Thenkabail, P.S., Nolte, C., 1996. Capabilities of Landsat-5 Thematic Mapper (TM) data in regional mapping and characterization of inland valley agroecosystems in West Africa. International Journal of Remote Sensing 17, 1505–1538.

Thenkabail, P.S., Nolte, C., 1995a. Regional characterization of inland valley agroecosystems in save, Bante, Bassila, and Parakou regions in South-central Republic of Benin. Inland Valley Characterization Report 1.

Thenkabail, P.S., Nolte, C., 1995b. Mapping and Characterising Inland Valley Agroecosystems of West and Central Africa: A Methodology Integrating Remote Sensing, Global Positioning System, and Ground-Truth Data in a Geographic Information Systems Framework. RCMD Monograph No.16, International Institute of Tropical Agriculture, Ibadan, Nigeria. p. 62.

Thenkabail, P.S., Nolte, C., 2000. Regional characterisation of inland valley agroecosystems in West and central Africa using high-resolution remotely sensed data. (Book Chapter # 8, pp. 77–99). In: John G. Lyon (ed) The Book Entitled: GIS Applications for Water Resources and Watershed Management. Taylor and Francis, London and New York, pp. 266.

Thenkabail, P.S., Nolte, C., Lyon, J.G., 2000a. Remote sensing and GIS modeling for selection of a benchmark research area in the inland valley agroecosystems of West and Central Africa. Photogrammetric Engineering and Remote Sensing 66, 755–768.

Thenkabail, P.S., Smith, R.B., De Pauw, E., 2000b. Hyperspectral Vegetation Indices and Their Relationships with Agricultural Crop Characteristics. Remote Sensing of Environment 71, 158–182.

Thenkabail, P.S., Stucky, N., Griscom, B.W., Ashton, M.S., Diels, J., van der Meer, B., Enclona, E., 2004. Biomass estimations and carbon stock calculations in the oil palm plantations of African derived savannas using IKONOS data. International Journal of Remote Sensing 25(23), 5447–5472. https://doi.org/10.1080/01431160412331291279.

Tiner, R.W., 2009. Global distribution of wetlands. In: Likens, G.E. (ed.) Encyclopedia of Inland Waters. Academic Press, Oxford, Cambridge, pp. 526–530. https://doi.org/10.1016/B978-012370626-3.00068-5.

Topp, S.N., Pavelsky, T.M., Jensen, D., Simard, M., Ross, M.R.V., 2020. Research trends in the use of remote sensing for inland water quality science: Moving towards multidisciplinary applications. Water 12, 169. https://doi.org/10.3390/w12010169.

Tschora, H., Cherubini, F., 2020. Co-benefits and trade-offs of agroforestry for climate change mitigation and other sustainability goals in West Africa. Global Ecology and Conservation 22, e00919.

Tucker, C.J. 1979. Red and photographic infrared linear combinations for monitoring vegetation. Remote Sensing of Environment 8(2), 127–150.

Wagner, W., Bloschl, G., Pampaloni, P., Calvet, J.-C., Bizzarri, B., Wigneron, J.-P., Kerr, Y., 2007. Operational readiness of microwave remote sensing of soil moisture for hydrologic applications. Nordic Hydrology 38, 1–20.

Wallace, C.S., Marsh, S., 2005. Characterizing the spatial structure of endangered species habitat using geostatistical analysis of IKONOS imagery. International Journal of Remote Sensing 26, 2607–2629.

WARDA, 2003. Strategic Plan 2003–2012. Bouaké, Côte d'Ivoire, WARDA The Africa Rice Center, p. 56.

WARDA, 2006. Medium Term Plan 2007–2009. Charting the Future of Rice in Africa. Africa Rice Center (WARDA), Cotonou, Republic of Benin.

Wdowinski, S., Kim, S.-W., Amelung, F., Dixon, T.H., Miralles-Wilhelm, F., Sonenshein, R., 2008. Space-based detection of wetlands' surface water level changes from L-band SAR interferometry. Remote Sensing of Environment 112, 681–696.

Wingard, G.L., Bernhardt, C.E., Wachnicka, A.H., 2017. The role of paleoecology in restoration and resource management—the past as a guide to future decision-making: Review and xample from the Greater Everglades Ecosystem, USA. Frontiers in Ecology and Evolution 5, 11.

Woodhouse, P., 2009. Technology, environment and the productivity problem in African agriculture: Comment on the world development report 2008. Journal of Agrarian Change 9, 263–276.

Wright, C., Gallant, A., 2007. Improved wetland remote sensing in Yellowstone National Park using classification trees to combine TM imagery and ancillary environmental data. Remote Sensing of Environment 107, 582–605.

Xia, N., Du, E., de Vries, W., 2024. Impacts of nitrogen deposition on soil methane uptake in global forests. In: Atmospheric Nitrogen Deposition to Global Forests. Academic Press, Cambridge, pp. 157–168.

Xie, Z., Hu, J., Kang, X., Duan, P., Li, S., 2021. Multilayer global spectral—spatial attention network for wetland hyperspectral image classification. IEEE Transactions on Geoscience and Remote Sensing 60, 1–13.

Xing, H., Niu, J., Feng, Y., Hou, D., Wang, Y., Wang, Z., 2023. A coastal wetlands mapping approach of Yellow River Delta with a hierarchical classification and optimal feature selection framework. Catena 223, 106897.

Yang, Z., Bai, J., Zhang, W., 2021. Mapping and assessment of wetland conditions by using remote sensing images and POI data. Ecological Indicators 127, 107485.

Zenna, N., Senthilkumar, K., Sie, M., 2017. Rice production in Africa. Rice Production Worldwide, 117–135.

Zhang, C., Xiao, X., Wang, X., Qin, Y., Doughty, R., Yang, X., Meng, C., Yao, Y., Dong, J., 2024. Mapping wetlands in Northeast China by using knowledge-based algorithms and microwave (PALSAR-2, Sentinel-1), optical (Sentinel-2, Landsat), and thermal (MODIS) images. Journal of Environmental Management 349, 119618.

Zhang, X., Liu, L., Zhao, T., Chen, X., Lin, S., Wang, J., Mi, J., Liu, W., 2022. GWL_FCS30: Global 30 m wetland map with fine classification system using multi-sourced and time-series remote sensing imagery in 2020. Earth System Science Data Discussions, 1–31.

Zhuang, Q., Melack, J.M., Zimov, S., Walter, K.M., Butenhoff, C.L., Khalil, M.A.K., 2009. Correction to "Global methane emissions from wetlands, rice paddies, and lakes". EOS Transactions 90, 92–92.

Part V

Water Use and Water Productivity

10 Remote Sensing of Evapotranspiration from Croplands

Trent W. Biggs, Pamela L. Nagler, Anderson Ruhoff, Triantafyllia Petsini, Michael Marshall, Stefanie Kagone, Gabriel B. Senay, George P. Petropoulos, Camila Abe, and Edward P. Glenn

ACRONYMS, ABBREVIATIONS AND VARIABLES

Variable	Description
a	Intercept in linear model relating T_R to T_1-T_2 (SEBAL, METRIC)
a_{ds}	Empirical coefficient in downscaling model (Kustas et al, 2003)
a_k	Coefficient in empirical coefficient method (Eq 8)
$a_{p,q}$	Coefficients in the T_R-VI-SVAT model (Carlson 2007)
α	Broadband, blue-sky albedo
α_{PT}	Priestley-Taylor coefficient
b	Slope of linear model relating T_R to T_1-T_2 (SEBAL, METRIC)
b_{ds}	Empirical coefficient in downscaling model (Kustas et al, 2003)
b_k	Coefficient in empirical coefficient method (Eq 8)
c	Temperature correction factor (SSEBop)
c_{ds}	Empirical coefficient in downscaling model (Kustas et al, 2003)
c_k	Coefficient in empirical coefficient method (Eq 8)
c_L	Mean potential stomatal conductance per unit leaf area (Mu et al, 2011)
C_p	Specific heat capacity of air
C_{rad}	Adjustment factor for sloped surfaces (Allen et al, 2007)
CWSI	crop water stress index = 1-ET_f
D	Vapor pressure deficit
d	Zero-plane displacement height ~ 2/3 h
dT	Difference between radiometric surface temperature (T_R) and air temperature (T_2).
e	Vapor pressure
e_{sat}	Saturated vapor pressure
ET	Actual evapotranspiration
ET_c	Potential evapotranspiration of a given crop
ET_f	Reference ET fraction
ET_o	Potential evapotranspiration of a grass reference crop
ET_{o24}	ET_o for a 24-hour period
$ET_{c\ adj}$	Actual ET, or ET_c under non-standard conditions in the FAO-56 method
EVI	Enhanced vegetation index
ε_o	Broad-band surface emissivity
-f	FANO constant (SSEBop)

f_c	Vegetation cover fraction
f_g	Green canopy fraction (Fisher et al, 2008)
f_M	Plant moisture constraint (Fisher et al, 2008)
FPAR	Photosynthetically active radiation
f_{SM}	Soil moisture constraint (Fisher et al, 2008)
f_T	Temperature constraint to ET (Fisher et al, 2008)
f_{wet}	Relative surface wetness (Fisher et al, 2008)
F_{wet}	Water cover fraction (Mu et al, 2011).
G	Ground heat flux
γ	Psychometric constant
γ^s	Surface psychrometric constant (Senay 2018)
H	Sensible heat flux
h	Vegetation height
η	Coefficient in empirical crop coefficient method
K_c	Crop coefficient in FAO-56 method
K_s	Soil moisture stress coefficient in FAO-56 method
LAI	Leaf area index
LST	Land surface temperature, equivalent to T_R
LW↑	Upwelling longwave radiation
LW↓	Downwelling longwave radiation
Λ	Evaporative fraction
Λ_{24}	Λ for 24-hour period
Λ_d	Λ for daylight hours
Λ_{op}	Λ at time of overpass
λ	Latent heat of vaporization
λE_I	Latent heat flux from evaporation from wet canopy leaf surfaces
λE_s	Latent heat flux from evaporation from the soil surface
λE_{SP}	Potential latent heat flux from soil evaporation (Mu et al, 2011)
λET	Latent heat flux
λET_c	Latent heat flux from transpiration
$m(D)$	Multiplier limiting stomatal conductance by D (Mu et al, 2007)
$m(Tmin)$	Multiplier limiting stomatal conductance by minimum air temperature (Mu et al, 2007)
Mo	Soil moisture
NDVI*	Normalized NDVI
NDVI	Net difference vegetation index
NDVIo	NDVI for bare soil
NDVIs	NDVI for dense vegetation
NDVImax	Maximum NDVI value
nRMSE	Normalized root mean squared error
Ω	Index of degree of clumping (ALEXI)
σ	Stefan-Boltzmann constant
r_a	Aerodynamic surface resistance (Penman-Monteith)
r_{a_s}	Aerodynamic resistance at the soil surface (Mu et al, 2011)
R_{ah}	Aerodynamic resistance to turbulent heat transport between z_1 and z_2 (SEBAL)
RH	Relative humidity
RMSD	Root mean squared difference
RMSE	Root mean squared error
R_n	Net radiation
R_{n24}	Net radiation over 24-hour period
R_{ns}	Net radiation at the soil surface

Rs	Heat exchange resistance of the soil surface
r_s	Resistance of the land surface or plant canopy to ET
r_{s_c}	Dry canopy resistance to transpiration (Mu et al, 2011)
r_{s_wetC}	Wet canopy resistance to evaporation
Rx	Bulk boundary layer resistance of the canopy in two-source energy balance models
ρ	Air density
s	Slope of the saturation vapor pressue versus temperature curve
SAVI	Soil-adjusted vegetation index
SW↓	Incoming shortwave radiation
T_1	Aerodynamic temperature of the evaporating surface at height z1
T_{1c}	Vegetation canopy temperature (ALEXI)
T_{1s}	Soil temperature (ALEXI)
T_2	Air temperature at height z_2
T_c	Theoretical temperature under cool/moist conditions (SSEBop)
T_h	Theoretical temperature under hot/dry conditions (SSEBop)
T_R	Radiometric surface temperature, equivalent to LST
$\widehat{T}_{Rhi}$	Predicted T_R at high spatial resolution (Kustas et al, 2003)
$\widehat{T}_{Rlow}(NDVI_{low})$	Predicted radiometric temperature using low resolution NDVI
$\widehat{T}_{Rlow}(NDVI_{hi})$	Predicted radiometric temperature using high-resolution NDVI (Kustas et al, 2003)
T_{Rmax}	Minimum T_R over vegetation
T_{Rmin}	Maximum T_R over bare soil
T_{scaled}	Scaled T_R
θ	View angle
VI	Vegetation index
VI_{max}	VI value when ET is maximum
VI_{min}	VI value for bare soil
z_1	Height above the ground surface of the evaporating surface, = $d+z_{0m}$
z_2	Height at which air temperature is measured (often 2 or 3 m)
z_{om}	Surface roughness for momentum transport, ~ 0.03h – 0.123h

ORGANIZATIONS, SATELLITE, AND MODEL ACRONYMS

ABL	Atmospheric boundary layer
AGRIMET	AGRicultural METeorological modeling system
ALEXI	Atmosphere-Land Exchange Inverse model (Anderson et al, 1997)
ASTER	Advanced Spaceborne Thermal Emission and Reflection Radiometer
AVHRR	Advanced Very High Resolution Radiometer
CERES	Clouds and Earth's Radiant Energy System
CESTEM	CubeSat Enabled Spatio-Temporal Enhancement Method
CHIME	Copernicus Hyperspectral Imaging Mission for the Environment
CHRIS	Compact High Resolution *Imaging* Spectrometer
CLM	Community Land Model
CONUS	Conterminous United States
CSIRO	Commonwealth Scientific and Industrial Research Organization, Australian Government
DAIS	Digital Airborne Imaging Spectrometer
DisALEXI	Disaggregation ALEXI model (Norman et al, 2003)
DSTV	diurnal surface temperature variation
DTD	Dual-temperature-difference
ECMWF	European Centre for Medium-range Weather Forecasts

ECOSTRESS	ECOsystem Spaceborne Thermal Radiometer Experiment on Space Station
EnMAP	Environmental Mapping and Analysis Program
EO	Earth observation
ERA	ECMWF atmospheric reanalysis
ESA	European Space Agency
FANO	Forcing and Normalizing Operation
FAO	Food and Agriculture Organization
FAO-56	FAO evapotranspiration model (Allen et al., 1998)
FIFE	First ISLSCP (International Satellite Land Surface Climatology Project) Field Experiment
FLUXNET	Global network of micrometeorological flux tower sites
GDAS	Global Data Assimilation System
GEE	Google Earth Engine
GG model	Granger and Gray (GG) model (Granger & Gray, 1989)
GLDAS	Global Land Data Assimilation System
GLEAM	Global Land-surface Evaporation Amsterdam Methodology
GMAO	Global Modeling and Assimilation Office
GOES	Geostationary Operational Environmental Satellites
HLS	Harmonized Landsat and Sentinel-2
LSA-SAF	Land Surface Analysis Satellite Applications Facility
LSM	Land surface model
MERRA	Modern Era Retrospective Analysis for Research and Applications
METRIC	Mapping EvapoTranspiration at high Resolution with Internalized Calibration
MMR	Modular Multispectral Radiometer
MOD16	MODIS ET product, input from the Terra satellite (1030am local time)
MOD43B3	MODIS albedo product
MODIS	Moderate Resolution Imaging Spectroradiometer
MSG	Meteosat Second Generation satellite
MYD16	MODIS ET product, input from the Aqua satellite (130pm local time)
NCEP-NCAR	National Centers for Environmental Prediction–National Center for Atmospheric Research
NOAA	National Oceanic and Atmospheric Administration
NWS	National Weather Service
PBMR	Push Broom Microwave Radiometer
PML-V2	Penman-Monteith-Leuning version 2 model (Zhang *et al.*, 2019)
PoLDER	Polarization and Directionality of Earth Reflectance instrument
PRISMA	Hyperspectral Precursor of the Application Mission
PT-JPL	Priestley-Taylor Jet Propulsion Laboratory model (Fisher et al, 2008)
RMSD	Root mean square difference
RTM	Radiative Transfer Model
SBG	Surface Biology and Geology mission, NASA
SEBAL	Surface energy balance algorithm (Bastiaanssen et al, 1998)
SEBS	Surface energy balance system (Su, 2002)
SEN2FLEX	SENtinel-2 and FLuorescence EXperiment
SEVIRI	Spinning Enhanced Visible and Infrared Imager
SGP	Southern Great Plains
Sim-ReSET	Simple Remote Sensing Evapotranspiration Model
SIMS	Satellite Irrigation Management Support from NASA
SMACEX	Soil Moisture–Atmosphere Coupling Experiment
SRB	Surface radiation budget
SSEB	Simplified Surface Energy Balance

S-SEBI	Simplified Surface Energy Balance Index
SSEBop	Operational Simplified Surface Energy Balance
STARFM	Spatial and Temporal Adaptive Reflectance Fusion Model
SVAT	Soil Vegetation Atmosphere Transfer model
TESSEL	Tiled ECMWF Surface Scheme for Exchange Processes over Land
TIMS	Tropospheric Infrared Mapping Spectrometers
TMS	Thematic Mapper Simulator
TRMM	Tropical Rainfall Measurement Mission
TSEB	Two source energy balance model
VIC	Variable infiltration capacity model
VIIRS	Visible Infrared Imaging Radiometer Suite
VMC	Vegetation and moisture coefficient, equivalent to ET_f
WaPOR	Water Productivity Open-access portal

10.1 INTRODUCTION

Climate change, population growth, dietary changes, and economic growth all contribute to crises in water resources (Heller *et al.*, 2021; Mbow *et al.*, 2019). Evapotranspiration (ET), defined as the loss of liquid water from the land surface to the atmosphere via evaporation from wet surfaces (moist soil, wet vegetation, or open water) and transpiration through the stomata of leaves, is the single most important term in global and river basin water balances after precipitation. ET exchanges mass and energy among the hydrosphere, biosphere, and atmosphere, playing a critical role in water cycles, the surface energy balance (Sellers *et al.*, 1996; Foley et al., 2005) and regional circulation patterns (Lee *et al.*, 2009, 2011). Quantitative information on ET is also important for understanding the processes that control ecosystem CO_2 exchange (Scott *et al.*, 2006) as well as the interactions between parameters in different ecosystem processes (Wever *et al.*, 2002). Better understanding of the physical processes controlling evapotranspiration (ET) from croplands is of paramount importance for the conservation of water resources (Petropoulos *et al.*, 2016; Foley et al., 2011). At field, watershed, and regional scales, determining ET and its accuracy is of crucial importance for water resources management (Bastiaanssen *et al.*, 2005; Petropoulos *et al.*, 2021). Measurements of ET are required for monitoring plant water requirements, plant growth, and productivity, as well as for irrigation management and deciding when to carry out cultivation procedures (e.g., Yang *et al.*, 2010; Shi and Liang, 2014; Deng *et al.*, 2019). ET has a dominant role in the hydrological cycle as well as in biogeochemical cycles such as the carbon cycle, all of which are changing with a changing climate (Yang *et al.*, 2023). ET is of key importance for the understanding and management of agriculture, water resources (Petropoulos *et al.*, 2016; Srivastava *et al.*, 2013a, 2015), and regional climate (Jung *et al.*, 2010; Srivastava *et al.*, 2015; Petropoulos and Hristopulos, 2020).

Several types of ground instrumentation can be used for measuring ET from croplands, including eddy flux correlation towers, atmometers, and lysimeters (Glenn *et al.*, 2007; Jarchow *et al.*, 2022; Petropoulos *et al.*, 2013). At continental and regional scales, ground-based techniques cannot estimate ET at high spatial resolution due to high costs and limited spatial coverage. Ground-based monitoring networks have been established that collect data for individual sites around the world (Dorigo *et al.*, 2021); a global inventory of eddy flux correlation towers (Baldocchi *et al.*, 2001; Pastorello *et al.*, 2017) reports data for ~1500 site-years at 212 unique sites, though most towers are in the United States and Europe, and there are no towers in some countries where knowledge of ET is critical to water management (Jung *et al.*, 2009).

Earth observation (EO) approaches, including imagery from aerial and satellite platforms, can help estimate ET at a variety of spatial and temporal scales. EO technology is the best alternative to ground-based measurements for obtaining estimates of ET at the spatiotemporal scales and accuracy levels required by many applications (Melesse et al., 2008; Tian et al., 2014). Previous reviews of the use of EO data for ET estimation are available (Kustas and Norman, 1996; Courault et al., 2005; Glenn et al., 2007, 2008a, 2010; Kalma et al., 2008; Verstraeten et al., 2008; González-Dugo et al., 2009; Li et al., 2009; Petropoulos, 2013; Liou and Kar, 2014; Zhang et al., 2016; Chen and Liu, 2020; García-Santos et al., 2022; FAO, 2023; Tran et al., 2023), including reviews on the use of uncrewed aerial systems (UAS) (Niu et al., 2020).

The present chapter provides a critical and systematic overview of different modeling approaches to estimate ET from EO data, with a focus on applications in agriculture. First, we review methods for estimating net radiation (R_n), which is required by all ET modeling techniques described in the chapter. In agricultural applications, special consideration is often required to calculate outgoing radiation at sufficiently high spatial and temporal resolution to capture spatial variability over heterogeneous land surfaces. We then discuss four major families of methods used to calculate ET using EO data: vegetation index, radiometric land surface temperature (T_R), scatterplot inversion, and land surface model (LSM). T_R is often called the land surface temperature (LST), which is essentially the atmospherically corrected surface temperature measured by the sensor (T_s). We use the term radiometric land surface temperature and the symbol T_R to emphasize that it is the temperature expression of thermal emissions detected at the sensor.

We use common mathematical symbols for each method to facilitate inter-comparison and to highlight similarities in the conceptual foundations of the various methods (please see the list of acronyms, abbreviations, and variables). Sufficient detail is provided to implement and compare some of the most commonly used algorithms, and suggestions for the application of each method are also given, discussing any special issues related to estimating ET in agricultural landscapes. The accuracy of the methods is then compared, special problems in application of the methods to crops are discussed, and future research directions are highlighted. The spatial resolution of various datasets is indicated by the size of one pixel on the ground (e.g., 30 m or 1 km). Several well-established algorithms are being implemented in cloud computing platforms, and ensembles of methods are being integrated in cloud-based platforms with domains covering the United States (Melton *et al.*, 2021). With automation and code-sharing platforms, individual researchers no longer have to write their own ET algorithms, and with cloud-based processing the user does not have to download imagery, making large-scale (continental to global) applications possible. Automated ET algorithms in cloud-based environments will allow for significant advances in algorithm inter-comparison, error documentation, algorithm improvement, and understanding of the spatiotemporal structure and changes in ET at continental to global scales. The chapter provides the theoretical and mathematical foundations of ET algorithms, allowing researchers to be informed users of automated algorithms.

10.2 OVERVIEW OF METHODS FOR ET CALCULATION USING REMOTE SENSING

EO sensors do not directly measure ET or its latent energy equivalent (λET), where λ is the latent heat of vaporization of water at a given temperature. Several algorithms have been developed for estimating ET using either space- or airborne systems, which can be broadly grouped into four basic categories (Table 10.1): (1) vegetation index, (2) radiometric land surface temperature; (3) triangle/trapezoid or scatterplot inversion; and (4) land surface model (LSM) methods. Table 10.1 summarizes the advantages and disadvantages of the first three; LSM methods are described separately.

Various terms are used to describe the water demand of the atmosphere and the actual use of water by crops (Allen *et al.*, 1998). Potential evapotranspiration (PET) is the amount of water lost to evaporation from the soil surface, wet canopy, and open water surfaces, and from transpiration from the canopy under well-watered conditions. PET is a function of R_n, temperature, wind velocity,

TABLE 10.1
Three Families of Methods to Estimate Evapotranspiration from Earth Observation Data, and Their Advantages, Disadvantages, Error, and Recommended Uses

Method of ETEstimation	Advantages	Disadvantages	Recommended for	Not recommended or Untested for	Error	Reference
1. Vegetation-based models						
Empirical	- Ease of implementation	- Requires ground data for calibration; does not directly estimate soil evaporation	- Small geographic regions with ground reference data, riparian and agricultural vegetation, desert, semi-arid rangelands, vegetated wetlands	Regional or global application without reference data for calibration	Mean error 0.12 for ETf and 5% error for annual ET from flux towers and soil moisture balance (Nagler et al., 2013) RSMD 10–30% compared to mean ET (Glenn et al., 2010) Mean error 4% compared with daily ET from eddy covariance towers (Samani et al., 2009); Mean error 15% compared with daily ET (Duchemin et al., 2006); RMSE 15% compared with daily ET (Hunsaker et al., 2007a, 2007b).	Power function: (Nagler et al., 2013); Extreme VI values: (Glenn et al., 2010); Beer-Lambert Law: (Nagler et al., 2013); CropCoeff-VI: (Bausch and Neale, 1987; Choudhury et al., 1994) Linear regression on NDVI, SAVI: (Samani et al., 2009; Campos et al., 2010); Desert: (Glenn et al., 2008b); Wetlands: Glenn et al., 2013
PT-JPL	- Relative ease of implementation; Minimal ground data requirements; Operational globally; Does not require calibration	- May underestimate soil evaporation from irrigated areas in dry climates	Global and regional rainfed systems	Needs further testing in irrigated agriculture with high soil evaporation	Mean error 13% compared with mean annual ET from eddy covariance towers (Fisher et al., 2008)	(Fisher et al., 2008); Alternate formulation: (Marshall et al., 2013)
MOD16	- Minimal ground data requirements; Operational globally; Does not require calibration	- May underestimate soil evaporation from irrigated areas in dry climates	Global and regional rainfed systems	Needs further testing in irrigated agriculture with high soil evaporation	RMSE ~20%, MAB 24–25%, compared with daily ET from eddy covariance towers (Mu et al., 2011); ~50% compared with ET at regional or national scale (Velpuri et al., 2013)	(Leuning et al., 2008; Mu et al., 2011, 2007, 2013; Nishida and Nemani, 2003)

(Continued)

TABLE 10.1 (Continued)
Three Families of Methods to Estimate Evapotranspiration from Earth Observation Data, and Their Advantages, Disadvantages, Error, and Recommended Uses

Method of ETEstimation	Advantages	Disadvantages	Recommended for	Not recommended or Untested for	Error	Reference
\multicolumn{7}{l}{2. Radiometric land surface temperature–based models}
Overall					RMSD < 50 W/m² and < 33 W/m² compared with instantaneous λET and H fluxes respectively (González-Dugo et al., 2009)	González-Dugo et al. (2009)
SEBAL, METRIC	- Minimal ground data; Accurate in semi-arid environments with irrigation; Accurate for wet soil and inundated surfaces	- Moderate complexity; Requires extremes of wet and dry to be present in a scene, and calibration for each image; Issues in merging across scenes	Routine application in irrigated agriculture in semi-arid and arid climates; Single scenes	Humid environments; Global applications; Data scarce regions	RMSE 15–20% compared to daily ET from eddy covariance towers (Allen et al., 2011a); RMSE ~5% compared to seasonal ET estimates (Bastiaanssen et al., 2005)	(Bastiaanssen et al., 2002, 1998; Allen et al. 2007)
ALEXI, DisALEXI	- Minimal ground data; Accurate in a variety of vegetation types and climates; Accurate for wet soil and inundated surfaces	High complexity of implementation; Requires two images for each daily ET estimate; Coarse resolution before downscaling	Regional and global applications, operational data product	Routine application for applied irrigation systems or management; Data scarce regions	RMSD 40 W/m², MAD 30 W/m², R² 0.77 (Anderson et al., 1997) RMSD 40–50 W/m² (Norman et al., 2003) MAD 15–20% for 30min avg, 10% daily, and ~5% seasonal (Anderson et al., 2013)	ALEXI: (Anderson et al., 1997); DisALEXI: (Norman et al., 2003);

Remote Sensing of Evapotranspiration from Croplands

SSEB, SSEBop	Minimal ground data; Ease of implementation; Operational application over large areas	Potentially inaccurate in regions with high spatial variability in albedo or elevation; Requires in situ meteorological data; Does not solve energy balance completely; Physical processes of ET are not fully represented	Regional and global applications, irrigated agriculture	Heterogeneous vegetation, mountainous regions; Regions with high albedo and high emissivity; Data scarce regions	Mean error < 30% compared to eddy covariance towers (Senay et al., 2007); Mean error ~60% compared with 60 eddy covariance towers pooled across CONUS (Velpuri et al., 2013)	SSEB: (Senay et al., 2007); SSEBop: (Senay et al., 2013);

3. Scatterplot or triangle methods

Overall	Ease of implementation, few parameters (except SVAT method)	Require extremes of wet and dry to be present in a scene; subjectivity in selecting wet/dry pixels; issues in merging across scenes for large areas	Routine application in irrigated agriculture in semi-arid and arid climates	Humid environments; Global applications		(Petropoulos et al., 2009)
T_R–VI methods	Computationally straightforward; Relative independence from site-specific tuning of model parameter	Assumes linear relationships between location in T_R–VI space and ET; Clouds, standing water, and sloping terrain need to be masked	Local to regional scale sites	Homogenous land cover	RSMD ~30% compared to daily ET (Jiang and Islam, 2001); RMSD 45 W/m^{-2}, bias of 5.6 W/m^{-2}, R^2 =0.86, compared to daily λET from eddy covariance towers (Nishida, 2003; Nishida and Nemani, 2003);	(Jiang and Islam, 2001; Nishida, 2003; Nishida and Nemani, 2003; Tang et al., 2010; Zhang et al., 2006)

(Continued)

TABLE 10.1 (CONTINUED)
THREE FAMILIES OF METHODS TO ESTIMATE EVAPOTRANSPIRATION FROM EARTH OBSERVATION DATA, AND THEIR ADVANTAGES, DISADVANTAGES, ERROR, AND RECOMMENDED USES

Method of ETEstimation	Advantages	Disadvantages	Recommended for	Not recommended or Untested for	Error	Reference
T_R-T_2 difference and VI scatterplot	Insensitive to absolute accuracy of T_R. Requires small number of in situ observations	Often applicable for homogenous areas (Moran et al., 1994)	Areas of partial vegetation cover		RMSD 29 W/m^{-2} compared to instantaneous λET (Moran et al., 1996); RMSD 59 W/m^{-2}, bias −42 W/m^{-2} compared to λET from eddy covariance towers (Jiang and Islam, 2003) RMSD 0.08 to 0.19 and R^2 0.4–0.7 in λ from MODIS and AVHRR (Venturini et al., 2004);	(Moran et al., 1994)
T_R-albedo scatterplot (moved up in table because it is above Day-Night Tr in text)	Requires small number of in situ observations; Realistic assumption (extreme T for the wet and dry conditions vary with changing surface reflectance)	Requirement to identify extreme points in the scatterplot domain	- Operational products in small geographic areas		Error 90 W/m^{-2} for instantaneous ET and 1mm/d daily ET compared to lysimeters (Gómez et al., 2005; Sobrino, et al., 2005) RMSD 64 W m^{-2}, R^2 0.85 compared to daily ET Zahira et al. (2009); RMSE 25–32% compared to mean daily ET from Bowen ratio tower (Bhattacharya et al., 2010);	S-SEBI: (Roerink et al., 2000)

Day-night T_R difference and VI scatterplot	Requires small number of in situ observations	Assumes three dominant land cover types; Requires both day- and nighttime observations	Areas with three dominant land cover types	Areas with mixed land cover types	Errors 2.8–3.9% compared to daily ET from lysimeters (Chen et al., 2002); RMSD 0.106, bias –0.002, R^2 0.61 in Λ (Wang et al., 2006)	(Chen et al., 2002; Wang et al., 2006)
Triangle-SVAT model	Non-linear interpretation of T_R/VI feature space; Potential for deriving additional parameters (soil surface moisture, daytime mean λET and H fluxes)	Large number of input parameters; Requires user expertise	Operational products; Global- and regional-scale sites	Homogenous land cover	RMSD ±10% for daily ET from FIFE/MONSOON (Gillies et al., 1997); Error ~ 15–50% for λET from in situ and airborne measurements (Brunsell and Gillies, 2003); RMSD 40 W/m^{-2} compared to λET from CarboEurope (Petropoulos and Carlson, 2011)	(Carlson, 2007)

humidity, and vegetation characteristics. PET can vary with leaf area, stage of development, photosynthetic pathway (C3 or C4), and rooting depth, so a second term, reference ET (ET$_o$) is defined as PET of a specific reference crop, often a well-watered grass with specific characteristics (see Section 10.2.2). Crop ET for a given vegetation or crop type without soil moisture limitation ("standard conditions") is indicated by the variable ET$_c$ (Allen *et al.*, 1998). Finally, actual ET is the amount of water that is lost via both evaporation from the soil surface and transpiration from a specific vegetation cover under actual field conditions, including limitations to ET caused by soil moisture stress, nutrient limitation, and pathogens. In Allen *et al.* (1998) actual ET is called ET$_c$ under non-standard conditions (ET$_{c\ adj}$). In this chapter we use the symbol ET and the term evapotranspiration to represent actual evapotranspiration of a given land surface (Glenn *et al.*, 2011a). The latent heat flux (λET, W m^{-2}) is the product of ET and λ. It is sometimes used instead of ET because it is a rate that can be expressed for a given instant, is reported by several field techniques, and is a key term in both the water and energy balance of the land surface. We use the terms ET or λET depending on the method being described.

10.2.1 Net Radiation

EO-based methods for estimating ET, including all methods reviewed in this chapter, depend on accurate determination of net radiation (R_n, W m^{-2}), which is calculated as:

$$R_n = (1-\alpha)SW\downarrow + (LW\downarrow - LW\uparrow) \tag{1}$$

where α is broadband blue-sky albedo (dimensionless), $SW\downarrow$ is incoming shortwave radiation, $LW\downarrow$ is downwelling longwave radiation, and $LW\uparrow$ is upwelling longwave radiation (all in W m^{-2}). For field-scale applications, R_n is often estimated with meteorological data alone using a variety of methods reviewed in several publications (e.g., Allen *et al.*, 1998).

R_n can be calculated from well-established approaches based on primarily EO data. Liang *et al.* (2013) and Liang *et al.* (2019) review methods for estimating the radiation budget from EO sensors, including operationally distributed products. $SW\downarrow$ and $LW\downarrow$ depend on atmospheric properties that can be estimated accurately with coarse-resolution datasets (10km—1°). α and $LW\uparrow$ depend on surface conditions, including reflectivity and temperature, which can be highly spatially variable. Next we review global datasets for $SW\downarrow$ and $LW\downarrow$ and other methods to estimate α and $LW\uparrow$.

10.2.1.1 Regional and Global Datasets for Net Radiation

At regional and global scales, R_n can be estimated with gridded data from surface climate reanalysis that assimilate remote sensing data or from EO data alone, with errors of ±10–20% compared to ground measurements (Bisht *et al.*, 2005; Bisht and Bras, 2010; Mira *et al.*, 2016). Gridded datasets used for regional- to global-scale R_n estimation can be separated into two spatiotemporal categories: 1979–present (1° spatial resolution or lower) and 2000–present (0.25° spatial resolution and higher). The higher spatial resolution post-2000 is due to the launch of the Moderate Resolution Imaging Spectroradiometer (MODIS) satellites and other EO systems that facilitate the downscaling of the surface energy budget (Gottschalck *et al.*, 2005). Liang *et al.* (2010) and Liang *et al.* (2013) provide good introductions to commonly used R_n datasets and associated uncertainties. For 1979–present, several coarse-resolution and downscaled sources exist. The most commonly used is the Global Energy and Water Cycle Experiment Surface Radiation Budget (SRB; NASA GEWEX Surface Radiation Budget, 2024) (Gupta *et al.*, 1999), which provides three-hourly shortwave and longwave radiation fluxes at 1° resolution. These data are generated primarily from the International Satellite Cloud Climatology Project (Schiffer and Rossow, 1983; Rossow and Schiffer, 1991, 1999) and Global Modeling and Assimilation Office (GMAO, http://gmao.gsfc.nasa.gov/) meteorology. The original dataset covering 1983–2007 has been expanded to cover from 1979 to the present, as part of the Modern Era-Retrospective Analysis for Research and Applications (MERRA; GMAO MERRA, 2024) (Rienecker *et al.*, 2011) dataset. MERRA is updated regularly with remote sensing and observed data and fed through a land surface catchment hydrology model, which provides additional outputs and further reduces inconsistencies. The Global Land Data Assimilation System (GLDAS) (Rodell *et al.*, 2004; NASA GLDAS, 2024) product is also at 1° resolution and assimilates atmospheric fields from the Global Data Assimilation System (GDAS; NOAA GDAS, 2024) maintained by the National Oceanic and Atmospheric Administration (NOAA), precipitation from the Climate Prediction Center Merged Analysis of Precipitation (NOAA CMAP, 2024), and observation-driven shortwave and longwave radiation using the Air Force Weather Agency's Agricultural Meteorological (AgriMet) modeling system (Idso, 1981; Shapiro, 1987; USBR AgriMet, 2024) to parameterize four land surface realizations: (1) Noah land surface model (Chen *et al.*, 1996, 1997), (2) community land model (Bonan, 1998), (3) mosaic (Koster and Suarez, 1996), and (4) variable infiltration capacity model (Liang *et al.*, 1994). The forcing data for GLDAS, like MERRA, are produced at three-hourly intervals at 1° resolution from 1948 to the present (Rienecker *et al.*, 2011). Although at much coarser spatial resolution, two other datasets are commonly used: the National Centers for Environmental Prediction—National Center for Atmospheric Research (NCEP-NCAR, 2024; Kalnay *et al.*, 1996) and the European Centre for Medium-range Weather Forecasts (ECMWF) interim reanalysis (Morcrette, 1991, 2002). NCEP-NCAR shortwave and longwave fluxes are available at six-hourly intervals from 1948 to the present at 2.5° resolution, while the ECMWF atmospheric reanalysis (ERA) shortwave and longwave flux is available at six-hourly intervals spanning 1979–present at 1.5° resolution (Bosilovich, 2008). A new version of ERA (ERA5) is globally available (ECMWF Reanalysis v5, 2024) with spatial resolution of 0.25° and 0.1° over land (ERA5-Land; ECMWF ERA5-Land, 2024) with temporal resolution down to one hour (Muñoz-Sabater *et al.*, 2021; Hersbach *et al.*, 2020). Sheffield *et al.* (2006) uses surface elevation to downscale the NCEP-NCAR to 1° resolution. Other downscaled R_n products are available on specific university sites (e.g., University of Southampton, 2024). Daymet (NASA ORNL Daymet, 2024) interpolates weather station data on solar radiation, vapor pressure, and maximum and minimum air temperature with a spatial resolution of 1 km for 1980–present over North America (Thornton *et al.*, 2022). R_n is not part of Daymet, so it must be estimated with other methods.

For regional estimation of R_n after 2000, EO data are incorporated directly or indirectly into some of the aforementioned reanalysis datasets. These methods involve several assumptions or ground-based estimates of α, T_R, and emissivity for model calibration (Bisht *et al.*, 2005). Since 2000, products from MODIS, including land surface temperature and emissivity (NASA MOD11, 2024) and α (NASA MOD43C3, 2024), have been combined and extrapolated from once-a-day measurements to daily values at a quasi-1km resolution. The Clouds and Earth's Radiant Energy System (CERES) initially aboard the Tropical Rainfall Measurement Mission (TRMM) platform from the National Aeronautics and Space Administration (NASA) and later placed on NASA's Terra and Aqua platforms, is a radiometer that collects solar-reflected, Earth-emitted, and total radiation to determine Earth's radiation budget. Data are available from 2000 to the present at three-hourly, monthly average, or monthly average by hour at 1° resolution (NASA CERES, 2024). Validation results from dos Santos Nascimento *et al.* (2019) indicated very good accuracy for the all CERES radiation components, with significant improvements between versions 3A and 4A. The Land Surface Analysis Satellite Applications Facility (LSA-SAF, 2024) has also developed a radiation budget for Africa and Europe using the Spinning Enhanced Visible and Infrared Imager (SEVIRI) radiometer on board the Meteosat Second Generation (MSG) satellite. The MSG-SEVIRI platform provides 30-minute 3km resolution α, TR (Martins *et al.*, 2019), emissivity, $SW\downarrow$, and $LW\downarrow$ information from 2018 to the present (EUMETSAT CM SAF, 2024). Satellite methods to estimate components of R_n are continually evolving and being evaluated against each other, including for shortwave (Ma and Pinker, 2012; Wang et al., 2019), longwave (Gui et al., 2010), and R_n (Bisht and Bras, 2010), and re-analysis datasets that incorporate satellite imagery will continue to be improved.

ET mapping often uses vegetation indices or T_R data that have a higher spatial resolution (e.g., 30m, 250m, 1km) than is available from the global grids of radiation, but not all parts of the radiation budget need to be downscaled in detail. Incoming shortwave ($SW \downarrow$) and incoming longwave ($LW \downarrow$) are determined primarily by atmospheric properties and are therefore often assumed homogeneous over a given cell in the global gridded products, so they can be taken directly from the gridded data, though some applications (e.g., MOD16; Mu et al., 2011) interpolate to the resolution of the other satellite imagery used to map vegetation indices or radiometric surface temperature in order to avoid abrupt changes at scene boundaries (Mu et al., 2011). MOD16 calculates net longwave as a function of grid-cell average air temperature, which is in turn taken from the global gridded dataset (MERRA) and interpolated to 1km using non-linear interpolation on the four nearest neighbors (Mu et al., 2007).

10.2.1.2 Outgoing Shortwave and Longwave at High Spatial Resolution

In contrast to incoming radiation, $SW\uparrow$ and $LW\uparrow$ depend strongly on land surface properties and so may exhibit significant spatial variation over short distances, particularly in agricultural areas with sharp boundaries in vegetation with different levels of soil moisture stress. Algorithms for calculating α from Landsat imagery are included in several ET estimation models, including the Mapping EvapoTranspiration at high Resolution with Internalized Calibration (METRIC) model (Allen et al., 2007), and an albedo product is directly available for MODIS (NASA MOD43B3, 2024). A review of the methods is also presented in Liang et al. (2013). $LW\uparrow$ can be calculated at high spatial resolution using TR from satellite imagery and an estimate of the surface emissivity. This approach is used in the METRIC (Allen et al., 2007) and Surface Energy Balance Algorithm (SEBAL; Bastiaanssen et al., 1998) models and in many other applications (Bisht et al., 2005; Tang and Li, 2008):

$$LW\uparrow = \varepsilon_o \sigma T_R^4 \tag{2}$$

where ε_o is the broad-band surface emissivity, σ is the Stefan-Boltzmann constant (5.67 x 10^{-8} W m^{-2} K^{-1}), and T_R is the radiometric surface temperature, which is often assumed equal to the surface temperature. Some remote sensing products of T_R include estimates of ε_o, including MODIS, Advanced Spaceborne Thermal Emission and Reflection Radiometer (ASTER), and Landsat collection 2, though there may be significant errors over heterogeneous landscapes (Liang et al., 2013).

10.2.1.3 Available Energy and the Ground Heat Flux

Available energy is the amount of R_n left over after accounting for the ground heat flux (G; $W\ m^{-2}$), and is calculated as $R_n - G$. G is usually close to zero over 24-hour, weekly, or ten-day periods and is assumed to be zero in several methods (e.g., Table 10.2), but G can be significant at the instant of satellite overpass and is particularly important for energy-based methods. Instantaneous G can account for up to 50% of R_n for sparse vegetation and can average 20–30% for normalized difference vegetation index (NDVI) values up to around 0.6 (Bastiaanssen et al., 1998), so it cannot be neglected unless there is a high canopy cover fraction. Instantaneous G can be calculated with a variety of algorithms (Murray and Verhoef, 2007), the simplest of which is to assume that G is a constant fraction of R_n, usually between 0.2 and 0.5 at midday (Choudhury, 1989), with specific model applications using constant values of, for example, 0.35 (Norman et al., 1995) or 0.31 (Anderson et al., 1997). Other models include those where G is a function of NDVI, which is applicable over vegetated areas but not water (Morse et al., 2000):

$$G = 0.30 * \left(1 - 0.98 NDVI^4\right) R_n \tag{3}$$

G can also be estimated by a more complicated empirical equation (Table 10.3) that describes heat transfer using T_R, α, and an extinction factor that describes attenuation of radiation through vegetation canopies using NDVI (Clothier *et al.*, 1986; Choudhury, 1989; Kustas and Daughtry, 1990; Van Oevelen, 1991). More detailed approaches incorporate soil properties (Murray and Verhoef, 2007). Using empirical equations, the error in *G* is often around 20–30% (Petropoulos, 2013). Over water bodies, *G* is usually larger and requires a different equation, often calibrated to local measurements (Morse *et al.*, 2000). The user is recommended to try a few different equations for *G* to test its sensitivity to the equation used.

10.2.2 Vegetation Index–Based Methods for ET Estimation

Vegetation index–based methods to estimate ET use an index of vegetation biomass or leaf area index to calculate crop ET and are reviewed in detail in Glenn *et al.* (2008a). These methods have been applied in both agricultural and natural land covers (reviewed in Glenn *et al.* 2011b) and in the drylands of Australia (reviewed in Glenn *et al.* 2011a). One of the most widely used global ET datasets, MOD16 (Nishida and Nemani, 2003; Mu *et al.*, 2007, 2011; Leuning *et al.*, 2008; Aguilar *et al.*, 2018; Faisol *et al.*, 2020), uses a variant of the Penman–Monteith equation (Mu *et al.*, 2011) to estimate ET from crop canopies and soil surfaces:

$$\lambda ET = \frac{s(R_n - G) + \rho C_p D / r_a}{s + \gamma \left(1 + \frac{r_s}{r_a}\right)} \quad (4)$$

where *s* (Pa K^{-1}) is the slope of the curve relating saturated vapor pressure (e_{sat} in Pa) to air temperature, ρ is air density (kg m^{-3}), C_p is the specific heat capacity of air (J kg^{-1} K^{-1}), D is the vapor pressure deficit (e_{sat}-e, where e is actual vapor pressure in Pa), r_a is the aerodynamic surface resistance (s m^{-1}), γ is the psychrometric constant (~0.066 kPa K^{-1} or as calculated in Mu *et al.*, 2007), and r_s is the resistance of the land surface or plant canopy to ET (s m^{-1}). R_n-G is for the 24-hour period; on a daily average *G* is usually close to zero. Several of the parameters (s, γ, ρ, C_p, e, e_{sat}) are determined from meteorological data or elevation and do not depend on satellite-derived vegetation characteristics. The meteorological inputs are taken from either local meteorological stations or gridded global meteorological datasets, and include air temperature, which is used to calculate e_{sat}; wind speed; and relative humidity, which is used to calculate e. The two main parameters that control ET for different vegetation types and different levels of soil moisture stress are r_a and, often more importantly, r_s. Reference ET (ET_o) is calculated with r_a and r_s parameters for a reference surface, for example, a grass 12 cm tall with r_s of 70 s m^{-1} and albedo 0.23 (Allen *et al.*, 1998). The MOD16 user guide was updated (Running *et al.*, 2017) and used to model agricultural field-scale ET over the continental USA (CONUS) (He *et al.*, 2019).

Alternatively, the Priestley–Taylor equation has been used in global models, particularly the Priestly–Taylor Jet Propulsion Laboratory (PT-JPL) model (Fisher *et al.*, 2008), and in several scatterplot methods (Jiang and Islam, 2001):

$$\lambda ET = \alpha_{PT} \frac{s}{s + \gamma} (R_n - G) \quad (5)$$

where α_{PT} (unitless) is an empirical coefficient. For open water and vegetation without soil moisture limitation, α_{PT} is set to 1.26, though adjustments may be applied in different environments. λET can be converted to ET in mm by dividing by the latent heat of vaporization (λ = 2.45 MJ kg^{-1}) (Allen *et al.*, 1998) and a time constant to convert from mm s^{-1} to mm day^{-1}.

Three basic approaches are used in the application of (4) or (5) to estimate actual ET: (1) empirical crop coefficient methods, (2) physically based coefficient methods, and (3) canopy resistance methods. Each of these three methods is reviewed next.

10.2.2.1 Empirical Vegetation Methods: Crop Coefficients
10.2.2.1.1 Calibration Methods

Crop coefficient methods calculate ET as the product of ET_o and a crop coefficient (K). In the original Food and Agriculture Organization model of crop ET (FAO-56) (Allen *et al.*, 1998; Allen, 2000), under conditions of soil moisture limitation, the coefficient describing the effect of crop type and growth stage (K_c) is multiplied by a coefficient quantifying the effect of soil moisture stress (K_s) (Allen *et al.*, 1998, Eq. 81). Most satellite methods estimate the product ($K_c K_s$), which is also sometimes called the reference evapotranspiration fraction (ET_f). To simplify notation, and because the method is often used for other vegetation types besides crops, here we use the term "reference ET fraction" and the symbol ET_f to represent $K_c K_s$ throughout the chapter. Some references refer to the crop coefficient derived from EO data as K_c or $K_{c\text{-}VI}$ (Glenn *et al.*, 2011b); here we call it ET_f to be consistent with recent publications (Nagler *et al.*, 2013), to highlight similarities with other methods that use ET_f (Allen *et al.*, 2007), and to reduce confusion with K_c from the original FAO-56 method, which is for conditions of no water stress and estimates potential ET for a given crop, and $ET_f = K_c K_s$, which estimates actual ET and is what EO-based crop coefficient methods determine.

In EO-based methods, the crop coefficient is modeled as a function of a vegetation index (VI) using several possible empirical equations (Eqs. 6–8), including as a power function (Nagler *et al.*, 2013):

$$ET_f = a_K VI^\eta \tag{6}$$

where a_K is an empirical coefficient determined by regression, and η is a coefficient that varies by the vegetation index. Alternatively, ET_f can be modeled by first normalizing VI (Glenn *et al.*, 2010):

$$ET_f = \left[1 - (VI_{max} - VI)/(VI_{max} - VI_{min})\right]^\eta \tag{7}$$

where VI_{max} is the VI where ET is at a maximum, and VI_{min} is the VI of bare soil ($VI = 0$). Other studies scaled VI as $(VI\text{-}VI_{min})/(VI_{max}\text{-}VI_{min})$ and called it VI^* (e.g., NDVI* in Groeneveld and Baugh, 2007; Groeneveld *et al.*, 2007). Nagler *et al.* (2005) scaled Enhanced Vegetation Index (EVI*) to determine water use of natural vegetation in riparian corridors. In a later study, ET was constrained using 0 x ET_o at EVI = 0.05 (e.g., bare soil or dormant vegetation) and 1.28 x ET_o at EVI = 0.973, which was the highest EVI value observed in agricultural and natural land covers (Nagler *et al.*, 2013). NDVI* and EVI* have been commonly used as primary inputs in calculating ET (Nagler *et al.*, 2009, 2013, 2014, 2020; Jarchow *et al.*, 2017, 2020; Abbasi *et al.*, 2023a).

EVI was developed for use with imagery from MODIS (Huete *et al.*, 2002; NASA MOD13, 2024). EVI was tested across other sensors and resolutions, when Jarchow *et al.* (2018) sampled both agricultural and riparian areas to compare EVI from MODIS to the Visible Infrared Imaging Radiometer Suite (VIIRS), Landsat 5 TM and Landsat 8 OLI Platforms. Landsat- and VIIRS-based EVI data were highly correlated with MODIS EVI over large areas with dense and fully transpiring agricultural vegetation; they report a very high correspondence between Landsat and MODIS EVI (R^2 = 0.97) for agricultural fields, but this strong relationship degraded over smaller, sparsely vegetated areas, (R^2 = 0.48) such as riparian vegetation. Nagler *et al.* (2020) found a much higher correlation between MODIS EVI and Landsat EVI (R^2 0.76–0.80) for the riparian reaches of the Colorado Delta in Mexico. They also found a higher correlation range between MODIS ET using EVI and Landsat ET using EVI (R^2 ranged between 0.87 and 0.90) for all of the riparian corridor in the Colorado Delta (Nagler *et al.*, 2020). They also tested the two-band EVI (EVI2, Jiang *et al.*, 2008), which showed high correlation (R^2 = 0.99) with EVI, and tested ET using NDVI* as well; EVI2 was selected for future use in arid regions (Nagler *et al.*, 2021, 2022a, 2022b).

In the ET_f equations (6) and (7), η is often close to 1 for some vegetation indices (EVI, SAVI) but may be less than 1 for NDVI due to NDVI's lack of sensitivity for leaf area indices greater than ~3 (Glenn *et al., 2011b). Equation (7) assumes evaporation is zero when VI equals VI_{min}, which may not be the case for wet soil, including irrigated fields at initial growth stages. Soil evaporation can be included by introducing a second coefficient K_e that is determined by modeling the soil drying curve is the single most important term in global and river basin water balances after precipitation or irrigation events, which is independent of remote sensing data (Glenn et al., 2010).

Other formulations of the ET_f-VI relationship include those derived from Beer–Lambert Law of absorption of light by a canopy, assuming a linear relationship between EVI and the leaf area index (LAI) (Nagler et al., 2013) (Figure 10.1):

$$ET_f = a_K \left(1 - exp(-b_K EVI)\right) - c_K \qquad (8)$$

where a_K, b_K, and c_K are the coefficients determined by regression of EVI against observed ET_f in pixels that have ground-level measurements of ET.

The coefficients in Eqs. 6–8 may vary by crop type, climate, or soil type, so implementation depends on the availability of ground data of ET from lysimeters, eddy flux correlation towers, sapflow measurements (Glenn et al., 2008b), or other methods to calibrate the ET_f-VI relationship. If data from lysimeters or towers are used for calibration, the equation estimates ET, and if data from sapflow measurements are used, the equation estimates transpiration only. The method assumes that all crops with identical VI have the same ET_f values. While the coefficients in the ET_f-VI relationship may be constant over several vegetation types (Nagler et al., 2009, 2013), they may vary under different climates, soil types, or soil moisture stress conditions, so the method is typically used to estimate ET over relatively small areas with available data. Regional ET_f-VI curves have been constructed for various locations in the western United States, Spain, and Australia, with an accuracy of within 5% of measured values on an annual basis (Nagler et al., 2013), suggesting the possibility of monitoring ET without field-by-field knowledge of cropping patterns. The spatial and temporal variability in the ET_f-VI relationships and the size region in which they can be applied with a given accuracy could be further documented.

The accuracy of empirical crop coefficient methods has been assessed using flux towers, soil water balances, and annual water balances. Nagler et al. (2013) reported a standard mean error in ET_f of 0.12 in the application of Eq. (8) to MODIS EVI data compared to flux towers (riparian

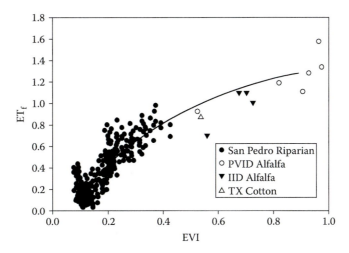

FIGURE 10.1 Reference evapotranspiration fraction (ET_f) from field measurements versus the enhanced vegetation index (EVI). The sites span a range of geographic locations (San Pedro River, AZ; Imperial Irrigation district, CA; Palo Verde irrigation district, CA; Texas). The line is the best fit regression line. Reprinted from Nagler et al. (2013) with permission from MDPI.

areas and irrigated alfalfa) and soil moisture balance (cotton). Glenn *et al.* (2010) reviewed numerous applications of empirical ET_f methods and reported root mean square deviation (RMSD) in the range of 10–30% of mean ET across several different biomes. The main disadvantage of the method is the meteorological data required to estimate ET_o and some ground-level measurements of ET to calibrate the ET_f-VI equation.

10.2.2.1.2 Application of the Crop Coefficient Approach to Agricultural Crops

Bausch and Neale (1987) and Neale *et al.* (1989) first established the validity of the empirical crop coefficient (ET_f) method at two experimental farm sites in Colorado. They grew irrigated corn in fields equipped with weighing lysimeters, and ET_f was calculated with alfalfa grown at the same facility as the reference crop (i.e., $ET_f = ET_{corn}/ET_{alfalfa}$). NDVI was measured with radiometers suspended over corn canopies. NDVI was strongly correlated with LAI and fractional vegetation cover, and ET_f derived from radiometric measurements closely tracked measured ET_f over the crop cycle. Choudhury *et al.* (1994) used a modeling approach to show the theoretical justification for replacing ET_f from ground-based measurements with ET_f estimated from a vegetation index, with ET_f from the vegetation index replacing canopy resistance terms in the Penman–Monteith equation. The ET_f method has since been successfully applied to a wide variety of crops, as outlined in the following examples.

Vineyards and orchards are difficult to model with normal crop coefficients due to differences in plant spacing and other crop variables among plantings. Campos *et al.* (2010) accurately predicted actual ET by simple linear regression equations in grape orchards in Spain. Samani *et al.* (2009) used NDVI from Landsat imagery to develop field-scale ET_f values for pecan orchards in the lower Rio Grande Valley, New Mexico, USA, with mean bias of 4% compared with estimates from a flux tower. Only 5% of fields were within the range of ET and ET_f set by expert opinion, indicating the potential for significant water savings.

Duchemin *et al.* (2006) used NDVI from Landsat images to map LAI and ET in irrigated wheat fields in Morocco, and reproduced ground measurements of ET within 15%. Wheat fields varied widely in ET, so irrigation efficiency could be improved by tailoring water applications to actual crop needs determined by satellite imagery. Gontia and Tiwari (2010) developed field-specific crop coefficients for a wheat field in West Bengal, India, using ET_f values determined from NDVI and SAVI from satellite sensors.

The United States Department of Agriculture (USDA) Agricultural Research Service (ARS) estimated ET from wheat in a desert irrigation district in Maricopa, Arizona, USA (Hunsaker *et al.*, 2005, 2007a, 2007b). They developed NDVI-derived ET_f values that tracked within 5% of measured ET_f for wheat grown in weighing lysimeters. Wheat was then grown in field plots for two seasons under stress and non-stress conditions. The NDVI method gave more accurate predictions of actual irrigation demands than the FAO-56 crop coefficient method (Allen *et al.*, 1998) under all treatment conditions with root mean square error (RMSE) of about 15% of measured water use with no bias across treatments. French *et al.* (2020) estimated the crop coefficient and crop ET for the same wheat fields using remotely sensed NDVI from Sentinel-2 and VENUS satellites with accurate estimates of cumulative, mid-season, and late season ET. Abbasi *et al.* (2023b) utilized the wheat data in French *et al.* (2020) to assess the performance of remotely sensed ET and K_c for the whole growing season (Abbasi *et al.*, 2023b), computed with the Modified 2-band Enhanced Vegetation Index (METEVI2) implemented in GEE for mapping ET in the Lower Colorado River Basin (Abbasi *et al.*, 2023a). Abbasi *et al.* (2021) translated and localized the Nagler *et al.* (2013) VI-based ET algorithm ("Nagler ET") using Landsat EVI and the 2-band EVI (EVI2, Jiang *et al.*, 2008) over croplands in the Zayandehrud River Basin in Iran. In Abbasi *et al.* (2023b), monthly ET variations for METEVI2 and OpenET models were comparable; METEVI2 had the lowest difference from the average ET with 17 mm underestimation, while the Satellite Irrigation Management Support (SIMS) model had the highest difference rate (82 mm) (Abbasi *et al.*, 2023b). METEVI2 is a cost-effective ET mapping tool in drylands (Abbasi *et al.*, 2023b).

Similar studies have been conducted with cotton in the southwestern USA (Hunsaker *et al.*, 2003) and Spain (González-Dugo and Mateos, 2008), where considerable water savings could be achieved by scheduling irrigations based on NDVI-derived crop models rather than ET_f values determined for crops grown under optimal conditions. ET_f methods have been developed for other crops, including potato (Jayanthi *et al.*, 2007), broccoli (El-Shikha *et al.*, 2007), sugarbeet (González-Dugo and Mateos, 2008), soybean (González-Dugo *et al.*, 2009), the oilseed crop camelina (Hunsaker *et al.*, 2011), sorghum (Singh and Irmak, 2009), and alfalfa (Singh and Irmak, 2009). All these studies reported positive results and pointed to the possibility of considerable water savings by replacing static FAO-56 crop coefficients with locally derived ET_f values. Guerschman *et al.* (2009) used EVI from MODIS data to estimate the parameter of the Priestly–Taylor equation and ET, validated with flux towers and water balance estimates for field locations in Australia. Guerschman *et al.* (2022) then developed and calibrated the MODIS ReScaled EvapoTranspiration (CMRSET) algorithm from the Commonwealth Scientific and Industrial Research Organisation (CSIRO), Australian government, for estimating ET from field to continent continental scales using imagery from Sentinel-2, Landsat, Visible Infrared Imaging Radiometer Suite (VIIRS), and MODIS.

Often the main interest is determining district-wide water demand or consumptive use, which requires estimating ET over mixed-crop areas. Choudhury *et al.* (1994) pointed out that the ET_f-VI relationship was not necessarily crop-specific and that the ET_f approach might be used over mixed crops without a serious loss of accuracy. Allen and Pereira (2009) found a reasonable agreement between measured ET_f and vegetation cover fraction (f_c) over a wide range of tree crops, and the relationship was improved by including plant height in the regression. Similar findings were reported by Trout *et al.* (2008) for tree and vegetable crops, with a close correlation between Landsat-derived f_c based on NDVI, and f_c measured on the ground over 30 fields with crops ranging from trees (almonds, pistachios) to vines (grapes) to row crops (onions, tomatoes, cantaloupes, watermelons, beans, peppers, garlic, and lettuce). The only aberrant crop was red lettuce, which has a low NDVI due to its reflection of red light. An operational ET-monitoring program was developed for California's irrigation districts based on Landsat-derived NDVI and ET_o from the California Irrigation Management Information System (CIMIS) network of micrometeorological stations (Johnson and Trout, 2012).

10.2.2.2 Physically Based Coefficient Methods: PT-JPL Model

For regional or global application, ground reference data are often not available at sufficient spatial density to support local calibration of the ET_f-VI relationship, so more physically based models that require minimum calibration have been developed. The Priestley–Taylor Jet Propulsion Lab (PT-JPL) model of Fisher *et al.* (2008) estimates ET directly from satellite imagery with minimal ground data requirements. PT-JPL is a crop coefficient method in that it estimates coefficients that are multiplied by ET_o to estimate ET, similar to the empirical ET_f method, but using physically based models to estimate the coefficients from satellite imagery and meteorological data. In the PT-JPL algorithm, λET is calculated as the sum of evaporation of water intercepted by the canopy (λE_I), evaporation from the soil surface (λE_s), and transpiration from the dry canopy (λE_c) (Table 10.2), where a "dry canopy" has no liquid water on the surface of the leaves. ET_f for transpiring vegetation is the product of four coefficients that account for variations in surface wetness (f_{wet}), green canopy fraction (f_g), a plant temperature constraint (f_T), and a plant moisture constraint (f_M):

$$\lambda ET_c = (1 - f_{wet}) f_g f_T f_M \alpha_{PT} \frac{s}{s+\gamma} (R_n - R_{ns}) \tag{9}$$

where R_{ns} is R_n at the soil surface and is a function of vegetation cover (Table 10.2). Alternate formulations use fractional vegetation cover without calculating R_{ns} separately (Marshall *et al.*, 2013). λE_I is calculated as:

$$\lambda E_I = f_{wet} \alpha_{PT} \frac{s}{s+\gamma} (R_n - R_{ns}) \tag{10}$$

Soil evaporation (λE_s) is calculated separately but also has coefficients related to surface wetness and soil moisture:

$$\lambda E_s = \left(f_{wet} + f_{SM}\left(1 - f_{wet}\right)\right)\alpha_{PT} \frac{s}{s+\gamma}(R_{ns} - G) \quad (11)$$

Several of the parameters in Table 10.2 may be adjusted according to local conditions (Fisher *et al.*, 2008).

Fisher *et al.* (2008) compared the PT-JPL model predictions to measurements at FLUXNET eddy covariance towers and reported an RMSE of 16 mm month^{-1}, and an error in annual ET of 12 mm yr^{-1} or 13% of the observed mean. The flux sites covered a range of biomes, including temperate C3/C4 crops, but did not include irrigated cropland, which might be expected to have higher error due to high evaporation from inundated and wet soil at the beginning of the growing season (Yilmaz *et al.*, 2014).

The PT-JPL model was originally implemented at 1° resolution (Fisher *et al.*, 2008), and subsequently with MODIS data at 500m resolution. The PT-JPL model has also been applied to the thermal infrared band of the Ecosystem Spaceborne Thermal Radiometer Experiment on Space Station (ECOSTRESS) sensor, which has four-day overpass frequency and 70m resolution, launched on June 29, 2018 (Fisher *et al.*, 2020). Errors in ET from PT-JPL ECOSTRESS were low (bias 8%, normalized RMSE [nRMSE] 6%).

10.2.2.3 Physically Based Vegetation Methods: Canopy Resistance and MOD16

Canopy resistance methods predict the resistance parameters in the Penman–Monteith equation (r_a, r_s in Eq. (4)) from satellite imagery. Examples include the Penman–Monteith–Leuning version 2 (PML-V2) model (Zhang *et al.*, 2019) and MOD16 (Mu *et al.*, 2011). Several resistance methods, including MOD16, are based on the model of Cleugh *et al.* (2007). MOD16 is vegetation-based in that the primary inputs driving ET for a given amount of R_n are derived from vegetation indices. The fraction of photosynthetically active radiation (FPAR) is used to determine the fraction of the surface covered by crop canopy (f_c) and soil (1-f_c). LAI is used to determine the dry canopy resistance to transpiration (r_{s_c}), the aerodynamic resistance (r_a), and wet canopy resistance to evaporation (r_{s_wetC}).

Dry canopy resistance to transpiration (r_{s_c}) in MOD16 is calculated using LAI, minimum air temperature, and vapor pressure deficit:

$$r_{s\_c} = \frac{1}{LAI c_L m(Tmin) m(D)} \quad (12)$$

where c_L is the mean potential stomatal conductance per unit leaf area, and $m(Tmin)$ and $m(D)$ are multipliers (range 0.1–1) that limit stomatal conductance by minimum air temperature ($Tmin$) and D. Equations for $m(Tmin)$ and $m(D)$ are given in Mu *et al.* (2007).

10.2.2.4 Vegetation-Based Methods and Soil Evaporation

In both the PT-JPL and MOD16 methods, ET increases with VI. Soil evaporation, including from both saturated and unsaturated surfaces, is assumed to increase with the fourth power of relative humidity (RH) (see Table 10.2, Figure 10.2). In MOD16, evaporation from saturated and unsaturated soils is calculated separately. The water cover fraction or fraction of the soil that is saturated at the surface (F_{wet}) is assumed zero in grid cells where relative humidity (RH) is less than 70% (0.7) (Mu *et al.*, 2011, Eq. 15):

$$F_{wet} = \begin{cases} 0, & RH < 0.7 \\ RH^4, & RH \geq 0.7 \end{cases} \quad (13)$$

Remote Sensing of Evapotranspiration from Croplands

TABLE 10.2
Inputs and Calculation Steps for the PT-JPL Model (Fisher et al., 2008)

Inputs	Units	Description	Source or Equation
1. R_n	W m^{-2}	24-hour mean net radiation	Eq (1)
2. r_{NIR}	-	Near infrared spectrum reflectance	Imagery
3. r_{VIS}	-	Visible spectrum reflectance	Imagery
4. T_{max}	°C	Maximum temperature	Meteorological data
5. e_a or RH	kPa	Water vapor pressure	Meteorological data
	—	Relative humidity	Meteorological data
Derived variables			
1. G	W m^{-2}	Ground heat flux	Assumed zero
2. SAVI	-	Soil adjusted vegetation index	$1.5(r_{NIR}-r_{VIS})/(r_{NIR}+r_{VIS}+0.5)$
3. NDVI	-	Net difference vegetation index	$(r_{NIR}-r_{VIS})/(r_{NIR}+r_{VIS})$
4. f_{APAR}	-	Fraction of PAR absorbed by green vegetation	m_1SAVI $+b_1$ $m_1 = 1.2*1.136$ $b_1 = 1.2*-0.04$
5. f_{IPAR}	-	Fraction of PAR intercepted by all vegetation cover	m_2NDVI $+b_2$ $m_2 = 1$, $b_2 = -0.05$
6. LAI	-	Leaf area index	$-\ln(1-f_{IPAR})/k_{PAR}$ $k_{PAR} = 0.5$
7. R_{ns}	W m^{-2}	Net radiation to the soil	$R_{ns} = R_n(\exp(-k_{Rn}\text{LAI}))$ $k_{Rn} = 0.6$
8. f_{wet}	-	Relative surface wetness	RH^4
9. f_g	-	Green canopy fraction	f_{APAR}/f_{IPAR}
10. T_{opt}	°C	Optimum plant growth temperature	T_{max} at max(PAR$f_{APAR}T_{max}$/VPD)
11. f_T	-	Plant temperature constraint	$\exp\left(-\left(\dfrac{T_{max}-T_{opt}}{T_{opt}}\right)^2\right)$
12. f_{SM}	-	Soil moisture constraint	RH$^{VPD/\beta}$ $\beta = 1.0$ kPa
13. λET_c	W m^{-2}	Transpiration from dry canopy	Eq (9)
14. λET_I	W m^{-2}	Evaporation from wet canopy	Eq (10)
15. λET_s	W m^{-2}	Soil evaporation	Eq (11)

In MOD16, potential evaporation from soils, both saturated and unsaturated, is:

$$\lambda E_{SP} = \frac{s(R_n - G) + \dfrac{\rho C_p D}{r_{a_s}}}{s + \gamma\left(1 + \dfrac{r_{tot}}{r_{a_s}}\right)} \quad (14)$$

where r_{tot} is the total aerodynamic resistance to vapor transport and r_{a_s} is the aerodynamic resistance at the soil surface (Mu et al., 2011). Total evaporation from both saturated and unsaturated soils, based on a rearrangement of Eq. 27 in Mu et al. (2011), is:

$$\lambda E_s = \lambda E_{SP}\left(F_{wet} + (1 - F_{wet})f_{SM}\right) \quad (15)$$

where f_{SM} is the same as f_{SM} in the PT-JPL method (Table 10.2). Soil evaporation is predicted to be very low in arid and semi-arid environments with low RH and high daytime temperature, since the model assumes that F_{wet} and soil evaporation are functions of the regional climate, with more soil evaporation in humid regions (Figure 10.2). For example, Eq. (15) predicts that at 40°C and RH = 0.5, soil evaporation is less than 0.1 of the potential soil evaporation due to low F_{wet}. While this assumption may be accurate in rainfed systems, it may be problematic in irrigated areas in arid and semi-arid climates, because F_{wet} and soil moisture may be high due to irrigation, even if the grid-cell average *RH* is low.

Recognizing the problems with MOD16 in irrigated areas and wetlands, Mu (2013) updated MOD16 in an application in the Nile River Delta, using T_R to calculate a revised surface resistance to ET. The MOD16 product was again revised in 2017 (Running *et al*., 2017; He *et al*., 2019), though the 2017 revised MOD16 algorithm retains the estimation of F_{wet} based on RH (Eq. 13) and underestimates ET during early growth stages of irrigated crops (He *et al*., 2019), consistent with other studies that used some of the same data in irrigated fields (Biggs *et al*., 2016). Brust *et al*. (2021) incorporated higher resolution meteorological data (4km vs. 50 x 70 km) and EO-based soil moisture data from the Soil Moisture Active Passive (SMAP) mission over the coterminous United States, replacing the equation for f_{SM} based on RH and vapor pressure deficit (Table 10.2) with the relative extractable soil water (REW) calculated from the time series of SMAP soil moisture operational product, version 4 Level 4 (Reichle *et al*., 2019; NASA SMAP, 2024). SMAP uses L-band brightness temperature, has 9km resolution and is available from March 31, 2015. The SMAP-based ET estimates were significantly improved compared to the default MOD16 algorithm, especially over croplands and arid regions where soil evaporation was a large fraction of ET: over cereal crops the mean bias decreased from −0.401 mm d^{-1} to −0.047 mm d^{-1} (Brust *et al*., 2021). Further improvements to the MOD16 soil evaporation algorithm promise to make MOD16 a much more accurate product over croplands; as of the date of publication, the SMAP-based improvements had not yet been incorporated into the global MOD16 product.

Mu *et al*. (2011) validated MOD16 against eddy flux covariance towers and reported mean absolute bias in daily ET of 0.31–0.40 mm day^{-1}, or 24–25% of observed daily ET when using GMAO MERRA as meteorological input. The flux towers were located in a range of biomes in the United States and at several sites in the Brazilian Amazon. The validation included two irrigated sites, both of which had higher error (1.2 mm day^{-1}, 72–76% of the observed mean) compared to the mean RMSE for all sites with flux tower data.

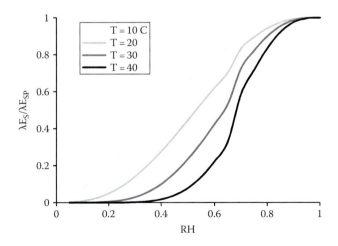

FIGURE 10.2 Soil moisture function in the MOD16 and PT-JPL models.

10.2.2.5 Comparison of Vegetation-Based Methods

Vegetation index methods can be run globally and continuously with remote sensing and surface climate reanalysis at low computational cost. The PT-JPL model was evaluated against several other vegetation-based approaches in the humid tropics, given its simplicity and strong dependence on R_n. In regions where light limits carbon assimilation, such as the humid tropics, R_n is the dominant control on λET (Fisher *et al.*, 2009). The PT-JPL model performed best overall, as Penman–Monteith (resistance-based) methods include more parameters, and therefore may be inherently more uncertain and more strongly coupled to the atmosphere. PT-JPL overpredicted ET, while MOD16 had lower bias but higher overall error (Chen *et al.*, 2014). The performance of these models has been evaluated primarily with eddy covariance flux tower data, and their performance can significantly degrade at larger spatial scales, due to the large uncertainties in surface climate reanalysis products, in particular RH (Vinukollu *et al.*, 2011a; Marshall *et al.*, 2013). The Abbasi *et al.* (2023a) VI-based ET method was validated with flux towers from several continents, performed very well in agricultural drylands, and was comparable to the results from the thermal methods available on OpenET (Melton *et al.*, 2021); future research could include validation on croplands in mesic climates.

Spatial resolution has important impacts on ET performance in croplands. For example, Bajgain *et al.* (2020) implemented the MOD16 algorithm using different spatial resolutions (30, 200, 500, and 1000 m); performance was highest for the 30m product. The most direct comparisons of various vegetation-based methods use the same input data at the same resolution. More detail on spatial resolution in ET estimates is provided in Section 10.4.1.

Forcing ET models with hyperspectral narrowbands (HNBs) has so far received little attention (Hank *et al.*, 2019). Hyperspectral sensors consist of hundreds of narrow (≤ 10nm) bands that distinguish biochemical processes and biophysical properties related to ET. Substitution of NDVI with hyperspectral vegetation indices (HVIs) in the VNIR and SWIR-based soil moisture derived from field spectroradiometers achieved 17% and 14% gains in transpiration and soil evaporation PT-JPL model performance, respectively (Marshall *et al.*, 2016). More reliable ET estimates with the two-source energy balance (TSEB) model were also obtained by estimating the vegetation fraction with an inverted-radiative transfer model and European Space Agency (ESA) Compact High Resolution Imaging Spectrometer (CHRIS) HNBs (Richter and Timmermans, 2009). The lack of research is due in part to the short supply of hyperspectral image data. ESA CHRIS and NASA Hyperion were the only two spaceborne hyperspectral missions in the past—neither of which was designed for global monitoring (Rast and Painter, 2019). The launch of the Italian Space Agency's PRISMA (Hyperspectral Precursor of the Application Mission) and German Space Agency's Environmental Mapping and Analysis Program (EnMAP) usher in a new era of Earth Observation missions with hyperspectral capability. The pioneering ESA project, HyRELIEF (ESA HyRELIEF, 2024) aims to improve ECOSTRESS ET by integrating PRISMA and EnMAP HNBs and hyperspectral vegetation indices (HVIs) into PT-JPL and TSEB via radiative transfer model (RTM) inversion. The project informs upcoming ESA Copernicus Hyperspectral Imaging Mission for the Environment (CHIME) and Surface Biology and Geology (SBG) missions, which will provide the first operational (Landsat-like) acquisition of HNBs from the global land surface.

10.2.3 Radiometric Land Surface Temperature Methods for ET Estimation

T_R-based methods for estimating ET are based on the consumption of energy by ET for vaporization, which reduces surface temperature (Su *et al.*, 2005). A subset of these methods is often called surface energy balance (SEB) methods, since they solve the energy balance equation using T_R to partition available energy (R_n-G) between the sensible and latent heat fluxes; here we use the more generic term radiometric land surface temperature (T_R) method.

T_R methods have been used as early as the 1970s, when Stone and Horton (1974) used a thermal scanner to estimate ET, and Verma *et al.* (1976) developed a resistance model with thermal imagery inputs. Since then, a variety of methods have been developed, including SEBAL (Bastiaanssen *et al.*, 1998, 2002), METRIC (Allen *et al.*, 2007), the Surface Energy Balance System (SEBS) (Su, 2002), ALEXI (Anderson *et al.*, 1997), DisALEXI (Norman *et al.*, 2003), and Operational Simplified Surface Energy Balance (SSEBop) (Senay *et al.*, 2013). Most methods in this category use T_R to estimate components of the energy balance, though some simplified methods (e.g., simplified surface energy balance, SSEB) use temperature directly without solving for the energy balance. Next we summarize the theoretical foundations of the energy balance methods, describe simplified approaches based on T_R, and highlight key differences in the most-used algorithms.

In energy-balance methods that use T_R, λET is computed as a residual of the energy balance equation:

$$\lambda ET = R_n - G - H \tag{16}$$

where G is soil heat flux, and H is the sensible heat flux (all in W m^{-2}). While there may be some energy exchange from photosynthesis, it is usually a small fraction of R_n, is not easily measured even by ground instrumentation (Wilson *et al.*, 2001), and is assumed to be zero (Meyers and Hollinger, 2004). In vegetation with high canopy cover, such as forests, energy exchange from photosynthesis can become high (7–15%), particularly over short time intervals (Meyers and Hollinger, 2004).

The most important term in the energy balance equation after R_n-G is H, which can be estimated using either one- or two-source energy balance models (Figure 10.3). One-source models, including SEBAL (Bastiaanssen *et al.*, 1998), METRIC (Allen *et al.*, 2007), and SEBS (Su, 2002), estimate ET from the surface as a whole. Two-source energy balance (TSEB) models separate ET into E from soil and ET from the vegetation canopy, which is sometimes further separated into evaporation of intercepted water from a wet canopy and transpiration from a dry canopy as in the PT-JPL and MOD16 models. TSEB models include the Atmosphere-Land Exchange Inverse (ALEXI; Anderson *et al.*, 1997) and distributed ALEXI (DisALEXI; Norman *et al.*, 2003). The separation into two sources results in two additional resistance variables to be estimated: R_x, the total boundary layer resistance of the canopy, and R_s, the sensible heat exchange resistance of the soil surface (Figure 10.3).

10.2.3.1 One-Source Models: SEBAL, METRIC, and SEBS

In one-source models (Figure 10.3) H is calculated as:

$$H = \frac{\rho C_p (T_1 - T_2)}{R_{ah}} \tag{17}$$

where T_1 is the aerodynamic temperature (K) of the evaporating surface at height z_1, which is the height of the zero-plane displacement (d) plus the surface roughness for momentum transport (z_{0m}); T_2 is the air temperature at height z_2, which is the height where air temperature is measured (usually 2 or 3 m above the evaporating surface); and R_{ah} is the aerodynamic resistance to turbulent heat transport from z_1 to z_2 (Figure 10.3). The model assumes that evaporating surfaces have a temperature equal to or hotter than the air above ($T_1 \geq T_2$), resulting in a positive value of H. The zero-plane displacement height (d) is the mean height where momentum is absorbed by the canopy, typically around 2/3 of the vegetation height (h). z_{0m} is a relatively small fraction of h (0.03h, 0.1h, or 0.123h in Morse *et al.*, 2000), and is around 0.03 m over grassland, 0.10–0.25 m over cropland, and 0.5–1.0 m over forest or shrubland. In practice, z_{0m} is estimated as a function of NDVI or with a land cover map (see Table 10.3).

There are two main uncertain variables in the calculation of *H* in Eq. (17). First, T_R may differ from the actual aerodynamic temperature (T_1). Note that neither of the air temperatures in Eq. (17) are directly sensed by the satellite, which estimates T_R based on thermal radiation reaching the sensor from the combined soil and canopy surfaces. The correspondence between T_R and air temperatures at either z_1 or z_2 varies by surface type, roughness, and crop canopy structure. Different

Remote Sensing of Evapotranspiration from Croplands

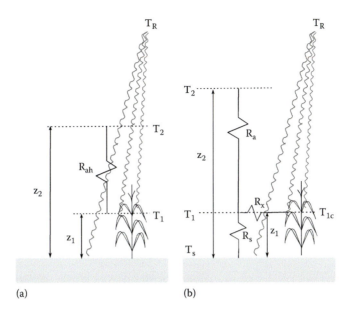

FIGURE 10.3 Schematic of one (left) and two-source (right) models for temperature-based ET calculations. T_R is the radiometric surface temperature detected at the satellite, and the gray lines indicate thermal radiation emission from the soil and canopy.

TABLE 10.3
Inputs and Calculation Steps for the SEBAL Model

Inputs	Description	Source
1. R_n	Instantaneous net radiation	Eq. 1
2. R_{n24}	24-hour mean net radiation	Eq. 1
3. NDVI	Normalized Difference Vegetation Index	Imagery
4. T_R	Radiometric Land Surface Temperature (K)	Imagery
5. U	Wind speed (m/s)	Wind speed from meteorological station or gridded data
6. z_{0m}	Surface roughness (m)	$z_{0m} = 0.123 * h$, where h is vegetation height in meters from land cover map or NDVI
7. Elevation	Surface elevation (m)	Digital Elevation Model (DEM)

Derived Variables	Description	Equation[a]
1. G	Ground heat flux (W/m2)	$G = R_n * \left(\frac{T_R - 273.15}{\alpha}\right)\left(0.0032\alpha + 0.0062\alpha^2\right)\left(1 - 0.978 NDVI^4\right)$ where α is daytime-average albedo[b]
2. U_x^*	Friction velocity at meteorological station (m/s)	$U_x^* = \dfrac{0.41 U}{\ln\left(\frac{z_2}{z_{0m} \text{ at station}}\right)}$
3. U_{200}	Wind speed at blending height (200m) above the meteorological station (m/s)	$U_{200} = U_x^* \dfrac{\ln\left(\frac{200}{z_{0m} \text{ at station}}\right)}{0.41}$
4. U^*	Initial value of friction velocity	$U^* = \dfrac{0.41 U_{200}}{\ln\left(\frac{200}{z_{0m}}\right)}$

(Continued)

TABLE 10.3 (*Continued*)
Inputs and Calculation Steps for the SEBAL Model

Inputs	Description	Source
5. R_{ah}	Initial value of aerodynamic resistance to heat transport	$R_{ah} = \frac{\ln\left(\frac{z_2}{z_1}\right)}{(0.41 U_*)}$
6. Dry pixel	The dry pixel selected for calibration	Selected from the image manually[a], or automatically using quantiles of NDVI and TR[c]
7. Wet pixel	The wet pixel is selected for calibration	Selected from the image manually[a], or automatically using quantiles of NDVI and TR[c]
8. T_R Dry, T_R Wet	T_R at dry and wet pixels	Imagery

Iteration loop starts here:

Inputs	Description	Source
9. T_2	Air temperature at height z2	$T_2 = T_R - dT$ Assume $T_2 = T_R$ for the first iteration
10. P_a	Air pressure (kPa)	$P_a = 101.3\left(\frac{T_2 - 0.0065z}{T_2}\right)^{5.26}$ where z is the elevation of the pixel or region (m)[d]
11. ρ_{air}	Air density (kg m−3)	$\rho_{air} = \frac{1000 P_a}{1.01 T_2 R}$ where R = 287 J kg^{-1} K^{-1}
12. a	Calibration coefficient a	$a = \frac{dT_{Dry}}{T_{RDry} - T_{RWet}}$ where $dT_{Dry} = \frac{H_{Dry} R_{ahDry}}{\rho_{air} C_p}$ and HDry = RnDry—GDry
13. b	Calibration coefficient b	$b = -aT_{RWet}$
14. dT	Temperature difference between z1 and z2 (K)	$dT = aT_R + b$
15. H	Sensible heat flux (W/m2)	$H = \frac{v_{air} C_p dT}{r_{ah}}$ where Cp = 1004 J kg^{-1} K^{-1}
16. L	Monin-Obhukov length (m), atmospheric stability condition	$L = -\frac{v_{air} C_p U_*^3 T_R}{0.41 g H}$ where g = 9.81 m s−2
17. Ψ_h	Stability correction for heat transport under unstable (L < 0), stable (L > 0), and neutral (L = 0) atmospheric conditions [e]	L < 0: $\psi_h = 2\ln\left(\frac{1+x_{z_2}^2}{2}\right)$ where $x_{z_2} = \left(1 - 16\left(\frac{z_2}{L}\right)\right)^{\frac{1}{4}}$ L > 0: $\psi_h = -5\left(\frac{z_2}{L}\right)$ L = 0: $\psi_h = 0$
18. Ψ_m	Stability correction for momentum transport under Unstable (L < 0), stable (L > 0), and neutral (L = 0) atmospheric conditions [e]	L < 0: $\psi_m = 2\ln\left(\frac{1+x_{200}}{2}\right) + \ln\left(\frac{1+x_{200}^2}{2}\right) - 2\arctan(x_{200}) + \frac{\pi}{2}$ where $x_{200} = \left(1 - 16\left(\frac{200}{L}\right)\right)^{\frac{1}{4}}$ L>0: $\psi_m = -5\left(\frac{z_2}{L}\right)4$ L=0: $\psi_m = 0$
19. U_*	Friction velocity with stability correction for momentum transport	$U_* = \frac{0.41 \times U_{200}}{\ln\left(\frac{200}{z_{0m}}\right) - \psi_m}$
20. R_{ah}	Aerodynamic resistance to heat transport with stability correction for heat transport	$R_{ah} = \frac{\ln\left(\frac{z_2}{z_1}\right) - \psi_h}{0.41 U_*}$

TABLE 10.3 (Continued)
Inputs and Calculation Steps for the SEBAL Model

Inputs	Description	Source

Iteration: Repeat steps 9–20 until changes in H are < 5%, usually ~ 5–10 iterations

21. Λ_{op} Evaporative fraction at overpass (dimensionless) $\Lambda_{op} = \frac{R_n - G - H}{R_n - G}$

22. ET_{24} 24-hour evapotranspiration (mm day^{-1}) $ET_{24} = \frac{86{,}400 \Lambda_{op} R_{n24}}{v_w \lambda}$

where 86,400 is s day^{-1}, ρw is density of water (kgm^{-3}), λ is latent heat of vaporization (J kg−1).

---

one-source models have different strategies for estimating the temperature difference between z_1 and z_2. Some models include an extra term in R_{ah}, while others such as SEBAL (Bastiaanssen *et al.*, 1998, 2002) and METRIC (Allen *et al.*, 2007) calibrate an empirical linear model relating T_1-T_2 to T_R. Second, R_{ah} has high spatial variability and may be difficult to predict.

In one-source models (SEBAL, METRIC, SEBS; Su, 2002), a linear equation predicts the difference between T_1 and T_2 as a function of T_R (Figure 10.4, Table 10.3):

$$T_1 - T_2 = a + bT_R \tag{18}$$

where a and b are empirical parameters determined from the imagery in a process called "internalized calibration" (Allen *et al.*, 2007). Field investigations suggest that Eq. (18) holds under a variety of conditions (Figure 10.4). In applications, a and b are determined from the wet and dry pixels only, with no field data on air temperatures for calibration and therefore no estimate of error of Eq. (18).

Combining Eq. (17) with Eq. (18) gives:

$$H = \frac{\rho_{air} C_p (a + bT_R)}{R_{ah}} \tag{19}$$

Identifying the parameters a, b, and R_{ah} requires calculating ρ_{air} and C_p (Table 10.3) and identifying some pixels where H and T_R are known. First, one pixel is selected that is "wet," where H is assumed to be 0 and λE is assumed to be equal to R_n-G, and another that is "dry," where H is assumed to be equal to R_n-G and λE is assumed to be 0 (Figure 10.4). An initial guess of R_{ah} is made for the image based on literature values by land cover type. An initial guess of $T_1 - T_2$ is made at the dry pixel by solving for it in Eq. (17), using the assumption $H = R_n - G$. Coefficients a and b in Eq. (18) are then determined from the observed T_R and estimated T_1-T_2 at the wet and dry pixels (Table 10.3). H is then calculated again using Eq. (19), this time accounting for unstable atmospheric conditions using the Monin-Obukhov (MO) equations (Table 10.3) (Bastiaanssen *et al.*, 1998, 2002; Allen *et al.*, 2011a). The values of a, b, and R_{ah} are then solved iteratively by updating the values of each until the result converges on H=0 for the wet pixel and $H = R_n - G$ for the dry pixel. The internal calibration of SEBAL and METRIC allows the estimation of ET without knowing either T_1 or T_2, which is an advantage in data-scarce regions.

Sensitivity analysis suggests that T_R at the hot and cold pixels is the most important control on H and λET estimates for a given image, followed by R_n at the hot pixel (Long *et al.*, 2011). Since H is assumed to be zero at the cold pixel, R_n at the cold pixel does not influence the resulting model parameters and calculated H. The criteria for pixel selection are important, but there is no generally agreed method for selecting them (Long *et al.*, 2011). Past applications of SEBAL and METRIC have used manual pixel selection, since user experience in the study area helps select the appropriate

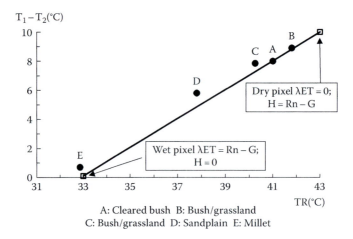

FIGURE 10.4 Example of the assumption of the linear relationship between radiometric temperature (T_R) and the temperature difference between heights z_1 and z_2. Data are from Bastiaanssen et al. (1998), and dry and wet pixels are added for illustration. Reprinted with permission from Elsevier.

pixels that represent typical field conditions in the image. The selection procedure can be automated, which reduces variability among users and allows for more rapid implementation (Kjaersgaard et al., 2009; Biggs et al., 2016). In the automation algorithm of Long et al. (2011), the dry pixel is the pixel with the highest T_R in the subset of pixels with specified land use (bare, urban, or dry cropland), and the wet pixel is the pixel with the lowest T_R, after screening for cloud contamination. Automation of pixel selection in METRIC (Allen et al., 2013) uses a combination of NDVI, T_R and α. Other semi-automated approaches simply select the highest and lowest T_R in a given image, using masks to exclude either clouds or non-representative land covers. The reasons for excluding certain land covers for the wet and dry pixel selection are often not explicit and vary by application. For the wet pixel, some studies advocate excluding water bodies since they have different aerodynamic properties than agricultural fields where ET is being estimated (Morse et al., 2000; Conrad et al., 2007), while others include water bodies, particularly if vegetated pixels have much higher temperatures than open water bodies. For dry pixels, some studies exclude urban environments (Conrad et al., 2007), while others include them (Long et al., 2011). The sensitivity of SEBAL ET to the quantiles used for calibration pixel selection can be assessed by varying the quantiles of both NDVI and T_R (Biggs et al., 2016). Kayser et al. (2022) and Bhattarai et al. (2017) also varied the quantiles used for pixel selection over a specified range. Kayser et al. (2022) observed a lower sensitivity of SEBAL to the meteorological inputs, suggesting that different quantiles may be used in the humid tropics to reduce bias. Bhattarai et al. (2017) search for hot and cold pixels within 7 x 7 Landsat pixel windows with only agricultural land cover and homogenous NDVI (coefficient of variation < 25%). Additional algorithms could be developed and tested for endmember pixel selection, with attention to how the optimal selection quantiles may differ with climate and land cover.

One-source models have the convenience of being relatively simple to use and are calibrated to wet and dry pixels, reducing the need for meteorological data. However, the calibration is performed on a single image, and the *a* and *b* parameters from Eq. (19) may only be valid for that image. While this may not be a problem for study areas the size of a single scene, areas that cover multiple scenes may suffer from problems of merging along scene boundaries (Bhattarai et al., 2019). The SSEBop model (Senay et al., 2013; Section 10.2.3.3) was designed to address scene boundary problems by estimating T_1-T_2 for each pixel under dry and wet conditions. de Andrade et al. (2024) developed a novel methodology for the application of the SEBAL model over very large study regions using MODIS or VIIRS imagery, overcoming the issues associated with domain size selection and scene boundaries. This new method employs a novel technique that normalizes T_R for latitude and elevation, enabling the application of contextual SEB models at continental scales.

Bastiaanssen *et al.* (2005), Kalma *et al.* (2008), and Teixeira *et al.* (2009) reported that errors in SEBAL are typically higher for smaller spatial and temporal scales and are often within the errors of measurements of the device used for validation (10–15%). Compared to field-scale measurements, one-source models have errors around 50 W m^{-2}, or a maximum error of around 15–30% for daily estimates (Kalma *et al.*, 2008), though the errors may vary with the spatial resolution of the input data. Errors over long time scales are typically lower (RMSE~5%) due to canceling out of daily errors (Bastiaanssen *et al.*, 2005).

Most validation sites, both for SEBAL/METRIC and for EO-based ET methods in general, are located in relatively large plots of homogeneous vegetation, which facilitates comparison with satellite imagery but may not assess accuracy well over heterogeneous landscapes. SEBAL and other energy balance models, for example, typically assume minimal advection of energy among pixels, which is likely valid over large homogenous vegetation but may not be valid in heterogeneous irrigated landscapes in semi-arid and arid climates. Advection may double the amount of ET in situations of extreme humidity gradients and high winds (Allen *et al.*, 2011b), which motivated the use of ET_o and ET_f in METRIC in place of R_n-G as used in SEBAL.

Water balance measurements have also been used to validate remotely sensed ET at the scales of individual fields, watersheds, or irrigation projects (Bastiaanssen *et al.*, 2002, 2005). Validation using water balances at the watershed scale is difficult in rainfed systems in arid and semi-arid environments, since streamflow as a percentage of precipitation is often within the error of ET estimated by any method. Water balance validation is more feasible in surface irrigated systems, where inflows and outflows are large relative to ET (Bastiaanssen *et al.*, 2002).

10.2.3.2 Two-Source Energy Balance Models (TSEB): ALEXI, DisALEXI

Two-source energy balance models (TSEB) account for the differences in aerodynamic resistance between soil and vegetation, which are lumped into a single resistance parameter in one-source models. Two-source models require estimation of the energy balance and therefore of T_1 and T_2 over vegetation and soil separately, and so cannot use internal calibration to wet and dry pixels. One popular two-source model, the ALEXI model (Anderson *et al.*, 1997), uses the TSEB of Norman *et al.* (1995). In ALEXI, T_1 of the soil (T_{1s}) and canopy (T_{1c}) are estimated using the vegetation cover fraction:

$$T_R = \left[f_c T_{1c}^4 + (1-f_c) T_{1s}^4 \right]^{1/4} \tag{20}$$

where f_c is the fractional vegetation cover at a given view angle, calculated as:

$$f_c = 1 - exp\left(\frac{-0.5 \Omega LAI}{cos(\theta)} \right) \tag{21}$$

where Ω is an index of the degree of clumping from the given view angle, and θ is the view angle.

T_2 is estimated using an atmospheric boundary layer (ABL) model (Anderson *et al.*, 2013) calibrated to the observed increase in temperature during the morning hours (from 1 to 1.5 hours after sunrise to before local noon), which is obtained from geostationary satellites such as the Geostationary Operational Environmental Satellites (GOES) (Anderson *et al.*, 2013). The ABL model used in ALEXI is relatively simple and can be programmed as a system of equations.

The spatial resolution of the ALEXI model is constrained by the resolution of geostationary satellites (5–10 km), so a different algorithm, DisALEXI (Norman *et al.*, 2003), uses higher-resolution imagery from MODIS (1km), Landsat (thermal band 100m; Anderson *et al.*, 2012b), or

ECOSTRESS (70m, Cawse-Nicholson et al., 2020) to generate high-resolution ET estimates using ALEXI. DisALEXI utilizes the temperature and wind speed at the blending height (~50m above the land surface) and downwelling short- and longwave radiation from ALEXI as input, assuming those four variables are spatially uniform at the resolution of the ALEXI model (usually 5–10 km). The high-resolution thermal imagery is then adjusted to the view angle of the GOES satellite to ensure consistency in the radiometric temperature. The angle-adjusted radiometric temperature, vegetation cover, and land use maps from the high-resolution imagery are then used to calculate R_n at high resolution, and the two-source model is run on each high-resolution pixel with the ALEXI-derived temperature at 50m as the upper boundary condition (Norman et al., 2003). The DisALEXI values are adjusted to match the mean ALEXI values by iteratively altering the air temperature map (T_2) until the aggregated DisALEXI values match the ALEXI ET values, ensuring consistency across scales.

Methods for fusing DisALEXI results with MODIS for daily resolution and with Landsat for high spatial resolution have been developed and tested over rainfed (Cammalleri et al., 2013) and irrigated areas (Cammalleri et al., 2014), based on a data fusion strategy for MODIS and Landsat (Gao et al., 2006). The Spatial and Temporal Adaptive Reflectance Fusion Model (STARFM; Cammalleri et al., 2013) finds the date with the highest correlation between Landsat- and MODIS-derived ET and uses that correlation structure to predict Landsat-scale ET on dates with only MODIS data (Figure 10.5). The use of MODIS improves the estimation of ET over the use of a simple spline function on ET using available Landsat dates. The results highlight the importance of having daily MODIS estimates for some locations where vegetation responds to changes in moisture (Figure 10.6). At some sites (N=162), little difference was observed between the interpolated Landsat ET and the MODIS-Landsat fusion product, but at other sites (N=161), ET increased rapidly and was higher than the interpolated Landsat values for a 15–20-day period following a rainfall event, showing the importance of high temporal resolution data in certain parts of the time series. DisALEXI has also incorporated ECOSTRESS thermal infrared data (four-day overpass frequency, 70m, Fisher et al., 2020) and fused it using STARFM with both MODIS (1km) and Landsat (30m resolution) to generate 30m data cubes with daily temporal resolution (Anderson et al., 2021). Errors in ET with continental-scale DisALEXI using ECOSTRESS data were 0.81 mm day^{-1}, similar to other studies using DisALEXI (Cawse-Nicholson et al., 2020). The high overpass frequency of ECOSTRESS reduces cloud cover issues substantially, by an average of 22%. Reductions in errors in ET from including ECOSTRESS were small when cloud-free Landsat data were available, but substantial (improvements of 65%) when ECOSTRESS provided imagery during periods when Landsat imagery was unavailable due to clouds (Anderson et al., 2021).

10.2.3.2.1 Other TSEB Models

Another TSEB model, the dual-temperature-difference (DTD) model (Norman et al., 2003), is based on the time rate of change in T_R and T_2, where the equations in ALEXI model (Anderson et al., 1997) are used to form a dual-difference ratio of radiometric and air temperatures. The H flux is then calculated from temporal measurements of T_2, T_R, and wind speed. An ABL model is not required in implementing the model, so the calculations can be made efficiently with minimal ground-based data. Agreement between the model-predicted λET and corresponding ground observations is ~50 W m^{-2} (Gowda et al., 2008).

Sun et al. (2009) proposed another TSEB model, the Simple Remote Sensing Evapotranspiration Model (Sim-ReSET). Compared to other TSEBs, in Sim-ReSET the aerodynamic resistance (R_{ah}) was calculated using a reference dry bare soil and canopy height, assuming homogeneous wind speed in the upper boundary layer. Unlike ALEXI, Sim-ReSET is based on a single image, and, like SEBAL and METRIC, is based on internal calibration to dry soil and wet vegetated pixels. λET from Sim-ReSET using MODIS imagery had a RMSD for instantaneous values of ~42 W m^{-2} and a MAE of 34 W m^{-2} compared with concurrent ground measurements from 12 experimental days. Error in the mean daily ET over a six-day period was lower (MAD=0.26 mm day^{-1},

Remote Sensing of Evapotranspiration from Croplands

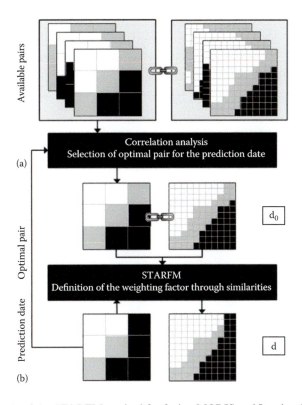

FIGURE 10.5 Schematic of the STARFM method for fusing MODIS and Landsat imagery for high spatial and temporal resolution of ET. Reprinted from Cammalleri *et al.* (2014), with permission from Elsevier.

RMSD= 0.30 mm day^{-1}). Sim-ReSET avoids the direct computation of the aerodynamic resistance, and all inputs can be estimated from remote sensing data alone.

10.2.3.3 Simplified T$_R$-Based Approaches: SSEB and SSEBop

Several other T$_R$-based methods have been developed, including the Simplified Surface Energy Balance (SSEB) model (Senay *et al.*, 2007). Similar to the crop coefficient methods in Section 10.2.2.1, SSEB calculates ET as the product of ET$_o$ (Eq. (4) or (5)) and ET_f (Table 10.4). The SSEB model assumes that ET_f for a given pixel can be estimated from the radiometric temperatures at the hot, cold, and observed pixels alone, without explicitly solving the energy balance equation:

$$ET_f = \frac{(T_h - T_R)}{(T_h - T_c)} \quad (22)$$

where T_h is T_R of the hot pixel and T_c is T_R of the cold pixel. Since the maximum ET of a crop with a high leaf area index may be higher than that of the reference grass, the ET$_o$ parameterized for the reference grass is multiplied by a correction factor, usually 1.2 (Senay *et al.*, 2007). Similar to other energy balance methods, the temperatures at the hot and cold pixel are derived from the image, assuming that pixels where ET = 0 and ET = 1.2ET$_o$ exist in the image.

While the original formulation of SSEB is easy to implement and produces estimates for regions with a uniform hydroclimate such as irrigated districts, an improvement was necessary to account for differences in T$_R$ caused by spatial variation in elevation and albedo (Figure 10.7). An enhanced version of SSEB was introduced by Senay *et al.* (2011) to adjust T$_R$ using a lapse rate correction before using it in Eq. (22). A similar lapse-rate adjustment is also performed in other temperature-based methods, including SEBAL and METRIC. A comparison between the enhanced SSEB and

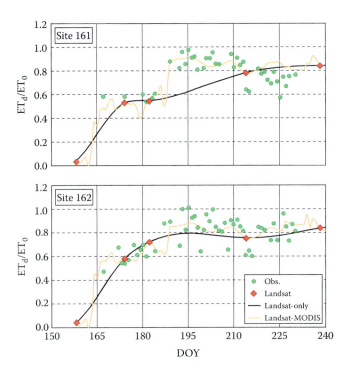

FIGURE 10.6 Time series of ET_d/ET_o estimated using Landsat alone and using MODIS downscaled to Landsat resolution using the STARFM algorithm over rainfed soybeans in Iowa. ET_d is daytime mean ET. Reprinted from Cammalleri *et al.* (2013), with permission from American Geophysical Union.

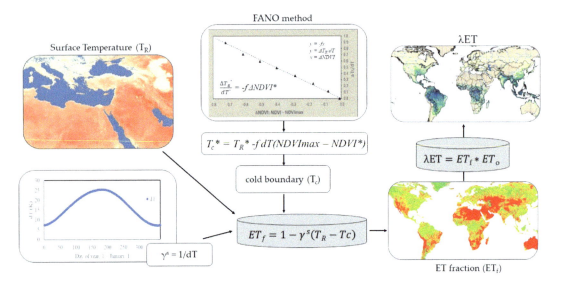

FIGURE 10.7 Flowchart of steps in the SSEBop algorithm.

METRIC showed a strong relationship at elevations below 2000 m ($R^2 = 0.91$) compared to elevations higher than 2000 m ($R^2 = 0.52$) in Central Idaho (Senay *et al.*, 2011). The original SSEB model also calculates ET_o using a fixed value for α (0.23). While SW↓, LW↓, and LW↑ can vary by pixel under the original SSEB method, using the fixed α ignores the impact of spatial variability in α and

TABLE 10.4
Inputs, Parameters, Steps, and Equations Used in the SSEBop Model (Senay et al. 2013)

No.	Inputs	Units	Description	Data Source
1	T_R	K	Radiometric surface temperature	Satellite imagery (Landsat, MODIS)
2	T_2	K	Air temperature at 2 m	PRISM
3	R_n	W/m²	Clear-sky net radiation	Allen et al. (1998)
4	R_{ah}	s/m	Aerodynamic resistance	constant value (110)
5	R	J/ kg K	Specific gas constant	constant value (287)
6	C_p	J/ kg K	Specific heat of air at constant pressure	constant value (1013)
7	ET_o	mm/day	Reference evapotranspiration	Penman–Monteith Equation
8	NDVI	-	Normalized Difference Vegetation Index	Satellite imagery (Landsat, MODIS)
9	Z	m	Surface elevation	Digital Elevation Model (DEM)
10	T_{R_cold}	K	T_R at a cold pixel	Satellite imagery (Landsat, MODIS)

Steps	Parameters	Units	Description	Equation
1	c	-	Temperature correction coefficient	$c = \dfrac{T_{R\_cold}}{T_2}$
2	P	kPa	Atmospheric pressure	$P = 101.3\left(\dfrac{293 - 0.0065 \times Z}{293}\right)^{5.26}$
3	T_{kv}	K	Virtual temperature	$T_{kv} = 1.01 T_2$
4	ρ_{air}	kg/m³	Density of air	$\rho_{air} = \dfrac{1000 P}{T_{kv} R}$
5	dT	K	Temperature difference	$dT = \dfrac{R_n R_{ah}}{\rho_a C_p}$
6	T_c	K	Cold reference temperature	$T_c = c T_2$
7	T_h	K	Hot reference temperature	$T_h = T_c + dT$
8	ET_f	-	Reference ET fraction	$ET_f = \dfrac{(T_h - T_R)}{dT}$
9	ET	mm	Actual ET	$ET = ET_f \times ET_o$

G on ET, which could result in overestimation of ET at pixels with high albedo, and overestimation of ET at pixels with high and positive G. An improved version of SSEB (Senay et al., 2011) adjusts ET_f by a factor that varies with NDVI, and ranges from 0.65 to 1.0 (Eq. 6 in Senay et al., 2011). The adjustment is designed to account for generally higher albedo and greater ground heat flux in pixels with low NDVI, though the relationship between the adjustment factor and NDVI may vary geographically and is not derivable from other measurements.

The SSEB model requires the selection of hot and cold pixels in the image, as in the SEBAL and METRIC models. This selection process generally inhibits operational application of such models over large areas and introduces problems along scene boundaries. Therefore, the SSEB model was updated and renamed to SSEBop (operational Simplified Surface Energy Balance) because of the operational capability to estimate ET without the manual selection of boundary conditions (Senay et al., 2013; Senay, 2018; Senay et al., 2023). SSEBop proposes that the hot and cold temperatures needed to estimate ET_f can be determined uniquely for each pixel, rather than as a lumped parameter for an entire raster image using the principle of satellite psychrometry (Senay, 2018):

$$ET_f = 1 - \gamma^s (T_R - T_c) \qquad (23)$$

where γ^s is a surface psychrometric constant over a dry-bare surface (°C^{-1} or K^{-1}), and T_c is the wet-bulb reference land surface temperature (K) limit. The constant 1 represents the ET_f value during maximum ET, i.e., when $T_R = T_c$.

Since the SSEBop model does not attempt to solve all terms of the surface energy balance, the surface psychrometric constant was introduced to determine the temperature difference (dT) between the desired location (dry limit or surface temperature over dry-bare soil) and the reference value (wet limit or canopy-level air temperature). dT is formulated in Senay (2018) as:

$$\gamma^s = 1/dT = (C_p \times \rho) / (R_n \times R_{ah}) \qquad (24)$$

where the variable definitions are the same as those in Eqs. (19) and (23). dT represents the inverse of the temperature difference ($T_1 - T_2$) when solving Eq. (17) and is calculated for each day and pixel by assuming G is zero for a dry-bare surface ($R_n = H$, $ET = 0$). R_{ah} is assumed to have a constant value of 180 s m^{-1} when R_n is based on gray-sky radiation (Senay et al., 2023). dT represents the radiometric or land surface temperature difference between a hypothetical dry-bare soil surface and a well-vegetated surface at each pixel to estimate the cooling effect of evaporation regardless of the actual cover condition. The wet-bulb reference surface temperature (T_c), which represents the cold boundary condition, is estimated as:

$$T_c = cT_2; c = \frac{T_R^* - fdT(NDVImax - NDVI^*)}{T_2^*} \qquad (25)$$

where T_c is the wet-bulb reference surface temperature; T_2 is the maximum daily air temperature obtained from a meteorological station or gridded meteorological data; and c is the temperature correction factor. The c factor is calculated using the Senay Approximation (numerator) of the innovative Forcing and Normalizing Operation (FANO) method (Eq. 7 in Senay et al., 2023), where T_R^* is the mean observed T_R over a chosen coarse grid; f is the FANO constant ($f = 1.25$); NDVImax is the NDVI value for a well-vegetated pixel (NDVImax = 0.9); NDVI* is the NDVI value over a chosen coarse grid; and T_2^* is the maximum daily air temperature over a chosen coarse grid. The * indicates that the original pixel size of the image is not used but instead an averaged value over a coarser grid (5-km), whereby the FANO approach removes outliers in T_R and NDVI. The FANO method determines the c factor and wet-bulb boundary condition for each pixel and a deterministic equation based on the inverse linear relationship of T_R and NDVI (Figure 10.4 in Senay et al., 2023). In earlier versions (Senay et al., 2013), a single c factor was established for the entire image instead of the current spatially distributed values. Implementation of the FANO equation was driven by over-/under-estimations of ET by SSEBop in different parts of the globe, e.g., underestimation in Western Europe (Senay et al., 2020). Also, the previous criterion of NDVI > 0.7 for the c factor estimation was not always available in the image, especially in arid and semi-arid regions or outside of the major growing seasons. Further, extrapolating T_c to an entire image for limited NDVI calibration points could introduce bias and errors in hydro-climatically complex regions. The NDVI threshold could also create widely varying T_c ranges because some images may only have a narrow range of NDVI that is closer to the threshold (NDVI = 0.7), while others may have T_c values derived from pixels with NDVI far higher than the 0.7 threshold (e.g., NDVI = 0.9). The FANO method is expected to apply to all hydro-climatic conditions for any set of remotely sensed data that includes surface temperature and NDVI parameters. Once γ^s and T_c are calculated, the ET_f can be estimated from the observed T_R using Eq. (23). ET_f ranges from 0 to 1, where 1 represents a wet or vegetated landscape and 0 represents a dry or less vegetated landscape. ET at a given pixel is then calculated by multiplying ET_f by ET_o. This approach simplifies the model and enables operational application over large areas with limited data requirements. Senay *et al*. (2023) provides more details on the method and implementation.

10.2.3.4 From Instantaneous to Daily ET

Several approaches may be used to calculate daily total ET from the instantaneous imagery (Crago, 1996; Chávez *et al*., 2008; Kalma *et al*., 2008; Petropoulos, 2013), though two methods are most commonly applied: the evaporative fraction method and the crop coefficient method. The evaporative fraction (Λ) approach uses the satellite-derived ET to calculate ET as a fraction of available energy (R_n-G):

$$\Lambda_{op} = \frac{\lambda E}{R_n - G} \quad (26)$$

The evaporative fraction at the time of overpass (Λ_{op}) is assumed equal to Λ for the daytime (Λ_d) or during a 24-hour period (Λ_{24}). Either daytime or 24-hour total ET is calculated as the product of Λ_{op} and the daytime net available energy (R_n–G) or 24-hour net radiation (R_{n24}), since G is assumed to be zero over 24 hours. The assumption that Λ_{op} equals Λ_d or Λ_{24} is justified by some field measurements (Jackson *et al*., 1983; Hall *et al*., 1992), though clouds can cause Λ to vary over time (Crago, 1996), and modeling studies suggest there may be diurnal variation in Λ, with minimum values during midday that can result in underestimation of the daily mean Λ of up to 20–40% when using overpass times between 11 am and 3 pm (Lhomme and Elguero, 1999; Gentine *et al*., 2007). Clouds reduce R_n-G, but typically H decreases more than λE, increasing Λ during cloudy periods, though the assumption of constant Λ over daytime hours is "surprisingly robust" (Crago, 1996). A correction factor that varies with the time of overpass and soil moisture has been developed (Gentine *et al*., 2007), though in practice, Λ is often assumed to be constant over a day.

Either daytime available energy $(R_n–G)_d$ or 24-hour total net radiation (R_{n24}) is used as the multiplier to calculate total ET from Λ_{op}. R_{n24} is most commonly used to estimate ET_{24} given the (near) zero G term, though Van Niel *et al*. (2011) caution that, in addition to the assumption that the evaporative fraction is constant over a 24-hour period, the use of R_{n24} also assumes that net available energy (R_n–G) and λET are zero or near zero at night (Van Niel *et al*., 2011). R_n–G is commonly negative at night due to longwave emission from the surface. While λET can also be negative at night, corresponding to condensation or dew formation, much of the negative available energy changes H rather than λET. λET can also be positive at night if sensible heat is advected onto a given location, which can occur where irrigated vegetation may have heat advected to it from surrounding drier surfaces. In an irrigated alfalfa plot, nighttime ET was > 7% of total daily ET (Tolk *et al*., 2006). Positive nighttime λET can result in significant underestimation of daily ET when using R_{n24}, of up to −24% to −38% when using the mid-morning value of Λ_{op} and lower errors when using mid-afternoon values (-5% to −21%). The main contributor to the error of using R_{n24} was the non-zero nighttime available energy flux, which was sometimes nearly equal to the daytime available energy in a wet forest site, and the error was smaller at a drier savanna site (Van Niel *et al*., 2011). More research would be helpful about how the magnitude of errors incurred by using R_{n24} instead of $(R_n–G)_d$ to calculate daily ET depend on season, climate, and vegetation.

Other methods, including the original METRIC model (Allen *et al*., 2007), use the crop coefficient approach, which calculates the ratio of actual to reference ET at the time of satellite overpass, then multiplies that fraction by reference ET for the day:

$$ET_{24} = C_{rad} * ET_f * ET_{o24} \quad (27)$$

where ET_{24} is ET over the 24-hour period, C_{rad} is an adjustment applied to sloped surfaces, ET_f is the ratio of actual to reference ET at the time of satellite overpass, and ET_{o24} is reference ET for the 24-hour period. C_{rad} is likely to be close to 1 for most crops, which are mostly grown on flat surfaces, but there may be local exceptions for agroforestry crops in mountainous terrain.

The crop coefficient method was advocated over the Λ method by Allen *et al.* (2007), who suggested that advection, which is not included in the Λ method, is important for heterogeneous irrigated landscapes and is accounted for by the Penman–Monteith equation. A review of field studies suggested that ET_f is relatively constant over a 24-hour period in irrigated plots (Romero, 2004, cited in Allen *et al.*, 2007). In one comparison study, the evaporative fraction method had a lower RMSE (7.0%) than the crop coefficient method (16.6%) (Chávez *et al.*, 2008), but the accuracies of each method likely change with meteorological conditions, vegetation, and soil moisture.

10.2.3.5 Comparison of Temperature-Based Methods

As energy balance methods gained popularity for their simplicity and accuracy in measuring energy fluxes across landscapes, the merits of one-source and two-source approaches were scrutinized. Timmermans *et al.* (2007) compared two common energy-based methods: one-source (SEBAL) and two-source (ALEXI). SEBAL accuracy was lower over hot, dry, heterogeneous terrain because of the difficulty in selecting a dry end-member pixel within the boundaries of the remote sensing image, which is then used to calibrate Eq. (18). ALEXI was less accurate in densely vegetated areas, where small changes in vegetation cover can significantly impact canopy temperature estimation. ALEXI and other TSEBs may perform better in conditions of either dense or sparse vegetation or extremes of soil moisture (Anderson *et al.*, 2013, p. 212), and field-scale comparisons suggest that TSEBs outperform one-source models (González-Dugo *et al.*, 2009), though both produce acceptable results.

González-Dugo *et al.* (2009) evaluated instantaneous λET fluxes derived from an empirical one-layer energy balance model (Chávez *et al.*, 2005), METRIC (Allen *et al.*, 2007), and the two-source model of Kustas and Norman (1999), an updated version of the Norman *et al.* (1995) TSEB model that forms the basis of ALEXI. RMSD was less than 50 W m^{-2} (λET) and 33 W m^{-2} (H) by all methods. The TSEB of Kustas and Norman (1999) had the closest agreement to the ground observations (RMSD 30 W m^{-2}, R^2 = 0.83), followed by METRIC (RMSD 42 W m^{-2}, R^2 = 0.70), and last by the empirical one-layer model (RMSD 50 W m^{-2}, R^2 = 0.70). González-Dugo *et al.* (2009) reported similar accuracy for the Λ method (Crago, 1996), the adjusted Λ method (Anderson *et al.*, 1997) and the reference evapotranspiration fraction (Doorenbos and Pruitt, 1977 as cited in González-Dugo *et al.*, 2009) (RMSD=0.74 mm day^{-1}, R^2 = 0.76). Senay *et al.* (2011) compared λET fractions derived from SSEB and METRIC models using Landsat images acquired for south-central Idaho during the growing season. SSEB compared well with METRIC output, and SSEB was more reliable over a wide elevation range (especially > 2000 m).

Several studies have compared SEBAL, METRIC, and some combination of S-SEBI, SEBS, and SSEBop (Acharya and Sharma, 2021; Wagle *et al.*, 2017; Singh and Senay, 2015; Liaqat and Choi, 2015). Several of those studies suggest that most T_R-based models overestimate ET under dry conditions (Acharya and Sharma, 2021; Wagle *et al.*, 2017). There is no consensus on the optimal model, though either METRIC or SEBAL are often among the highest-performing models, attributed in part to internal calibration.

10.2.4 SCATTERPLOT-BASED METHODS FOR ET ESTIMATION

Scatterplot inversion methods plot T_R versus VI, the boundaries of which often approximate a triangle (Carlson, 2007; Maltese *et al.*, 2015; Kasim and Usman, 2016) or trapezoid (Gillies *et al.*, 1997; Carlson, 2007; Maltese *et al.*, 2015) (Figure 10.8). The position of a pixel in the T_R-VI or T_R-α space

Remote Sensing of Evapotranspiration from Croplands 349

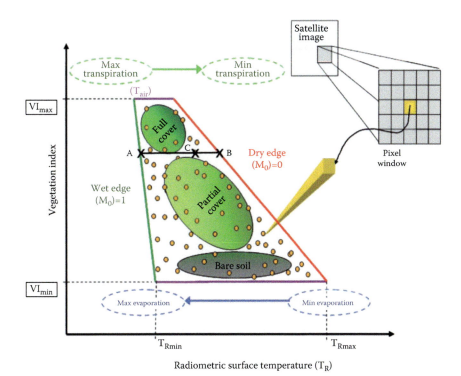

FIGURE 10.8 Example of a scatterplot of T_R vs. VI, indicating wet and dry edges. Reprinted from Petropoulos *et al.* (2009) with permission from SAGE Publications Ltd.

relative to the boundaries of the observed scatterplot determine the reference evapotranspiration fraction (ET_f) as in empirical methods with vegetation index (Section 10.2.2.1) and the evaporation fraction (EF) as in T_R-based methods (Section 10.2.3). Petropoulos *et al.* (2018) review T_R-VI scatterplot methods for ET and soil moisture retrieval.

In Figure 10.8, each yellow circle defines the measurements at a single pixel. The shape of the group of points in the T_R-VI feature space is triangular or trapezoidal if pixels with clouds or standing water are masked. This shape is a result of the insensitivity of T_R to water content for areas with dense vegetation (Petropoulos and Hristopulos, 2020). The plot has four boundary points defined by the "dry edge" on the right side and the "wet edge" on the left side. The "dry" or "warm edge" represents limited water content in and evaporation from the surface soil and defines the warmest pixels for different amounts of bare soil and vegetation cover. The "wet" or "cold edge" represents the maximum surface soil water content and defines the coldest pixels for different amounts of vegetation cover. Bare ground pixels are along the base of the triangle, while full vegetation cover is at the top. The strongest evaporative cooling is represented by pixels with the minimum T_R (point A), while the weakest cooling, by those with the maximum T_R (point B), where ET is low. NDVI quantifies vegetation cover in the triangle method, while the trapezoidal method uses the fractional vegetation cover. The ratio of CB and AB distances defines ET_f (Carlson, 2020; Petropoulos and Hristopulos, 2020). T_R-VI scatterplot methods have advantages over other methods that use satellite data to retrieve surface soil moisture (Singh *et al.*, 2022), including not being dependent on external meteorological and surface parameters and better results with heterogeneous soils (Petropoulos *et al.*, 2018).

Scatterplot methods include four groups based on the variables used in the scatterplot: (1) T_R-VI scatterplots; (2) Surface-to-air temperature difference vs. VI scatterplots; (3) T_R-*a* scatterplots; (4) day-night temperature difference vs. VI scatterplots; and (5) methods that couple the T_R-VI scatterplot with a Soil Vegetation Atmosphere Transfer (SVAT) model. Each of the groups of methods is reviewed in the following sections (Table 10.1).

10.2.4.1 T$_R$-VI Scatterplot Methods

Several approaches use the T$_R$-VI triangle scatterplot. Jiang and Islam (1999, 2001) use the Priestley–Taylor equation, where α$_{PT}$ is 1.26 times the ratio of the CB to AB distance (Figure 10.8), with RMSD of ET equal to 30% of the observed mean (Jiang and Islam, 2001). This technique assumes that extremes occur along the dry (ET = 0) and wet (ET = ETo) edges of the triangle. Nishida (2003) and Nishida and Nemani (2003) addressed this problem by estimating the evaporative fraction (Λ) with MODIS data for vegetation and soil separately, where ET from vegetation is calculated from a combination of the Penman–Monteith equation and the complementary relationship between potential ET and actual ET, while the triangle method calculates the soil evaporation. Λ was computed every eight days for a range of climates and biomes. The results showed good agreement for specific Ameriflux locations (RMSD = 45.1 W m^{-2}, bias = 5.6 W m^{-2}, R^2 = 0.86). The determination of wet and dry edges is essential for the application of the T$_R$-VI methods, which can be automated (Zhang et al., 2006).

A simplified triangle method has been proposed by Carlson and Petropoulos (2019) that calculates the surface evapotranspiration fraction (EF) and surface soil water content (M$_0$). VI is replaced by the vegetation cover fraction (F$_c$), defined as (Gillies and Carlson, 1995):

$$F_c = \left(\frac{NDVI - NDVI_O}{NDVI_s - NDVI_O} \right)^2 \tag{28}$$

where NDVI$_s$ and NDVI$_o$ are the NDVI for dense vegetation (s) and bare ground (o), respectively. Note that F$_c$ is the same as NDVI* used in other VI-based methods. The scaled surface temperature (T$_{scaled}$) is calculated as:

$$T_{scaled} = \frac{T_R - T_{Rmin}}{T_{Rmax} - T_{Rmin}} \tag{29}$$

where T$_{Rmin}$ and T$_{Rmax}$ are the minimum and maximum temperature, for wet vegetation and dry bare soil, respectively, defined by the T$_R$-VI scatter plot. M$_o$ and EF are calculated as:

$$M_0 = 1 - \frac{T_{scaled(pixel)}}{T_{scaled(dry\,edge)}} \tag{30}$$

$$EF = EF_{soil}(1 - F_c) + F_c(EF_{veg}) = M_O(1 - F_c) + F_c \tag{31}$$

where EF$_{soil}$ is the ratio between soil evaporation and net radiation (R$_n$), and EF$_{veg}$ is the transpiration fraction for vegetation. T$_{scaled(dry\,edge)}$ is T$_{scaled}$ at the dry edge of the triangle for the F$_c$ of a given pixel. Eqs. (30) and (31) assume a linear variation between T$_{scaled}$ and M$_0$ or EF. For conditions where a pixel has both vegetation and bare soil, the canopy EF is taken as the weighted value of EF for the vegetation fraction of the pixel (EF$_{veg}$ = 1, by definition) (Petropoulos et al., 2021).

The method is relatively easy to apply and does not require a land surface model or atmospheric or ancillary surface data, Fuzzo et al. (2019) used the simplified method for a specific crop. Petropoulos et al. (2020) tested the accuracy of the simplified triangle model using Sentinel-3 data in a Mediterranean environment. There was a reasonable agreement with field measurements as well as seasonal variation and differences among land cover types were largely consistent with M$_0$ and EF. Petropoulos et al. (2021) used the method with UAS data in a Mediterranean environment with very good accuracy (RMSE = 0.053 for EF), sometimes better than other studies in the same region. The use of UAS and the simplified triangle method is well suited for applications related to routine EF and M$_0$ monitoring in agricultural environments.

Remote Sensing of Evapotranspiration from Croplands

10.2.4.2 Surface-to-Air Temperature Difference and VI Scatterplot Methods

Surface-air temperature difference methods replace T_R with the difference between the surface temperature and the air surface (dT = T_R-T_2). λET is estimated from the vegetation index/temperature (VIT) trapezoid (Moran *et al.*, 1994) in areas with partial vegetation cover based on the Penman–Monteith (PM) equation (4) and the crop water stress index (CWSI = 1-ET_f) (Jackson *et al.*, 1981). The PM equation is inverted following Jackson *et al.* (1981) to estimate T_R at the four vertices of the dT-VI trapezoid, and 1-ET_f is calculated as the ratio of the difference in temperature between a given pixel and the dry temperature (CB in Figure 10.8) to the difference in temperature between the dry edge and wet edge at the pixel's NDVI value (AB in Figure 10.8). Inversion of the PM equation to determine T_R at the four corners of the trapezoid avoids the requirement that there be a pixel in the image where ET = ET_o and another where ET = 0. Moran *et al.* (1994) validated the model over both agricultural and semi-arid grasslands in Arizona, USA. Moran *et al.* (1996) used Landsat TM data to estimate instantaneous λET for grassland sites (Arizona, USA, RMSD of 29 W m^{-2}) with consistent overestimation of λET.

Jiang and Islam (2003) modified the Jiang and Islam (1999) method, using T_R-T_2 in place of T_R and F_c instead of NDVI. λET was predicted with a RMSD and bias of 58.6 and −42.4 W m^{-2} respectively. Venturini *et al.* (2004) validated the method utilizing the same sensors for a region over South Florida, with RMSD in Λ of 0.08–0.19, and R^2 0.4–0.7. Stisen *et al.* (2008) combined the method of Jiang and Islam (2003) with thermal inertia information obtained from the geostationary SEVIRI sensor to estimate the evaporative fraction (Λ) for a semi-arid region in South Senegal in Africa, with a close agreement for both Λ (RMSD of 0.13 and R^2 of 0.63) and instantaneous λET (RMSD of 41.45 W m^{-2} and R^2 of 0.66). Shu *et al.* (2011) validated the Stisen *et al.* (2008) method using observations from the Fengyun-2C (NSMC FY-2 Series, 2024) and MODIS satellites over a subtropical region in the North China Plain, with R^2 of 0.73 and RMSD of 0.92 mm day^{-1} for daily ET, and R^2 of 0.55 and RMSE of 0.14 for Λ. The Jiang and Islam (2003) method does not depend on the absolute accuracy of the T_R measures, since T_R-T_2 equal to zero in their technique always represents the cold edge of the triangle space where Λ equals zero. Jiang and Islam (2003) assumed a linear variation in Λ across the F_c-(T_R-T_2) feature space, while Stisen *et al.* (2008) assumed non-linear relationships.

10.2.4.3 T_R-Albedo Scatterplot Methods

Other triangle methods are based on a scatterplot of T_R and α. The Simplified SEBI (S-SEBI) calculates Λ using the same equation that SSEB uses to calculate ET_f (22), but in S-SEBI, T_h and T_c are linear functions of α, where the linear function coefficients are determined from the boundaries of the T_R-α plot (Roerink *et al.*, 2000). Gómez *et al.* (2005) extended S-SEBI to the retrieval of daily evapotranspiration (λET) from high spatial resolution data (20m) from PoLDER (Polarization and Directionality of Earth Reflectance) airborne instrument and a thermal infrared camera (20 m), with an error of 90 W m^{-2} (instantaneous λET) and 1 mm day^{-1} (daily total λET). Sobrino *et al.* (2005) used S-SEBI and high spatial resolution airborne images from the Digital Airborne Imaging Spectrometer (DAIS) over agricultural areas in Spain, with accuracy in daily ET prediction better than 1 mm day^{-1}. Sobrino *et al.* (2007, 2008) adapted S-SEBI to the low spatial resolution AVHRR data and during the Sentinel-2 and Fluorescence Experiment (SEN2FLEX, field measurements, airborne data) project for similar sites in Spain, with RMSD of ~1 mm day^{-1} in the estimation of daily λET compared to measurements from lysimeters. García *et al.* (2007) evaluated three operative models for estimating the "non-evaporative" fraction (NEF) as an indicator of the surface water deficit in a semi-arid area of southeast Spain. Zahira *et al.* (2009) monitored the drought status in an Algerian forest using S-SEBI with the visible, near infrared, and thermal infrared bands of Landsat ETM+ imagery, with R^2 of 0.85 and a RMSD of 64 W m^{-2}. Bhattacharya *et al.* (2010) calculated Λ from the T_R-α plot, validated over an agricultural region in India from the Indian geostationary meteorological satellite Kalapana-1, with RMSE in daily ET of 25–32%, and eight-day RMSD of 26% (0.45 mm day^{-1}) with r = 0.8. The technique of Bhattacharya *et al.* (2010) does not need any ground observations, but assumes uniform atmospheric conditions, and, like SEBAL and METRIC, requires wet and dry pixels in the image.

10.2.4.4 Day-Night Temperature Difference and VI Scatterplot Methods

The difference between daytime and nighttime T_R (diurnal surface temperature change or DSTV) versus VI is another form of the triangle method (Chen et al., 2002), based on the observed relationship between DSTV, soil moisture, and thermal inertia (Van de Griend et al., 1985; Engman and Gurney, 1991). A simple linear mixture model is used to determine the fractional contributions of vegetation, dry soil surfaces, and wet soil surfaces to the observed values of NDVI and T_R. The vegetation and moisture coefficient (VMC), which is the same as ET_f in the crop coefficient methods, was calculated at each pixel as the sum of VMC for each of the three surface types weighted by the fraction contribution at the given pixel (Chen et al., 2002). Chen et al. (2002) implemented the algorithm with AVHRR data for a wetland in South Florida, USA, with mean bias of 2.8–23.9% and RMSD of 3.08–5.74 mm day^{-1}. The Chen et al. (2002) method requires a small number of ground parameters. Limitations include the assumption of only three land cover types in the mixture modeling, the need for two T_R images to calculate the DSTV, and the assumption that ET is the same for all dense vegetation, similar to Jiang and Islam (2001). The method also requires identification of homogenous and large areas of the three distinct land cover types.

The use of DSTV and NDVI in place of daily T_R was a modification of the method of Jiang and Islam (2003) by Wang et al. (2006). Wang et al. (2006) used MODIS 1km Aqua and Terra products to map DSTV-NDVI, and estimated Λ as a function of air temperature and the Priestley–Taylor parameter αPT, which was compared with observations taken over 16 days in 2004, with RMSD of 0.106, bias = 0.002, and R^2 of 0.61. The approach is relatively simple and requires a small number of input parameters for its application.

10.2.4.5 Coupling T_R-VI Scatterplots with SVAT Models

λET and H (both instantaneous and averaged over the day) and soil moisture availability (M$_o$) can be calculated using the T_R-VI scatter plot using a SVAT model (Petropoulos et al., 2018). The SVAT model is parameterized using geographical location, biophysical characteristics (vegetation, soil profile) and atmospheric data. The SVAT model is then iterated and calibrated until the modeled extreme values of F$_c$ and T_R match the satellite data. The simulated model values are aligned with the observed T$_{scaled}$ by intersecting the "dry edge" with two extreme values (NDVI$_0$, NDVI$_s$). The SVAT-calculated T$_{scaled}$ along the "dry edge" defined by NDVI$_0$ and NDVI$_s$, are assumed to agree with the modeled values for soil moisture (M$_0$) equal to zero. The model continues to iterate, varying all possible values of F$_c$ and M$_0$, from 0% to 100% and 0–1, respectively, for all theoretical combinations of these two parameters. The result is a table where the SVAT model output values are retrieved for the given satellite overpass time. A sequence of nonlinear equations is derived by relating F$_c$ and T_R to each of the other variables of interest (H, λET, and Λ) from the model. The nonlinear equations are empirical third-order polynomial equations between F$_c$, T_R, and the output variables:

$$ET = \sum_{p=0}^{3}\sum_{q=0}^{3} a_{p,q}\left(T_{scaled}\right)^p F_c^q \qquad (33)$$

where a$_{p,q}$ are coefficients and the exponents p and q range from 0 to 3. Gillies and Carlson (1995) first applied the technique using AVHRR images in the United Kingdom. Carlson et al. (1995) estimated daily ET for a site in Pennsylvania, USA, and validated the results using ground-based measurements from the Push Broom Microwave Radiometer (PBMR) and the NS001 instrument (30m spatial resolution). Gillies et al. (1997) validated the method using high-resolution airborne data from the NS001 instrument and field observations from the First International Satellite Land Surface Climatology Project (ISLSCP) Field Experiment (FIFE; Vernekar et al., 2003) and Monsoon 90 (Kustas et al., 1991) campaigns. The RMSD was ±10% for ET (Gillies et al., 1997). Brunsell and

Gillies (2003) used data from the Thematic Mapper Simulator (TMS) and Tropospheric Infrared Mapping Spectrometers (TIMS) airborne (12m) and coarse AVHRR (1km) radiometers and the triangle method. λET for the high-resolution airborne data was within ~15% error, but results from the satellite data were in poor agreement with both the observations and the airborne data (50% difference for λET). Petropoulos and Carlson (2011) evaluated the triangle-SVAT method at several CarboEurope sites using ASTER data. Closer agreements with the ground observations were generally found when comparisons were limited to cloud-free days at flat terrain sites (RMSD in λET of 27 W m^{-2}). Petropoulos and Hristopulos (2020) combined satellite data from AATSR and ground-based data from CarboEurope to calculate λET, H, and M_0 using the SimSphere model (Gillies et al., 1997).

Triangle-SVAT methods combine the detailed physical properties embedded in the SVAT model with spectral information and its spatial variation. The triangle-SVAT method does not assume a linear relationship between the location in T_R-VI space and ET, which may give more reliable ET estimates in heterogeneous surfaces. The triangle-SVAT method can also estimate daily ET without assuming constant daily Λ (Brunsell and Gillies, 2003). Difficulties include the large number of input parameters and required user expertise (Cayrol et al., 2000).

In sum, triangle methods are an option for some users, especially for relatively quick estimates where other operational products are not available. Remaining challenges include the need for evaluation of its accuracy for a wider range of environmental and ecological conditions (Carlson, 2020). For SVAT model approaches, sensitivity analysis of the model can help reduce subjectivity in parameterization (Petropoulos et al., 2020; Singh et al., 2022).

10.2.5 LAND SURFACE MODELS AND REANALYSIS FOR ET ESTIMATION

Land surface models (LSM) track the water and energy balance, including soil moisture, runoff, and land surface fluxes, while reanalysis assimilates observational data to reconstruct land surface fluxes and atmospheric conditions over historical periods. Many LSMs and reanalysis products have been developed; examples include the Community Land Model (CLM), including adaptations for agriculture (Levis et al., 2012), the Noah model series (Niu et al., 2011; Srivastava et al., 2013b), the variable infiltration capacity (VIC) model (Wood et al., 1992), and the Global Land Surface Evaporation: Amsterdam Methodology (GLEAM; Martens et al., 2017). LSMs assimilate EO data to different extents, with some using remotely sensed land cover for a single year (e.g., ERA-5) or multiple years (e.g., CLM, Shi et al., 2013; Shi and Liang, 2014), and others directly assimilating time-varying EO data (e.g., GLEAM). GLEAM uses EO data to estimate vegetation cover fraction and microwave vegetation optical depth to calculate transpiration stress (Martens et al., 2017). EO data on LAI has been assimilated into the Noah Multiparameterization (Noah-MP) model (Kumar et al., 2019). LSA-SAF provides a λET operational product over Europe, northern Africa, southern Africa, and South America based on imagery from the second-generation Meteosat angularly stabilized satellite (MSG-2) and the Spinning Enhanced Visible and Infrared Imager (SEVIRI) radiometer, with a temporal resolution of 30 minutes (Ghilain et al., 2014). SEVIRI is a multi-spectral radiometer with a spatial resolution of 3 km at nadir (1 km for the high-resolution visible channels) and 12 spectral channels from visible to TIR. MSG-SEVIRI products use a simplified version of the SVAT TESSEL (Tiled ECMWF Surface Scheme for Exchange Processes over Land) model (Viterbo and Beljaars, 1995; Van den Hurk et al., 2000). The algorithm assimilates real-time data from meteorological satellites (Gellens-Meulenberghs et al., 2007), and EO inputs including daily albedo (Geiger et al., 2008a; Carrer et al., 2010) and half-hourly shortwave (Geiger et al., 2008b) and longwave fluxes (Ineichen et al., 2009). The spatial resolution of the daily product is 1 km and with a time lag of one day and a quality flag. The retrieval accuracy of ET is ~25% if ET is greater than 0.4 mm h^{-1} and 0.1 mm h^{-1} otherwise (Ghilain et al., 2011). Several studies have validated both instantaneous and daily ET from MSG-SEVIRI (Sepulcre-Cantó et al., 2014; Petropoulos et al., 2016). Potential uncertainties are due to algorithm standardization, surface heterogeneity, land cover classification, and auxiliary input parameters. There is good agreement with ground-based measurements of selected sites (Petropoulos et al., 2015, 2016), with RMSD 0.065–0.107 mm h^{-1}. Additional documentation would be helpful of the accuracy of the MSG-SEVIRI ET product. ETLook (Bastiaanssen et al., 2012) downscales microwave-based soil moisture estimates to 1 km and uses the Penman–Monteith equation to estimate the ET using albedo, vegetation cover, and LAI from MODIS. ETLook is operationally available for Africa and the Middle East as the FAO Water Productivity open-access product (FAO WaPOR, 2024). Sample accuracies include RMSE of 1.2 mm day^{-1} for 14 flux towers in Africa (Blatchford et al., 2020).

LSMs have the advantage of regional to global coverage with high temporal resolution under all-weather conditions, but often have coarse spatial resolution (25–100 km), though LSMs are achieving higher resolutions for specific study areas (1 km Rouf et al., 2021), including for irrigated areas (Gibson et al., 2017). LSMs may not be able to simulate irrigation effectively, especially since irrigation volumes are often difficult to quantify over large areas. LSMs therefore may not be useful for evaluating ET from individual fields, especially irrigated fields, but can be used for assessments of regional rainfed cropland responses to climate variability and change. We anticipate that LSMs will increasingly incorporate EO data, facilitate evaluation of products more exclusively forced by EO data, and allow for scenarios including assessment of climate change's impact on crop ET and production.

10.2.6 Seasonal ET Estimates and Cloud Cover Issues

All four families of methods for estimating ET reviewed in this chapter (vegetation-based, temperature- or energy-based, scatterplot-based, and land surface models) produce daily maps of ET that can be temporally interpolated to estimate seasonal ET, which is often the main output of concern to water managers and agriculturalists. Interpolation is necessary because the satellite platforms that generate high-resolution imagery often have long overpass return periods (e.g., Landsat at two weeks), and because of clouds, which compromise the generation of daily ET maps even when daily scenes are potentially available (e.g., MODIS). Cloud cover is less of a problem in arid and some semi-arid climates, and VI-based methods have been utilized for both seasonal and annual ET estimation, and for monitoring changes in ET time series (Nagler et al., 2020, 2021, 2022a, 2022b). Cloud cover is a significant obstacle for determining season-total ET in semi-humid and humid climates. In Mediterranean climates, the main growing season in summer corresponds to cloud-free conditions, and satellite methods work well for determining seasonal ET from daily values. In other locations, where the growing season coincides with the wet season, such as monsoon-dominated areas, cloud-free imagery is often not available during the main crop growing season, and EO-based methods for ET estimation may need to be supplemented with other modeling approaches like the FAO-56 method (Allen et al., 1998) or more complex models.

Microwaves can penetrate cloud cover, and significant progress has been made in microwave-based methods for monitoring agricultural areas (Liu et al., 2020). Active L-band microwaves can penetrate typical cropland vegetation canopies regardless of cloud cover, weather, and daytime acquisition conditions (Petropoulos et al., 2015). Passive microwave sensors acquire microwaves over a wider range of frequencies but are limited by coarse spatial resolution at 25–50 km (Liu and Yang, 2022). However, use of microwave sensors has been rather limited particularly in ET retrieval, and very few studies are available in the literature exploring their use for this purpose (e.g., Liu et al., 2018; Walker et al., 2018, 2019; Liu and Yang, 2022). Instead, microwave data have been used for calculating intermediate parameters required for use in ET models (Reyes-González et al., 2019; Da Rocha et al., 2020; Gan et al., 2021). Some of the parameters estimated by microwave data include soil moisture, vegetation optical depth, land surface temperature, leaf area index, surface roughness, net surface shortwave radiation, and gross primary production (Walker et al., 2019; Stankevich et al., 2017; Duan et al., 2020; Bousquet et al., 2021). Some operational ET algorithms (e.g., GLEAM; ETlook) rely on microwave sensors and downscale the coarse resolution for applications in agriculture (Martens et al., 2017; Bastiaanssen et al., 2012). Microwave applications may be particularly important in agricultural areas where the rainy season coincides with the cropping season, which is the case in many rainfed regions.

Three methods to interpolate ET maps for days without cloud-free imagery are (1) the evaporative fraction (Λ) method, (2) the crop coefficient (ET_f) method, and (3) the simulation model method (Long and Singh, 2010). The Λ and crop coefficient methods for estimating seasonal ET are very similar to the methods for generating daily estimates from instantaneous estimates, but there are special problems with cloud cover when interpolating over longer time scales that are discussed further in this section.

The Λ method, which is also used to calculate daily ET from instantaneous values of the evaporative fraction (Λ_{op}) (Section 10.2.3.5), calculates ET for a date without cloud-free imagery as the product of R_n for the day without cloud-free imagery and Λ_{op} for the date with cloud-free imagery. The Λ method assumes that Λ_{op} is constant or varies linearly between the dates of cloud-free imagery, which is more likely to be violated over several days than for a single day, especially if cloud cover changes significantly. Farah *et al.* (2004) found that Λ does not vary with cloud cover over short (weekly) time intervals over woodland and grassland in central Kenya. The method may produce accurate ET values over periods of 5–10 days (Farah *et al.*, 2004), but over many areas, cloudy conditions persist much longer. Λ likely increases during cloudy periods due to a larger proportionate reduction in H than in λET (Van Niel *et al.*, 2011).

The crop coefficient approach calculates the ratio of actual to reference ET ($=ET_f$) on the day of satellite imagery, then multiplies that fraction by ET_o for each day without imagery (Long and Singh, 2010):

$$ET_{period} = \sum_{i=d1}^{n} ET_f ET_{oi} \qquad (34)$$

where ET_f is for the 24-hour period on the date with satellite-derived ET, and ET_{oi} is reference ET on day i, which does not have imagery, *d1* is the beginning day without ET data, and *n* is the number of consecutive days without image-derived ET estimates. ET_f can be assumed constant between images or could be linearly varied between available ET images. Allen *et al.* (2007) suggest that one ET image per month is sufficient to estimate seasonal total ET, though this may not be the case under conditions of rapidly varying soil moisture conditions or surface saturation, as might be expected in irrigated areas. The ET_f method has been applied using the METRIC model (Allen *et al.*, 2007) and in northern China (Li *et al.*, 2008).

Simulation model methods use satellite-based ET on clear days to calibrate a SVAT or LSM, which is then run for all days, including cloudy days. Simplified models of the relationship between meteorological conditions and ET, such as the Granger and Gray (GG) model (Granger and Gray, 1989), have been used with SEBAL to estimate ET during cloudy periods (Long and Singh, 2010). The GG model uses the complementary relationship between actual and potential ET to estimate actual ET when imagery are not available. The GG model depends on the availability of meteorological data at a comparable spatial resolution to the observed heterogeneity in ET. Other simple models of soil moisture stress have been developed that use ET derived from remote sensing to estimate model states on clear days and extrapolate those state variable values to dates with clouds (Anderson *et al.*, 2007).

10.3 ET METHODS INTERCOMPARISON STUDIES

Each of the four families of methods used to estimate ET has different strengths and weaknesses (Table 10.1). Temperature-based methods were developed to estimate ET from irrigated agriculture, but are often sensitive to how they are calibrated, and sometimes depend on the existence of extreme values of ET in the image. Vegetation index–based methods were developed for global application, with a focus on rainfed systems, and may have lower accuracy in irrigated systems where ET may be decoupled from a vegetation index, particularly on the shoulders of the growing season (Biggs *et al.*, 2016). Scatterplot methods incorporate both temperature and vegetation, but usually require internal calibration and, like some one-source energy-based methods, often depend on extremes of ET to be present in the image. Studies that compare vegetation-, temperature-, and scatterplot-based methods together are not common; here we review some examples.

Rafn *et al.* (2008) compared an NDVI-based empirical coefficient method with METRIC over mixed crops in an Idaho irrigation district. The empirical coefficient method was fast, easy, and less costly to employ than METRIC. González-Dugo and Mateos (2008) found that NDVI-derived ET_f combined with ET_o measurements predicted ET of corn and soybean crops as well as thermal-band methods in Central Iowa, although it over-estimated ET of corn during a dry-down period. Vinukollu *et al.* (2011b) compared three different models: Surface Energy Balance System (SEBS; Su, 2002), MOD16 (Mu *et al.*, 2007), and PT-JPL (Fisher *et al.*, 2008) using MODIS-Aqua satellite augmented by AVHRR data for vegetation characterization at three spatial scales: (1) eddy covariance flux tower data; (2) a basin-scale water balance; (3) global scale using a hydrologic model driven by surface climate reanalysis. For towers where soil evaporation plays an important role following precipitation events, SEBS and PT-JPL showed the highest and similar correlations with observations, though large differences occurred during the primary growing season. Correlations between λET measured at towers in densely vegetated areas, such as evergreen and deciduous broadleaf forests, were highest for PT-JPL. At the basin scale, the performance of each model was similar. At the global scale, the vegetation and energy-based ET methods tended to underestimate simulated soil moisture storage in water-limited (arid) regions of the world. Ruhoff *et al.* (2022) intercompared eight ET products, validated with the water balance in 50 river basins across South America. Performance was best in regions with highly seasonal precipitation, and weakest in humid regions, including the central Amazon.

Ershadi *et al.* (2014) evaluated the PT-JPL model against SEBS, the 2011 updated MOD16 (Mu *et al.*, 2011), and a complementary approach (Advection-Aridity model) against ET observed at FLUXNET towers. The PT-JPL model had the closest correlation with the FLUXNET-estimated ET, followed closely by SEBS. The PT-JPL model did particularly well in densely vegetated areas and was comparable to SEBS over croplands and grasslands. On a seasonal basis, all models, except the PT-JPL model, exhibited strong seasonality. The poor performance of MOD16 and SEBS in densely vegetated areas was attributed to the uncertainties that arise from a large number of model input requirements and complexity, including the sensitivity of ET estimates to resistance parameters. All models did poorly over shrublands and evergreen needle leaf forests, reflecting the difficulty of NDVI in capturing vegetation dynamics for these land cover types. Ershadi *et al.* (2014) recommend, for regional to global studies, an ensemble of models weighted by the success of contributing models for each land cover type.

Velpuri *et al.* (2013) compared SSEBop ET (Senay *et al.*, 2013) with point and gridded flux tower observations and water balance ET, gridded FLUXNET ET (Jung *et al.*, 2011) and MOD16 ET (Mu *et al.*, 2011) over the conterminous United States (CONUS). At the 60 FLUXNET towers, both monthly SSEBop and MOD16 showed overall comparable annual accuracies with mean errors in the order of 30–60%. SSEBop showed lower RMSE than MOD16 for grassland, irrigated cropland, and forest classes, while MOD16 performed better than SSEBop in rainfed croplands, shrublands, and woody savanna classes. At basin scales, both MOD16 and SSEBop ET matched the accuracies of the global gridded FLUXNET ET. MOD16 effectively reproduced basin scale ET (up to 25% uncertainty) compared to CONUS-wide point-based ET (up to 50–60% uncertainty) illustrating the potential for MODIS ET products for basin-scale ET estimation. The apparent CONUS-wide uncertainties (up to 50–60%) for monthly MODIS ET represented errors from several FLUXNET stations. The uncertainty for individual stations was lower: 20% (MOD16, Mu *et al.*, 2011), 30% (SSEBop, Senay *et al.*, 2013), and as low as 10% for individual stations (Singh *et al.*, 2013). Despite an apparent high level of uncertainty at the CONUS-scale, the monthly SSEBop ET products can be useful for localized applications.

Choi et al. (2009) compared three models (TSEB based on ALEXI, METRIC, and the T_R-VI method) for estimating spatially distributed λET over a region in Iowa, USA, using Landsat TM/ETM+ imagery and ancillary observations from the Soil Moisture Atmosphere Coupling Experiment (SMACEX) 2002 field experiment (Kustas et al., 2005). TSEB and METRIC yielded similar and reasonable agreement with measured λET and H fluxes, with RMSD of 50–75 W m^{-2}, whereas for the T_R-VI method RMSD was over 100 W m^{-2} (3.8 mm day^{-1}). TSEB and METRIC agreed at the point scale, though the gridded λET had significant discrepancies that correlated with vegetation density.

Intercomparisons of different EO-based ET methods is greatly facilitated by automated methods, many of which are currently operationally implemented in cloud computing frameworks. For western and central parts of the continental United States the OpenET project (Melton et al., 2021) integrates six ET algorithms, including geeSEBAL (Laipelt et al., 2021; Kayser et al., 2022), METRIC, DisALEXI, SSEBop, SIMS (Melton et al., 2012), and PT-JPL. OpenET, was developed to increase ET estimates accuracy at field scale (30 m) for the western United States (Melton et al., 2021), with expected international expansion in the following years. Errors in automated and large-scale implementations of ET algorithms are typically similar to the errors from manual calibration on individual scenes. At daily time scale, the ensemble mean of OpenET products yielded a mean absolute error (MAE) of 0.74 mm day^{-1} (22% of the mean) and RMSE of 0.96 (28% of the mean) compared to eddy flux correlation towers for cropland sites (Melton et al., 2021). Another accuracy assessment (Volk et al., 2024) documented errors over croplands of 15.8 mm month^{-1} (17%), mean bias of −5.3 mm month^{-1} (6%) and R^2 of 0.9 when compared to ground measurements from towers. Abbasi et al. (2023a) compared vegetation-based methods (METEVI2, PT_JPL) with temperature-based methods (SEBAL, METRIC, DisALEXI) in an irrigated wheat field at 30m resolution. Correlation coefficients (r) between models were greater than 0.91 for all model pairs and were greater than 0.95 in 86% of model pairs. Mean bias ranged from +8% for Earth Engine METRIC (eeMETRIC) and SSEBop to −13% for SIMS, with an ensemble average 7% and smallest error of 3% (METEVI2). The use of multiple models to estimate cropland ET, as in the OpenET project, allows users to assess model uncertainties and achieve higher accuracy when compared to the use of single models, facilitating comparison of different models and helping refine automation methods.

10.4 SPECIAL PROBLEMS IN CROPPED AREAS

10.4.1 Landscape Heterogeneity and Spatial Disaggregation

Estimation of ET from croplands using remote sensing is particularly challenging in heterogeneous landscapes where agricultural plots are small (Figure 10.9). In India, for example, there are large areas of homogeneous irrigated cropping in canal-irrigated systems, but more than 50% of the irrigated area is supplied from groundwater wells, which are typically individually owned bore wells supplying small plots (< 1ha). Small groundwater-irrigated plots are often topographically organized, occurring mostly near stream channels where the water table is shallow, resulting in narrow bands of irrigation (Figure 10.9), which requires mapping irrigated areas as fractional cover of 1km MODIS pixels (Biggs et al., 2006). Most globally available and operational ET datasets are at a resolution of 500m (MOD16: USGS MOD16A2 v061, 2024) or coarser, which is significantly larger than irrigated patches in many areas, though some newer products are available at higher resolution (30m) (Table 10.5). Even in the United States, where agricultural fields are large, 1km resolution can be too coarse to resolve individual fields and to map ET differences by crop (Kustas et al., 2004). Kustas et al. (2004) documented that 250m resolution was necessary to resolve differences in ET among agricultural fields in Iowa, and that the variance in estimated ET decreases with increasing pixel size. Townshend and Justice (1988) estimated that land cover change mapping requires a spatial resolution of 250–500m. Irrigated landscapes might

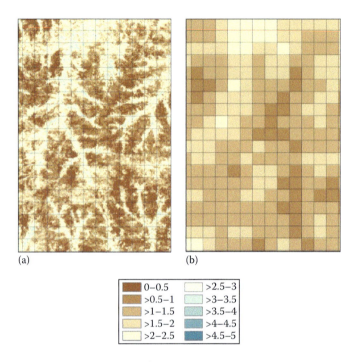

FIGURE 10.9 Maps of SEBAL ET (mm day^{-1}) in a groundwater-irrigated area in southern India at 30 m (left) and aggregated to 1 km (right). The grid in both panels represents 1km pixels. Based on data from Ahmad *et al.* (2006).

be expected to require finer resolution than 250m (He *et al.*, 2019), especially in regions where plots are small.

Global methods (PT-JPL, MOD16) were designed to estimate ET over large spatial scales, often as input to community land surface models, rather than to assess crop-specific ET. In heterogeneous irrigated landscapes in semi-arid climates, extreme spatial variability in soil moisture and ET means that extremes of low and high ET may not occur in any 1km pixel, which significantly reduces ET estimates in the 1km aggregated average. This could result in an underestimation of ET from irrigated cropland if no further disaggregation technique were applied. High-resolution imagery (< 100m, e.g., Landsat 30m) is only available at two weeks or greater temporal resolution, which limits its application in areas with high cloud cover and dynamic land cover, though observations from a suite of satellites (Landsat 7–9, Sentinel-2) improve the possible temporal resolution. The Harmonized Landsat and Sentinel-2 (HLS) record (Claverie *et al.*, 2018) assembles and harmonizes data from various platforms, which increases the temporal resolution of ET mapping. Some efforts have focused on combining imagery from different platforms to generate high-resolution maps of seasonal ET, such as the 30m resolution version of MOD16 (He *et al.*, 2019) and fusion of 70m ECOSTRESS (four-day revisit) with Landsat (100m) (Anderson *et al.*, 2021). Here we review a few select studies to illustrate the potential for cross-platform downscaling.

Thermal imagery typically has a coarser resolution than visible and near-infrared (VNIR) bands for a given spectroradiometer. For the MODIS sensor, VNIR bands are available at 250m resolution, while the thermal band is at 1km. For Landsat, VNIR are at 30m, while the thermal infrared band is 120m but has been resampled to either 60m (before February 25, 2010) or 30m (after February 25, 2010). The higher-resolution bands can be used to sharpen the coarse-resolution thermal band using the inverse relationship between T_R and NDVI (Kustas *et al.*, 2003; Agam *et al.*, 2007). The first step is to coarsen the NDVI image to the resolution of the thermal image. Both the mean NDVI and

TABLE 10.5
Summary of Operational ET Products

Name	Method	Satellite	Spatial Resolution	Spatial Extent	Temporal Resolution	Publisher	References
SEVIRI	LSM	SEVIRI MSG-2	1–3km	Europe, North Africa, South Africa, South America	30 min.	EUMETSAT	Algorithm Theoretical Basis Document MSG-2 Evapotranspiration Product, 2008
MOD16	VI	MODIS	1km	Global	8-day, month, annual	NASA	Mu et al. (2011)
OpenET	ALEXI/DisALEXI, eeMETRIC, geeSEBAL, PT-JPL, SIMS, SSEBop	Various	30m–1km	US	Daily	NASA	Melton et al. (2022)
FAO WaPOR	ETLook	Various, MODIS	30–250m	Africa, Middle East	Daily 10-day	FAO	FAO (2020)
PML ET v2	PM-VI	MODIS	500m	Global	8-day	GEE	Zhang et al., (2019)
SSEBop	T_R	MODIS	1km	Global	10-day	USGS	Senay et al. (2020)
METRIC-EEFLUX	T_R	Landsat	30m	Global	14-day	GEE	Allen et al. (2015)

the coefficient of variation (CV) of the high-resolution NDVI are calculated at the low resolution. Then, a percentage of pixels with the lowest CV with each of several bins of NDVI (e.g., 0–0.25, 0.2–0.5, > 0.5) are selected to parameterize a function relating low-resolution T_R to low-resolution NDVI, which is used to predict T_R at low resolution:

$$\hat{T}_{Rlow}(NDVI_{low}) = a_{ds} + b_{ds}NDVI_{low} + c_{ds}NDVI_{low}^2 \tag{35}$$

where a_{ds}, b_{ds}, and c_{ds} are empirical coefficients determined through least squares regression, and the ds subscript designates downscaling to differentiate the coefficients from the a and b parameters of the SEBAL algorithm. A linear equation could also be used, and the form of the equation depends on the observed T_R-NDVI relationship. While Eq. (35) can be used directly to predict T_R at high spatial resolution using high-resolution NDVI, the correlation between T_R and NDVI may break down for pixels with low NDVI, including irrigated areas in either the beginning or end of the growing season, where high soil moisture or standing water, and therefore low T_R, can co-occur with low vegetation cover (Figure 10.7). Kustas *et al.* (2003) proposed an additional correction using the residual of (35):

$$\hat{T}_{R_{hi}} = \hat{T}_{R_{low}}(NDVI_{hi}) + (T_{R_{low}} - \hat{T}_{Rlow}(NDVI_{low})) \tag{36}$$

where $\hat{T}_{R_{hi}}$ is the predicted radiometric temperature at high spatial resolution, $\hat{T}_{R_{low}}(NDVI_{hi})$ is the predicted radiometric temperature using high-resolution NDVI and (35), T_{Rlow} is the observed radiometric temperature at low resolution, and $\hat{T}_{Rlow}(NDVI_{low})$ is the predicted radiometric temperature using low-resolution NDVI (32). Based on observed scatterplots of NDVI and T_R (Figure 10.7), this correction will be minimal for pixels with high NDVI and largest for pixels with low NDVI.

More complex algorithms that use all visible and NIR bands to downscale T_R in a moving window may be more successful in irrigated landscapes (Gao *et al.*, 2012). Other algorithms have been developed to downscale T_R by fusing MODIS and Landsat, including those that use artificial neural network (ANN) models (Bindhu *et al.*, 2013). One downscaling method specific to the ALEXI model, DisALEXI, was discussed in Section 10.2.3.2. DisALEXI ensures consistency in aggregated ET across scales by adjusting the high-resolution estimates to match the low-resolution average. ET estimated using algorithms forced by imagery with different resolutions may disagree. SEBS-ET from Landsat or ASTER differed markedly from SEBS-ET derived from MODIS, even when the Landsat and ASTER ET were aggregated to MODIS resolution (McCabe and Wood, 2006) (Figure 10.10), likely due to non-linear averaging of important inputs to the energy balance equations. The spatial average ET from all three image sources matched to within 10–15%, suggesting that the low-resolution MODIS imagery were useful for watershed-scale estimates of ET (50 km^2), but the MODIS data underestimated the variability present in the landscape. McCabe and Wood (2006) conclude that MODIS data are sufficient for estimating ET at watershed scales but are likely not accurate for estimating crop ET at resolutions that resolve ET from individual fields.

The Landsat archive can provide historical imagery at 30m resolution back to the late 1970s and has been used in many ET applications (Norman *et al.*, 2003; Ahmad *et al.*, 2006; Glenn *et al.*, 2011b; Kjaersgaard *et al.*, 2011; Anderson *et al.*, 2012a, 2012b; Allen *et al.*, 2013), though problems with cloud cover remain, particularly where the rainy season coincides with the cropping season, and Landsat imagery may not be sufficient to capture high-quality data in areas with dynamic land cover. ASTER generates data with 15m resolution in the visible and 90m in the thermal infrared bands, but the image footprints and times of acquisition are often irregular, complicating its use for seasonal ET estimation (Er-Raki *et al.*, 2008; Galleguillos *et al.*, 2011). Data from new remote sensing platforms could prove very useful for mapping ET at spatial and temporal resolutions that more closely approximate actual variability in agricultural landscapes. Sentinel-2 and -3 and the Planet constellation of satellites can provide high-resolution data of vegetation indices at high temporal frequency. Sentinel-2 provides data on visible, near infrared, and shortwave infrared wavelengths at

Remote Sensing of Evapotranspiration from Croplands

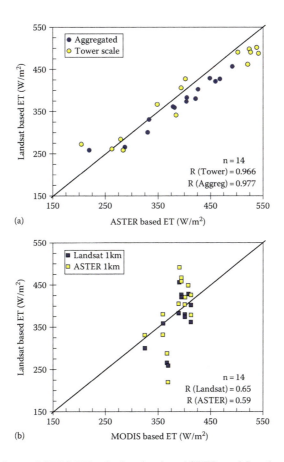

FIGURE 10.10 Comparison of SEBS-ET calculated using ASTER and Landsat data (top) and MODIS, ASTER, and Landsat (bottom). In the top panel, tower scale refers to individual pixels of Landsat or ASTER, while aggregated are the Landsat and ASTER SEBS-ET aggregated to MODIS resolution (1km). Reprinted from McCabe and Wood (2006) with permission from Elsevier.

10 or 20m with a revisit time of five days at the equator and two to three days at mid-latitudes and has been used for vegetation-based ET models (Vanino *et al.*, 2018; Elfarkh *et al.*, 2023). Sentinel-3 provides thermal imagery at 1km resolution with revisit times of approximately one day at the equator, though the temporal and spatial resolution is similar to existing MODIS data, so the additional gains for temperature-based methods may come in downscaling the 1km data using Sentinel-2 data (e.g., Guzinski and Nieto, 2019). Planet data (CubeSat) are available in RGB and near-infrared (NIR) bands at high spatial (~3–4m) and temporal (quasi daily) resolution but have the added challenge that each image is taken by different sensors that are not inter-calibrated, complicating the use of multiple Planet images over time. Landsat surface reflectance data can be used to constrain the mean reflectance of Planet pixels, a process known as CubeSat Enabled Spatio-Temporal Enhancement Method (CESTEM) (Houborg and McCabe, 2018) as applied for ET estimation using PT-JPL by Aragon *et al.* (2018, 2021).

Additional sensors have been developed that allow for EO-based ET estimates at high spatial resolution with high overpass frequency. ECOSTRESS ET, with data available starting in 2018, has 70m resolution and four-day overpass (Fisher *et al.*, 2020). ECOSTRESS is mounted on the International Space Station, so coverage is limited to 52° N and S latitude, and samples different times of day on each overpass (Hook and Hulley, 2019), which has the advantage of providing pseudo-diurnal profiles of ET but can make images difficult to compare if land surface conditions change rapidly.

10.4.2 Model Complexity, Equifinality, and Sources of Error in ET Models

The EO-based estimates of ET presented in this chapter all rely on models to predict ET from EO data, and so have problems similar to other applications of models to estimate hydrological processes, including equifinality and parameter uncertainty. Equifinality arises in models that have many variables in the determining equations but few actual data to populate the equations (Franks et al., 1997; Medlyn et al., 2005; Beven, 2006a). Models are frequently calibrated using approximated or assumed values for unmeasured variables. Models with different assumptions and levels of complexity can converge on the same output values, despite different process representations. Equifinality in hydrological models complicates the choice of a model for a given application, makes the use of models for hypothesis testing difficult, and results in uncertain prediction for times or places that have not been used for calibration (Beven, 2006a).

In remote sensing of ET, equifinality can occur in process-based models in mixed natural and agricultural areas, for which few spatially distributed ground data are available. Remote sensing data usually consist only of radiance values from two or three bands (VNIR), and an imperfect estimate of T_R from thermal bands. The determining equations generally have numerous variables, including fractional cover, LAI, roughness lengths for mass and momentum transfer, albedo, emissivity, net radiation, and ground heat flux (e.g., Bastiaanssen et al., 1998; Kustas and Norman, 1999). All of these are related in some way to vegetation cover, so they are often estimated by the use of vegetation indices. Glenn et al. (2008a) gave examples where five different plant stands in a mixed agricultural and riparian environment had the same NDVI values but differed markedly in plant heights (and presumably therefore roughness lengths), fractional cover, and LAI. VIs cannot uniquely determine separate biophysical variables; rather, they give an integrated measure of canopy "greenness" (Baret and Buis, 2008; Glenn et al., 2008a). Equifinality may occur in surface energy balance and VI methods that parameterize numerous variables with limited remote sensing data. For example, the SEBAL algorithm calibrates the aerodynamic resistance and the coefficients in the T_R and air temperature (T_2) relationship at the wet and dry pixels, but other representations of the T_R-T_2 relationship may result in different predictions at other locations in the image. McCabe et al. (2005) explicitly include the impact of different model parameterizations on ET estimated by calibrating a hydrological model with remotely sensed data. Explicit treatments of uncertainty in ET estimation could be produced in future efforts.

A problem related to equifinality is parameter uncertainty. In the PT-JPL model, for example, several parameters are based on field studies in specific locations, with unknown applicability in other regions. Over-simplification of the process representation may require additional ground-level data or calibration of the simplified model to a specific location, as in the empirical crop coefficient methods. This complicates the use of simplified methods in global applications or in other regions with limited ground-level data for input or calibration. Overall, a balance has to be struck between model complexity and applicability to unmeasured locations; the ideal model is the simplest one that provides adequate fit to observed data given the limitations in input data availability. There is significant scope for future research in the level of complexity needed for different scales of application, particularly at regional and global scales.

Another problem with application of EO data to estimate ET is error and uncertainty in the ground data used for calibration and validation. Ground measurements usually come from eddy covariance flux tower measurements, with typical errors of 15–30% when compared to lysimeters or other highly accurate measurements of ET (Allen et al., 2011b). Eddy covariance data have "energy closure" errors, where the sum of measured $\lambda E + H$ does not equal measured $R_n - G$. λE and H are usually increased to force closure, preserving the $\lambda E:H$ ratio of the measured data (Twine et al., 2000), but usually there is no way to check if this correction improves the ET estimates. Scott (2010) compared eddy covariance results at three flux tower sites in a semiarid rangeland, at which precipitation, infiltration, and runoff were also measured. Uncorrected eddy covariance data gave ET values close to precipitation minus runoff and infiltration, where data corrected to force closure overestimated ET at each site by 10–20%.

A further problem is the mismatch between scales of measurement, especially when advection of atmospheric moisture between dry and wet areas violates the one-dimensional assumption of several

EO-based ET models. Evett *et al.* (2012) compared surface energy balance components used to calculate ET at scales ranging from weighing lysimeters (4.7 m diameter) to small plots to whole fields (several hundred ha) captured by aerial and satellite imagery. Even with the best equipment and expertise, it was difficult to measure ET accurately using flux towers. Advection led to underestimates of ET by towers compared to lysimeters even after correcting for energy closure. Evett *et al.* (2012) urged caution in interpreting ET data from semi-arid environments with advective conditions, especially those with mixes of irrigated and dryland crops and native vegetation. In dryland areas in Spain (Morillas *et al.*, 2013) a two-source energy balance model had errors of up to 90% in estimating λET compared to eddy covariance because λET was estimated as a residual and was a small component of the overall energy budget. Both energy balance and vegetation-based remote sensing methods overestimated ET of salt-stressed shrubs by 50% or more in a riparian corridor surrounded by dry uplands (Glenn *et al.*, 2013).

There has been a proliferation of remote sensing ET methods, most of which have uncertainty or errors of 10–30% (Glenn *et al.*, 2007, 2011b) compared to ground measurements on a daily or monthly basis with lower error (e.g., 5–10%) on a seasonal or annual basis (Table 10.1) (Allen *et al.*, 2011a). Comparison studies often do not identify a clear choice of methods due to the problem of equifinality and errors and uncertainties in ground data (e.g., González-Dugo *et al.*, 2009). Simple methods tend to perform as well as more complex methods (e.g., Jiang *et al.*, 2004; Abbasi *et al.*, 2023b), but simple methods often depend on calibration to ground-level data and may not be applicable outside of the area of calibration. Reducing the error and uncertainty in remote sensing estimates of ET depends in part on improving ground methods for measuring ET. Medlyn *et al.* (2005) and Beven (2006b) recommended more rigorous sensitivity analyses of ET models and explicit representation of uncertainty in model estimates. Medlyn *et al.* (2005) concluded that simplistic comparisons of ET models with eddy covariance data could lead to errors due to problems of equifinality, insensitivity, and uncertainty in both the models and the ground data. Their main focus was on SVAT models, but their conclusions can also be applied to remote sensing methods for estimating ET (Glenn *et al.*, 2008a). Future research could calculate pixel-wise estimates of the uncertainty in ET estimated from satellite imagery that accompany any map of ET, calculated using either a range of plausible parameter values or an intercomparison of models with different assumptions. Cloud-based implementations of multiple ET models will greatly facilitate such inter-comparisons.

10.5 EO-BASED OPERATIONAL PRODUCTS AVAILABLE FOR ET

Rapid advances have been made to estimate ET from EO data in operational products (Table 10.5). Some operational products are available globally, and others for specific regions (MSG-SEVIRI, OpenET, WaPOR). Although some ET operational products have been available for a few decades, a comprehensive accuracy assessment is lacking. The recent availability of EO datasets from more sophisticated sensors (such as MSG-3 SEVIRI, Sentinel-3, ECOSTRESS, continuation of Landsat mission), provides an opportunity to develop new operational ET products. Continuity will be critical for porting algorithms across platforms and for assembling comparable datasets over time; for example, the MODIS suite has provided data for more than 20 years, and continuity with other sensors, in particular VIIRS, is being planned (Román *et al.*, 2024).

10.6 CONCLUSIONS

This chapter provided a critical overview of methods for estimating ET from EO platforms, with a focus on croplands. The chapter used consistent mathematical symbols across all methods, facilitating a systematic intercomparison of multiple techniques. With a single comparison of multiple methods in one text, practitioners and researchers can more clearly compare different methods, identify how their particular model choice could be extended to include other methods, and identify the assumptions, strengths, and weaknesses of each family of methods. Many different methods are available with different degrees of complexity, utilizing EO data acquired from different platforms. Each method has different strengths and limitations related to their practical implementation and varying accuracy depending on the dataset and land cover. Validation studies have confirmed the potential for regional- and global-scale ET mapping at 500 m to 1 km, with recent advances in 30 m products with operational implementation (Table 10.5). Several global models are vegetation-based methods and may not perform well in irrigated areas, where soil evaporation may occur in dry environments, but modifications are being incorporated to improve ET in such conditions (e.g., Zhang et al., 2021). T_R-based methods, especially those requiring internal calibration, have been widely tested in irrigated environments but face challenges in application for geographic scales larger than a single satellite image, though progress has been made in applying temperature methods at continental scales (de Andrade et al., 2024). We encourage further intercomparison of the different EO-based modeling schemes for deriving ET in croplands, especially for operational models being applied at regional and continental scales (Table 10.5).

Most studies that evaluate different techniques to predict ET have been based on direct comparisons between predicted fluxes and corresponding in situ measurements. Other modeling approaches, such as uncertainty or sensitivity analysis, have so far been little incorporated in such studies, despite their importance for any all-inclusive model validation/verification (Petropoulos et al., 2013). Explicit estimation of uncertainty in ET, including pixel-wise estimates of model error, would be a significant advance.

More validation studies for operationally distributed products could be conducted in different ecosystems globally. Such studies, if conducted in a systematic way following an acceptable protocol, will identify issues in the algorithmic design of these products that will improve our capability to operationally estimate ET from EO sensors. More work could be directed toward the development of schemes for the temporal interpolation of the instantaneous ET estimates, as well as of downscaling approaches of ET to the resolution of individual fields. Since no model performs optimally in all land covers, regional or global applications could consider using different models for different land cover types. Further intercomparison of vegetation, temperature-based, and land surface model methods, and further research on downscaling to the resolution of individual fields would help further refine model selection for a given time and location.

ACKNOWLEDGMENTS

We wish to acknowledge the passing of Dr. Edward P. Glenn in 2017 and his important contributions to this chapter in the last edition. His review articles on the state of evapotranspiration science through ground and remote sensing methods continue to be cited by academics from around the world and by myriad practitioners who seek to understand water accounting through evapotranspiration estimation in both natural ecosystems and agricultural lands. Thanks to John Kimball, Mutlu Ozdogan, Geneva Chong, and Prasad S. Thenkabail for their helpful comments. The participation of George P. Petropoulos was financially supported by the LISTEN-EO project funded by the framework of H.F.R.I called "Basic research Financing (Horizontal support of all Sciences)" under the National Recovery and Resilience Plan "Greece 2.0" funded by the European Union—NextGenerationEU (H.F.R.I. Project Number: 15898). Any use of trade, firm, or product names is for descriptive purposes only and does not imply endorsement by the US government.

REFERENCES

Abbasi, N., Nouri, H., Didan, K., Barreto-Muñoz, A., Chavoshi Borujeni, S., Salemi, H., Opp, C., Siebert, S., Nagler, P. 2021. Estimating actual evapotranspiration over croplands using vegetation index methods and dynamic harvested area. Remote Sensing 13(24), 5167.

Abbasi, N., Nouri, H., Didan, K., Barreto-Muñoz, A., Chavoshi Borujeni, S., Opp, C., Nagler, P., Thenkabail, P.S., Siebert, S. 2023a. Mapping vegetation index-derived actual evapotranspiration across croplands using the google earth engine platform. Remote Sensing 15(4), 1017.

Abbasi, N., Nouri, H., Nagler, P., Didan, K., Chavoshi Borujeni, S., Barreto-Muñoz, A., Opp, C., Siebert, S., 2023b. Crop water use dynamics over arid and semi-arid croplands in the lower Colorado River Basin. European Journal of Remote Sensing 56, 2259244. doi:10.1080/22797254.2023.2259244.

Acharya, B., Sharma, V., 2021. Comparison of satellite driven surface energy balance models in estimating crop evapotranspiration in semi-arid to arid inter-mountain region. Remote Sensing 13, 1822. doi:10.3390/rs13091822.

Agam, N., Kustas, W.P., Anderson, M.C., Li, F., Colaizzi, P.D. 2007. Utility of thermal sharpening over Texas high plains irrigated agricultural fields. Journal of Geophysical Research 112, D19110. doi:10.1029/2007JD008407.

Aguilar, A., Flores, H.E.M., Crespo, G., Marin, M., Campos, I., Calera, A. 2018. Performance assessment of MOD16 in evapotranspiration evaluation in Northwestern Mexico. Water 10(7), 901. doi:10.3390/w10070901.

Ahmad, M., Biggs, T.W., Turral, H., Scott, C.A. 2006. Application of SEBAL approach to map the agricultural water use patterns in the data scarce Krishna River Basin of India. Water Science and Technology 53, 83–90.

Allen, R.G. 2000. Using the FAO-56 dual crop coefficient method over an irrigated region as part of an evapotranspiration intercomparison study. Journal of Hydrology 229, 27–41.

Allen, R.G., Burnett, B., Kramber, W., Huntington, J., Kjaersgaard, J., Kilic, A., Kelly, C., Trezza, R. 2013. Automated calibration of the METRIC-landsat evapotranspiration process. JAWRA Journal of the American Water Resources Association 49, 563–576. doi:10.1111/jawr.12056.

Allen, R.G., Irmak, A., Trezza, R., Hendrickx, J.M.H., Bastiaanssen, W., Kjaersgaard, J. 2011a. Satellite-based ET estimation in agriculture using SEBAL and METRIC. Hydrological Processes 25, 4011–4027. doi:10.1002/hyp.8408.

Allen, R.G., Morton, C., Kamble, B., Kilic, A., Huntington, J., Thau, D., Gorelick, N., Erickson, T., Moore, R., Trezza, R., Ratcliffe, I., Robison, C. 2015. EEFlux: A landsat-based evapotranspiration mapping tool on the Google Earth Engine, in: 2015 ASABE/IA Irrigation Symposium: Emerging Technologies for Sustainable Irrigation—A Tribute to the Career of Terry Howell, Sr. Conference Proceedings. American Society of Agricultural and Biological Engineers, pp. 1–11. doi:10.13031/irrig.20152143511.

Allen, R.G., Pereira, L.S. 2009. Estimating crop coefficients from fraction of ground cover and height. Irrigation Science 28, 17–34.

Allen, R.G., Pereira, L.S., Howell, T.A., Jensen, M.E. 2011b. Evapotranspiration information reporting: I. Factors governing measurement accuracy. Agricultural Water Management 98, 899–920.

Allen, R.G., Pereira, L.S., Raes, D., Smith, M. 1998. Crop evapotranspiration-guidelines for computing crop water requirements-FAO Irrigation and drainage paper 56. Rome.

Allen, R.G., Tasumi, M., Trezza, R. 2007. Satellite-based energy balance for Mapping Evapotranspiration with Internalized Calibration (METRIC)—model. Journal of Irrigation and Drainage Engineering 133, 380–394. doi:10.1061/(ASCE)0733-9437(2007)133:4(380).

Anderson, M.C., Allen, R.G., Morse, A., Kustas, W.P. 2012a. Use of Landsat thermal imagery in monitoring evapotranspiration and managing water resources. Remote Sensing of Environment 122, 50–65. doi:10.1016/j.rse.2011.08.025.

Anderson, M.C., Kustas, W.P., Alfieri, J.G., Gao, F., Hain, C., Prueger, J.H., Evett, S., Colaizzi, P., Howell, T., Chávez, J.L., 2012b. Mapping daily evapotranspiration at Landsat spatial scales during the BEAREX'08 field campaign. Advances in Water Resources 50, 162–177. doi:10.1016/j.advwatres.2012.06.005.

Anderson, M.C., Kustas, W.P., Hain, C.R. 2013. Mapping surface fluxes and moisture conditions from field to global scales using ALEXI/DisALEXI. In: Remote sensing of energy fluxes and soil moisture content (G.P. Petropoulos, ed.), pp. 207–232. Taylor and Francis, New York, NY.

Anderson, M.C., Norman, J.M., Diak, G.R., Kustas, W.P., Mecikalski, J.R. 1997. A two-source time-integrated model for estimating surface fluxes using thermal infrared remote sensing. Remote Sensing of Environment 60, 195–216. doi:10.1016/S0034-4257(96)00215-5.

Anderson, M.C., Norman, J.M., Mecikalski, J.R., Otkin, J.A., Kustas, W.P. 2007. A climatological study of evapotranspiration and moisture stress across the continental United States based on thermal remote sensing: 1. Model formulation. Journal of Geophysical Research: Atmospheres 112, D10117. doi:10.1029/2006JD007506.

Anderson, M.C., Yang, Y., Xue, J., Knipper, K.R., Yang, Y., Gao, F., Hain, C.R., Kustas, W.P., Cawse-Nicholson, K., Hulley, G., Fisher, J.B., Alfieri, J.G., Meyers, T.P., Prueger, J., Baldocchi, D.D., Rey-Sanchez, C. 2021. Interoperability of ECOSTRESS and Landsat for mapping evapotranspiration time series at sub-field scales. Remote Sensing of Environment 252, 112189. doi:10.1016/j.rse.2020.112189.

Aragon, B., Houborg, R., Tu, K., Fisher, J.B., McCabe, M. 2018. CubeSats enable high spatiotemporal retrievals of crop-water use for precision agriculture. Remote Sensing 10(12), 1867. doi:10.3390/rs10121867.

Aragon, B., Ziliani, M.G., Houborg, R., Franz, T.E., McCabe, M.F. 2021. CubeSats deliver new insights into agricultural water use at daily and 3 m resolutions. Scientific Reports 11, 12131. doi:10.1038/s41598-021-91646-w.

Bajgain, R., Xiao, X., Wagle, P., Kimball, J., Brust, C., Basara, J., Gowda, P., Starks, P., Neel, J., 2020. Comparing evapotranspiration products of different temporal and spatial scales in native and managed prairie pastures. Remote Sensing 13, 82. doi:10.3390/rs13010082.

Baldocchi, D., Falge, E., Gu, L., Olson, R., Hollinger, D., Running, S., Anthoni, P., Bernhofer, C., Davis, K., Evans, R. 2001. FLUXNET: A new tool to study the temporal and spatial variability of ecosystem-scale carbon dioxide, water vapor, and energy flux densities. Bulletin of the American Meteorological Society 82, 2415–2434.

Baret, F., Buis, S. 2008. Estimating canopy characteristics from remote sensing observations: Review of methods and associated problems. In: Advances in land remote sensing, pp. 173–201. Springer, Dordrecht. https://doi.org/10.1007/978-1-4020-6450-0_7

Bastiaanssen, W.G.M., Ahmad, M.D., Chemin, Y. 2002. Satellite surveillance of evaporative depletion across the Indus Basin. Water Resources Research 38, 1273, doi:10.1029/2001WR000386.

Bastiaanssen, W.G.M., Cheema, M.J.M., Immerzeel, W.W., Miltenburg, I.J., Pelgrum, H. 2012. Surface energy balance and actual evapotranspiration of the transboundary Indus Basin estimated from satellite measurements and the ETLook model. Water Resources Research 48, n/a-n/a. doi:10.1029/2011wr010482.

Bastiaanssen, W.G.M., Menenti, M., Feddes, R.A., Holtslag, A.A.M. 1998. A remote sensing surface energy balance algorithm for land (SEBAL): 1. Formulation. Journal of Hydrology 212–213, 198–212.

Bastiaanssen, W.G.M., Noordman, E.J.M., Pelgrum, H., Davids, G., Thoreson, B.P., Allen, R.G. 2005. SEBAL model with remotely sensed data to improve water-resources management under actual field conditions. Journal of irrigation and drainage engineering 131, 85–93.

Bausch, W.C., Neale, C.M.U. 1987. Crop coefficients derived from reflected canopy radiation: A concept. American Society of Agricultural Engineers 30, 703–709.

Beven, K. 2006a. A manifesto for the equifinality thesis. Journal of Hydrology 320, 18–36.

Beven, K. 2006b. On undermining the science? Hydrological Processes 20, 3141–3146.

Bhattacharya, B.K., Mallick, K., Patel, N.K., Parihar, J.S. 2010. Regional clear sky evapotranspiration over agricultural land using remote sensing data from Indian geostationary meteorological satellite. Journal of Hydrology 387, 65–80.

Bhattarai, N., Mallick, K., Jain, M. 2019. Sensitivity of four contextual remote sensing based surface energy balance models to spatial domain. International Archives Photogrammetry Remote Sensing Spatial Information Sciences XLII-3/W6, 3–7. doi:10.5194/isprs-archives-XLII-3-W6-3-2019.

Bhattarai, N., Quackenbush, L.J., Im, J., Shaw, S.B. 2017. A new optimized algorithm for automating endmember pixel selection in the SEBAL and METRIC models. Remote Sensing of Environment 196, 178–192. doi:10.1016/j.rse.2017.05.009.

Biggs, T.W., Marshall, M., Messina, A. 2016. Mapping daily and seasonal evapotranspiration from irrigated crops using global climate grids and satellite imagery: Automation and methods comparison. Water Resources Research 52, 7311–7326. doi:10.1002/2016WR019107.

Biggs, T.W., Thenkabail, P.S., Gumma, M.K., Scott, C.A., Parthasaradhi, G.R., Turral, H.N. 2006. Irrigated area mapping in heterogeneous landscapes with MODIS time series, ground truth and census data, Krishna Basin, India. International Journal of Remote Sensing 27, 4245–4266. doi:10.1080/01431160600851801.

Bindhu, V.M., Narasimhan, B., Sudheer, K.P. 2013. Development and verification of a non-linear disaggregation method (NL-DisTrad) to downscale MODIS land surface temperature to the spatial scale of Landsat thermal data to estimate evapotranspiration. Remote Sensing of Environment 135, 118–129. doi:10.1016/j.rse.2013.03.023.

Bisht, G., Bras, R.L. 2010. Estimation of net radiation from the MODIS data under all sky conditions: Southern Great Plains case study. Remote Sensing of Environment 114, 1522–1534. doi:10.1016/j.rse.2010.02.007.

Bisht, G., Venturini, V., Islam, S., Jiang, L. 2005. Estimation of the net radiation using MODIS (Moderate Resolution Imaging Spectroradiometer) data for clear sky days. Remote Sensing of Environment 97, 52–67.

Blatchford, M.L., Mannaerts, C.M., Njuki, S.M., Nouri, H., Zeng, Y., Pelgrum, H., Wonink, S., Karimi, P. 2020. Evaluation of WaPOR V2 evapotranspiration products across Africa. Hydrological Processes 34, 3200–3221. doi:10.1002/hyp.13791.

Bonan, G.B. 1998. The land surface climatology of the NCAR land surface model coupled to the NCAR community climate model. Journal of Climate 11, 1307–1326.

Bosilovich, M. 2008. NASA's modern era retrospective-analysis for research and applications: Integrating Earth observations. Earthzine. [Online] 26.

Bousquet, E., Mialon, A., Rodríguez-Fernández, N., Prigent, C., Wagner, F., Kerr, Y.H. 2021. Influence of surface water variations on VOD and biomass estimates from passive microwave sensors. Remote Sensing of Environment, 257, 112345. doi:10.1016/j.rse.2021.112345.

Brunsell, N.A., Gillies, R.R. 2003. Scale issues in land—atmosphere interactions: Implications for remote sensing of the surface energy balance. Agricultural and Forest Meteorology 117, 203–221.

Brust, C., Kimball, J.S., Maneta, M.P., Jencso, K., He, M., Reichle, R.H. 2021. Using SMAP Level-4 soil moisture to constrain MOD16 evapotranspiration over the contiguous USA. Remote Sensing of Environment 255, 112277. doi:10.1016/j.rse.2020.112277.

Cammalleri, C., Anderson, M.C., Gao, F., Hain, C.R., Kustas, W.P. 2013. A data fusion approach for mapping daily evapotranspiration at field scale. Water Resources Research 49, 4672–4686. doi:10.1002/wrcr.20349.

Cammalleri, C., Anderson, M.C., Gao, F., Hain, C.R., Kustas, W.P. 2014. Mapping daily evapotranspiration at field scales over rainfed and irrigated agricultural areas using remote sensing data fusion. Agricultural and Forest Meteorology 186, 1–11. doi:10.1016/j.agrformet.2013.11.001.

Campos, I., Neale, C.M.U., Calera, A., Balbontín, C., González-Piqueras, J. 2010. Assessing satellite-based basal crop coefficients for irrigated grapes (Vitis vinifera L.). Agricultural Water Management 98, 45–54. doi:10.1016/j.agwat.2010.07.011.

Carlson, T. 2007. An overview of the "triangle method" for estimating surface evapotranspiration and soil moisture from satellite imagery. Sensors 7, 1612–1629.

Carlson, T.N. 2020. A brief analysis of the triangle method and a proposal for its operational implementation. Remote Sensing 12(22), 3832. doi:10.3390/rs12223832.

Carlson, T.N., Gillies, R.R., Schmugge, T.J. 1995. An interpretation of methodologies for indirect measurement of soil water content. Agricultural and Forest Meteorology 77, 191–205. doi:10.1016/0168-1923(95)02261-U.

Carlson, T.N., Petropoulos, G.P. 2019. A new method for estimating of evapotranspiration and surface soil moisture from optical and thermal infrared measurements: The simplified triangle. International Journal of Remote Sensing 40(20), 7716–7729. doi:10.1080/01431161.2019.1601288.

Carrer, D., Roujean, J., Meurey, C. 2010. Comparing operational MSG/SEVIRI Land Surface Albedo products from Land SAF with ground measurements and MODIS. IEEE Transactions on Geoscience and Remote Sensing 48(4), 1714–1728. doi:10.1109/tgrs.2009.2034530.

Cawse-Nicholson, K., Braverman, A., Kang, E.L., Li, M., Johnson, M., Halverson, G., Anderson, M., Hain, C., Gunson, M., Hook, S. 2020. Sensitivity and uncertainty quantification for the ECOSTRESS evapotranspiration algorithm—DisALEXI. International Journal of Applied Earth Observation and Geoinformation 89, 102088. doi:10.1016/j.jag.2020.102088.

Cayrol, P., Chehbouni, A., Kergoat, L., Dedieu, G., Mordelet, P., Nouvellon, Y. 2000. Grassland modeling and monitoring with SPOT-4 VEGETATION instrument during the 1997–1999 SALSA experiment. Agricultural and Forest Meteorology 105(1–3), 91–115. doi:10.1016/s01681923(00)00191-x.

Chávez, J.L., Neale, C.M.U., Hipps, L.E., Prueger, J.H., Kustas, W.P. 2005. Comparing aircraft-based remotely sensed energy balance fluxes with eddy covariance tower data using heat flux source area functions. Journal of Hydrometeorology 6, 923–940.

Chávez, J.L., Neale, C.M.U., Prueger, J., Kustas, W. 2008. Daily evapotranspiration estimates from extrapolating instantaneous airborne remote sensing ET values. Irrigation Science 27, 67–81. doi:10.1007/s00271-008-0122-3.

Chen, F., Janjić, Z., Mitchell, K. 1997. Impact of atmospheric surface-layer parameterizations in the new land-surface scheme of the NCEP mesoscale Eta model. Boundary-Layer Meteorology 85, 391–421.

Chen, F., Mitchell, K., Schaake, J., Xue, Y., Pan, H., Koren, V., Duan, Q.Y., Ek, M., Betts, A. 1996. Modeling of land surface evaporation by four schemes and comparison with FIFE observations. Journal of Geophysical Research: Atmospheres (1984–2012) 101, 7251–7268.

Chen, J.-H., Kan, C.-E., Tan, C.-H., Shih, S.-F. 2002. Use of spectral information for wetland evapotranspiration assessment. Agricultural Water Management 55, 239–248.

Chen, J.M., Liu, J., 2020. Evolution of evapotranspiration models using thermal and shortwave remote sensing data. Remote Sensing of Environment 237, 111594. doi:10.1016/j.rse.2019.111594.

Chen, Y., Xia, J., Liang, S., Feng, J., Fisher, J.B., Li, X., Li, X., Liu, S., Ma, Z., Miyata, A., Mu, Q., Sun, L., Tang, J., Wang, K., Wen, J., Xue, Y., Yu, G., Zha, T., Zhang, L., Zhang, Q., Zhao, T., Zhao, L., Yuan, W. 2014. Comparison of satellite-based evapotranspiration models over terrestrial ecosystems in China. Remote Sensing of Environment 140, 279–293. doi:10.1016/j.rse.2013.08.045.

Choi, M., Kustas, W.P., Anderson, M.C., Allen, R.G., Li, F., Kjaersgaard, J.H. 2009. An intercomparison of three remote sensing-based surface energy balance algorithms over a corn and soybean production region (Iowa, US) during SMACEX. Agricultural and Forest Meteorology 149, 2082–2097.

Choudhury, B.J. 1989. Estimating evaporation and carbon assimilation using infrared temperature data. In: Theory and applications of optical remote sensing (G. Asrar, ed.), pp. 628–690. Wiley-Interscience, New York.

Choudhury, B.J., Ahmed, N.U., Idso, S.B., Reginato, R.J., Daughtry, C.S.T. 1994. Relations between evaporation coefficients and vegetation indices studied by model simulations. Remote Sensing of Environment 50, 1–17.

Claverie, M., Ju, J., Masek, J.G., Dungan, J.L., Vermote, E.F., Roger, J.-C., Skakun, S.V., Justice, C., 2018. The Harmonized Landsat and Sentinel-2 surface reflectance data set. Remote Sensing of Environment 219, 145–161. doi:10.1016/j.rse.2018.09.002.

Cleugh, H.A., Leuning, R., Mu, Q., Running, S.W. 2007. Regional evaporation estimates from flux tower and MODIS satellite data. Remote Sensing of Environment 106, 285–304.

Clothier, B.E., Clawson, K.L., Pinter, P.J., Moran, M.S., Reginato, R.J., Jackson, R.D. 1986. Estimation of soil heat flux from net radiation during the growth of alfalfa. Agricultural and Forest Meteorology 37, 319–329. doi:10.1016/0168-1923(86)90069-9.

Conrad, C., Dech, S., Hafeez, M., Lamers, J., Martius, C., Strunz, G. 2007. Mapping and assessing water use in a Central Asian irrigation system by utilizing MODIS remote sensing products. Irrigation and Drainage Systems 21, 197–218. doi:10.1007/s10795-007-9029-z.

Courault, D., Seguin, B., Olioso, A. 2005. Review on estimation of evapotranspiration from remote sensing data: From empirical to numerical modeling approaches. Irrigation and Drainage Systems 19, 223–249.

Crago, R.D. 1996. Conservation and variability of the evaporative fraction during the daytime. Journal of Hydrology 180, 173–194. doi:10.1016/0022-1694(95)02903-6.

Da Rocha, N.S., Käfer, P.S., Skokovic, D., Veeck, G.P., Diaz, L.R., Kaiser, E.A., Carvalho, C.M., Cruz, R.C., Sobrino, J.A., Roberti, D.R., Rolim, S.B.A. 2020. The influence of land surface temperature in evapotranspiration estimated by the S-SEBI model. Atmosphere 11(10), 1059. doi:10.3390/atmos11101059.

de Andrade, B.C., Laipelt, L., Fleischmann, A., Huntington, J., Morton, C., Melton, F., Erickson, T., Roberti, D.R., De Arruda Souza, V., Biudes, M., Gomes Machado, N., Antonio Costa Dos Santos, C., Cosio, E.G., Ruhoff, A. 2024. geeSEBAL-MODIS: Continental-scale evapotranspiration based on the surface energy balance for South America. ISPRS Journal of Photogrammetry and Remote Sensing 207, 141–163. doi:10.1016/j.isprsjprs.2023.12.001.

Deng, K.A.K., Lamine, S., Pavlides, A., Petropoulos, G.P., Bao, Y., Srivastava, P.K., Guan, Y. 2019. Large scale operational soil moisture mapping from passive MW radiometry: SMOS product evaluation in Europe and USA. International Journal of Applied Earth Observation and Geoinformation 80, 206–217. doi:10.1016/j.jag.2019.04.015.

Doorenbos, J., Pruitt, W.O. 1977. Guidelines for Predicting Crop Water Requirements. FAO Irrigation and Drainage Paper No. 24 (second rev. ed.), FAO, Rome, Italy.

Dorigo, W., Himmelbauer, I., Aberer, D., Schremmer, L., Petrakovic, I., Zappa, L., Preimesberger, W., Xaver, A., Annor, F., Ardö, J., Baldocchi, D., Bitelli, M., Blöschl, G., Bogena, H., Brocca, L., Calvet, J., Camarero, J.J., Capello, G., Choi, M., . . . Sabia, R. 2021. The international soil moisture network: Serving earth system science for over a decade. Hydrology and Earth System Sciences 25(11), 5749–5804. doi:10.5194/hess-25-5749-2021.

dos Santos Nascimento, G.S., Ruhoff, A., Cavalcanti, J.R., da Motta Marques, D., Roberti, D.R., da Rocha, H.R., Munar, A.M., Fragoso, C.R., de Oliveira, M.B.L. 2019. Assessing CERES surface radiation components for tropical and subtropical biomes. IEEE Journal of Selected Topics in Applied Earth Observations and Remote Sensing 12(10), 3826–3840.

Duan, S., Han, X., Huang, C., Li, Z., Wu, H., Ye, Q., Gao, M., Leng, P. 2020. Land surface temperature retrieval from passive microwave satellite observations: State-of-the-art and future directions. Remote Sensing 12(16), 2573. doi:10.3390/rs12162573.

Duchemin, B., Hadria, R., Erraki, S., Boulet, G., Maisongrande, P., Chehbouni, A., Escadafal, R., Ezzahar, J., Hoedjes, J.C.B., Kharrou, M.H. 2006. Monitoring wheat phenology and irrigation in Central Morocco: On the use of relationships between evapotranspiration, crops coefficients, leaf area index and remotely-sensed vegetation indices. Agricultural Water Management 79, 1–27.

ECMWF ERA5-Land. 2024. https://cds.climate.copernicus.eu/cdsapp#!/dataset/reanalysis-era5-land?tab=overview, access date April 13, 2024.

ECMWF Reanalysis v5. 2024. www.ecmwf.int/en/forecasts/dataset/ecmwf-reanalysis-v5, access date April 13, 2024.

Elfarkh, J., Johansen, K., El Hajj, M.M., Almashharawi, S.K., McCabe, M.F. 2023. Evapotranspiration, gross primary productivity and water use efficiency over a high-density olive orchard using ground and satellite based data. Agricultural Water Management 287, 108423. doi:10.1016/j.agwat.2023.108423.

El-Shikha, D.M., Waller, P., Hunsaker, D., Clarke, T., Barnes, E. 2007. Ground-based remote sensing for assessing water and nitrogen status of broccoli. Agricultural Water Management 92, 183–193.

Engman, E.T., Gurney, R.J. 1991. Recent advances and future implications of remote sensing for hydrologic modeling. In: Recent advances in the modeling of hydrologic systems (D.S. Bowles, ed.), pp. 471–495. Kluwer Academic Publishers, Dordrecht, Netherlands.

Er-Raki, S., Chehbouni, A., Hoedjes, J., Ezzahar, J., Duchemin, B., Jacob, F. 2008. Improvement of FAO-56 method for olive orchards through sequential assimilation of thermal infrared-based estimates of ET. Agricultural Water Management 95, 309–321.

Ershadi, A., McCabe, M.F., Evans, J.P., Chaney, N.W., Wood, E.F. 2014. Multi-site evaluation of terrestrial evaporation models using FLUXNET data. Agricultural and Forest Meteorology 187, 46–61. doi:10.1016/j.agrformet.2013.11.008.

ESA HyRELIEF. 2024. https://eo4society.esa.int/projects/hyrelief/, access date April 13, 2024.

EUMETSAT CM SAF. 2024. https://wui.cmsaf.eu/safira/action/viewHome, access date April 13, 2024.

Evett, S.R., Kustas, W.P., Gowda, P.H., Anderson, M.C., Prueger, J.H., Howell, T.A. 2012. Overview of the ushland Evapotranspiration and Agricultural Remote sensing EXperiment 2008 (BEAREX08): A field experiment evaluating methods for quantifying ET at multiple scales. Advances in Water Resources 50, 4–19.

Faisol, A., Indarto, I., Novita, E., Budiyono, B. 2020. An evaluation of MODIS global evapotranspiration product (MOD16A2) as terrestrial evapotranspiration in East Java—Indonesia. IOP Conference Series: Earth and Environmental Science 485(1), 012002. doi:10.1088/1755-1315/485/1/012002.

FAO. 2020. WaPOR database methodology, version 2. https://doi.org/10.4060/ca9894en, access date April 13, 2024.

FAO. 2023. Remote sensing determination of evapotranspiration: Algorithms, strengths, weaknesses, uncertainty and best fit-for-purpose. Cairo. doi:10.4060/cc8150en.

FAO WaPOR. 2024. https://data.apps.fao.org/wapor/?lang=en, access date April 13, 2024.

Farah, H.O., Bastiaanssen, W.G.M., Feddes, R.A. 2004. Evaluation of the temporal variability of the evaporative fraction in a tropical watershed. International Journal of Applied Earth Observation and Geoinformation 5, 129–140.

Fisher, J.B., Lee, B., Purdy, A.J., Halverson, G.H., Dohlen, M.B., Cawse-Nicholson, K., Wang, A., Anderson, R.G., Aragon, B., Arain, M.A., Baldocchi, D.D., Baker, J.M., Barral, H., Bernacchi, C.J., Bernhofer, C., Biraud, S.C., Bohrer, G., Brunsell, N., Cappelaere, B., Castro-Contreras, S., Chun, J., Conrad, B.J., Cremonese, E., Demarty, J., Desai, A.R., De Ligne, A., Foltýnová, L., Goulden, M.L., Griffis, T.J., Grünwald, T., Johnson, M.S., Kang, M., Kelbe, D., Kowalska, N., Lim, J., Maïnassara, I., McCabe, M.F., Missik, J.E.C., Mohanty, B.P., Moore, C.E., Morillas, L., Morrison, R., Munger, J.W., Posse, G., Richardson, A.D., Russell, E.S., Ryu, Y., Sanchez-Azofeifa, A., Schmidt, M., Schwartz, E., Sharp, I., Šigut, L., Tang, Y., Hulley, G., Anderson, M., Hain, C., French, A., Wood, E., Hook, S. 2020. ECOSTRESS: NASA's next generation mission to measure evapotranspiration from the international space station. Water Resources Research 56, e2019WR026058. doi:10.1029/2019WR026058.

Fisher, J.B., Malhi, Y., Bonal, D., Da Rocha, H.R., De Araújo, A.C., Gamo, M., Goulden, M.L., Hirano, T., Huete, A.R., Kondo, H., Kumagai, T., Loescher, H.W., Miller, S., Nobre, A.D., Nouvellon, Y., Oberbauer, S.F., Panuthai, S., Roupsard, O., Saleska, S., Tanaka, K., Tanaka, N., Tu, K.P., Von Randow, C. 2009. The land—atmosphere water flux in the tropics. Global Change Biology 15, 2694–2714. doi:10.1111/j.1365-2486.2008.01813.x.

Fisher, J.B., Tu, K.P., Baldocchi, D.D. 2008. AVHRR and ISLSCP-II data, validated at 16 FLUXNET sites. Remote Sensing of Environment 112, 901–919.

Foley, J.A., DeFries, R., Asner, G.P., Barford, C., Bonan, G., Carpenter, S.R., Chapin, F.S., Coe, M.T., Daily, G.C., Gibbs, H.K., Helkowski, J.H., Holloway, T., Howard, E.A., Kucharik, C.J., Monfreda, C., Patz, J.A., Prentice, I.C., Ramankutty, N., Snyder, P.K. 2005. Global consequences of land use. Science 309, 570–574. doi:10.1126/science.1111772.

Foley, J.A., Ramankutty, N., Brauman, K.A., Cassidy, E.S., Gerber, J.S., Johnston, M., Mueller, N.D., O'Connell, C., Ray, D.K., West, P.C., Balzer, C., Bennett, E.M., Carpenter, S.R., Hill, J., Monfreda, C., Polasky, S., Rockström, J., Sheehan, J., Siebert, S., Tilman, D., Zaks, D.P.M. 2011. Solutions for a cultivated planet. Nature 478, 337–42. doi:10.1038/nature10452.

Franks, S.W., Beven, K.J., Quinn, P.F., Wright, I.R. 1997. On the sensitivity of the soil-vegetation-atmosphere transfer (SVAT) schemes: Equifinality and the problem of robust calibration. Agricultural and Forest Meteorology 86, 63–75.

French, A.N., Hunsaker, D.J., Sanchez, C.A., Saber, M., Gonzalez, J.R., Anderson, R., 2020. Satellite-based NDVI crop coefficients and evapotranspiration with eddy covariance validation for multiple durum wheat fields in the US Southwest. Agricultural Water Management 239, 106266. doi:10.1016/j.agwat.2020.106266.

Fuzzo, D.F.S., Carlson, T.N., Kourgialas, N.N., Petropoulos, G.P. 2019. Coupling remote sensing with a water balance model for soybean yield predictions over large areas. Earth Science Informatics 13(2), 345–359. doi:10.1007/s12145-019-00424-w.

Galleguillos, M., Jacob, F., Prévot, L., French, A., Lagacherie, P. 2011. Comparison of two temperature differencing methods to estimate daily evapotranspiration over a Mediterranean vineyard watershed from ASTER data. Remote Sensing of Environment 115, 1326–1340.

Gan, G., Zhao, X., Fan, X., Xie, H., Jin, W., Zhou, H., Cui, Y., Liu, Y. 2021. Estimating the gross primary production and evapotranspiration of rice paddy fields in the Sub-Tropical region of China using a Remotely-Sensed based Water-Carbon coupled model. Remote Sensing 13(17), 3470. doi:10.3390/rs13173470.

Gao, F., Kustas, W.P., Anderson, M.C. 2012. A data mining approach for sharpening thermal satellite imagery over land. Remote Sensing 4, 3287–3319.

Gao, F., Masek, J., Schwaller, M., Hall, F. 2006. On the blending of the Landsat and MODIS surface reflectance: Predicting daily Landsat surface reflectance. Geoscience and Remote Sensing, IEEE Transactions on 44, 2207–2218.

García, M., Villagarcía, L., Contreras, S., Domingo, F., Puigdefábregas, J. 2007. Comparison of three operative models for estimating the surface water deficit using ASTER reflective and thermal data. Sensors 7, 860–883.

García-Santos, V., Sánchez, J., Cuxart, J. 2022. Evapotranspiration acquired with remote sensing thermal-based algorithms: A State-of-the-art review. Remote Sensing 14, 3440. doi:10.3390/rs14143440.

Geiger, B., Carrer, D., Franchistéguy, L., Roujean, J., Meurey, C. 2008a. Land surface albedo derived on a daily basis from Meteosat second generation observations. IEEE Transactions on Geoscience and Remote Sensing 46(11), 3841–3856. doi:10.1109/tgrs.2008.2001798.

Geiger, B., Meurey, C., Lajas, D., Franchistéguy, L., Carrer, D., Roujean, J. 2008b. Near real-time provision of downwelling shortwave radiation estimates derived from satellite observations. Meteorological Applications 15(3), 411–420. doi:10.1002/met.84.

Gellens-Meulenberghs, F., Arboleda, A., Ghilain, N. 2007. Towards a continuous monitoring of evapotranspiration based on MSG data. IAHS-AISH Publication 228–234. www.cabdirect.org/abstracts/20093105756.html.

Gentine, P., Entekhabi, D., Chehbouni, A., Boulet, G., Duchemin, B. 2007. Analysis of evaporative fraction diurnal behaviour. Agricultural and Forest Meteorology 143, 13–29. doi:10.1016/j.agrformet.2006.11.002.

Ghilain, N., Arboleda, A., Gellens-Meulenberghs, F. 2011. Evapotranspiration modelling at large scale using near-real time MSG SEVIRI derived data. Hydrology Earth System Sciences 15, 771–786. doi:10.5194/hess-15-771-2011.

Ghilain, N., De Roo, F., Gellens-Meulenberghs, F. 2014. Evapotranspiration monitoring with Meteosat Second Generation satellites: Improvement opportunities from moderate spatial resolution satellites for vegetation. International Journal of Remote Sensing 35(7), 2654–2670. doi:10.1080/01431161.2014.883093.

Gibson, J., Franz, T.E., Wang, T., Gates, J., Grassini, P., Yang, H., Eisenhauer, D., 2017. A case study of field-scale maize irrigation patterns in western Nebraska: Implications for water managers and recommendations for hyper-resolution land surface modeling. Hydrology Earth System Sciences 21, 1051–1062. doi:10.5194/hess-21-1051-2017.

Gillies, R.R., Carlson, T.N. 1995. Thermal remote sensing of surface soil water content with partial vegetation cover for incorporation into climate models. Journal of Applied Meteorology 34, 745–756.

Gillies, R.R., Kustas, W.P., Humes, K.S. 1997. A verification of the 'triangle' method for obtaining surface soil water content and energy fluxes from remote measurements of the Normalized Difference Vegetation Index (NDVI) and surface radiant temperature. International Journal of Remote Sensing 18, 3145–3166.

Glenn, E.P., Doody, T.M., Guerschman, J.P., Huete, A.R., King, E.A., McVicar, T.R., Van Dijk, A.I., Van Niel, T.G., Yebra, M., Zhang, Y. 2011a. Actual evapotranspiration estimation by ground and remote sensing methods: The Australian experience. Hydrological Processes 25(26), 4103–4116. doi:10.1002/hyp.8391.

Glenn, E.P., Huete, A.R., Nagler, P.L., Hirschboeck, K.K., Brown, P. 2007. Integrating remote sensing and ground methods to estimate evapotranspiration. Critical Reviews in Plant Sciences 26, 139–168.

Glenn, E.P., Huete, A.R., Nagler, P.L., Nelson, S.G. 2008a. Relationship between remotely-sensed vegetation indices, canopy attributes and plant physiological processes: What vegetation indices can and cannot tell us about the landscape. Sensors 8, 2136–2160.

Glenn, E.P., Morino, K., Didan, K., Jordan, F., Carroll, K., Nagler, P.L., Hultine, K., Sheader, L., Waugh, J. 2008b. Scaling sap flux measurements of grazed and ungrazed shrub communities with fine and coarse-resolution remote sensing. Ecohydrology 1, 316–329.

Glenn, E.P., Nagler, P.L., Huete, A.R. 2010. Vegetation index methods for estimating evapotranspiration by remote sensing. Surveys in Geophysics 31, 531–555.

Glenn, E.P., Nagler, P.L., Morino, K., Hultine, K.R. 2013. Phreatophytes under stress: Transpiration and stomatal conductance of saltcedar (Tamarix spp.) in a high-salinity environment. Plant and Soil 371, 655–672.

Glenn, E.P., Neale, C.M.U., Hunsaker, D.J., Nagler, P.L. 2011b. Vegetation index-based crop coefficients to estimate evapotranspiration by remote sensing in agricultural and natural ecosystems. Hydrological Processes 25(26), 4050–4062. doi:10.1002/hyp.8392.

GMAO MERRA. 2024. https://gmao.gsfc.nasa.gov/reanalysis/MERRA-2/, access date April 13, 2024.

Gómez, M., Olioso, A., Sobrino, J.A., Jacob, F. 2005. Retrieval of evapotranspiration over the Alpilles/ReSeDA experimental site using airborne POLDER sensor and a thermal camera. Remote Sensing of Environment 96, 399–408.

Gontia, N.K., Tiwari, K.N. 2010. Estimation of crop coefficient and evapotranspiration of wheat *Triticum aestivum* in an irrigation command using remote sensing and GIS. Water Resources Management 24, 1399–1414.

González-Dugo, M.P., Mateos, L. 2008. Spectral vegetation indices for benchmarking water productivity of irrigated cotton and sugarbeet crops. Agricultural Water Management 95, 48–58.

González-Dugo, M.P., Neale, C.M.U., Mateos, L., Kustas, W.P., Prueger, J.H., Anderson, M.C. and Li, F. 2009. A comparison of operational remote sensing-based models for estimating crop evapotranspiration. Agricultural and Forest Meteorology 149, 1843–1853.

Gottschalck, J., Meng, J., Rodell, M., Houser, P. 2005. Analysis of multiple precipitation products and preliminary assessment of their impact on global land data assimilation system land surface states. Journal of Hydrometeorology 6, 573–598. doi:10.1175/JHM437.1.

Gowda, P., Chavez, J., Colaizzi, P., Evett, S., Howell, T., Tolk, J. 2008. ET mapping for agricultural water management: Present status and challenges. Irrigation Science 26, 223–237. doi:10.1007/s00271-007-0088-6.

Granger, R.J., Gray, D.M. 1989. Evaporation from natural nonsaturated surfaces. Journal of Hydrology 111, 21–29.

Groeneveld, D.P., Baugh, W.M. 2007. Correcting satellite data to detect vegetation signal for eco-hydrologic analyses. Journal of Hydrology 344, 135–145. doi:10.1016/j.jhydrol.2007.07.001.

Groeneveld, D.P., Baugh, W.M., Sanderson, J.S., Cooper, D.J. 2007. Annual groundwater evapotranspiration mapped from single satellite scenes. Journal of Hydrology 344(1–2), 146–156. doi:10.1016/j.jhydrol.2007.07.002.

Guerschman, J.P., McVicar, T.R., Vleeshower, J., Van Niel, T.G., Peña-Arancibia, J.L., Chen, Y. 2022. Estimating actual evapotranspiration at field-to-continent scales by calibrating the CMRSET algorithm with MODIS, VIIRS, Landsat and Sentinel-2 data. Journal of Hydrology 605, 127318. doi:10.1016/j.jhydrol.2021.12.

Guerschman, J.P., van Dijk, A.I.J.M., Mattersdorf, G., Beringer, J., Hutley, L.B., Leuning, R., Pipunic, R.C., Sherman, B.S. 2009. Scaling of potential evapotranspiration with MODIS data reproduces flux observations and catchment water balance observations across Australia. Journal Hydrology 369, 107–119. doi:10.1016/j.jhydrol.2009.02.013.

Gui, S., Liang, S., Li, L. 2010. Evaluation of satellite-estimated surface longwave radiation using ground-based observations. Journal Geophysical Research 115, D18214. doi:10.1029/2009JD013635.

Gupta, S.K., Ritchey, N.A., Wilber, A.C., Whitlock, C.H., Gibson, G.G., Stackhouse, P.W.J. 1999. A climatology of surface radiation budget derived from satellite data. Journal of Climate 12, 2691–2710.

Guzinski, R., Nieto, H. 2019. Evaluating the feasibility of using Sentinel-2 and Sentinel-3 satellites for high-resolution evapotranspiration estimations. Remote Sensing of Environment 221, 157–172. doi:10.1016/j.rse.2018.11.019.

Hall, F.G., Huemmrich, K.F., Goetz, S.J., Sellers, P.J., Nickeson, J.E. 1992. Satellite remote sensing of surface energy balance: Success, failures, and unresolved issues in FIFE. Journal of Geophysical Research: Atmospheres 97, 19061–19089. doi:10.1029/92JD02189.

Hank, T.B., Berger, K., Bach, H., Clevers, J.G.P.W., Gitelson, A., Zarco-Tejada, P., Mauser, W. 2019. Spaceborne imaging spectroscopy for sustainable agriculture: Contributions and challenges. Surveys in Geophysics 40, 515–551. doi:10.1007/s10712-018-9492-0.

He, M., Kimball, J.S., Yi, Y., Running, S.W., Guan, K., Moreno, A., Wu, X., Maneta, M. 2019. Satellite data-driven modeling of field scale evapotranspiration in croplands using the MOD16 algorithm framework. Remote Sensing of Environment 230, 111201. doi:10.1016/j.rse.2019.05.020.

Heller, M.C., Willits-Smith, A., Mahon, T., Keoleian, G.A., Rose, D. 2021. Individual US diets show wide variation in water scarcity footprints. Nature Food 2, 255–263. https://doi.org/10.1038/s43016-021-00256-2.

Hersbach, H., Bell, B., Berrisford, P., Hirahara, S., Horányi, A., Muñoz-Sabater, J., Nicolas, J., Peubey, C., Radu, R., Schepers, D., Simmons, A. 2020. The ERA5 global reanalysis. Quarterly Journal of the Royal Meteorological Society 146(730), 1999–2049.

Hook, S.J., Hulley, G. 2019. ECOSTRESS cloud mask daily L2 global 70 m V001. doi:10.5067/ECOSTRESS/ECO2CLD.001.

Houborg, R., McCabe, M.F. 2018. A Cubesat Enabled Spatio-Temporal Enhancement Method (CESTEM) utilizing planet, landsat and MODIS data. Remote Sensing of Environment 209, 211–226. doi:10.1016/j.rse.2018.02.067.

Huete, A., Didan, K., Miura, T., Rodriguez, E.P., Gao, X., Ferreira, L.G. 2002. Overview of the radiometric and biophysical performance of the MODIS vegetation indices. Remote Sensing of Environment 83(1–2), 195–213. doi:10.1016/S0034-4257(02)00096-2.

Hunsaker, D.J., Fitzgerald, G.J., French, A.N., Clarke, T.R., Ottman, M.J., Pinter, P.J., Jr. 2007a. Wheat irrigation management using multispectral crop coefficients: I. Crop evapotranspiration prediction. Transactions of the American Society of Agricultural and Biological Engineers 50, 2017–2033.

Hunsaker, D.J., Fitzgerald, G.J., French, A.N., Clarke, T.R., Ottman, M.J., Pinter, P.J., Jr. 2007b. Wheat irrigation management using multispectral crop coefficients: II. Irrigation scheduling performance, grain yield, and water use efficiency. Transactions of the American Society of Agricultural and Biological Engineers 50, 2035–2050.

Hunsaker, D.J., French, A.N., Clarke, T.R., El-Shikha, D.M. 2011. Water use, crop coefficients, and irrigation management criteria for camelina production in arid regions. Irrigation Science 29, 27–43. doi:10.1007/s00271-010-0213-9.

Hunsaker, D.J., Pinter, P.J., Barnes, E.M., Kimball, B.A. 2003. Estimating cotton evapotranspiration crop coefficients with a multispectral vegetation index. Irrigation Science 22, 95–104. doi:10.1007/s00271-003-0074-6.

Hunsaker, D.J., Pinter, P.J., Jr., Kimball, B.A. 2005. Wheat basal crop coefficients determined by normalized difference vegetation index. Irrigation Science 24, 1–14.

Idso, S.B. 1981. A set of equations for full spectrum and 8-to 14-µm and 10.5-to 12.5-µm thermal radiation from cloudless skies. Water Resources Research 17, 295–304.

Ineichen, P., Barroso, C.S., Geiger, B., Hollmann, R., Marsouin, A., Mueller, R. 2009. Satellite application facilities irradiance products: Hourly time step comparison and validation over Europe. International Journal of Remote Sensing 30(21), 5549–5571. doi:10.1080/01431160802680560.

Jackson, R.D., Hatfield, J.L., Reginato, R.J., Idso, S.B., Pinter, P.J., Jr. 1983. Estimation of daily evapotranspiration from one time-of-day measurements. Agricultural Water Management 7, 351–362.

Jackson, R.D., Idso, S.B., Reginato, R.J., Pinter, P.J. 1981. Canopy temperature as a crop water stress indicator. Water Resources Research 17, 1133–1138.

Jarchow, C.J., Didan, K., Barreto-Muñoz, A., Nagler, P.L., Glenn, E.P. 2018. Application and comparison of the MODIS-derived enhanced vegetation index to VIIRS, landsat 5 TM and landsat 8 OLI platforms: A case study in the arid colorado river delta, Mexico. Sensors 18(5), 1546.

Jarchow, C.J., Nagler, P.L., Glenn, E.P. 2017. Greenup and evapotranspiration following the Minute 319 pulse flow to Mexico: An analysis using Landsat 8 Normalized Difference Vegetation Index (NDVI) data. Ecological Engineering 106, 776–783. doi:10.1016/j.ecoleng.2016.08.007.

Jarchow, C.J., Waugh, W.J., Didan, K., Barreto-Muñoz, A., Herrmann, S., Nagler, P.L. 2020. Vegetation-groundwater dynamics at a former uranium mill site following invasion of a biocontrol agent: A time series analysis of Landsat normalized difference vegetation index data. Hydrological Processes 34(12), 2739–2749. doi:10.1002/hyp.13772.

Jarchow, CJ, Waugh, W.J., Nagler, P.L. 2022. Calibration of an evapotranspiration algorithm in a semiarid sagebrush steppe using a 3-ha lysimeter and Landsat normalized difference vegetation index data. Ecohydrology 15(3), e2413. doi:10.1002/eco.2413.

Jayanthi, H., Neale, C.M.U., Wright, J.L. 2007. Development and validation of canopy reflectance-based crop coefficient for potato. Agricultural Water Management 88, 235–246. doi:10.1016/j.agwat.2006.10.020.

Jiang, L., Islam, S. 1999. A methodology for estimation of surface evapotranspiration over large areas using remote sensing observations. Geophysical Research Letters 26, 2773–2776.

Jiang, L., Islam, S. 2001. Estimation of surface evaporation map over southern great plains using remote sensing data. Water Resources Research 37, 329–340.

Jiang, L., Islam, S. 2003. An intercomparison of regional latent heat flux estimation using remote sensing data. International Journal of Remote Sensing 24, 2221–2236. doi:10.1080/01431160210154821.

Jiang, L., Islam, S., Carlson, T.N. 2004. Uncertainties in latent heat flux measurement and estimation: Implications for using a simplified approach with remote sensing data. Canadian Journal of Remote Sensing 30, 769–787.

Jiang, Z., Huete, A.R., Didan, K., Miura, T. 2008. Development of a two-band enhanced vegetation index without a blue band. Remote sensing of Environment 112(10), 3833–3845. doi:10.1016/j.rse.2008.06.006.

Johnson, L.F., Trout, T.J. 2012. Satellite NDVI assisted monitoring of vegetable crop evapotranspiration in California's San Joaquin Valley. Remote Sensing 4, 439–455.

Jung, M., Reichstein, M., Bondeau, A. 2009. Towards global empirical upscaling of FLUXNET eddy covariance observations: Validation of a model tree ensemble approach using a biosphere model. Biogeosciences 6, 2001–2013. doi:10.5194/bg-6-2001-2009.

Jung, M., Reichstein, M., Ciais, P., Seneviratne, S.I., Sheffield, J., Goulden, M.L., Bonan, G.B., Cescatti, A., Chen, J., De Jeu, R., Dolman, A.J., Eugster, W., Gerten, D., Gianelle, D., Gobron, N., Heinke, J., Kimball, J.S., Law, B.E., Montagnani, L., . . . Zhang, K. 2010. Recent decline in the global land evapotranspiration trend due to limited moisture supply. Nature, 467, 951–954. doi:10.1038/nature09396.

Jung, M., Reichstein, M., Margolis, H.A., Cescatti, A., Richardson, A.D., Arain, M.A., Arneth, A., Bernhofer, C., Bonal, D., Chen, J., Gianelle, D., Gobron, N., Kiely, G., Kutsch, W., Lasslop, G., Law, B.E., Lindroth, A., Merbold, L., Montagnani, L., Moors, E.J., Papale, D., Sottocornola, M., Vaccari, F., Williams, C. 2011. Global patterns of land-atmosphere fluxes of carbon dioxide, latent heat, and sensible heat derived from eddy covariance, satellite, and meteorological observations. Journal of Geophysical Research 116, G00J07. doi:10.1029/2010JG001566.

Kalma, J.D., McVicar, T.R., McCabe, M.F. 2008. Estimating land surface evaporation: A review of methods using remotely sensed surface temperature data. Surveys in Geophysics 29, 421–469.

Kalnay, E., Kanamitsu, M., Kistler, R., Collins, W., Deaven, D., Gandin, L., Iredell, M., Saha, S., White, G., Woollen, J., Zhu, Y., Chelliah, M., Ebisuzaki, W., Higgins, W., Janowiak, J., Mo, K.C., Ropelewski, C., Wang, J., Leetmaa, A., Reynolds, R., Jenne, R., Joseph, D. 1996. The NCEP/NCAR 40-year reanalysis project. Bulletin of the American Meteorological Society 77, 437–472.

Kasim, A.A., Usman, A.A. 2016. Triangle method for estimating soil surface wetness from satellite imagery in Allahabad district, Uttar Pradesh, India. Journal of Geoscience and Environment Protection 4(1), 84–92. doi:10.4236/gep.2016.41010.

Kayser, R.H., Ruhoff, A., Laipelt, L., Kich, E.D.M., Roberti, D.R., Souza, V.D.A., Rubert, G.C.D., Collischonn, W., Neale, C.M.U., 2022. Assessing geeSEBAL automated calibration and meteorological reanalysis uncertainties to estimate evapotranspiration in subtropical humid climates. Agricultural and Forest Meteorology 314, 108775. doi:10.1016/j.agrformet.2021.108775.

Kjaersgaard, J., Allen, R.G., García, M., Kramber, W., Trezza, R. 2009. Automated selection of anchor pixels for landsat based evapotranspiration estimation. In: World Environmental and Water Resources Congress 2009, pp. 1–11. American Society of Civil Engineers. doi:10.1061/41036(342)442.

Kjaersgaard, J., Allen, R.G., Irmak, A. 2011. Improved methods for estimating monthly and growing season ET using METRIC applied to moderate resolution satellite imagery. Hydrological Processes 25, 4028–4036. doi:10.1002/hyp.8394.

Koster, R.D., Suarez, M.J. 1996. Energy and water balance calculations in the Mosaic LSM. NASA Technical Memoranum 104606, 59.

Kumar, S.V., Mocko, D.M., Wang, S., Peters-Lidard, C.D., Borak, J. 2019. Assimilation of remotely sensed leaf area index into the Noah-MP land surface model: Impacts on water and carbon fluxes and states over the continental United States. Journal of Hydrometeorology 20(7), 1359–1377. doi:10.1175/JHM-D-18-0237.1.

Kustas, W.P., Daughtry, C. 1990. Estimation of the soil heat flux/net radiation ratio from spectral data. Agricultural and Forest Meteorology 49, 205–223. doi:10.1016/0168-1923(90)90033-3.

Kustas, W.P., Hatfield, J.L., Prueger, J.H. 2005. The Soil Moisture—Atmosphere Coupling Experiment (SMACEX): Background, hydrometeorological conditions, and preliminary findings. Journal of Hydrometeorology 6, 791–804.

Kustas, W.P., Jackson, T.J., Schmugge, T.J., Parry, R., Goodrich, D.C., Amer, S.A., Bach, L.B., Keefer, T.O., Weltz, M.A., Moran, M.S. 1991. An interdisciplinary field study of the energy and water fluxes in the atmospheric-biosphere system over semiarid rangelands: Description and some preliminary results. Bulletin of the American Meteorological Society 72, 1683–1705.

Kustas, W.P., Li, F., Jackson, T.J., Prueger, J.H., MacPherson, J.I., Wolde, M. 2004. Effects of remote sensing pixel resolution on modeled energy flux variability of croplands in Iowa. Remote Sensing of Environment 92, 535–547.

Kustas, W.P., Norman, J.M. 1996. Use of remote sensing for evapotranspiration monitoring over land surfaces. Hydrological Sciences Journal 41, 495–516. doi:10.1080/02626669609491522.

Kustas, W.P., Norman, J.M. 1999. Evaluation of soil and vegetation heat flux predictions using a simple two-source model with radiometric temperatures for partial canopy cover. Agricultural and Forest Meteorology 94, 13–29.

Kustas, W.P., Norman, J.M., Anderson, M.C., French, A.N. 2003. Estimating subpixel surface temperatures and energy fluxes from the vegetation index—radiometric temperature relationship. Remote Sensing of Environment 85, 429–440. doi:10.1016/S0034-4257(03)00036-1.

Laipelt, L., Kayser, R.H.B., Fleischmann, A.S., Ruhoff, A., Bastiaanssen, W., Erickson, T.A., Melton, F., 2021. Long-term monitoring of evapotranspiration using the SEBAL algorithm and Google Earth Engine cloud computing. ISPRS Journal of Photogrammetry and Remote Sensing 178, 81–96. doi:10.1016/j.isprsjprs.2021.05.018.

Lee, E., Chase, T.N., Rajagopalan, B., Barry, R.G., Biggs, T.W., Lawrence, P.J. 2009. Effects of irrigation and vegetation activity on early Indian summer monsoon variability. International Journal of Climatology 29, 573–581.

Lee, E., Sacks, W.J., Chase, T.N., Foley, J.A. 2011. Simulated impacts of irrigation on the atmospheric circulation over Asia. Journal of Geophysical Research: Atmospheres (1984–2012) 116, D08114. doi:10.1029/2010JD014740.

Leuning, R., Zhang, Y.Q., Rajaud, A., Cleugh, H., Tu, K. 2008. A simple surface conductance model to estimate regional evaporation using MODIS leaf area index and the Penman-Monteith equation. Water Resources Research 44, W10419. doi:10.1029/2007WR006562.

Levis, S., Bonan, G.B., Kluzek, E., Thornton, P.E., Jones, A., Sacks, W.J., Kucharik, C.J. 2012. Interactive Crop Management in the Community Earth System Model (CESM1): Seasonal influences on land: Atmosphere fluxes. Journal of Climate 25, 4839–4859. doi:10.1175/JCLI-D-11-00446.1.

Lhomme, J.-P., Elguero, E. 1999. Examination of evaporative fraction diurnal behaviour using a soil-vegetation model coupled with a mixed-layer model. Hydrology and Earth System Sciences 3, 259–270.

Li, H., Zheng, L., Lei, Y., Li, C., Liu, Z., Zhang, S. 2008. Estimation of water consumption and crop water productivity of winter wheat in North China Plain using remote sensing technology. Agricultural Water Management 95, 1271–1278.

Li, Z., Tang, R., Wan, Z., Bi, Y., Zhou, C., Tang, B., Yan, G., Zhang, X. 2009. A review of current methodologies for regional evapotranspiration estimation from remotely sensed data. Sensors 9(5), 3801–3853. doi:10.3390/s90503801.

Liang, S., Wang, D., He, T., Yu, Y. 2019. Remote sensing of earth's energy budget: Synthesis and review. International Journal of Digital Earth 12, 737–780. doi:10.1080/17538947.2019.1597189.

Liang, S., Wang, K., Zhang, X., Wild, M. 2010. Review on estimation of land surface radiation and energy budgets from ground measurement, remote sensing and model simulations. IEEE Journal of Selected Topics in Applied Earth Observations and Remote Sensing 3, 225–240.

Liang, S., Zhang, X., He, T., Cheng, J., Wang, D. 2013. Remote sensing of the land surface radiation budget. In: Remote Sensing of Energy Fluxes and Soil Moisture Content (G.P. Petropoulos, ed.), pp. 121–162. Taylor and Francis, New York.

Liang, X., Lettenmaier, D.P., Wood, E.F., Burges, S.J. 1994. A simple hydrologically based model of land surface water and energy fluxes for general circulation models. Journal of Geophysical Research 99, 14, 414–415, 428.

Liaqat, U.W., Choi, M. 2015. Surface energy fluxes in the Northeast Asia ecosystem: SEBS and METRIC models using Landsat satellite images. Agricultural and Forest Meteorology 214–215, 60–79. doi:10.1016/j.agrformet.2015.08.245.

Liou, Y.A., Kar, S. 2014. Evapotranspiration estimation with remote sensing and various surface energy balance algorithms: A review. Energies 7(5), 2821–2849. doi:10.3390/en7052821.

Liu, M., Tang, R., Li, Z.L., Mao, H., Zhou, F., Yan, G. 2018. Estimation of annual averaged evapotranspiration by using passive microwave observations. IGARSS 2018–2018 IEEE International Geoscience and Remote Sensing Symposium, Valencia, Spain, 2018, pp. 791–794. doi:10.1109/IGARSS.2018.8518728.

Liu, R., Wen, J., Wang, X., Wang, Z., Liu, Y., Zhang, M. 2020. Estimates of daily evapotranspiration in the source region of the Yellow River combining Visible/Near-Infrared and microwave remote sensing. Remote Sensing 13(1), 53. doi:10.3390/rs13010053.

Liu, Y., Yang, Y. 2022. Advances in the quality of global soil moisture products: A review. Remote Sensing 14(15), 3741. doi:10.3390/rs14153741.

Long, D., Singh, V.P. 2010. Integration of the GG model with SEBAL to produce time series of evapotranspiration of high spatial resolution at watershed scales. Journal of Geophysical Research 115, D21128. doi:10.1029/2010jd014092.

Long, D., Singh, V.P., Li, Z.-L. 2011. How sensitive is SEBAL to changes in input variables, domain size and satellite sensor? Journal Geophysical Research 116, D21107. doi:10.1029/2011jd016542.

LSA-SAF. 2024. www.eumetsat.int/lsa-saf, access date April 13, 2024.

Ma, Y., Pinker, R.T. 2012. Modeling shortwave radiative fluxes from satellites. Journal of Geophysical Research 117, D23202. doi:10.1029/2012JD018332.

Maltese, A., Capodici, F., La Loggia, G. 2015. Soil water content assessment: Critical issues concerning the operational application of the triangle method. Sensors 15(3), 6699–6718. doi:10.3390/s150306699.

Marshall, M., Thenkabail, P., Biggs, T., Post, K. 2016. Hyperspectral narrowband and multispectral broadband indices for remote sensing of crop evapotranspiration and its components (transpiration and soil evaporation). Agriculture and Forest Meteorology 218, 122–134. doi:10.1016/j.agrformet.2015.12.025.

Marshall, M., Tu, K., Funk, C., Michaelsen, J., Williams, P., Williams, C., Ardö, J., Boucher, M., Cappelaere, B., de Grandcourt, A., Nickless, A., Nouvellon, Y., Scholes, R., Kutsch, W. 2013. Improving operational land surface model canopy evapotranspiration in Africa using a direct remote sensing approach. Hydrology and Earth System Sciences 17, 1079–1091. doi:10.5194/hess-17-1079-2013.

Martens, B., Miralles, D.G., Lievens, H., Van Der Schalie, R., De Jeu, R.A.M., Fernández-Prieto, D., Beck, H.E., Dorigo, W.A., Verhoest, N.E.C. 2017. GLEAM v3: Satellite-based land evaporation and root-zone soil moisture. Geoscientific Model Development 10, 1903–1925. doi:10.5194/gmd-10-1903-2017.

Martins, J.P.A., Trigo, I.F., Ghilain, N., Jimenez, C., Göttsche, F.-M., Ermida, S.L., Olesen, F.-S., Gellens-Meulenberghs, F., Arboleda, A. 2019. An all-weather land surface temperature product based on MSG/SEVIRI observations. Remote Sensing 11, 3044. doi:10.3390/rs11243044.

Mbow, C., Rosenzweig, C., Barioni, L.G., Benton, T.G., Herrero, M., Krishnapillai, M., Liwenga, E., Pradhan, P., Rivera-Ferre, M.G., Sapkota, T., Tubiello, F.N., Xu, Y. 2019. Food Security. In: Climate Change and Land: An IPCC special report on climate change, desertification, land degradation, sustainable land management, food security, and greenhouse gas fluxes in terrestrial ecosystems. www.ipcc.ch/srccl/chapter/chapter-5/.

McCabe, M.F., Kalma, J.D., Franks, S.W. 2005. Spatial and temporal patterns of land surface fluxes from remotely sensed surface temperatures within an uncertainty modelling framework. Hydrology and Earth System Sciences 9, 467–480.

McCabe, M.F., Wood, E.F. 2006. Scale influences on the remote estimation of evapotranspiration using multiple satellite sensors. Remote Sensing of Environment 105, 271–285.

Medlyn, B.E., Robinson, A.P., Clement, R., McMurtrie, R.E. 2005. On the validation of models of forest CO2 exchange using eddy covariance data: Some perils and pitfalls. Tree Physiology 25, 839–857.

Melesse, A.M., Frank, A., Nangia, V., Hanson, J. 2008. Analysis of energy fluxes and land surface parameters in a grassland ecosystem: A remote sensing perspective. International Journal of Remote Sensing 29, 3325–3341.

Melton, F.S., Huntington, J., Grimm, R., Herring, J., Hall, M., Rollison, D., Erickson, T., Allen, R., Anderson, M., Fisher, J.B., Kilic, A., Senay, G.B., Volk, J., Hain, C., Johnson, L., Ruhoff, A., Blankenau, P., Bromley, M., Carrara, W., Daudert, B., Doherty, C., Dunkerly, C., Friedrichs, M., Guzman, A., Halverson, G., Hansen, J., Harding, J., Kang, Y., Ketchum, D., Minor, B., Morton, C., Ortega-Salazar, S., Ott, T., Ozdogan, M., ReVelle, P.M., Schull, M., Wang, C., Yang, Y., Anderson, R.G. 2021. OpenET: Filling a critical data gap in water management for the Western United States. Journal American Water Resources Association 58, 971–994. doi:10.1111/1752-1688.12956.

Melton, F.S., Johnson, L.F., Lund, C.P., Pierce, L.L., Michaelis, A.R., Hiatt, S.H., Guzman, A., Adhikari, D.D., Purdy, A.J., Rosevelt, C., Votava, P., Trout, T.J., Temesgen, B., Frame, K., Sheffner, E.J., Nemani, R.R. 2012. Satellite irrigation management support with the terrestrial observation and prediction system: A framework for integration of satellite and surface observations to support improvements in agricultural water resource management. IEEE Journal of Selected Topics in Applied Earth Observations Remote and Sensing 5, 1709–1721. doi:10.1109/JSTARS.2012.2214474.

Meyers, T.P., Hollinger, S.E. 2004. An assessment of storage terms in the surface energy balance of maize and soybean. Agricultural and Forest Meteorology 125, 105–115.

Mira, M., Olioso, A., Gallego-Elvira, B., Courault, D., Garrigues, S., Marloie, O., Hagolle, O., Guillevic, P., Boulet, G. 2016. Uncertainty assessment of surface net radiation derived from Landsat images. Remote Sensing of Environment 175, 251–270. doi:10.1016/j.rse.2015.12.054.

Moran, M.S., Clarke, T.R., Inoue, Y., Vidal, A. 1994. Estimating crop water deficit using the relation between surface-air temperature and spectral vegetation index. Remote Sensing of Environment 49, 246–263.

Moran, M.S., Rahman, A.F., Washburne, J.C., Goodrich, D.C., Weltz, M.A., Kustas, W.P. 1996. Combining the Penman-Monteith equation with measurements of surface temperature and reflectance to estimate evaporation rates of semiarid grassland. Agricultural and Forest Meteorology 80, 87–109.

Morcrette, J. 1991. Radiation and cloud radiative properties in the European Centre for Medium Range Weather Forecasts forecasting system. Journal of Geophysical Research: Atmospheres (1984–2012) 96, 9121–9132.

Morcrette, J.-J. 2002. The surface downward longwave radiation in the ECMWF forecast system. Journal of Climate 15, 1875–1892.

Morillas, L., Leuning, R., Villagarcía, L., García, M., Serrano-Ortiz, P. and Domingo, F. 2013. Improving evapotranspiration estimates in Mediterranean drylands: The role of soil evaporation. Water Resources Research 49, 6572–6586. doi:10.1002/wrcr.20468.

Morse, A., Tasumi, M., Allen, R.G., Kramber, W.J. 2000. Application of the SEBAL methodology for estimating consumptive use of water and streamflow depletion in the Bear River Basin of Idaho through remote sensing. Final Report, Idaho Department of Water Resources, Boise, ID, December 15.

Mu, Q.M. 2013. MODIS 1-km2 Terrestrial Evapotranspiration (ET) Product for the Nile Basin Algorithm Theoretical Basis Document. University of Montana, Missoula, MT.

Mu, Q.M., Heinsch, F.A., Zhao, M., Running, S.W. 2007. Development of a global evapotranspiration algorithm based on MODIS and global meteorology data. Remote Sensing of Environment 111, 519–536.

Mu, Q.M., Zhao, M.S., Running, S.W. 2011. Improvements to a MODIS global terrestrial evapotranspiration algorithm. Remote Sensing of Environment 115, 1781–1800.

Mu, Q.M., Zhao, M.S., Running, S.W. 2013. MOD16 1-km2 terrestrial evapotranspiration (ET) product for the Nile Basin; algorithm theoretical basis document. Numerical Terradynamic Simulation Group, College of Forestry and Conservation, University of Montana, Missoula, MT.

Muñoz-Sabater, J., Dutra, E., Agustí-Panareda, A., Albergel, C., Arduini, G., Balsamo, G., Boussetta, S., Choulga, M., Harrigan, S., Hersbach, H., Martens, B. 2021. ERA5-Land: A state-of-the-art global reanalysis dataset for land applications. Earth System Science Data 13(9), 4349–4383.

Murray, T., Verhoef, A. 2007. Moving towards a more mechanistic approach in the determination of soil heat flux from remote measurements: I. A universal approach to calculate thermal inertia. Agricultural and Forest Meteorology 147, 80–87. doi:10.1016/j.agrformet.2007.07.004.

Nagler, P.L., Barreto-Muñoz, A., Chavoshi Borujeni, S., Jarchow, C.J., Gómez-Sapiens, M.M., Nouri, H., Herrmann, S.M., Didan, K. 2020. Ecohydrological responses to surface flow across borders: Two decades of changes in vegetation greenness and water use in the riparian corridor of the Colorado River delta. Hydrological Processes 34(25), 4851–4883. doi:10.1002/hyp.8392.

Nagler, P.L., Barreto-Muñoz, A., Chavoshi Borujeni, S., Nouri, H., Jarchow, C.J., Didan, K. 2021. Riparian area changes in greenness and water use on the lower Colorado river in the USA from 2000 to 2020. Remote Sensing 13(7), 1332. doi:10.3390/rs13071332.

Nagler, P.L., Barreto-Muñoz, A., Sall, I., Lurtz, M.R., Didan, K. 2022b. Riparian plant evapotranspiration and consumptive use for selected areas of the little Colorado River watershed on the Navajo nation. Remote Sensing 15(1), 52. doi:10.3390/rs15010052.

Nagler, P.L., Glenn, E.P., Nguyen, U., Scott, R.L., Doody, T. 2013. Estimating riparian and agricultural actual evapotranspiration by reference evapotranspiration and MODIS enhanced vegetation index. Remote Sensing 5, 3849–3871. doi:10.3390/rs5083849.

Nagler, P.L., Morino, K., Murray, R.S., Osterberg, J., Glenn, E.P. 2009. An empirical algorithm for estimating agricultural and riparian evapotranspiration using MODIS Enhanced Vegetation Index and ground measurements of ET: I. Description of method. Remote Sensing 1, 1273–1297.

Nagler, P.L., Pearlstein, S., Glenn, E.P., Brown, T.B., Bateman, H.L., Bean, D.W., Hultine, K.R. 2014. Rapid dispersal of saltcedar (*Tamarix* spp.) biocontrol beetles (*Diorhabda carinulata*) on a desert river detected by phenocams, MODIS imagery and ground observations. Remote Sensing of Environment 140, 206–219. doi:10.1016/j.rse.2013.08.017.

Nagler, P.L., Sall, I., Barreto-Muñoz, A., Gómez-Sapiens, M., Nouri, H., Chavoshi Borujeni, S., Didan, K. 2022a. Effect of restoration on plant greenness and water use in relation to drought in the riparian corridor of the Colorado River delta. JAWRA Journal of the American Water Resources Association 58(5), 746–784. doi:10.1111/1752-1688.13036.

Nagler, P.L., Scott, R.L., Westenburg, C., Cleverly, J.R., Glenn, E.P., Huete, A.R. 2005. Evapotranspiration on western US rivers estimated using the Enhanced Vegetation Index from MODIS and data from eddy covariance and Bowen ratio flux towers. Remote Sensing of Environment 97(3), 337–351. doi:10.1016/j.rse.2005.05.011.

NASA CERES. 2024. https://ceres.larc.nasa.gov/, access date April 13, 2024.
NASA GEWEX Surface Radiation Budget. 2024. https://science.larc.nasa.gov/gewex-srb/?doing_wp_cron=1713021086.8910689353942871093750, access date April 13, 2024.
NASA GLDAS. 2024. https://ldas.gsfc.nasa.gov/gldas, access date April 13, 2024.
NASA MOD11. 2024. https://modis.gsfc.nasa.gov/data/dataprod/mod11.php, access date April 13, 2024.
NASA MOD13. 2024. https://modis.gsfc.nasa.gov/data/dataprod/mod13.php, access date April 13, 2024.
NASA MOD43B3. 2024. https://ladsweb.modaps.eosdis.nasa.gov/filespec/MODIS/4/MOD43B3, access date April 13, 2024.
NASA MOD43C3. 2024. https://lpdaac.usgs.gov/products/mcd43c3v006/, access date April 13, 2024.
NASA ORNL Daymet. 2024. https://daymet.ornl.gov/, access date April 13, 2024.
NASA SMAP. 2024. https://nsidc.org/data/smap, access date April 13, 2024.
NCEP-NCAR. 2024. https://psl.noaa.gov/data/gridded/data.ncep.reanalysis.html, access date April 13, 2024.
Neale, C.M.U., Bausch, W.C., Heermann, D.F. 1989. Development of reflectance-based crop coefficients for corn. Transactions of the American Society of Agricultural Engineers 32, 1891–1899.
Nishida, K. 2003. An operational remote sensing algorithm of land surface evaporation. Journal of Geophysical Research: Atmospheres (1984–2012) 108, 4270. doi:10.1029/2002JD002062.
Nishida, K., Nemani, R. 2003. Development of an evapotranspiration index from Aqua/MODIS for monitoring surface moisture status. IEEE Transactions on Geoscience and Remote Sensing 41, 493–501.
Niu, G.-Y., Yang, Z.-L., Mitchell, K.E., Chen, F., Ek, M.B., Barlage, M., Kumar, A., Manning, K., Niyogi, D., Rosero, E., Tewari, M., Xia, Y. 2011. The community Noah land surface model with multiparameterization options (Noah-MP): 1. Model description and evaluation with local-scale measurements. Journal of Geophysical Research: Atmospheres 116. doi:10.1029/2010JD015139.
Niu, H., Hollenbeck, D., Zhao, T., Wang, D., Chen, Y., 2020. Evapotranspiration estimation with small UAVs in precision agriculture. Sensors 20, 6427. doi:10.3390/s20226427.
NOAA CMAP. 2024. www.cpc.ncep.noaa.gov/products/global_precip/html/wpage.cmap.shtml, access date April 13, 2024.
NOAA GDAS. 2024. www.ncei.noaa.gov/products/weather-climate-models/global-data-assimilation, access date April 13, 2024.
Norman, J.M., Anderson, M.C., Kustas, W.P., French, A.N., Mecikalski, J., Torn, R., Diak, G.R., Schmugge, T.J., Tanner, B.C.W. 2003. Remote sensing of surface energy fluxes at 10 1 -m pixel resolutions. Water Resources Research 39, 1221. doi:10.1029/2002WR001775.
Norman, J.M., Kustas, W.P., Humes, K.S. 1995. Source approach for estimating soil and vegetation energy fluxes in observations of directional radiometric surface temperature. Agricultural and Forest Meteorology 77, 263–293.
NSMC FY-2 Series. 2024. www.nsmc.org.cn/nsmc/en/satellite/FY2.html, access date April 13, 2024.
Pastorello, G., Papale, D., Chu, H., Trotta, C., Agarwal, D., Canfora, E., Baldocchi, D., Torn, M. 2017. A new data set to keep a sharper eye on land-air exchanges. Eos. doi:10.1029/2017EO071597.
Petropoulos, G.P. 2013. Remote sensing of surface turbulent energy fluxes. In: Remote sensing of energy fluxes and soil moisture content (G.P. Petropoulos, ed.), pp. 49–84. Taylor and Francis, New York.
Petropoulos, G.P., Carlson, T.N. 2011. Retrievals of turbulent heat fluxes and surface soil water content by remote sensing. In: Advances in environmental remote sensing: Sensors, algorithms, and applications (Q. Weng, ed.), p. 469. CRC Press, London.
Petropoulos, G.P., Carlson, T.N., Griffiths, H.M. 2013. Turbulent fluxes of heat and moisture at the Earth's land surfade: Importance, controlling parameters, and conventional measurement techniques. In: Remote sensing of energy fluxes and soil moisture content (G.P. Petropoulos, ed.), pp. 3–28. CRC Press, New York.
Petropoulos, G.P., Carlson, T.N, Wooster, M.J., Islam, S. 2009. A review of Ts/VI remote sensing based methods for the retrieval of land surface fluxes and soil surface moisture content. Progress in Physical Geography 33(2), 1–27.
Petropoulos, G.P., Hristopulos, D.T. 2020. Retrievals of key biophysical parameters at mesoscale from the Ts/VI scatterplot domain. Geocarto International 37(8), 2385–2405. doi:10.1080/10106049.2020.1821099.
Petropoulos, G.P., Ireland, G., Barrett, B. 2015. Surface soil moisture retrievals from remote sensing: Current status, products and future trends. Physics and Chemistry of the Earth, Parts a/B/C 83–84, 36–56. doi:10.1016/j.pce.2015.02.009.
Petropoulos, G.P., Ireland, G., Lamine, S., Griffiths, H., Ghilain, N., Anagnostopoulos, V., North, M., Srivastava, P.K., Georgopoulou, H. 2016. Operational evapotranspiration estimates from SEVIRI in support of sustainable water management. International Journal of Applied Earth Observation and Geoinformation 49, 175–187. doi:10.1016/j.jag.2016.02.006.

Petropoulos, G.P., Maltese, A., Carlson, T.N., Pavlides, A., Hristopulos, D.T., Capodici, F., Chalkias, C., Dardanelli, G., Manfreda, S. 2021. Exploring the use of Unmanned Aerial Vehicles (UAVs) with the simplified 'triangle' technique for soil water content and evaporative fraction retrievals in a Mediterranean setting. International Journal of Remote Sensing 42(5), 1623–1642. doi:10.1080/01431161.2020.1841319.

Petropoulos, G.P., Şandric, I., Hristopulos, D.T., Carlson, T.N. 2020. Evaporative fluxes and surface soil moisture retrievals in a mediterranean setting from sentinel-3 and the "simplified triangle". Remote Sensing 12(19), 3192. doi:10.3390/rs12193192.

Petropoulos, G.P., Srivastava, P.K., Ferentinos, K., Hristopulos, D. 2018. Evaluating the capabilities of optical/TIR imaging sensing systems for quantifying soil water content. Geocarto International 35(5), 494–511. doi:10.1080/10106049.2018.1520926.

Rafn, E.B., Contor, B., Ames, D.P. 2008. Evaluation of a method for estimating irrigated crop-evapotranspiration coefficients from remotely sensed data in Idaho. Journal of Irrigation and Drainage Engineering 134, 722–729.

Rast, M., Painter, T.H. 2019. Earth observation imaging spectroscopy for terrestrial systems: An overview of its history, techniques, and applications of its missions. Surv. Geophys. 40, 303–331. doi:10.1007/s10712-019-09517-z.

Reichle, R.H., Liu, Q., Koster, R.D., Crow, W.T., De Lannoy, G.J.M., Kimball, J.S., Ardizzone, J.V., Bosch, D., Colliander, A., Cosh, M., Kolassa, J., Mahanama, S.P., Prueger, J., Starks, P., Walker, J.P. 2019. Version 4 of the SMAP level-4 soil moisture algorithm and data product. Journal of Advances in Modeling Earth Systems 11, 3106–3130. do:10.1029/2019MS001729.

Reyes-González, A., Kjaersgaard, J., Trooien, T.P., Sánchez, D.G.R., Duarte, J.I.S., Rangel, P.P., Fortis-Hernández, M. 2019. Comparison of leaf area index, surface temperature, and actual evapotranspiration estimated using the METRIC model and in situ measurements. Sensors 19(8), 1857. doi:10.3390/s19081857.

Richter, K., Timmermans, W.J. 2009. Physically based retrieval of crop characteristics for improved water use estimates. Hydrology and Earth System Sciences 13, 663–674. doi:10.5194/hess-13-663-2009.

Rienecker, M.M., Suarez, M.J., Gelaro, R., Todling, R., Bacmeister, J., Liu, E., Bosilovich, M.G., Schubert, S.D., Takacs, L., Kim, G.-K., Bloom, S., Chen, J., Collins, D., Conaty, A., da Silva, A., Gu, W., Joiner, J., Koster, R.D., Lucchesi, R., Molod, A. 2011. MERRA: NASA's modern-era retrospective analysis for research and applications. Journal of Climate 24, 3624–3648.

Rodell, M., Houser, P.R., Jambor, U., Gottschalck, J., Mitchell, K., Meng, C.-J., Arsenault, K., Cosgrove, B., Radakovich, J., Bosilovich, M., Entin*, J.K., Walker, J.P., Lohmann, D., Toll, D. 2004. The global land data assimilation system. Bulletin of the American Meteorological Society 85, 381–394. doi:10.1175/BAMS-85-3-381.

Roerink, G., Su, Z., Menenti, M. 2000. S-SEBI: A simple remote sensing algorithm to estimate the surface energy balance. Physics and Chemistry of the Earth, Part B: Hydrology, Oceans and Atmosphere 25, 147–157. doi:10.1016/S1464-1909(99)00128-8.

Román, M.O., Justice, C., Paynter, I., Boucher, P.B., Devadiga, S., Endsley, A., Erb, A., Friedl, M., Gao, H., Giglio, L., Gray, J.M., Hall, D., Hulley, G., Kimball, J., Knyazikhin, Y., Lyapustin, A., Myneni, R.B., Noojipady, P., Pu, J., Riggs, G., Sarkar, S., Schaaf, C., Shah, D., Tran, K.H., Vermote, E., Wang, D., Wang, Z., Wu, A., Ye, Y., Shen, Y., Zhang, S., Zhang, S., Zhang, X., Zhao, M., Davidson, C., Wolfe, R. 2024. Continuity between NASA MODIS Collection 6.1 and VIIRS Collection 2 land products. Remote Sensing of Environment 302, 113963. doi:10.1016/j.rse.2023.113963.

Romero, M.G. 2004. Daily evapotranspiration estimation by means of evaporative fraction and reference evapotranspiration fraction: Daily evapotranspiration estimation by means of evaporative fraction and reference evapotranspiration fraction, Utah State University, Department of Biological and Irrigation Engineering, Logan, UT.

Rossow, W.B., Schiffer, R.A. 1991. ISCCP cloud data products. Bulletin of the American Meteorological Society 72, 2–20.

Rossow, W.B., Schiffer, R.A. 1999. Advances in understanding clouds from ISCCP. Bulletin of the American Meteorological Society 80, 2261–2287.

Rouf, T., Maggioni, V., Mei, Y., Houser, P. 2021. Towards hyper-resolution land-surface modeling of surface and root zone soil moisture. Journal of Hydrology 594, 125945. doi:10.1016/j.jhydrol.2020.125945.

Ruhoff, A., De Andrade, B.C., Laipelt, L., Fleischmann, A.S., Siqueira, V.A., Moreira, A.A., Barbedo, R., Cyganski, G.L., Fernandez, G.M.R., Brêda, J.P.L.F., Paiva, R.C.D.D., Meller, A., Teixeira, A.D.A., Araújo, A.A., Fuckner, M.A., Biggs, T., 2022. Global evapotranspiration datasets assessment using water balance in South America. Remote Sensing 14, 2526. doi:10.3390/rs14112526.

Running, S.W., Mu, Q., Zhao, M., Moreno, A. 2017. MODIS global terrestrial evapotranspiration (ET) product (NASA MOD16A2/A3) NASA earth observing system MODIS land algorithm. NASA, Washington, DC, USA. May 5.

Samani, Z., Bawazir, A.S., Bleiweiss, M., Skaggs, R., Longworth, J., Tran, V.D., Pinon, A. 2009. Using remote sensing to evaluate the spatial variability of evapotranspiration and crop coefficient in the lower Rio Grande Valley, New Mexico. Irrigation Science 28, 93–100. doi:10.1007/s00271-009-0178-8.

Schiffer, R.A., Rossow, W.B. 1983. The International Satellite Cloud Climatology Project (ISCCP): The first project of the world climate research programme. American Meteorological Society, Bulletin 64, 779–784.

Scott, R.L. 2010. Using watershed water balance to evaluate the accuracy of eddy covariance evaporation measurements for three semiarid ecosystems. Agricultural and Forest Meteorology 150, 219–225.

Scott, R.L., Huxman, T.E., Cable, W.L., Emmerich, W.E. 2006. Partitioning of evapotranspiration and its relation to carbon dioxide exchange in a Chihuahuan Desert shrubland. Hydrological Processes 20, 3227–3243.

Sellers, P.J., Randall, D.A., Collatz, G.J., Berry, J.A., Field, C.B., Dazlich, D.A., Zhang, C., Collelo, G.D., Bounoua, L. 1996. A revised land surface parameterization (SiB2) for atmospheric GCMs: Part I: Model formulation. Journal of Climate 9, 676–705.

Senay, G.B., 2018. Satellite psychrometric formulation of the operational simplified surface energy balance (SSEBop) model for quantifying and mapping evapotranspiration. Applied Engineering in Agriculture 34, 555–566. doi:10.13031/aea.12614.

Senay, G.B., Bohms, S., Singh, R.K., Gowda, P.H., Velpuri, N.M., Alemu, H., Verdin, J.P. 2013. Operational evapotranspiration mapping using remote sensing and weather datasets: A new parameterization for the SSEB approach. JAWRA Journal of the American Water Resources Association 49, 577–591. doi:10.1111/jawr.12057.

Senay, G.B., Budde, M.E., Verdin, J.P. 2011. Enhancing the Simplified Surface Energy Balance (SSEB) approach for estimating landscape ET: Validation with the METRIC model. Agricultural Water Management 98, 606–618.

Senay, G.B., Budde, M.E., Verdin, J.P., Melesse, A. 2007. A coupled remote sensing and simplified surface energy balance approach to estimate actual evapotranspiration from irrigated fields. Sensors 7, 979–1000.

Senay, G.B., Kagone, S., Velpuri, N.M., 2020. Operational global actual evapotranspiration: Development, evaluation, and dissemination. Sensors 20, 1915. doi:10.3390/s20071915.

Senay, G.B., Parrish, G.E.L., Schauer, M., Friedrichs, M., Khand, K., Boiko, O., Kagone, S., Dittmeier, R., Arab, S., Ji, L. 2023. Improving the operational simplified surface energy balance evapotranspiration model using the forcing and normalizing operation. Remote Sensing 15, 260. doi:10.3390/rs15010260.

Sepulcre-Cantó, G., Vogt, J., Arboleda, A., Antofie, T. 2014. Assessment of the EUMETSAT LSA-SAF evapotranspiration product for drought monitoring in Europe. International Journal of Applied Earth Observation and Geoinformation 30, 190–202. doi:10.1016/j.jag.2014.01.021.

Shapiro, R. 1987. A simple model for the calculation of the flux of direct and diffuse solar radiation through the atmosphere. Hanscom Air Force Base, MA.

Sheffield, J., Goteti, G., Wood, E.F. 2006. Development of a 50-year high-resolution global dataset of meteorological forcings for land surface modeling. Journal of Climate 19, 3088–3111. doi:10.1175/JCLI3790.1.

Shi, Q., Liang, S. 2014. Surface-sensible and latent heat fluxes over the Tibetan Plateau from ground measurements, reanalysis, and satellite data. Atmospheric Chemistry and Physics 14(11), 5659–5677. doi:10.5194/acp-14-5659-2014.

Shi, X., Mao, J., Thornton, P.E., Huang, M. 2013. Spatiotemporal patterns of evapotranspiration in response to multiple environmental factors simulated by the community land model. Environmental Research Letters 8: 024012. doi:10.1088/1748-9326/8/2/024012.

Shu, Y., Stisen, S., Jensen, K.H., Sandholt, I. 2011. Estimation of regional evapotranspiration over the North China Plain using geostationary satellite data. International Journal of Applied Earth Observation and Geoinformation 13, 192–206.

Singh, R.K., Irmak, A. 2009. Estimation of crop coefficients using satellite remote sensing. Journal of Irrigation and Drainage Engineering 135, 597–608.

Singh, R.K., Senay, G. 2015. Comparison of four different energy balance models for estimating evapotranspiration in the midwestern United States. Water 8, 9. doi:10.3390/w8010009.

Singh, R.K., Senay, G.B., Velpuri, N.M., Bohms, S., Scott, R.L., Verdin, J.P. 2013. Actual evapotranspiration (water use) assessment of the Colorado River basin at the landsat resolution using the operational simplified surface energy balance model. Remote Sensing 6, 233–256.

Singh, R.K., Srivastava, P.K., Petropoulos, G.P., Shukla, S., Prasad, R. 2022. Improvement of the "triangle method" for soil moisture retrieval using ECOSTRESS and Sentinel-2: Results over a heterogeneous agricultural field in Northern India. Water 14(19), 3179. doi:10.3390/w14193179.

Sobrino, J.A., Gómez, M., Jiménez-Muñoz, J.C., Olioso, A. 2007. Application of a simple algorithm to estimate daily evapotranspiration from NOAA—AVHRR images for the Iberian Peninsula. Remote Sensing of Environment 110, 139–148.

Sobrino, J.A., Gómez, M., Jiménez-Muñoz, J.C., Olioso, A., Chehbouni, G. 2005. A simple algorithm to estimate evapotranspiration from DAIS data: Application to the DAISEX campaigns. Journal of hydrology 315, 117–125.

Sobrino, J.A., Jiménez-Muñoz, J.C., Sòria, G., Gómez, M., Ortiz, A.B., Romaguera, M., Zaragoza, M., Julien, Y., Cuenca, J., Atitar, M. 2008. Thermal remote sensing in the framework of the SEN2FLEX project: Field measurements, airborne data and applications. International Journal of Remote Sensing 29, 4961–4991.

Srivastava, P.K., Han, D., Ramirez, M.A., Islam, T. 2013a. Machine learning techniques for downscaling SMOS satellite soil moisture using MODIS land surface temperature for hydrological application. Water Resources Management 27, 3127–3144.

Srivastava, P.K., Han, D., Rico-Ramirez, M.A., Al-Shrafany, D., Islam, T. 2013b. Data fusion techniques for improving soil moisture deficit using SMOS satellite and WRF-NOAH land Surface model. Water Resources Management. doi:10.1007/s11269-013-0452-7.

Srivastava, P.K., Han, D., Islam, T., Petropoulos, G.P., Gupta, M., Dai, Q. 2015. Seasonal evaluation of evapotranspiration fluxes from MODIS satellite and mesoscale model downscaled global reanalysis datasets. Theoretical and Applied Climatology 1–14. doi:10.1007/ s00704-015-1430-1.

Stankevich, S.A., Kozlova, A.A., Piestova, I.O., Lubskyi, M.S. 2017. Leaf area index estimation of forest using sentinel-1 C-band SAR data. IEEE Microwaves, Radar and Remote Sensing Symposium (MRRS), Kiev, Ukraine, pp. 253–256, doi:10.1109/MRRS.2017.8075075.

Stisen, S., Sandholt, I., Nørgaard, A., Fensholt, R., Jensen, K.H. 2008. Combining the triangle method with thermal inertia to estimate regional evapotranspiration: Applied to MSG-SEVIRI data in the Senegal River basin. Remote Sensing of Environment 112, 1242–1255.

Stone, L.R., Horton, M.L. 1974. Estimating evapotranspiration using canopy temperatures: Field evaluation. Agronomy Journal 66, 450–454.

Su, H., McCabe, M.F., Wood, E.F., Su, Z., Prueger, J.H. 2005. Modeling evapotranspiration during SMACEX: Comparing two approaches for local- and regional-scale prediction. Journal of Hydrometeorology 6, 910–922.

Su, Z. 2002. The Surface Energy Balance System (SEBS) for estimation of turbulent heat fluxes. Hydrology and Earth System Sciences Discussions 6, 85–100.

Sun, Z., Wang, Q., Matsushita, B., Fukushima, T., Ouyang, Z., Watanabe, M. 2009. Development of a Simple Remote Sensing EvapoTranspiration model (Sim-ReSET): Algorithm and model test. Journal of Hydrology 376, 476–485.

Tang, B., Li, Z.-L. 2008. Estimation of instantaneous net surface longwave radiation from MODIS cloud-free data. Remote Sensing of Environment 112, 3482–3492.

Tang, R., Li, Z.-L., Tang, B. 2010. An application of the Ts-VI triangle method with enhanced edges determination for evapotranspiration estimation from MODIS data in arid and semi-arid regions: Implementation and validation. Remote Sensing of Environment 114, 540–551.

Teixeira, A.H.D.C., Bastiaanssen, W.G.M., Ahmad, M.D., Bos, M.G. 2009. Reviewing SEBAL input parameters for assessing evapotranspiration and water productivity for the Low-Middle São Francisco River basin, Brazil: Part A: Calibration and validation. Agricultural and Forest Meteorology 149, 462–476.

Thornton, M.M., Shrestha, R., Wei, Y., Thornton, P.E., Kao, S.-C., Wilson, B.E. 2022. Daymet: Daily surface weather data on a 1-km grid for North America, Version 4 R1 0 MB. doi:10.3334/ORNLDAAC/2129.

Tian, F., Qiu, G., Lü, Y., Yang, Y., Xiong, Y. 2014. Use of high-resolution thermal infrared remote sensing and "three-temperature model" for transpiration monitoring in arid inland river catchment. Journal of Hydrology, 515, 307–315. doi:10.1016/j.jhydrol.2014.04.056.

Timmermans, W.J., Kustas, W.P., Anderson, M.C., French, A.N. 2007. An intercomparison of the surface energy balance algorithm for land (SEBAL) and the two-source energy balance (TSEB) modeling schemes. Remote Sensing of Environment 108, 369–384.

Tolk, J.A., Howell, T.A., Evett, S.R. 2006. Nighttime evapotranspiration from alfalfa and cotton in a semiarid climate. Agronomy Journal 98, 730–736.

Townshend, J.R., Justice, C.O. 1988. Selecting the spatial resolution of satellite sensors required for global monitoring of land transformations. International Journal of Remote Sensing 9, 187–236. doi:10.1080/01431168808954847.

Tran, B.N., van der Kwast, J., Seyoum, S., Uijlenhoet, R., Jewitt, G., Mul, M. 2023. Uncertainty assessment of satellite remote sensing-based evapotranspiration estimates: A systematic review of methods and gaps. EGUsphere, pp. 1–40.

Trout, T.J., Johnson, L.F., Gartung, J. 2008. Remote sensing of canopy cover in horticultural crops. Hortscience 43, 333–337.

Twine, T.E., Kustas, W.P., Norman, J.M., Cook, D.R., Houser, P., Meyers, T.P., Prueger, J.H., Starks, P.J., Wesely, M.L. 2000. Correcting eddy-covariance flux underestimates over a grassland. Agricultural and Forest Meteorology 103, 279–300.

University of Southampton. 2024. Hydrology @ University of Southampton. https://hydrology.soton.ac.uk/, access date April 13, 2024.

USBR AgriMet. 2024. www.usbr.gov/main/agrihydro.html, access date April 13, 2024.

USGS MOD16A2 v061. 2024. https://lpdaac.usgs.gov/products/mod16a2v061/, access date April 13, 2024.

Van de Griend, A.A., Camillo, P.J. and Gurney, R.J. 1985. Discrimination of soil physical parameters, thermal inertia, and soil moisture from diurnal surface temperature fluctuations. Water Resources Research 21, 997–1009.

Van den Hurk, B.J.J.M., Viterbo, P., Beljaars, A.C.M., Betts, A.K. 2000. Offline validation of the ERA40 surface scheme, ECMWF Technical Memorandum No. 295, p. 41.

Vanino, S., Nino, P., De Michele, C., Bolognesi, S.F., Guido D'Urso, G., Claudia Di Bene, C., Pennelli, B., et al. 2018. Capability of sentinel-2 data for estimating maximum evapotranspiration and irrigation requirements for tomato crop in Central Italy. Remote Sensing of Environment 215, 452–470. doi:10.1016/j.rse.2018.06.035.

Van Niel, T.G., McVicar, T.R., Roderick, M.L., van Dijk, A.I.J.M., Renzullo, L.J., van Gorsel, E. 2011. Correcting for systematic error in satellite-derived latent heat flux due to assumptions in temporal scaling: Assessment from flux tower observations. Journal of Hydrology 409, 140–148. doi:10.1016/j.jhydrol.2011.08.011.

Van Oevelen, P.J. 1991. Determination of the available energy for evapotranspiration with remote sensing. Determination of the available energy for evapotranspiration with remote sensing, Wageningen University, The Netherlands.

Velpuri, N.M., Senay, G.B., Singh, R.K., Bohms, S., Verdin, J.P. 2013. A comprehensive evaluation of two MODIS evapotranspiration products over the conterminous United States: Using point and gridded FLUXNET and water balance ET. Remote Sensing of Environment 139, 35–49. doi:10.1016/j.rse.2013.07.013.

Venturini, V., Bisht, G., Islam, S., Jiang, L. 2004. Comparison of evaporative fractions estimated from AVHRR and MODIS sensors over South Florida. Remote Sensing of Environment 93, 77–86. doi:10.1016/j.rse.2004.06.020.

Verma, S.B., Rosenberg, N.J., Blad, B.L., Baradas, M.W. 1976. Resistance-energy balance method for predicting evapotranspiration: Determination of boundary layer resistance and evaluation of error effects. Agronomy Journal 68, 776–782. doi:10.2134/agronj1976.00021962006800050023x.

Vernekar, K.G., Sinha, S., Sadani, L.K., Sivaramakrishnan, S., Parasnis, S.S., Mohan, B., Dharmaraj, S., Patil, M.N., Pillai, J.S., Murthy, B.S., Debaje, S.B., Bagavathsingh, A. 2003. An overview of the land surface processes experiment (Laspex) over a semi-arid region of India. Boundary-Layer Meteorology 106, 561–572.

Verstraeten, W.W., Veroustraete, F., Feyen, J. 2008. Assessment of evapotranspiration and soil moisture content across different scales of observation. Sensors 8, 70–117.

Vinukollu, R.K., Meynadier, R., Sheffield, J., Wood, E.F. 2011a. Multi-model, multi-sensor estimates of global evapotranspiration: Climatology, uncertainties and trends. Hydrological Processes 25, 3993–4010. doi:10.1002/hyp.8393.

Vinukollu, R.K., Wood, E.F., Ferguson, C.R., Fisher, J.B. 2011b. Global estimates of evapotranspiration for climate studies using multi-sensor remote sensing data: Evaluation of three process-based approaches. Remote Sensing of Environment 115, 801–823.

Viterbo, P., Beljaars, A. 1995. An improved land surface parameterization scheme in the ECMWF model and its validation. Journal of Climate 8(11), 2716–2748. doi:10.1175/1520-0442(1995)008.

Volk, J.M., Huntington, J.L., Melton, F.S., Allen, R., Anderson, M., Fisher, J.B., Kilic, A., Ruhoff, A., Senay, G.B., Minor, B., Morton, C. 2024. Assessing the accuracy of OpenET satellite-based evapotranspiration data to support water resource and land management applications. Nature Water, 1–13.

Wagle, P., Bhattarai, N., Gowda, P.H., Kakani, V.G., 2017. Performance of five surface energy balance models for estimating daily evapotranspiration in high biomass sorghum. ISPRS Journal of Photogrammetry and Remote Sensing 128, 192–203. doi:10.1016/j.isprsjprs.2017.03.022.

Walker, E., García, G.A., Venturini, V. 2018. Actual evapotranspiration estimation over flat lands using soil moisture products from SMAP mission. Revista de Teledetección 52, 17–26. doi:10.4995/raet.2018.10566.

Walker, E., García, G.A., Venturini, V. 2019. Evapotranspiration estimation using SMAP soil moisture products and bouchet complementary evapotranspiration over Southern Great Plains. Journal of Arid Environments 163, 34–40. doi:10.1016/j.jaridenv.2019.01.002.

Wang, K., Li, Z., Cribb, M. 2006. Estimation of evaporative fraction from a combination of day and night land surface temperatures and NDVI: A new method to determine the Priestley—Taylor parameter. Remote Sensing of Environment 102, 293–305.

Wang, Y., Jiang, B., Liang, S., Wang, D., He, T., Wang, Q., Zhao, X., Xu, J. 2019. Surface Shortwave Net Radiation Estimation from Landsat TM/ETM+ Data Using Four Machine Learning Algorithms. Remote Sensing 11(23), 2847. doi:10.3390/rs11232847.

Wever, L.A., Flanagan, L.B., Carlson, P.J., 2002. Seasonal and interannual variation in evapotranspiration, energy balance and surface conductance in a northern temperate grassland. Agricultural and Forest Meteorology 112, 31–49. doi:10.1016/S0168-1923(02)00041-2.

Wilson, K.B., Hanson, P.J., Mulholland, P.J., Baldocchi, D.D., Wullschleger, S.D. 2001. A comparison of methods for determining forest evapotranspiration and its components: Sap-flow, soil water budget, eddy covariance and catchment water balance. Agricultural and Forest Meteorology 106, 153–168.

Wood, E.F., Lettenmaier, D.P., Zartarian, V.G., 1992. A land-surface hydrology parameterization with subgrid variability for general circulation models. Journal Geophysical Research 97. doi:10.1029/91jd01786.

Yang, Y., Fang, J., Fay, P.A., Bell, J.E., Ji, C., 2010. Rain use efficiency across a precipitation gradient on the Tibetan Plateau. Geophysical Research Letters 37, L15702. doi:10.1029/2010gl043920.

Yang, Y., Roderick, M.L., Guo, H., Miralles, D.G., Zhang, L., Fatichi, S., Luo, X., Zhang, Y., McVicar, T.R., Tu, Z., Keenan, T.F., Fisher, J.B., Gan, R., Zhang, X., Piao, S., Zhang, B., Yang, D. 2023. Evapotranspiration on a greening Earth. Nature Reviews Earth Environment. doi:10.1038/s43017-023-00464-3.

Yilmaz, M.T., Anderson, M.C., Zaitchik, B., Hain, C.R., Crow, W.T., Ozdogan, M., Chun, J.A., Evans, J. 2014. Comparison of prognostic and diagnostic surface flux modeling approaches over the Nile River basin. Water Resources Research 50, 386–408. doi:10.1002/2013WR014194.

Zahira, S., Abderrahmane, H., Mederbal, K., Frederic, D. 2009. Mapping latent heat flux in the western forest covered regions of Algeria using remote sensing data and a spatialized model. Remote Sensing 1, 795–817.

Zhang, K., Kimball, J.S., Running, S.W., 2016. A review of remote sensing based actual evapotranspiration estimation. WIREs Water 3, 834–853. doi:10.1002/wat2.1168.

Zhang, L., Marshall, M., Nelson, A., Vrieling, A., 2021. A global assessment of PT-JPL soil evaporation in agroecosystems with optical, thermal, and microwave satellite data. Agricultural and Forest Meteorology 306, 108455. doi:10.1016/j.agrformet.2021.108455.

Zhang, Y., Kong, D., Gan, R., Chiew, F.H.S., McVicar, T.R., Zhang, Q., Yang, Y. 2019. Coupled estimation of 500 m and 8-day resolution global evapotranspiration and gross primary production in 2002–2017. Remote Sensing of Environment 222, 165–182. doi:10.1016/j.rse.2018.12.031.

Zhang, Y., Liu, C., Lei, Y., Tang, Y., Yu, Q., Shen, Y., Sun, H. 2006. An integrated algorithm for estimating regional latent heat flux and daily evapotranspiration. International Journal of Remote Sensing 27, 129–152.

11 Modeling and Monitoring Water Productivity by Using Geotechnologies
In Some Brazilian Agroecosystems

Antônio Heriberto de Castro Teixeira, Janice Freitas Leivas, Celina Maki Takemura, Edson Patto Pacheco, Edlene Aparecida Monteiro Garçon, Inajá Francisco de Sousa, André Quintão de Almeida, Prasad S. Thenkabail, and Ana Flávia Maria Santos

ACRONYMS AND DEFINITIONS

BIO	Biomass Production
DOY	Day of the Year
ET	Evapotranspiration
FAO	Food and Agricultural Organization
GS	Growing Season
HI	Harvest Index
IC	Irrigated Crops
LUE	Light Use Efficiency
MG	Minas Gerais state
MODIS	Moderate-Resolution Imaging Spectroradiometer
NDVI	Normalized Difference Vegetation Index
NIR	Near-Infrared
NV	Natural Vegetation
PAR	Photosynthetically Active Radiation
PM	Penman–Monteith
RUE	Radiation Use Efficiency
SAFER	Simple Algorithm for Evapotranspiration Retrieving
SD	Standard Deviations
SUREAL	Surface Resistance Algorithm
WP	Water Productivity

11.1 INTRODUCTION

Observations and modeling have been demonstrated evidence of changes in the climatic systems worldwide, which are connected to human activities (IPCC, 2023). In addition to these climate changes, the expansion of agriculture over the natural ecosystems has caused large alterations in the river flows and ground tables, what can, together with bad agricultural managements, contribute to environmental degradation, affecting the energy, water, and carbon fluxes between the surfaces and the low atmosphere (Akhtar et al., 2021; Oliveira et al., 2022; Silva et al., 2023).

To provide sustainable water resources, water managers may observe the water accounting approach, which recognizes the various water users of a hydrological basin and the water flows in terms of net water production or net water consumption, i.e., water productivity—WP (Cai et al., 2002). Understanding the effects of climate and land use changes upon WP is critical for ecological restoration (Yang et al., 2016; Zhang and Zhang, 2019), demanding large-scale studies to support sustainable explorations of the water resources (Almeida et al., 2023a, 2023b; Araujo et al., 2019; Santos et al., 2020). Geotechnologies can be also used to detect anomalies on the water productivity components for specific periods by using long-term satellite and weather data (Beguería et al., 2014; Bento et al., 2018; Gouveia et al., 2017; Vicente-Serrano et al., 2015, 2018; Yang et al., 2014; Zhang and Zhang, 2019; Zhang et al., 2021; Teixeira et al., 2021a).

Quantifying energy, water, and carbon balance components by using remote sensing together with gridded weather data in mixed agroecosystems, is a suitable way to assess WP, giving support for the rational management of water resources. After accounting for all the radiation balance components, the net radiation (R_n) is the difference between incoming and outgoing radiation for both short and long wavelengths, then R_n is partitioned into latent (λE), sensible (H), and ground (G) heat fluxes. Acquiring λE deserves highlighting, because it represents the energy for ET, which is the main use of water resources by well-watered vegetation and is also related to BIO (Teixeira et al., 2020a, 2020b, 2021a, 2021b, 2021c, 2023).

For WP assessments in this chapter, distinctions are made between reference (ET_0), potential (ET_p), and actual (ET) evapotranspiration (Allen et al., 1998). ET_0 is considered the water flux from a reference surface, not a shortage of water, which in standard practice is a hypothetical grass surface with specific characteristics. ET_p refers to the water flux from vegetation growing in large fields under optimum root-zone moisture, excellent management, and environmental conditions, achieving full productivity. ET is the real water flux occurring from vegetation in a specific situation involving all environmental conditions. Due to sub-optimal managements and environmental constraints that affect plant growth and limit the water fluxes, ET is frequently smaller than ET_p (Teixeira et al., 2015).

On the one hand, although ET is related to BIO, increasing its rates means less water availability for ecological and human uses. On the other hand, H magnitude may indicate surface warming or cooling effects (Bhattarai et al., 2017; Teixeira et al., 2021b). Replacing natural vegetation with crops may produce carbon sinks, affecting BIO (Ceschia et al., 2010), while water scarcity increases vegetation mortality rates and changes the ecosystem's species compositions (Zhao and Running, 2010). Monitoring these WP components along the years is essential for ecological restoration and to assess the dimension of environmental impacts (Yang et al., 2016; Zhang and Zhang, 2019; Zhang et al., 2021).

WP may be defined as the ratio of the net benefits from crop, forestry, fishery, livestock, and mixed agricultural systems to the amount of water required to produce those benefits (Molden et al., 2007). In the current chapter, water productivity based on biomass production is considered the ratio of BIO by ET (WP_{BIO}), and the water productivity based on actual yield (Ya) is the ratio of Ya by ET (WP_{Ya}). There are many ways to raise WP_{BIO} and WP_{Ya} for both irrigated and rainfed crops (Adak et al., 2013, Almeida et al., 2023a; Molden et al., 2007; Teixeira et al., 2019, 2021c; Yuan et al., 2013).

Field methods for ET and BIO measurements provide values for specific sites and are not suitable for large-scale WP assessments. The spatial variability of root-zones moisture may be significant, and these variations are caused by different amounts of precipitation, seepage, flooding, irrigation, hydraulic characteristics of soils, vegetation types and densities, while the temporal variations on ET can be ascribed to weather conditions and vegetation development. Such difficulties stimulated remote sensing together with weather data as valuable tools in the determination of WP components in distinct Brazilian agroecosystems (Silva et al., 2019; Santos et al., 2020; Jardim et al., 2022; Rampazo et al., 2020; Teixeira et al. 2020a, 2020b, 2021a, 2021b, 2021c, 2023).

For WP assessments through geotechnologies, algorithms have been elaborated and validated at different spatial and temporal sales (Allen et al., 2007; Bastiaanssen et al., 1998; Teixeira, 2010). They are very useful to evaluate the impacts from human activities upon natural resources, due to their efficiencies to account the WP components at different spatial and temporal resolutions. Because of its operationality, the Penman–Monteith (PM) equation has been inserted in some algorithms (Cleugh et al., 2007; Consoli and Vanella, 2014; Consoli et al., 2016; Nagler et al., 2013; Olivera-Guerra et al., 2018), which when applied with gridded weather data, are suitable for using with low spatial resolution satellite images (Mateos et al., 2013; Teixeira et al., 2023; Vanella et al., 2019).

Considering operational aspects of the Penman–Monteith equation, the SAFER (Simple Algorithm for Evapotranspiration Retrieving) algorithm was developed by using simultaneous field and remote sensing measurements in northeast Brazil to estimate the WP components (Teixeira et al., 2008; Teixeira, 2010). The reason for the SAFER's choice for the WP assessments in this chapter is that, besides its applicability, other important advantage, regarding other algorithms, is that in its newest version it is possible to carry out these assessments with and without the thermal bands, being possible to use only the visible and near infrared bands, more easily available (Consoli and Vanella, 2014). For example, the thermal bands of the MODIS sensor having a spatial resolution of 1 km means that the images should cover more mixed surface types, when comparing with the 250-m spatial resolution of its red and near infrared bands (Teixeira et al., 2023).

Retrieving ET from SAFER together with estimations of the available energy allows BIO estimations for WP assessments on large scales. The energy captured in photosynthesis, depending on the radiation use efficiency (ε), is in part represented by BIO, which is a key indicator for any agroecosystem (Wu et al., 2010), and its values, as for ET, are also highly variable in both space and time (Adak et al., 2013; Teixeira et al., 2023). In water-limited environments, the challenge is to make improvements in BIO through optimized management practices (Adak et al., 2013).

The slope of the linear regression between BIO and cumulative Photosynthetically Active Radiation (PAR) intercepted by canopy has been used to determine ε (Ceotto and Castelli, 2002; Tesfaye et al., 2006). However, the Radiation Use Efficiency (RUE) model proposed by Monteith (1972) to estimate BIO, based on the absorbed photosynthetically active radiation (PAR_{abs}) and canopy development, can be applied together with remote sensing data (Almeida et al., 2023a; Bastiaanssen and Ali, 2003; Claverie et al., 2012; Nyolei et al., 2019; Rampazo et al., 2020; Teixeira et al., 2023; Zhao et al., 2005). The model proposes a direct proportional relationship between BIO and PAR_{abs}, which is variable throughout the year and during the crop growing periods (Tesfaye et al., 2006).

Although uncertainties arise in connection with ε values, due to their spatiotemporal variability (Zhao et al., 2005), the Monteith RUE model accuracy has been considered acceptable for large-scale applications with different remote sensing data. These data are efficient for BIO estimates because they provide spatial and temporal information on the location and state of water and vegetation conditions, overcoming the lack of extensive observations and/or measurements over large areas (Wu et al., 2010; Ahamed et al., 2011, Teixeira et al., 2023).

This chapter highlights the combination of the newest version of the SAFER algorithm and the Monteith RUE model, with applications in agroecosystems inside some Brazilian biomes. This is done to demonstrate that remote sensing measurements, together with weather data, can be used for water productivity assessments on different spatial and temporal scales, to support the rational water resources management. A third model for the surface resistance to water fluxes (r_s), SUREAL (Surface Resistance Algorithm), is used to classify the vegetation into irrigated crops and natural ecosystems (Teixeira, 2010; Teixeira et al., 2013) to retrieve the incremental values of ET and BIO, resulted from the replacement of natural vegetation by irrigated crops.

Following the introduction, the study regions, data set, and the steps for modeling are described. Water productivity assessments are made using remote sensing parameters at different spatial and temporal resolutions, involving natural vegetation and agricultural crops, under both irrigation and rainfed conditions, for distinct agroecosystems inside the Brazilian biomes. The successful applications carried out in Brazil may encourage replications of the methods in other countries through simple calibrations of the modeling equations.

11.2 STUDY REGIONS AND DATA SET

Figure 11.1 shows the study regions at the scales of agricultural growing regions and crops within the Brazilian biomes Atlantic Forest, Caatinga, and Cerrado, for water productivity assessments.

In Figure 11.1a, the agricultural growing regions and crops assessed for WP studies are depicted in Brazil, namely in the biomes Cerrado, Caatinga, and Atlantic Forest. Regarding the agricultural growing regions, Petrolina/Juazeiro is fully within the Caatinga biome, while the North of Minas Gerais is in a transition zone between the Caatinga and Cerrado biomes. All Fruit Circuit agricultural areas are within the Atlantic Forest biome.

Caatinga and Cerrado face frequent droughts, and their natural species develop resilience with increasing aridity under these dry conditions (Almagro et al., 2017; Azevedo et al., 2020; Jardim et al., 2022; Sano et al., 2019; Santos et al., 2014), and Atlantic Forest has a humid tropical climate (Ribeiro et al., 2009), but contrasting microclimates among natural and anthropized areas (Souza et al., 2020). However, all these agricultural regions are facing the coupled effects of climate and land-use changes (Teixeira et al., 2023).

The Petrolina/Juazeiro pole involves the counties of Petrolina (Pet) in Pernambuco state and Juazeiro (Jua), Bahia state, both in northeast Brazil (Figure 11.1b). Intensification of irrigated crops has caused widespread replacement of the Caatinga species, affecting the water productivity components. The visible, near infrared (NIR) and thermal bands from MODIS images, six for 2010 and nine for 2011, together with a net of 14 automatic weather stations, were used for water productivity assessments in this region, following field calibrations and spatial and temporal interpolations (Teixeira et al., 2013).

The North of Minas Gerais pole, southeast Brazil (Figure 11.1c), comprises the smaller counties of Matias Cardoso (MC), Jaíba (JAI), Varzelândia (VA), Verdelândia (VE), Pai Pedro (PP),

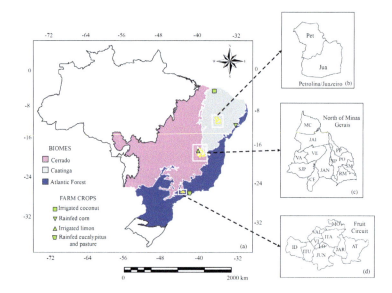

FIGURE 11.1 Study regions for water productivity assessments in the respective Brazilian biomes involving the agricultural growing regions and farm crops.

* Agricultural growing regions: Juazeiro/Petrolina involving the counties of Petrolina and Juazeiro, the North of Minas Gerais, and Fruit Circuit of São Paulo
* Irrigated crops: limon and coconut
* Rainfed crops: corn, eucalyptus, and pasture

Nova Porteirinha (NP), Porteirinha (PO), Janaúba (JAN), São João da Ponte (SJP), Riacho dos Machado (RM), Serranópolis de Minas (SM), and Capitão Eneas (CE). In this agricultural growing region intensification of irrigated crops has also caused widespread replacement of the Caatinga and Cerrado natural species. In this case, the WP assessments were carried out by using the visible, NIR, and thermal bands from the Landsat 8 images, together with a net of 12 automatic weather stations during the year 2015 (Teixeira et al., 2018).

The Fruit Circuit pole is inside São Paulo state, southeast Brazil (Figure 11.1d). It comprises the smaller counties of Indaiatuba (ID), Itupeva (ITU), Valinhos (VAL), Vinhedo (VI), Louveira (LO), Jundiaí (JUN), Morungaba (MO), Itatiba (ITA), Jarinu (JAR), and Atibaia (AT). In this agricultural growing region fruit crops are replacing the natural species from the Atlantic Forest biome. In this case, the WP assessments were carried out by using the visible and IR bands from the MOD13Q1 reflectance product, together with a net of nine automatic weather stations involving the long-term period from 2012 to 2016 (Teixeira et al., 2020a).

According to the green polygons in Figure 11.1a, considering crop scales, WP assessments were carried out under irrigation (limon and coconut) and rainfed (corn, eucalyptus, and pasture) conditions.

Irrigated limon crops were in a commercial farm, County of Jaíba, within the Caatinga biome, northern Minas Gerais state (MG), southeast Brazil (green up-pointed triangle in Figure 11.1a). A weather station was installed close to the limon areas, and the data used together with the visible and NIR bands from Landsat 8 images, along the year 2015. Another WP study under irrigation conditions was carried out in a commercial farm of dwarf coconut, in the Caatinga biome but in Camocim County, coastal zone of Ceará state, northeast Brazil (green square in Figure 11.1a). A weather station was also installed close to the coconut areas, and the data used together with the visible and NIR bands from Landsat 8 images, during the year 2016.

Land use masks were used for rainfed eucalyptus and pasture crops, interspaced by natural vegetation were in the Atlantic Forest biome, inside the eastern side of São Paulo state, southeast Brazil (green trapeze in Figure 11.1a). The red and near infrared bands from the MODIS MOD13Q1 reflectance product were used together with a net of seven automatic weather stations classifying these agroecosystems during the year 2015. Rainfed irrigated corn crop was within the Caatinga biome, Sergipe state, northeast Brazil (green down-pointed triangle in Figure 11.1a). In this case, it used the green, red, red edge, and NIR bands of aerial images from camera Sequoia on board a drone for the main corn crop stages in the year 2017.

Table 11.1 presents the characteristics of the sensors on board satellites and drones, together with the respective bands of wavelengths.

The input weather data for WP modeling were incident global solar radiation (R_G); air temperature (T_a), air relative humidity (RH), and wind speed at 2m height (u_2), for calculations of reference evapotranspiration (ET_0) using the standard FAO-56 method (Allen et al., 1998); actual evapotranspiration (ET), applying the SAFER algorithm (Teixeira, 2010); and biomass production (BIO), using the RUE (Radiation Use Efficient) model (Monteith, 1972, 1977). For agricultural growing regions and land use masks, gridded weather data were layered with the remote sensing parameters, contributing to a better spatial characterization of the water productivity components (Teixeira et al., 2023).

11.3 MODELING WATER PRODUCTIVITY COMPONENTS

Figure 11.2 shows the flowchart for modeling the large-scale water productivity components by applying the SAFER algorithm and the RUE model using remote sensing parameters and gridded weather data.

Regarding the Landsat 8 images, following Figure 11.2, the spectral radiances for the visible and near infrared spectral ranges (L_B) are computed from their digital numbers (DN_B):

$$L_B = a + bDN_B \tag{1}$$

TABLE 11.1

Characteristics of the Sensors Used, on Board Satellites and Drones, with Respective Bands of Wavelengths. Used for Water Productivity Assessments in Brazil

Satellite/Drone	Spatial Resolutions	Spectral Bands (μm)	Sensor Swatch	Temporal Resolution
Landsat 8	Visible/NIR: 30 m Thermal: 100 m	B1: 0.43–0.45 B2: 0.45–0.51 B3: 0.53–0.59 B4: 0.64–0.69 B5: 0.85–0.88 B6: 1.57–1.65 B7: 2.110–2.290 B10: 10.60–11.19 B11: 11.50–12.50	185 km	16 days
MODIS	Visible/NIR: 250 m Thermal: 1000 m	B1: 0.62–0.67 B2: 0.84–0.87 B31: 10.78–11.28 B32: 11.77–12.27	2330 km	1 to 2 days
Drone (Sequoia)	Visible/NIR: 4 cm	B1: 0.53–0.57 B2: 0.64–0.68 B3: 0.73–0.74 B4: 0.77–0.82	32 m (at 40 m height)	–

Landsat 8: Thermal bands B10 and B11; MODIS: Thermal bands B31 and B32

where L_B is in W m^{-2} sr^{-1} μm^{-1}, and a and b are regression coefficients given in the metadata file (Teixeira et al., 2021b).

In general, the reflectance for each satellite band (ρ_B) is calculated as:

$$\rho_B = \frac{L_B \pi d^2}{R_{a_B} \cos \varphi} \quad (2)$$

where d is the relative earth-sun distance; R_{a_B} is the mean solar irradiance at the top of the atmospheric irradiance for each band (W m^{-2} μm^{-1}), and j the solar zenith angle.

The broadband planetary albedo (α_p) is computed as the total sum of the ρ_B values, according to the weights for each band (w_B):

$$\alpha_p = \sum_{B_1}^{B_n} w_B \rho_B \quad (3)$$

for Landsat 8, bands 1 to 7; for MODIS images, bands 1 and 2; and for Sequoia camera, bands 1 to 4 are considered in this chapter (Teixeira et al., 2021b, 2021c, 2023).

NDVI is a land cover indicator acquired from remote sensing measurements as:

$$\text{NDVI} = \frac{\rho_{NIR} - \rho_{RED}}{\rho_{NIR} + \rho_{RED}} \quad (4)$$

where ρ_{NIR} and ρ_{RED} are the reflectances over the ranges of wavelengths in the near infrared (NIR) and red (RED) regions of the solar spectrum, respectively (Almeida et al., 2023a, 2023b; Teixeira et al., 2023).

Modeling and Monitoring Water Productivity

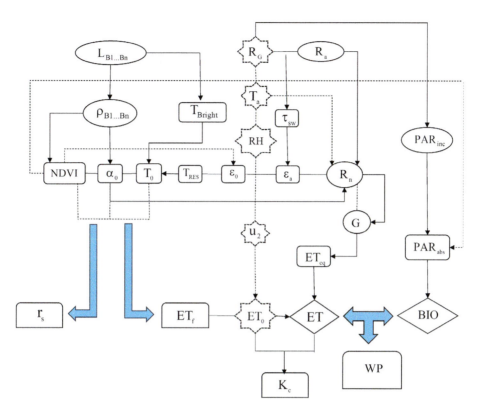

FIGURE 11.2 Flowchart for modeling the water productivity components, by using remote sensing parameters together with gridded weather data applying the SAFER algorithm and the LUE model. Dashed polygonal-shaped boxes are data from the weather stations.

Note: $L_{B1...Bn}$—spectral radiance from band 1 to n; $\rho_{B1...Bn}$—spectral reflectance from band 1 to n; NDVI—Normalized Difference Vegetation Index; α_0—Surface albedo; T_{Bright}—Brightness temperature; T_{RES}—Residual temperature; T_0—Surface temperature; r_s—Surface resistance; ET_f—Evapotranspiration fraction; R_G—Incident global solar radiation at the surfaces; R_a—Incident solar radiation at the top of the atmosphere; τ_{sw}—shortwave atmospheric transmissivity; T_a—Air temperature; RH—Relative humidity; u_2—Wind speed at 2m height; ET_0—Reference evapotranspiration; ε_0—Surface emissivity; ε_a—Atmospheric emissivity; R_n—Net radiation; G—Ground heat flux; ET_{eq}—Equilibrium evapotranspiration; ET—Actual evapotranspiration; K_c—Crop coefficient; PAR_{inc}—Incident photosynthetically active radiation; PAR_{abs}—Absorbed photosynthetically active radiation; BIO—Biomass production; WP—Water productivity.
*Dashed polygons mean data from weather stations.

Two ways to estimate the surface temperature (T_0) are demonstrated: through the brightness temperature (T_{bright}) applying the Plank's low to the thermal bands at the satellite overpass timescale; and by the residual temperature (T_{RES}), without the thermal bands, using the Stefan–Boltzmann low for the atmospheric and surface emitted radiations in the radiation balance at daily timescale.

The brightness temperature (T_{bright}), following the Planck's low, is estimated as:

$$T_{bright} = \frac{K_2}{\ln\left(\frac{K_1}{L_b} + 1\right)} \quad (5)$$

where L_b is the uncorrected radiance from the thermal bands, and K_1 and K_2 are the conversion coefficients. For Landsat 8 the bands 10 and 11 are used, while for the MODIS images the thermal bands 31 and 32 are considered.

The results for both α_p and T_bright need to be corrected atmospherically to acquire the surface values of surface albedo (α_0) and T_0 at the satellite overpass time, which is carried out through simple linear regressions from satellite estimations of α_p and field measurements of α_0; and satellite estimations of T_bright and field measurements of T_0 (Teixeira, 2010; Teixeira et al., 2010, 2013).

Considering the residual method to estimate T_0 as a residue in the radiation balance, with R_G and T_a measured at the weather stations, and R_a astronomically calculated, R_n can be estimated using the Slob equation (de Bruin, 1987):

$$R_n = (1-\alpha_0)R_G - a_L \tau_{sw} \qquad (6)$$

where τ_{sw} is the shortwave atmospheric transmissivity (R_G/R_a) and a_L is a regression coefficient upscaled throughout the T_a pixel values (Teixeira et al., 2008). $a_L = a_T T_a + b_T$ (7) where the regression coefficients a_T and b_T in Brazil, involving strong thermo-hydrological conditions in the Caatinga biome, were 6.8 and −40, respectively, but may be adjusted for specific environmental conditions (Teixeira, 2010; Teixeira et al., 2008).

The atmospheric emissivity (ε_A) is calculated according to Almeida et al. (2023a, 2023b) and Teixeira et al. (2023):

$$\varepsilon_A = a_A \left(\ln \tau_{sw} \right)^{b_A} \qquad (8)$$

where a_A and b_A are regression coefficients, which are reported as 0.94 and 0.11, found in Brazil, involving strong thermo-hydrological conditions, but may be calibrated using field radiation balance measurements for specific environmental environments (Teixeira, 2010; Teixeira et al., 2008).

Surface emissivity (ε_0) is estimated according to Rampazo et al. (2020) and Silva et al. (2019):

$$\varepsilon_0 = a_0 \ln \text{NDVI} + b_0 \qquad (9)$$

where a_0 and b_0 are regression coefficients, which were reported as 0.06 and 1.00 in Brazil, involving strong thermo-hydrological conditions, but may be acquired for specific environmental environments by simultaneous field radiation balance and remote sensing NDVI measurements (Teixeira, 2010; Teixeira et al., 2008).

Using the residual method, according to the physical principle of the Stefan–Boltzmann low, T_{RES} is estimated as (Almeida et al., 2023a, 2023b; Teixeira et al., 2023):

$$T_{RES} = \frac{\sqrt[4]{R_G(1-\alpha_0) + \sigma \varepsilon_a T_a^4 - R_n}}{\sigma \varepsilon_0} \qquad (10)$$

where σ = 5.67 10⁻⁸ W m⁻² K⁻⁴ is the Stefan--Boltzmann constant. In the case of using the SAFER algorithm without the thermal band, T_{RES} is considered as T_0 at daily time scale.

To acquire ET, the ratio of ET to ET_0 (the evapotranspiration fraction—ET_f) at satellite overpass (with thermal bands) or daily (without thermal bands) scales, is modeled (Araujo et al., 2019; Dehziari and Sanaienejad, 2019; Safre et al., 2022; Teixeira et al., 2021a, 2021b, 2023; Venâncio et al., 2021):

$$ET_f = \exp\left[a_{sf} + b_{sf} \left(\frac{T_0}{\alpha_0 \text{NDVI}} \right) \right] \qquad (11)$$

where a_{sf} and b_{sf} are regression coefficients, which were 1.90 and −0.008 involving distinct environmental conditions in Brazil, being possible to be calibrated using field measurements for ET and ET_0 and the remote sensing parameters α_0, NDVI, and T_0 in contrasting hydrological surfaces (Venâncio et al., 2021; Safre et al., 2022).

Eq. 9 does not work for water bodies or mixtures of land and water (NDVI < 0); thus, under these circumstances, the concept of equilibrium evapotranspiration—ET_{eq} (Raupasch, 2001) is used in the SAFER algorithm:

$$ET_{eq} = 0.035\left(\frac{\Delta\,(R_n - G)}{\Delta + \gamma}\right) \quad (12)$$

where Δ is the inclination of the curve relating to the saturation vapor pressure (e_s) and T_a, γ is the psychrometric constant, and the ground heat flux (G) is estimated according to:

$$\frac{G}{R_n} = a_G \exp(b_G \alpha_0) \quad (13)$$

where a_G and b_G are regression coefficients, 3.98 and −25.47, respectively, in Brazil, involving strong thermo-hydrological conditions, but may be calibrated using simultaneous field energy balance measurements for other environmental conditions (Teixeira, 2010; Teixeira et al., 2008).

Throughout conditional functions applied to the NDVI values, daily ET rates were acquired as:

$$ET = ET_f ET_0 \text{ or } ET_{eq} \quad (14)$$

where ET_0 is calculated using daily gridded weather data on R_G, T_a, RH, and u_2 (Teixeira et al., 2023).

For BIO estimations, the Monteith RUE model is applied introducing the root-zone moisture effect through ET_f (Almeida et al., 2023a, 2023b; Araujo et al., 2019; Rampazo et al., 2020; Teixeira et al., 2023):

$$BIO = \varepsilon_{max} ET_f PAR_{abs}\, 0.864 \quad (15)$$

where ε_{max} is the maximum radiation use efficiency, which for the majority of C3 crops in the Brazilian biomes was considered 2.45 g MJ^{-1}, and 0.864 is the unit conversion factor.

The incident photosynthetically active radiation (PAR_{inc}) may be measured, but it can be also estimated from R_G:

$$PAR_{inc} = a_{pi} RG \quad (16)$$

where the regression coefficient a_{pi} in Brazil was 0.44 (Teixeira et al., 2008), but it can be acquired for specific environmental conditions from simultaneous measurements of R_G and PAR_{inc}.

The fraction of the absorbed PAR_{inc} (f_{PAR}) is calculated as:

$$f_{PAR} = a_{pa} NDVI + b_{pa} \quad (17)$$

where a_{pa} and b_{pa} are regression coefficients, 1.257 and −0.161, respectively, found in mixed crops (Bastiaanssen and Ali, 2003), but they may be calibrated for specific environmental conditions using field measurements of PAR above and below the canopies, together with NDVI values.

PAR_{abs} is estimated from PAR_{inc} and f_{PAR}:

$$PAR_{abs} = f_{PAR} PAR_{inc} \quad (18)$$

For the separation of irrigated crops and natural vegetation, the SUREAL is applied during the naturally driest period of a year, considering threshold limits for the surface resistance (r_s) to water fluxes (Teixeira et al., 2013):

$$r_s = \exp\left[a_r\left(\frac{T_0}{\alpha_0}\right)(1 - NDVI) + b_r\right]\Bbbk \quad (19)$$

where a_r and b_r are regression coefficients, found to be 0.04 and 2.72, respectively, for agroecosystems in Brazil, but they can be calibrated having all the variables of the Penman–Monteith equation measured in the field together with simultaneous remote sensing measurements.

Water productivity (WP) according to Teixeira et al. (2019, 2021c, 2023) is considered as:

$$WP_{BIO,Ya} = \frac{BIO \text{ or } Ya}{ET} \quad (20)$$

where Ya is the actual yield.

The ET_f pixel values without water deficits are used for estimations of the crop coefficient (K_c) as a function of the accumulated degree-days (DD_{ac}):

$$K_c = a_d DD_{ac}^2 + b_d DD_{ac} + c_d \quad (21)$$

where a_d, b_d, and c_d are the crop-specific regression coefficients (Teixeira et al., 2021b).

11.4 AGRICULTURAL GROWING REGION SCALE

11.4.1 Petrolina/Juazeiro Pole

Within the Brazilian Caatinga biome (see Figures 11.1a,b), the Petrolina/Juazeiro pole represents an important agricultural growing region. Figure 11.3 presents the spatial variation of the ET monthly values in the mixed agroecosystems limited by the Petrolina (Pet) and Juazeiro (Jua) counties, northeast Brazil, throughout the year 2011, by using the red (B1), infrared (B2), and thermal bands (B31 and B32) of the MODIS sensor (see Table 11.1).

Monthly spatial and temporal ET variations throughout the year is evident, mainly when comparing the wettest period from February to April, with the driest one between August and October. The maximum ET rates were observed in April, with averages of 60 and 45 mm month^{-1} for Pet and Jua, respectively. However, the highest pixel values, reaching 200 mm month^{-1}, were from November to December for both counties, representing well-irrigated crops. Intermediate ET values in natural vegetation (Caatinga) occurred just after the rainy period, from May to June, because

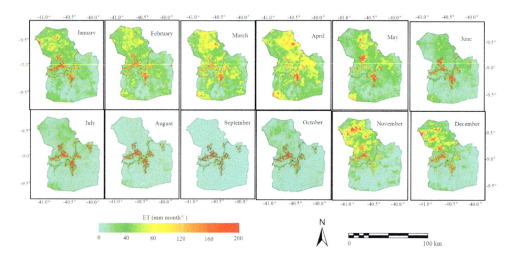

FIGURE 11.3 Spatial distribution for the monthly values of actual evapotranspiration (ET) from the mixed agroecosystems of the agricultural growing region within the Caatinga biome, limited by the Petrolina (Pet) and Juazeiro (Jua) counties, northeast Brazil, during the year 2011, by using the red (B1), infrared (B2), and thermal bands (B31 and B32) of the MODIS sensor.

previous rainfalls kept their species wet and green. During this time of year, the hydrological large-scale uniformity promotes ET rates from natural vegetation like those from irrigated crops or even higher on some occasions.

Because the largest fractions of the available energy are used as sensitive heat fluxes during the driest period of the year, between August and October, natural vegetation presented the lowest ET pixel values (bluish pixels), while the irrigated fields showed the highest ones (reddish pixels). Stomata of the Caatinga species close under these conditions, limiting transpiration and photosynthesis, while, in general, irrigation intervals in crops are short (daily irrigation), with a uniform water supply, reducing the heat losses to the atmosphere.

Applying the SUREAL model (Eq. 19), it was possible to separate the results for the WP components into irrigated crops (IC) and natural vegetation (NV). Table 11.2 shows the ET monthly average values and standard deviations (SD) for IC and NV inside the Caatinga biome, limited by the Petrolina (Pet) and Juazeiro (Jua) counties, northeast Brazil, during the year 2011, by using the red (B1), infrared (B2), and thermal bands (B31 and B32) of the MODIS sensor.

Regarding IC, differences between the counties are clear, with November presenting the highest crop ET rates in Pet, while in Jua they happened in April. In NV, the largest ET values were in April, while the lowest ones were in September, for both counties. IC consumed 555 mm yr^{-1} more water than NV in Pet, while in Jua, this extra consumption was 545 mm yr^{-1}.

Considering both counties, the lowest SD values for both IC and NV were during the climatically driest period of the year, from July to October. Under rainfed or irrigation conditions, the plants are strongly sensitive to the spatial distribution of precipitation and root-zone moisture (Claverie et al., 2012; Zhang et al., 2021; Teixeira et al., 2023). During the climatically driest periods, the higher ET

TABLE 11.2

Monthly Average Values and Standard Deviations (SD) of Actual Evapotranspiration (ET) from Irrigated Crops (IC) and Natural Vegetation (NV) within the Caatinga Biome, Limited by the Petrolina and Juazeiro Counties, Respectively, Pernambuco (PE) and Bahia (BA) States, northeast Brazil, during the Year 2011, by Using the Red (B1), Infrared (B2), and Thermal Bands (B31 and B32) of the MODIS Sensor

Months/Year	Petrolina-PE ET (mm month^{-1}) IC	Petrolina-PE ET (mm month^{-1}) NV	Juazeiro-BA ET (mm month^{-1}) IC	Juazeiro-BA ET (mm month^{-1}) NV
1	95 ± 41	42 ± 20	69 ± 38	17 ± 17
2	87 ± 32	44 ± 19	69 ± 33	24 ± 20
3	94 ± 34	52 ± 24	82 ± 39	34 ± 26
4	87 ± 28	58 ± 25	87 ± 36	42 ± 26
5	74 ± 37	37 ± 21	67 ± 34	26 ± 19
6	67 ± 30	23 ± 12	61 ± 31	15 ± 11
7	66 ± 27	15 ± 8	64 ± 36	9 ± 9
8	69 ± 34	5 ± 6	59 ± 41	6 ± 8
9	50 ± 30	2 ± 4	37 ± 33	2 ± 4
10	59 ± 32	12 ± 8	39 ± 33	4 ± 7
11	97 ± 42	53 ± 31	52 ± 36	9 ± 13
12	91 ± 42	43 ± 27	54 ± 38	8 ± 13
Year	938 ± 351	385 ± 141	739 ± 328	194 ± 141

*IC—Irrigated crops; NV—Natural vegetation

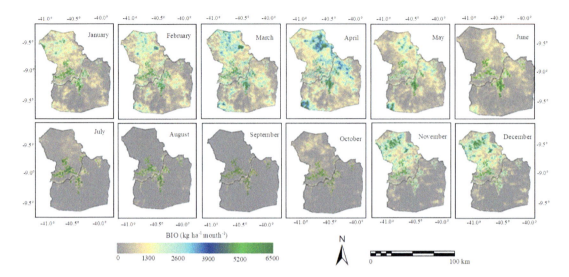

FIGURE 11.4 Spatial distribution for the monthly values of biomass production (BIO) from the mixed agroecosystems of the agricultural growing region within the Caatinga biome, limited by Petrolina (Pet) and Juazeiro (BA) counties, northeast Brazil, during the year 2011, by using the red (B1), infrared (B2), and thermal bands (B31 and B32) of the MODIS sensor.

and SD values for IC are mainly caused by different levels of fertilization, crop stages, and irrigation (Safre et al., 2022; Teixeira et al., 2021b, 2021c; Wu et al., 2010).

Figure 11.4 presents the spatial variation of the BIO monthly values in the mixed agroecosystems within the biome Caatinga, limited by the Petrolina (Pet) and Juazeiro (Jua) counties, northeast Brazil, throughout the year 2011, by using the red (B1), infrared (B2), and thermal bands (B31 and B32) of the MODIS sensor.

As there is a relation between water consumption and BIO (Yuan et al., 2013), the spatial and temporal variations of BIO are also strong throughout the year, for both IC and NV agroecosystems. The highest values are during the rainy season from February to April, and the lowest ones are in the driest period, between July and September. The maximum BIO rates occurred in April, with averages of 1,817 and 1,306 kg ha^{-1} month^{-1} for Pet and Jua, respectively. During April, it was noticed that several areas presented rates higher than 4,000 kg ha^{-1} month^{-1}, including both NV and IC. The lowest BIO happened in September for both counties, with corresponding mean pixel values of 142 and 92 kg ha^{-1} month^{-1}, respectively.

Rainfalls from January to April provide enough water storage in the root zones of the Caatinga species to maintain their growth, while from June to September, IC present larger BIO rates than those for NV, as with absence of rains, root-zone moisture of IC is close to the field capacity due to the general daily irrigation, making sure crops are well visible on BIO maps. Similarly, irrigated corn crop showed twice the BIO values when comparing with natural alpine meadow in the Heihe River Basin (Wang et al., 2012), irrigation being considered the main reason for these marked differences. The same effects on BIO values between NV and IC were observed by Teixeira et al. (2018) under the semi-arid conditions of the agricultural growing region of North of Minas Gerais state, southeast Brazil.

Table 11.3 shows the BIO monthly average values and standard deviations (SD) for IC and NV inside the Caatinga biome, limited by Petrolina (Pet) and Juazeiro (Jua) counties, northeast Brazil, during the year 2011, by using the red (B1), infrared (B2), and thermal bands (B31 and B32) of the MODIS sensor.

In general, Pet presented higher BIO than Jua for both NV and IC. However, for IC, this difference was only 33%, while for NV it was double. For all agroecosystems, the period with the highest

TABLE 11.3
Monthly Average Values and Standard Deviations (SD) of Biomass Production (ET) from Irrigated Crops (IC) and Natural Vegetation (NV) within the Caatinga Biome, Limited by the Petrolina and Juazeiro Counties, Respectively, Pernambuco (PE) and Bahia (BA) States, Northeast Brazil, during the Year 2011, by Using the Red (B1), Infrared (B2), and Thermal Bands (B31 and B32) of the MODIS Sensor

Months/Year	Petrolina-PE BIO (t ha^{-1}) IC	Petrolina-PE BIO (t ha^{-1}) NV	Juazeiro-BA BIO (t ha^{-1}) IC	Juazeiro-BA BIO (t ha^{-1}) NV
January	2.8 ± 1.7	1.1 ± 0.7	1.9 ± 1.5	0.4 ± 0.6
February	2.8 ± 1.4	1.2 ± 0.8	2.2 ± 1.4	0.6 ± 0.7
March	3.2 ± 1.6	1.5 ± 1.0	2.4 ± 1.6	0.8 ± 0.8
April	2.9 ± 1.3	1.7 ± 1.1	2.9 ± 1.7	1.2 ± 1.0
May	2.4 ± 1.4	0.9 ± 0.8	2.4 ± 1.7	0.7 ± 0.8
June	2.0 ± 1.1	0.4 ± 0.5	2.0 ± 1.5	0.4 ± 0.4
July	1.9 ± 1.1	0.1 ± 0.3	0.9 ± 0.8	0.5 ± 0.1
August	1.9 ± 1.4	0.0 ± 0.1	1.7 ± 1.7	0.1 ± 0.2
September	1.3 ± 1.2	0.0 ± 0.1	1.0 ± 1.2	0.0 ± 0.1
October	1.6 ± 1.2	0.2 ± 0.4	0.9 ± 1.2	0.1 ± 0.2
November	2.9 ± 1.7	1.4 ± 1.1	1.3 ± 1.3	0.2 ± 0.4
December	2.6 ± 1.7	1.0 ± 1.0	1.4 ± 1.4	0.2 ± 0.4
Year	28.5 ± 14.4	9.9 ± 4.9	21.3 ± 13.4	4.6 ± 4.7

*IC—Irrigated crops; NV—Natural vegetation

BIO was from March to April. In IC, the lowest average values were from September to October, while in natural vegetation, they occurred between August and September. The annual incremental BIO rates (differences between IC and NV) were 18.6 and 16.7 t ha^{-1} yr^{-1}, on average, for Pet and Jua, respectively. As in the case of ET, the lowest SD values occurred during the driest period of the year, independently of the agroecosystem. However, Pet presented slightly higher BIO spatial variation in comparison with Juazeiro.

Figure 11.5 presents the monthly average values and standard deviations (SD) for water productivity based on biomass production (WP$_{BIO}$) considering irrigated crops (IC) and natural vegetation (NV) in Petrolina (Figure 11.5a) and Juazeiro (Figure 11.5b) counties, during the year 2011, by using the red (B1), infrared (B2), and thermal bands (B31 and B32) of the MODIS sensor.

Analyzing only IC, WP$_{BIO}$ differences between the counties can be also identified. The period from March to April presented the highest values in Pet, while in Jua they happened in May. In NV, the largest WP$_{BIO}$ values were in April and the lowest ones occurred in October for both counties. The WP$_{BIO}$ values from IC in Jua were 86% of those for Pet, while this percentage was 74% for NV, evidencing the effect of the differences in the root-zone moisture conditions between the counties.

Considering the annual incremental WP$_{BIO}$, it was an average 1.0 kg m^{-3} for both counties, with the spatial variation in Jua being slightly higher than that for Pet. Although the lowest SD values were during the driest period, as they were for ET and BIO, the monthly WP$_{BIO}$ spatial differences were smaller, showing compensation when considering the SD values for BIO and ET together in Eq. 20.

Multiplying BIO by harvest index (HI) makes it possible to estimate the water productivity based on actual yield—WP$_{Ya}$ (Teixeira and Bassoi, 2009; Teixeira, 2012). HI values were found to be around 0.60 for vineyards and 0.80 for mango orchards under these conditions in the Petrolina/

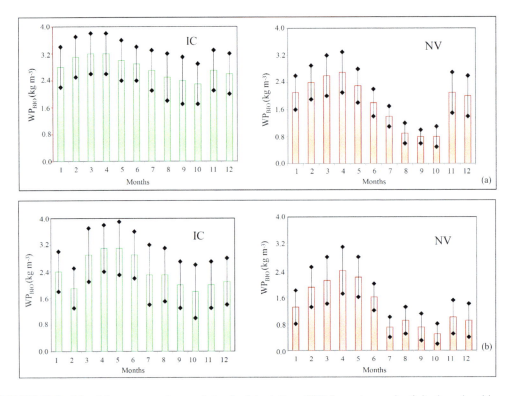

FIGURE 11.5 Monthly average values and standard deviations (SD) for water productivity based on biomass production (WP$_{BIO}$) considering irrigated crops (IC) and natural vegetation (NV) in Petrolina (a) and Juazeiro (b) counties, during the year 2011, by using the red (B1), infrared (B2), and thermal bands (B31 and B32) of the MODIS sensor.

Juazeiro agricultural growing region (Teixeira et al., 2009). In Southwest France, HI values were 0.25 for rainfed sunflower and 0.48 for irrigated corn (Claverie et al., 2012), while for wheat, an average of 0.35 for wheat was reported because of several experiments around the word (Zwart et al., 2010).

11.4.2 NORTH OF MINAS GERAIS POLE

Figure 11.6 shows the spatial distribution and the average daily pixel values together with standard deviations (SD) for the ET (Figure 11.6a) and BIO (Figure 11.6b) in irrigated crops (IC) and natural vegetation (NV), within the transition zone of Caatinga and Cerrado biomes, in the North of Minas Gerais agricultural growing region, southeast Brazil, under different hydrological conditions during the year 2015, in terms of day of the year (DOY), by using the visible and near infrared bands (B1 to B7), and the thermal bands (B10 and B11) from the Landsat 8 sensor.

The ET (Figure 11.6a) and BIO (Figure 11.6b) spatial and temporal variations are strong, mainly when comparing the wettest conditions (represented by the image of DOY 019—January 19) with the driest ones (represented by the image of DOY 259—September 16), where the pixels with high ET and BIO values represent irrigated crops (IC). The maximum ET and BIO rates occurred during the rainy period, when rainfall favored the NV species, while besides the natural water supply, in the IC areas, the plants are continuously beneficed by supplementary irrigation.

The lowest ET and SD values for IC were soon after the rainy period, conditions represented by the image of DOY 163 (June 12), while for BIO, they were during the climatically driest conditions,

Modeling and Monitoring Water Productivity 397

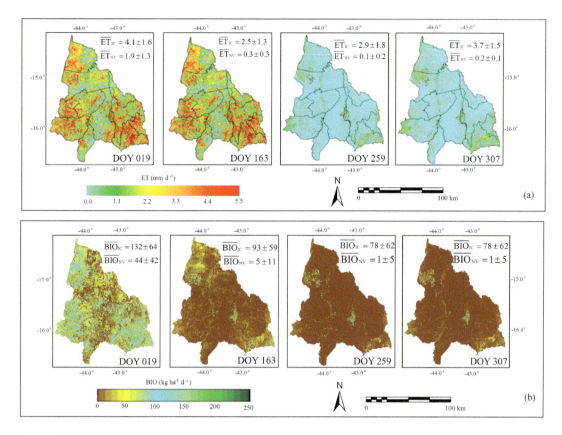

FIGURE 11.6 Spatial distribution of the daily values for the water productivity parameters, under different hydrological conditions during the year 2015 in terms of day of the year (DOY), by using the visible and near infrared bands (B1 to B7) and the thermal bands (B10 and B11) of the Landsat 8 sensor in the North of Minas Gerais agricultural growing region, southeast Brazil. (a) Actual Evapotranspiration (ET) and (b) Biomass Production (BIO). Overbars means averages showed together with standard deviations (SD). The subscripts IC and NV mean Irrigated Crops and Natural Vegetation, respectively.

represented by the image of DOY 259 (September 16), but with the lowest SD in November (DOY 307). Considering NV, the highest ET and BIO values occurred during the rainy period, represented by the image of DOY 019 (January 19), while the lowest ones were during the climatically driest period (DOY 259, September 16), because of the low root-zone moisture conditions promoting short vegetative development of the NV species. Under these last conditions, the native plants are in dormancy stage, closing stomata that limit both transpiration and photosynthesis, and in general, crops are regularly daily irrigated, increasing ET and BIO (Pereira et al., 2020).

The average pixel values for ET and BIO, in IC, ranged respectively from 2.5 ± 1.3 to 4.1 ± 1.6 mm d^{-1} and from 78 ± 62 to 132 ± 64 kg ha^{-1} d^{-1}. The corresponding ranges for NV were 0.1 ± 0.2 to 1.9 ± 1.3 mm d^{-1} and de 1 ± 1 to 44 ± 42 kg ha^{-1} d^{-1}. Leivas et al. (2016), using MODIS products, reported maximum ET values of 3.5 ± 1.0 mm d^{-1} in the Jaíba irrigation perimeter located in the North of Minas. In the Petrolina/Juazeiro agricultural growing region, Teixeira et al. (2013), also with MODIS images, found BIO maximums of 100 kg ha^{-1} d^{-1} and 46 kg ha^{-1} d^{-1} for the IC and NV agroecosystems, respectively. These differences may be related, in part, to different spatial resolutions between satellite images.

While in the second half of the year the ET and BIO values were progressively declining, reaching mean pixel values close to zero in November (DOY 307) in the NV ecosystem, in IC they were always above 2.0 mm d⁻¹ and 75 kg ha d⁻¹, respectively. At the annual scale, the incremental rates, resulting from the replacement of NV by IC were 2.7 mm d⁻¹ and 83 kg ha d⁻¹, for ET and BIO, respectively.

The largest ET and BIO values were for the Jaíba (JAI) and Matias Cardoso (MC) counties (see Figures 11.1 and 11.6) because of the irrigation water availability from the Jaíba irrigation perimeter, with the water source being from the São Francisco River. Highlights for this agricultural growing region are also for Nova Porteirinha (NP) and Janaúba (JAN) counties, inside the Gorotuba irrigation perimeter, but in this last case, the dam Bico da Pedra is the water source. These irrigation perimeters concentrate mainly irrigated fruit crops and sugar cane.

Figure 11.7 shows the spatial distribution and the average daily values together with standard deviations (SD) for the water productivity based on biomass production (WP$_{BIO}$), for irrigated crops (IC) and natural vegetation (IC), within the transition zone of Caatinga and Cerrado biomes, in the North of Minas Gerais, Southeast Brazil, during the year 2015. The images are in terms of day of the year (DOY), by using the visible and near infrared bands (B1 to B7), and the thermal bands (B10 and B11) of the Landsat 8 sensor.

Regarding WP$_{BIO}$, on the one hand, the highest rates and spatial variations for IC were in June (representative image of DOY 163), conditions of high crop root-zone moisture conditions soon after the rainy period, when previous rainfall was added to irrigation water. On the other hand, during the rainy period (represented by the image of DOY 019) the highest WP$_{BIO}$ values for NV happened. The spatial variations indicated different root-zone moisture and vegetation conditions in NV species and heterogeneity on crop stages in IC. More uniformity on WP$_{BIO}$ values were for NV, evidenced by the lower SD values when compared with those for IC.

The WP$_{BIO}$ seasonal values for IC ranged from 2.2 ± 0.8 to 3.3 ± 0.9 kg m⁻³, while the corresponding range for the NV was between 0.6 ± 0.3 and 1.8 ± 0.8 kg m⁻³. When multiplied by HI they become the water productivity based on actual yield (WP$_{Ya}$). Reported HI values were around 0.60 and 0.80 for vineyards and mango orchard surrounded by Caatinga species under the semi-arid conditions of northeast Brazil, retrieving WP$_{Ya}$ values of 2.8 and 3.4 kg m⁻³ (Teixeira et al., 2012). The WP$_{BIO}$ maximums of the North of Minas Gerais when multiplied by these HI values are lower, the probable reason being the water allocation restriction for irrigation perimeters during the drought events of 2015.

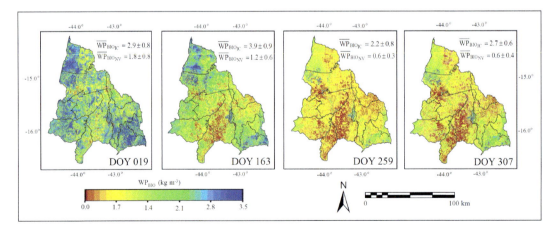

FIGURE 11.7 Spatial distribution of the daily values for the water productivity based on biomass production (WP$_{BIO}$), for irrigated crops (IC) and natural vegetation (IC), within the transition zone of Caatinga and Cerrado biomes, in the North of Minas Gerais, southeast Brazil, during the year 2015, in terms of day of the year (DOY), by using the visible and near infrared bands (B1 to B7) and the thermal bands (B10 and B11) of the Landsat 8 sensor.

Modeling and Monitoring Water Productivity 399

These previous WP assessments carried out within the Petrolina/Juazeiro and North of Minas Gerais agricultural growing regions, were done using the thermal bands of MODIS images (spatial resolution of 1 km) and Landsat 8 (spatial resolution of 120 m). With the newer version of the SAFER algorithm, it is possible to acquire ET and BIO without thermal bands, increasing the spatial resolution for 250 m in case of the MODIS images and 30 m for Landsat 8 images, allowing better classification of the IC and NV agroecosystems. Teixeira et al. (2015) have reported these scale problems, mainly when using MODIS images with the thermal bands in mixed IC and NV, to separate the specific results for each agroecosystem. Thus, in the following sections, the residual method for retrieving the surface temperature (T_{RES}) is considered (see Eq. 10).

11.4.3 Fruit Circuit Pole

The WP components in the Fruit Circuit pole within the Atlantic Forest biome, São Paulo state, Southeast Brazil (Figure 11.1d), were assessed at quarter timescales (Q) during a year. The spatial distribution, average values, and the standard deviations (SD) for ET and BIO quarterly values, involving the long-term period from 2002 to 2016, are shown in Figure 11.8.

The ET spatial variations along the year are evident (Figure 11.8a), confirming the suitability of applying the SAFER algorithm and the RUE model to MODIS reflectance products together with gridded weather data at a 250m spatial resolution, taking as reference the Fruit Circuit agricultural growing region. The hydrological contrasts are well-noticed when comparing the first (Q1—January to March) and third (Q3—July to September) quarters of the year because of the root-zone moisture decline between the previous rainy season in Q1 and the period when the rains come back in Q3. However, the lowest atmospheric demand (< ET_0) in the middle of the year somewhat limits ET rates, even under high root-zone moisture conditions.

Teixeira et al. (2016) applied the SAFER algorithm to MODIS images and found ET values ranging from 0.6 to 4.0 mm d^{-1} for sugarcane during four general crop stages in the northeastern region of the state of São Paulo, which is within the spatial values shown in Figure 11.8a. The construction of ET maps using historical data set is important to detect anomalies for specific periods along a

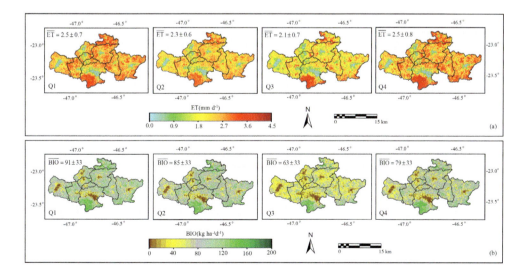

FIGURE 11.8 Spatial distribution, averages, and standard deviations (SD) of the quarterly pixel values of actual evapotranspiration—ET (a) and biomass production—BIO (b), for the long-term period 2002–2016, in the Fruit Circuit area within the Atlantic Forest biome, São Paulo, southeast Brazil. Q1, first quarter; Q2, second quarter; Q3, third quarter; Q4, fourth quarter.

particular year (Vicente-Serrano et al., 2018). Franco et al. (2014) applied the SAFER algorithm to Landsat images in the northwestern region of the state of São Paulo and detected an ET increase of 153% from 1997 to 2010, in mixed agroecosystems composed by pasture, sugarcane, irrigated crops, and natural vegetation. These issues are crucial for an accurate assessment of the water scarcity impacts on vegetation activities in different years to understand the response of irrigated crops and natural vegetation to anomalies (Zhang and Zhang, 2019).

The BIO spatial variations are also evident (Figure 11.8b), with the contrasts mainly noticed when comparing the moistest quarters Q1 (January—March) and Q2 (April—June), when the pixel values reached 200 kg ha^{-1} d^{-1}, with the driest one (Q3—July to September), which shows several zero-pixel values.

In the northeastern region of the state of São Paulo, Teixeira et al. (2016) reported BIO ranging from 20 to 200 kg ha^{-1} d^{-1}, throughout the sugar cane crop stages. These values are like the maximum ones depicted in Figure 11.8b for vegetated surfaces. Nuñez et al. (2017) found annual average BIO values ranging from 3 to 78 kg ha^{-1} d^{-1} in different ecosystems of the northwestern region of São Paulo state, following the rainfall dynamics of the region. As for ET, BIO maps involving long-term climate conditions are also important to detect anomalies during specific years. Silva et al. (2018) reported BIO declines in the aquifers of São Paulo during the years 2014 and 2015, with consequent impacts on groundwater recharges. The maximum BIO rates during the first half of the year in the Fruit Circuit areas, following the root-zone moisture conditions, are in accordance with studies in other regions under similar rainfall patterns, where the maximum NDVI values occur from Q1 to Q2—January to June (Bento et al., 2018; Vicente-Serrano et al., 2015, 2018).

Figure 11.9 presents the spatial distribution, average values, and standard deviations (SD) of the quarterly values of water productivity based on biomass production (WP$_{BIO}$) in the Fruit Circuit areas, within the Atlantic Forest biome, São Paulo state, southeast Brazil, for the long-term period from 2002 to 2016.

Although the BIO values are higher in Q1 (January to March) than in Q2 (April to June), the lower ET rates in this last quarter promote higher WP$_{BIO}$, while even though Q3 (July to September) presenting the lowest ET rates among all quarters, this did not increase WP$_{BIO}$, due to declines on BIO, which drop the WP$_{BIO}$ values in the second half of the year.

Franco et al. (2016) applied the SAFER algorithm with Landsat 8 images and weather data to assess the WP$_{BIO}$ dynamics, from 2013 to 2014, in the northwestern region of the state of São Paulo, reporting an average value of 4.9 kg m^{-3}. The reason for the higher values from their study, in comparison with those for the Fruit Circuit agricultural growing region, is the specificity of distinct agroecosystems. However, Nuñez et al. (2017) evaluated WP$_{BIO}$ also using Landsat 8 images in the same region of Franco et al. (2016) and detected a mean value of 3.5 kg m^{-3}, like the ones obtained for good root-zone moisture conditions from Figure 11.9.

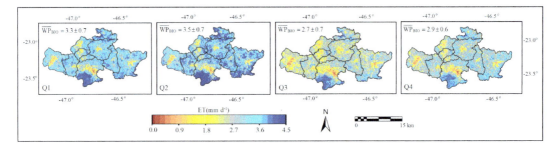

FIGURE 11.9 Spatial distribution, averages, and standard deviations (SD) of the quarterly pixel values for water productivity based on biomass production (WP$_{BIO}$) (2002–2016), in the Fruit Circuit area, within the Atlantic Forest biome, São Paulo, southeast Brazil. Q1, first quarter (January to March); Q2, second quarter (April to June); Q3, third quarter (July to September); Q4, fourth quarter (October to December).

Modeling and Monitoring Water Productivity

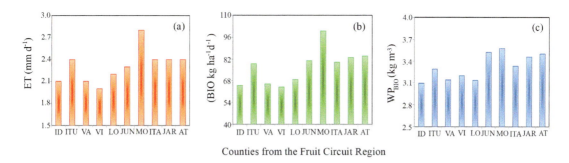

FIGURE 11.10 Average annual pixel values of the water productivity components during the period from 2002 to 2016 for each county of the Fruit Circuit region: Indaiatuba (ID), Itupeva (ITU), Valinhos (VA), Vinhedo (VI), Louveira (LO), Jundiaí (JUN), Itatiba (ITA), Morungaba (MO), Jarinu (JAR), and Atibaia (AT). ET, actual evapotranspiration; BIO, biomass production; WP_{BIO}, water productivity based on BIO.

Separating the counties within the Fruit Circuit agricultural growing region, within the Atlantic Forest biome, São Paulo state, southeast Brazil, Figure 11.10 shows the average annual values of the water productivity components for each of them, regarding the long-term period from 2002 to 2016.

Although all counties are well-supplied with rainfall water and attend the atmospheric demand in most situations throughout the year, the one with the highest annual values for both ET (2.8 mm d^{-1}) and BIO (100 kg ha^{-1} d^{-1}), is Morungaba (MO), promoting a WP_{BIO} value of 3.6 kg m^{-3}, the best option for rainfed agriculture expansion, whereas the lowest both ET (2.0 mm d^{-1}) and BIO (64 kg ha^{-1} d^{-1}) are for Vinhedo (VI), but still resulting in a high WP_{BIO} of 3.3 kg m^{-3}, however, evidencing that for this last county supplementary irrigation should be beneficial in some critical crop stages.

11.5 CROP SCALE

11.5.1 Irrigated Crops

11.5.1.1 Irrigated Limon Crop

Overlapping satellite crossings gave the opportunity of 27 Landsat 8 image acquisitions for irrigated limon water productivity assessments (up-pointed green triangle in Figure 11.1a), covering all orchard phenological stages inside six commercial farms during the year 2015. The crop was irrigated by drip (Santa Fé and Saara farms), micro sprinkler (Esperança, Yamada, Santa Fé, Marazul, and Tropical farms), and pivot systems (Yamada farm).

Figure 11.11 shows the spatial distribution, averages, and standard deviations (SD) of the ET quarterly values for the lemon orchards irrigated by drip, micro sprinkler, and pivot systems on these commercial farms.

Considering all analyzed irrigation systems, the highest ET rates occurred in Q1 (January to March) and Q2 (April to June), above 3.3 mm d^{-1} for micro sprinkler and pivot irrigation systems. The average ET quarterly values ranged respectively from 2.0 to 3.9 mm d^{-1}, for drip (Q3—April to June) and pivot (Q1—January March) irrigation systems. At the annual scale, ET averaged 2.6, 3.1, and 3.7 mm d^{-1}, under drip, micro sprinkler, and pivot irrigation systems, respectively.

The ET range encompassed the values from 2.0 to 3.8 mm d^{-1}, reported for olive orchards (Ramírez-Cuesta et al., 2019), and drip irrigated orange (0.7–3.9 mm d^{-1}) (Consoli and Papa, 2013), both studies under the Mediterranean semi-arid conditions. The ET rates were most affected by variations on root-zone moisture levels, which in turn depended on the weather conditions but also the type of irrigation system, which affects ET partitions into transpiration and soil evaporation (Fandino et al., 2012; Longo-Minnolo et al., 2020; Rosa et al., 2016).

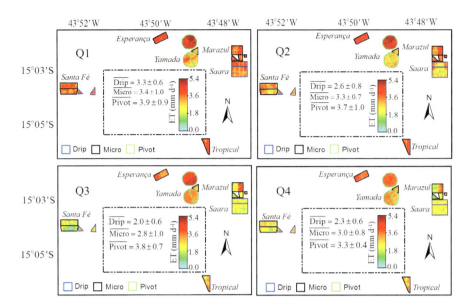

FIGURE 11.11 Spatial distribution, averages, and standard deviations (SD) for the actual evapotranspiration (ET) quarterly values. Blue, black, and green contour lines represent lemon orchards irrigated by drip, micro sprinklers, and pivot irrigation systems, respectively, inside the assessed commercial farms. Q1—First quarter (January to March); Q2—Second quarter (April to June); Q3—Third quarter (July to September); Q4—Fourth quarter (October to December).

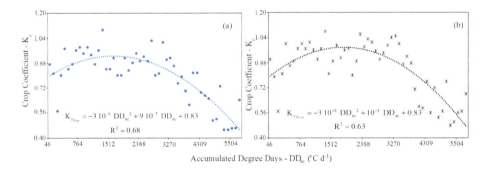

FIGURE 11.12 Models for estimating single lemon crop coefficients (K_c) in the Caatinga biome, northern Minas Gerais (MG) state, southeast Brazil. (a) Drip irrigated orchards (subscript Drip); (b) Micro sprinkler irrigated orchards (subscript Micro).

Because the ET results showed that lemon orchards under pivot irrigation systems use too much water, these systems are not recommended under the water scarcity scenarios of the Brazilian semi-arid region. Figure 11.12 shows the models to estimate crop coefficient (K_c) as a function of accumulated degree-days (DD_{ac}), for the recommended drip (Figure 11.11a) and micro sprinkler (Figure 11.11b) irrigation systems under these conditions.

Similarly, Rallo et al. (2017), to infer the phenological stages of drip irrigated orange orchards, used a site-specific polynomial equation to determine K_c as a function of the measured canopy fractional cover, to determine water requirements under the semi-arid conditions of Spain. Points above the curves from Figure 11.12 indicate high contributions of soil evaporation while the ones below represent conditions of water fluxes lower than those for potential conditions.

For drip irrigation systems, the modeled K_c values ranged from 0.37 to 0.90, averaging 0.78, while the corresponding ranges and average for the micro sprinkler irrigation systems were between 0.43 and 0.92, with a mean of 0.81. The maximum K_c values occurred during the transition from Flowering to Fruit Growth crop stages (CS), between March and April, and from September to October, respectively; while the minimum ones were from November to December, during the CS of harvest peaks.

The maximum single K_c values represented by the peak of the polynomial curves in Figure 11.12 (0.90 and 0.92, for drip and micro sprinkler irrigation systems, respectively) were higher than the FAO single K_c ones tabulated for citrus by Allen et al. (1998). One of the reasons of the lower values for drip irrigation systems in comparison with the micro sprinkler ones, should be its higher partition into transpiration. Longo-Minnolo et al. (2020), analyzing irrigation strategies in citrus orchards from remote sensing and meteorological data in Southern Italy, reported K_c values ranging from 0.56 to 0.67 for drip-irrigated oranges, varying according to treatments as indicators of transpiration activities, also below the FAO tabled values.

Differences on K_c values from the recommended FAO tabulated values were also found by Consoli and Papa (2013) from field measurements in an irrigated orange orchard under the Mediterranean semi-arid conditions, resulting in a K_c range from 0.20 to 1.10, averaging 0.68. Also, for irrigated orange orchard, K_c values averaging 0.91 and 0.75 were reported, during the summer and winter seasons in Japan, respectively (Yang et al., 2003). However, in São Paulo state, southeast Brazil, Junior et al. (2008) found K_c values from 0.82 to 1.18 from lysimeter measurements in an irrigated lemon orchard, like the range of the K_c values for those in the North of Minas Gerais agricultural growing region.

Actual yield (Ya) data together with the ET results, allowed water productivity assessments based on Ya (Eq. 20), for each studied lemon producer farm. The WP_{Ya} ranged from 1.6 to 3.6 kg m^{-3}. The highest values were for Saara farm, with 100% of its lemon cropped area being drip irrigated, while the lowest ones were for tropical farm, under only micro sprinkler irrigation system (see Figure 11.11).

Although with large WP_{Ya} range in all assessed farms, it was noticed rooms for water management improvements in limon crop with irrigation strategies. Panigrahi et al. (2012) reported a WP_{Ya} increasing from 2.3 to 2.9 kg m^{-3} proportionally to water stress level, in a drip irrigated orange orchard from central India. Improvements on WP_{Ya} with deficit irrigation were also reported in a semi-arid region of Spain (García-Tejero et al., 2010, 2019). Considering the irrigation water applied ranging from 45% to 100% of the reference evapotranspiration (ET$_0$) in the semi-arid region of Iran, Jamshidi et al. (2020) reported respective WP_{Ya} values for drip irrigated oranges from 2.0 to 3.0 kg m^{-3}. These last authors recommended the treatment of irrigation water applied at 60% or 70% of ET$_0$ as the best option for minimizing the amount of water use while still maintaining the benefit of a good yield.

According to the water productivity assessments in limon orchards discussed in the current chapter, techniques about the use of regulated deficit irrigation strategies are recommended in some phenological stages of the crop are encouraged, aiming to maintain or improve yield, while promoting water savings. Applications of the SAFER algorithm without the satellite thermal band improving spatial resolutions can contribute to these studies, being of great value for both researchers and advisors (Cancela et al., 2019). In addition, the algorithm implementation can also contribute to minimize conflicts among farmers and other water users, under the water scarcity conditions in semi-arid environments.

11.5.1.2 Irrigated Coconut Crop

Figure 11.13 shows the spatial distributions and averages of quarterly (Q) daily values for ET (Figure 11.13a) and BIO (Figure 11.13b) of irrigated dwarf coconut crop, involving different palm ages during the year 2016, inside the Caatinga biome, in the Camocim County, coastal zone of the Ceará state, northeast Brazil (green square of Figure 11.1a).

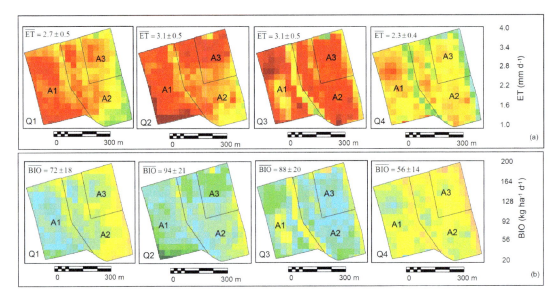

FIGURE 11.13 Spatial distributions and averages of the quarterly (Q) daily values of the water productivity parameters for the irrigated dwarf coconut during 2016, inside the Caatinga biome, in the Camocim County, coastal zone of Ceará state, northeast Brazil, with palms aging three (A1), four (A2), and five (A3) years. ET—Actual evapotranspiration and BIO—Biomass production.

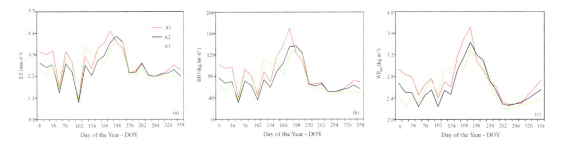

FIGURE 11.14 Tendencies of the daily mean pixel values for actual evapotranspiration (ET), biomass production (BIO), and water productivity based on BIO (WP_{BIO}), in terms of day of the year (DOY), inside the areas with three- (A1), four- (A2), and five- (A3) year-old dwarf coconut palms, inside the Caatinga biome, Camocim County, coastal zone of Ceará state, northeast Brazil.

Clearly one can see spatial differences on ET and BIO pixel values in the A1 (three-year-old palms), A2 (four-year-old palms), and A3 (five-year-old palms) dwarf coconut areas, accordingly to the period of the year. The largest BIO values, with averages above 90 kg ha^{-1} d^{-1}, in the second quarter (Q2—April to June) promoted the maximum WP_{BIO} of 3.0 kg m^{-3}, at the end of the rainy period, conditions of high root-zone moisture, but for both coconut palms and inter-row weeds. The lowest BIO rates in Q4 (October—December), with mean pixel below 60 kg ha^{-1} d^{-1}, the climatically driest period of the year, retrieved the lowest WP_{BIO} of 2.4 kg m^{-3}, when coconut palms were covering only 50% of the soil and receiving localized irrigation water, while inter-row weeds dried out.

Figure 11.14 presents the tendencies of the ET, BIO, and WP_{BIO} daily average pixel values, in terms of day of the year (DOY), for the dwarf coconut areas, with palms three (A1), four (A2), and five (A3) years old, during the year 2016, in terms of DOY (see also the green square from Figure 11.1).

The largest ET values occurred from May to July (DOY 150–214), reaching mean rates of 4.5 mm d^{-1} for the three-year-old palms (A1 area). During this peak period, the maximum rates in the A2 and A3 areas, four- and five-year-old palms, were 4.3 and 4.1 mm d^{-1}, respectively. The lowest ET values, considering all coconut areas, were around 1.0 mm d^{-1}, happening in April (DOY 102), at the end of the rainy period, while the annual averages ranged from 2.6 to 2.9 mm d^{-1} among the different palm ages.

From the soil water balance approach, in Sergipe state, northeast Brazil, Azevedo et al. (2006) found mean ET values from 2.5 to 3.2 mm d^{-1} for six-year-old dwarf coconut palms, varying according to the amount of irrigation water applied. Also in Sergipe and for dwarf coconut, Sousa et al. (2011), from lysimetric measurements, reported an average value of 3.9 mm d^{-1} in a range from 2.0 to 5.6 mm d^{-1}. In the coastal zone of Ceará state, northeast Brazil, Miranda et al. (2007), using the soil water balance method, found maximum ET of 5.0 mm d^{-1} for a three-year-old dwarf coconut, averaging 3.9 mm d^{-1} when the crop was aged four years. Carr (2011) presented a summary of average ET for mature dwarf coconut ranging from 1.7 to 5.0 mm d^{-1}. According to this last author, a typical coconut ET rate should be 3.0–3.5 mm d^{-1}. One of the reasons for the ET differences between the results from these field measurements and remote sensing values from Figure 11.14 is that for the satellite image cases, they represent average pixel values.

As there is a relation between ET and BIO, maximum values for both occurred at the same periods, from May to July (DOY 150–214), with BIO reaching daily rates of 170 kg ha^{-1} d^{-1} in A1 (three-year-old palms). The highest BIO rates in A2 (four-year-old palms) and A3 (five-year-old palms) were 137 and 132 kg ha^{-1} d^{-1}, respectively. However, the lowest BIO, around 36 kg ha^{-1} d^{-1}, considering all coconut areas, occurred in February (DOY 054), inside the rainy period, because of declining solar radiation levels that limited photosynthetic activities. The mean annual BIO ranged from 73 to 84 kg ha^{-1} d^{-1} among all palm ages. Under different agroclimatic zones in India, simulated average BIO values ranged from 93 to 100 kg ha^{-1} d^{-1} for irrigated coconuts (Naresh Kumar et al., 2008), a little higher than those for the Brazilian northeast region.

Regarding WP$_{BIO}$, the maximum values coincided with those for ET and BIO, from May to July (DOY 150–214), reaching 3.7 kg m^{-3} in A1 (three-year-old palms). During this period, those for A2 (four-year-old palms) and A3 (five-year-old palms) were 3.4 and 3.3 kg m^{-3}, respectively. The lowest values, around 2.3 kg m^{-3}, occurred from September to November (DOY 262–310) for all palm ages. Nuñez et al. (2017), applying the SAFER algorithm and the RUE model to Landsat 8 images in the northwestern side of São Paulo state, southeast Brazil, found average WP$_{BIO}$ of 3.4 kg m^{-3} for perennial crops, like the dwarf coconut results in the coastal areas of northeast Brazil.

The analyses of the root-zone moisture tendencies, along the year 2016, were carried out by the evapotranspiration fraction—ET$_f$, considering the different palm ages, through the average pixel values in A1, A2, and A3 dwarf coconut areas. However, the irrigation water amounts (mm) were monitored by hydrometer only in the A2 (four-year-old palms) area. Figure 11.15 shows the tendencies for ET$_f$ in the dwarf coconut areas, with palms aging three (A1), four (A2), and five (A3) years.

Despite the similar WP$_{BIO}$ and WP$_{Ya}$ tendencies among all dwarf coconut areas, the ET$_f$ average values ranged between 0.46 in October (DOY 294) in A3 and 1.08 in June (DOY 166–182) in A1, areas with, respectively, five and three-year-old palms. The highest ET$_f$ values can be considered the coconut crop coefficient—K$_c$ (Mokhtari et al., 2019). According to Carr (2011), The K$_c$ values for mature coconut crop range from 0.50 to 1.02, with some evidence of seasonal variability; however, according to the author, for immature palms they should be lower.

Excepting from September 2 (DOY 246) to November 5 (DOY 310), when ET$_f$ values dropped below 0.50, the palm root-zones were well water supplied, with the annual ET$_f$ averages from 0.69 (A2 and A3 areas) to 0.76 (A1 area), respectively. From the similarity on ET$_f$ values outside the rainy period, disregarding palm ages, it could be assumed no strong differences on irrigation water management among the dwarf coconut areas. The lower ET$_f$ values from September to November (DOY 230–310) mean a gap between ET and the atmospheric demand, what should be unfavorable for coconut yield (Madurapperuma et al., 2009).

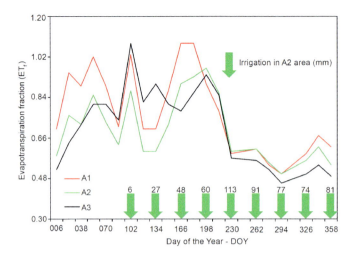

FIGURE 11.15 Tendencies for the evaporation fraction (ET_f) in the dwarf coconut areas, with palms aging three (A1), four (A2), and five (A3) years. Green arrows show the 30-day accumulated irrigation (mm) in the four-year-old palms inside the A2 area.

At the end of the rainy period, previously accumulated irrigation of 134 mm, between May 16 and June 16, was not enough to compensate the water consumed and/or percolated, pointing out water stress conditions in the palm root zones. Miranda et al. (2004) reported that more than 80% of the absorptive dwarf coconut roots are located until 0.60 m soil depth in the coastal zone of Ceará state, northeast Brazil.

It must be considered that during the first half of the year, all cropped areas were covered by the dwarf coconut palms and inter-row vegetation, due to rainfall. After this period, however, the invasive plants dried out, due to the localized irrigation, and, because of the large spaces between palms, the soil cover and moist fractions by the irrigation system, estimated from field measurements, were only 0.45 and 0.32, respectively, which reduced ET while the atmospheric demand increased, hence dropping ET_f. According to Roupsard et al. (2006), the ET partition into transpiration (T) and evaporation (E) between coconut palms and inter-row vegetation is considerably important to understand the water requirements and competition mechanisms. These authors, separating T and E measurements by the sap flow method in coconut crop under tropical conditions, found T representing 68% of ET, for a 75% of soil covered by the palms, much larger than that for the coconut crop in northeast Brazil.

Li et al. (2016) reported that WP_{Ya} is more affected by agronomy practices than by climatic factors, and its improvements should be based on advanced technologies involving crop and water management. The results in northeast Brazil brought attention to controlling inter-row invasive species during the rainy period, for instance, by using mulching, concentrating the water consumption only for the dwarf coconut palms, while reducing the invasive species. Jayakumar et al. (2017), through a field experiment in Coimbatore, India, reported the benefits of polythene mulching on hybrid coconut, which increased palm height, canopy development, spathe length, number of inflorescences, number of bunches/palm/year and number of fruits/bunches, raising WP_{Ya} when compared with the control treatment without mulch. In addition, mulching can help root-zone moisture conservation during the following climatically drier periods after the rainy season, minimizing crop water stress (Carr, 2011; Resende et al., 2015).

11.5.2 Rainfed Crops

11.5.2.1 Rainfed Eucalyptus and Pasture Crops

By using crop land use masks, Figure 11.16 presents the spatial distribution of the actual daily ET values for some representative MODIS 16-day periods in the eucalyptus and pasture crops,

Modeling and Monitoring Water Productivity 407

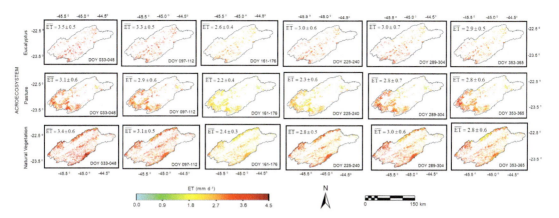

FIGURE 11.16 Spatial distribution of the daily actual evapotranspiration (ET) values for some representative 16-day periods in the eucalyptus and pasture, together with those for interspaced natural vegetation, during the year 2015, located at the Atlantic Forest biome, in the eastern side of the São Paulo state, southeast Brazil, in terms of day of the year (DOY). Over bars mean averages shown together with standard deviations (SD).

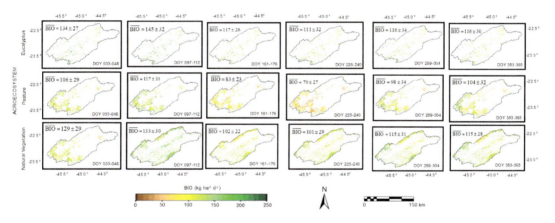

FIGURE 11.17 Spatial distribution of the daily biomass production (BIO) values for some representative 16-day periods in the eucalyptus and pasture crops, together with those for the interspaced natural vegetation (NV), during the year 2015, inside the Atlantic Forest biome, located on the eastern side of São Paulo state, southeast Brazil, in terms of day of the year (DOY). Over bars mean averages shown together with standard deviations.

interspaced by natural vegetation, during the year 2015, located at the Atlantic Forest biome in the eastern side of São Paulo state, southeast Brazil (green trapeze in Figure 11.1a), in terms of DOY.

More variation on the ET pixel values for pasture are noticed when compared with those for eucalyptus and natural vegetation, according to the standard deviations (SD). The lowest ET rates, below 2.5 mm d^{-1} in pasture and natural Vegetation, occurred in the middle of the year (June, DOY 161–176), while the highest rates were for eucalyptus, reaching a mean pixel value of 3.5 mm d^{-1} in February (DOY 033–048), during the rainiest period of the year.

The annual ET values were 1017, 886, and 967 mm yr^{-1}, with SD representing 18%, 22%, and 19% of the average value for eucalyptus, pasture, and natural vegetation agroecosystems, respectively. On one hand, besides the lowest ET rates, pasture presented also the largest SD. On the other hand, eucalyptus is highlighted with the highest ET rates but with the largest uniformity by its lowest SD.

Figure 11.17 shows the spatial distribution of the daily BIO values for some of the representative MODIS 16-day periods in eucalyptus, pasture, and natural vegetation, during the year 2015, inside the Atlantic Forest biome, located on the eastern side of São Paulo state, southeast Brazil, in terms of DOY.

Comparing Figures 11.17 and 11.18, one can see the relations between ET and BIO for all three agroecosystems; however, the period with the highest BIO is in April (DOY 097–112), after the rainiest season, with an average of 145 kg ha^{-1} d^{-1} in eucalyptus, while the lowest ones, around 70 kg ha^{-1} d^{-1} were in August (DOY 225–245) for pasture. High rainfall amounts together with large solar radiation levels increase vegetation growth at the start of the year, promoting large BIO rates.

The annual BIO values were 41, 32, and 38 t ha^{-1} yr^{-1} for eucalyptus, pasture, and natural vegetation, respectively. The BIO spatial variations followed those for ET, with the lowest SD in June (DOY 161–176) and the highest in October (DOY 289–304). At the annual scale SD represented 26% of the average values for eucalyptus and natural vegetation, while for pasture this fraction was 31%.

Figure 11.18 presents the average pixel values of WP$_{BIO}$ (Figure 11.18a) and the evapotranspiration fraction—ET$_f$ (Figure 11.18b), for the 16-day MODIS images periods, during the year 2015, in the eucalyptus and pasture crops, and those for the interspaced natural vegetation, inside the Atlantic Forest biome, located on the eastern side of São Paulo state, southeast Brazil, in terms of DOY.

The 16-day average WP$_{BIO}$ for eucalyptus, pasture, and natural vegetation (Figure 11.18a) ranged, respectively, from 3.5 to 4.5 kg m^{-3}; 2.7 to 3.9 kg m^{-3}; and 3.3 to 4.3 kg m^{-3}. The highest values for all agroecosystems happened from April to June (DOY 097–176), while the lowest ones were from August to September (DOY 225–257). At the annual scale, the WP$_{BIO}$ values for the eucalyptus were 3% and 15% higher than those for the natural vegetation and pasture, respectively.

On the one hand, the highest ET rates for eucalyptus should be a negative aspect; however, it presented the best WP$_{BIO}$, what is a positive issue, under conditions of climate and land use changes, together with water scarcity. Considering the importance of eucalyptus for paper and wood production, the high WP$_{BIO}$ values under rainfed conditions, with sustainable crop management, may compensate the negative effects of its fast expansion over natural vegetation in the Atlantic Forest biome in southeast Brazil.

According to Figure 11.18b, the ET$_f$ values were always above 0.50, indicating good moisture conditions for all agroecosystems, with absence of water stress in their root zones. The three ET$_f$ curves were symmetric with the highest values, happening from DOY 113 (end of April) to 241 (end of August), reaching 1.38 at the end of June, for eucalyptus. Although the ET$_f$ curve pictured in Figure 11.18b is for rainfed agroecosystems, under optimum root-zone moisture conditions they correspond to K$_c$ values and can be used in rational irrigation management for other environmental conditions, as well as for agroclimatic aptitude zoning.

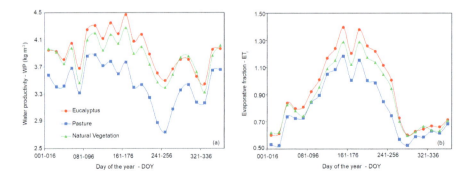

FIGURE 11.18 Average pixel values for water productivity based on BIO—WP$_{BIO}$ (a), and evapotranspiration fraction—ET$_f$ (b), for the 16-day period MODIS images during the year 2015, in eucalyptus and pasture crops, and those for the interspaced natural vegetation, inside the Atlantic Forest biome, located on the eastern side of São Paulo state, southeast Brazil, in terms of ay of the year (DOY).

Zhang et al. (2012), studying a temperate desert steppe in Inner Mongolia, China, reported seasonal ET_f variations from mean daily values of 0.16 to maximum of 0.75, lower than those for the studied agroecosystems in southeast Brazil. However, Lu et al. (2011), in the same Chinese region, found ET_f higher than 1.00 for six different ecosystems, while Sumner and Jacobs (2005) reported ET_f values between 0.47 and 0.92 in a non-irrigated pasture site in Florida, USA. These last values are like those from Figure 11.18b for several periods throughout the year. Zhou and Zhou (2009) concluded that air temperature, air humidity, and available energy were the most important variables for the ET_f variations in a reed marsh in the northeast China. In the Brazilian Atlantic Forest biome, however, the most important reason for the highest ET_f values were the previous rainy seasons making the soil moistier in the subsequent period, but these values are also dependent on the stomatal regulation and plant adaptation to water scarcity conditions (Mata-González et al., 2005).

Considering all the analyzed agroecosystems within the Atlantic Forest biome, on the eastern side of São Paulo state, southeast Brazil, on average, the ET rates accounted for 86% of those for the reference grass ($K_c = 0.86$). Eucalyptus had the highest ET_f (0.92 at the annual scale), when comparing to those for natural vegetation (0.88 at the annual scale) and pasture (0.79 at the annual scale). These results indicated that the eucalyptus plants have greater ability to conserve soil moisture in their root zones, increasing water productivity.

11.5.2.2 Rainfed Corn Crop

The WP assessments in rainfed corn crop with drone were carried out in the Caatinga biome, within Sergipe state, northeast Brazil (down-pointed green triangle in Figure 11.1a). Figure 11.19 shows the corn plots cover fertilized under different nitrogen (N) levels and sources, Nitrate—Nt and Urea—Ur (Figure 11.19a), together with spatial distributions, average pixel values, and standard deviations (SD), for the daily corn actual evapotranspiration—ET (Figure 11.19b) and biomass production—BIO (Figure 11.19c), considering the four analyzed crop stages, during the year 2017.

The effect of the cover fertilizing under different N levels and sources on water productivity components were assessed through drone flights at V6 (plants with six leaves), V10 (plants with ten leaves), PF (pre-flowering), and FF (full flowering) stages. With grain yield data, it was possible to quantify WP in terms of both BIO and actual yield (Ya) at the growing season (GS) timescale, at a 4cm spatial resolution in the plots with 1m of width by 6m of length for each treatment.

According to Figure 11.19b, there was a strong increase on ET rates along the four analyzed crop stages for all N cover fertilizing levels and sources, with mean pixel values ranging from 2.2 to 4.4 mm d^{-1} and SD representing, respectively, 59% and 33% of the averages, for the V6 (July 11) and PF (August 04) crop stages (CI). However, besides CI, the ET and SD values are also related to root-zone moisture, represented by ET_f (Eq. 11), which ranged from 0.59 (V6) to 1.03 (FF).

As for ET, spatial and temporal variations on BIO pixel values in the images from Figure 11.19c are clearly perceived, but in this case the effects of canopy development for all N cover fertilizing treatments are more noticeable. Comparing the representative images of the V6 and FF stages, BIO rates increased 4.4 fold. However, differently from ET, whose rates are also influenced by soil evaporation at good soil moisture conditions, BIO spatial variations rapidly decreased as the canopies developed, with SD values dropping from 97% in V6 to 40% of the average BIO pixel values already in the V10 stage.

Table 11.4 presents ET average and SD values, considering the N cover fertilizing levels and sources for each of the N treatments (0 to 250 kg ha^{-1} of N via Nitrate [Nt] and Urea [Ur]).

In the V6 vegetative stage (July 11), the highest ET rates were for N cover fertilizing treatments 200Nt and 200Ur, 20% above of those from the control one (0NrUr). For all corn plots, ET for the Ur source was only 8% higher than those for Nt. All treatments were well-rainfall-water supplied in this CS, under low soil cover, but with ET_f averaging 0.56 and 0.60 for Nt and Ur, respectively. During the V10 vegetative stage (July 21), the plots at 100 kg ha^{-1} from Nt (100Nt) and 200 kg ha^{-1} from Ur (200Ur), presented the highest ET rates, above 7% of those for 0NtUr; however, ET rates for Ur were only 2% higher than those for Nt. During this CS, the ET_f values

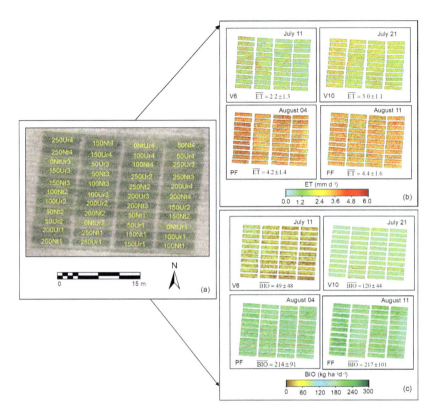

FIGURE 11.19 Corn plots cover fertilized with different nitrogen (N) levels from nitrate (Nt) and urea (Ur) sources, at 0, 50, 100, 150, 200, 250 kg ha⁻¹ (a); and spatial distributions, average pixel values, and standard deviations (SD), for the daily corn actual evapotranspiration—ET (b) and biomass production—BIO (c).

* V6—Vegetative stage with six leaves per plant (July 11), V10—vegetative stage with ten leaves per plant (July 21), PF—reproductive stage pre-flowering (August 04), and reproductive stage FF—full flowering (August 11).

averaged 0.97 and 0.99 for Nt and Ur, respectively, indicating good water availability in the root-zone, at increasing soil cover conditions.

In the PF reproductive stage (August 4), the highest ET rates happened at 50 kg ha⁻¹ for Ur (50Ur) and 100 kg ha⁻¹ for Nt (100Nt), but only 2% and 3% higher, respectively, than those for the control treatment (0NtUr). The average ET rates for Ur were only 4% higher than those for Nt. The ET$_f$ values in this CS averaged 1.00 and 1.04 for Nt and Ur, respectively. During the FF reproductive stage (August 11), the highest ET rates were for 0NtUr and 150Ur. Regarding all corn plots, the average pixel values for Ur were only 2% higher than those for Nt. The ET$_f$ values during this CS averaged 1.03 and 1.04 for Nt and Ur, respectively, again indicating good water storage in the root-zone at the highest soil cover conditions.

ET rates were affected by crop development, but there were no large differences among N cover fertilizing levels and sources inside each specific CS, which were most affected by variations on root-zone moisture levels, which in turn depend on the weather conditions and the ET partitions into transpiration (T) and soil evaporation (E), making it difficult to distinguish the effects of N treatments inside a CS, because of the alternated magnitudes of T and E according to soil cover (Consoli and Vanella, 2014; Fandino et al., 2012; Longo-Minnolo et al., 2020; Rosa et al., 2016).

Table 11.5 presents BIO average and SD values, considering the N cover fertilizing levels and sources for each of the N treatment (0 to 250 kg ha⁻¹ of N via Nitrate [Nt] and Urea [Ur]).

In the V6 vegetative stage (July 11), the highest BIO values were for N cover fertilizing of 100 kg ha⁻¹ for both Nt and Ur sources, 130% of those for the 0NrUr control treatment. However, there were

TABLE 11.4

Average Pixel Values and Standard Deviations (SD) of Actual Evapotranspiration (ET), Considering the N Cover Fertilizing Levels (0 to 250 kg ha⁻¹) and Sources (Nitrate and Urea) for the Analyzed Phenological Stages (V6, V10, PF, and FF), inside the Caatinga Biome, in the County of Nossa Senhora das Dores, State of Sergipe, Northeast Brazil

		Actual Evapotranspiration—ET (mm d⁻¹)						
N levels Date	CS¹	0Nt²	50Nt	100Nt	150Nt	200Nt	250Nt	Mean
July 11	V6	1.96±1.23	1.97±1.24	2.08±1.28	1.98±1.31	2.33±1.30	2.12±1.31	2.07±1.28
July 21	V10	2.88±1.21	2.90±1.20	3.06±1.10	2.92±1.15	2.88±1.06	2.90±1.05	2.92±1.13
August 4	PF	4.26±1.21	4.10±1.48	4.12±1.45	4.10±1.37	4.07±1.31	4.04±1.32	4.12±1.46
August 11	FF	4.54±1.76	4.39±1.58	4.35±1.63	4.44±1.57	4.37±1.39	4.35±1.51	4.41±1.57
Mean	-	3.41±1.35	3.34±1.38	3.40±1.37	3.36±1.35	3.41±1.27	3.35±1.30	3.38±1.36

		Actual Evapotranspiration—ET (mm d⁻¹)						
N levels Date	CS	0Ur³	50Ur	100Ur	150Ur	200Ur	250Ur	Mean
July 11	V6	1.96±1.23	2.19±1.28	2.33±1.30	2.27±1.26	2.32±1.32	2.30±1.32	2.23±1.28
July 21	V10	2.88±1.21	2.86±1.19	3.00±1.13	3.07±1.13	3.08±1.10	2.91±1.00	2.97±1.13
August 4	PF	4.26±1.21	4.35±1.51	4.29±1.44	4.32±1.33	4.27±1.33	4.13±1.38	4.27±1.47
August 11	FF	4.54±1.76	4.44±1.61	4.41±1.74	4.53±1.58	4.51±1.62	4.46±1.36	4.48±1.61
Mean	-	3.41±1.35	3.46±1.40	3.51±1.40	3.55±1.33	3.55±1.34	3.45±1.27	3.49±1.37

¹CS: Crop stages; ²Nt: Nitrate; ³Ur: Urea. V6—Vegetative stage with six leaves per plant, V10—Vegetative stage with ten leaves per plant, PF—Reproductive pre-flowering, and FF—Reproductive full flowering.

---

no large differences for N levels from 150 to 200 kg ha⁻¹ from both N sources. For this CS, BIO rates for Ur were only 2% higher than those for Nt. The low solar radiation levels and canopy development reduced transpiration during this CS, conditions not favorable for BIO. During the V10 vegetative stage, the corn plots at 150 kg ha⁻¹ of N cover fertilizing, for both sources (150Nt and 150Ur) presented the highest BIO rates, above, respectively, 28% and 23% of those for 0NtUr. However, the still low solar radiation levels somewhat limited BIO rates, even at increasing soil cover.

In the PF reproductive stage, the highest BIO values for Nt happened already at 50 kg ha⁻¹ (50Nt), while for Ur, this was at 250 kg ha⁻¹ (250Ur), above, respectively, 20% and 10% of the 0NtUr control treatment. The average BIO for Nt was 4% above that for Ur, with increasing solar radiation levels, promoting BIO pixel values above 200 kg ha d⁻¹ for all non-zero N cover fertilizing levels. During the FF reproductive stage, the highest BIO rates were for both 50Nt and 50Ur treatments, above, respectively, 22% and 12% of 0NtUr. The Nt source promoted BIO values 3% higher than those for Ur, under good root-zone moisture levels, with a continuous increase on solar radiation interception as a consequence of crop development.

The FF reproductive stage presenting the maximum BIO values agrees with Taghvaeian et al. (2012) and Zhang et al. (2019), who found maximum rates when the canopies were fully covering the soil. In addition, the solar radiation levels increase in August raised the photosynthetic activity and together with crop development favoring transpiration brought BIO to maximum values (Yang et al., 2019). According to Kang et al. (2002) and Driscoll et al. (2006), transpiration promotes high levels of photosynthetic activity, increases BIO under good root-zone moisture levels, and agrees with the results in northeast Brazil.

From Tables 11.4 and 11.5, it is noticed that N cover fertilizing did not significantly affect the ET rates, while BIO followed the development of leaf areas with increases on transpiration rates, ET

TABLE 11.5

Average Pixel Values and Standard Deviations (SD) of Biomass Production (BIO), Considering the N Cover Fertilizing Levels (0 to 250 kg ha⁻¹) and Sources (Nitrate and Urea) for the Analyzed Phenological Stages (V6, V10, PF, and FF), inside the Caatinga Biome, in the County of Nossa Senhora das Dores, State of Sergipe, Northeast Brazil

		Biomass Production—BIO (kg ha⁻¹ d⁻¹)						
N Levels								
Dates	**CS[1]**	**0Nt[2]**	**50Nt**	**100Nt**	**150Nt**	**200Nt**	**250Nt**	**Mean**
July 7	V6	40±41	51±45	52±47	50±49	50±52	44±49	48±47
July 21	V10	105±64	118±67	115±65	134±66	120±61	121±61	119±64
August 4	PF	192±22	230±102	221±100	222±94	215±92	217±93	216±100
August 11	FF	200±111	244±101	206±105	215±101	228±88	222±96	219±100
Mean	-	134±60	161±79	149±79	155±78	153±73	151±75	151±78

		Biomass Production—BIO (kg ha–1 d–1)						
N Levels								
Dates	**CS**	**0Ur[3]**	**50Ur**	**100Ur**	**150Ur**	**200Ur**	**250Ur**	**Mean**
July 7	V6	40±41	50±46	51±49	50±46	51±49	50±51	49±47
July 21	V10	105±64	117±67	117±65	129±66	121±63	121±58	118±64
August 4	PF	192±22	210±103	210±98	218±92	209±92	211±95	208±100
August 11	FF	200±111	224±103	205±110	215±102	211±102	215±87	212±103
Mean	-	134±60a	150±80b	146±81b	153±77b	148±77b	149±73	147±79

[1]CS: Crop stage; [2]Nt: Nitrate; [3]Ur: Urea. V6—Vegetative stage with six leaves per plant, V10—Vegetative stage with ten leaves per plant, PF—Reproductive stage pre-flowering, and FF—Reproductive stage full flowering

being much related with its partition into transpiration and soil evaporation (Longo-Minnolo et al., 2020), which in turn is also dependent on soil cover (Rosa et al., 2016). Campos et al. (2018) and Twohey et al. (2019), from remote sensing measurements, confirm high correlations between BIO and transpiration in both irrigated and rainfed corn crop, as the soil evaporation does not contribute to BIO.

Crossing Tables 11.4 and 11.5, WP$_{BIO}$ was assessed for the four analyzed corn CS. In the V6 vegetative stage, the highest WP$_{BIO}$ values, around 2.6 kg m⁻³, were for N cover fertilizing treatments at 50 kg ha⁻¹ for both N sources, Nt and Ur, respectively, more than 27% and 12% above 0NtUr. The reasons for these highest values were the lowest ET rates (mean of 2.0 mm d⁻¹), as BIO ones were high, averaging 51 kg ha⁻¹ d⁻¹ (see Tables 11.4 and 11.5). During the V10 vegetative stage, the plots with N cover fertilizing at 150 kg ha⁻¹ presented the highest WP$_{BIO}$ rates for both Nt and Ur sources, around 4.6 kg m⁻³ and 4.2 kg m⁻³, above, respectively, 26% and 15% of 0NtUr. The main reason was the highest BIO, with respective averages of 134 and 129 kg ha⁻¹ d⁻¹, as for ET there were no significant differences among treatments, around 3.0 mm d⁻¹ (Tables 11.4 and 11.5).

In the PF reproductive stage, the highest WP$_{BIO}$ values were for the 50Nt (5.6 kg m⁻³) and 250Ur (5.1 kg m⁻³), 24% and 13% higher than those for 0NtUr, respectively. The main reason was high BIO (above 210 kg ha⁻¹ d⁻¹, Table 11.5) at ET rates around 4.0 mm d⁻¹ (Table 11.4). This CS was characterized as one of the highest WP$_{BIO}$, because of the coupled effect of large root-zone moisture, together with also high solar radiation interception, favoring BIO. In the FF reproductive stage, the highest WP$_{BIO}$ values were for N cover fertilizing level at 50 kg ha⁻¹ for both Nt and Ur sources, respectively, 5.6 and 5.0 kg m⁻³, more than 27% and 14% of those for 0NtUr. The BIO rates, also above 210 kg ha⁻¹ d⁻¹ (Table 11.5), at ET rates around 4.4 mm d⁻¹ (Table 11.4) for this CS, were not

Modeling and Monitoring Water Productivity   413

so different from the previous ones, under also high root-zone moisture and radiation interception levels.

There were distinctions on WP_{BIO} values among N cover fertilizing treatments due to root-zone moisture, crop development, and solar radiation levels. These variations together with soil stains and germination failures sometimes made it difficult to understand the effects of N cover fertilizing treatments on ET and BIO pixel variations, and then on WP_{BIO}, inside a CS. However, considering the four repetitions for each N cover fertilizing treatments, it was clearly noticed that the N cover at 150 kg ha^{-1} for both Nt and Ur sources should be the best option for having good corn yields while promoting water saving, avoiding N leaching to the ground water.

Figure 11.20 shows the growing season (GS) values for ET, BIO, and Y_a, considering all N cover fertilizing levels and sources of the studied rainfed corn crop.

The highest GS ET value was for the 250Ur treatment (394.6 mm GS^{-1}), but only 0.4 mm GS^{-1} above that for the control (0NtUr) (Figure 11.20a). According to Figure 11.20b, the largest GS Y_a and GS BIO values were for the N cover fertilizing level of 150 kg ha^{-1} for both N sources (respective averages 9.8 and 17.7 t ha^{-1} GS^{-1}), but not significantly different from those for 200 and 250 kg ha^{-1} (only 1–2% above). The mean GS values for Y_a and BIO were 8.77 and 8.74 t ha^{-1} GS^{-1}, and 17.32 and 17.28 t ha^{-1} GS^{-1}, respectively, for Nt and Ur N sources.

Considering the harvest index (HI) as the ratio of Y_a by BIO, it ranged from 0.37 to 0.55 for the respective treatments without N cover fertilizing (0NtUr) and those for N cover applications at 150 kg ha^{-1} N for both N sources, averaging 0.50 and 0.51, for Nt and Ur, respectively. Nyolei et al. (2019) reported average corn ET, Y_a, and BIO respective values of 331 mm, 3.2 t ha^{-1}, and 6.8 t ha^{-1}, yielding an average HI of 0.47, using Sentinel and Landsat 8 images in the northeastern part of Tanzania. However, this last study involved different corn growth stages and pixel contaminations with other crops.

The GS WP_{BIO} and GS WP_{Ya} values for nitrate (subscript Nt) and urea (subscript Ur) N sources are presented in Table 11.6.

The highest BIO and Y_a values together with low ET rates for Nt source favored slightly higher WP_{BIO} and WP_{Ya} values when comparing with those for Ur source. The top values were for N cover applications at 150 kg ha^{-1}, with WP_{BIO} being 9% and 8% higher and WP_{Ya} 61% and 60% above those for the 0NtUr control treatment, regarding Nt and Ur, respectively. However, these values stabilized after this N cover fertilizing level for both N sources. The average WP_{Ya} is higher than that found by Nyolei et al. (2019), with an average of 0.97 kg m^{-3} in Tanzania, but the authors attributed

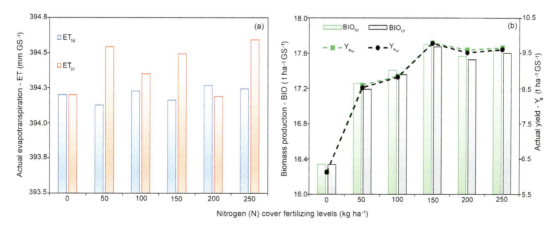

FIGURE 11.20 Growing season (GS) values for the water productivity (WP) components. Actual evapotranspiration—ET (a); biomass production—BIO and actual yield—Y_a (b), for nitrogen (N) cover fertilizing levels (0 to 250 kg ha^{-1}) and sources from nitrate (subscript Nt) and urea (subscript Ur).

TABLE 11.6
Average Pixel Values for the Water Productivity Based on Biomass Production—BIO (WP$_{BIO}$) and Based on Actual Yield—Ya (WP$_{Ya}$), Considering Cover Fertilizing with Different Nitrogen (N) Levels and Sources for the Rainfed Corn Growing Season (GS)

	Water Productivity Based on Biomass Production— WP$_{BIO}$ (kg m^{-3})						
N levels							
N sources	0	50	100	150	200	250	Mean
Nitrate—Nt	4.14	4.39	4.43	4.51	4.47	4.48	4.40
Urea—Ur	4.14	4.36	4.40	4.48	4.45	4.46	4.38
Mean	4.14	4.38	4.42	4.50	4.46	4.47	4.39
	Water Productivity Based on Actual Yield—WP$_{Ya}$ (kg m^{-3})						
N levels	0	50	100	150	200	250	Mean
N sources							
Nitrate—Nt	1.55	2.18	2.25	2.49	2.43	2.45	2.23
Urea—Ur	1.55	2.16	2.24	2.48	2.41	2.43	2.21
Mean	1.55	2.17	2.25	2.49	2.42	2.44	2.22

the low values to poor yields obtained in the mid- and lowlands area of their studied areas in the catchment. However, Teixeira et al. (2014b), by using MODIS images, reported an average value of 2.10 kg m^{-3} for rainfed corn crop in the state of Mato Grosso, central west Brazil, while for pivot irrigated corn, Teixeira et al. (2014a) using Landsat 8 images, found WP$_{Ya}$ average of 2.00 kg m^{-3} in the state of São Paulo, southeast Brazil.

The stabilization of WP$_{Ya}$ at N cover fertilizing at 150 kg ha^{-1} means that farmers applying N above this level will lose money and increase the risk of N leaching into the groundwater. However, the advantage of the price of kg of Ur being 63% of that for Nt is summed with its lower N leaching rates, promoting fewer environmental problems. However, besides these differences, other issues are important to consider in the rainfed corn water productivity, such as the overall production costs.

11.6 CONCLUSIONS AND POLICY IMPLICATIONS

Remote sensing models have been applied to demonstrate the feasibility of assessing water variables on large scales. Surface albedo, surface temperature, and NDVI from satellite images, together with weather data, are the input parameters for the modeling. This combination allowed the large-scale water productivity assessments from different agroecosystems in Brazil. The analyses may contribute to better understandings of the dynamics of biophysical parameters, important issues for the appraisal of climate and land-use change impacts, key information when federal and municipal governments plan expansions of agricultural areas with rational criteria.

It was demonstrated that the large-scale water productivity components can be analyzed from instantaneous remote sensing measurements in the visible and infrared bands, with and without the thermal spectrum. This was possible by modeling the ratio of the actual to the reference evapotranspiration, in conjunction with the availability of daily weather variables, at satellite overpass and daily timescales. These approaches provide large-scale temporal information on vegetation growth rates as well as plant responses to dynamic weather under irrigation and rainfed conditions, being useful for monitoring biophysical parameters. In the specific case of Brazil, this is particularly relevant because there is rainfall scarcity, mainly in semi-arid regions, and the analyses from the current study therefore become highly relevant to the application of concepts of evidence-based decision-making on the sustainable water resources use.

The available tools tested here can be operationally implemented to monitor the intensification of agriculture and the adverse impact on downstream water users in changing environments. Although the tools were tested in Brazilian agroecosystems, they can be used in other parts of the world, with probably only the need to adjust the modeling equation regression coefficients.

REFERENCES

Adak, T., Kumar, G., Chakravarty, N.V.K., Katiyar, R.K., & Deshmukh, P.S. 2013. Biomass and biomass water use efficiency in oilseed crop (Brassica Jnceae L.) under semi-arid microenvironments. Biomass and Bioenergy, 51, 154–162.

Ahamed, T., Tian, L., Zhang, Y., & Ting, K.C. 2011. A review of remote sensing methods for biomass feedstock production. Biomass and Bioenergy, 35, 2455–2469.

Akhtar, N., Syakir Ishak, M.I., Bhawani, S.A., & Umar, K. 2021. Various natural and anthropogenic factors responsible for water quality degradation: A review. Water, 13, 2660.

Allen, R.G., Pereira, L.S., Raes, D., & Smith, M. 1998. Crop evapotranspiration: Guidelines for computing crop water requirements, Food and Agriculture Organization of the United Nations: Rome, Italy.

Allen, R.G., Tasumi, M., Morse, A., Trezza, R., Wright, J.L., Bastiaanssen, W.G.M., Kramber, W., Lorite, I., & Robison, C.W. 2007. Satellite-based energy balance for mapping evapotranspiration with internalized calibration (METRIC)—Applications. Journal of Irrigation and Drainage Engineering, 133, 395–406.

Almagro, A., Oliveira, P.T.S., & Nearing, M.A. 2017. Projected climate change impacts in rainfall erosivity over Brazil. Scientific Reports, 7, 8130.

Almeida, S.L.H.D., Souza, J.B.C., Nogueira, S.F., Pezzopane, J.R.M., Teixeira, A.H.D.C., Bosi, C., Adami, M., Zerbato, C., Bernardi, C.C., Bayma, G., & Silva, P.R. 2023a. Forage mass estimation in silvopastoral and full sun systems: Evaluation through proximal remote sensing applied to the SAFER model. Remote Sensing, 15, 815.

Almeida, S.L.H.D., Souza, J.B.C., Pilon, C., Teixeira, A.H.D.C., Santos, A.F.D., Sysskind, M.N., Vellidis, G., & Silva, R.P.D., 2023b. Performance of the SAFER model in estimating peanut maturation. European Journal of Agronomy, 147, 126844.

Araujo, L.M., Teixeira, A.H.D.C., & Bassoi, L.H. 2019. Evapotranspiration and biomass modelling in the Pontal Sul Irrigation Scheme. International Journal of Remote Sensing, 41, 1–13.

Azevedo, G.B.D., Rezende, A.V., Azevedo, G.T.O.S., Miguel, E.P., Aquino, F.G., Bruzinga, J.S.C., Oliveira, L.S.C.D., Pereira, R.S., & Teodoro, P.E. 2020. Woody biomass accumulation in a Cerrado of Central Brazil monitored for 27 years after the implementation of silvicultural systems. Forest Ecology and Management, 455, 117718.

Azevedo, P.V., Sousa, I.F., Silva, B.B.D., & Silva, V.D.P.R. 2006. Water-use efficiency of dwarf-green coconut (Cocos nucifera L.) orchards in northeast Brazil. Agricultural Water Managent, 84, 259–264.

Bastiaanssen, W.G.M. & Ali, S. 2003. A new crop yield forecasting model based on satellite measurements applied across the Indus Basin, Pakistan. Agriculture, Ecosystems & Environment, 94, 32–340.

Bastiaanssen, W.G.M., Menenti, M., Feddes, R.A., Roerink, G.J., & Holtslag, A.A.M. 1998. A remote sensing surface energy balance algorithm for land (SEBAL) 1: Formulation. Journal of Hydrology, 212–213, 198–212.

Beguería, S., Vicente-Serrano, S.M., Reig, F., & Latorre, B., 2014. Standardized precipitation evapotranspiration index (SPEI) revisited: Parameter fitting, evapotranspiration models, tools, datasets and drought monitoring. International Journal Climatology, 34, 3001–3023.

Bento, V.A., Gouveia, C.M., Dacamara, C.C., Trigo, I.F. 2018. A climatological assessment of drought impact on vegetation health index. Agricultural and Forest Meteorology, 259, 286–295.

Bhattarai, N., Wagle, P., Gowda, P.H., & Kakani, V.G. 2017. Utility of remote sensing-based surface energy balance models to track water stress in rain-fed switchgrass under dry and wet conditions. ISPRS Journal Photogrammetry and Remote Sensing, 113, 128–141.

Cai, X., Mckinney, C., & Lasdon, S. 2002. A framework for sustainable analysis in water resources management and application to the Syr Darya Basin. Water Resources Research, 38, 21–1/14.

Campos, I., Neale, C.M.U., Arkebauer, T.J., Suyker, A.E., & Gonçalves, I.Z. 2018. Water productivity and crop yield: A simplified remote sensing driven operational approach. Agricultural and Forest Meteorology, 249, 501–511.

Cancela, J.J., Gonzalez, X.P., Vilanova, M., & Miras-Avalos, J.M. 2019. Water management using drones and satellites in agriculture. Water, 11, 874.

Carr, M.K.V. 2011. The water relations and irrigation requirements of coconut (cocos nucifera): A review. Experimental Agriculture, 47, 27–51.

Ceotto, E. & Castelli, F. 2002. Radiation use efficiency in flue-cured tobacco (<cite lang="en">Nicotiana tabacum</cite> L.): Response to nitrogen supply, climate variability and sink limitations. Field Crops Research, 74, 117–130.

Ceschia, E., Beziat, P., Dejoux, J.F., Aubinet, M., Bernhofer, C., Bodson, B., Carrara, A., Cellier, P., Di Tommasi, P., Elbers, J.A., Eugster, W., Grünwald, T., Jacobs, C.M.J., Jans, W.W.P., Jones, M.J., Kutsch, W., Lanigan, G., Magliulo, E., Marloie, O., Moors, E.J., Moureaux, C., Olioso, A., Osborne, B., Sanz, M.J., Saunders, M., Smith, P., Soegaard, H., & Wattenbach, M. 2010. Management effects on net ecosystem carbon and GHG budgets at European crop sites. Agriculture, Ecosystems & Environment, 139, 363–383.

Claverie, M., Demarez, V., Duchemin, B., Hagolle, O., Ducrot, D., Marais-Sicre, C., Dejuoux, J.-F., Huc, M., Keravec, P., Béziat, P., Fieuzal, R., Ceschia, E., & Dedieu, G. 2012. Maize and sunflower biomass estimation in southwest France using spatial and temporal resolution remote sensing data. Remote Sensing of Environment, 124, 884–857.

Cleugh, H.A., Leuning, R., Mu, Q., & Running, S.W. 2007. Regional evaporation estimates from flux tower and MODIS satellite data. Remote Sensing of Environment, 106, 285–304.

Consoli, S., Licciardello, F., Vanella, D., Pasotti, L., Villani, G., & Tomei, F. 2016. Testing the water balance model CRITERIA using TDR measurements, micrometeorological data, and satellite-based information. Agricural Water Management, 170, 68–80.

Consoli, S. & Papa, R., 2013. Corrected surface energy balance to measure and model the evapotranspiration of irrigated orange orchards in semi-arid Mediterranean conditions. Irrigation Science, 31, 1159–1171.

Consoli, S. & Vanella, D., 2014. Comparisons of satellite-based models for estimating evapotranspiration fluxes. Journal of Hydrology, 513, 475–489.

de Bruin, H.A.R. 1987. From Penman to Makkink. In: Hooghart, J.C. (Ed.), Proceedings and information: TNO committee on hydrological sciences, vol. 39, Gravenhage, The Netherlands, pp. 5–31.

Dehziari, S.A. & Sanaienejad, S.H. 2019. Energy balance quantification using Landsat 8 images and SAFER algorithm in Mashhad, Razavi Khorasan, Iran. Journal of Applied Remote Sensing, 13, 014528.

Driscoll, S.P., Prins, A., Olmos, E., Kunert, K.J., & Foyer, C.H. 2006. Specification of adaxial and abaxial stomata, epidermal structure and photosynthesis to CO2 enrichment in maize leaves. Journal of Experimental Botany, 57, 381–390.

Fandino, M., Cancela, J.J., Rey, B.J., Martínez, E.M., Rosa, R.G., & Pereira, L.S. 2012. Using the dual-Kc approach to model evapotranspiration of Albarino vineyards (Vitis vinifera L. cv. Albarino) with consideration of active ground cover. Agricultural Water Management, 112, 75–87.

Franco, R.A.M., Hernandez, F.B.T., & Teixeira, A.H.D.C. 2014. Water productivity of different land uses in watersheds assessed from satellite imagery Landsat 5 Thematic Mapper. Proceedings of SPIE, 9239, 92392E-1–92392E-7.

Franco, R.A.M., Hernandez, F.B.T., Teixeira, A.H.D.C., Leivas, J.F., Nuñez, D.N.C., & Neale, C.M.U. 2016. Water productivity mapping using Landsat 8 satellite together with weather stations. Proceedings of SPIE, 9998, 99981H-1–99981H-12.

García-Tejero, I.F., Durán Zuazo, V.H., & Rubio Casal, A.E. 2019. Deficit-irrigation strategies to enhance the water productivity in orange trees in semi-arid environments. Jornal of Agriculture Food and Development, 5, 43–51.

García-Tejero, I.F., Romero-Vicente, R., Jimenez-Bocanegra, J.A., Martínez-García, G., Duran-Zuazo, V.H., & Muriel-Fernandez, J.L. 2010. Response of citrus trees to deficit irrigation during different phenological periods in relation to yield, fruit quality, and water productivity. Agricultural Water Management, 97, 689–699.

Gouveia, C., Trigo, R.M., Beguería, S., & Vicente-Serrano, S.M. 2017. Drought impacts on vegetation activity in the Mediterranean region: An assessment using remote sensing data and multi-scale drought indicators. Global and Planetary Change, 151, 15–27.

IPCC. 2023. Climate change 2023: Synthesis report: A report of the intergovernmental panel on climate change: Contribution of working groups I, II and III to the sixth assessment report of the intergovernmental panel on climate change. In: Lee, H. & Romero, J. (Eds.), IPCC, Geneva, Switzerland, pp. 35–115.

Jamshidi, S., Zand-Parsa, S., Kamgar-Haghighi, A.A., Shahsavar, A.R., & Niyogi, D. 2020. Evapotranspiration, crop coefficients, and physiological responses of citrus trees in semi-arid climatic conditions. Agricultural Water Management, 227, 105838.

Jardim, A.M.R.F., Júnior, G.N.A., da Silva, M.V., dos Santos, A., da Silva, J.L.B., Pandorfi, H., Oliveira-Júnior, J.F.D., Teixeira, A.H.D.C., Teodoro, P.E., de Lima, J.L.P.M., Junior, C.A.S., Souza, L.S.B., Silva, E.A., & Silva, T.G.F.S. 2022. Using remote sensing to quantify the joint effects of climate and land use/land cover changes on the caatinga Biome of Northeast Brazilian. Remote Sensing, 14, 1911.

Jayakumar, M., Janapriya, S., & Surendran, U. 2017. Effect of drip fertigation and polythene mulching on growth and productivity of coconut (Cocos nucifera L.), water, nutrient use efficiency and economic benefits. Agricultural Water Management, 182, 87–93.

Junior, C.R.A.B., Folegatti, M.V., Rocha, F.J., & Atarassi, R.T. 2008. Coeficiente de cultura da lima ácida Tahiti no outono-inverno determinado por lisimetria de pesagem em Piracicaba-SP. Engenharia Agrícola, 28, 691–698.

Kang, S., Zhang, F., Hu, X., & Zhang, J. 2002. Benefits of CO2 enrichment on crop plants are modified by soil water status. Plant and Soil, 238, 69–77.

Leivas, J.F., Teixeira, A.H.D.C., Bayma-Silva, G., Ronquim, C.C., & Reis, J.B.R.D.S. 2016. Biophysical indicators based on satellite images in an irrigated area at the São Francisco River basin, Brazil. Proceedings of SPIE, 9998, 99981N-99981N.

Li, X., Zhang, X., Niu, J., Tong, L., Kang, S., Du, T., Li, S., & Ding, R. 2016. Irrigation water productivity is more influenced by agronomic practice factors than by climatic factors in Hexi Corridor, Northwest China. Science Reports, 6, 1–10.

Longo-Minnolo, G., Vanella, D., Consoli, S., Intrigliolo, D.S., & Ramirez-Cuesta, J.M. 2020. Integrating forecast meteorological data into the ArcDualKc model for estimating spatially distributed evapotranspiration rates of a citrus orchard. Agricultural Water Management, 231, 105967.

Lu, N., Chen, S., Wilske, B., Sun, G., & Chen, J. 2011. Evapotranspiration and soil water relationships in a range of disturbed and undisturbed ecosystems in the semi-arid Inner Mongolia, China. Journal of Plant Ecology, 4, 49–60.

Madurapperuma, W.S., de Costa, W.A.J.M., Sangakkara, U.R., & Jayescara, C. 2009. Estimation of water use of mature coconut (Cocos nucifera L.) cultivars (CRIC 60 and CRIC 65) grown in the low country intermediate zone using the compensation heat pulse method (CHPM). Journal of the National Science Foundation of Sri Lanka, 37, 175–186.

Mata-González, R., McLendon, T., & Martin, D.W. 2005. The inappropriate use of crop transpiration coefficients (Kc) to estimate evapotranspiration in arid ecosystems: A review. Arid Land Research and Management, 19, 285–295.

Mateos, L., González-Dugo, M.P., Testi, L., & Villalobos, F.J. 2013. Monitoring evapotranspiration of irrigated crops using crop coefficients derived from time series of satellite images: I. Method validation. Agricultural Water Management, 125, 81–91.

Miranda, F.R., Gomes, A.R.M., de Oliveira, C.H.C., Montenegro, A.A.T., & Bezerra, F.M.L. 2007. Evapotranspiração e coeficientes de cultivo do coqueiro anão-verde na região litorânea do Ceará. Revista Ciência Agronômica, 38, 129–135.

Miranda, F.R., Montenegro, A.A.T., Lima, R.N., Rosseti, A.G., & Freitas, J.A.D. 2004. Distribuição do sistema radicular de plantas jovens de coqueiro-anão sob diferentes frequências de irrigação. Revista Ciência Agronômica, 35, 309–318.

Mokhtari, A., Noory, H., Pourshakouri, F., Haghighatmehr, P., Afrasiabian, Y., Razavi, M., Fereydooni, F., & Naeni, A.S. 2019. Calculating potential evapotranspiration and single crop coefficient based on energy balance equation using Landsat 8 and Sentinel-2. ISPRS Journal Photogrammetry and Remote Sensing, 154, 231–245.

Molden, D., Frenken, K., Barker, R., de Fraiture, C., Bancy, M., Svendsen, M., Sadoff, C., & Finlayson, C.M. 2007. Trends in water and agricultural development. In: Chapter 2 in water for food, water for life: A comprehensive assessment of water management in agriculture, International Water Management Institute, London, Earthscan, Colombo.

Monteith, J.L. 1972. Solar radiation and productivity in tropical ecosystems. Journal of Applied Ecology, 9, 747–766.

Monteith, J.L. 1977. Climate and efficiency of crop production in Britain. Philosophical Transactions of the Royal Society B, 281, 277–294.

Nagler, P.L., Glenn, E.P., Nguyen, U., Scott, R.L., & Doody, T. 2013. Estimating riparian and agricultural actual evapotranspiration by reference evapotranspiration and MODIS enhanced vegetation index. Remote Sensing, 5, 3849–3871.

Naresh Kumar, S., Kasturi Bai, K.V., Rajagopal, V., & Aaggarwal, P.K. 2008. Simulating coconut growth, development and yield with the InfoCrop-coconut model. Tree Physiology, 28, 1049–1058.

Nuñez, D.C., Hernandez, F.B.T., Teixeira, A.H.D.C., Franco, R.A.M., & Leivas, J.F.L. 2017. Water productivity using SAFER—simple algorithm for evapotranspiration retrieving in watershed. Revista Brasileira de Engenharia Agrícola e Ambiental, 21, 524–529.

Nyolei, D., Nsaali, M., Minaya, V., van Griensven, A., Mbilinyi, B., Diels, J., Hessels, T., & Kahimba, F. 2019. High resolution mapping of agricultural water productivity using SEBAL in a cultivated African catchment, Tanzania. Physics and Chemistry of the Earth, 112, 36–39.

Oliveira, M.L., Santos, C.A.C.D., Oliveira, G.D., Silva, M.T., Silva, B.B.D., Cunha, J.E.D.B.L., Ruhoff, A., & Santos, C.A.G. 2022. Remote sensing-based assessment of land degradation and drought impacts over terrestrial ecosystems in Northeastern Brazil. Science of the Total Environment, 835, 155490.

Olivera-Guerra, L., Merlin, O., Er-Raki, S., Khabba, S., & Escorihuela, M.J. 2018. Estimating the water budget components of irrigated crops: Combining the FAO-56 dual crop coefficient with surface temperature and vegetation index data. Agricultural Water Management, 208, 120–131.

Panigrahi, P., Srivastava, A.K., & Huchche, A.D. 2012. Effects of drip irrigation regimes and basin irrigation on Nagpur mandarin agronomical and physiological performance. Agricultural Water Management, 104, 79–88.

Pereira, M.P.S., Mendes, K.R., Justino, F.J., Couto, F., Silva, A.S.D., Silva, D.F.D., & Malhado, A.C.M. 2020. Brazilian dry forest (Caatinga) response to multiple ENSO: The role of Atlantic and Pacific Ocean. Science of the Total Environment, 705, 135717.

Rallo, G., Gonzalez-Altozano, P., Manzano-Juarez, J., & Provenzano, G. 2017. Using field measurements and FAO-56 model to assess the eco-physiological response of citrus orchards under regulated deficit irrigation. Agricultural Water Management, 180, 136–147.

Ramírez-Cuesta, J.M., Allen, R.G., Zarco-Tejada, P.J., Kilic, A., Santos, C., & Lorite, I.J. 2019. Impact of the spatial resolution on the energy balance components on an open-canopy olive orchard. International Journal Applied Earth Observation and Geoinformation, 74, 88–102.

Rampazo, N.A.M., Picoli, M.C.A., Teixeira, A.H.D.C., & Cavaleiro, C.K.N. 2020. Water consumption modeling by coupling MODIS images and agrometeorological data for sugarcane crops. Sugar Tech, 23, 524–535.

Raupasch, M.R. 2001. Combination theory and equilibrium evaporation. Quarterly Journal of Royal Meteorology Society, 127, 1149–1181.

Resende, R.S., Santos, H.R., de Amorim, J.R.A., Souza, G., & Meneses, T.N. 2015. Efeito da cobertura morta no padrão de distribuição da água em microaspersão. Revista Brasileira de Agricultura Irrigada, 9, 278–286.

Ribeiro, M.C., Metzger, J.P., Martensen, A.C., Ponzoni, F.J., & Hirota, M.M. 2009. The Brazilian Atlantic Forest: How much is left, and how is the remaining forest distributed? Implications for conservation. Biological Conservation, 142, 1141–1153.

Rosa, R.D., Ramos, T.B., & Pereira, L.S. 2016. The dual Kc approach to assess maize and sweet sorghum transpiration and soil evaporation under saline conditions: Application of the SIMDualKc model. Agricultural Water Management, 177, 77–94.

Roupsard, O., Bonnefond, J.M., Irvine, M., Berbigier, P., Nouvellon, Y., Dauzat, J., Taga, S., Hamel, O., Jourdan, C., Saint-André, L., Mialet-Serra, I., Labouisse, J.P., Epron, D., Joffre, R., Braconnier, S., Rouzière, A., Navarro, M., & Bouillet, J.P. 2006. Partitioning energy and evapotranspiration above and below a tropical palm canopy. Agricultural and Forest Meteorology, 139, 252–268.

Safre, A.L.S., Nassar, A., Torres-Rua, A., Aboutalebi, M., Saad, J.C.C., Manzione, R.L., Teixeira, A.H. D.C., Prueger, J.H., McKee, L.G., Alfieri, J.G., Hipps, L.E., Nieto, H., White, W.A., Alsina, M.D.M., Sanchez, L., Kustas, W.P., Dokoozlian, N., Gao, F., & Anderson, M.C. 2022. Performance of Sentinel-2 SAFER ET model for daily and seasonal estimation of grapevine water consumption. Irrigation Science, 40, 635–654.

Sano, E.E., Rodrigues, A.A., Martins, E.S., Bettiol, G.M., Bustamante, M.M.C., Bezerra, A.S., Couto, A.F., Vasconcelos, V., Schüler, J., & Bolfe, E.L. 2019. Cerrado ecoregions: A spatial framework to assess and prioritize Brazilian savanna environmental diversity for conservation. Journal of Environmental Management, 232, 818–828.

Santos, J.E.O., Cunha, F.F., Filgueiras, R., Silva, G.H., Teixeira, A.H.D.C., Silva, F.C.S., & Sediyama, G.C. 2020. Performance of SAFER evapotranspiration using missing meteorological data. Agricultural Water Management, 233, 1–8.

Santos, M.G., Oliveira, M.T., & Figueiredo, K.V. 2014. Caatinga, the Brazilian dry tropical forest: Can it tolerate climate changes? Theoretical and Experimental Plant Physiology, 26, 83–99.

Silva, C.O.F., Manzione, R.L., & Teixeira, A.H.D.C. 2018. Modelagem espacial da evapotranspiração e produtividade hídrica na porção paulista do afloramento do aquífero Guarani entre 2013 e 2015. Holos Environment, 18, 126–140.

Silva, C.O.F., Teixeira, A.H.D.C., & Manzione, R.L. 2019. An R package for spatial modelling of energy balance and actual evapotranspiration using satellite images and agrometeorological data. Environmental Modelling & Software, 120, 104497.

Silva, J.L.B.D., Moura, G.B.D.A., Silva, M.V.D., Oliveira-Júnior, J.F.D., Jardim, A.M.R.F., Refati, D.C., Lima, R.C.C., Carvalho, A.A.D., Ferreira, M.B., Brito, J.I.B.B., Guedesf, R.V.S., Lopes, P.M.O., Nobrega, R.S., Pandorfi, H., Bezerra, A.C., Batista, P.H.D., Jesus, F.L.F., Sanches, A.C., & Santos, R.C. 2023. Environmental degradation of vegetation cover and water bodies in the semiarid region of the Brazilian Northeast via cloud geoprocessing techniques applied to orbital data. Journal of South American Earth Sciences, 121, 104164.

Sousa, I.F., Netto, A.O.A., Campeche, L.F.M.S., Barros, A.C., da Silva, V.P.S., & Azevedo, P.V. 2011. Lisímetro de pesagem de grande porte. Parte II: Consumo hídrico do coqueiro anão verde irrigado. Revista Brasileira de Engenharia Agrícola e Ambiental, 15, 526–532.

Souza, C.M., Jr, Z., Shimbo, J., Rosa, M.R., Parente, L.L., Alencar, A., Rudorff, B.F.T., Hasenack, H., Matsumoto, M.G., Ferreira, L., Souza-Filho, P.W.M., de Oliveira, S.W., Rocha, W.F., Fonseca, A.V., Marques, C.B., Diniz, C.G., Costa, D., Monteiro, D., Rosa, E.R., Vélez-Martin, E., Weber, E.J., Lenti, F.E.B., Paternost, F.F., Pareyn, F.G.C., Siqueira, J.V., Viera, J.L., Neto, L.C.F., Saraiva, M.M., Sales, M.H., Salgado, M.P.G., Vasconcelos, R., Galano, S., Mesquita, V.V., & Azevedo, T. 2020. Reconstructing three decades of land use and land cover changes in Brazilian biomes with Landsat archive and Earth Engine. Remote Sensing, 12, 2735.

Sumner, D.M. & Jacobs, J. 2005. Utility of Penman-Monteith, Priestley-Taylor, reference evapotranspiration, and pan evaporation methods to estimate pasture evapotranspiration. Journal of Hydrology, 308, 81–104.

Taghvaeian, S., Chavez, J.L., & Hansen, N.C. 2012. Infrared thermometry to estimate crop water stress index and water use of irrigated maize in northeastern Colorado. Remote Sensing, 4, 3619–3637.

Teixeira, A.H.D.C. 2010. Determining regional actual evapotranspiration of irrigated and natural vegetation in the São Francisco River basin (Brazil) using remote sensing and Penman-Monteith equation. Remote Sensing, 2, 1287–1319.

Teixeira, A.H.D.C. 2012. Modelling water productivity components in the Low-Middle São Francisco River basin, Brazil. In: Bilibio, B., Hensel, O., & Selbach, J. (Org.), Sustainable water management in the tropics and subtropics and case studies in Brazil. 1st ed., vol. 3, University of Kassel, Kassel, Germany, pp. 1077–1100.

Teixeira, A.H.D.C. & Bassoi, L.H. 2009. Crop water productivity in semi-arid regions: From field to large scales. Annals of Arid Zone, 48, 1–13.

Teixeira, A.H.D.C., Bastiaanssen, W.G.M., Ahmad, M.U.D., & Bos, M.G. 2009. Reviewing SEBAL input parameters for assessing evapotranspiration and water productivity for the Low-Middle São Francisco River basin, Brazil Part B: Application to the large scale. Agricultural and Forest Meteorology, 149, 477–490.

Teixeira, A.H.D.C., Bastiaanssen, W.G.M., Ahmad, M.D., Moura, M.S.B., & Bos, M.G. 2008. Analysis of energy fluxes and vegetation-atmosphere parameters in irrigated and natural ecosystems of semi-arid Brazil. Journal of Hydrology, 362, 110–127.

Teixeira, A.H.D.C., Hernandez, F.B.T., Andrade, R.G., Leivas, J.F., Victoria, D.C., & Bolfe, E.L. 2014a. Irrigation performance assessments for corn crop with Landsat images in the São Paulo state, Brazil. Water Resources and Irrigation Management, 3, 91–100.

Teixeira, A.H.D.C., Hernandez, F.B.T., Scherer-Warren, M., Andrade, R.G., Victoria, D.D.C., Bolfe, E.L., Thenkabail, P.S., & Franco, R.A.M. 2015. Water productivity studies from earth observation data: Characterization, modeling, and mapping water use and water productivity In: Tenkabail, P. (Ed.), Remote sensing of water resources, disasters, and urban studies, 1st ed., vol. 3, Taylor and Francis, Boca Raton, Florida, pp. 101–126.

Teixeira, A.H.D.C., Leivas, J.F., Garçon, E.A.M., Takeura, C.M., Quartaroli, C.F., & Alvarez, I.A. 2020a. Modeling large-scale biometeorological indices to monitor agricultural-growing areas: Applications in the fruit circuit region, São Paulo, Brazil. International Journal of Biometeorology, 1, 1–14.

Teixeira, A.H.D.C., Leivas, J.F., Pacheco, E.P., Garçon, E.A.M., & Takemura, C.M., 2021a. Biophysical characterization and monitoring large-Scale water and vegetation anomalies by remote sensing in the agricultural growing areas of the Brazilian semi-arid region. In: Pandey, P.C. & Sharma, L.K. (Eds.), Advances in remote sensing for natural resource monitoring, 1st ed., vol. 1, Wiley Online Library, New Jersey, pp. 94–109.

Teixeira, A.H.D.C., Leivas, J.F., Ronquim, C.C., & Victoria, D.D.C. 2016. Sugarcane water productivity assessments in the São Paulo state, Brazil. International Journal of Remote Senings Applications, 6, 84–95.

Teixeira, A.H.D.C., Leivas, J.F., Struiving, T.B., Reis, J.B.R.S., & Simão, F.R. 2021b. Energy balance and irrigation performance assessments in lemon orchards by applying the SAFER algorithm to Landsat 8 images. Agricultural Water Management, 247, 1–9.

Teixeira, A.HD.C., Leivas J.F., Takemura, C.M., Bayma, G., Garçon, E.A.M., Sousa, I.D.F., Farias, F.D.J., & Silva, C.O.F. 2023. Remote sensing environmental indicators for monitoring spatial and temporal dynamics of weather and vegetation conditions: Applications for Brazilian biomes. Environment Monitoring Assessments, 195, 944.

Teixeira, A.H.D.C., Miranda, F.R.D., Leivas, J.F., Pacheco, E.P., & Garçon, E.A.M. 2019. Water productivity assessments for dwarf coconut by using Landsat 8 images and agrometeorological data. ISPRS Journal of Photogrammetry and Remote Sensing, 155, 150–158.

Teixeira, A.H.D.C., Pacheco, E.P., Silva, C.O.F., Dampieri, M.G., & Leivas, J.F. 2021c. SAFER applications for water productivity assessments with aerial camera onboard a remotely piloted aircraft (RPA): A rainfed corn study in Northeast Brazil. Remote Sensing Applicatios: Society and Environment, 22, 100514.

Teixeira, A.H.D.C., Scherer-Warren, M., Hernandez, F.B.T., Andrade, R.G., & Leivas, J.F. 2013. Large-scale water productivity assessments with MODIS images in a changing semi-arid environment: A Brazilian case study. Remote Sensing, 5, 5783–5804.

Teixeira, A.H.D.C., Simão, F.R., Leivas, J.F., Gomide, R.L., Reis, J.B.R.S., Kobayashi, M.K., & Oliveira, F.G. 2018. Water productivity modeling by remote sensing in the semiarid region of Minas Gerais State, Brazil. In: Arman, H. & Yuksel, I. (Eds.), Arid environments and sustainability, 1st ed., InTech, Londres, pp. 94–108.

Teixeira, A.H.D.C., Takemura, C.M., Leivas, J.F., Pacheco, E.P., Silva, G.B., & Garçon, E.A.M. 2020b. Water productivity monitoring by using geotechnological tools in contrasting social and environmental conditions: Applications in the São Francisco River basin, Brazil. Remote Sensing Applicatios: Society and Environment, 18, 1–9.

Teixeira, A.H.D.C., Victoria, D.D.C., Andrade, R.G., Leivas, J.F., Bolfe, E.L., & Cruz, C.R. 2014b. Coupling MODIS images and agrometeorological data for agricultural water productivity analyses in the Mato Grosso state, Brazil. Proceedings of SPIE, 9239, 92390W-1–92390W-14.

Tesfaye, K., Walker, S., & Tsubo, M. 2006. Radiation interception and radiation use efficiency of three grain legumes under water deficit conditions in a semi-arid environment. European Journal of Agronomy, 25, 60–70.

Twohey, R.J., Roberts, L.M., & Studer, A.J. 2019. Leaf stable carbon isotope composition reflects transpiration efficiency in Zea mays. Plant Journal, 97, 475–484.

Vanella, D., Ramírez-Cuesta, J.M., Intrigliolo, D.S., & Consoli, S. 2019. Combining electrical resistivity tomography and satellite images for improving evapotranspiration estimates of Citrus orchards. Remote Sensing, 11 (4), 373.

Venâncio, L.P., Mantovani, E.C., Amaral, C.H.D., Neale, C.M.U., Filgueiras, R., Gonçalves, I.Z., & Cunha, F.F.D. 2021. Evapotranspiration mapping of commercial corn fields in Brazil using SAFER algorithm. Scientia Agricola, 78, 1–12.

Vicente-Serrano, S.M., Cabello, D., Tomás-Burguera, M., Martín-Hernández, N., Beguería, S., Azorin-Molina, C., & Kenawy, A. 2015. Drought variability and land degradation in semiarid regions: Assessment using remote sensing data and drought indices (1982–2011). Remote Sensing 7, 4391–4423.

Vicente-Serrano, S.M., Miralles, D.G., Domínguez-Castrom, F., Azorin-Molina, C., El Kenawy, A., McVicar, T.R., Tomás-Burguera, M., Beguería, S., Maneta, M., & Peña-Gallardo, M. 2018. Global assessment of the standardized evapotranspiration deficit index (SEDI) for drought analysis and monitoring. Journal of Climate, 31, 5371–5393.

Wang, X., Ma, M., Huang, G., Veroustraete, F., Zhang, Z., Song, Y., & Tan, J. 2012. Vegetation primary production estimation at maize and alpine meadow over the Heihe River Basin, China. International Journal of Applied Earth Observation Geoinformation, 17, 94–101.

Wu, C., Munger, J.W., Niu, Z., & Kuanga, D. 2010. Comparison of multiple models for estimating gross primary production using MODIS and eddy covariance data in Havard Forest. Remote Sensing of Environment, 114, 2925–2939.

Yang, H.B., Qi, J., Xu, X.Y., Yang, D.W., & Lv, H.F. 2014, The regional variation in climate elasticity and climate contribution to runoff across China. Journal of Hydrology, 517, 607–616.

Yang, L.S., Yano, T.M.A., & Li, S. 2003. Evapotranspiration of orange trees in greenhouse lysimeters. Irrigation Science, 21, 145–149.

Yang, Y., Guan, H., Batelaan, O., McVicar, T.R., Long, D., Piao, S., Liang, W., Liu, B., Jin, Z., & Simmons, C.T. 2016. Contrasting responses of water use efficiency to drought across global terrestrial ecosystems. Science Reports, 6, 1–8.

Yang, Y., Xu, W., Hou, P., Liu, G., Liu, W., Wang, Y., Zhao, R., Ming, B., Xie, R., Wang, K., Zhao, R., Ming, B., Xie, R., Wang, K., & Li, S. 2019. Improving maize grain yield by matching maize growth and solar radiation. Science Reports, 9, 1–11.

Yuan, M., Zhang, L., Gou, F., Su, Z., Spiertz, J.H.J., & Werf, W. van der. 2013. Assessment of crop growth and water productivity for five C3 species in the semi-arid Inner Mongolia. Agricultural Water Management, 122, 28–38.

Zhang, F., Zhou, G., Wang, Y., Yan, F., & Christer Nilsson, C. 2012. Evapotranspiration and crop coefficient for a temperate desert steppe ecosystem using eddy covariance in Inner Mongolia, China. Hydrological Processes, 26, 379–386.

Zhang, G., Su, X., Singh, V.P., & Ayantobo, O. 2021. Appraising standardized moisture anomaly index (SZI) in drought projection across China under CMIP6 forcing scenarios. Journal of Hydrology: Regional Studies, 37, 100898.

Zhang, L., Niu, Y., Zhang, H., Han, W., Li, G., Tang, J., & Peng, X. 2019. Maize canopy temperature extracted from UAV thermal and RGB imagery and its application in water stress monitoring. Frontiers in Plant Science, 10, 1–18.

Zhang, X. & Zhang, B. 2019. The responses of natural vegetation dynamics to drought during the growing season across China. Journal of Hydrology, 574, 706–714.

Zhao, M., Heinsch, F.A., Nemani, R.R., & Running, S.W. 2005. Improving of the MODIS terrestrial gross and net primary production global dataset. Remote Sensing of Environment. 95, 164–176.

Zhao, M. & Running, S.W. 2010. Drought-induced reduction in global terrestrial net primary production from 2000 through 2009. Science, 329, 940–943.

Zhou, L. & Zhou, G. 2009. Measurement and modeling of evapotranspiration over a reed (Phragmitesaustralis) marsh in Northeast China. Journal of Hydrology, 372, 41–47.

Zwart, S.J., Bastiaanssen, W.G.M., Fraiture de, F., & Molden, D.J. 2010. WATPRO: A remote sensing-based model for mapping water productivity of wheat. Agricultural Water Management, 97, 1628–1636.

Part VI

Snow and Ice

12 Remote Sensing Mapping and Modeling of Snow Cover Parameters and Applications

Hongjie Xie, Tiangang Liang, Xianwei Wang, Guoqing Zhang, Xiaodong Huang, and Xiongxin Xiao

ACRONYMS AND DEFINITIONS

AMSR-E	Advanced Microwave Scanning Radiometer—Earth Observing System
ANN	Artificial Neural Network
APHRODITE	Asian Precipitation-Highly Resolved Observational Data Integration Towards the Evaluation of Water Resources
AVHRR	Advanced Very High Resolution Radiometer
BME	Bayesian Maximum Entropy
BP-ANN	back propagation artificial neural network
DDF	Degree Day Factor
DEM	Digital Elevation Model
DNB	Day/Night Band
EOS	Earth Observing System
GDP	Gross Domestic Product
GOES	Geostationary Observational Environmental Satellite
HMRF	Hidden Markov random field
ICESat	Instrument aboard the Ice, Cloud, and land Elevation
IMS	Ice Mapping System
MAE	Mean Absolute Error
MODIS	Moderate Resolution Imaging Spectroradiometer
NASA	National Aeronautics and Space Administration
NDSI	Normalized Difference Of Snow Index
NDVI	Normalized Difference Vegetation Index
NME	Negative Mean Error
NOHRSC	National Operational Hydrologic Remote Sensing Center
NS	Nash–Sutcliffe
NSCI	Normalized Snow Cover Index
NSIDC	National Snow and Ice Data Center
OK	Ordinary Kriging
PME	Positive Mean Error
RF	Random Forests
RMSE	Root Mean Square Error
SA	Simple Averaging
SCA	Snow-Covered Area
SCD	Snow Cover Duration
SCE	Snow Cover Extent

SCED	Snow Cover End Dates
SCI	Snow Cover Index
SCOD	Snow Cover Onset Dates
SK	Simple Kriging
SRM	Snowmelt Runoff Model
SRTM	Shuttle Radar Topography Mission
SSL	Snow Season Length
SSM/I	Special Sensor Microwave/Imager
SWE	Snow Water Equivalent
TAC	Terra and Aqua MODIS combination
TP	Tibetan Plateau
VIIRS	Visible Infrared Imaging Radiometer Suite

12.1 INTRODUCTION

Snowpack or snow-covered area (SCA) is an important component of the hydrologic cycle, especially in mountainous basins where the majority of water originates from snowmelt. Basins with snowmelt water such as the Western USA and the Tibetan Plateau (TP) and its surrounding areas, are a source of major water resources for agriculture, residential, industry, and many other needs (Immerzeel et al., 2010, 2020; Musselman et al., 2021; Qin et al., 2020). Due to its high albedo, snow cover is one of the key variables impacting the Earth's energy balance. Snow is also an excellent insulator with its low thermal conductivity, having effect on the underground permafrost thermal regime and potentially regulating the emission of carbon dioxide from carbon-laden soil by microbial decomposition (Jan & Painter, 2020; Park et al., 2015). Therefore, accurate monitoring of surface snow cover properties (i.e., snow covered area, snow depth, and/or snow water equivalent) has been a major task for the cryosphere remote sensing community. Using its unique high spectral signature, snow can be relatively easy to separate from other types of land cover (Brubaker et al., 2005; Simic et al., 2004). However, the accuracy of snow cover mapping techniques is affected by variables such as the sensor's spectral resolution, snow depth on the ground, cloud cover, and forest canopy (Hall et al., 1998; Vikhamar and Solberg, 2002). The importance of snow cover products as a substitute for ground stations/sensors is especially underscored in large areas where the ground station density is sparse or the terrain is inaccessible due to the remoteness of the region and its large spatial extent.

In situ measurements of snow depth and snow water equivalent (SWE) can provide good snow depth observations at localized points and can be used as ground truth data to validate remote sensing products for snow area extent (Zhou et al., 2005; Wang et al., 2008; Xiao et al., 2018, 2020). However, these stations are sparsely distributed and are usually located near open urban areas or at lower elevations, which result in considerable uncertainties of snow cover conditions in the mountainous areas or higher elevation regions. Furthermore, these *in situ* stations provide point measurements of snow depth rather than the extent of snow cover. In some instances, airborne measurements are used for snow cover mapping, but these measurements are limited in time and space. The aforementioned requirements highlight the need for satellite-based snow cover products that provide a near-continuous mapping of snow, both in space and time.

Snow cover has been derived from various satellite remote sensing sensors such as the Landsat series, Moderate Resolution Imaging Spectroradiometer (MODIS), the Sentinel series, Special Sensor Microwave/Imager (SSM/I), Advanced Microwave Scanning Radiometer—Earth Observing System (AMSR-E), Geostationary Observational Environmental Satellite (GOES), and Advanced Very High Resolution Radiometer (AVHRR). The last two sensors have been used by the United States National Weather Service at the National Operational Hydrologic Remote Sensing Center (NOHRSC) in a physically based snow model to produce daily gridded snow cover maps for the continental USA at the spatial resolution of 1km. This product has been used as a reference for snow cover mapping in the US region and has been used to validate other satellite

SCA products, such as the daily MODIS snow cover at 500m resolution (Maurer et al., 2003; Klein and Barnett, 2003). Another widely known satellite-based SCA product, which is produced by the Interactive Multisensor Snow and Ice Mapping System (IMS) combining optical and passive microwave satellite remote sensing data, is global snow cover extent data at a daily temporal resolution and at three spatial resolutions (1km, 4km, and 24km) (Romanov, 2017; Romanov & Tarpley, 2007). This cloud-free snow cover extent product has a better spatial resolution than passive microwave remote sensing–derived snow cover extent data (such as AMSR-E) (Brubaker et al., 2005) and has demonstrated compatible snow mapping accuracy with MODIS (Mazari et al., 2013). Different from traditional optical sensors, Visible Infrared Imaging Radiometer Suite (VIIRS) Day/Night Band (DNB) nighttime light satellite observations have been innovatively investigated to determine snow cover to address the snow cover detection issue in polar regions and high-latitude regions (Huang et al., 2022a; Stopic & Dias, 2023). This kind of snow cover product is a promising piece of auxiliary data to fill the MODIS snow cover data gaps due to cloud contamination (Chen et al., 2023; Huang et al., 2022a; Liu et al., 2023).

It is noted that MODIS sensors have worked for over 20 years. As a new generation, VIIRS sensors are expected to replace and prolong the data collection of the aging MODIS sensors in the visible and infrared spectra. Some studies started to perform the consistent evaluation of snow cover determination when both VIIRS and MODIS sensors are available (Liu et al., 2022; Rittger et al., 2020a; Zhang et al., 2020). In this chapter, we primarily introduce the application and modeling of MODIS snow cover products. MODIS mounted on the Earth Observing System (EOS) Terra and Aqua platforms provides the most recent and advanced SCA products for global coverage (Hall et al., 2002).

12.2 PRINCIPLES OF MODIS SNOW COVER MAPPING AND STANDARD PRODUCTS

Snow has high reflectance in the visible wavelength (0.3–0.7 μm) and low reflectance in the middle infrared wavelength (1.4 μm and longer) (Figure 12.1) (Wang et al., 2014; Xiao et al., 2022a), and a high albedo up to 0.9 (Picard et al., 2020). This outstanding reflectance/characteristic makes it possible to use optical remote sensing data to seperate snow elements based on the radiative response of different spectral bands to solar radiation. However, there are several challenges and limitations using optical data to detect surface snow cover. First, optical remote sensing cannot provide valid surface reflectance information under polar night conditions. Second, electromagnetic radiation within the visible spectral range cannot pass through thick cloud to observe land surface information. Another cloud-associated problem is that thin cloud has similar high reflectance in visible spectral band as snow, easily causing confusion classification results for thin cloud and snow, while the consideration of reflectance feature in the near and mid-infrared spectral bands partly alleviate the misclassification for snow and cloud (Frei et al., 2012). Third, vegetations, particularly the dense forest canopies, obstructing the satellite's viewing of ground snow's spectral signal at nadir perspective, make it difficult to accurately detect snow cover under vegetation (Klein et al., 1998; Nolin, 2004; Rittger et al., 2020b; Xiao et al., 2022a). Therefore, the associate uncertainties of snow cover detection vary with viewable gap fraction of vegetation and satellite viewing geometry (Rittger et al., 2020b; Xiao et al., 2022a). Last, surface heterogeneity, such as in mountainous areas or in the polar regions with wetlands and lakes, and pixels scratch in along-track and cross-track at a non-nadir scan angle, making it difficult to accurately map snow cover using moderate-resolution imagery without high-resolution (tens m) land surface data (Frei and Lee, 2010; Dozier et al., 2008; Margulis et al., 2019).

12.2.1 BINARY/FRACTIONAL SNOW COVER MAPPING

Similar to the normalized difference of vegetation index (NDVI) for vegetation mapping, normalized difference of snow index (NDSI) was developed to distinguish snow from no snow (Dozier, 1989; Hall et al., 1995). It was first tested using Landast images: band 2 (0.52–0.60 μm) and band

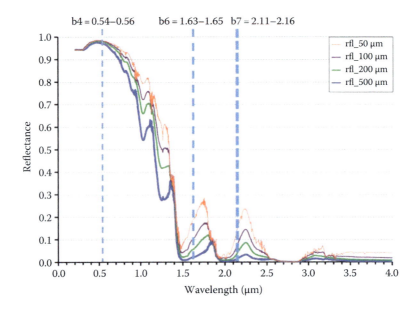

FIGURE 12.1 Snow model predicted spectral reflectance of dry snow surface under clear sky at the Summit station (72.5794N, 38.5042W) of Greenland. Snow density is 250.0 kg.m^{-3}, and snow grain radii are 50, 100, 200, and 500 μm from top to bottom, respectively. The three dashed lines from left to right are the wavelength positions of MODIS band 4, band 6, and band 7, respectively, and the width of the line is proportional to the width of the wavelength (Adapted from Wang et al., 2014).

5 (1.55–1.75 μm), with compatable wavelengths from MODIS, band 4 (0.545–0.565 μm), and band 6 (1.628–1.652 μm) for Terra MODIS (eqn. 1) or band 7 (2.105–2.155 μm) for Aqua MODIS, due to the non-functional detector in Aqua band 6. The snow-mapping algorithm (SNOMAP) tested by Landsat images showed 98% accuracy in identifying snow pixels, with snow covered by 60% or more (Hall et al., 1995).

$$NDSI = \frac{Band4 - band6}{band4 + band6} \quad (1)$$

In the standard MODIS snow cover products, four masks are also used in classifying a certain NDSI value as snow or no snow, including a dense forest stands mask, a thermal mask, a cloud mask, and an ocean and inland water mask. For MODIS images, a pixel in a non-dense forested region is identified as snow when its NDSI value is larger than or equal to 0.4, the reflectance value of band 2 (0.841–0.876 μm) is larger than 0.11, and the reflectance value of band 4 is larger than 0.1. The latter rule prevents pixels containing very dark targets, such as black spruce forests, from being marked as snow. The detailed algorithm and processing steps were documented in several sources (Hall et al., 2002; Riggs et al., 2006).

The snow binary classification only tells whether a pixel is covered by snow or is snow-free, while a given pixel with less than 50% of snow cover area should not be mapped as snow-free. The fractional snow cover was then introduced to denote the fraction of snow (0–100%) within a pixel. The standard MODIS fractional snow cover (F_{SCA}) estimation method was based on the empirical relationships (eqns. 2, 3) (Rittger et al., 2013; Salomonson and Appel, 2004; 2006) and was used to generate Collection 5 MODIS fractional snow cover product (Hall and Riggs, 2007). However, it should be noted that Collection 6 and the following versions, the noticeable change in the MODIS snow products is that it no longer provide fractional snow cover layer and binary snow-covered area layer, but NDSI snow cover layer (Riggs and Hall, 2015).

Terra MODIS: $F_{SCA} = -0.01 + 1.45 \times NDSI$ (2)

Aqua MODIS: $F_{SCA} = -0.64 + 1.91 \times NDSI$ (3)

Apart from the standard MODIS fractional snow cover estimation algorithm, spectral mixing analysis is a commonly used approach for estimating fractional snow cover from MODIS satellite observations. The spectral mixing approach generally selects the spectral information of multiple endmembers, such as vegetation, soil, and snow, from the spectral database of laboratory or in situ measurement (Painter et al., 2009) or the "pure" pixels of satellite images (Shi, 2012). A representative spectral mixing analysis algorithm, MODSCAG (MODIS Snow Covered-Area and Grain size) retrieval algorithm, was developed to retrieve fractional snow cover and snow grain size based on a spectral mixing analysis approach using MODIS spectral reflectance data (Nolin et al., 1993; Dozier and Painter, 2004; Painter et al., 2009; Sirguey et al., 2009). This algorithm uses the relative shape of the snow spectrum, which is sensitive to the spectral reflectance of the snow fraction. Thus, MODSCAG allows the snow spectral reflectance to vary pixel by pixel and can address the spatial heterogeneity that characterizes snow and its albedo in mountainous and patchy snow regions. Validation and comparison experiments show that the MODSCAG algorithm is more accurate to characterize fractional snow cover than the MODIS standard fractional snow cover algorithm (eqns. 2, 3) based on NDSI (Rittger et al., 2013). To obtain accurate estimation through spectral mixing analysis, a significant limitation arises when acquiring endmembers' spectral from a spectral database, as it demands significant time and computational sources. In addition, spectral databases may fail to accurately depict surface features within real-world environments under varying solar illumination and atmospheric conditions. In response to these challenges, an automated endmember selection method was introduced to extract the necessary endmembers from MODIS images, based on the identification of "pure" pixels through multiple indices (Shi, 2012). This efficient methodology has undergone validation and assessment in numerous studies (Hao et al., 2019; Pan et al., 2022; Wang et al., 2017; Zhu and Shi, 2018).

In addition to conventional formula-based approaches, machine learning methods, also called black-box method, have gained significant attention in the field of environmental remote sensing. They can offer distinct advantages in enhancing the precision of snow cover parameter estimation from satellite remote sensing data (Dobreva and Klein, 2011; Kuter et al., 2018; Xiao et al., 2022b, 2018). An increasing number of machine learning methods have been applied in fractional snow cover estimation in recent decades, such as artificial neural network (ANN), support vector machine, multivariate adaptive regression splines, random forest, and extremely randomized trees (Czyzowska-Wisniewski et al., 2015; Kuter, 2021; Kuter et al., 2018; Xiao et al., 2022b). Regarding specific areas, for instance, mountainous areas and forest coverage areas, some studies utilized machine learning methods to show a very promising performance in estimating fractional snow cover area from satellite reflectance images with lower uncertainties and higher improvements, as compared to traditional function-based methods (Czyzowska-Wisniewski et al., 2015; Dobreva and Klein, 2011; Xiao et al., 2022a)

12.2.2 MODIS STANDARD SNOW COVER PRODUCTS AND ACCURACY

Two MODIS sensors on board the Terra and Aqua satellites as a part of NASA's EOS were launched on December 18, 1999, and May 4, 2002, respectively, with the aim of providing global monitoring of atmospheric, land, and ocean processes. Terra overpasses the equator at around 10:30 am (10:30 pm) local time and Aqua at around 1:30 pm (1:30 am). MODIS standard snow cover products are produced as a series (see Figure 12.2), beginning with a swath (scene) product at a nominal pixel spatial resolution of 500 m with nominal swath coverage of 2330 × 2030 km. The multiple swath observations at 500 m resolution of snow cover (MOD10_L2) are then

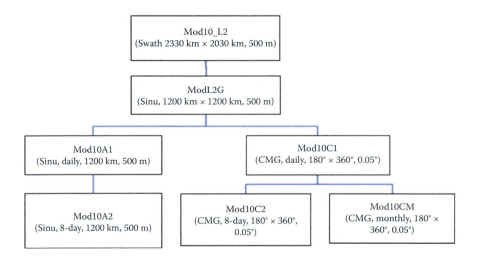

FIGURE 12.2 Terra/Aqua MODIS standard snow cover products. Aqua MODIS snow cover product's names begin with MYD (Adapted from Wang et al., 2014).

projected onto a sinusoidal gridded tile (1200 × 1200 km) of MOD10L2G, which is further processed as a sinusoidal 500 m grid of daily (MOD10A1) and eight-day (MOD10A2) composite tile (1200 × 1200 km) products, or 0.05° global Climate Modeling Grid (CMG) daily product (MOD10C1), eight-day product (MOD10C2), and monthly product (MOD10CM) (Hall et al., 2002; Riggs et al., 2006). In Collection 6 and subsequent series, the snow cover mapping algorithm of MODIS standard snow products has been updated to address and alleviate errors in snow cover determination, both in terms of commission and omission errors (Masson et al., 2018; Riggs and Hall, 2015; Sulla-Menashe and Friedl, 2018). Compared to Collection 5 and previous versions, a significant modification in this context pertains to the NDSI calculation for Collection 6 of the Aqua MODIS snow product. To enhance the scientific usability of the Aqua MODIS band 6 data for snow cover detection, the Quantitative Image Restoration algorithm (Gladkova et al., 2012) has been integrated, enabling the application of consistent snow cover algorithms for both Terra and Aqua sensors (Riggs and Hall, 2015).

MODIS standard snow cover products (both daily and eight-day) have high snow accuracy (over 90%) in clear-sky conditions (Hall et al., 1998; Hall and Riggs, 2007; Klein and Barnett, 2003; Liang et al., 2008b; Riggs et al., 2006; Wang et al., 2008; Wang and Xie, 2009; Wang et al., 2009; Zhou et al., 2005), but only 30–50% in all-sky conditions (Gao et al., 2010b; Xie et al., 2009). In spite of the replacement of Aqua band 6 with the Aqua band 7 due to the non-functional detector of band 6, both Collection 5 snow cover products (MOD10A1 and MYD10A1) have high agreement in snow cover classification for cloud free based on the validation analysis in northern Xinjiang, China (Wang et al., 2009). Based on an extensive dataset comprising 1.5 million observations obtained from more than 800 Snow Telemetry (SNOTEL) stations across the United States, a comprehensive evaluation of Collection 5 MODIS snow cover production demonstrated their high accuracy, both in binary and fractional representations. The assessment also revealed that several factors influence the accuracy of snow cover classification, with the primary factor being variation in land cover types, followed by consideration of viewing geometry factor (Coll and Li, 2018). With regard to Collection 6 MODIS NDSI snow cover data, numerous studies have undertaken similar validation analyses on regional and continental scales. These studies consistently report an improvement in the accuracy of Collection 6 snow cover products for mapping binary snow cover under diverse conditions, as compared to that of Collection 5 (Dong et al., 2014; Huang et al., 2018; Xiao et al., 2022a, 2021).

12.3 IMPROVED DAILY AND FLEXIBLE MULTIDAY COMBINATIONS OF SNOW COVER MAPPING

As mentioned, MODIS snow cover mapping suffers cloud contaminations. Obtaining cloud-free or even cloud percentage less than 10% of MODIS images remains a huge challenge. Various approaches were proposed to reduce cloud obscuration and maximize snow cover identification by improving the quality of MODIS cloud mask (Riggs and Hall, 2003), or separating cloud-masked pixels into snow or land (snow-free), through spatial-temporal or multi-sensor combinations. The solution aimed at improving the quality of the MODIS cloud mask have inherent limitations, given the perpetual and unchanging presence of clouds in the natural environment. Therefore, filling cloud pixels through spatial-temporal or multi-sensor combinations has been a hot topic in the last two decades, since cloud moves. Spatial approaches aim to replace cloud pixels with the majority of non-cloud pixels in the eight neighborhood pixels (Parajka and Blöschl, 2008; Tong et al., 2009; Zhao and Fernandes, 2009). This can effectively reduce cloud coverage of daily MODIS snow products by ~7%. But this method has limited capability to handle the areas covered with massive cloud (i.e., all neighborhood pixels are cloud-covered).

Temporal approaches merge daily or multi-day MODIS snow cover products to minimize cloud coverage and maximize snow coverage, by the sacrifice of temporal resolution. Daily combination of Terra and Aqua MODIS, whose overpass times differ by three hours, can reduce ~10–20% of daily cloud cover (Parajka and Blöschl, 2008; Wang et al., 2009; Xie et al., 2009; Yang et al., 2006). The MODIS eight-day Snow Cover Product (MOD10A2 or MYD10A2) is the representative of fixed-day combination products. Using the same algorithm, some user-defined multi-day snow cover products are produced in the fixed temporal windows (Liang et al., 2008b; Parajka and Blöschl, 2008; Yang et al., 2006; Yu et al., 2012). Based on a predefined maximum cloud coverage threshold, such as 10%, other multi-day composite products with flexible starting and ending dates were produced (Wang et al., 2009; Xie et al., 2009). These products have higher temporal resolution (average two to three days per image) and relatively low cloud coverage, but ignoring some special cases. For example, when weather conditions remain overcast for over a week, this kind of method may result in a composite product of over eight days or even several weeks. Hall et al. (2010) developed a new method to fill cloud gap based on the most recent cloud-free observations for each cloud pixel. This gap-filling strategy is a very useful and dynamic method that uses all the nearest non-cloud observations. However, one disadvantage of this kind of method is that it does not control the cloud percentage for the entire image area.

A multi-sensor combination approach can generate cloud-free binary snow cover maps by merging MODIS and AMSR-E snow products but at the expense of spatial resolution. This approach takes the advantage of both high spatial resolution of optical sensors and cloud penetration of passive microwave sensors (Foster et al., 2007; Gao et al., 2010b; Hall and Riggs, 2007; Liang et al., 2008a). The snow identification accuracy of the MODIS Terra and AMSR-E blended daily binary snow cover products is 85.6%, which is much higher than the 30.7% of MODIS daily products in all-sky conditions (Liang et al., 2008a).

While both spatial and temporal approaches effectively contribute to cloud elimination with high classification accuracy, most methods struggle to achieve complete cloud removal. Consequently, an increasing number of researchers have turned their attention to developing and implementing spatiotemporal combination approaches or multi-sensor combination approaches (Li et al., 2019). A novel spatiotemporal gap-filling technique, named as Hidden Markov random field (HMRF)–based spatiotemporal modeling, was introduced to address data gaps in MODIS daily snow cover images (Huang et al., 2018). This approach leverages spectral information, spatial and temporal contextual information, and associated environmental information. Validation experiments have demonstrated that this technique enhances snow cover mapping performance and significantly reduces data gaps from 32% to less than 1%. This modified technique has been applied in generating snow cover extent products based on AVHRR snow cover product (Hao et al., 2021) and

MODIS snow cover product (Hao et al., 2022), with applications across China further confirming the efficiency of the HMRF-based spatiotemporal modeling technique in snow cover mapping, achieving high accuracy (> 85%) (Hao et al., 2022, 2021; Huang et al., 2022b).

While numerous data gap-filling methods exist for binary snow cover products, relatively few researchers have dedicated their efforts in improving the coverage of valid snow or land (snow-free) information. Mathematical function provides a good solution for filling data gap in fractional snow cover or NDSI data. For instance, Dozier et al. (2008) employed a space-time cube method to fill gaps in fractional snow cover products, considering both temporal and spatial information and applying a series of rules to eliminate cloud pixels in MODIS snow cover data. Furthermore, based on the Terra and Aqua MODIS combination (TAC) data over the Tibetan Plateau, a spatiotemporal fusion framework, consisting of Gaussian kernel function for spatiotemporal interpolation and error correction, was applied to remove cloud pixels and fill data gaps in MODIS Collection 6 NDSI snow cover products. Extensive validation experiments across the entire region of China (Jing et al., 2019, 2022) have also validated this method's capability to fill data gaps.

Here we present three typical and improved snow cover extent products, based on daily combination of Terra and Aqua MODIS (TAC), that are currently widely used in the snow remote sensing community for various applications.

12.3.1 Flexible Multiday Combination of Snow Cover Mapping

Flexible multi-day combination approach is controlled by two thresholds, namely, maximum cloud percentage (P) and maximum composite days (N), user-defined parameters that can be assigned depending on the application (Gao et al., 2010a). In the analysis of the influence of maximum composite days, Coll and Li (2018) conducted a statistical assessment covering the entire United States. Their findings indicated that increasing the length of the combination window, represented by maximum composite days, led to the removal of a higher number of cloud pixels but resulted in a slight reduction in the overall accuracy of snow cover mapping. Previous study usually defined the P as 10% and N as eight days (Gao et al., 2010a). This approach first calculates the cloud percentage (P) of the first input image, a daily combination of Terra and Aqua MODIS (TAC). If it is less than the threshold P, the image directly outputs; otherwise, it is combined with the second input image, and the cloud percentage of the study area in the combined image is updated. The process does not stop until either the cloud percentage is less than or equal to the threshold P, or the number of composite days equals the threshold N. The resultant image from this process is a flexible multi-day combination of snow cover map.

Figure 12.3 shows an example of cloud cover decrease and snow cover increase for daily Terra-Aqua combination on October 10, 2002, and flexible multi-day combination from October 7 to 10, 2002, in the Nam Co drainage basin in the TP region. The cloud cover decreased greatly from 27.19% for the origin MODIS Terra and 21.67% for the MODIS Aqua snow cover products to 1.87% for the flexible multi-day combination snow cover mapping, while the snow cover percentage improved from ~45% to ~65%.

12.3.2 Daily Cloud-Free Snow Cover and Snow Water Equivalent Mapping

Although the cloud cover is significantly reduced by combination of daily Terra and Aqua MODIS, it still presents in the snow cover mapping even after flexible multiday combination. Gao et al. (2010b) presented a method that combines the daily AMSR-E snow water equivalent (SWE) data with the daily combined Terra and Aqua (TAC) snow cover map to produce the cloud-free snow cover extent (MAC-SC) and the SWE (MAC-SWE) maps with 500 m spatial resolution (Figure 12.4), through redistributing the 25 km grid cell of AMSR-E SWE product into pixels (500 m) covered only with snow. Eqn. (4) is used to calculate the SWE of every sub-pixel within one AMSR-E pixel (25 km).

Snow Cover Parameters and Applications 433

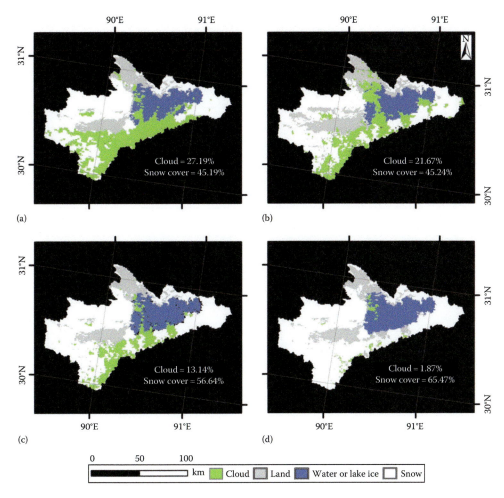

FIGURE 12.3 Cloud and snow cover changes after daily and multiday combinations of MODIS images in the Nam Co basin of the Tibetan Plateau: (A) MODIS Terra; (B) MODIS Aqua; (C) Daily Terra-Aqua combination; (D) Flexible multi-day combination of October 07–10, 2002 (Adapted from Zhang et al., 2012).

$$SWE_n = \frac{SWE_o \times 2 \times 2500}{N_{snow}} \quad (4)$$

where SWE_n is the SWE value (mm) of the new MAC-SWE product; SWE_o is the SWE value (mm) of the AMSR-E product; N_{snow} is the number of snow sub-pixels (500 m) within one AMSR-E snow pixel (25 km); 2 is the scaling factor. For example, the SWE_o value of one AMSR-E pixel is 3 mm that should be 6 mm as AMSR-E SWE values was scaled by a factor of 2 into the HDF-EOS format file archived by the National Snow and Ice Data Center (NSIDC). Within the 2,500 sub-pixels of one AMSR-E pixel, if we suppose 625 sub-pixels are snow, the SWE_n of the new MAC product should come from these 625 snow sub-pixels and therefore should be 24 mm. This means that the remaining 1875 sub-pixels have neither SC value nor SWE value.

Figure 12.5 is an example to demonstrate the outstanding capability in removing cloud obscuration of MODIS SC and enhancing spatial resolution of AMSR-E SWE products.

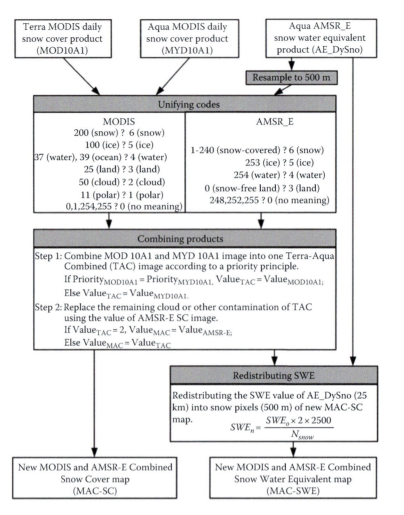

FIGURE 12.4 Flow chart of MODIS and AMSR-E combination (MAC) method for producing daily cloud-free snow cover (MAC-SC) and snow water equivalent (MAC-SWE) at 500 m cell size. (Adapted from Gao et al., 2010b).

Image B (TAC-SC) is the combination of MOD10A1 and MYD10A1 on October 15, 2006, which still presents considerable cloud pixels. Image E (MAC-SC), in which cloud cover pixels were completely removed, is the combination of the TAC-SC and the derived AMSR-E SC from SWE data. However, there are some important points to mention. (1) Scattered cloud cover areas (less than 25 km × 25 km) in Image B are removed in Image E and the resolution of snow or land (snow-free) indeed increased to a 500 m spatial resolution in those regions. (2) Massive and continuous cloud cover in Image B is simply replaced by the snow or land (snow-free) of the AMSR-E SC product. Although those cloud pixels are removed, the spatial resolution of the snow or land (snow-free) pixels remain at 25 km. For example, the cloud pixels in the upper left corner of Image B are now simply AMSR-E SC in Image E, while the actual resolution of snow or land (snow-free) pixels does not change to 500 m. (3) Ice cover (25 km resolution) in AMSR-E (Images C and D) is increased to 500 m resolution in Image E.

Based on Image E (MAC-SC), Image F (MAC-SWE) is the SWE redistribution result of Image D (AMSR-E SWE), indicating the 25 km SWE in Image D is converted to a 500 m resolution in

Snow Cover Parameters and Applications

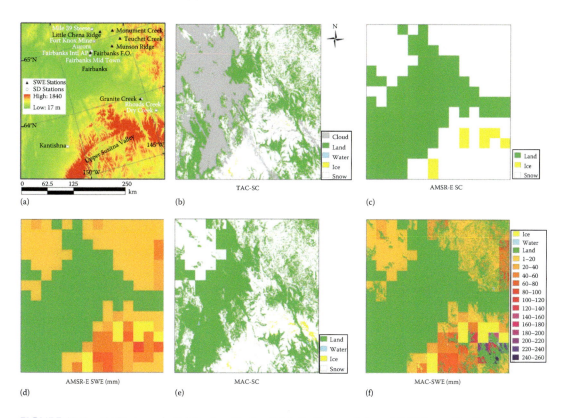

FIGURE 12.5 (A) Test area in Fairbanks, Alaska, (B) daily Terra and Aqua MODIS combination snow cover (TAC-SC), (C) AMSR-E snow cover, (D) AMSR-E snow water equivalent (SWE), (E) MODIS TAC-SC and AMSR-E combined snow cover (MAC-SC), and (F) redistributed SWE, for the October 15, 2006 (Adapted from Gao et al., 2010b).

Image F. However, this is true only for those pixels that have a spatial resolution of 500 m in Image E. For example, for the pixels in the upper left corner of Image E still keep a 25 km spatial resolution in Image F. It is very clear that the SWE values in the upper right have a true 500 m spatial resolution due to the true 500 m SC of Image E, while most portions of the upper left of Image F has the false 500 m SWE due to the false 500 m SC in Image E.

The evaluation against *in situ* snow observations (Gao et al., 2010b) showed that overall accuracy in snow cover identification in all sky conditions increase from 31% (MYD10A1) and 45% (MOD10A1) to 49% (TAC-SC) and 86% (MAC-SC). A similar combination experiment between Terra MODIS and AMSR-E snow cover product was tested in a previous study (Liang et al., 2008a) in northern Xinjiang, China, against 20 *in situ* snow observation stations. This assessment reported the overall accuracy of the combined snow cover product of Terra MODIS (no Aqua MODIS involved) and AMSR-E snow cover products is 75%, which is much higher than the 34% of MOD10A1 alone. However, this accuracy is lower than 86% found in Gao et al. (2010b), partly due to the consideration of the Aqua MODIS in the latter study to remove cloud pixels. The monthly comparison between MODIS and AMSR-E snow cover products indicated that good correspondences were found in the land (snow-free) or snow stable period. However, the AMSR-E SWE maps were not as accurate as the MODIS snow cover maps during the snow accumulating and melting period (the early fall and the later spring) due to the coarse resolution of AMSR-E snow cover product. It was also found that for pixels with discrete cloud (less than 25 km) in the MODIS TAC-SC, the new daily 500 m MAC SC and SWE products indeed improve the spatial resolution of those pixels to 500 m. For massive cloud cover (larger than 25 km), the real resolution of those pixels in the MAC products kept 25 km, even in 500 m pixel scale. Despite these limitations, the resultant new daily and cloud-free MAC SC and SWE maps are great supplements to NASA's current standard MODIS and AMSR-E snow products and are suitable for hydrological and meteorological modeling on a daily basis.

12.3.3 Improved Daily Cloud-Free Snow Cover Mapping

As indicated in Section 12.3.2, directly replacing the massive clouds with the snow or land (snow-free) information of AMSR-E snow cover products does remove those cloud pixels in the daily Terra and Aqua combination image. But this procedure does not really enhance the spatial resolution since patchy snow within a 25 km AMSR-E pixel was determined as land (snow-free) due to its coarse pixel resolution. Wang et al. (2015) developed an improved algorithm that is capable to greatly fill cloud pixels in MODIS snow products, before combining with AMSR-E. The algorithm is shown in Figure 12.6, through four steps: (1) combining MOD10A1 and MYD10A1 to obtain the combined snow image (MOYD) (Xie et al., 2009); (2) performing adjacent temporal reduction (ATR) using three days' images (previous, current, and next day); (3) applying SNOWL method to further separate the remaining cloud into snow or cloud based on elevation and snow information (Parajka et al., 2010); and (4) replacing the remaining cloud pixel with the snow or land (snow-free) information of AMSR-E SWE data, i.e., the final combination of MODIS and AMSE-E snow data

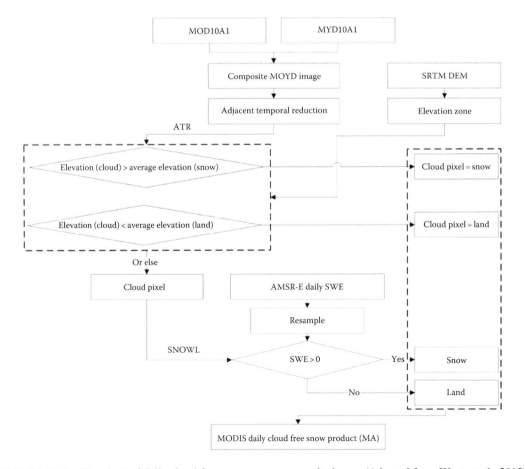

FIGURE 12.6 Flowchart of daily cloud-free snow cover composite image (Adapted from Wang et al., 2015).

TABLE 12.1
Snow classification accuracy and errors for SCA images from 2003 to 2010 (Adapted from Wang et al., 2014a)

							Clear sky		All sky	
SCA image	S-S	S-L	S-C	L-L	L-S	L-C	Snow accuracy (%)	Overall accuracy (%)	Snow accuracy (%)	Overall accuracy (%)
MOD10A1	1936	448	3855	139735	797	102290	81.21	99.13	31.03	56.88
MYD10A1	1452	613	4326	122914	1486	114580	70.31	98.34	22.72	50.68
MOYD	2552	546	2733	179275	2372	63011	82.38	98.42	43.77	72.59
ATR	3710	663	2183	200816	2461	41204	84.84	98.50	56.59	81.47
SNOWL	4670	781	1736	215291	2558	26068	85.67	98.50	64.98	87.60
MA	4979	875	0	245364	3199	0	85.05	98.40	85.05	98.40

Note: S, L, and C are respectively for snow, land, and cloud; the first letter is for climate station and the second letter for satellite. For example, S-S means that 'snow' seen from climate stations is also seen as 'snow' from satellite; L-S means that 'land' seen from climate stations is seen as snow from satellite.

(MA). The final cloud-free products achieved high snow mapping and overall accuracies (~85% and ~98%, respectively), which is much higher than those of existing daily snow cover products in all sky conditions, and are very closer to or even slightly higher than those in clear sky conditions of the daily MODIS products based on the validation results over the Tibetan Plateau (Table 12.1) (Wang et al., 2015).

12.4 SNOW COVER PHENOLOGY PARAMETERS

Based on flexible multiday combination of snow cover or daily cloud-free snow cover maps, four snow cover phenology parameters, i.e., snow cover index (SCI), snow cover duration (SCD), snow cover onset dates (SCOD), and snow cover end dates (SCED), in each hydrological year can be further derived to examine the spatiotemporal variations of snow cover (Gao et al., 2011; Wang and Xie, 2009). The SCI (km^2·day) contains both snow cover extent and duration for one hydrological year and can be used to study annual snow cover condition. To intercompare snow cover condition for multiple basins, a normalized SCI (NSCI, unit km·day/ km^2) can be used (Zhang et al., 2012). The NSCI is defined as the SCI divided by the corresponding basin's area and indicates the average snow cover days in each basin. Figure 12.7 shows an example of SCOD, SCED, and SCD maps for the US Pacific Northwestern for the 2006–2008 hydrological years (Gao et al., 2011). Three SCD maps have very similar patterns even though their corresponding SCOD and SCED maps change a lot from year to year or from area to area. When compared SCD maps with elevation distribution (Figure 12.7D), it is clear that SCD variation has high correlation with elevation. SCD less than 90 days (greenish) is in the area where elevation is lower than 1000 m. In the middle 1000–2000 m areas, most of SCDs are in the range of 90–150 days (yellowish). At the high elevation areas, SCDs are larger than 150 days, and even larger than 240 days in mountaintop areas (reddish). At a few mountainous peak areas above 3000 m, SCD even exceeded 270 days (deep red) especially in 2008.

Figure 12.8 shows an example of NSCI variations for the four basins (Nam Co, Selin Co, CedoCaka, and YamzhogYumco) in the TP region (Zhang et al., 2012). Generally, the Nam Co basin shows the highest NSCI (i.e., the greatest snow cover condition) among the four basins in all years. The HY2003 shows the highest NSCI for the Nam Co basin. The smallest NSCI for the

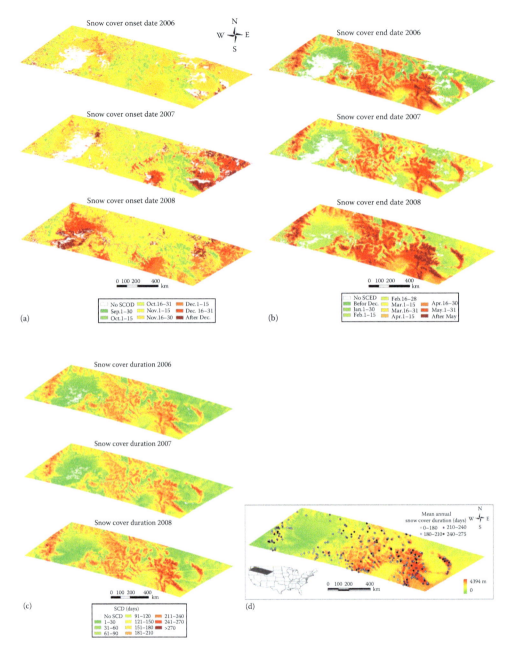

FIGURE 12.7 Flexible multiple combination of MODIS snow cover derived (A) snow cover onset date (SCOD), (B) snow cover end date (SCED), and (C) snow cover duration (SCD) of (D) the US Pacific Northwest for the 2006–2008 hydrological years. Dots on the DEM map (D) are 244 SNOTEL stations with the mean annual SCDs in four different color groups (Adapted from Gao et al., 2011).

Nam Co basin occurs in HY2010. For CedoCaka, Selin Co, and YamzhogYumco, the largest NSCI was in HY2007. HY2001 and HY2010 show similar small NSCI for CedoCaka, Selin Co, and YamzhogYumco basins, while the smallest NSCI for the CedoCaka basin was in HY2004.

Over the past decade, an increasing number of snow cover datasets have been used for the extraction of snow cover phenology parameters, which have emerged as a new and effective

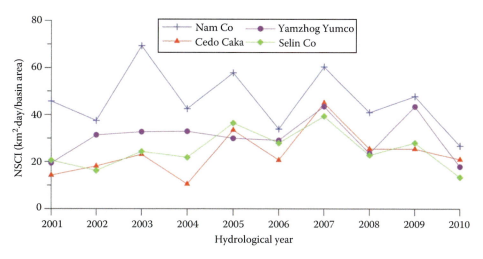

FIGURE 12.8 Normalized snow cover index (NSCI) for the basins CedoCaka, Selin Co, Nam Co and YamzhogYumco in the Tibetan Plateau during HY2001–2010 (Adapted from Zhang et al., 2012).

approach for characterizing the variation in snow cover and understanding its impact on climate. For example, Northern Hemisphere snow depth data, because of its freedom from no cloud obscuration and independent from sunlight, has been proposed to calculate parameters such as SCOD, SCED, snow season length (SSL), and SCD over the period from 1988 to 2018 (Yue et al., 2022). Furthermore, a gapless snow cover extent product, created by combining MODIS Collection 6 NDSI products, AMSR-E SWE products, and data from the Interactive Multisensor Snow and Ice Mapping System (IMS), have been employed to calculate four snow cover phenology parameters—SCAI, SCOD, SCED, and SCD, for exploring spatiotemporal variations in snow cover across the Eurasia region from 2000 to 2016 (Sun et al., 2020). Similar methodologies, although with different data combination strategies, was also applied in other studies (Wang et al., 2018; Xu et al., 2023; Zhao et al., 2022). The results of these analyses reveal significant spatial heterogeneity in the variations of these four snow cover phenology parameters. Generally, most regions exhibited delayed SCOD, an advance in SCED, and a shortening of SCD (SSL). However, regional trends are relying on temperature, elevations, and other climatic factors.

To elucidate the factors influencing the variations of snow cover phenology, many researchers have investigated the contributions of meteorological factors, geographical location, topography, vegetation greenness, and atmospheric pollution factors (Peng et al., 2013; Xu et al., 2023; Zhao et al., 2022). Conversely, the profound changes in snow cover inevitably impact vegetation growth, necessitating further research to bridge this knowledge gap (Wang et al., 2018; Xu et al., 2022).

12.5 SNOW COVER AS A WATER RESOURCE FOR LAKE-LEVEL AND WATERSHED ANALYSES

Snow plays an important role in the energy and water balance of drainage basins in alpine regions. Contribution of snow melt to runoff is one of the most important water resources in mountainous regions in addition to rainfall and melting glaciers. With the improved snow cover mapping based on the MODIS daily snow cover products, we can now examine the time series snow cover change as a water resource change for basin and watershed managements. Here we present two application examples: (1) snow cover as a water resource for lake-level change and (2) snowmelt and runoff for watershed analysis.

12.5.1 Snow Cover over Lake Basin for Lake-Level Analysis

Snow over the Tibetan Plateau (TP) greatly influences the water availability of several major Asian rivers, such as the Yellow, Yangtze, Indus, Ganges, Brahmaputra, Irrawaddy, Salween, and Mekong rivers (Barnett et al., 2005; Immerzeel et al., 2009). Discharge from these rivers sustains the lives of more than 1 billion people living both in the region and downstream (Barnett et al., 2005). The TP has undergone warming in the past three decades (Liu and Chen, 2000). The glaciers over the TP showed accelerated retreating because of the warmer climate, except for in the Karakoram (Bolch et al., 2010; 2012; Ding et al., 2006; Gardelle et al., 2012; Yao et al., 2007, 2010, 2012). Lake's water level studies using NASA's ICESat laser altimetry data have shown that over 70% of the lakes in the TP saw lake-level increase in the past years (2003–2009) (Phan et al., 2012; Song et al., 2013; Zhang et al., 2011a, 2013). Did snow cover change play a role in the lake-level increase?

Based on the flexible multiday combination of MODIS snow cover products, Zhang et al. (2012) examined four lake basins over the Tibetan Plateau for snow cover dynamic change (Figure 12.9). The relation between lake-level change and combined snow cover, precipitation, and pan evaporation during hydrological years 2001–2010 (September through August) were examined (Figure 12.10).

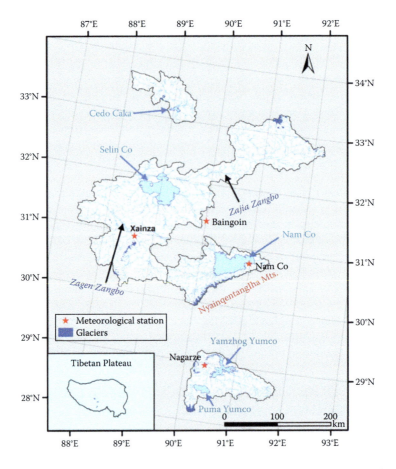

FIGURE 12.9 Location of lakes CedoCaka, Selin Co, Nam Co, and YamzhogYumco (also in the inset map) and their corresponding drainage basins over the Tibetan Plateau. Streams and boundaries of lake basins are delineated from SRTM DEM and glacier coverage from GLIMS at www.glims.org. Four available meteorological stations are also denoted, with no station nearby the Cedo Cake basin (Adapted from Zhang et al., 2012).

Snow Cover Parameters and Applications

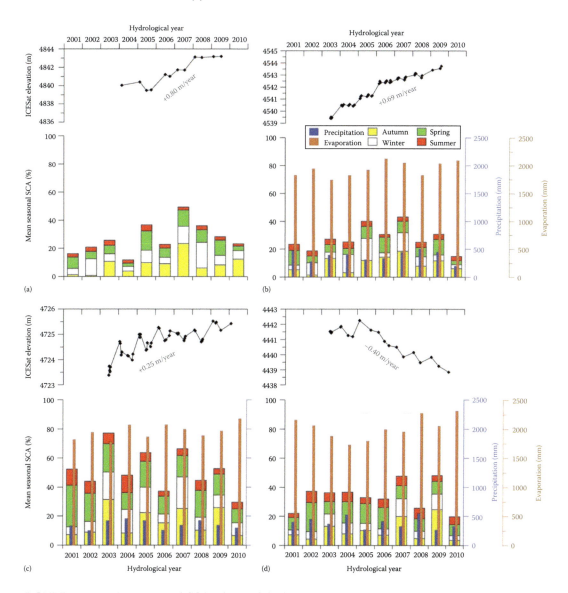

FIGURE 12.10 Mean seasonal SCA (%), precipitation, and pan evaporation (observed at nearby stations) in the CedoCaka basin (A), Selin Co basin (B), Nam Co basin (C), and YamzhogYumco basin (D) during HY2001–2010, lake-level changes derived from ICESat data for each basin during HY2003–2009 are also included (Adapted from Zhang et al., 2012).

The lake CedoCaka had the largest lake-level increase (+0.80 m/year) among all examined lakes in Zhang et al. (2011a). The lake Selin Co had the second largest lake-level increase (+0.69 m/year). The lake YamzhogYumco, however, had the fastest drop in water level (−0.40 m/year) among all examined lakes in Zhang et al. (2011a).

As shown in Figures 12.9 and 12.10, except the CedoCaka lake basin, all three lake basins have nearby weather stations. However, for the CedoCaka lake, the correlation coefficient between SCA in the basin and lake-level change is 0.94, indicating that snow cover variation played an important role in the lake-level increase. This means that snow cover alone can explain 88% of lake-level change, although it is not significant at the 95% confidence level (p = 0.063). This means other parameters such as precipitation, evaporation, and glacier melt also played a certain role for the lake-level change. For the Selin Co lake, the precipitation, evaporation, and snow cover together can explain 98% of lake-level change, and it is significant at the 95% (p = 0.029). This means glacier melts in the basin had almost no impact to the lake-level change. For the Namco lake, the combined precipitation, evaporation, and snow cover can explain 76% of lake-level change, although it is statistically insignificant at the 95% confidence level (p > 0.05). This suggests the glacier melts from the Nyainqentanglha Mts. also played additional role for the lake-level increase. For the YamzhogYumco lake, the combined precipitation, evaporation, and snow cover can explain 76% of the lake-level decrease, although it is statistically insignificant at the 95% confidence level (p > 0.05). It is noticed that the operation of a power station nearby the YamzhogYumco lake would have played an additional role for the lake's water decrease (Zhang et al., 2012).

12.5.2 Quantitative Water Resource Assessment Using Snowmelt Runoff Model

Snow cover, in particular seasonal snowpack in mountainous areas, plays an important role in maintaining a global water balance (Goodison et al., 1999; Immerzeel et al., 2010). The monitoring of snowmelt runoff in the snow- and/or glacier-fed basins is important for efficient flood forecasting and water resources management practices, such as effective irrigation, controlled reservoir levels, and the generation of hydropower.

Several models have been developed to simulate snowmelt discharge, such as the degree-day Snowmelt Runoff Model (SRM) (Martinec, 1975), and energy balance models (Blöschl et al., 1991a, 1991b). However, physically based distributed models are often impractical to use in data-sparse mountainous regions such as the TP. Owing to the requirement of only minimal temperature, precipitation, and snow cover data as inputs, and only nominal calibration procedures, the SRM appears to be the most ideal model for use in data-sparse regions.

Zhang et al. (2014) did a SRM simulation on Lake Qinghai (Figure 12.11) and compared the simulation performance of two different snow cover products: MODIS standard eight-day snow

FIGURE 12.11 Lake Qinghai basin and Buha watershed delineated from SRTM DEM, including glacier coverage within this basin and beyond. The locations of meteorological stations Tianjun and Gangcha, hydrologic station Buha are denoted. The map inset shows the outline of Lake Qinghai and the large rivers on the Tibetan Plateau (TP), and the direction of the Westerlies and Asian summer monsoon (ASM) that influence the lake (Adapted from Zhang et al., 2014).

cover product (MOD10A2) and the flexible multiday combined MODIS snow cover product (MODISMC). Lake Qinghai is situated near the northeastern margin of the TP. The water level of this lake decreased overall by ~3.7 m from 1959 to 2004, but then increased by ~1 m between 2004 and 2009 (Zhang et al., 2011b). There is a long record of gauge runoff observations from the Buha watershed (the largest catchment area within the Lake Qinghai basin) in comparison with the other lake basins within the TP.

12.5.2.1 Datasets and Methodology

Besides the two MODIS snow cover datasets: MOD10A2 and MODISMC, for the period of HY2003–2009, daily observed data were also used, including temperature and precipitation at the Tianjun meteorological station, and streamflow of the Buha watershed at the Buha hydrologic station, and pan evaporation (E_{pan}) at the Gangcha meteorological station (the closest station to Lake Qinghai). The recently released APHRODITE APHRO_MA_V1101 product (available at www.chikyu.ac.jp/precip/) for the period HY2003–2007 was also used. The APHRODITE-precipitation product is created by the Asian Precipitation-Highly Resolved Observational Data Integration Towards the Evaluation of Water Resources (APHRODITE) project (Yatagai et al., 2012).

To derive the depletion curves of the SRM, short-lived snow cover in summer should not be used (Martinec et al., 2008) and were replaced from a linear interpolation based on adjacent days.

The daily simulated streamflows, as well as the discriminated runoff from snowmelt and rainfall in the Buha watershed, were calculated according to Eqn. (5):

$$Q_{n+1} = \underbrace{C_{sn} \cdot a_n (T_n + \Delta T_n) S_n \cdot A \cdot 0.116 (1 - k_{n+1})}_{snowmelt\ runoff}$$
$$+ \underbrace{C_{rn} P_n . A \cdot 0.116 (1 - k_{n+1})}_{rain\ runoff} \qquad (5)$$
$$+ \underbrace{(Qs_n + Qr_n) k_{n+1}}_{runoff\ contribution\ from\ the\ previous\ day}$$

where Q is the mean daily discharge (m³/s); C is the snow (C_s) or rain (C_r) runoff coefficient; a, the degree day factor (DDF) (cm °C⁻¹d⁻¹); $T + \Delta T$ are the degree-days (°C d); S is the ratio of the snow covered area to the total area; P is the measured precipitation on that day (cm); A is the area of basin or elevation zone (km²); k is the recession coefficient; and n is the sequence of days during the simulation period. The factor of 0.116 converts data from cm km²/day to m³/s.

The accuracy of the SRM was evaluated using the Nash–Sutcliffe determination coefficient (R^2) and the volume difference (D_v). The model was run using both basin-wide simulations and zone-wise simulations for the Buha watershed during the period HY2003–2009. The parameters were calibrated with data from HY2003–2005 and validated with data from HY2006–2007.

12.5.2.2 Results

The runoff simulations were conducted using measured precipitation and snow products for basin-wide applications: MODISMC-MB and MOD10A2-MB, and also with APHRODITE-precipitation and snow products: MODISMC-AB and MOD10A2-AB. The average Nash–Sutcliffe (NS) determination coefficients are very similar between 0.74 and 0.75 for MODISMC-MB (Figure 12.12A), MOD10A2-MB (Figure 12.12B), MODISMC-AB (Figure 12.12C), and MOD10A2-AB (Figure 12.12D). The mean absolute value of volume difference (D_v%), however, is lower when MODISMC is used vs. MODIS10A2, i.e., 10.12% vs. 16.55% (measured precipitation used), or 12.20% vs. 19.65% (APHRODITE-precipitation used). These results indicate that MODISMC performs better in basin-wide simulation than MOD10A2. For all the different precipitation and snow cover used for basin-wide simulations, the correlation of determination is ~0.76 (Figure 12.12A–D).

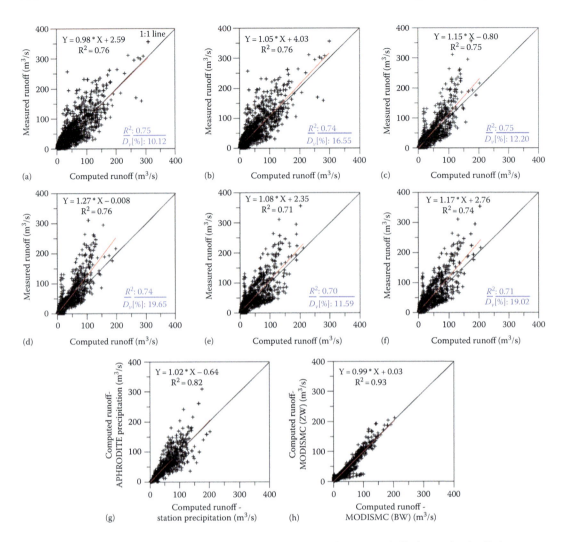

FIGURE 12.12 Relationship between measured discharge and computed discharge in the Buha watershed from HY2003 to HY2009: (A) MODISMC-MB (Measured precipitation for basin-wide simulation); (B) MOD10A2-MB; (C) MODISMC-AB (APHRODITE-precipitation for basin-wide simulation); (D) MOD10A2-AB; (E) MODISMC-AZ (APHRODITE-precipitation for zone-wise simulation); (F) MOD10A2-AZ. Two computed discharges: (G) MODISMC-AB and MODISMC-MB, and (H) MODISMC-AB and MODISMC-AZ. The blue number in the figures is the mean Nush–Sutcliffe coefficient of determination (R^2) (range), and the mean absolute value of volume difference (D_v) between the simulated and measured runoff (Adapted from Zhang et al., 2014).

Zone-wise runoff simulations were further conducted in the Buha watershed. The mean NS coefficient and absolute value of volume difference were 0.70 and 11.59%, respectively, for MODISMC-AZ data (Figure 12.12E). For the MOD10A2-AZ simulation, the mean NS coefficient and absolute value are 0.71 and 19.02%, respectively (Figure 12.12F). The results show that simulations with MODISMC data achieved lower volume differences than MOD10A2 data, although with smaller NS values. Figure 12.12G shows that there is a high correlation between computed runoff using measured precipitation and runoff using APHRODITE-precipitation for basin-wide simulations with the MODISMC snow product during the periods HY2003–2007 ($R^2 = 0.82$). In addition, runoff modeling from basin-wide applications has a high correlation with runoff from zone-wise simulations ($R^2 = 0.93$) (Figure 12.12H).

Figure 12.13 shows the daily runoff contribution fraction from snowmelt and rainfall in the Buha watershed. The average annual estimated runoff from snowmelt is 13% (range 10–17%) of the total runoff, which indicates that rainfall is a major source of discharge to this lake.

A quantitative assessment (Table 12.2) showed that the lake-level changes are highly correlated with single variables: R_m ($R^2 = 0.95$), R_s ($R^2 = 0.77$), R_s-rainfall ($R^2 = 0.75$), and SCA (zone B of Buha watershed, $R^2 = 0.68$) at the 95% confidence level, and multivariates: R_s–rain + P_L ($R^2 = 0.91$)

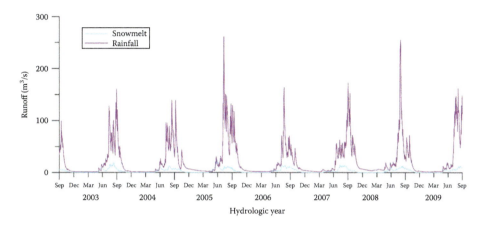

FIGURE 12.13 Daily runoff contribution fraction from snowmelt and rainfall from HY2003–2009 using MODISMC-MB (Adapted from Zhang et al., 2014).

TABLE 12.2

The coefficient of determination (R^2) between annual differences of in situ lake-level changes, measured runoff (R_m), and simulated runoff (R_s), for MODISMC-MB simulation in the Buha watershed, precipitation changes over the lake surface (P_L), and pan evaporation (E_{pan}) changes measured at the Gangcha station, and SCA changes in the Buha watershed (SCA$_B$ and Zones A, B, C, D) and Qinghai Lake basin (SCA$_Q$) from HY2003 to HY2009 (Adapted from Zhang et al., 2014).

Variable	Lake level change	
R_m	0.95	0.001
R_s	0.77	0.02
R_s-rain	0.75	0.03
P_L	0.64	0.06
R_s-rain + P_L	0.91	0.03
R_s-snow + P_L	0.64	0.22
$R_m + P_L + E_{pan}$	0.98	0.02
$R_s + P_L + E_{pan}$	0.95	0.08
R_s-rain + P_L + E_{pan}	0.95	0.08
SCA$_B$ (Zone A, B, C, D)	0.60 (0.48, 0.68, 0.59, 0.38)	0.07 (0.13, 0.04, 0.07, 0.19)
SCA$_Q$	0.60	0.07
SCA$_B$ + P_L + E_{pan}	0.94	0.09
SCA$_Q$ + P_L + E_{pan}	0.85	0.22

and $R_m + P_L + E_{pan}$ ($R^2 = 0.98$) at the 95% confidence level. These results suggest that rainfall-derived runoff and precipitation over the lake's surface are major contributors to lake-level changes in comparison to that of snowmelt-derived runoff. This differs from the four lake basins at inner TP examined that snow cover played important role for lake-level change (Section 12.5.1).

12.5.2.3 Summary

Accurate snowmelt streamflow simulations and forecasts are critical for solving water resources management issues pertaining to irrigation, recreation, flood control, and hydroelectric power generation. The flexible multiday combination of MODIS snow cover (MODISMC) product performs better than the standard eight-day product (MOD10A2), with a slightly higher Nush–Sutcliffe coefficient for basin-wide streamflow simulation, and overall smaller volume differences for basin- and zone-wide simulations. The two to three days combination seems to give better snow cover product and then better runoff simulation. This study demonstrated that the SRM is a suitable hydrologic model for use in estimating snow melt runoff in the TP and that precipitation in the Qinghai Lake basin plays a dominant role in lake-level variations.

12.6 SCALING EFFECT IN SNOW COVER MAPPING

The scaling effects hold great significance in the field of snow cover remote sensing (Armstrong and Brodzik, 2001). Snow cover in the Tibetan Plateau exhibits pronounced spatial heterogeneity, with its distribution being fragmented and shallow. However, remotely sensed snow depth products have a coarse spatial resolution of ten to tens of kilometers. In general, the true observation value of snow depth is typically obtained on a point scale. There are large errors and uncertainties in constructing and validating remotely sensed snow depth inversion models using field-based snow depth measurement data because the spatial scale of ground observation points and remote sensing pixels does not match (Justice et al., 2000). Additionally, the coarse spatial resolution of snow depth products further leads to the presence of mixed pixels, seriously impeding the capture of detailed information about snow depth changes in mountainous areas. Spatial scale conversion is currently considered to be an important method that can overcome the scale effect, encompassing upscaling and downscaling. Next, the two scale conversion methods are introduced through examples.

12.6.1 Upscaling

Input data for upscaling methods generally include topographic factors that affect the spatial distribution of snow depth, such as slope, aspect, and elevation (Kour et al., 2016). Point-to-surface upscaling methods encompass various algorithms, including simple averaging (SA), simple kriging (SK), ordinary kriging (OK), Bayesian maximum entropy (BME), and random forests (RF) (Wu et al., 2019). Hou et al. (2022) analyzed the application of these above methods in snow depth upscaling fields. Their results demonstrated that snow depth has consistent spatial distribution characteristics among the OK, SK, and BME methods. The upscaling results were primarily influenced by the spatial distribution of observed snow depth points. The region with the maximum and minimum snow depth were both closely related to the corresponding observed maximum and minimum values of observed snow depth. The upscaling snow depth showed obvious "island" characteristics (Figure 12.14). Because the covariates, such as elevation, slope, and aspect were used in the RF upscaling algorithm, the spatial distribution characteristics of the observed snow depth have less influence on upscaling snow depth, and some snow depth variation characteristics can be exhibited for regions lacking observation points.

Table 12.3 provides a summary of the root mean square error (RMSE) and mean absolute error (MAE) for different upscaling methods. The RMSE statistics reveal that the observation scale significantly impacts the upscaling results. The upscaling results of the 500-m snow sample are significantly better than that of the 25-km snow sample. All the upscaling methods consistently exhibit a decrease

Snow Cover Parameters and Applications

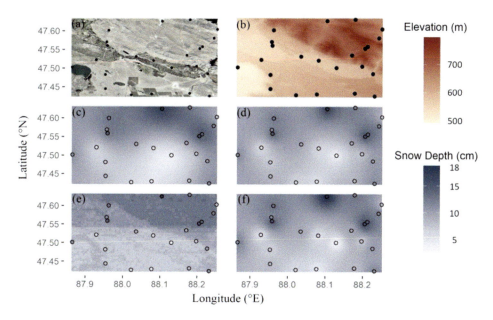

FIGURE 12.14 Upscaling results of (c) OK, (d) SK, (e) RF, and (f) BME for a 25 km² plot; the satellite image and DEM for the snow plot are shown in (a) and (b), respectively. Note: OK: ordinary kriging, SK: simple kriging, RF: random forests, and BME: Bayesian maximum entropy (Adapted from Hou et al., 2022).

TABLE 12.3
The RMSE and MAE for upscaling results at different snow samples (unit: cm). Note: SA: simple averaging, OK: ordinary Kriging, SK: simple Kriging, BME: Bayesian maximum entropy, and RF: random forests. (Adapted from Hou et al., 2022)

Index	Snow samples	SA	OK	SK	BME	RF
RMSE	25 km	4.10	3.48	3.42	3.48	2.01
	500 m	2.98	2.95	2.68	2.41	2.11
	Average	3.54	3.22	3.05	2.95	2.08
MAE	25 km	3.08	1.50	1.87	0.32	2.01
	500 m	2.25	1.81	2.13	0.97	2.11
	Average	2.67	1.66	2.00	0.65	2.08

in RMSE with smaller sample size. The MAE statistics demonstrate that BME yields the smallest errors. The MAE values are 0.32 cm and 0.97 cm for sample sizes of 25 km and 500 m, respectively, with an average value of 0.65 cm. In contrast, the SA exhibits the largest MAE, with a mean value of 2.67 cm. The OK demonstrates superior upscaling performance compared to the SK, while the RF shows slightly worse MAE performance than the OK and the SK. In addition, the observation scale has minimal impact on the MAE. The SA exhibits the largest error, with RMSE of 3.54 cm.

Figure 12.15 depicts the validation analysis of the upscaling results utilizing the cross-validation method. In the 500-m snow sample, the overall disparity among the five upscaling methods is minimal. The BME exhibits the highest overall performance, with a median RMSE of 1.91 cm, followed by the SK, OK, SA, and RF. The best stability performance is also the BME, with RMSE of only 1.87

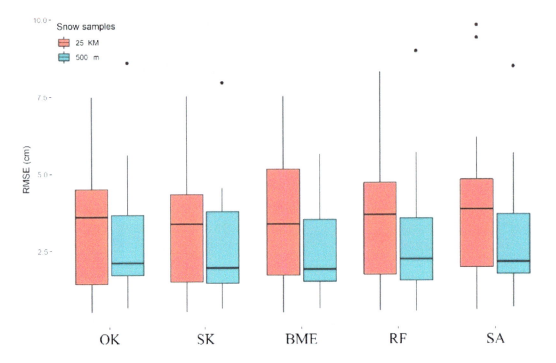

FIGURE 12.15 Boxplot of RMSE for two types of snow samples in upscaling models. Note: OK: ordinary kriging, SK: simple kriging, BME: Bayesian maximum entropy, RF: random forests, and SA: simple averaging (Adapted from Hou et al., 2022).

cm and then following the SA, OK, RF, and SK. These findings affirm that the BME stands as the optimal upscaling approach for the 500-m snow sample. Conversely, in the 25-km snow sample, SK achieves the smallest median RMSE (3.4 cm) and optimal stability (2.83 cm). It indicates that the SK method works best for the 25-km snow sample. The observation scale significantly impacts both the overall level and stability of the upscaling results. The results of this study reveal that the upscaling results for the 500-m sample size are better than those of the 25-km sample size. This disparity arises due to the smaller range and lower spatial heterogeneity present in the 500-m sample size, where the observation points exhibit greater representativeness during the upscaling process.

Numerous factors, including the upscaling methods, topography, and the spatial distribution characteristics of the snow sample, have a more pronounced effect on the snow depth upscaling effect. Different upscaling methods perform best under specific conditions. The SA is suitable for areas with low spatial heterogeneity and a well-distributed network of observation sites. This method provides representative upscaling results for relatively homogeneous surfaces with minimal vegetation and larger areas. However, the SA method tends to produce larger errors in the complex topography areas. For the mountainous areas with a wide range of observations, large variations in snow depth and complex terrain, BME shows better performance in upscaling. This is because the BME is able to kernel-smooth the data during the upscaling process, eliminating the effect of sampling data bias on the upscaling results. Additionally, increasing the number of observation points within the sample can effectively improve the accuracy of the upscaling results.

12.6.2 Downscaling

Downscaling methods is a crucial approach to address the issue of image element mixing caused by coarse resolution. These methods can integrate the strengths of passive microwave remote sensing snow depth products, optical remote sensing snow cover products, and other environmental factors (such as topographic and geographic factors) that influence the spatial distribution of snow cover (e.g., fractional snow cover, snow depth, and snow water equivalent) (Wang et al., 2019; Wei et al., 2021; Rittger et al., 2021). There is a strong correlation between the snow-covered days (SCD) and the spatial distribution of snow water equivalent or snow depth (Mhawej et al., 2014). For alpine regions where snow depth observation sites are scarce, the probabilistic modeling of the SCD spatial distribution is an important downscaling method to generate high-resolution snow depth products. A spatial–temporal downscaling method was developed to produce a daily 0.05° snow depth product from 25 km daily snow depth data for the Tibetan Plateau by taking advantage of snow cover distribution information (i.e., SCD) from MODIS -day cloud-free (500 m) snow cover products (Yan et al., 2022). The multifactor regression model is also a very effective downscaling method for snow depth (Pulwicki et al., 2018; Wang et al., 2019). However, it is crucial to note that too many input variables may result in high correlation among them, significantly impacting the model and its accuracy (Pulwicki et al., 2018). Machine learning algorithms, such as RF (Meloche et al., 2022), artificial neural networks (Tedesco et al., 2004), support vector machines (Liang et al., 2015), and deep learning (Zhu et al., 2021), have also been employed to generate the downscaling snow depth or snow water equivalent products. While these algorithms demonstrate favorable model validation performance, there remains challenges in unraveling the underlying physical mechanisms and resolving model uncertainties (Meloche et al., 2022).

Eqn. 6 presents the downscaling algorithm employed to generate daily snow depth data at a 500 m spatial resolution ($SD_{500\,m}$) for the period 1980–2020. Initially, the snow cover dataset is used to determine whether the surface is covered by snow. In other words, if the snow cover extent (SCE) is equal to 0, the $SD_{500\,m}$ is set as 0. For areas where the SCE value is non-0, two algorithms are employed to obtain $SD_{500\,m}$. In the areas below 4000 m, multiple linear regression is utilized, while $SD_{500\,m}$ in areas above 4000 m are determined based on the spatial distribution probability of the SCD.

$$\begin{cases} if & SCE=0, \ SD_{500\,m}=0 \\ else & \begin{cases} if & elevation \leq 4000, SD_{500\,m}=0.0659*SD_{25\,km}+0.0077*SCD \\ & \qquad +0.0168*longitude-0.0453*latitude+0.1199 \\ else & SD_{500\,m}=\dfrac{SD_{25\,km}*SCD*2500}{SDT} \end{cases} \end{cases} \quad (6)$$

where $SD_{500\,m}$ is the downscaling snow depth to be produced, $SD_{25\,km}$ is the original passive microwave snow depth data product, SCD is the snow-covered days in the Julian year, SDT is the sum of the SCD within one 25 km resolution pixel, and the number 2500 is the number of pixels for 500 m resolution within a 25 km resolution pixel.

The accuracy of the $SD_{500\,m}$ and the $SD_{25\,km}$ are evaluated using ground-based snow depth data (Table 12.4). The $SD_{500\,m}$ exhibits higher accuracy compared to the $SD_{25\,km}$, as evidenced by lower values of RMSE, MAE, positive mean error (PME), and negative mean error (NME). The PME and NME are employed to evaluate whether the $SD_{25\,km}$ or the $SD_{500\,m}$ overestimated or underestimated snow depth against the ground-observed snow depth, respectively (Wang et al., 2019; Liu et al., 2020). The result showed that this downscaling model effectively reduces the overestimation (33.4%) snow depth in the Tibetan Plateau against the $SD_{25\,km}$.

This downscaling model uses different snow cover data products before and after 2002, with the AVHRR snow cover dataset at 5 km spatial resolution for 1980–2002 (Hao et al., 2021) and the MODIS snow cover dataset at 500 m spatial resolution for 2003–2020 (Qiu et al., 2021). In addition, various downscaling algorithms were employed in the area above and below 4000 m. Figure 12.16 presents the results of the accuracy evaluation of the $SD_{500\,m}$ compared to the $SD_{25\,km}$ against the

TABLE 12.4

The accuracy evaluation of the SD$_{500\,m}$ and the SD$_{25\,km}$ against the ground-observed snow depth (unit: cm/d).

Condition	Data Type	Total	Year 1980–2002	Year 2003–2017	Elevation ≤4000 m	Elevation >4000 m
RMSE	SD$_{500\,m}$	**2.75**	3.26	1.98	2.47	3.53
	SD$_{25\,km}$	4.16	4.79	3.26	4.26	3.77
MAE	SD$_{500\,m}$	**0.54**	0.74	0.29	0.41	0.97
	SD$_{25\,km}$	1.60	1.96	1.18	1.60	1.61
PME	SD$_{500\,m}$	**3.31**	3.42	1.64	1.23	7.77
	SD$_{25\,km}$	4.97	5.52	4.18	5.02	4.78
NME	SD$_{500\,m}$	**−3.44**	−3.43	−3.48	−3.60	−3.02
	SD$_{25\,km}$	−3.93	−3.99	−3.80	−4.04	−3.67

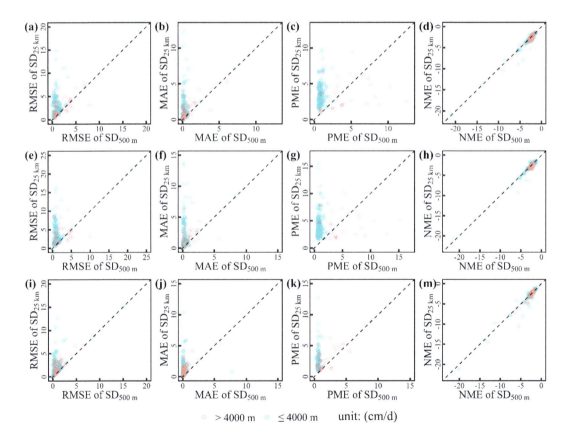

FIGURE 12.16 The accuracy evaluation of the SD$_{500\,m}$ and SD$_{25\,km}$ against the ground-observed snow depth in the period of 1980–2017 (a–d), 1980–2002 (e–h) and 2003–2017 (i–m), respectively.

ground-based observed snow depth. It was found that higher-resolution snow cover datasets, and more appropriate downscaling algorithms are crucial for constructing highly accurate downscaling models.

SD$_{pan}$ (Pan et al., 2022), SD$_{zhang}$ (Yan et al., 2021), and SD$_{che}$ (Che et al., 2021) are three widely used snow depth products downloaded from the Tibetan Plateau Science Data Center (https://data.tpdc.ac.cn/home). These snow depth products were generated based on satellite remote sensing products, such as AMSR-2, AMSR-E, NHSD, and GlobSnow using a process model, a downscaling model of snow cover distribution probabilities, or machine learning algorithms. The spatial resolution of these products ranges from 5 km to 25 km. When compared with the ground-based snow depth measurements, it is found that the accuracy of the new SD$_{500m}$ product is better than the aforementioned three widely used snow depth products, mainly in terms of lower RMSE and MAE (Figure 12.17).

12.7 SNOW-CAUSED LIVESTOCK DISASTERS IN PASTORAL AREA: RISK AND WARNING

In this section we present a case study of using the daily MODIS cloud-free snow cover products (the MAC-SC of Figure 12.4) and other parameters for application on snow-caused livestock disasters. In the pastoral areas of China, natural disasters resulting in a large amount of livestock deaths caused by continuous snowfall in winter and spring is called snow-caused livestock disaster or snow disaster (Liang et al., 2007). Snow has less impact on animal husbandry in developed countries due to better infrastructure in grassland and livestock industry. However, in the pastoral areas of China, a great deal of snowfall often leads to grassland being buried and transportation being disrupted, resulting in a large number of deaths of livestock due to lower temperature and lack of forage stock. Snow disasters generally begin in October and end in April of the following year in the Tibetan Plateau (Wen, 2007). Spatially, snow disasters mainly occur in high-elevation and high-latitude areas, as well as in rich natural grasslands, especially in Inner Mongolia, Xinjiang, Qinghai, Tibet, and other places (Hao et al., 2006). This has a severe influence on the sustainable development of grassland animal husbandry.

Early warning and risk assessment are the two most important yet difficult issues in the study of snow-caused livestock disasters. Early warning of snow disasters involves many factors, such as grassland, snow, weather, livestock, society, and economy, of which many have strong temporal and spatial heterogeneity. The accuracy of warning results is not only closely related to weather forecasting information but also connected with regional environment conditions and capability of disaster prevention. Therefore, the premise for an operational snow disaster warning system is to establish snow disaster warning and risk assessment models based on a long-term series of regional snow disaster monitoring databases using remote sensing, ground-based observation data and weather prediction data in pastoral areas. Furthermore, studying early warning mechanisms of snow disasters is extremely important in both theory and operational practice for improving the ability to prevent disaster and minimize loss in pastoral areas. Therefore, it is important to construct accurate risk assessments of snow-caused disasters and pre-disaster early warning systems to prevent and reduce these disasters.

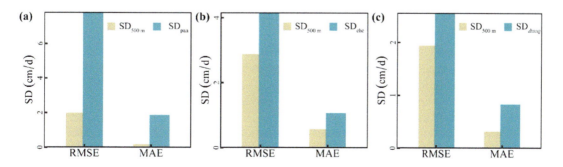

FIGURE 12.17 The RMSE and MAE values of the new SD$_{500\,m}$ and three widely used SD products (a, b, c).

12.7.1 Factors of Snow Disaster Early Warning and Risk Assessment

Snow disaster warning and risk assessment mainly involve factors such as terrain, grassland, snow, weather, livestock, and social economy and are detailed here:

1. Terrain and grassland factors. These factors mainly include the topography conditions (e.g., elevation, slope, aspect), grassland distribution, and herbage growing status (e.g., herbage yield, cover, and height of different grassland types).
2. Snow and weather factors. These mainly include (1) winter (from October to January of the next year) and spring (from February to May) snow disaster probabilities based on multiple-year data (e.g., 50 years) on an administrative unit basis (e.g., township, county); (2) snow depth and number of snow-covered days, snow water equivalent (SWE) products, the rate of snow-covered area, the rate of snow-covered grassland area, grassland burial index (i.e., the ratio of snow depth to grass height) in the unit; and (3) observed and forecasted meteorological conditions during a warning period (e.g., daily minimum, maximum, and average temperatures, precipitation, wind speed, and snow depth observed and predicted at climatic stations) and continuous days of low temperature (e.g., below 0°C, below −10°C, or below −20°C).
3. Livestock factors. Livestock is the main hazard-affected body. The related factors mainly include the number of livestock at the beginning of a year, number of livestock by the end of a year, actual livestock stocking rate, and fraction of small livestock.
4. Social economy factors. These factors mainly include (1) social and economic factors consisting of population, per-capita gross domestic product (GDP), per-capita farming income, density of road network, and spatial distribution of residential areas; (2) forage stock and shed rate of livestock; and (3) per livestock GDP (i.e., the ratio of regional GDP to the number of livestock by the end of a year).

12.7.2 Models for Early Warning and Qualitative Risk Assessment of Snow Disasters

Snow disaster warning and risk assessment have been studied for a long time. Many studies have been focused on snow disaster monitoring methods (Nakai et al., 2012), spatiotemporal analyses of the disaster-causing factors (Su et al., 2011), snow-caused disaster risk assessment (Liang et al., 2007; Romanov et al., 2002; Gao et al., 2017), loss evaluation of post-disaster (i.e., the research for the pastoral areas after snow disasters have occurred, and have resulted in loss of livestock) (Nakamura and Shindo, 2001), snow disaster and avalanche mapping, as well as their relations to climate change (Bocchiola et al., 2008; Delparte et al., 2008; Hendrikx et al., 2005; Hirashima et al., 2008; Jones and Jamieson, 2001; Lato et al., 2012; Martelloni et al., 2013; Mercogliano et al., 2013; Williamson et al., 2002). Tachiiri et al. (2008) evaluated the economic loss of snow disasters in arid inland pastoral areas of Mongolia using a tree-based regression model with parameters including livestock mortality rate of current year, grassland NDVI (Normalized Difference Vegetation Index), SWE (Snow Water Equivalent), and livestock numbers and mortality rates of previous years. In addition, Tominaga et al. (2011) predicted snow cover area and potential extent of snow disasters in build-up environments by combining a meteorological model and a computational fluid dynamics model. Nakai et al. (2012) established a snow disaster early warning system using meteorological factors (i.e., precipitation, wind speed, temperature), which would predict avalanche potential, visibility in blowing snow, and snow conditions on roads.

Due to a lack of operational models and information systems for real-time warning of snow disasters in pastoral areas in developing countries (Liang et al., 2007), there has been much difficulty in snow disaster warning and risk assessments. As for snow disasters in pastoral areas in China, the emphasis has been placed on monitoring the change of snow distribution and livestock loss evaluation of post-disasters. Early warning of snow disaster requires developing a model that can be used to quantify the potential damages caused by heavy snowfall events in a particular period and area. In 2006, the Chinese government issued a national standard for grading snow disasters in pastoral areas (GB/T 20482–2006) (General Administration of Quality Supervision of China, 2006). According to the standard, Zhou et al. (2006) analyzed the potential conditions of snow disasters based on evaluating the vulnerability of hazard-affected bodies and dynamic change of precipitation in the natural environment. Liu et al. (2008) did a preliminary study on an indicating system for early warning and hazard assessment models of snow disasters in pastoral areas of northern Xinjiang utilizing livestock mortality rate as a factor of disaster assessment. Zhang et al. (2008) proposed several indicators and methods for quantification of those indicators for early warning of snow disaster in the pastoral areas of northern Qinghai province. Wang et al. (2013) developed a model and a warning standard for early warning on a county basis using the chosen 45 typical cases of snow disasters and proposed a method of qualitative risk assessment of snow disasters at 500 m resolution for pastoral areas in the TP. Gao et al. (2017) developed and validated a snow disaster model for early warning using the back propagation artificial neural network (BP-ANN) machine learning method at a 500 m spatial resolution based on a risk analysis and the factors that influence snow disasters in pastoral areas of Qinghai Province (Figure 12.18). The results show that the BP-ANN model for an early warning of snow disaster is a practicable predictive method with an overall accuracy of 80% (Table 12.5). This model has quite a few advantages over previously published models, such as it is raster-based, has a high resolution, and has an ideal capacity of generalization and prediction.

According to the methodology for early warning and risk evaluation of snow disasters, the map of potential snow disaster risk intensity (such as Figure 12.19) and snow disaster grades (such as Figures 12.20 and 12.21) at 500 m pixel scale resolution can be made for decision-making of relieving and preventing a snow disaster (Gao et al., 2017).

It is well known that the degree and extent of snow disasters are not only related to the infrastructure (e.g., roads, communications and sheds for livestock) of disaster resistance and forage stock in a pastoral area, but also closely related to disaster rescue efforts. With the development of high-resolution

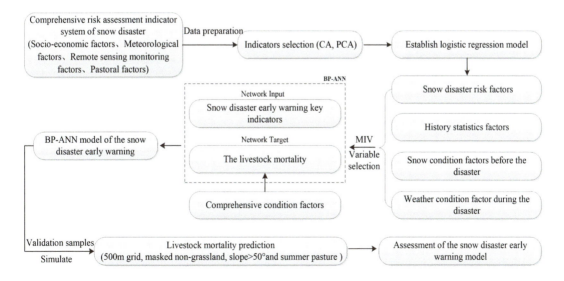

FIGURE 12.18 Flow chart of the snow disaster early warning model (Adapted from Gao et al., 2017).

TABLE 12.5
Accuracy assessment of the simulated livestock mortality from the snow disaster early warning model. (Adapted from Gao et al., 2017)

Cases	In Mid-February 2008			In Late February 2015	
Disaster areas	Guoluo Prefecture	Yushu Prefecture	Huangnan Prefecture	Dulan county	Wulan county
Quantity of livestock at the beginning of the year (10^4)	200.6	270.3	212.5	23.6	144.3
Number of livestock deaths (10^4)	116.7	181.5	40.6	1.4	0.6
Actual livestock mortality (%)	0.0582	0.0671	0.0191	0.0593	0.0416
Simulated livestock mortality (%)	0.051	0.0414	0.0148	0.0709	0.0462
Average accuracy (%)		76			85
Overall accuracy (%)			80		

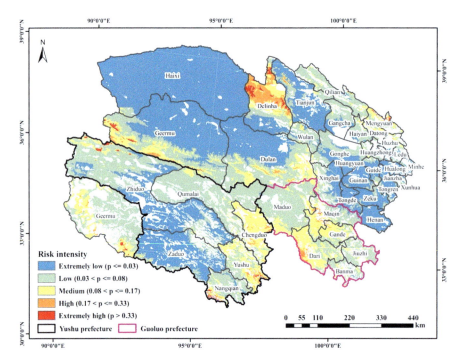

FIGURE 12.19 Potential snow disaster risk at 500 m pixel scale resolution in Qinghai (white spaces are non-grassland areas and slopes >50°) (Adapted from Gao et al., 2017).

Earth observation systems, simulations of snow disasters for early warning have improved from analysis of a single hazard event to comprehensive evaluations of various concurrent or connected disasters. The occurrence of snow disasters was often not just the result of one heavy snowfall or temperature drop, but also affected by continuous snowfall and other factors, climatic indicators and snow cover condition being the main factors. However, the developed models for the early warning of snow

Snow Cover Parameters and Applications 455

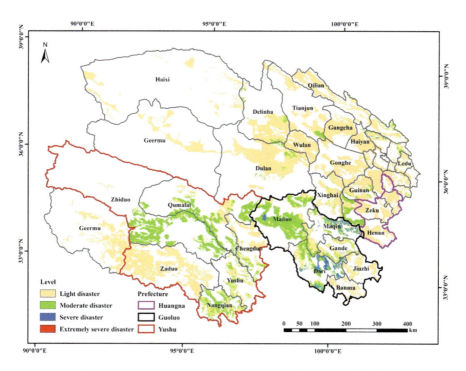

FIGURE 12.20 The simulation results of a snow disaster in mid-February 2008 in the winter and spring pastures in Qinghai Province (white spaces are non-grassland ages, slope > 50°, and summer grazing grassland) (Adapted from Gao et al., 2017).

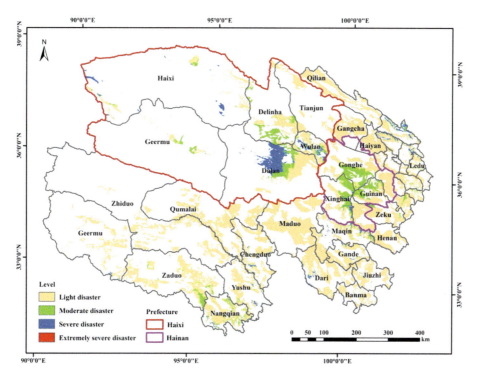

FIGURE 12.21 The simulation results of a snow disaster in late February 2015 in the winter and spring pastures in Qinghai Province (white spaces are non-grassland areas, slope > 50°, and summer grazing grassland) (Adapted from Gao et al., 2017).

disasters and risk assessment in pastoral areas in China are completely based on existing and available datasets, without considering any interference or assistance of human beings such as disaster relief by supplying a great amount of forage from other regions. Uncertainties of the warning model require verification, and several indicators and parameters in the models and classification standards for early warning of snow disaster and risk assessment still need to be further improved in the future.

12.8 CONCLUSIONS

Snow cover is an important fact for regional climate change studies, agriculture, and water source management. Satellite-based snow measurements have revolutionized the monitoring of spatiotemporal variation of snow cover in complex natural conditions at regional and global scales. This chaper introduces the principle of optical remote sensing snow cover detection and the MODIS standard snow cover products, summarizes the recent efforts and improvements on reducing cloud contamination and increasing snow mapping accuracy in all-sky conditions, and finally illustrates the spatiotemporal variatons of snow cover and its applications in the TP and a few other places.

Overall, MODIS standard snow cover products have high accuracy in clear-sky observations and are widely applied in all kinds of studies around the world. In contrast, MODSCAG uses the relative shape of the snow's spectrum and is more accurate to characterize fractional snow cover than the MODIS standard snow cover products (Rittger et al., 2013). However, the impact of this improvement for fractional snow cover on the total water resource budget is limited. Meanwhile, cloud blockage is a severe obstacle to the wide application of MODIS daily snow cover products. Many efforts have been conducted to mitigate the negative effects of cloud cover.

Benefiting from twice-daily observations of Terra and Aqua MODIS sensors and the cloud-free microwave AMSR-E/Aqua obvervations, the major improvements presented here include (1) the daily Terra and Aqua combination, (2) the flexible multiday combination, (3) the daily cloud-free snow cover processing, and (4) the daily redistribution of 25km SWE to 500m pixel size.

Based on the improved snow cover mapping, four specific snow cover parameters are introduced, namely, snow cover index (SCI), snow-covered duration/days (SCD) map, snow cover onset dates (SCOD) map, and snow cover end dates (SCED) map, one each per year or per hydrological year. While an SCD map gives the overall spatial distribution of snow cover duration, both SCOD and SCED maps provide the spatial distribution of the specific dates when the snow cover starts and when the snow cover melts away at the pixel scale of 500 m. The SCI ($km^2 \cdot day$) contains both snow cover extent and duration for a year and can be used to study the snow cover condition on a yearly basis. Together, SCD, SCI, SCOD, and SCED provide important information on the snow cover conditions and can be applied in any region of interest.

Based on those snow cover parameters, we present four examples of application: (1) snow cover as a water resource for study of lake-level change, (2) snow cover as an input for snowmelt and runoff modeling to seprarate snow and rainfall contribution to total runoff, (3) snow cover scaling effects, and (4) snow cover for modeling assessment and prediction of snow-caused disasters in pastoral area.

Overall, remotely sensed snow cover and parameters could be critical information for local government, such as land use planning, agriculture, livestock, and water resource management, to mitigate snow-caused disasters and to plan for agriculture and industry water use, etc. Long-term availability of MODIS/VIIRS type for producing such snow cover datasets is key to study the connection between snow cover variations and climate change. For example, possible application for this data in the future is to study the relation between SCI and El Niño-Southern Oscillation (ENSO). Additional questions can also be addressed through the type of data presented. Does global warming reduce the mean snow cover days for a region? Does global warming have potential to forward shift or backward shift the mean snow cover onset date and/or mean snow cover melting date? How does the spring sand-storm or dust soil impact the melting of snow cover?

ACKNOWLEDGMENTS

This work was in part supported by the National Natural Science Foundation of China (31228021, 31372367, and 41301063) and the University of Texas at San Antonio. Two new authors (Drs. Xiaodong Huang and Xiongxin Xiao) were invited in this version to make sure we included most of the newly updated literature and improvements in remote sensing of snow cover products and applications. Many figures and tables are adapted from published journal papers co-authored with many of our colleagues and graduate students, for which they are sincerely acknowledged. In particular, we would like to thank Drs. Ying Ma and Jinglong Gao for their contribution to this new version. We also would like to thank Dr. Prasad S. Thenkabail, Editor-in-Chief of the series, for the opportunity to update the chapter.

REFERENCES

Armstrong, R., Brodzik, M., 2001. Recent Northern Hemisphere snow extent: A comparison of data derived from visible and microwave satellite sensors. Geophys. Res. Lett., 28(19), 3673–3676.

Barnett, T.P., Adam, J.C., Lettenmaier, D.P., 2005. Potential impacts of a warming climate on water availability in snow-dominated regions. Nat., 438(7066), 303–309.

Blöschl, G., Gutknecht, D., Kirnbauer, R., 1991a. Distributed snowmelt simulations in an alpine catchment: 2. parameter study and model predictions. Water Resour. Res., 27(12), 3181–3188.

Blöschl, G., Kirnbauer, R., Gutknecht, D., 1991b. Distributed snowmelt simulations in an alpine catchment: 1. model evaluation on the basis of snow cover patterns. Water Resour. Res., 27(12), 3171–3179.

Bocchiola, D., Janetti, E.B., Gorni, E., Marty, C., Sovilla, B., 2008. Regional evaluation of three day snow depth for avalanche hazard mapping in Switzerland. Nat. Hazards Earth Syst. Sci., 8(4), 685–705.

Bolch, T. et al., 2010. A glacier inventory for the western Nyainqentanglha range and Nam Co basin, Tibet, and glacier changes 1976–2009. The Cryosphere, 4(3), 419–433.

Bolch, T. et al., 2012. The state and fate of Himalayan glaciers. Sci., 336(6079), 310–314.

Brubaker, K.L., Pinker, R.T., Deviatova, E., 2005. Evaluation and comparison of MODIS and IMS snow-cover estimates for the continental United States using station data. J. Hydrometeorol., 6(6), 1002–1017.

Che, T., Hu, Y., Dai, L., Xiao, L., 2021. Long-term series of daily snow depth dataset over the Northern Hemisphere based on machine learning (1980–2019). A Big Earth Data Platform for Three Poles. https://data.tpdc.ac.cn/en/data/5576dba9-1b15-4ab0-a387-bac9bad80daf.

Chen, B., Zhang, X., Ren, M., Chen, X., Cheng, J., 2023. Snow cover mapping based on SNPP-VIIRS day/night band: A case study in Xinjiang. China. Remote Sens., 15, 3004. https://doi.org/10.3390/rs15123004.

Coll, J., Li, X., 2018. Comprehensive accuracy assessment of MODIS daily snow cover products and gap filling methods. ISPRS J. Photogramm. Remote Sens., 144, 435–452. https://doi.org/10.1016/j.isprsjprs.2018.08.004.

Czyzowska-Wisniewski, E.H., van Leeuwen, W.J.D., Hirschboeck, K.K., Marsh, S.E., Wisniewski, W.T., 2015. Fractional snow cover estimation in complex alpine-forested environments using an artificial neural network. Remote Sens. Environ., 156, 403–417. https://doi.org/10.1016/j.rse.2014.09.026.

Delparte, D., Jamieson, B., Waters, N., 2008. Statistical runout modeling of snow avalanches using GIS in Glacier National Park, Canada. Cold Reg. Sci. Technol., 54(3), 183–192.

Ding, Y., Liu, S., Li, J., Shangguan, D., 2006. The retreat of glaciers in response to recent climate warming in western China. Ann. Glaciol., 43, 97–105.

Dobreva, I.D., Klein, A.G., 2011. Fractional snow cover mapping through artificial neural network analysis of MODIS surface reflectance. Remote Sens. Environ., 115, 3355–3366. https://doi.org/10.1016/j.rse.2011.07.018.

Dong, J., Ek, M., Hall, D., Peters-Lidard, C., Cosgrove, B., Miller, J., Riggs, G., Xia, Y., 2014. Using air temperature to quantitatively predict the MODIS fractional snow cover retrieval errors over the continental United States. J. Hydrometeorol., 15, 551–562. https://doi.org/10.1175/JHM-D-13-060.1.

Dozier, J., 1989. Spectral signature of alpine snow cover from the landsat thematic mapper. Remote Sens. Environ., 28, 9–22.

Dozier, J., Painter, T.H., 2004. Multispectral and hyperspectral remote sensing of alpine snow properties. Annu. Rev. Earth Planet. Sci., 32, 465–494.

Dozier, J., Painter, T.H., Rittger, K., Frew, J.E., 2008. Time-space continuity of daily maps of fractional snow cover and albedo from MODIS. Adv. Water Resour., 31, 1515–1526. https://doi.org/10.1016/j.advwatres.2008.08.011.

Foster, J. et al., 2007. Blended Visible, Passive Microwave and Scatterometer Global Snow Products, Proceedings of the 64th Eastern Snow Conference, May 29th–June 1st, 2007 St. John's, Newfoundland, Canada.

Frei, A., Lee, S., 2010. A comparison of optical-band based snow extent products during spring over North America. Remote Sens. Environ., 114(9), 1940–1948.

Frei, A. et al., 2012. A review of global satellite-derived snow products. Adv. Space Res., 50(8), 1007–1029.

Gao, J., Huang, X.D., Ma, X.F., et al., 2017. Snow disaster early warning in pastoral areas of Qinghai Province. China. Remote Sens., 9, 475.

Gao, Y., Xie, H., Yao, T., Xue, C., 2010a. Integrated assessment on multi-temporal and multi-sensor combinations for reducing cloud obscuration of MODIS snow cover products of the Pacific Northwest USA. Remote Sens. Environ., 114(8), 1662–1675.

Gao, Y., Xie, H., Lu, N., Yao, T., Liang, T., 2010b. Toward advanced daily cloud-free snow cover and snow water equivalent products from Terra-Aqua MODIS and Aqua AMSR-E measurements. J. Hydrol., 385(1–4), 23–35.

Gao, Y., Xie, H., Yao, T., 2011. Developing snow cover parameters maps from MODIS, AMSR-E, and blended snow products. Photogramm. Eng. Remote Sens., 77(4), 351–361.

Gardelle, J., Berthier, E., Arnaud, Y., 2012. Slight mass gain of Karakoram glaciers in the early twenty-first century. Nat. Geosci., 322–325.

General Administration of Quality Supervision of China, Inspection and Quarantine of the People's Republic of China, Standardization Administration of the People's Republic of China, 2006. Chinese national standard, in: Grade of Pastoral Area Snow Disaster. China Standards Press, Beijing, pp. 1–7 (in Chinese).

Gladkova, I., Grossberg, M., Bonev, G., Romanov, P., Shahriar, F., 2012. Increasing the accuracy of MODIS/Aqua snow product using quantitative image restoration technique. IEEE Geosci. Remote Sens. Lett., 9, 740–743. https://doi.org/10.1109/LGRS.2011.2180505.

Goodison, B.E., Brown, R.D., Crane, R.G., 1999. Cryospheric systems, in: King, M.D. (Ed.), EOS Science Plan: The State of Science in the EOS Program, pp. 261–307. https://eospso.nasa.gov/sites/default/files/publications/SciencePlan.pdf.

Hall, D.K., Riggs, G.A., Salomonson, V.V., 1995. Development of methods for mapping global snow cover using moderate resolution imaging spectroradiometer data. Remote Sens. Environ., 54(2), 127–140.

Hall, D.K., Foster, J.L., Verbyla, D.L., Klein, A.G., Benson, C.S., 1998. Assessment of snow-cover mapping accuracy in a variety of vegetation-cover densities in Central Alaska. Remote Sens. Environ., 66(2), 129–137.

Hall, D.K., Riggs, G.A., Salomonson, V.V., DiGirolamo, N.E., Bayr, K.J., 2002. MODIS snow-cover products. Remote Sens. Environ., 83(1–2), 181–194.

Hall, D.K., Riggs, G.A., 2007. Accuracy assessment of the MODIS snow products. Hydrol. Process., 21(12), 1534–1547.

Hall, D.K., Riggs, G.A., Foster, J.L., Kumar, S.V., 2010. Development and evaluation of a cloud-gap-filled MODIS daily snow-cover product. Remote Sens. Environ., 114(3), 496–503.

Hao, L., Gao, J.M., Yang, C.Y., 2006. Snow disaster system of grassland animal husbandry and control countermeasures. Pratacult. Sci., 23, 48–54. (In Chinese).

Hao, S., Jiang, L., Shi, J., Wang, G., Liu, X., 2019. Assessment of MODIS-based fractional snow cover products over the Tibetan Plateau. IEEE J. Sel. Top. Appl. Earth Obs. Remote Sens., 12, 533–548. https://doi.org/10.1109/JSTARS.2018.2879666.

Hao, X., Huang, G., Che, T., Ji, W., Sun, X., Zhao, Q., Zhao, H., Wang, J., Li, H., Yang, Q., 2021. The NIEER AVHRR snow cover extent product over China—a long-term daily snow record for regional climate research. Earth Syst. Sci. Data, 13, 4711–4726. https://doi.org/10.5194/essd-13-4711-2021.

Hao, X., Huang, G., Zheng, Z., Sun, X., Ji, W., Zhao, H., Wang, J., Li, H., Wang, X., 2022. Development and validation of a new MODIS snow-cover-extent product over China. Hydrol. Earth Syst. Sci., 26, 1937–1952. https://doi.org/10.5194/hess-26-1937-2022.

Hendrikx, J., Owens, I., Carran, W., Carran, A., 2005. Avalanche activity in an extreme maritime climate: The application of classification trees for forecasting. Cold Reg. Sci. Technol., 43(1), 104–116.

Hirashima, H., Nishimura, K., Yamaguchi, S., Sato, A., Lehning, M., 2008. Avalanche forecasting in a heavy snowfall area using the snowpack model. Cold Reg. Sci. Technol., 51(2), 191–203.

Hou, Y., Huang, X., Zhao, L., 2022. Point-to-surface upscaling algorithms for snow depth ground observations. Remote Sens., 14(19), 4840.

Huang, Y., Liu, H., Yu, B., Wu, J., Kang, E.L., Xu, M., Wang, S., Klein, A., Chen, Y., 2018. Improving MODIS snow products with a HMRF-based spatio-temporal modeling technique in the Upper Rio Grande Basin. Remote Sens. Environ., 204, 568–582. https://doi.org/10.1016/j.rse.2017.10.001.

Huang, Y., Song, Z., Yang, H., Yu, B., Liu, H., Che, T., Chen, J., Wu, J., Shu, S., Peng, X., Zheng, Z., Xu, J., 2022a. Snow cover detection in mid-latitude mountainous and polar regions using nighttime light data. Remote Sens. Environ., 268, 112766. https://doi.org/10.1016/j.rse.2021.112766.

Huang, Y., Xu, Jiahui, Xu, Jingyi, Zhao, Y., Yu, B., Liu, H., Wang, S., Xu, W., Wu, J., Zheng, Z., 2022b. HMRFS-TP: Long-term daily gap-free snow cover products over the Tibetan Plateau from 2002 to 2021 based on hidden Markov random field model. Earth Syst. Sci. Data, 14, 4445–4462. https://doi.org/10.5194/essd-14-4445-2022.

Immerzeel, W.W., Beek, L.P.H., Bierkens, M.F.P., 2010. Climate change will affect the Asian water towers. Sci., 328(5984), 1382–1385.

Immerzeel, W.W., Droogers, P., de Jong, S.M., Bierkens, M.F.P., 2009. Large-scale monitoring of snow cover and runoff simulation in Himalayan river basins using remote sensing. Remote Sens. Environ., 113(1), 40–49.

Immerzeel, W.W., Lutz, A.F., Andrade, M., Bahl, A., Biemans, H., Bolch, T., Hyde, S., Brumby, S., Davies, B.J., Elmore, A.C., Emmer, A., Feng, M., Fernández, A., Haritashya, U., Kargel, J.S., Koppes, M., Kraaijenbrink, P.D.A., Kulkarni, A.V., Mayewski, P.A., Nepal, S., Pacheco, P., Painter, T.H., Pellicciotti, F., Rajaram, H., Rupper, S., Sinisalo, A., Shrestha, A.B., Viviroli, D., Wada, Y., Xiao, C., Yao, T., Baillie, J.E.M., 2020. Importance and vulnerability of the world's water towers. Nat., 577, 364–369. https://doi.org/10.1038/s41586-019-1822-y.

Immerzeel, W.W., Van Beek, L.P.H., Bierkens, M.F.P., 2010. Climate change will affect the asian water towers. Sci., 328, 1382–1385. https://doi.org/10.1126/science.1183188.

Jan, A., Painter, S.L., 2020. Permafrost thermal conditions are sensitive to shifts in snow timing. Environ. Res. Lett., 15, 084026. https://doi.org/10.1088/1748-9326/AB8EC4.

Jing, Y., Li, X., Shen, H., 2022. STAR NDSI collection: A cloud-free MODIS NDSI dataset (2001–2020) for China. Earth Syst. Sci. Data, 14, 3137–3156. https://doi.org/10.5194/essd-14-3137-2022.

Jing, Y., Shen, H., Li, X., Guan, X., 2019. A two-stage fusion framework to generate a spatio—temporally continuous MODIS NDSI product over the Tibetan Plateau. Remote Sens., 11, 2261. https://doi.org/10.3390/rs11192261.

Jones, A.S., Jamieson, B., 2001. Meteorological forecasting variables associated with skier-triggered dry slab avalanches. Cold Reg. Sci. Technol., 33(2), 223–236.

Justice, C., Belward, A., Morisette, J., Lewis, P., Privette, J., Baret, F., 2000. Developments in the 'validation' of satellite sensor products for the study of the land surface. Int. J. Remote. Sens., 21(17), 3383–3390.

Klein, A.G., Barnett, A.C., 2003. Validation of daily MODIS snow cover maps of the upper Rio Grande river basin for the 2000–2001 snow year. Remote Sens. Environ., 86(2), 162–176.

Klein, A.G., Hall, D.K., Riggs, G.A., 1998. Improving snow cover mapping in forests through the use of a canopy reflectance model. Hydrol. Process., 12(10–11), 1723–1744.

Kour, R., Patel, N., Krishna, A., 2016. Effects of terrain attributes on snow-cover dynamics in parts of Chenab basin, western Himalayas. Hydrol. Sci. J., 61(10), 1861–1876.

Kuter, S., 2021. Completing the machine learning saga in fractional snow cover estimation from MODIS Terra reflectance data: Random forests versus support vector regression. Remote Sens. Environ., 255, 112294. https://doi.org/10.1016/j.rse.2021.112294.

Kuter, S., Akyurek, Z., Weber, G.-W., 2018. Retrieval of fractional snow covered area from MODIS data by multivariate adaptive regression splines. Remote Sens. Environ., 205, 236–252. https://doi.org/10.1016/j.rse.2017.11.021.

Lato, M., Frauenfelder, R., Bühler, Y., Gauthier, D., Lanorte, A., 2012. Automated detection of snow avalanche deposits: Segmentation and classification of optical remote sensing imagery. Nat. Hazards Earth Syst. Sci., 12(9), 2893–2906.

Li, X., Jing, Y., Shen, H., Zhang, L., 2019. The recent developments in cloud removal approaches of MODIS snow cover product. Hydrol. Earth Syst. Sci., 23, 2401–2416. https://doi.org/10.5194/hess-23-2401-2019.

Liang, J., Liu X., Huang K., Li X., Shi X., Chen Y., Li J., 2015. Improved snow depth retrieval by integrating microwave brightness temperature and visible/infrared reflectance. Remote Sens. Environ., 156, 500–509.

Liang, T., Liu, X., Wu, C., Guo, Z., Huang, X., 2007. An evaluation approach for snow disasters in the pastoral areas of northern Xinjiang, PR China. N.Z. J. Agric. Res., 50(3), 369–380.

Liang, T. et al., 2008a. Toward improved daily snow cover mapping with advanced combination of MODIS and AMSR-E measurements. Remote Sens. Environ., 112(10), 3750–3761.

Liang, T. et al., 2008b. An application of MODIS data to snow cover monitoring in a pastoral area: A case study in Northern Xinjiang, China. Remote Sens. Environ., 112(4), 1514–1526.

Liu, A., Che, T., Huang, X., Dai, L., Wang, J., Deng, J., 2022. Effect of cloud mask on the consistency of snow cover products from MODIS and VIIRS. Remote Sens., 14, 6134. https://doi.org/10.3390/rs14236134.

Liu, C.Y., Huang, X.D., Li, X.B., Liang, T.G., 2020. MODIS fractional snow cover mapping using machine learning technology in a mountainous area. Remote Sens., 12(6), 962. https://doi.org/10.3390/rs12060962.

Liu, D., Shen, Y., Wang, Y., Wang, Z., Mo, Z., Zhang, Q., 2023. Monitoring the spatiotemporal dynamics of arctic winter snow/ice with moonlight remote sensing: Systematic evaluation in Svalbard. Remote Sens., 15, 1255(15), 1255. https://doi.org/10.3390/RS15051255.

Liu, X., Liang, T., Guo, Z., Zhang, X., 2008. Early warning and risk assessment of snow disaster in pastoral area of northern Xinjiang. Ying yong sheng tai xue bao= The Journal of Applied Ecology/Zhongguo sheng tai xue xue hui, Zhongguo ke xue yuan Shenyang ying yong sheng tai yan jiu suo zhu ban, 19(1), 133–138.

Liu, X.D., Chen, B.D., 2000. Climatic warming in the Tibetan Plateau during recent decades. Int. J. Climatol., 20(14), 1729–1742.

Margulis, S.A., Liu, Y., Baldo, E., 2019. A joint landsat- and MODIS-based reanalysis approach for mid-latitude montane seasonal snow characterization. Front. Earth Sci., 7, 1–23. https://doi.org/10.3389/feart.2019.00272.

Martelloni, G., Segoni, S., Lagomarsino, D., Fanti, R., Catani, F., 2013. Snow accumulation/melting model (SAMM) for integrated use in regional scale landslide early warning systems. Hydrol. Earth Syst. Sci., 17, 1229–1240. https://doi.org/10.5194/hess-17-1229-2013.

Martinec, J., 1975. Snowmelt runoff model for river flow forecasts. Nord. Hydrol., 6(3), 145–154.

Masson, T., Dumont, M., Mura, M., Sirguey, P., Gascoin, S., Dedieu, J.-P., Chanussot, J., 2018. An assessment of existing methodologies to retrieve snow cover fraction from MODIS data. Remote Sens., 10, 619. https://doi.org/10.3390/rs10040619.

Martinec, J., Rango, A., Roberts, R., 2008. Snowmelt Runoff Model (SRM) User's Manul, USDA Jornada Experimental Range, New Mexico State University, Las Cruces, NM 88003, U.S.A.

Maurer, E.P., Rhoads, J.D., Dubayah, R.O., Lettenmaier, D.P., 2003. Evaluation of the snow-covered area data product from MODIS. Hydrol. Process., 17(1), 59–71.

Mazari, N., Tekeli, A.E., Xie, H., Sharif, H.O., El Hassan, A.A., 2013. Assessment of ice mapping system and moderate resolution imaging spectroradiometer snow cover maps over Colorado Plateau. J. Appl. Remote Sens., 7(1), 073540–073540.

Meloche, J., Langlois, A., Rutter, N., McLennan, D., Royer, A., Billecocq, P., Ponomarenko, S., 2022. High-resolution snow depth prediction using Random Forest algorithm with topographic parameters: A case study in the Greiner watershed, Nunavut. Hydrol. Process., 36(3), e14546.

Mercogliano, P., Segoni, S., Rossi, G., Sikorsky, B., Tofani, V., Schiano, P., Catani, F., Casagli, N., 2013. Brief communication "A prototype forecasting chain for rainfall induced shallow landslides". Nat. Hazards Earth Syst. Sci., 13, 771–777. https://doi.org/10.5194/nhess-13-771-2013.

Mhawej, M., Faour, G., Fayad, A., Shaban, A., 2014. Towards an enhanced method to map snow cover areas and derive snow-water equivalent in Lebanon. J. Hydrol., 513, 274–282.

Musselman, K.N., Addor, N., Vano, J.A., Molotch, N.P., 2021. Winter melt trends portend widespread declines in snow water resources. Nat. Clim. Chang., 11, 418–424. https://doi.org/10.1038/s41558-021-01014-9.

Nakai, S., Sato, T., Sato, A., Hirashima, H., Nemoto, M., Motoyoshi, H., Lwamoto, K.A., 2012. Snow Disaster Forecasting System (SDFS) constructed from field observations and laboratory experiments. Cold Reg. Sci. Technol., 70, 53–61.

Nakamura, M., Shindo, N., 2001. Effects of snow cover on the social and foraging behavior of the great tit Parus major. Ecol. Res., 16(2), 301–308.

Nolin, A.W., 2004. Towards retrieval of forest cover density over snow from the Multi-angle Imaging SpectroRadiometer (MISR). Hydrol. Process., 18(18), 3623–3636.

Nolin, A.W., Dozier, J., Mertes, L.A., 1993. Mapping alpine snow using a spectral mixture modelling technique. Ann. Glaciol., 17, 121–124.

Painter, T.H. et al., 2009. Retrieval of subpixel snow covered area, grain size, and albedo from MODIS. Remote Sens. Environ., 113(4), 868–879.

Pan, F., Jiang, L., Zheng, Z., Wang, G., Cui, H., Zhou, X., Huang, J., 2022. Retrieval of fractional snow cover over High Mountain Asia using 1 km and 5 km AVHRR/2 with simulated mid-infrared reflective band. Remote Sens., 14, 3303. https://doi.org/10.3390/rs14143303.

Pan, J., Yang, J., Jiang, L., Xiong, C., Pan, F., Shi, J., Gao, X., 2022. Physical snow process model supported global snow depth product retrieved from the passive microwave AMSR2 sensor (2013–2020). National Tibetan Plateau Data Center.. https://data.tpdc.ac.cn/en/data/c3b7d80c-a43b-4b0b-a09e-a38f837fb921.

Parajka, J., Blöschl, G., 2008. Spatio-temporal combination of MODIS images-potential for snow cover mapping. Water Resour. Res., 44(3): W03406.

Parajka, J., Pepe, M., Rampini, A., Rossi, S., Blöschl, G., 2010. A regional snow-line method for estimating snow cover from MODIS during cloud cover. J. Hydrol., 381(3–4), 203–212.

Park, H., Fedorov, A.N., Zheleznyak, M.N., Konstantinov, P.Y., Walsh, J.E., 2015. Effect of snow cover on pan-Arctic permafrost thermal regimes. Clim. Dyn., 44, 2873–2895. https://doi.org/10.1007/S00382-014-2356-5/FIGURES/16.

Peng, S., Piao, S., Ciais, P., Friedlingstein, P., Zhou, L., Wang, T., 2013. Change in snow phenology and its potential feedback to temperature in the Northern Hemisphere over the last three decades. Environ. Res. Lett., 8, 014008. https://doi.org/10.1088/1748-9326/8/1/014008.

Phan, V.H., Lindenbergh, R., Menenti, M., 2012. ICESat derived elevation changes of Tibetan lakes between 2003 and 2009. Int. J. Appl. Earth Obs. Geoinf., 17, 12–22.

Picard, G., Dumont, M., Lamare, M., Tuzet, F., Larue, F., Pirazzini, R., Arnaud, L., 2020. Spectral albedo measurements over snow-covered slopes: Theory and slope effect corrections. Cryosphere, 14, 1497–1517. https://doi.org/10.5194/tc-14-1497-2020.

Pulwicki, A., Flowers, G., Radić, V., Bingham, D., 2018. Estimating winter balance and its uncertainty from direct measurements of snow depth and density on alpine glaciers. J. Glaciol., 64(247), 781–795.

Qin, Y., Abatzoglou, J.T., Siebert, S., Huning, L.S., AghaKouchak, A., Mankin, J.S., Hong, C., Tong, D., Davis, S.J., Mueller, N.D., 2020. Agricultural risks from changing snowmelt. Nat. Clim. Chang., 10, 459–465. https://doi.org/10.1038/s41558-020-0746-8.

Qiu, Y.B., Liu, L.J., Shi, L.J., 2021. MODIS daily cloud-free snow cover product over the Tibetan Plateau. Science Data Bank. https://doi.org/10.11922/sciencedb.55.

Riggs, G., Hall, D., 2015. MODIS Snow Products Collection 6 User Guide. NSIDC User Guid. Ser. https://landweb.modaps.eosdis.nasa.gov/data/userguide/MODIS-snow-user-guide-C6.pdf.

Riggs, G.A., Hall, D.K., 2003. Reduction of Cloud Obscuration in the MODIS Snow Data Product. Proceedings of the 60th Eastern Snow Conference, Sherbrooke, Québec, 4–6, June, 2003 (pp. 205–212).

Riggs, G.A., Hall, D.K., Salomonson, V.V., 2006. MODIS snow products user guide to collection 5. Online article, retrieved on January 2, 2007, www.modis-snow-ice.gsfc.nasa.gov/userguides.html.

Rittger, K., Bormann, K.J., Bair, E.H., Dozier, J., Painter, T.H., 2020a. Evaluation of VIIRS and MODIS snow covered fraction in High Mountain Asia using Landsat 8. Front. Remote Sens., 2, 1–15. https://doi.org/10.3389/frsen.2021.647154.

Rittger, K., Krock, M., Kleiber, W., Bair, E.H., Brodzik, M.J., Stephenson, T.S., Rajagopalan, B.R., Bormann, K.J., Painter, T.H., 2021. Multi-sensor fusion using random forests for daily fractional snow cover at 30m. Remote Sens. Environ., 264, 112608. https://doi.org/10.1016/j.rse.2021.112608.

Rittger, K., Raleigh, M.S., Dozier, J., Hill, A.F., Lutz, J.A., Painter, T.H., 2020b. Canopy adjustment and improved cloud detection for remotely sensed snow cover mapping. Water Resour. Res., 56, 1–20. https://doi.org/10.1029/2019WR024914.

Rittger, K., Painter, T.H., Dozier, J., 2013. Assessment of methods for mapping snow cover from MODIS. Adv. Water Res., 51(0), 367–380.

Romanov, P., 2017. Global multisensor automated satellite-based snow and ice mapping system (GMASI) for cryosphere monitoring. Remote Sens. Environ., 196, 42–55. https://doi.org/10.1016/j.rse.2017.04.023.

Romanov, P., Gutman, G., Csiszar, I., 2002. Satellite-derived snow cover maps for north America: Accuracy assessment. Adv. Space Res., 30(11), 2455–2460.

Romanov, P., Tarpley, D., 2007. Enhanced algorithm for estimating snow depth from geostationary satellites. Remote Sens. Environ., 108, 97–110. https://doi.org/10.1016/j.rse.2006.11.013.

Salomonson, V., Appel, I., 2004. Estimating fractional snow cover from MODIS using the normalized difference snow index. Remote Sens. Environ., 89(3), 351–360.

Salomonson, V., Appel, I., 2006. Development of the Aqua MODIS NDSI fractional snow cover algorithm and validation results. IEEE Trans. Geosci. Remote Sens., 44(7), 1747–1756.

Shi, J., 2012. An automatic algorithm on estimating sub-pixel snow cover from MODIS. Quat. Sci., 32, 6–15.

Simic, A., Fernandes, R., Brown, R., Romanov, P., Park, W., 2004. Validation of VEGETATION, MODIS, and GOES + SSM/I snow-cover products over Canada based on surface snow depth observations. Hydrol. Process., 18(6), 1089–1104.

Sirguey, P., Mathieu, R., Arnaud, Y., 2009. Subpixel monitoring of the seasonal snow cover with MODIS at 250 m spatial resolution in the Southern Alps of New Zealand: Methodology and accuracy assessment. Remote Sens. Environ., 113(1), 160–181.

Song, C., Huang, B., Ke, L., 2013. Modeling and analysis of lake water storage changes on the Tibetan Plateau using multi-mission satellite data. Remote Sens. Environ., 135(0), 25–35.

Stopic, R., Dias, E., 2023. Examining thresholding and factors impacting snow cover detection using nighttime images. Remote Sens., 15, 868. https://doi.org/10.3390/rs15040868.

Su, W., Zhang, X.D., Wang, Z., Su, X.H., Huang, J.X., Yang, S.Q., Liu, S.C., 2011. Analyzing disaster-forming environments and the spatial distribution of flood disasters and snow disasters that occurred in China from 1949 to 2000. Math. Comput. Model., 2011, 54, 1069–1078.

Sulla-Menashe, D., Friedl, M.A., 2018. User Guide to Collection 6 MODIS Land Cover (MCD12Q1 and MCD12C1) Product. https://lpdaac.usgs.gov/documents/101/MCD12_User_Guide_V6.pdf.

Sun, Y., Zhang, T., Liu, Y., Zhao, W., Huang, X., 2020. Assessing snow phenology over the large part of eurasia using satellite bservations from 2000 to 2016. Remote Sens., 12, 2060. https://doi.org/10.3390/rs12122060.

Tachiiri, K., Shinoda, M., Klinkenberg, B., Morinaga, Y., 2008. Assessing mongolian snow disaster risk using livestock and satellite data. J. Arid. Environ., 72(12), 2251–2263.

Tedesco, M., Pulliainen, J., Takala, M., Hallikainen, M., Pampaloni, P., 2004. Artificial neural network-based techniques for the retrieval of SWE and snow depth from SSM/I data. Remote Sens. Environ., 90(1), 76–85.

Tominaga, Y. et al., 2011. Development of a system for predicting snow distribution in built-up environments: Combining a mesoscale meteorological model and a CFD model. J. Wind Eng. Ind. Aerodyn., 99(4), 460–468.

Tong, J., Déry, S.J., Jackson, P.L., 2009. Interrelationships between MODIS/Terra remotely sensed snow cover and the hydrometeorology of the Quesnel River Basin, British Columbia, Canada. Hydrol. Earth Syst. Sci. Discuss., 6(3), 3687–3723.

Vikhamar, D., Solberg, R., 2002. Subpixel mapping of snow cover in forests by optical remote sensing. Remote Sens. Environ., 84(1), 69–82.

Wang, G., Jiang, L., Wu, S., Shi, J., Hao, S., Liu, X., 2017. Fractional snow cover mapping from FY-2 VISSR imagery of China. Remote Sens., 9, 983. https://doi.org/10.3390/rs9100983.

Wang, X., Wu, C., Peng, D., Gonsamo, A., Liu, Z., 2018. Snow cover phenology affects alpine vegetation growth dynamics on the Tibetan Plateau: Satellite observed evidence, impacts of different biomes, and climate drivers. Agric. For. Meteorol., 256–257, 61–74. https://doi.org/10.1016/j.agrformet.2018.03.004.

Wang, W., Huang, X., Liang, T., Xie, H., Chen, M., 2015. Spatio-tempoeral change of snow cover and its response to climate over Tibetan Plateau based on an improved daily cloud-free snow cover products. Remote Sens., 7(1), 169–194. https://doi.org/10.3390/rs70100169.

Wang, W., Liang, T., Huang, X., Feng, Q., Xie, H., Liu, X., Chen, M., Wang, X., 2013. Early warning of snow-caused disasters in pastoral areas on the Tibetan Plateau. Nat. Hazards Earth Syst. Sci., 13, 1411–1425.

Wang, X., Xie, H., 2009. New methods for studying the spatiotemporal variation of snow cover based on combination products of MODIS Terra and Aqua. J. Hydrol., 371(1–4), 192–200.

Wang, X., Xie, H., Liang, T., 2008. Evaluation of MODIS snow cover and cloud mask and its application in Northern Xinjiang, China. Remote Sens. Environ., 112(4), 1497–1513.

Wang, X., Xie, H., Liang, T., Huang, X., 2009. Comparison and validation of MODIS standard and new combination of Terra and Aqua snow cover products in northern Xinjiang, China. Hydrol. Process., 23(3), 419–429.

Wang, X., Xie, H., Liang, T., 2014. Spatiotemporal variation of snow cover from space in Northern Xinjiang, in: Yaning Chen. (Ed.), Book Chapter 6 of Water Resources Research in Northwest China. Springer, https://doi.org/10.1007/978-94-017-8016-2.

Wang, Y., Huang, X., Wang, J.S., Zhou, M., Liang, T., 2019. AMSR2 snow depth downscaling algorithm based on a multifactor approach over the Tibetan Plateau, China. Remote Sens. Environ., 231, 111268.

Wei, P., Zhang, T., Zhou, X., Yi, G., Li, J., Wang, N., Wen, B., 2021. Reconstruction of snow depth data at moderate spatial resolution (1 km) from remotely sensed snow data and multiple optimized environmental factors: A case study over the Qinghai-Tibetan Plateau. Remote Sens., 13(4), 657.

Wen, K.G., 2007. Meteorological Disaster Chinese Ceremony: Qinghai. China Meteorological Press, Beijing, China.

Williamson, R.A., Hertzfeld, H.R., Cordes, J., Logsdon, J.M., 2002. The socioeconomic benefits of Earth science and applications research: Reducing the risks and costs of natural disasters in the USA. Space Policy, 18, 57–65.

Wu, X., Xiao, Q., Wen, J.G., You, D., Hueni A., 2019. Advances in quantitative remote sensing product validation: Overview and current status. Earth-Sci. Rev., 196, 102875.

Xiao, X., He, T., Liang, S., Liu, X., Ma, Y., Liang, S., Chen, X., 2022a. Estimating fractional snow cover in vegetated environments using MODIS surface reflectance data. Int. J. Appl. Earth Obs. Geoinf., 114, 103030. https://doi.org/10.1016/j.jag.2022.103030.

Xiao, X., He, T., Liang, S., Zhao, T., 2022b. Improving fractional snow cover retrieval from passive microwave data using a radiative transfer model and machine learning method. IEEE Trans. Geosci. Remote Sens., 60, 1–15. https://doi.org/10.1109/TGRS.2021.3128524.

Xiao, X., Liang, S., He, T., Wu, D., Pei, C., Gong, J., 2021. Estimating fractional snow cover from passive microwave brightness temperature data using MODIS snow cover product over North America. Cryosph., 15, 835–861. https://doi.org/10.5194/tc-15-835-2021.

Xiao, X., Zhang, T., Zhong, X., Li, X., 2020. Spatiotemporal variation of snow depth in the northern hemisphere from 1992 to 2016. Remote Sens., 12, 2728. https://doi.org/10.3390/rs12172728.

Xiao, X., Zhang, T., Zhong, X., Shao, W., Li, X., 2018. Support vector regression snow-depth retrieval algorithm using passive microwave remote sensing data. Remote Sens. Environ., 210, 48–64. https://doi.org/10.1016/j.rse.2018.03.008.

Xie, H., Wang, X., Liang, T., 2009. Development and assessment of combined Terra and Aqua snow cover products in Colorado Plateau, USA and northern Xinjiang, China. J. Appl. Remote Sens., 3, 033559.

Xu, J., Tang, Y., Dong, L., Wang, S., Yu, B., Wu, J., 2023. Temperature-dominated spatiotemporal variability in phenology factors on the Tibetan Plateau from 2002 to 2021 snow. Cryosph. Discuss. https://doi.org/10.5194/tc-2023-135.

Xu, J., Tang, Y., Xu, J., Shu, S., Yu, B., Wu, J., Huang, Y., 2022. Impact of snow cover phenology on the vegetation green-up date on the Tibetan Plateau. Remote Sens., 14. https://doi.org/10.3390/rs14163909.

Yan D., Ma N., Zhang Y., 2021. A daily, 0.05° snow depth dataset for Tibetan Plateau (2000–2018). National Tibetan Plateau Data Center. https://data.tpdc.ac.cn/en/data/0515ce19-5a69-4f86-822b-330aa11e2a28/.

Yan, D., Ma, N., Zhang, Y., 2022. Development of a fine-resolution snow depth product based on the snow cover probability for the Tibetan Plateau: Validation and spatial—temporal analyses. J. Hydrol., 604, 127027. https://doi.org/10.1016/j.jhydrol.2021.127027.

Yang, W. et al., 2006. Analysis of leaf area index products from combination of MODIS Terra and Aqua data. Remote Sens. Environ., 104(3), 297–312.

Yao, T., Pu, J., Lu, A., Wang, Y., Yu, W., 2007. Recent glacial retreat and its impact on hydrological processes on the tibetan plateau, China, and sorrounding regions. Arct. Antarct. Alp. Res., 39(4), 642–650.

Yao, T. et al., 2010. Glacial distribution and mass balance in the Yarlung Zangbo River and its influence on lakes. Chin. Sci. Bull., 55(20), 2072–2078.

Yao, T. et al., 2012. Different glacier status with atmospheric circulations in Tibetan Plateau and surroundings. Nat. Clim. Change, 2(9), 663–667.

Yatagai, A. et al., 2012. APHRODITE: Constructing a long-term daily gridded precipitation dataset for Asia based on a dense network of rain gauges. Bull. Am. Meteorol. Soc., 93(9), 1401–1415.

Yu, H. et al., 2012. A new approach of dynamic monitoring of 5-day snow cover extent and snow depth based on MODIS and AMSR-E data from Northern Xinjiang region. Hydrol. Process., 26, 3052–3061.

Yue, S., Che, T., Dai, L., Xiao, L., Deng, J., 2022. Characteristics of snow depth and snow phenology in the high latitudes and high altitudes of the Northern Hemisphere from 1988 to 2018. Remote Sens., 14, 5057. https://doi.org/10.3390/rs14195057.

Zhang, G. et al., 2008. Study on warning indicator system of snow disaster and risk management in headwaters region. Pratacult. Sci., 26, 144–150.

Zhang, G., Xie, H., Duan, S., Tian, M., Yi, D., 2011a. Water level variation of Lake Qinghai from satellite and in situ measurements under climate change. J. Appl. Remote Sens., 5, 053532.

Zhang, G., Xie, H., Kang, S., Yi, D., Ackley, S.F., 2011b. Monitoring lake level changes on the Tibetan Plateau using ICESat altimetry data (2003–2009). Remote Sens. Environ., 115(7), 1733–1742.

Zhang, G., Xie, H., Yao, T., Liang, T., Kang, S., 2012. Snow cover dynamics of four lake basins over Tibetan Plateau using time series MODIS data (2001–2010). Water Resour. Res., 48(10): W10529.

Zhang, G., Xie, H., Yao, T., Li, H., Duan, S., 2014. Quantitative water resources assessment of Qinghai Lake basin using Snowmelt Runoff Model (SRM). J. Hydrol., 519, 976–987. https://doi.org/10.1016/j.jhydrol.2014.08.022.

Zhang, G., Yao, T., Xie, H., Kang, S., Lei, Y., 2013. Increased mass over the Tibetan Plateau: From lakes or glaciers? Geophys. Res. Lett., 40(10), 2125–2130.

Zhang, H., Zhang, F., Che, T., Wang, S., 2020. Comparative evaluation of VIIRS daily snow cover product with MODIS for snow detection in China based on ground observations. Sci. Total Environ., 724, 138156. https://doi.org/10.1016/j.scitotenv.2020.138156.

Zhao, H., Fernandes, R., 2009. Daily snow cover estimation from advanced very high resolution radiometer polar pathfinder data over Northern Hemisphere land surfaces during 1982–2004. J. Geophys. Res., 114(D5): D05113.

Zhao, Q., Hao, X., Wang, J., Luo, S., Shao, D., Li, H., Feng, T., Zhao, H., 2022. Snow cover phenology change and response to climate in China during 2000–2020. Remote Sens., 14, 3936. https://doi.org/10.3390/rs14163936.

Zhou, B., Shen, F., Li, S., 2006. A synthetical forcasting model of snow disaster in Qinghai-Tibet Plateau. Meteorol., 9, 106–110.

Zhou, X., Xie, H., Hendrickx, J.M.H., 2005. Statistical evaluation of remotely sensed snow-cover products with constraints from streamflow and SNOTEL measurements. Remote Sens. Environ., 94(2), 214–231.

Zhu, J., Shi, J., 2018. An algorithm for subpixel snow mapping: Extraction of a fractional snow-covered area based on ten-day composited AVHRR/2 data of the Qinghai-Tibet Plateau. IEEE Geosci. Remote Sens. Mag., 6, 86–98. https://doi.org/10.1109/MGRS.2018.2850963.

Zhu, L., Zhang, Y., Wang, J., Tian, W., Liu, Q., Ma, G., Kan, X., Chu, Y., 2021. Downscaling snow depth mapping by fusion of microwave and optical remote-sensing data based on deep learning. Remote Sens., 13(4), 584.

Part VII

Summary and Synthesis for Volume V

13 Remote Sensing Handbook, Volume V
Water Resources: Hydrology, Floods, Snow and Ice, Wetlands, and Water Productivity

Prasad S. Thenkabail

ACRONYMS AND DEFINITIONS

AGB	Aboveground Biomass
ALEXI	Atmosphere-Land Exchange Inverse
ALOS	Advanced Land Observing Satellite
AMSR-E	Advanced Microwave Scanning Radiometer—Earth Observing System
ANN	Artificial Neural Network
ASAR	Advanced Synthetic Aperture Radar on Board ENVISAT
ASTER	Advanced Spaceborne Thermal Emission and Reflection Radiometer
AVHRR	Advanced Very High-Resolution Radiometer
BGB	Belowground Biomass
BRT	Boosted Regression Tree
CAM	Cropland Area Mapping
CNES	The Centre national d'études spatiales, or the National Center of Space Studies
CNN	Convolution Neural Network
COSMO-SkyMed	Constellation of small Satellites for Mediterranean basin Observation
CPM	Crop Productivity Mapping
CryoSat	Europe's Satellite to Study Ice
CWP	Crop Water Productivity
DA	data Assimilation
DEM	Digital Elevation Model
DisALEXI	Disaggregated ALEXI
DMSP	Defense Meteorological Satellite
DSS	Decision Support System
EGSMA	Egyptian Geological Survey and Mining Authority
EL	Elevation
ENVISAT	Environmental Satellite
EOS	Earth Observing System
ERS	European Remote Sensing Satellites
ERS2-SAR	European Remote Sensing Satellite2 Synthetic Aperture radar
ESTARFM	Enhanced Spatial and Temporal Adaptive Reflectance Fusion Model
ET	Evapotranspiration
ETM+	Enhanced Thematic mapper Plus

DOI: 10.1201/9781003541400-20

EVI	Enhanced Vegetation Index
FAO	Food and Agricultural Organization
FRM	flood Risk Mapping
GBM	Green Biomass
GEE	Google Earth Engine
GEO	Geostationary Earth Orbital
GIS	Geographic Information Systems
GMS	Geostationary Meteorological Satellite
GOES	Geostationary Operational Environmental Satellite
GPS	Global Positioning Systems
GRACE	Gravity Recovery and Climate Experiment
HI	Harvest Index
HNB	Hyperspectral Narrowband
HVI	Hyperspectral Vegetation Indices
ICESat	Instrument aboard the Ice, Cloud, and land Elevation
IKONOS	A commercial earth observation satellite typically collecting sub-meter to 5 m data
InSAR	Interferometric SAR
IRS	Indian Remote Sensing Satellites
ISODATA	Iterative Self-organizing Data Analysis Technique
IVs	Inland Valleys
JAXA	Japan Aerospace Exploration Agency
JERS	Japanese Earth Resources Satellite
LAI	Leaf Area Index
LEO	Low Earth Orbital
LiDAR	Light Detection and Ranging
LST	Land Surface Temperature
LU	Land Use
LULC	Land Use and Land Cover
M1DP	Maximum One-Day Precipitation
MaxEnt	Maximum Entropy
MERIS	Medium-Resolution Imaging Spectrometer
METRIC	Mapping Evapotranspiration at high Resolution and with Internalized Calibration
MLA	Machine Learning Algorithms
MOD16 MODIS	Global Evapotranspiration Dataset
MODIS	Moderate Resolution Imaging Spectroradiometer
MSS	Multi-Spectral Scanner
NASA	National Aeronautics and Space Administration
NDSI	Normalized Difference Snow Index
NDVI	Normalized Difference Vegetation Index
NDWI	Normalized Difference Water Index
NIR	Near-Infrared
NOAA	National Oceanic and Atmospheric Administration
OBIA	object-based image analysis
PALSAR	Phased Array type L-band Synthetic Aperture Radar
PCA	Principal Component Analysis
PERSIANN	Precipitation Estimation from Remotely Sensed Information using Artificial Neural Networks
PMW	Passive Microwave
PolSAR	polarimetric SAR

PT-JPL	Priestley–Taylor Jet Propulsion Laboratory
RADARSAT	RADAR Satellite
RD	River Distance
RF	Random Forest
RISAT	Radar Imaging Satellite
RS	Remote Sensing
SA	Slope Aspect
SAR	Synthetic Aperture Radar
SCD	Snow-Covered Duration/Days
SCED	Snow Cover End Dates
SCI	Snow Cover Index
SCOD	Snow Cover Onset Dates
SEASAT	First satellite designed for remote sensing of the Earth's oceans with synthetic aperture radar (SAR)
SEB	Surface Energy Balance
SEBAL	Surface Energy Balance Algorithm for Land
SEBS	Surface Energy Balance System
SIR	Spaceborne Imaging Radar
SL	Slope
SMMR	Scanning Multi-channel Microwave Radiometer
SPI	Stream Power Index
SPOT	Satellite Pour l'Observation de la Terre, French Earth Observing Satellites
SR	Surface Reflectance
SRTM	Shuttle Radar Topography Mission
SSEB	Simplified Surface Energy Balance
SSMI/S	Special Sensor Microwave Imager
ST	Soil Texture
SVM	Support Vector Machines
SWIR	Shortwave Infrared
SWOT	Surface Water and Ocean Topography
TerraSAR-X	A radar Earth observation satellite, with its phased array synthetic aperture radar
TLS	Terrestrial Laser Scans
TMI	TRMM Microwave Imager
TRMM	Tropical Rainfall Measuring Mission
TWI	Topographic Wetness Index
VF	Vegetation Fraction
VHRI	Very High-Resolution Imagery
WF	Water Footprint
WP	Water Productivity
WPM	Water Productivity Mapping
WRD	Waterway and River Density
WUM	Water Use Mapping

This chapter provides a brief summary of all 12 chapters appearing in this volume of the *Remote Sensing Handbook*. Volume V has a focus on hydrology, water resources, ice, wetlands, and crop water productivity. The chapters are broadly classified into: (1) geomorphology, (2) hydrology and water resources, (3) floods, (4) wetlands, (5) crop water use and crop water productivity, and (6) snow and ice. Under each of these topics, one or more chapters provide comprehensive coverage. For example, there are four chapters under wetlands. The summary in this chapter provides a "window view" of what exists in each chapter as well as provides inter-linkages to various chapters. You can read the summary chapter in three ways: (1) before reading the chapters to get an overview, (2) after reading all the chapters to re-cap and refresh major highlights, and (3) do both (1) and (2) to capture the totality of all the chapters. The chapter summaries also help the reader establish inter-linkages that exist between chapters, in a nutshell. An overview of the organization and structure of the chapters is provided in Figure 13.0.

13.1 GEOMORPHOLOGICAL STUDIES OF REMOTE SENSING STUDIES

Geomorphology is the study of landforms and the processes that shape them (Smith and Pain, 2006; Rao, 2002). Remote sensing offers a synoptic view of large areas to study landform characteristics, their classification, and their changes over space and time. Geomorphological features include such features as land topography and lithology, land use and vegetation, coastal landforms, river systems, and tidal deltas. Such and other information derived from remote sensing will help enrich geomorphological maps (e.g., Figure 13.1). Chapter 1, by Dr. James B. Campbell and Dr. Lynn M. Resler, presents geomorphological studies addressing topics such as:

1. Alpine and polar periglacial environments
2. Glacial geomorphology
3. Mass wasting
4. Fluvial landforms
5. Floodplain analysis
6. Channel migration
7. Stream bank retreat
8. Coastal geomorphology
9. Aeolian landforms
10. Biogeomorphology

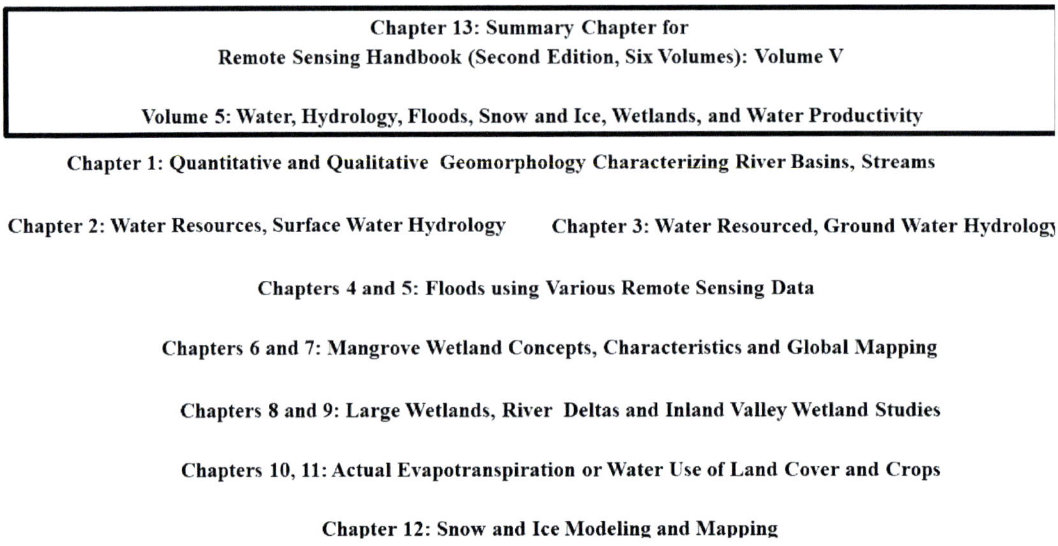

Chapter 13: Summary Chapter for
Remote Sensing Handbook (Second Edition, Six Volumes): Volume V

Volume 5: Water, Hydrology, Floods, Snow and Ice, Wetlands, and Water Productivity

Chapter 1: Quantitative and Qualitative Geomorphology Characterizing River Basins, Streams

Chapter 2: Water Resources, Surface Water Hydrology   Chapter 3: Water Resourced, Ground Water Hydrology

Chapters 4 and 5: Floods using Various Remote Sensing Data

Chapters 6 and 7: Mangrove Wetland Concepts, Characteristics and Global Mapping

Chapters 8 and 9: Large Wetlands, River Deltas and Inland Valley Wetland Studies

Chapters 10, 11: Actual Evapotranspiration or Water Use of Land Cover and Crops

Chapter 12: Snow and Ice Modeling and Mapping

FIGURE 13.0 Overview of the chapters in Volume V of the *Remote Sensing Handbook* (Second Edition).

Some of the important lessons we learn from Chapter 1 include:

- **Landslides:** studied through three types of information gathered from different remote sensing approaches: (1) historical inventories of landslides using, for example, Landsat or IRS imagery; (2) site-specific detection of mass wasting events using radar imagery; and (3) examination of active sites using ground-based LiDAR.
- **Mass wasting and change detection:** often focus on before and after an event, using a multi-sensor approach and depending on image availability.
- **Terrain deformation:** studied using interferometric SAR (InSAR), employing phase differences derived from two or more SAR images to identify changes in local topography over time.
- **Terrain deformation:** studied using interferometric SAR (InSAR), employing phase differences derived from two or more SAR images to identify changes in local topography over time.
- **Characterization of steep periglacial high mountain faces:** requires very high spatial resolution imagery (< 5 m)
- **Inaccessible or difficult to access areas,** such as arctic environments and remote mountainous areas, are best studied using remotely sensed data.
- **Fluvial processes and floodplain analysis:** Changes in river morphology, river meanders, and floodplain characteristics, including land use in these landscapes, are studied using multi-date and multi-sensor imagery.
- **Bio-geomorphological studies** using remote sensing include fluvial geomorphology and hydrology, vegetation cover, and land use\land cover changes.
- **Terrain studies with and without vegetation:** LiDAR data offers the ability to observe the terrain surface free of vegetation by comparing LiDAR pulses reflected off the top of the canopy with the corresponding LiDAR data reflected from below the canopy.
- **Contributing factors:** In many of the aforementioned studies, a number of contributing factors such as vegetation, land use, hydrology, and terrain are required. These contributing factors can be obtained from one or more types of remote sensing data from optical, radar, LiDAR, and thermal platforms.

Drs. Campbell and Resler highlight the importance of LiDAR as a key to multiple geomorphological studies, with significant contributions from digital elevation model (DEM) data, a wide array of remote sensing data from various platforms, and integration of these data with GPS in a GIS framework.

Availability of high-resolution (30m or better) DEMs are helping increase precision and accuracies of various geomorphologic parameters. Rai et al. (2018) demonstrated a use of ASTER DEM to extract the vital hydrological parameters such as stream order, stream length, bifurcation ratio, drainage density, drainage frequency, drainage texture, form factor, circularity ratio, and elongation ratio of Kosi River Basin (India) and demonstrated their value in water management programs, site selection of water-harvesting structures, and in flood management activities. Remotely sensed data in a GIS environment is effective in evaluating drainage morphometric parameters and in exploring the relationship between the drainage morphometric, landform/basin properties and land uses (Temitope and Oyedotun, 2022). Increasingly automated approaches are used to extract morphometric parameters, as demonstrated by Monegaglia et al. (2018) and Shahrood et al. (2020) using Landsat data to extract: (1) river channel centerline, (2) river channel migration\meandering, and (3) sediment transport and quantification. Increasingly quantitative geomorphological parameters are derived on the cloud using platforms such as the Google Earth Engine (GEE) (Gorelick et al., 2017; Boothroyd et al., 2020; Pu et al., 2021; Velastegui-Montoya et al., 2023).

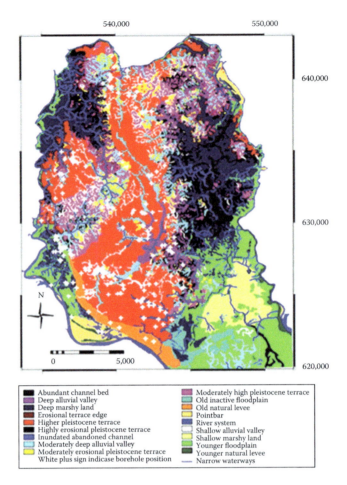

FIGURE 13.1 Geomorphology accompanied with landfill thickness in Dhaka city area derived using a spatially enhanced (or fused) image of Indian Remote Sensing (IRS)-1D PAN data and Landsat Enhanced Thematic mapper plus (ETM+) bands 5, 4, and 3, acquired February 2000 and 2002, respectively. In addition, data was also obtained from topographic maps (Source: Kamal and Midorikawa, 2004).

13.2 HYDROLOGICAL STUDIES USING MULTI-SENSOR REMOTE SENSING

Remote sensing has played a key role in better understanding and characterizing the hydrological cycle. A wide array of hydrological parameters can be consistently characterized and mapped using a set of remote sensing sensors gathering data in radar, optical, or thermal wavelengths. Some of these parameters, such as evapotranspiration and basin characteristics (e.g., topography, river morphology, vegetation, and land use) can be mapped with a great degree of accuracy. Some other hydrological parameters, such as ground water and soil moisture, have a greater degree of uncertainty, mainly as a result of an insufficient number of sensors of adequate resolution (e.g., may have anywhere between 5 and 25 km spatial resolution). Chapter 2 by Dr. Sadiq I. Khan et al. provides us a comprehensive assessment of the key hydrological parameters (precipitation, evapotranspiration, soil moisture, and ground water) monitored and mapped by various sensors and their strengths and limitations.

The key hydrological parameters required in hydrological studies through the satellite sensors are briefly presented next (e.g., Figure 13.2):

13.2.1 Precipitation

Precipitation retrieval from satellite data comes from either visible and NIR bands of Geostationary Earth Orbital (GEO) satellites gathered every 15–30 minutes and/or passive/active microwave images from Low Earth Orbital (LEO) satellites but acquired with much lower sampling frequency. GEO data is only indirectly related to surface rainfall, whereas passive microwave (PMW) sensors on LEO satellites give more direct sensing of rain clouds. PERSIANN (Precipitation Estimation from Remotely Sensed Information using Artificial Neural Networks) (Sorooshian et al., 2011) utilized both GEO and LEO data for rainfall estimation at high temporal resolution (e.g., three-hourly [60S-60N] and six-hourly (50S-50N). These data are available as 0.25° (~27 km) grid. Tropical Rainfall Measuring Mission (TRMM; e.g., Figure 13.2), GEOS-8,10, Geostationary Meteorological Satellite (GMS), Defense Meteorological Satellite (DMSP), and National Oceanic and Atmospheric Administration (NOAA), are some of the other satellite systems gathering precipitation data. Data are gathered in visible and NIR passive and active microwave sensors.

13.2.2 Evapotranspiration

Evapotranspiration (ET) is one of the most widely measured hydrological parameters from remote sensing at various scales. The most widely used technique for actual ET (water use; mm/day) computation is through surface energy balance (SEB) modeling. Surface energy balance models require short wave (optical) bands and thermal bands. So, data from sensors that have both optical and thermal bands such as Landsat-5,7,8 (60–120 m), ASTER (90 m), Terra\Aqua MODIS (1020 m), and NOAA AVHRR (1000 m) are all used in ET modeling. The most widely used SEB models are:

- Surface Energy Balance Algorithm for Land (SEBAL)
- Mapping Evapotranspiration at high Resolution and with Internalized Calibration (METRIC)
- Surface Energy Balance System (SEBS)

13.2.3 Soil Moisture

Soil moisture (e.g., Figure 13.2) is important in: (1) agricultural crop management (e.g., irrigation scheduling, droughts), (2) understanding ET flux, (3) understanding water and heat energy between the land surface and the atmosphere through evaporation and plant transpiration; and (4) amount of precipitation runoff. Soil moisture is mostly measured passive remote sensing in the microwave region with best frequencies of about 6 GHz (C-band), but also in 1 and 3 GHz (L-band) frequencies. Soil moisture is measured by sensors such as Advanced Microwave Scanning Radiometer for the Earth observing system (AMSR-E; e.g., Figure 13.2) on board Aqua with 12-channel, six-frequency passive microwave radiometer acquiring data in 6 x 4 km (89 GHz), 14 x 8 km (36.5 GHz), and 74 x 43 km (6.9 GHz); DMSP SSM/I in 19.3–85.5 GHz in resolutions of 15 x 13 km (85.5 GHz) and 37 x 28 (37 GHz), Tropical Rainfall Measuring Mission (TRMM) Microwave Imager (TMI) observations at 10.65 GHz with 25 km; and Scanning Multi-channel Microwave Radiometer (SMMR) on board Nimbus-7 in 6.6–37 GHz and in resolutions 27 x 18 (37 GHz) and 148 x 95 (6.6 GHz). As is clear from these sensor characteristics, the main problem with soil moisture sensors is the very coarse resolution.

13.2.4 Groundwater

Ground water is very difficult to map and often done by indirect means. Early studies used lineaments. The twin Gravity Recovery and Climate Experiment (GRACE) satellites measure Earth's gravity field and its data is used to establish terrestrial water storage variation by detecting gravitational anomalies (Rodell, 2006).

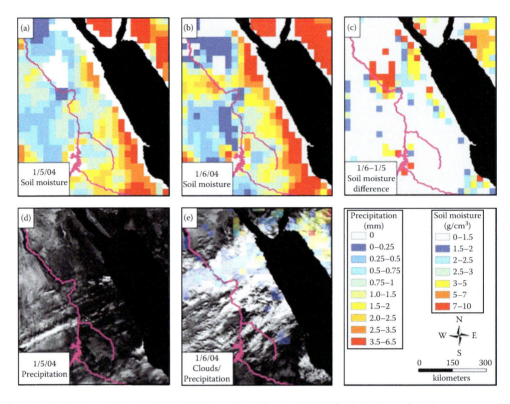

FIGURE 13.2 Validation of Tropical Rainfall Measuring Mission (TRMM)–derived precipitation event over the Eastern Desert of Egypt on 1/6/04. (a) Soil moisture content extracted from Advanced Microwave Scanning Radiometer for EOS (AMSR-E) acquired on 1/5/04. (b) Soil moisture on 1/6/04. (c) Soil moisture difference image: (1/6/04 image minus 1/5/04 image). (d) AVHRR image showing minimal cloud coverage on 1/5/04. (e) Advanced Very High-Resolution Radiometer (AVHRR) image acquired on 1/6/04 showing extensive cloud coverage (white areas). Also shown on Figure 271e is TRMM-derived precipitation (colored areas) (Source: Milewski et al., 2009).

13.2.5 Surface Water

Surface water changes can be measured using AMSR-E (Khaki, 2023). GRACE data can be used to assess surface-water storage (Alshehri and Mohamed, 2023). NASA-CNES, with contributions from Canada Space Agency and UK space Agency, launched Surface Water and Ocean Topography (SWOT) mission in 2022 to study surface water (Cacal et al., 2023).

13.2.6 Snow, Ice, Glaciers

Snow, ice, and glaciers play a key role in Earth's energy and water balance. Many rivers have significant to large proportions of the flow from snowmelt runoff. Satellite sensors such as the AVHRR, MODIS, GOES, and a number of other sensors have been used for snow cover mapping over many years. In addition to microwave sensors such as conventional SSMI/S, and AMSR-E, more recent missions and sensors such as Radarsat, ENVISAT, MERIS, ICEsat-2, and CryoSat have played an important role in studies pertaining to snow, ice, and glaciers (Gabarró et al., 2023).

13.2.7 Basin Characteristics

Water balance is often accounted at varying watershed and river basin scales. This requires us to obtain basin characteristics such as topography, geology, soils, vegetation, and land use, geomorphological characteristics (e.g., river length, stream orders, and stream elevation). All of these properties can be studied using a wide array of optical and radar remote sensing.

Chapter 2 by Dr. Sadiq I. Khan presents the main water budget parameters discussed earlier except the basin characteristics.

The use of multi-sensor remote sensing and other data along with advanced methods are the key to accurately studying various hydrological parameters. For example, lake water assessments are best achieved using multi-sensor remote sensing data through the study of glacier surface elevation using TanDEM-X, seasonal snow cover area changes using MODIS snow cover products, and snow depth using passive microwave remote sensing (Zhang et al., 2020). Remote sensing when integrated with topographic and geomorphologic datasets and machine learning algorithms along with field-based techniques (e.g., electrical resistivity and ground-penetrating radar) are key to mapping shallow groundwater occurrences in arid lands (Sahour et al., 2022). Field-based methods are accurate but hard to scale. Remote sensing helps to scale. Other data like DEMs will help increase accuracies. Combining these data in machine learning and cloud computing platforms helps speed computing over very large areas. Multi-sensor data assimilation (DA) on river discharge estimations showed (Wu et al., 2022): (1) GRACE-DA improved the spatial pattern of discharge over snow-dominated river basins and (2) on top of MODIS and AMSR-E DA, GRACE DA shows an added value in simulating inter-seasonal and interannual variability in discharge. The benefits of single and multi-sensor approaches in long-term water monitoring of temporary water bodies, against extensive ground truth data were evaluated for all available observations from Landsat 7, Landsat 8, Sentinel-2, and MODIS over 1999–2019 in the Google Earth Engine platform by Gorelick et al. (2017), Velastegui-Montoya et al. (2023), and Ogilvie et al. (2020). They made some interesting inferences: (1) There is a need to integrate coarser MODIS observations with Landsat time series before 2013 when there was a single Landsat acquiring data; (2) from 2013, with two Landsats (Landsat 7 and Landsat 8) data became sufficient, and integrating MODIS observations degrades performance marginally; and (3) combining Landsat and Sentinel-2 yields modest improvements after 2015. These results clearly imply the need for intelligent use of multi-sensor data. More data is better, but more data is not always the best. What is needed is optimal data to meet the needs of the goal. Overall, hydrological studies can be extensive, ranging from river basin land use\/and cover, geomorphology, rainfall-runoff, droughts, floods, and a host of other studies like evapotranspiration studies and water productivity studies. The user is required to have thorough understanding of multi-sensor data for gathering the most accurate information through models, maps, and statistics.

13.3 GROUNDWATER STUDIES USING REMOTE SENSING

Groundwater accounts for ~30% of the world's freshwater resources, and the current withdrawal rates in the estimated range of 982 km^3/year forms the largest extraction of any raw material on Planet Earth (Margat and van der Gun, 2013). Three countries: India (251 km^3/year), the USA, and China (112 km^3/year each) account for nearly 50% of this extraction. Fortunately, groundwater is replenishable and a renewable water resource. Nevertheless, many parts of the world suffer from overexploitation that is unsustainable. About 60% of groundwater withdrawn worldwide is used for agriculture; the rest is almost equally divided between the domestic and industrial sectors (Siebert et al., 2010). In many parts of the world, especially in rural areas, groundwater is the primary source of drinking water and often the safest, given the extent of surface water pollution (Dubey et al., 2023; Mounirou et al., 2023).

Groundwater is subterranean. Hence there is no real direct measure of groundwater from remote sensing, but several indirect measures exist, such as through the use of several thematic maps (e.g., Figure 13.3) in a spatial modeling, to determine potential groundwater zones. These thematic maps are derived from remote sensing and non–remote sensing sources and provide a wide array of information of an area. When this information is integrated in a spatial modeling framework, it is possible to derive potential groundwater zones. Remote sensing is used as a: (1) reconnaissance tool, (2) indirect indicator requiring terrain image interpretation of various natures, and (3) input data in the GIS spatial modeling framework, which in turn will help establish groundwater potential zones or groundwater stress areas or in supporting groundwater management. Long-wave radar remote sensing data can penetrate ground under ideal conditions and can at times detect groundwater at depths of a few meters. Thermal imagery can detect temperature anomalies in water bodies, thus, for example, helping identify areas of discharge into lakes and rivers from groundwater. Optical and radar imagery are interpreted for various geomorphological, geological, terrain, and vegetation characteristics that help determine the potential areas of groundwater source or stress. Most of these interpretations require considerable skill of an expert groundwater scientist. In Chapter 3 on groundwater, Dr. Santhosh Kumar Seelan shows us several approaches and methods used for groundwater detection, groundwater stress areas, and groundwater management issues using remote sensing. These are summarized here:

Lineaments and lithology: Remote sensing plays an important and powerful role in geological structural analysis such as lineament detection and mapping. Lineaments are often the best indicators of groundwater presence, especially in hardrock landscapes (e.g., crystalline and basaltic terrains) and are mapped using image enhancement and filtering techniques using optical or radar imagery. Wells located on the lineament zones produced as much as 14 times better yield than wells located away from the lineaments (Opoku et al., 2024; Echogdali et al., 2023).

Quantitative geomorphologic parameters: Drainage density and stream lengths have a bearing on recharge conditions and permeability of the rocks. These are studied using optical as well as radar data. One common indicator is lesser the drainage density, greater the groundwater potential. Fossil drainage in deserts and paleo (old) drainage, for example, can be detected using radar imagery, which can penetrate up to 3 m and yield far more groundwater than elsewhere (Abdelkareem et al., 2020; Abotalib et al., 2021).

Vegetation, land use, croplands (e.g., irrigated or rainfed) information: These helps support water balance budgets and are studied using optical, radar, or thermal data. For example, actual evapotranspiration (ET) or water use by crops and natural landscape is assessed using surface energy balance modeling where thermal and optical remote sensing plays a key role. Vegetation associated with fault zones helps detect near-surface groundwater. Fine-grained clayey soils generally tend to permit less recharge and more runoffs, while coarse-grained sandy soils allow more recharge to groundwater.

Groundwater discharges into lakes and rivers are mapped based on temperature differences of lake or river water (higher temperature) compared to groundwater discharge into lakes (cooler water). In case of two adjacent streams, the one fed by a hot spring appears brighter.

Groundwater-fed irrigation in many cases has a center-pivot (e.g., in much of the southern Ogallala system) and can easily be detected, and the water use pattern modeled using surface energy balance models in which both thermal and optical remote sensing are key inputs.

Hydrogeological studies based on faults in sedimentary rocks: These are often indicative of groundwater presence mapped using either optical or radar data.

Unconsolidated terrains such as alluvial plains, deltas, and other Pleistocene deposits: These morphological units often hold large quantities of groundwater. These landscapes are best mapped using optical or radar remote sensing and provide a good reconnaissance of the area for further investigations.

Landforms of various natures where groundwater potential exists can be mapped using optical, radar, or thermal remote sensing and DEM data. These landforms include weathered rocks, landform, and soil associations, karst terrain (in karst depressions using DEM data), and volcanoes (e.g., groundwater discharge in lower volcanoes).

Total water storage change quantification studies are conducted using GRACE satellite microgravity measurements (Rodell et al., 2006). Regional, national, and river basin scale total water storage changes are studied using the GRACE satellite measurements.

Groundwater modeling inputs: Remote sensing data can provide useful inputs to groundwater modeling. For example, in areas of heavy irrigated agriculture, information such as the area cropped and the type of crops will be useful inputs to groundwater numerical models.

Spatial modeling for groundwater assessment and management: This involves a wide array of spatial data such as geology, geomorphology, terrain, soils, and vegetation that will help make a decision on groundwater potential or stress and will help groundwater management decide on aquifer recharge areas.

The readers with deep interest in groundwater remote sensing are also encouraged to read the works of Ray (1960), Sabins (1987), and Meijerink et al. (2007).

The advantage remote sensing has is its ability to quantify and provide reliable, objective, and accurate information of various sources of water (e.g., groundwater, surface water) and its uses (e.g., crop water use or actual evapotranspiration, green water for rainfed agriculture, blue water for irrigated agriculture). Groundwater is often the only source or significant supplemental source of water in many places and is often studied by combined use of remote sensing (RS), and geographic information system (GIS). For example, nine groundwater occurrence and movement controlling parameters (i.e., lithology, rainfall, geomorphology, slope, drainage density, soil, land use/land cover, distance to river, and lineament density) were derived from remote sensing and\or ancillary data and transformed into raster data using ArcGIS software (Mussa et al., 2020) for identification of groundwater potential zones of the Shatt Al-Arab Basin in Iran (Allafta et al., 2021). Similarly, groundwater control factors derived from remote sensing data by Lee et al. (2020) include nine topographic factors, two hydrological factors, forest type, soil material, land use, and two geological factors. Indeed, mapping groundwater potential mapping is one of the major applications of remote sensing, especially in dry areas (e.g., Andualem and and Demeke, 2019; Alshehri et al., 2020) and urban areas (Kalhor and Emaminejad, 2019). Increasingly, advanced machine learning algorithms (MLAs) such as deep boosting, logistic model trees, boosted regression trees, k-nearest neighbors, and random forest (Kamali et al., 2020) are used in mapping groundwater potential zones or locations.

13.4 FLOOD STUDIES BY INTEGRATING REMOTE SENSING IN HYDROLOGICAL MODELS AND OTHER DATA

Chapter 4 by Dr. Allan S. Arnesen shows the advantages of flood studies when SAR remote sensing data is combined with: (1) ancillary data and (2) hydrological models. First, they demonstrate the use of SAR data and algorithms to map flood extent. Second, they integrate SAR remote sensing–derived information in hydrological models. Third, flood forecasting to minimize the socioeconomic impacts of floods.

First, Chapter 4 demonstrates that in flood mapping, SAR data offers the advantage of looking through clouds, identifying alter below trees and other vegetation, and helping study aquatic vegetation (e.g., submerged rice fields, mangrove forests). The following critical information of SAR data for flood mapping is established by Dr. Arnesen in Chapter 4:

1. Three scattering mechanisms are dominant in floodplain areas: double bounce, volumetric (or canopy), and surface (or specular).
2. The longer the SAR wavelength, the higher the radiation penetration in the canopy will be, and high backscattering values are expected due to double bounce occurrence.
3. Shorter wavelengths (X- and C-bands) have reduced penetration at the canopy and volumetric and surface backscattering mechanisms are predominant.
4. Co-polarization configuration (HH and VV) data is preferable to cross-polarization (HV or VH) to identify folded forest.
5. Flood mapping includes such approaches as: (1) thresholding backscatter values and (2) binary (flood versus no-flood) algorithms based on backscatter value in decibel (dB). Both supervised and unsupervised classification methods are applied. However, pixel-based

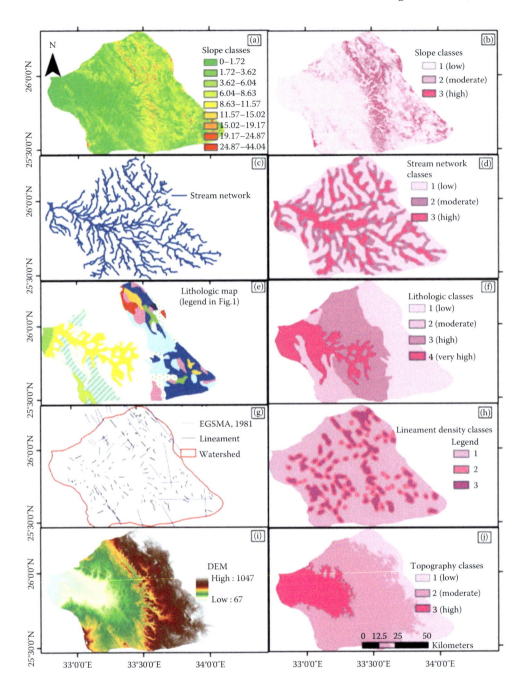

FIGURE 13.3 Example of some of the thematic maps used in determining groundwater potential zones through spatial modeling. Some of the thematic maps include: (a) slope map derived from Shuttle Radar Topography Mission (SRTM) 90 m, (b) slope classes map, (c) stream network map, (d) stream density map, (e) simplified geology map modified after Egyptian Geological Survey and Mining Authority (EGSMA, 1981) Geologic Map of Egypt, Scale 1:2,000,000, (f) geologic map assigned weight factor depending on capability for holding water, (g) lineament map, (h) lineament density map, (i) Digital Elevation Model of the study area, and (j) digital elevation model (DEM) classified map assigned weight factor depending on water infiltration (Source: Abdalla, 2012).

classification approaches provide low accuracies. Many times as low as 30% due to speckle effects in SAR data.
6. Object-oriented classification methodologies involving segmentation procedure and fuzzy logic classifiers using TerraSAR-X and COSMO-SkyMed SARs have shown significantly improved classification accuracies of 80% or above (Guimarães et al., 2020).

Second, Chapter 4 established the advantage of integrating remote sensing with hydrological models, allowing for the determination of various aspects of flood that include extent, stage, frequency, and different scenarios of storms. Remote sensing data (e.g., water-level data from ENVISAT altimeter, land cover from SAR) can help substantially improve the performance of hydrological models by providing a number of parameters such as land cover maps, vegetation type maps, and drainage network information needed in the hydrological models. In addition to remote sensing, the use of parameters derived from other ancillary data (e.g., SRTM-derived DEM, TRMM precipitation, AVHRR skin temperature) will further enhance the performance of hydrological models. Flood simulations will help flood risk zone mapping, flood prevention, and preparedness.

Third, Chapter 4 shows an approach to real-time flood forecasting that uses hydrological models fed by SAR remote sensing data, other remote sensing data, and ancillary data. The hydrological models use several types of data (e.g., Figure 13.4), such as land use, soil type, soil moisture, stream/river base flow, rainfall amount/intensity, snowpack characterization, DEM data, and drainage density/frequency. Most of these are best derived using one or the other remote sensing data.

Modern flood studies integrate multi-sensor remote sensing with Geographical Information System (GIS) data and work in cloud platforms using machine learning, deep learning, and artificial intelligence. Farhadi and Najafzadeh (2021) demonstrate flood risk mapping (FRM) using the web-based Google Earth Engine (GEE) platform (Gorelick et al., 2017, Velastegui-Montoya et al., 2023) with the aid of Landsat 8 satellite imagery, the Shuttle Radar Topography Mission (SRTM) Digital Elevation Model (DEM) to derive 11 risk indices (elevation [El], slope [Sl], slope aspect [SA], land use [LU], Normalized Difference Vegetation Index [NDVI]) Normalized Difference Water Index [NDWI], Topographic Wetness Index [TWI], river distance [RD], waterway and river density [WRD], soil texture [ST], and maximum one-day precipitation [M1DP]). Flood inundation probability estimates were performed using geotagged photographs sourced from social media, optical remote sensing and high-resolution terrain mapping, and by developing a Bayesian statistical model (Rosser et al., 2017). Machine learning models such as random forest (RF) model and a Bayesian generalized linear model (GLMbayes) (Avand et al., 2021) along with remote sensing, and GIS data are often used to investigate flood probability due to various causes, such as due to effects of changing climates and land uses (e.g., Avand et al., 2021). Hydrological models that predict floods have many parameters, such as land cover, snow cover, and quantitative geomorphological information coming from remote sensing data. The difference between machine learning–based methods and hydrological models is that machine learning methods are data-driven and mainly depend on the training data to produce accurate results, while hydrological models are knowledge-based, which implies that the human experts already feed them the knowledge to make flood-related decisions (Munawar et al., 2022). Multiscale and multi-sensor application for flooded area mapping using free satellite data from low- to medium-high-resolution from both the SAR (Sentinel-1, COSMO-Skymed) and multispectral sensors (MODIS, Sentinel-2) (Giordan et al., 2018), whereas damage assessments require sub-meter to 5 m very high spatial resolution imagery. Merging Landsat, Sentinel-2, and Sentinel-1 increased detection of extreme hydrological events such as floods from 7% for single sensors to up to 66% for a multi-sensor product (Munasinghe et al., 2023).

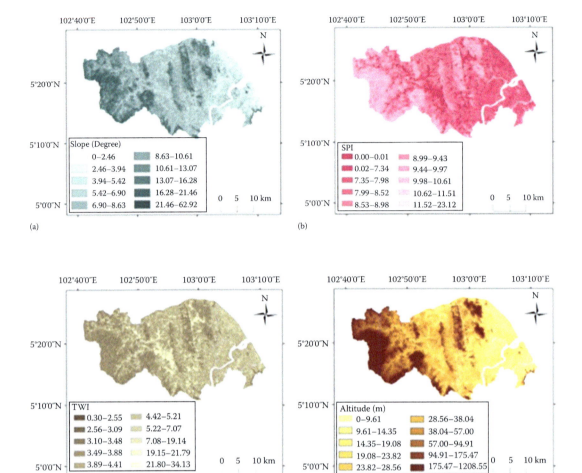

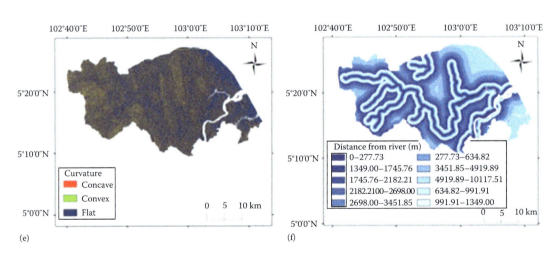

FIGURE 13.4 Some input thematic layers in flood susceptibility mapping shown for Terengganu, Malaysia: (a) slope, (b) stream power index (SPI), (c) topographic wetness index (TWI), (d) altitude, (e) curvature, (f) distance form river, (g) geology, (h) rainfall, (i) land use and land cover (LULC), and (j) soil type (Source: Tehrany et al., 2014).

Remote Sensing Handbook, Volume V   481

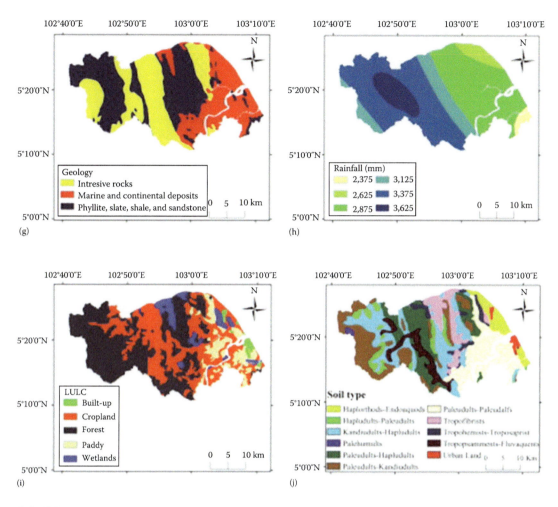

FIGURE 13.4 (Continued)

13.5 FLOOD STUDIES USING SAR REMOTE SENSING

Every year floods affect nearly 100 million people, with an average of nearly 7000 people killed and causing ~$14 billion in damages (Rentschler et al., 2022; Taguchi et al., 2022; Yu et al., 2022). Great floods in the past have caused immense damage to human life, such as the 1991 floods of Bangladesh (139,000 deaths), the 1938 floods of China (870,000 deaths), and the 1530 flood of the Netherlands (400,000 deaths) (Alcamo et al., 2000). Apart from looking at floods from a disaster point of view, they are also natural phenomenons that are often the lifeline of several billions living along the river courses that bring water for agriculture, fertile soil from elsewhere, and replenish groundwater. Floods should also be looked as "cleansing" the river systems of pollution. Flood water when harnessed can serve humanity during lean flow periods. So, there are many reasons to study floods.

There are numerous aspects of floods about which remote sensing provides useful information. Some of the following are discussed in Chapter 5 by Martinis et al.:

1. Flood inundation studies that include before-, during-, and after-flood events
2. Establishing flood-prone areas
3. Assessing areas of damage and areas benefited by floods
4. Determining storage volumes in reservoirs as a result of flooding
5. River morphological studies and changes in river meandering as a result of flooding

6. Flood prediction studies that include snowmelt runoff studies, rainfall, and soil moisture
7. Flood stage (which in turn relates to discharge)
8. Flood sedimentation studies. This last aspect is not covered in the chapter but is important for readers to further explore

Both active and passive remote sensing is widely used in flood studies. Passive sensors are preferred when these data are available during the flood events since they often offer better spatial resolution than SAR data and the multiple spectral characteristics of optical sensors allow us to study most of the flood parameters listed earlier. Passive sensors include data from sensors such as MODIS, ASTER, Landsat, and a suite of very high resolution imagery. Using these data is always a trade-off between spatial resolution, the large area coverage, and frequency of coverage (e.g., MODIS has coarse resolution of 250–1000m, but has daily global coverage). Floods often require large area coverage, such as the entire river basin, to understand, model, and map the flood phenomenon rather than to look at just the areas flooded.

The microwave region (1 mm–1 m wavelength) or 300 GHz–300 MHz (frequency) is specifically beneficial to overcome clouds. Chapter 6 by Martinis et al. focuses on flood studies from SAR data. Since most floods occur when clouds are at maximum, SAR data becomes very useful. The microwave region of the electromagnetic spectrum can be divided into the following bands: P-band (30–100 cm wavelength, 0.3–1 GHz frequency); L-band (15–30 cm, 1–2 GHz); S-band (8–15 cm, 2–4 GHz); C-band (4–8 cm, 4–8 GHz); X-band (2.5–4 cm, 8–12 GHz); Ku**-band (1.7–2.5 cm, 12–18 GHz); K-band (1.1–1.7 cm, 18–27 GHz); Ka**-band (0.75–1.1 cm, 27–40 GHz); V-band (0.4–0.75 cm, 4–75 GHz); W-band (0.27–0.4 cm, 75–100 GHz); and mm-band (< 0.27 cm, 110–200 GHz). L, C, X, and Ka radar bands are especially used in flood studies, with C being most common and L being used to map inundation under vegetation. Spaceborne SAR data includes data gathered from such sensors (Chapter 5 by Martinis et al. as well as Chapter 4): TerraSAR-X/TanDEM-X, COSMO-SkyMed, X- (SIR-C/X-SAR, SRTM), C- (ERS-1/2 AMI, Envisat ASAR, RADARSAT-1, RISAT-1, SIR-C/X-SAR), and L-band domain (SEASAT-1, JERS-1, ALOS PALSAR, SIR-A/B/C/X-SAR), ERS-1, ERS-2, JERS-1,2, RADARSAT-1,2, ENVISAT-1, Sentinel-1, SMMR, SSM/I, DMSP, and radar altimetry. Nevertheless, poor spatial resolutions of most of the SAR sensors limit their use. Since 2007, the successful launch of the platforms TerraSAR-X/TanDEM-X, COSMO-SkyMed, and RARARSAT-2 marks a new generation of civil SAR systems suitable for detailed flood mapping purposes up to a resolution of 0.24 m (TerraSAR-X Staring Spotlight mode).

Martinis et al. (Chapter 5) helps us understand the interactions between SAR signals and water bodies, including their strengths and limitation in flood mapping. For example, they establish factors that overestimate flooding as: (1) shadowing effects behind vertical objects such as vegetation, topography, and anthropogenic structures; (2) smooth natural surface features such as sand dunes, slat and clay pans, and bare ground; and (3) smooth anthropogenic features such as streets, airstrips, and heavy rain cells. They also enumerate the underestimating factors of flooding such as: (1) volume scattering of partially submerged vegetation; (2) double bounce scattering of partially submerged vegetation; (3) anthropogenic features on the water surface such as ships and debris; (4) roughening of the water surface by wind, heavy rain, or high flow velocity; and (5) layover effect on vertical objects such as topography and urban structures. This theoretical basis provides many insights (Chapter 5), such as:

1. Water monitoring using X-band SAR (e.g., Figure 13.5) is considered more suitable than using longer wavelengths, such as C- and L-bands.
2. HH polarization provides best discrimination between water and non-water terrain.
3. Superiority of cross-polarization HV over like polarization VV in monitoring roughened water surfaces.
4. Partially submerged vegetation may cause a backscatter increase over water bodies.
5. L-band SAR are more effective in detecting water under forest canopies.

6. Rice crop in early to mid-phases increase backscatter, whereas in the ripening phase backscatter decreases.
7. Likelihood to detect flooding in urban areas generally increases with decreasing incident angle.

Chapter 5 provides an exhaustive picture of flood mapping using SAR data (e.g., Figure 13.5) through visual as well as digital image analysis. They discuss in detail the three flood mapping approaches (Amitrano et al., 2024; Martinis et al., 2015):

- Semi-automatic object-based flood detection (RaMaFlood)
- Automatic pixel-based water detection (WaMaPro)
- Fully automatic pixel-based flood detection (TerraSAR-X Flood Service)

These are comprehensive and thorough, clearly illustrated with case studies.

The remote sensing technologies used for flood prediction can be divided into three types: multispectral, radar, and light detection and ranging (LIDAR) (Munawar et al., 2022). SAR images are used to penetrate cloud and haze and provide excellent discrimination of flooded areas relative to surrounding non-flooded areas due to distinct differences in radar coefficients. Fast flood extent mapping and monitoring with SAR change detection using the Google Earth Engine (GEE) (Gorelick et al., 2017; Velastegui-Montoya et al., 2023) has shown high agreement between the SAR and optical imagery (77%–80%), with SAR providing the benefit of under-cloud detection (Hamidi et al., 2023). To identify flooded areas, the combination of Sentinel-2 optical images, SAR Sentinel-1, and DEM, with ML algorithms provide more reliable results than if we used only Sentinel-2 or Sentinel-1 for binary classification (water and non-water) (Soria-Ruiz et al., 2022). Munawar et al. (2022) recommend use of: (1) machine learning (e.g., CNN, AI, KNN, Bayesian linear, SVM) for flood forecasting, (2) image segmentation to determine water fluctuation levels to predict flooding likelihood in urban areas, and (3) hydrological models for flood forecasting using remotely sensed data. Overall, all studies recommend a combined use of SAR and optical data in flood studies.

13.6 REMOTE SENSING OF MANGROVE FORESTS

Mangroves occupy ~0.1% of the terrestrial Earth but are unique and irreplaceable ecosystems with very high primary productivity, most carbon-rich biomes (Figure 13.6) containing an average of 937 tC ha^{-1} (Alongi, 2014); rich and unique flora and fauna housing many endangered species; supporting shoreline protection (e.g., during tsunamis); sustaining livelihoods through food, fodder, and several renewable resources (e.g., firewood, furniture); and encouraging coastal tourism. Even though mangroves account for approximately 1% (13.5 Gt year 1) of carbon sequestration by the world's forests (due to their low percentage of area), but as coastal habitats they account for 14% of carbon sequestration by the global ocean (Alongi, 2014). As per the United Nation's Food and Agricultural Organization (FAO) data, mangroves once covered more than 20 million hectares (Spalding et al., 1997) of sheltered tropical and subtropical coastlines, but it is disappearing at 1–2% per year (Duke et al., 2007). Mangrove wetlands are encroached for multiple purposes that include shrimp farming, fish farming, agriculture, deforestation, diversion of fresh water, and coastal development for urbanization and other purposes.

In Chapter 6, Wang et al. provide a good overview and definitions of the mangroves and lay out methods and approaches to best monitor them using remote sensing. A unique feature of the chapter is the various spatial scales at which mangroves are studied and the appropriate remote sensing data used for the same. At the leaf level, hyperspectral data and hyperspectral vegetation indices (HVIs) play a critical role. Since hyperspectral narrowband (HNB) data is acquired as near-continuous mode along the electromagnetic spectrum, hyperspectral vegetation indices (HVIs) help derive specific biophysical and biochemical properties of the leaf level accurately (Thenkabail et al., 2014,

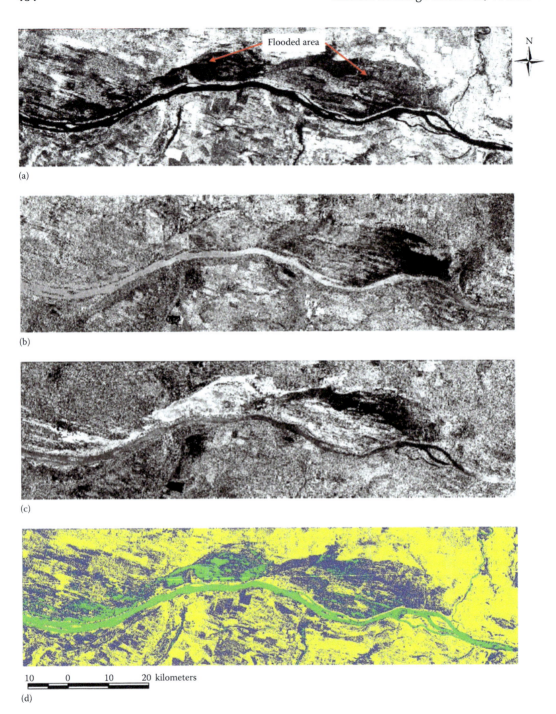

FIGURE 13.5 Plots of: (a) principal component 1 (PC1) that explains 39.81% of the total variance, (b) PC2 that explains 15.47% of the total variance, and (c) PC3 that explains 12.79% of the total variance, of the principal component analysis (PCA) of seven European Remote Sensing Satellite2 Synthetic Aperture radar (ERS2-SAR) images of original pixel values for the RCB reach of the Danube River at Romania (using seven ERS2 SAR images), where the dark color represents permanent water (Danube River) and the gray color represents part of the flooded area. Light gray and white likely represent the dry land; and (d) the classification of PC1 of seven ERS2-SAR images by the Isodata classifier (green: permanent water; blue: flooded area; yellow: dry land) (Source: Gan et al., 2012).

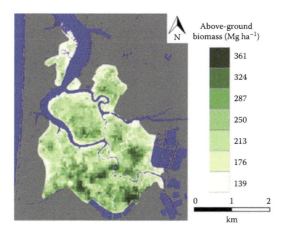

FIGURE 13.6 Map of mangrove biomass produced using high-resolution GeoEye-1 satellite imagery, medium resolution ASTER satellite elevation data, field-based biomass data, and a support vector machine regression model. The aboveground biomass (AGB) for the 151-ha Kamphuan mangrove, as computed by the equation AGB = 0.16 * elevation + 0.27 * (band 1/band 2)—0.11 * band 2 + 0.41 * band 4−0.03 and was 250 ± 53.4 Mg ha−1 with the highest carbon stocks located in the mangrove interior and the landward edge (Figure 13.6). Using the 0.38 ratio of belowground biomass (BGB) to AGB yielded an estimated BGB of 95 Mg ha−1. Combined AGB and BGB at the site was 345 ± 72.5 Mg ha−1. Using the 0.45 conversion factor between biomass and carbon stock, the estimated above- and belowground carbon biomass was 113 and 42.8 Mg C ha−1, respectively (Source: Jachowski et al., 2013).

2004). At the individual tree level, it is important to have sufficient spatial resolution, such as data from sub-meter to 5 m very high-resolution imagery (VHRI) data from sensors such as WorldView or LiDAR data from uncrewed aerial vehicles. The VHRI data helps gather individual tree crown data and their biomass. At the forest canopy level, a wide range of remote sensing data can be used. For example, to map land cover types, Landsat and sentinel data are ideal. For species types, a combination of hyperspectral and hyperspatial data will be ideal. For forest biomass and carbon assessments (Figure 13.6), multi-sensor data that integrated Landsat and WorldView type data with terrestrial laser scans (TLS) data will be ideal. At the ecosystem level, various mangrove studies can be conducted, such as tidal-driven nutrient exchange, biodiversity studies, biomass and carbon assessments, and coastal area protection studies. These are described in the chapter along with the type of remote sensing data used to study them.

Wang et al. (2019) conducted the first chronological review of mangrove remote sensing studies in six decades (1956–2018). What is interesting here is the evolution of mangrove studies using remote sensing with the evolution of satellite remote sensing. In this regard, Wang et al. (2019) outlined mangrove studies chronologically as follows: (1) mangrove extent before 1989; (2) mangrove LAI during 1990–1999; (3) mangrove species classification, structure and biomass modeling, carbon stock estimation, and health and condition during 2000–2009; and (4) carbon fluxes, ecohydrology, and the impact of climate change on mangroves during 2010–2018. Each period had a set of new satellites and sensors that facilitated these advances in studies. For example, global carbon emissions are monitored from space, by three pioneering satellites: NASA's Orbiting Carbon Observatory-2 (OCO-2), which was launched in 2014 and measures CO_2, Japan's Greenhouse Gases Observing Satellite (GOSAT), which was launched in 2009 and observes CO2 and methane, and China's TanSat (Wang et al., 2019). Currently, all of the aforementioned mangrove characteristics are studies using a wide array of satellite sensors such as Landsat series, Sentinel series, MODIS, WorldView series, GeoEye series, Planet Labs Doves and Super Doves, LiDAR, SAR, radar, and various hyperspectral sensors. Normalized Difference Vegetation Index (NDVI) was found to be the most widely applied index in mangroves, used in 82% of the studies reviewed, followed by the Enhanced Vegetation Index (EVI), used in 28% of the studies (Tran et al., 2022). Giri (2023) studied mangrove health conditions and dynamics of the world using the Moderate Resolution Imaging Spectroradiometer (MODIS) NDVI dataset during 2000–2018. Mangroves are overwhelmingly in tropical and subtropical regions, with cloud cover issues for most of the year. Hence, a combination of multispectral and SAR images is preferred.

13.7 MANGROVE WETLANDS OF THE WORLD

In Chapter 7, Dr. Chandra Giri discusses the rich importance and many uses of mangrove wetlands. The chapter shows us how the mangrove wetlands are mapped using remote sensing. Chapter 7 defines mangroves for mapping and monitoring purposes using remote sensing and enumerates on scale/resolution issues of various remote sensing data in mangrove mapping and monitoring. The chapter answers questions such as: areal extent of mangrove forests, where they are located, their change over space and time, causes of those changes, and potential areas for regeneration. Chapter 7 does this by using Landsat data for 1975, 1990, 2000, and 2005. The chapter presents and discusses various classification and change detection approaches and methods. It also shows mangrove species discrimination methods and approaches. The challenges and opportunities in using very high spatial resolution data in mangrove wetland mapping have also been presented.

More recently, mangroves of the world have been mapped at 10m resolution (Figure 13.7) by Jia et al. (2023) using Sentinel-2 data that estimated 145,068 km² mangrove forests in 2020 with the highest in Asia (39.2%) and amongst the countries highest in Indonesia followed by Brazil and Australia. The most popular application used in the remote sensing of mangroves is the mapping of mangrove and non-mangrove areas and conducting time-series change analysis (Giri, 2023). Accurately mapping global mangrove aboveground biomass (AGB) will help us understand how

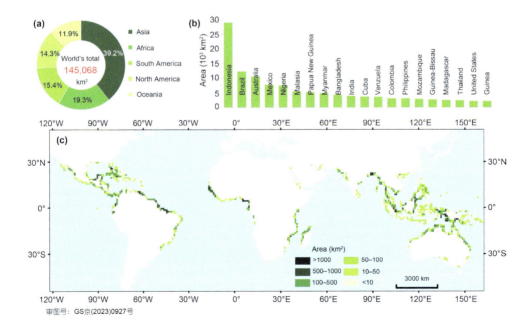

FIGURE 13.7 Areal extent and distribution of global mangrove forests in 2020. (a) Area and proportion of mangrove forests on each continent. (b) Areal extents of mangrove forests in the top 20 mangrove-rich countries. (c) Distributions of mangrove forests summarized in each decimal degree square (Source: Jia et al., 2023).

mangrove ecosystems are affected by the impacts of climatic change and human activities (Hu et al., 2020). Bunting et al. (2022) used L-band Synthetic Aperture Radar (SAR) global mosaic datasets from the Japan Aerospace Exploration Agency (JAXA) for 11 epochs from 1996 to 2020 to develop a long-term time series of global mangrove extent and change. They determined that, globally, there was 3.4% reduction in mangroves over the 24-year period. Mangroves reduced from 152,604 km^2 in 1996 to 147,359 km^2 in 2020. Global mangrove forest mapping is of critical importance. However, Chowdhury and Hafsa (2022) and Giri (2023) found that the four global mangrove maps have little consensus on the location of mangrove range limits. So, more thorough studies of regional, local mangrove forest characteristics are significant. This is now possible with freely available, well-calibrated harmonized Landsat and Sentinel Data (Masek et al., 2022; Masek et al., 2021). For example, time-series archival Landsat data along with detailed GIS data at the local/regional level is now allowing multi-decadal land cover change analysis such as illustrated over the Sundarbans Mangrove Forest of Bangladesh (Chowdhury and Hafsa, 2022), which is the largest mangrove forest in the world.

13.8 WETLAND MODELING AND MAPPING METHODS USING MULTI-SENSOR REMOTE SENSING

Wetlands are areas where water is the primary factor controlling the environment and the associated plant and animal life (Ramsar). Article 1.1 of the Ramsar convention defines wetlands as: "areas of marsh, fen, peatland or water, whether natural or artificial, permanent or temporary, with water that is static or flowing, fresh, brackish or salt, including areas of marine water the depth of which at low tide does not exceed six metres" (Gell et al., 2023; Gardner et al., 2023). Wetlands are found throughout the world, except Antarctica, and occupy anywhere between 7 and 9 million km^2 area (~4–6% of the terrestrial area). They are major habitats for a diverse group of plants and animals. Since wetlands constitute a mix of water, land, and vegetation and are of various sizes, from very large areas such as the flood plains along major rivers and river deltas (e.g., Figure 13.8) to tiny streams, every type of remote sensing data acquired in various spectral, spatial, temporal, and radiometric resolutions and in various portions of the electromagnetic spectrum (e.g., optical, thermal, radar) have been used to study them. Chapter 8 by Dr. Deepak R. Mishra et al. starts by providing an overview of the evolution of wetland remote sensing. This reveals that various widely used remote sensing techniques such as supervised and unsupervised clustering algorithms for wetland classification, vegetation indices for modeling wetland biophysical and biochemical quantities, as well as advanced image processing techniques such as object-based image analysis (OBIA) and data fusion involving multiple sensor data, such as the LiDAR and hyperspectral sensors are used to quantify, model, map, and monitor wetlands. They provide three studies to illustrate the value of various aspects of wetland studies using widely varying remotely sensed data:

1. LiDAR (1 m, 1 band, acquired at 1047 nm) and hyperspectral data (1 m, 63 bands, over 400–980 nm) fusion for accurate habitat mapping and elevation mapping in salt marsh environments (Chen et al., 2022).
2. Biophysical and biochemical quantification studies of the salt marsh habitats of the entire Georgia coast using the MODIS 250\500 m time-series data (Tang et al., 2022).
3. Estimation of aboveground and belowground biomass and foliar nitrogen (N) using satellite data and hybrid modeling in the freshwater marsh through hyperspectral and multispectral data (Hemati et al., 2024)

Some of the highlights of this comprehensive study by Dr. Deepak R. Mishra et al. in Chapter 8 are:

1. Hyperspectral data was able to classify nine distinct classes of salt marshes with overall accuracy of ~90% and producer and user accuracies of individual species ~80%.

2. Decision trees incorporating LiDAR-derived DEM and hyperspectral narrowband NDVI significantly improved classification accuracies.
3. Vegetation classification by fusing LiDAR and hyperspectral data offered significantly higher classification accuracies, especially for the producer and user accuracies.
4. LiDAR tends to overestimate saltmarsh elevations due to poor penetration of dense vegetation
5. When there are multiple species within a hyperspectral pixel, species separation becomes more complicated. This indicated hyperspectral data needs to be supported by hyperspatial pixel size that can capture the species spectra for better classification of the species.
6. Species separability needs to consider phenology, biomass productivity, and biochemical compositions (e.g., N) for better understanding and separation.
7. When highly accurate vegetation classification maps are available (e.g., using hyperspectral data), the DEMs of wetland obtained from LiDAR can be substantially improved.
8. Biophysical variables such as canopy chlorophyll (Chl) content, green leaf area index (LAI) (a ratio of green foliage area vs. ground area), green vegetation fraction (VF) (percent green canopy cover), and aboveground green biomass (GBM) can be modeled using MODIS and Landsat data with reasonable consistency over large areas.
9. There were substantial difficulties in determining the aboveground and belowground biomass of freshwater marsh species, with the best hyperspectral and multispectral models explaining up to 56% variability in data. This is mainly due to mixed pixel signatures making it difficult to acquire data of specific species. Uncertainties in data lead to uncertainty in results.

Global wetland maps have huge discrepancies. Hu et al. (2017) found estimates of global wetland area ranging from 0.54 to 21.26 million km². This is not surprising given that wetland definitions in

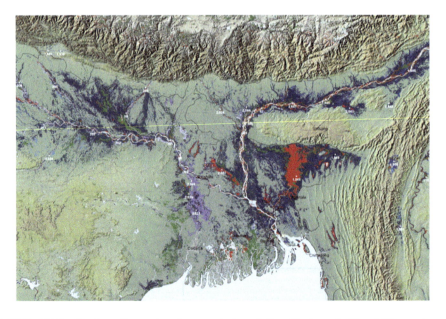

FIGURE 13.8 Wetlands along the major rivers and river deltas. Dartmouth Flood Observatory (DFO)–produced spatial flood coverage (1999–2009) depicting large wetlands along the Ganges-Brahmaputra Rivers (India, Bangladesh, China), and their deltas are characterized with data from three remote sensing systems (AMSR-E, SRTM, and MODIS). Colors represent different years, with more recent years overlying earlier flooded areas. Often, flooded areas reoccur from year to year (Source: Syvitski et al., 2012).

mapping varies widely and one should use the Ramsar definitions (Gell et al., 2023; Gardner et al., 2023) for consistency and inter-comparability. The most recent high-resolution estimate of global wetland area is in excess of 12.1 × 106 km², of which 54% is permanently inundated and 46% is temporarily inundated, and 92.8% of continental wetland area is inland, while only 7.2% is coastal (Davidson et al., 2018). They also determined that the largest wetland areas are in Asia (31.8%), North America (27.1%), Latin America and the Caribbean (Neotropics; 15.8%), with smaller areas in Europe (12.5%), Africa (9.9%), and Oceania (2.9%). Experimental results by Rezaee et al. (2018) demonstrated that using machine learning convolution neural network (CNN) was superior to random forest (RF) for complex wetland mapping even by incorporating the small number of input features (i.e., three features) for CNN compared to RF (i.e., eight features). Similar advantages of using CNN over traditional machine learning methods (i.e., support vector machine, linear discriminant analysis, K-nearest neighborhood, canonical correlation forests, and AdaBoost.M1) were determined by Günen (2022) in mapping wetland water areas.

13.9 INLAND VALLEY WETLAND CHARACTERIZATION AND MAPPING

Inland valleys (IVs) are lowland ecosystems spread across landscapes. Unlike large wetlands that occur along the higher order streams and in the river deltas, IVs occur along the lower order (e.g., 1st to 4th order) streams (e.g., Figure 13.9). Inland valleys offer an extensive, fairly unexploited potential for agricultural production due to rich soils and significantly higher water and moisture availability when compared with adjoining uplands. As discussed in Chapter 9 by Dr. Murali Krishna Gumma et al., IV wetlands have high potential for growing agricultural crops due to their: (1) easy access to river water, (2) significantly longer duration of adequate soil moisture to grow crops when compared with adjoining uplands, and (3) rich soils (depth and fertility) (Akumu et al., 2018). Even though Ramsar convention standards (Gell et al., 2023; Gardner et al., 2023) are widely followed in wetland mapping, IV wetlands are not covered in their definitions (Mandishona and Knight, 2022) and most mapping exercises ignore or only partially map them. This is one of the causes of huge differences in wetland areas mapped by different studies.

Chapter 9 focuses on characterizing and mapping IV wetlands of Africa, given highly unexploited IVs found throughout Africa and the need to understand, model, map and then prioritize the use and conservation of wetlands. The wetlands of Africa are increasingly considered "hot spots" for agricultural development and for expediting Africa's Green and Blue Revolutions. Currently, these IV wetlands are un-utilized or highly under-utilized in Africa despite their rich soils and abundant water availability as a result of: (1) limited road access to these wetlands and (2) prevailing diseases such as malaria, trypanosomiasis (sleeping sickness), and onchocerciasis (river blindness). However, the utilization of IV wetlands for agriculture is becoming unavoidable in African countries due to increasing pressure for food from a ballooning human population and difficulty finding arable land with access to water resources (Akpoti et al., 2022; Vogels et al., 2019). Given these pressures, it is critical to recognize the important functions of wetlands, which include holding about 20% of all carbon on Earth and providing habitat for unique and increasingly rare flora and fauna. Therefore, any proposed use of wetlands for agriculture has to be carefully weighed against the social benefits provided by the inherent ecological services of conserved and intact wetlands. In order to address the issue of the development vs. conservation of wetlands, Chapter 9 provides a structured approach to utilizing multi-sensor remote sensing data as well as various secondary data to develop methods and approaches of mapping IVs leading to informed decision-making by local, regional, national, and global stakeholders. The chapter follows the earlier, pioneering work of Thenkabail and Nolte (1995) and Thenkabail et al. (2000a, 2000b).

Chapter 9 first demonstrates how to identify, delineate, map, and characterize wetlands over large areas using data fusion involving satellite sensor data (e.g., Landsat, SPOT, MODIS Terra/Aqua, JERS SAR, IKONOS/QuickBird), secondary data (SRTM, FAO soils, precipitation), and in situ data. Second, the chapter shows methods and approaches to develop a decision support system

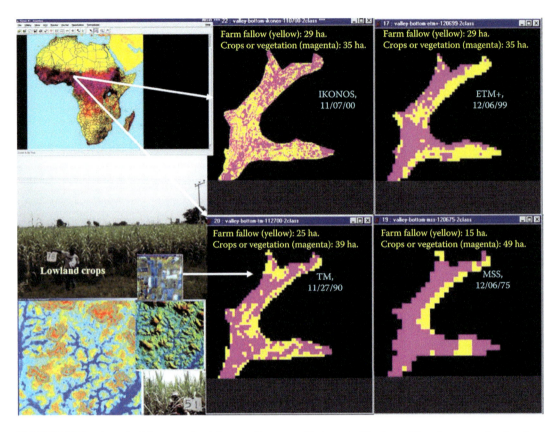

FIGURE 13.9 A typical lower-order inland valley near Kaduna in Northern Nigeria mapped using four satellite sensor data types: 4 m IKONOS, 30 m Landsat ETM+, 30 m Landsat TM, and 56 m Landsat MSS. In each case, two classes—farmland fallow and cultivated farmlands—were mapped. It is clear from these images that the spatial detail and precision seen in 4 m IKONOS is significantly higher than any other imagery. The lower left bottom images show 4 m DEM image derived from stereo pairs of IKONOS showing distinct lowland inland valleys (deep blue) as opposed to rest of the landscape. The photos (#78, #51) are two typical farms in the inland valleys of Northern Nigeria.

(DSS) through spatial modeling to perform land suitability analysis in order to determine which of the IV wetland areas are best suited for: (1) agricultural development or (2) preservation. Chapter 9 demonstrates the differences in detail and accuracies mapped using widely varying resolution (e.g., Figure 13.9) of remote sensing data.

Small inland wetlands are difficult to map and require a combination of data and methods. Géant et al. (2023) combined optical and SAR images and by integrating topographic, hydrological, and vegetation indices into the four most used ML classifiers (artificial neural network [ANN], random forest [RF], boosted regression tree [BRT], and maximum entropy [MaxEnt]) and found that RF exhibited the most accurate predictions in mapping small inland wetlands from non-wetlands, achieving an overall classification accuracy of 95.67%.

13.10 ACTUAL EVAPOTRANSPIRATION (WATER USE) OF CROPLANDS FROM REMOTE SENSING

The actual water consumed by the crops to grow food is referred to as actual evapotranspiration (ET_a expressed in mm/day). Currently, based on different estimates, anywhere between 4500 and

7500 km³/yr of water is consumed by 1.5–1.7 billion hectares of global croplands (Mekonnen and Gerbens-Leenes, 2020; Thenkabail, 2010; Siebert et al., 2006). This consumption is expected to increase by 2050 to 12,000–13,500 km³/yr (Droogers et al., 2010). Since agriculture accounts for about 80% of all human water use, determining and managing crop water use is of great importance. Remote sensing is the most powerful data source for consistent and accurate estimates of ET_a (water use) from a single farm to very large areas routinely and repeatedly. For example, Figure 13.10 shows ten-day cumulative ET_a or water use in mm of irrigated crops derived using multiple sensor time-series imagery.

Chapter 10 by Dr. Trent W. Biggs et al. provides systematic steps on methods of estimating ET_a using remote sensing data. They categorize the ET_a methods using remote sensing into three distinct groups:

1. Vegetation-based methods include empirical crop coefficient methods that require ground-level reference data, and process-based approaches that have minimal ground-level data requirements. Process-based approaches include the Priestley–Taylor Jet Propulsion Laboratory (PT-JPL) and MODIS global Evapotranspiration Data Set (MOD16) methods (Tang et al., 2024; Ling et al., 2022). Vegetation-based methods may have difficulty predicting evaporation from wet or inundated soil, which complicates their use in some types of irrigated agriculture. However, there is significant uncertainty that is yet to be resolved in these methods and approaches, requiring further studies.
2. Temperature- or energy-based methods include one-source models (Surface Energy Balance Algorithm for Land [SEBAL], Mapping Evapotranspiration at high Resolution with Internalized Calibration [METRIC]), two-source models (Atmosphere-Land Exchange Inverse [ALEXI], Disaggregated ALEXI [DisALEXI]), and simplified methods (Simplified Surface Energy Balance [SSEB], and operational SSEB [SSEB$_{op}$]). SEBAL and METRIC are most useful in irrigated areas in semi-arid and arid landscapes where extremes of ET_a are present in the image. Two-source models are more complex but are often more accurate than one-source models, particularly over areas with partial vegetation cover. SSEB is simpler to understand and apply and often has comparative performance with other one source models.
3. Scatterplot or triangle-trapezoidal methods use plots of surface temperature or albedo versus a vegetation index. This simple approach may provide comparable results with surface energy balance models.

For large regions or global applications, the authors recommend the use of an ensemble of models, since no one model performs the best across land cover types. However, the uncertainty in ET_a models is still very high, especially when the models are applied over large areas.

Each model has its own intricacies and complexities involving parameters derived from remote sensing data as well as some from meteorological data, as discussed in detail in various sections and sub-sections of Chapter 10. In its most simplistic form ET_a or the water used by crops (expressed m³/ha or mm/m²) is derived by (Platonov et al., 2008):

- Determining the ET fraction (e.g., Landsat ETM+ thermal data);
- Calculating the reference ET (e.g., using Penman–Monteith equations; Tang et al., 2024; Ling et al., 2022; and Mekonnen and Gerbens-Leenes, 2020); and
- Computing ET_a by multiplying ET fraction with reference ET.

The ET fraction (ET_f), or evaporative fraction, is the ratio of ET_a over reference ET (ET_0). The Simplified Surface Energy Balance (SSEB) model, for example, calculates ET_f based on the assumption that the latent heat flux (ET_a) varies linearly between the land surface temperature (LST) of "hot" and "cold" pixels (Platonov et al., 2008):

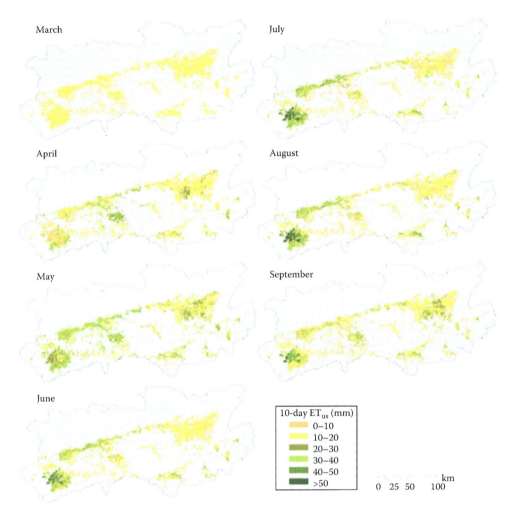

FIGURE 13.10 Actual evapotranspiration (water use) of irrigated crops. Maps of cumulative ET_{us} for the first ten-day of each 2009 campaign month over the Guadalquivir River basin irrigated area. When the ETs from different months are added, we will get cumulative ET_{act} (water use) of these irrigated crops for the entire season. Note: ET_{us} (with the subscript "us" meaning unstressed) to explicitly acknowledge that the reduction in plant transpiration due to stomata closure or water vapor pressure deficit is ignored. However, the effect of water stress on plant growth is reflected through the vegetation index, thus this indirect effect of water stress is accounted for in ET_{us} (Source: González-Dugo et al., 2013).

$$ET_f = (T_{hot} - T)/(T_{hot} - T_{cold}) \qquad (1)$$

where ET_f is the fraction of ET (dimensionless); T is the land surface temperature (LST) of any pixel; T_{hot} and T_{cold} are the LST of "hot" and "cold" pixels, respectively, the LST expressed in degree Kelvin or Celsius. The "hot" and the "cold" pixels are selected inside the irrigated fields of the investigated area for each image.

Reference ET (ET_0) is typically calculated using various methods (e.g., Priestley–Taylor, Blaney–Criddle, Hargreaves, Penman, Penman–Monteith) using various meteorological data. $ET_a = ET_f * ET_o$.

The surface energy balance models (e.g., SEBAL, METRIC) compute ET_a on a pixel-by-pixel basis for the instantaneous time of the satellite image, as the residual amount of energy remaining from the classical energy balance:

$$\lambda ET = Rn - G - H \tag{2}$$

where λET is latent heat flux (the energy used for evapotranspiration), Rn is net radiation at the surface, G is soil heat flux, and H is sensible heat flux to the air. All fluxes are in W m^{-2} day^{-1} units.

ET (mm day^{-1}) is calculated from latent heat flux by dividing it by the latent heat of water vaporization (λ) (Platonov et al., 2008).

Readers should note that the previous equation only covers one-source energy-based methods (and not PT-JPL or MOD16). Please refer to Chapter 10 for equations pertaining to PT-JPL or MOD16.

The daily or, in the least, every few days measurement of crop water use (actual evapotranspiration) data is desired for efficient and active agricultural crop monitoring. This is increasingly becoming possible with data such as: (1) enhanced spatial and temporal adaptive reflectance fusion model (ESTARFM) derived data that fuses Moderate Resolution Imaging Spectroradiometer (MODIS) data with Landsat Enhanced Thematic Mapper Plus (ETM+) for estimation of daily evapotranspiration and irrigation water efficiency at Landsat-like scale (Ma et al., 2018), and (2) fusion of Landsat-8,9 and Sentinel-2A,2B (S2) surface reflectance (SR) products already available in GEE (Gorelick et al., 2017; Velastegui-Montoya et al., 2023), and NASA's Harmonized Landsat Sentinel-2 (HLS) Landsat product (HLSL30) for 2013–present and HLS Sentinel-2 product (HLSS30) for 2015–present, that together have two to three days of global coverage (Masek et al., 2021, 2022) at nominal 30m resolution. These high spatial and temporal resolution data enable the study of crop biophysical, biochemical, plant health, plant stress, plant structural quantities, plant water, and plant water productivity throughout the growing season. Such data, along with climate data from meteorological stations, provides the ideal data required for farmers and decision-makers. Crop water use is increasingly monitored by an ensemble of six ET models (ALEXI/DisALEXI, eeMETRIC, geeSEBAL, PT-JPL, SIMS, and SSEBop) that have now become the gold standard in actual evapotranspiration modeling using remote sensing and fills a critical gap in water management (Melton et al., 2022).

13.11 MODELING WATER PRODUCTIVITY STUDIES FROM EARTH OBSERVATION SYSTEMS

Globally, ~80% (4500–7500 km^{3}yr^{-1}) of all human water use goes toward agriculture to produce food from ~1.5 billion hectares of existing irrigated and rainfed croplands (Thenkabail et al., 2010). Also, competition for water from multiple sectors (e.g., urban, recreation, environmental flows, and industries) is rising steeply, making such large quantities of agricultural land and water use untenable. On the other hand, global population is increasing and expected to reach 9.2–10 billion by the year 2050 from its current 7.2 billion, further increasing demand for food and nutrition, and associated increases in land and water allocations (Dorling, 2021). However, it is widely perceived that neither the increase allocation of land nor increased allocations of water is practical. Indeed, it is highly likely that the rising demand for food and nutrition needs to be met by decreasing land and water for growing food. Also, the Green Revolution (productivity increases per unit of land) era of the last 50 years has come to an end. Thereby, there are several attempts for increased food production through smart technologies and scientific advancement that are, hitherto, rarely explored. Improved water productivity (WP) of croplands around the world is considered the most promising opportunity of all the available measures for increasing food productivity.

Crop water productivity (CWP; kgm^{-3}; productivity per unit of water, crop per drop, biomass per evapotranspiration, or crop yield per evapotranspiration) studies are best conducted by integrating multi-sensor remote sensing (RS) data with surface energy balance modeling (evapotranspiration – ET modeling for water use by crops), agro-meteorological data, water withdrawal data, and biophysical and yield data in a geographical information system (GIS). Crop water productivity (CWP) is a performance indicator to monitor and evaluate water use efficiency in agriculture (Blatchford et al.,

2019). Methods and protocols for crop water productivity (CWP) typically involve five broad steps: (1) cropland area mapping (CAM, ha); (2) crop productivity mapping (CPM, kgm^{-2}; yield per unit of land) through biophysical modeling; (3) water use mapping (WUM, m^{3}ha^{-1}) through surface energy balance modeling (ET modeling) or through water balance from hydrological models; (4) water productivity mapping (WPM, kgm^{-3}) through a simple ratio of CPM and WUM; and (5) developing a spatial decision support system (DSS) for government agencies and farmers to determine where, how, and by how much water can be saved through improved water productivity.

The steps for water productivity modeling are well illustrated in Chapter 11 by Dr. Antônio Heriberto de Castro Teixeira et al. by taking rainfed and irrigated crops in study areas of Brazil. They have focused their research on physical and economic values of crop water productivity. The physical crop water productivity or CWP [kgm^{-3}; is the ratio of crop biomass or yield (kgm^{-2}) to the amount of water used (m^3 m^{-2}); in Chapter 11 water productivity is defined as biomass/ET$_{actual}$, which after applying a harvest index (HI) becomes the CWP]. The economic water productivity ($m^{-3}) relates the economic benefits per unit of water used. Water productivity (yield per unit of water per drop; kg m^{-3}) is the inverse of water footprint (WF) and is illustrated (Figure 13.11) for several main global crops by Mekonnen and Hoekstra (2014). Raising WP would mean reducing WF

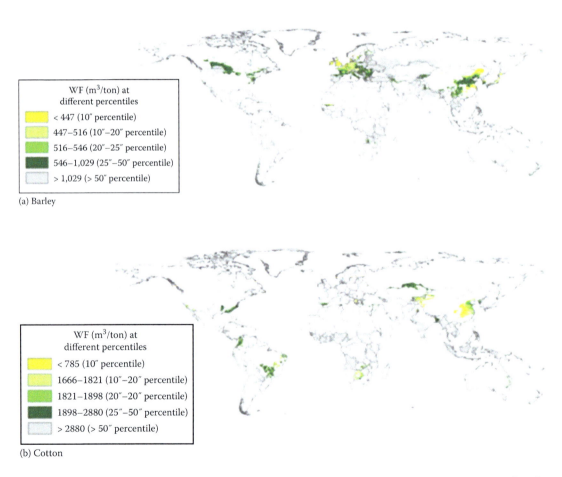

FIGURE 13.11 Spatial distribution of the green-blue water footprint (WF) of selected crops (in m^{3}ton^{-1}), classified based on the WFs at the different production percentiles. Water productivity (productivity per unit of water or crop per drop; kg m^{-3}) is inverse of WF. Raising water productivity in agriculture, that is, reducing the water footprint (WF) per unit of production, will contribute to reducing the pressure on the limited global freshwater resources (Source: Mekonnen and Hoekstra, 2014).

Remote Sensing Handbook, Volume V

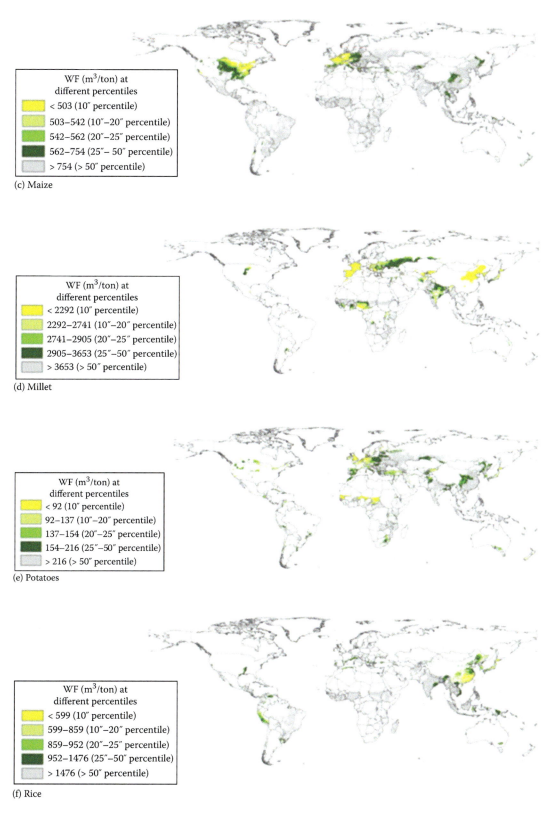

FIGURE 13.11 (Continued)

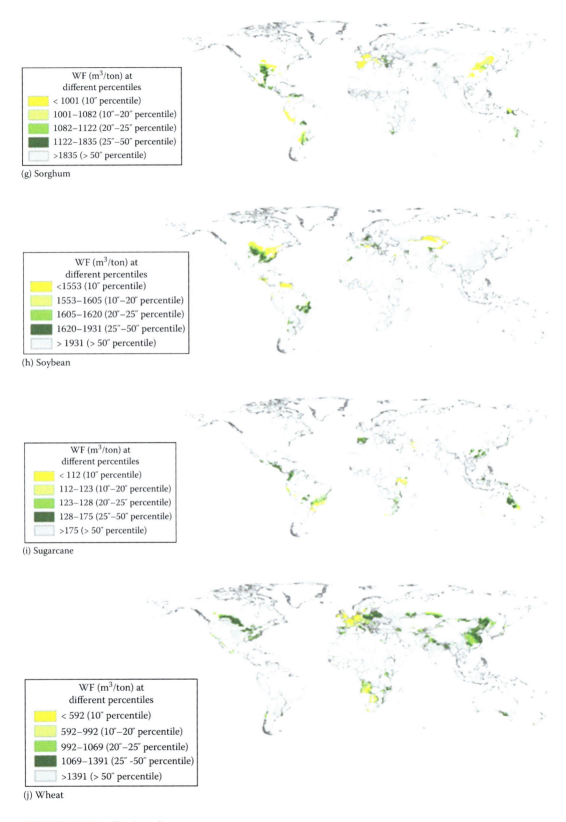

FIGURE 13.11 (Continued)

as illustrated. Remote sensing data of various resolutions can be used to study WP at different spatial scales (e.g., global, continental, regional, local). The accuracy and precision of WP studies will depend on the resolution (e.g., spatial, spectral, radiometric, and temporal) of the imagery used. Chapter 11 shows studies conducted using Landsat (30–120 m) and MODIS (250–1000 m) imagery.

Foley et al. (2020) provided meta-analysis of irrigated agricultural crop water productivity (CWP) of the world's three leading crops—wheat, corn, and rice—based on three decades of remote sensing– and nonremote-sensing-based studies. This study established the volume of water that can be saved for each crop in each country when there is an increase in CWP by 10%, 20%, and 30%. Cheng et al. (2022) developed at 1km resolution of crop yield and CWP for maize and wheat across China, based on the multiple remotely sensed indicators and random forest algorithm. Ghorbanpour et al. (2022) developed a cloud-based model within the Google Earth Engine (GEE) (Gorelick et al., 2017; Velastegui-Montoya et al., 2023) based on Landsat 7 and 8 satellite imagery to facilitate WP mapping at regional scales (30m resolution) and analyzing the state of the water use efficiency and productivity of the agricultural sector as a means of benchmarking its WP and defining local gaps and targets at spatiotemporal scales. Foley et al. (2023) developed crop water productivity from Cloud-Based Landsat to help assess California's water savings. Nevertheless, the CWP using remote sensing are still evolving (Cheng et al., 2023), and robust models with high levels of accuracies are yet to be achieved. For example, the percentage error of CWP from in situ methods ranges from 7% to 67%, depending on method and scale, whereas the error of CWP from remote sensing ranges from 7% to 22%, based on the highest reported performing remote sensing products (Blatchford et al., 2019). However, when considering the entire breadth of reported crop yield and ET_a accuracy, the achievable errors propagate to CWP ranges of 74–108% (Blatchford et al., 2019). This is not surprising given complexities such as remote sensing data resolutions, still-evolving model developments relating remote sensing data to crop biophysical and biochemical quantities, accuracies of field data itself, and uncertainties in methods and approaches in CWP modeling.

13.12 REMOTE SENSING OF SNOW COVER AND ITS APPLICATIONS

Snowfall is an important component of the hydrological cycle and has major impacts on global water resources. Unlike rainfall, snowfall often has significant lag time on when it melts and flows into rivers or recharges groundwater. About 48 million km^2 of the Earth's surface is covered by snow, of which ~98% is in the Northern Hemisphere. Snow is the largest single component of the cryosphere (ice, snow, glaciers, and permafrost) (Bishop et al., 2004). Snow cover is very dynamic, with as much as 50% of the Northern Hemisphere land surface that can be snow-covered at times during winter.

Visible wavelengths (400–700 nm) are best to study snow. This is because snow has very high albedo (80–90% reflectivity in 400–700 nm) relative to just 10–30% for vegetation and soils. Clouds, typically, have about 10% less albedo in 400–700 nm compared to fresh snow. However, older snow can have as low as 40% albedo. In NIR bands the contrast between snow and no-snow is poor. Cloud and snow are best discriminated in wavebands beyond 1500 nm. In this region clouds have high reflectance (around 60%), whereas snow < 10%. Active microwave or SAR is especially good in differentiating base soil, wet snow, and melting snow. Frequencies versus backscatter coefficient plots show this discrimination. Thermal data is not very useful, except in the case of detecting snow/land boundaries. This is because in order to measure snow temperature spectral emissivity and related properties of liquid water content and grain size need to be known. Any sensor with visible and SWIR bands can be used for snow studies. Some of the satellite sensors used in snow studies include: AMSR-E, MODIS Aqua/Terra (e.g., Figure 13.12), GOES, NOAA AVHRR, SSM/I, and SMMR.

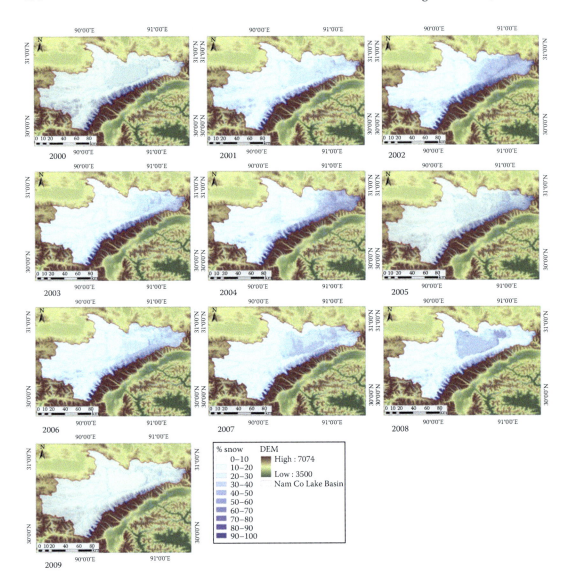

FIGURE 13.12 Monitoring changes of snow cover, lake, and vegetation phenology in Nam Co Lake Basin (Tibetan Plateau) using remote SENSING (2000–2009). Annual snow cover from 2000 to 2009 based on MODIS/Terra Snow Cover 8-Day L3 Global 500m Grid (MOD10A2) snow cover time series from March 2000 to December 2009. The values show the percentage of time that a pixel was snow covered throughout a year during the study period (2000–2009). The decreasing trend over time of persistent snow cover and intra-annual pattern of decreasing frequency of cover from SE (higher elevations along watershed divide) to NW can be observed. Note: MOD10A2 data set contains data fields for maximum snow cover extent over an eight-day period (Source: Zhang et al., 2013).

Remote sensing is used to measure snow parameters such as (e.g., Figure 13.12):

- Snow cover
- Snow depth
- Snow water equivalent

Normalized Difference Snow Index (NDSI) uses data from visible and far infrared or SWIR. In Landsat thematic mapper, for example, TM band 2 (530–610 nm) and TM band 5 (1570–1780 nm) are used to compute NDSI. If NDSI ≥ 0.4 the area is classified as snow.

Chapter 12 by Dr. Hongjie Xie et al. presents snow mapping approaches and methods using MODIS Terra/Aqua data as well as AMSR-E. The NDSI is calculated using MODIS band 4 (545–565 nm) and band 5 (1628–1652 nm) or band 7 (2105–2155 nm). The standard MODIS snow product, four masks are also used in classifying NDSI value as snow or no snow. These masks are dense forest stand mask, thermal mask, cloud mask, and ocean and inland water mask. MODIS Terra/Aqua also provides the fractional snow cover within pixel.

Some of the facts highlighted in Chapter 12 by Dr. Hongjie Xie et al. include:

1. MODIS standard snow cover products provide accuracies of > 90% (clear-sky conditions) and 30–50% (all-sky conditions).
2. Combining Terra and Aqua MODIS data (acquired in three-hour gap daily) can reduce cloud cover by ~10–20%.
3. By merging MODIS (500 m) and AMSR-E (25 km), cloud-free snow cover maps are often achieved, but by compromising spatial resolution. Accuracies of snow cover maps can be above 85% by combining these two data.
4. Flexible multi-day combination of snow mapping is controlled by two thresholds: maximum cloud percentage (P) and maximum composite days (N). Based on this approach, four snow cover parameters are calculated to examine spatio-temporal variations of snow cover during the hydrologial year. These four parameters are: snow cover index (SCI), snow-covered duration/days (SCD) map, snow cover onset dates (SCOD) map, and snow cover end dates (SCED) map.
5. Snow products are used in snowmelt runoff models to determine discharges into lakes and reservoirs. These computations are made with high level of certainty (e.g., < 90% variability explained compared to measured values), especially when multivariate analysis is performed.
6. Landsat NDSI can map snow with accuracy as high as 98% when snow cover > 60%.
7. Snow covers duration/day (SCD) maps, and snowmelt runoff predictions assessments to predict floods and disasters have also been illustrated.

Snow is a critical part of the hydrological cycle, a major part of the cryosphere, and is the water tower for many lakes and rivers. Snow cover assessments are critical in changing climate with melting glaciers, unseasonal snow, varying snow depth within and between seasons, increasing human activity (e.g., building roads, settlements, tourism), and rising temperature affecting snow extent and depth. Due to global daily coverage, the Moderate-Resolution Imaging Spectroradiometer (MODIS) Normalized Difference Snow Index (NDSI) is widely used for snow extent mapping. Chapter 12 by Dr. Hongjie Xie et al. developed the novel algorithm MODSAT-NDSI to harness the strengths of both coarse and finer spatial resolution imagery by fusing MODIS and Landsat NDSI to produce daily 30 m snow cover maps with an overall accuracy of 90% (Mityók et al., 2018). Optical remote sensing data are very effective in snow cover mapping, but the availability of data will depend on solar illumination conditions. SAR images over snow-covered areas are specifically valuable to see through the clouds in mountainous areas. In addition to widely applied backscattering-based methods, many new approaches based on interferometric SAR (InSAR) and polarimetric SAR (PolSAR) have been developed since the launch of ERS-1 in 1991 to monitor snow cover under both dry and wet snow conditions (Tsai et al., 2019).

ACKNOWLEDGMENTS

I would like to thank the lead authors and co-authors of each of the chapters for providing their insights and edits of my chapter summaries. Any use of trade, firm, or product names is for descriptive purposes only and does not imply endorsement by the US government.

REFERENCES

Abdalla, F. 2012. Mapping of groundwater prospective zones using remote sensing and GIS techniques: A case study from the Central Eastern Desert, Egypt, Journal of African Earth Sciences, Volume 70, 27 July, Pages 8–17, ISSN 1464–343X, http://dx.doi.org/10.1016/j.jafrearsci.2012.05.003.

Abdelkareem, M., Abdalla, F., Mohamed, S.Y., El-Baz, F. 2020. Mapping paleohydrologic features in the arid areas of Saudi Arabia using remote-sensing data, Water, Volume 12, Page 417, https://doi.org/10.3390/w12020417.

Abotalib, A.Z., Heggy, E., Bastawesy, M.E., Ismail, E., Gad, A., Attwa, M. 2021. Groundwater mounding: A diagnostic feature for mapping aquifer connectivity in hyper-arid deserts, Science of The Total Environment, Volume 801, Page 149760, ISSN 0048–9697, https://doi.org/10.1016/j.scitotenv.2021.149760. (www.sciencedirect.com/science/article/pii/S004896972104835X).

Akpoti, K., Groen, T., Dossou-Yovo, E., Kabo-bah, A.T., Zwart, S.J. 2022. Climate change-induced reduction in agricultural land suitability of West-Africa's inland valley landscapes, Agricultural Systems, Volume 200, Page 103429, ISSN 0308–521X, https://doi.org/10.1016/j.agsy.2022.103429. (www.sciencedirect.com/science/article/pii/S0308521X22000658).

Akumu, C., Henry, J., Gala, T., Dennis, S., Reddy, C., Tegegne, F., Haile, S., Archer, R. 2018. Inland wetlands mapping and vulnerability assessment using an integrated geographic information system and remote sensing techniques. Global Journal of Environmental Science and Management, Volume 4, Issue 4, Pages 387–400, https://doi.org/10.22034/gjesm.2018.04.001.

Alcamo, J. Henrichs, T., Rosch, T. 2000. World Water in 2025. Global world commission on water for the 21st century. Report A0001. Center for sity of Kassel, Kurt Wolters Strasse 3, 34109, Kassel, Germany.

Allafta, H., Opp, C., Patra, S. 2021. Identification of groundwater potential zones using remote sensing and GIS techniques: A case study of the Shatt Al-Arab Basin, Remote Sensing, Volume 13, Issue 1, Pages 112, https://doi.org/10.3390/rs13010112.

Alongi, D.M. 2014. Carbon sequestration in mangrove forests, Carbon Management, Volume 3, Issue 3, Pages 313–322.

Alshehri, F., Mohamed, A. 2023. Analysis of groundwater storage fluctuations using GRACE and remote sensing data in Wadi As-Sirhan, Northern Saudi Arabia, Water, Volume 15, Page 282, https://doi.org/10.3390/w15020282.

Alshehri, F., Sultan, M., Karki, S., Alwagdani, E., Alsefry, S., Alharbi, H., Sahour, H., Sturchio, N. 2020. Mapping the distribution of shallow groundwater occurrences using remote sensing-based statistical modeling over Southwest Saudi Arabia. Remote Sensing, Volume 12, Issue 9, Page 1361, https://doi.org/10.3390/rs12091361.

Amitrano, D., Di Martino, G., Di Simone, A. 2024. Imperatore, P. Flood detection with SAR: A review of techniques and datasets, Remote Sens, Volume 16, Page 656, https://doi.org/10.3390/rs16040656.

Andualem, T.G., Demeke, G. G. 2019. Groundwater potential assessment using GIS and remote sensing: A case study of Guna tana landscape, upper blue Nile Basin, Ethiopia, Journal of Hydrology: Regional Studies, Volume 24, Page 100610, ISSN 2214–5818, https://doi.org/10.1016/j.ejrh.2019.100610. (www.sciencedirect.com/science/article/pii/S2214581818302428).

Avand, M., Moradi, H., Lasboyee, M.R. 2021. Using machine learning models, remote sensing, and GIS to investigate the effects of changing climates and land uses on flood probability, Journal of Hydrology, Volume 595, Page 125663, ISSN 0022–1694, https://doi.org/10.1016/j.jhydrol.2020.125663. (www.sciencedirect.com/science/article/pii/S0022169420311240).

Blatchford, M.L., Mannaerts, C.M., Zeng, Y., Nouri, H., Karimi, P. 2019. Status of accuracy in remotely sensed and in-situ agricultural water productivity estimates: A review, Remote Sensing of Environment, Volume 234, Page 111413, ISSN 0034–4257, https://doi.org/10.1016/j.rse.2019.111413. (www.sciencedirect.com/science/article/pii/S0034425719304328).

Boothroyd, R.J., Williams, R.D., Hoey, T.B., Barrett, B., Prasojo, O.A. 2020. Applications of Google Earth Engine in fluvial geomorphology for detecting river channel change. WIREs Water, Volume 8, Issue 1, Page e21496, https://doi.org/10.1002/wat2.1496.

Bunting, P., Rosenqvist, A., Hilarides, L., Lucas, R.M., Thomas, N., Tadono, T., Worthington, T.A., Spalding, M., Murray, N.J., Rebelo, L.-M. 2022. Global mangrove extent change 1996–2020: Global mangrove watch version 3.0, Remote Sensing, Volume 14, Issue 15, Page 3657, https://doi.org/10.3390/rs14153657.

Cacal, J.C., Taboada, E.B., Mehboob, M.S. 2023. Strategic implementation of integrated water resource management in selected areas of Palawan: SWOT-AHP Method, Sustainability, Volume 15, Page 2922, https://doi.org/10.3390/su15042922.

Chen, C., Ma, Y., Ren, G., Wang, J. 2022. Aboveground biomass of salt-marsh vegetation in coastal wetlands: Sample expansion of in situ hyperspectral and Sentinel-2 data using a generative adversarial network, Remote Sensing of Environment, Volume 270, Page 112885, ISSN 0034–4257, https://doi.org/10.1016/j.rse.2021.112885. (www.sciencedirect.com/science/article/pii/S0034425721006052).

Cheng, M., Jiao, X., Shi, L., et al. 2022. High-resolution crop yield and water productivity dataset generated using random forest and remote sensing, Science Data, Volume 9, Page 641, https://doi.org/10.1038/s41597-022-01761-0.

Cheng, M., Yin, D., Wu, W., Cui, N., Nie, C., Shi, L., Liu, S., Yu, X., Bai, Y., Liu, Y., Zhu, Y., Jin, X. 2023. A review of remote sensing estimation of crop water productivity: Definition, methodology, scale, and evaluation, International Journal of Remote Sensing, Volume 44, Issue 16, Pages 5033–5068, https://doi.org/10.1080/01431161.2023.2240523.

Chowdhury, S., Hafsa, B. 2022. Multi-decadal land cover change analysis over Sundarbans Mangrove Forest of Bangladesh: A GIS and remote sensing based approach, Global Ecology and Conservation, Volume 37, Page e02151, ISSN 2351–9894, https://doi.org/10.1016/j.gecco.2022.e02151. (www.sciencedirect.com/science/article/pii/S2351989422001536).

Davidson, N.C., Fluet-Chouinard, E., Finlayson, C.M. 2018. Global extent and distribution of wetlands: Trends and issues, Marine and Freshwater Research, Volume 69, Pages 620–627, https://doi.org/10.1071/MF17019.

Dorling, D. 2021. 7: World population prospects at the UN: Our numbers are not our problem? In The struggle for social sustainability, Bristol, UK: Policy Press. Retrieved May 4, 2024, from https://doi.org/10.51952/9781447356127.ch007.

Dubey, A.K., Gupta, V., Gupta, S.N., Dev, R. 2023. Sustainable drinking water system for rural areas powered by non-conventional energy sources, International Conference on Computational Intelligence and Sustainable Engineering Solutions (CISES), Greater Noida, India, Pages 1052–1057, http://dx.doi.org/10.1109/CISES58720.2023.10183472.

Duke, N.C., Meynecke, J.O., Dittmann, S., Ellison, A.M., Anger, K., Berger, U., Cannicci, S., Diele, K., Ewel, K.C., Field, C.D., Koedam, N., Lee, S.Y., Marchand, C., Nordhaus, I., Dahdouh-Guebas, F. (Kavanagh, E., Ed.). 2007. A world without mangroves?, Letters Science, Volume 317, Page 41.

Echogdali, F.Z., Boutaleb, S., Abioui, M., Aadraoui, M., Bendarma, A., Kpan, R.B., Ikirri, M., El Mekkaoui, M., Essoussi, S., El Ayady, H., et al. 2023. Spatial mapping of groundwater potentiality applying geometric average and fractal models: A sustainable approach, Water, Volume 15, Page 336, https://doi.org/10.3390/w15020336.

Farhadi, H., Najafzadeh, M. 2021. Flood risk mapping by remote sensing data and random forest technique, Water, Volume 13, Issue 21, Page 3115, https://doi.org/10.3390/w13213115.

Foley, D.J., Thenkabail, P.S., Aneece, I.P., Teluguntla, P.G., Oliphant, A.J. 2020. A meta-analysis of global crop water productivity of three leading world crops (wheat, corn, and rice) in the irrigated areas over three decades, International Journal of Digital Earth, Volume 13, Issue 8, Pages 939–975, https://doi.org/10.1080/17538947.2019.1651912.

Foley, D., Thenkabail, P., Oliphant, A., Aneece, I., Teluguntla, P. 2023. Crop water productivity from cloud-based landsat helps assess California's water savings, Remote Sensing, Volume 15, Issue 19, Page 4894, https://doi.org/10.3390/rs15194894.

Gabarró, C., Hughes, N., Wilkinson, J., Bertino, L., Bracher, A., Diehl, T., Dierking, W., Gonzalez-Gambau, V., Lavergne, T., Madurell, T., Malnes, E., Wagner, P.M. 2023. Improving satellite-based monitoring of the polar regions: Identification of research and capacity gaps, Front Remote Sens, Volume 4, Page 952091, https://doi.org/10.3389/frsen.2023.952091.

Gan, T.Y., Zunic, F., Kuo, C.C., Strobl, T. 2012. Flood mapping of Danube River at Romania using single and multi-date ERS2-SAR images, International Journal of Applied Earth Observation and Geoinformation, Volume 18, August, Pages 69–81, ISSN 0303–2434, http://dx.doi.org/10.1016/j.jag.2012.01.012.

Gardner, R.C., Okuno, E., Pritchard, D. 2023. 2—Ramsar convention governance and processes at the international level, Gell, P.A., Davidson, N.C., Finlayson, C.M. (Eds.), Ramsar Wetlands: Elsevier, Pages 37–67, ISBN 9780128178034, https://doi.org/10.1016/B978-0-12-817803-4.00003-6. (www.sciencedirect.com/science/article/pii/B9780128178034000036).

Géant, C.B., Gustave, M.N. Schmitz, S. 2023. Mapping small inland wetlands in the South-Kivu province by integrating optical and SAR data with statistical models for accurate distribution assessment, Scientific Reports, Volume 13, Page 17626 (2023). https://doi.org/10.1038/s41598-023-43292-7.

Gell, P.A., Finlayson, C.M., Davidson, N.C. 2023. 1—An introduction to the Ramsar Convention on Wetlands, Gell, P.A., Davidson, N.C., Finlayson, C.M. (Eds.), Ramsar Wetlands: Elsevier, Pages 1–36, ISBN 9780128178034, https://doi.org/10.1016/B978-0-12-817803-4.00018-8. (www.sciencedirect.com/science/article/pii/B9780128178034000188).

Ghorbanpour, A.K., Kisekka, I., Afshar, A., Hessels, T., Taraghi, M., Hessari, B., Tourian, M.J., Duan, Z. 2022. Crop water productivity mapping and benchmarking using remote sensing and google earth engine cloud computing, Remote Sensing, Volume 14, Issue 19, Page 4934, https://doi.org/10.3390/rs14194934.

Giordan, D., Notti, D., Villa, A., Zucca, F., Calò, F., Pepe, A., Dutto, F., Pari, P., Baldo, M., Allasia, P. 2018. Low cost, multiscale and multi-sensor application for flooded area mapping, Natural Hazards and Earth System Sciences, Volume 18, Pages 1493–1516, https://doi.org/10.5194/nhess-18-1493-2018.

Giri, C. 2023. Frontiers in global mangrove forest monitoring, Remote Sensing, Volume 15, Issue 15, Page 3852, https://doi.org/10.3390/rs15153852.

González-Dugo, M.P., Escuin, S., Cano, F., Cifuentes, V., Padilla, F.L.M., Tirado, J.L., Oyonarte, N., Fernández, P., Mateos, L. 2013. Monitoring evapotranspiration of irrigated crops using crop coefficients derived from time series of satellite images: II. Application on basin scale, Agricultural Water Management, Volume 125, July, Pages 92–104, ISSN 0378–3774, http://dx.doi.org/10.1016/j.agwat.2013.03.024.

Gorelick, N., Hancher, M., Dixon, M., Ilyushchenko, S., Thau, D., Moore, R. 2017. Google earth engine: Planetary-scale geospatial analysis for everyone, Remote Sensing of Environment, Volume 202, Pages 18–27, ISSN 0034–4257, https://doi.org/10.1016/j.rse.2017.06.031. (www.sciencedirect.com/science/article/pii/S0034425717302900).

Guimarães, U.S., Galo, M.D.L.B.T., Narvaes, I.D.S., Silva, A.D.Q.D. 2020. Cosmo-SkyMed and TerraSAR-X datasets for geomorphological mapping in the eastern of Marajó Island, Amazon coast, Geomorphology, Volume 350, Page 106934, ISSN 0169–555X, https://doi.org/10.1016/j.geomorph.2019.106934. (www.sciencedirect.com/science/article/pii/S0169555X19304258).

Günen, M.A. 2022. Performance comparison of deep learning and machine learning methods in determining wetland water areas using EuroSAT dataset, Environmental Science and Pollution Research, Volume 29, Pages 21092–21106, https://doi.org/10.1007/s11356-021-17177-z.

Hall, D.K. Robinson, D.A. in press. Global snow cover. In Satellite image atlas of glaciers of the world, Williams, R.S., Jr., Ferrigno, J.G. (Eds.), USGS Professional Paper 1386.

Hamidi, E., Peter, B.G., Muñoz, D.F., Moftakhari H., Moradkhani, H. 2023. Fast flood extent monitoring with SAR change detection using google earth engine, IEEE Transactions on Geoscience and Remote Sensing, Volume 61, Pages 1–19, Art No. 4201419, https://doi.org/10.1109/TGRS.2023.3240097.

Hemati, M., Mahdianpari, M., Shiri, H., Mohammadimanesh, F. 2024. Integrating SAR and optical data for aboveground biomass estimation of coastal Wetlands using machine learning: Multi-scale approach, Remote Sens, Volume 16, Page 831, https://doi.org/10.3390/rs16050831.

Hu, S., Niu, Z. Chen, Y. 2017. Global Wetland datasets: A Review, Wetlands, Volume 37, Page 807–817, https://doi.org/10.1007/s13157-017-0927-z.

Hu, T., Zhang, Y.Y., Su, Y., Zheng, Y., Lin, G., Guo, Q. 2020. Mapping the global mangrove forest aboveground biomass using multisource remote sensing data, Remote Sensing, Volume 12, Issue 10, Page 1690, https://doi.org/10.3390/rs12101690.

Jachowski, N.R.A., Quak, M.S.Y., Friess, D.A., Duangnamon, D., Webb, E.L., Ziegler, A.D. 2013. Mangrove biomass estimation in Southwest Thailand using machine learning, Applied Geography, Volume 45, December, Pages 311–321, ISSN 0143–6228, http://dx.doi.org/10.1016/j.apgeog.2013.09.024.

Jia, M., Wang, Z., Mao, D., Ren, C., Song, K., Zhao, C., Wang, C., Xiao, X., Wang, Y. 2023. Mapping global distribution of mangrove forests at 10-m resolution, Science Bulletin, Volume 68, Issue 12, Pages 1306–1316, ISSN 2095–9273, https://doi.org/10.1016/j.scib.2023.05.004. (www.sciencedirect.com/science/article/pii/S2095927323003110).

Kalhor, K., Emaminejad, N. 2019. Sustainable development in cities: Studying the relationship between groundwater level and urbanization using remote sensing data, Groundwater for Sustainable Development, Volume 9, Page 100243, ISSN 2352–801X, https://doi.org/10.1016/j.gsd.2019.100243. (www.sciencedirect.com/science/article/pii/S2352801X19300712).

Kamal, A.S.M.M., Midorikawa, S. 2004. GIS-based geomorphological mapping using remote sensing data and supplementary geoinformation: A case study of the Dhaka city area, Bangladesh, International Journal of Applied Earth Observation and Geoinformation, Volume 6, Issue 2, December, Pages 111–125, ISSN 0303–2434, http://dx.doi.org/10.1016/j.jag.2004.08.003.

Kamali, M., Ehsan, S.A.N., Hashemi, H., Berndtsson, R. 2020. Application of advanced machine learning algorithms to assess groundwater potential using remote sensing-derived data, Remote Sensing, Volume 12, Issue 17, Page 2742, https://doi.org/10.3390/rs121727.

Khaki, M. 2023. Land surface model calibration using satellite remote sensing data, Sensors, Volume 23, Page 1848, https://doi.org/10.3390/s23041848.

Lee, S., Hyun, Y., Lee, S., Lee, M.-J. 2020. Groundwater potential mapping using remote sensing and GIS-based machine learning techniques, Remote Sensing, Volume 12, Issue 7, Page 1200, https://doi.org/10.3390/rs12071200.

Ling, M., Yang, Y., Xu, C., Yu, L., Xia, Q., Guo, X. 2022. Temporal and spatial variation characteristics of actual evapotranspiration in the Yiluo River Basin based on the priestley: Taylor jet propulsion laboratory model, Applied Sciences, Volume 12, Page 9784, https://doi.org/10.3390/app12199784.

Ma, Y., Liu, S., Song, L., Xu, Z., Liu, Y., Xu, T., Zhu, Z. 2018. Estimation of daily evapotranspiration and irrigation water efficiency at a Landsat-like scale for an arid irrigation area using multi-source remote sensing data, Remote Sensing of Environment, Volume 216, Pages 715–734, ISSN 0034–4257, https://doi.org/10.1016/j.rse.2018.07.019. (www.sciencedirect.com/science/article/pii/S0034425718303523).

Mandishona, E., Knight, J. 2022. Inland wetlands in Africa: A review of their typologies and ecosystem services, Progress in Physical Geography: Earth and Environment, Volume 46, Issue 4, Pages 547–565, https://doi.org/10.1177/03091333221075328.

Margat, J., van der Gun, J. 2013. Groundwater around the World, Balkema: CRC Press.

Martinis, S., Kersten, J., Twele, A. 2015. A fully automated TerraSAR-X based flood service, ISPRS Journal of Photogrammetry and Remote Sensing, Volume 104, Pages 203–212, ISSN 0924–2716, https://doi.org/10.1016/j.isprsjprs.2014.07.014. (www.sciencedirect.com/science/article/pii/S0924271614001981).

Masek, J.G., Ju, J., Claverie, M., Skakun, S., Roger, J.C., Vermote, E., Franch, B., Yin, Z., Dungan. J.L. 2022. Harmonized Landsat Sentinel-2 (HLS) Product User Guide Product Version 2.0.

Masek, J., Ju, J., Roger, J., Skakun, S., Vermote, E., Claverie, M., Dungan, J., Yin, Z., Freitag, B., Justice, C. 2021. HLS Sentinel-2 MSI surface reflectance daily global 30m v2.0., distributed by NASA EOSDIS Land Processes DAAC, https://doi.org/10.5067/HLS/HLSS30.002.

Meijerink, A.M.J., Bannert, D., Batelaan, O., Lubczynski, W., Pointet, T. 2007. Remote sensing applications to groundwater. IHP-VI, Series on Groundwater No. 16. Published by the United Nations Educational, Scientific, and Cultural Organization, UNESCO. 7, Place de Fontenoy, 75352 Paris 07 SP (France). Composed by Marina Rubio, 93200 Saint-Denis. Printed by UNESCO. Page 312.

Mekonnen, M.M., Hoekstra, A.Y. 2014. Water footprint benchmarks for crop production: A first global assessment, Ecological Indicators, Volume 46, November, Pages 214–223, ISSN 1470–160X, http://dx.doi.org/10.1016/j.ecolind.2014.06.013.

Mekonnen, M.M., Gerbens-Leenes, W. 2020. The water footprint of global food production, Water, Volume 12, Page 2696, https://doi.org/10.3390/w12102696.

Melton, F., Huntington, J., Grimm, R., Herring, J., Hall, M., Rollison, D., Erickson, T., Allen, R., Anderson, M., Fisher, J., Kilic, A., Senay, G., Volk, J., Hain, C., Johnson, L., Ruhoff, A., Blankenau, P., Bromley, M., Carrara, W., Daudert, B., Doherty, C., Dunkerly, C., Friedrichs, M.K., Guzman, A., Halverson, G., Hansen, J., Harding, J., Kang, Y., Ketchum, D.C., Minor, B., Morton, C., Ortega-Salazar, S., Ott, T., Ozdogan, M., Revelle, P., Schull, M., Wang, C., Yang, Y., Anderson, R.G. 2022. OpenET: Filling a critical data gap in water management for the western United States, JAWRA Journal of the American Water Resources Association, Voume 58, Issue 6, Pages 971–994, https://doi.org/10.1111/1752-1688.12956.

Milewski, A., Sultan, M., Yan, E., Becker, R., Abdeldayem, A., Soliman, F., Gelil, K.A. 2009. A remote sensing solution for estimating runoff and recharge in arid environments, Journal of Hydrology, Volume 373, Issues 1–2, 30 June, Pages 1–14, ISSN 0022–1694, http://dx.doi.org/10.1016/j.jhydrol.2009.04.002.

Mityók, Z.K., Bolton, D.K., Coops, N.C., Berman, E.E., Senger, S. 2018. Snow cover mapped daily at 30 meters resolution using a fusion of multi-temporal MODIS NDSI data and Landsat surface reflectance, Canadian Journal of Remote Sensing, Volume 44, Issue 5, Pages 413–434, https://doi.org/10.1080/07038992.2018.1538775.

Monegaglia, F., Zolezzi, G., Güneralp, I., Henshaw, A.J., Tubino, M. 2018. Automated extraction of meandering river morphodynamics from multitemporal remotely sensed data, Environmental Modelling & Software, Volume 105, Pages 171–186, ISSN 1364–8152, https://doi.org/10.1016/j.envsoft.2018.03.028. (www.sciencedirect.com/science/article/pii/S1364815217309118).

Mounirou, L.A., Sawadogo, B., Yanogo, H., Yonaba, R., Zorom, M., Faye, M.D., Kafando, M.B., Biaou, A.C., Koïta, M., Karambiri, H. 2023. Estimation of the actual specific consumption in drinking water supply systems in Burkina Faso (West Africa): Potential implications for infrastructure sizing, Water, Volume 15, Page 3423, https://doi.org/10.3390/w15193423.

Munasinghe, D., Frasson, R.P.D.M., David, C.H., et al. 2023. A multi-sensor approach for increased measurements of floods and their societal impacts from space, Commun Earth Environ, Volume 4, Page 462, https://doi.org/10.1038/s43247-023-01129-1.

Munawar, H.S., Hammad, A.W.A., Waller, S.T. 2022. Remote sensing methods for flood prediction: A review, Sensors, Volume 22, Issue 3, Page 960, https://doi.org/10.3390/s22030960.

Ogilvie, A., Poussin, J.-C., Bader, J.-C., Bayo, F., Bodian, A., Dacosta, H., Dia, D., Diop, L., Martin, D., Sambou, S. 2020. Combining multi-sensor satellite imagery to improve long-term monitoring of temporary surface water bodies in the senegal river floodplain, Remote Sensing, Volume 12, Issue 19, Page 3157, https://doi.org/10.3390/rs12193157.

Opoku, P.A., Shu, L., Amoako-Nimako, G.K. 2024. Assessment of groundwater potential zones by integrating hydrogeological data, Geographic Information Systems, Remote Sensing, and Analytical Hierarchical Process Techniques in the Jinan Karst Spring Basin of China, Water, Volume 16, Page 566, https://doi.org/10.3390/w160405.

Platonov, A., Thenkabail, P.S., Biradar, C.M., Cai, X., Gumma, M., Dheeravath, V., Cohen, Y., Alchanatis, V., Goldshlager, N., Ben-Dor, E., Vithanage, J., Manthrithilake, H., Kendjabaev, S., Isaev, S. 2008. Water Productivity Mapping (WPM) using landsat ETM+ data for the irrigated croplands of the Syrdarya River Basin in Central Asia, Sensors, Volume 8, Pages 8156–8180.

Pu, G., Quackenbush, L.J., Stehman, S.V. 2021. Using google earth engine to assess temporal and spatial changes in river geomorphology and riparian vegetation. Journal of the American Water Resources Association (JAWR), Volume 57, Issue 5, Pages 789–806, https://doi.org/10.1111/1752-1688.12950.

Rai, P.K., Chandel, R.S., Mishra, V.N., et al. 2018. Hydrological inferences through morphometric analysis of lower Kosi river basin of India for water resource management based on remote sensing data, Applied Water Sciences, Volume 8, Page 15, https://doi.org/10.1007/s13201-018-0660-7.

Rao, D. 2002. Remote sensing application in geomorphology, Tropical Ecology, Volume 43, Pages 49–59.

Ray, R.G. 1960. Aerial photographs in geologic interpretation, USGS Professional Paper Issue 373.

Rentschler, J., Salhab, M., Jafino, B.A. 2022. Flood exposure and poverty in 188 countries, Nature Communications, Volume 13, Page 3527, https://doi.org/10.1038/s41467-022-30727-4.

Rezaee, M., Mahdianpari, M., Zhang, Y. Salehi. B. 2018. Deep convolutional neural network for complex wetland classification using optical remote sensing imagery, IEEE Journal of Selected Topics in Applied Earth Observations and Remote Sensing, Volume 11, Issue 9, September, Pages 3030–3039, https://doi.org/10.1109/JSTARS.2018.2846178.

Rodell, M., Chen, J., Kato, H., Famiglietti, J., Nigro, J., Wilson, C. 2006. Estimating ground water storage changes in the Mississippi River basin (USA) using GRACE, Hydrogeology Journal, https://doi.org/10.1007/s10040-006-0103-7.

Rosser, J.F., Leibovici, D.G., Jackson, M.J. 2017. Rapid flood inundation mapping using social media, remote sensing and topographic data, Nature Hazards, Volume 87, Pages 103–120, https://doi.org/10.1007/s11069-017-2755-0.

Sabins, F.F. 1987. Remote sensing: Principles and interpretation, 2nd. edn, San Francisco, USA: W. H. Freeman.

Sahour, H., Sultan, M., Abdellatif, B., Emil, M., Abotalib, A.Z., Abdelmohsen, K., Vazifedan, M., Mohammad, A.T., Hassan, S.M., Metwalli, M.R., El Bastawesy, M. 2022. Identification of shallow groundwater in arid lands using multi-sensor remote sensing data and machine learning algorithms, Journal of Hydrology, Volume 614, Part A, Page 128509, ISSN 0022–1694, https://doi.org/10.1016/j.jhydrol.2022.128509. (www.sciencedirect.com/science/article/pii/S0022169422010794).

Shahrood, A.J., Menberu, M.W., Darabi, H., Rahmati, O., Rossi, P.M., Kløve, B., Haghighi, A.T. 2020. RiMARS: An automated river morphodynamics analysis method based on remote sensing multispectral datasets, Science of The Total Environment, Volume 719, Pages 137336, ISSN 0048–9697, https://doi.org/10.1016/j.scitotenv.2020.137336. (www.sciencedirect.com/science/article/pii/S0048969720308469).

Siebert, S., et al. 2010. Groundwater use for irrigation, Hydrology and Earth Systems Science, Volume 14, Pages 1863–1880, (www.hydrol-earth-syst-sci.net/14/1863/2010/doi:10.5194/hess-14-1863-2010).

Siebert, S., Hoogeveen, J., Frenken, K. 2006. Irrigation in Africa, Europe and Latin America: Update of the digital global map of irrigation areas to version 4, Frankfurt Hydrology Paper 05, Frankfurt am Main, Germany and Rome, Italy: Institute of Physical Geography, University of Frankfurt, Page 134.

Smith, M., Pain, C. 2006. Applications of remote sensing in geomorphology, Progress in Physical Geography, Volume 33, Pages 568–582.

Soria-Ruiz, J., Fernandez-Ordoñez, Y.M., Ambrosio-Ambrosio, J.P., Escalona-Maurice, M.J., Medina-García, G., Sotelo-Ruiz, E.D., Ramirez-Guzman, M.E. 2022. Flooded extent and depth analysis using optical and SAR remote sensing with machine learning algorithms, Atmosphere, Volume 13, Issue 11, Page 1852, https://doi.org/10.3390/atmos13111852.

Sorooshian, S., AghaKouchak, A., Arkin, P., Eylander, J., Foufoula-Georgiou, E., Harmon, R., Hendrickx, J. 2011. Advanced concepts on remote sensing of precipitation at multiple scales, Bulletin of the American Meteorological Society, Volume 92, Issue 10, Pages 1353–1357, https://doi.org/10.1175/2011BAMS3158.1.

Spalding, M., Blasco, F., Field, C. 1997. World mangrove atlas, Okinawa, Japan: International Society for Mangrove Ecosystems, Page 178.

Syvitski, J.P.M., Overeem, I., Brakenridge, G.R., Hannon, M. 2012. Floods, floodplains, delta plains: A satellite imaging approach, Sedimentary Geology, Volumes 267–268, 1 August, Pages 1–14, ISSN 0037-0738, http://dx.doi.org/10.1016/j.sedgeo.2012.05.014.

Taguchi, R., Tanoue, M., Yamazaki, D., Hirabayashi, Y. 2022. Global-scale assessment of economic losses caused by flood-related business interruption, Water, Volume 14, Page 967, https://doi.org/10.3390/w14060967.

Tang, R., Peng, Z., Liu, M., Li, Z., Jiang, Y., Hu, Y., Huang, L., Wang, Y., Wang, J., Jia, L., Zheng, C., Zhang, Y., Zhang, K., Yao, Y., Chen, X., Xiong, Y., Zeng, Z., Fisher, J.B. 2024. Spatial-temporal patterns of land surface evapotranspiration from global products, Remote Sensing of Environment, Volume 304, Page 114066, ISSN 0034-4257, https://doi.org/10.1016/j.rse.2024.114066. (www.sciencedirect.com/science/article/pii/S0034425724000774).

Tang, Y.-N., Ma, J., Xu, J.-X., Wu, W.-B., Wang, Y.-C., Guo, H.-Q. 2022. Assessing the impacts of tidal creeks on the spatial patterns of coastal salt marsh vegetation and its aboveground biomass, Remote Sens, Volume 14, Page 1839, https://doi.org/10.3390/rs14081839.

Tehrany, M.S., Pradhan, B., Jebur, M.N. 2014. Flood susceptibility mapping using a novel ensemble weights-of-evidence and support vector machine models in GIS, Journal of Hydrology, Volume 512, 6 May, Pages 332–343, ISSN 0022-1694, http://dx.doi.org/10.1016/j.jhydrol.2014.03.008.

Temitope, D., Oyedotun, T. 2022. Quantitative assessment of the drainage morphometric characteristics of Chaohu Lake Basin from SRTM DEM Data: A GIS-based approach, Geology, Ecology, and Landscapes, Volume 6, Issue 3, Pages 174–187, http://dx.doi.org/10.1080/24749508.2020.1812147.

Thenkabail, P.S., Enclona, E.A., Ashton, M.S., Legg, C., Jean De Dieu, M. 2004. Hyperion, IKONOS, ALI, and ETM+ sensors in the study of African rainforests, Remote Sensing of Environment, Volume 90, Pages 23–43.

Thenkabail, P.S., Gumma, M.K., Teluguntla, P., Mohammed, I.A., 2014. Hyperspectral remote sensing of vegetation and agricultural crops: Highlight article, Photogrammetric Engineering and Remote Sensing, Volume 80, Issue 4, Pages 697–709.

Thenkabail, P.S., Hanjra, M.A., Dheeravath, V., Gumma, M.A. 2010. A holistic view of global croplands and their water use for ensuring global food security in the 21st century through advanced remote sensing and non-remote sensing approaches, Remote Sensing Open Access Journal, Volume 2, Issue 1, Pages 211–261. http://dx.doi.org/10.3390/rs2010211. (www.mdpi.com/2072-4292/2/1/211).

Thenkabail, P.S., Nolte, C. 1995. Mapping and characterising inland valley agroecosystems of West and Central Africa: A methodology integrating remote sensing, global positioning system, and ground-truth data in a geographic information systems framework, RCMD Monograph No.16, Ibadan, Nigeria: International Institute of Tropical Agriculture, p. 62.

Thenkabail, P.S., Nolte, C., Lyon, J.G. 2000a. Remote sensing and GIS modeling for selection of benchmark research area in the inland valley agroecosystems of West and Central Africa, Photogrammetric Engineering and Remote Sensing, Africa Applications Special Issue, Volume 66, Issue 6, Pages 755–768.

Thenkabail, P.S., Smith, R.B., De-Pauw, E. 2000b. Hyperspectral vegetation indices for determining agricultural crop characteristics, Remote Sensing of Environment, Volume 71, Pages 158–182.

Tran, T.V., Reef, R., Zhu, X. 2022. A review of spectral indices for mangrove remote sensing, Remote Sensing, Volume 14, Issue 19, Pages 4868, https://doi.org/10.3390/rs14194868.

Tsai, Y.-L.S., Dietz, A., Oppelt, N., Kuenzer, C. 2019. Remote sensing of snow cover using spaceborne SAR: A review. Remote Sensing, Volume 11, Issue 12, Page 1456, https://doi.org/10.3390/rs11121456.

Velastegui-Montoya, A., Montalván-Burbano, N., Carrión-Mero, P., Rivera-Torres, H., Sadeck, L., Adami, M. 2023. Google earth engine: A global analysis and future trends, Remote Sens, Volume 15, Page 3675, https://doi.org/10.3390/rs15143675.

Vogels, M.F.A., de Jong, S.M., Sterk, G., Douma, H., Addink, E.A. 2019. Spatio-temporal patterns of smallholder irrigated agriculture in the horn of Africa using GEOBIA and sentinel-2 imagery, Remote Sens, Volume 11, Page 143, https://doi.org/10.3390/rs11020143.

Wang, L., Jia, M., Yin, D., Tian, J. 2019. A review of remote sensing for mangrove forests: 1956–2018, Remote Sensing of Environment, Volume 231, Page 111223, ISSN 0034–4257, https://doi.org/10.1016/j.rse.2019.111223. (www.sciencedirect.com/science/article/pii/S0034425719302421).

Wu, W.Y., Yang, Z.L., Zhao, L., Lin, P. 2022. The impact of multi-sensor land data assimilation on river discharge estimation, Remote Sensing of Environment, Volume 279, Page 113138, ISSN 0034–4257, https://doi.org/10.1016/j.rse.2022.113138. (www.sciencedirect.com/science/article/pii/S0034425722002528).

Yu, Q., Wang, Y., Li, N. 2022. Extreme flood disasters: Comprehensive impact and assessment, Water, Volume 14, Page 1211, https://doi.org/10.3390/w14081211.

Zhang, B., Wu, Y., Lei, L., Li, J., Liu, L., Chen, D., Wang, J. 2013. Monitoring changes of snow cover, lake and vegetation phenology in Nam Co Lake Basin (Tibetan Plateau) using remote SENSING (2000–2009), Journal of Great Lakes Research, Volume 39, Issue 2, June, Pages 224–233, ISSN 0380–1330, http://dx.doi.org/10.1016/j.jglr.2013.03.009.

Zhang, G., Chen, W., Li, G., Yang, W., Yi, S., Luo, W. 2020. Lake water and glacier mass gains in the northwestern Tibetan Plateau observed from multi-sensor remote sensing data: Implication of an enhanced hydrological cycle, Remote Sensing of Environment, Volume 237, Page 111554, ISSN 0034–4257, https://doi.org/10.1016/j.rse.2019.111554. (www.sciencedirect.com/science/article/pii/S0034425719305747).

Index

Note: Page numbers in *italics* indicate figures, and page numbers in **bold** indicate tables in the text

A

Absorbed Photosynthetically Active Radiation (APAR), 180, *180*
active contour models (ACMs), 149
active sensors, 115, 217
active systems, 6, 8
administrative floodplain mapping, 19
Advanced Baseline Imager (ABI), 46
Advanced Land Observing Satellite (ALOS), 117, 122, 138
Advanced Microwave Scanning Radiometer-Earth Observing System (AMSR-E), 40, 41, 52, 426, 431–436, 439, 451, 474
advanced neural network, 221
Advanced Spaceborne Thermal Emission and Reflection Radiometer (ASTER), 13, 164, 199, 206, 281, 326, 353, 360–361, *361*, 473
Advanced Synthetic Aperture Radar (ASAR), 138–139, 147–148, 150–151, 156–157, *158–160*, 162, 168
Advanced Very High Resolution Radiometer (AVHRR), 41, 73, 127–128, 227, 426, 431, 449, 473, 474, 479
aeolian deposits, 85
aeolian landforms, 24–26
aerial photograph/image, 5–7, *11*, 12–13, *14*, 15, 19, 21, 23, 25, *25*, 26, 66–69, 71, 75–76, 86, 100, *101*, 197–198, 206, 208, 387
　advantage in data acquisition time, 218
　advantage in spatial resolution, 218
　analysis of, 6
　commercial satellite corporations, 7
　floodplain limits, 19
　of fluvial landscape, 18
　for generation of DEMS, 13
　geomorphological inquiry, 5
　groundwater targeting, 66–69
　of Gulf Shores, 23
　limitations, 218
　mangroves, 197–198
　rock glacier movement, *14*
　sequential, *11*, 18, 21, 23
　study of glaciated terrain, 13
Aerial Photographs in Geologic Interpretation and Mapping (Ray), 6
Aerospace Center (DLR), Germany, 42, 139, 151
agricultural growing region scale, 392–401
　Fruit Circuit pole, 399–401
　North of Minas Gerais pole, 396–399
　Petrolina/Juazeiro pole, 392–396
Airborne LiDAR Bathymetry (ALB), 22–23
Air Force Weather Agency's Agricultural Meteorological (AgriMet) modeling system, 325
alluvial aquifers, 80–84
　alluvial fans/piedmont deposits, 80–81, *81*
　alluvial plains, 82–83, *83*
　deltaic terrains, 83–84
　exploration parameters in, 80–81
　valley fills, 81–82
alluvial deposits, defined, 80
alluvial fans, 80–81, *81*
　arid region parameters, 81
　humid region parameters, 80–81
alluvial plains, 82–83, *83*
along-track interferometry, 8, *8*
Amazon River, 118, 122, *123–124*, 125
American Society of Photogrammetry, 69
apparent optical properties (AOPs), 222
arid region parameters, 81
artificial neural network (ANN), 148, 221, 360, 429, 490
Atmosphere–Land Exchange Inverse (ALEXI), 336, 341–343, 348, 360
atmospheric boundary layer (ABL) model, 341, 342
atmospheric emissivity, 390
Atmospheric Radiation Measurements (ARM), 47, *49*
automatic classification algorithms, 222
automatic pixel-based water detection, 156–162
　methodology, 156–158
　results, 159–162
available energy, 326–327
avulsion, defined, 20
Ayeyarwady Delta, 202, *203*

B

backscattering variation, 116–118, *117*
Basalt, 89–93, *91*
Bayesian models, 220
belowground biomass, 243–254
　methods, 246–247
　　experimental design, 246
　　spectral reflectance data collection, 246
　　spectral reflectance model development, 246–247
　　study sites, 246
　overview, 243–246
　results, 247–253
　　aboveground biomass and spectral reflectance, 251–252
　　correlations among foliar N, 251–252
　　hyper- and multispectral reflectance, 247–248
　　performance, 250
　　predicting root, 249–250
　　prediction, 249, 252–253
Bernoulli Effect of winds, 24
Bhabar belt, 81
binary/fractional mapping, SC, 427–429
biodiversity
　in African continent, 136
　floral and faunal, 209
　mangrove forests, 185–186
　in Mekong Delta in Vietnam, 136
biogeomorphology, 26–27
biophysical data collection, *240*

507

biophysical parameter inversion, 221
biophysical parameter mapping, 253
    green biomass (GBM), 239
    leaf area index (LAI), 239–241
    leaf chlorophyll content (LCC), 239
    methods, 239–242
    model calibration and validation, 241
    monthly composite products, 241–242
    overview, 238–239
    phenology extraction, 241–242
    results, 242–243
    satellite data, 241
    using multispectral satellite data, 238–243
    vegetation fraction (VF), 239
Blue Revolution, WCA, 270, 279, 302, 489
broadband planetary albedo, 388
broad flood plains, *82*

C

California Irrigation Management Information System (CIMIS) network, 331
canopy height model (CHM), 181
carbonate rocks, 87–88
carbon budget, wetlands, 279
carbon conservation, 187
case studies, 150–168
    automatic pixel-based water detection, 156–162
    fully automatic pixel-based flood detection, 162–168
    Gagnoa, Cote d'IvVoire, 287–294, **290–292**
    Ghana, 287–294
    groundwater, 100–102
    groundwater targeting, 100–102
    IV wetland, 287–294
    semi-automatic object-based flood detection, 150–156
    wetlands, 226–254
CBERS-02B imagery, 228
Central Ground Water Board, 100
central volcano, groundwater in, *91*
change detection, 7, 10, 12, 188, 199–200, 220–221, 471
    flood, 148
    multi-temporal change detection methods, 222
    SAR-based water detection, 148
    wetlands, 220–221
change vector analysis (CVA), 221
channel migration, 20
classification and regression trees (CART), 221
classification hierarchy, *123*
Climate Prediction Center morphing algorithm (CMORPH), 40, 45
cloud cover issues, ET estimation, 354–355
cloud-free snow cover mapping
    daily, 432–436
    improved daily, 436–437
    snow water equivalent mapping and, 432–436
Clouds and Earth's Radiant Energy System (CERES), 325
coarse spatial resolution, 122, 218, 354, 446
coastal area protection, 187–188
coastal erosion, 7, 21, *24*, 204
coastal geomorphology, 21–24
    Airborne LiDAR Bathymetry (ALB), 22–23
    coastal changeand retreat, 23–24
    LiDAR surveys, *22*
    multispectral bathymetry, 22
    overview, 21
    suspended sediment concentration (SSC), 23
Coastal Resources Division (CRD), Georgia, 242
Coastal Service Center (CSC) of NOAA, 24
coconut crop, 403–406
community land model (CLM), 325, 353
Compact High Resolution Imaging Spectrometer (CHRIS), 335
complex auto-logistic regression, 222
conditional random field, 220
Congaree River, South Carolina, 26
Consortium of Spatial Information (CSI) network, 279
Constellation of small Satellites for Mediterranean basin Observation (COSMO-SkyMed), 115, 138, 145, 148, 151, 153, 218
Consultative Group on International Agricultural Research (CGIAR), 279
convolutional neural network (CNN), 221, 489
corn crop, 409–414
correlation analysis, wetlands vegetation, 230
coupling T_R–VI scatterplots with SVAT models, 352–353
cross-track interferometry, *8*
crown diameter (CD), 181
crystalline terrains, profile of, *94*

D

daily evapotranspiration, 347–348
Damming effect, 97, *98*
data/derivative maps, *78*
decision support system (DSS), 279, 489, 494
deep learning (DL), 180, 221, 449, 479
Defense Meteorological Satellite Program (DMSP), 128, 473
Deformation, Ecosystem Structure, and Dynamics of Ice (DESDynI), **42**
deltaic terrains, *83*, 83–84
Department of Space, 100
depiction of wetlands, *280*
diameter at breast height (DBH), 187
digital elevation models (DEMs), 12–13, 18, 40, 70, 125–126, 149, 163, 165, 167, 206, 281
digital orthophoto quarter quads (DOQQs), 19
digital terrain models (DTMs), 11, 12
dikes, 97–99, *98*
disaggregated ALEXI (DisALEXI), 341–343
disasters, 114–115, 136–137, 196, 206–207, 253, 451–456, **454**, *455*, 481
dissolved organic carbon (DOC), 196
dolomite, 87–88
double bounce scattering, 116
downscaling methods, SC, 448–451
drainage density, 6
drop size distribution (DSD), 43
droughts, 54–55, 102, 227, 242–243, 351, 386, 475
dry smooth soil (DS), 122
dual-fiber system, 223
Dual-frequency Precipitation Radar (DPR), 45
dual-temperature-difference (DTD) model, 342

Index

E

Earth observation (EO), 57, 153, 157, 217, 317–318, **319–323**, 493–497
   in disaster management, 136
   for IV wetlands, 302
   missions, 39
   open-source data, 221
   satellite sensors, 225
   time-series, 283
Earth Observation Center (EOC), 139
Earth Observing System (EOS), 40
Earth Resources Observation and Science (EROS), 52, 197
ecosystem level, 183–188. *see also under* mangrove forests
Ecosystem Spaceborne Thermal Radiometer Experiment on Space Station (ECOSTRESS) sensor, 332, 342, 361
Eddy Covariance (EC), 182
electromagnetic radiation, 117, 185–186
electromagnetic sensing, 41
electromagnetic spectrum, 40
El Niño, 226–227, 237
emergent macrophyte (EM), 122
energy flux process, *47*
engineering intelligence, 5
enhanced ellipsoid corrected (EEC), 162
Enhanced Thematic Mapper Plus (ETM+), 70, 72, 284
Environmental Satellite (ENVISAT), 26
ERS-1, *138*, 217, 499
ERS-2, *138*, 217
eskers, 84
ET Mapping Algorithm (ETMA), 40
eucalyptus crops, 406–409
European Centre for Medium-range Weather Forecasts (ECMWF), 325
European Remote Sensing Satellites (ERS), 70, 148
European Space Agency (ESA), 41, 136, 219
evaporation fraction (EF), 51, 349, *406*
evapotranspiration (ET), 39, 40, 46–52, 313–364
   defined, 317
   Earth observation (EO), **319–323**
   EO-based operational products for, 363
   estimation methods for, 40–41, *51*, 318–355
      land surface models (LSM), 353–354
      net radiation, 324–327
      radiometric land surface temperature methods for, 335–348
      scatterplot-based methods for, 348–353
      seasonal estimates and cloud cover issues, 354–355
      vegetation-based methods, 327–335
   global, 50–52
   intercomparison studies, 355–357
   overview, 317–318
   problems in cropped areas, 357–363
      landscape heterogeneity, 357–361
      model complexity, equiynality, and errors in, 362–363
      spatial disaggregation, 357–361
   regional local/field-scale studies, 47, 50
   SEB, 46

exploration parameters, 80–81, 96–98
exploratory drilling, 68
Extreme Ultra-Violet and X-Ray Irradiance Sensors (EXIS), 46

F

False Color Composite (FCC), 70–71, 93
Feature-oriented Principal Components Selection, 70
field-of-view (FOV), 223
field-plot data, IV wetland, 281
floating Mmcrophyte (FM), 122
flood, 113–130, 135–169
   case studies, 150–168
      RAMAFLOOD, 150–156
      TFS, 162–168
      WAMAPRO, 156–162
   defined, 114
   forecasting, 127–129
      preparedness phase, 127
      prevention phase, 127–128
      structure of, 128
   overview, 114–115, 136–139
   SAR-based water detection, **146**, 146–150
      change detection, 148
      contextual classification, 148–149
      integration of auxiliary data, 149–150
      thresholding, 147–148
      visual interpretation, 147
   SAR interaction/water bodies, 139–146
      partially submerged vegetation, 143–145
      rough open water, 143
      smooth open water, 141–143
      urban flooding, 145–146
   satellite observations of, 127–128
   socioeconomic impacts of, 115
   studies, remote sensing in, 115–129
      forecasting, 127–129
      GIS and hydrological models, 123–127
      mapping with SAR and ancillary data, 116–123
flooded forest (FF), *117*, 122
flood extent variation, *123–124*
flood mapping, 116–123, **119–121**
   accuracies/errors/uncertainties of, 122–123
   characteristics, 116–117
   methodologies, 118–122
floodplain analysis, 18–19, *20*
floodplain mapping, 18–19
flood risk zone identification, 127
fluvial landforms, 18–21
   channel migration, 20
   floodplain analysis, 18–19, *20*
   LiDAR imagery, 18
   overview, 18
   streambank retreat (SR), 21
Food and Agriculture Organization of the United Nations (FAO), 196, 201
food security
   in Africa, 268, 269, 279, 302
   blue revolution, 270
   and economic development, 279
   green revolution, 270

needs of fastest-growing continent, 268
population growth and, 269
forest level, 181–183
Fractal Net Evolution Approach (FNEA), 153–154
fractured hard rocks, 96–98. *see also* hard rocks
exploration parameters in, 97–98
groundwater potential in, 97
types of, 97
FROM-GLC, 218
Fruit Circuit pole, 399–401
fully automatic pixel-based flood detection, 162–168
methodology, 162–166
results, 166–168

G

Gaussian Lapp filter, 180
GEE cloud platform, 221
generalized additive modeling, 11
Geocoded Incidence Angle Mask (GIM), 163
geodiversity mapping, 27
Geographic Information Systems (GIS), 10, 17, 20, 23, 28, 74, 78, *78*, 78–79, *78–79*, 115, 123–127, 129–130, 151, 157, 162, 284, 302, 477, 479, 487, 493
geomorphological studies, 3–28, **4**, *28*
aeolian landforms, 24–26
aircraft and satellite systems, 6
alpine and polar periglacial environments, 10–12
biogeomorphology, 26–27
coastal geomorphology, 21–24
fluvial landforms, 18–21
glacial geomorphology, 12–14
historical perspective, 5–10
mass wasting, 14–18
overview, 3–5
quantitative strategies for, 6
remote sensing and, 27–28
spatial and temporal scales, *9*, 9–10
Geostationary Earth Orbital (GEO), 44, **48**, 473
Geostationary Lightning Mapper (GLM), 46
Geostationary Operational Environmental Satellite-R Series (GOES-R), 40, 46
Geostationary Operational Environmental Satellites (GOES), 40, 46, 57, 128, 341–342, 425
glacial geomorphology, 12–14
opportunities and challenges, 13–14
overview, 12–13
glacial terrains, 84–85
GLC_FCS30_2020, 219
Global Data Assimilation System (GDAS), 325
Global Digital Elevation Model Version 2 (GDEM V2), 164–165, 167
Global Flood Detection System (GFDS), 129
Global Flood Monitoring System (GFMS), 129
global flood monitoring tools, **129**
Global Inventory Modeling and Mapping Studies (GIMMS), 226, 227
Global Lakes and Wetlands Database (GLWD), 219
Global Land Data Assimilation (GLDAS), 325
Global Land Survey (GLS), 200–201
Global Land Survey 2005 (GLS2005) Landsat, 282
Global Mangrove Watch (GMW), 184, 219

Global Modeling and Assimilation Office (GMAO), 324–325
global positioning system (GPS), 7, 17, 18, 27–28, *28*, 471
Global Potential Wetland Distribution Dataset (GPWD), 220
Global Precipitation Climatology Project (GPCP), 45
global precipitation estimation, 44–45
Global Precipitation Measurement (GPM), **42**, 45, 57, 129
Global Reservoir and Dam Database (GRanD), 219
GlobCover, 219
GlobeLand30, 218–219
Google Earth Engine (GEE), 217, 294
Google Earth imagery, *24*
GPM Microwave Imager (GMI), 45
graphical user interface (GUI), 150
gravity recovery and climate experiment (GRACE), 42, 56, *56*, 75, 473, 474–476
green biomass (GBM), 238, 239
green revolution, WCA, 270, 279, 302, 493
Gross Primary Productivity (GPP), 182–183
ground ellipsoid corrected (GEC), 162
ground heat flux, 326–327
groundwater, 39, 41, 56–57, *67*, 68–81, *79*, 81, 83–103, *90–91*, *358*, 473, 475–477, 481, 497
assessments, 67–68
aerial photo satellite, 68
exploratory drilling, 68
hydrogeological/geophysical, 68
medium-resolution satellite, 67
prospects map, *79*
stress, 75–77
factors causing, 76
indicators of, 76–77
groundwater targeting, 65–102
case studies, 100–102
exploration, 66, *67*
hard rock terrain, 94–99
fractured hard rocks, 96–98
remote sensing parameters, 98–99
weathered hard rocks, 95–96
indicators, **67**
information extraction, 68–79
groundwater recharge/discharge, 73–74
groundwater stress, 75–77
groundwater use, 74–75
hydrogeomorphology, 70–72
overview, 65–66
potential, 95–96
role of remote sensing, 66–68
semi-consolidated/consolidated sedimentary rock, 86–89
carbonate rocks, 87
remote sensing parameters, 88–89
sandstone-shale aquifers, 87
sandstone-shale/carbonate rocks, 87–88
unconsolidated rock terrain, 79–86
aeolian deposits, 85
alluvial aquifers, 80–84
exploration parameters, 85–86
glacial terrains, 84–85
volcanic terrain, 89–93
exploration parameters, 92–93
remote sensing parameters, 93
typical profiles in, 90–91

Index

H

hard rock terrain, 94–99, *96. see also under* groundwater targeting
height above nearest drainage index (HAND), 150
HH polarization, 117, 145, 151
humid region parameters, 80–81
hydrogeological/geophysical, 68
hydrogeomorphology, 70–72, *71–72*
hydrological cycle, 40
hydrological models, 123–127
Hydrologic Modeling System (HMS), 127
hydrologic variables, 39
hyperspectral reflectance, 178, *179*

I

ice, *8*, 11, *11*, 13, 54–55, 84, 219, 467–499
Ice Mapping System (IMS), 427
IKONOS, 186, 197, 218
image histogram thresholding, 222
image texture algorithms, 222
incident photosynthetically active radiation (PARinc), 391
Indian Remote Sensing Satellites (IRS), 70–71, 76, 89, 197, 199
Individual Tree Crowns (ITC), 180–181
individual tree detection and delineation (ITDD), 181
individual tree level, 180–181
Indus River, *20*
Information Technology for Humanitarian Assistance, Cooperation and Action (ITHACA), 136
inherent optical properties (IOPs), 222
inland valley (IV) wetlands, 267–302, *269, 280*
   accuracies, error and uncertainties, 302
   agriculture *vs.* conservation, 295–301
   agroecological and soil zones, **282**, *285*
   characterization and classification of, **271–273**, 287–294
   cloud computing, 294
   defined, 279–280
   delineation and mapping methods, **271–273**, 282–287, *284*
      automated methods for, 283–284, **286**
      existing methods, 282–283
      semi-automated methods for, 284–287
   ecoregional approach, 279–280, 281
   field-plot data on, 281
   overview, 268–270, 279
   remote sensing data for, 280–281
   satellite sensor data, **274–278**
   spatial data layers and classes, **297–301**
   spatial data weights models, 295–301
   study areas, 279–280, 281
inland waters, 55, 143, 222, 428
in situ data, 126, 199, 279
instantaneous evapotranspiration, 347–348
Instrument aboard the Ice, Cloud, and land Elevation (ICESat), 42, 440, 441
integrated approach, 77–79
integrated Grassland Observing Site (iGOS), 53, *54*
integrate spatial-contextual, 148–149
integration of auxiliary data, 149–150
Interactive Multisensor Snow, 427
International Institute for Tropical Agriculture (IITA), 279
irrigated crops, 401–406
   irrigated coconut crop, 403–406
   irrigated limon crop, 401–403
Iterative Self-Organizing Data Analysis Techniques (ISODATA) Field, 221

J

Japan Aerospace Exploration Agency (JAXA), 487
Japanese Earth Resources Satellite (JERS), 118, 122, 126, 138, 217
Japanese Earth Resources Satellite Synthetic Aperture Radar (JERS SAR), 282–283
Joint Research Centre Global Surface Water Survey and Mapping map, 219

K

Kalangi river, 100–101
kame terraces, 84
karst, defined, 87
kernel methods, 220
kettle holes, 84
k-means clustering, 221
k-nearest neighbor (KNN), 221, 477, 489

L

lacunarity technique, 205, *205*
Land Processes Distributed Active Archive Center (LP DAAC), 52
Landsat, 23, 25–26, 69–77, 89, 102, 127, *186*, *199*, 199–202, 204–209, 218–219, 221–223, *291*, *292*, 326, 330–331, 351, 357–358, 471, 472, 485–487
   false color image, *51*
   for geomorphic features, 6–7, 18
   in India, 73
   Missouri River floodplain, *19*
   moderate-resolution satellite data, 198
   pre-processed and classified, 202
   to study of dune fields, 25
   Terra and Aqua missions, 40
   use as opposed to aerial photography, 13
   of Yellow River Delta in China, 20
Landsat 1, 74
Landsat 5, 50, 219, 328
Landsat 7, 11, 70, 72, 74, 206, 245–248, **248**, *248*, *249*, 250, 252, 475, 497
Landsat 8, 206, 252, 254, 328, 387–389, 396, *397–398*, 398–401, 405, 413–414, 479
Landsat imagery, 6
Landsat TM, 12, 26, 72–74, 89, 101–102, 218
landscape heterogeneity, and ET estimation, 357–361
landslide studies, 16–17
   applications of remote sensing, 16–17
   ground-based laser scans, 17
   inventory strategy, 17
   by LiDAR imagery, 16, *16*
   monitoring critical sites, 17
   role of SAR, 15, *15*
Land Surface Analysis Satellite Applications Facility (LSA-SAF), 325
land surface models (LSM), 318, 353–354

land surface temperature (LST), 51, 318, 491–492
Land Surface Water Index (LSWI), 220
land use/land cover (LULC), 50, 184, 281, 287
land use/land cover classification system, 288, **289**, 291
land-water surface discrimination, 222
La Niña, 226–227, 237
Large Format Camera (LFC), 69–70
lava plateau, groundwater in, 90
L-band SAR sensors, 144
leaf area index (LAI), 144, 182, 239–241
leaf chlorophyll content (LCC), 239
leaf level, 178–180
length of the growing period (LGP) method, 281
Light Detection and Ranging (LiDAR), 25, 6–7, 9, 12–13, 17–18, 22, 22–23, 25, 25–26, 149, 168, 178, 181, 186, 188, 206, 208, 217, 471, 487–488
    fluvial landforms, 18
    for food zone mapping, 9
    landslide studies by, 16, 16
    point cloud from, 181
Light Use Efficiency (LUE), 180, 180, 253, 389
limestone, 87–88, 88
limon crop, 401–403
lineaments, 69–70, 98–99
LISFLOOD-FP model, 126
lithology, 69–70
loess, 85
Low Earth Orbital (LEO), 44

M

machine learning (ML) methods, 148, 182, 211, 221, 429
Magnetometer (MAG), 46
mangrove forests, 176–188
    aerial view of, 183
    ecosystem level, 183–188
        biodiversity, 185–186
        carbon conservation, 187
        coastal area protection, 187–188
        tidal-driven nutrient exchange, 184–185
    forest level, 181–183
    functional indicators of, 177, 177–178
    individual tree level, 180–181
    leaf level, 178–180
    overview, 176–178
    photosynthetic capacity of, 182–183, 187
Mangrove Recognition Index (MRI), 184
mangroves mapping, 218
mangrove wetlands, 195–210, **196**, 197, **198**
    defined, 197
    Global Land Survey (GLS), 200–201
    high-resolution, 207–208
    for mapping/monitoring, 197, 201–205
    methods of, 198–200
    natural disasters, 206–207
    overview, 195–197
    remote sensing data, 197
    scale issues, 198
    species discrimination, 205
Mapping EvapoTranspiration at high Resolution with Internalized Calibration (METRIC) model, 326, 336–341
Marena Oklahoma In Situ Sensor Testbed (MOISST), 53

marker-controlled watershed segmentation (MCWS), 181
Markov random field (MRF), 149, 220
Mask R-CNN, 221
mass wasting, 14–18
    defined, 14
    landslide studies, 16–17
    role of SAR in landslide analysis, 15, 15
meandering, 82–83
mean original reflectance, 179
medium-resolution satellite, 67
Mekong Delta, 156, 159, 160–161
microwave energy, 217
Microwave Imaging Radiometer with Aperture Synthesis (MIRAS), 54
microwave radiation, 116, 116
microwave remote sensing, 41
mid-century revolution, 6
military demobilization, 5
minimum distance (MD), 221
minimum mapping unit (MMU), 166
Missouri River floodplain, 19
M/M-ET estimation algorithm, 49
model calibration and validation, 241
model complexity, equifinality, and errors in evapotranspiration (ET), 362–363
Modelo de Grandes Bacias-Instituto de Pesquisas Hidráulicas (MGB-IPH), 125–126
Moderate Resolution Imaging Spectroradiometer (MODIS), 47, 51–52, 74, 122, 197, 218, 241–242, 243, 283–284, 343–344, 385–389, 392, **393**, 394, **395**, 396, 408, 426–456
    -based mapping protocol, 226
    Global Evapotranspiration Dataset (MOD16), 51–52
Modern Era-Retrospective Analysis for Research and Applications (MERRA), 325
Modified Normalized Difference Water Index (MNDWI), 220
Monitoring, Reporting, and Verication (MRV), 209
Monteith RUE model, 385, 387, 389
monthly composite products, 241–242
moraines, defined, 84
multi-scale hydrological studies, 37–57
    evapotranspiration estimation, 46–52
        global, 50–52
        regional local/field-scale studies, 47, 50
        SEB, 46
    overview, 38–40
    precipitation estimation, 41–46
        global, 44–45
        missions on, 45–46
        NEXRAD, 43–44
        radar principles of, 41–43
    remote sensing advances in, 40–41
    soil moisture estimation, 52–55
        global, 54
        regional local/ field-scale studies, 52–53
        satellite mission on, 54–55
multi-sensor data
    evapotranspiration estimation on, 46–52
    precipitation estimation on, 41–46
    radar, 41–43
    soil moisture estimation, 52–55
multi-sensor joint observation, 225–226

Index

multispectral bathymetry, 22
Multi-Spectral Scanner (MSS), *71*, 77
multi-temporal change detection methods, 222
Murray Global Intertidal Change (MGIC), 220

N

National Aeronautics and Space Administration (NASA), **42**, **129**, 241, 325
National Centers for Environmental Prediction (NCEP), 128
National Centre of Space Research (CNES), France, **42**, 136
National Environmental Satellite, Data, and Information Service (NESDIS), **45**, **48**
National Geospatial-Intelligence Agency (NGA), 208
National Land Cover Database, 184
National Oceanic and Atmospheric Administration (NOAA), 46, 73, 127–128, 325
  Climate Prediction Center Merged Analysis of Precipitation (CMAP NOAA), 325
National Oceanic and Atmospheric Administration Advanced Very High Resolution Radiometer (NOAA AVHRR), 73, 218
National Remote Sensing Agency, 71, 76–77
Natural Resources Conservation Service (NCRS), 129
net radiation, 324–327
  available energy and ground heat flux, 326–327
  calculation, 324
  outgoing shortwave/longwave at high spatial resolution, 326
  regional and global datasets for, 324–326
neural network (NN), 221
Next Generation Radar (NEXRAD), 43–44, *44*
Noah Multiparameterization (Noah-MP) model, 353
non-flooded forest (NFF), *117*, 122
Normalized Difference Vegetation Index (NDVI), 73, 182, 206–207, *207*, 220, 360, 388, 390–391, 414
Normalized Difference Vegetation Index product (NDVI3g), 226, 227
Normalized Difference Water Index (NDWI), 220
normalized snow cover index (NSCI), *439*
North of Minas Gerais pole, 396–399
NPOESS Preparatory Project (NPP), 57
numerical models, **125**
Numerical Terradynamic Simulation Group (NTSG), 51

O

object-based image analysis (OBIA), 17, 221
Ogallala aquifer system, 75
Oklahoma Mesonet, 53. *see also* Marena Oklahoma In Situ Sensor Testbed (MOISST)
one-dimensional (1D), 126
Open Geospatial Consortium (OGC), 166
operational land imager (OLI), 206
ordinary least squares linear regression method (OLS), 229–230
outgoing shortwave/longwave radiation, 326

P

palaeochannels, 100–101, *101*
Palmer Drought Severity Index (PDSI), 55, *56*, 243, *244*
partial least squares (PLS) regression, 222, 225, 245
partially submerged vegetation, 143–145
particle size distribution (PSD), 43
passive remote sensors, 218
passive sensors, 218
pasture crops, 406–409
periglacial environments, 10–12
  change analysis, 11–12
  opportunitiesand challenges, 12
  overview, 10–11
permafrost degradation, 10
persistent scatterer interferometry (PSI), 8, 15, 17
Petrolina/Juazeiro pole, 392–396
Phased Array type L-band Synthetic Aperture Radar (PALSAR), 117, 122, 138, 197
phenology extraction, 241–242
Photochemical Reflectance Index (PRI), 182–183
photogrammetrically derived topographic data (PDTD), 19
photosynthetic capacity, 182–183, 187
physics, principles of, 39
piedmont deposits, 80–81, *81*
  arid region parameters, 81
  humid region parameters, 80–81
Planck's law, 389
point cloud, *181*
polarization, 142–143, 145
Polar Operational Environmental Satellites (POES), 128
potential evapotranspiration (PET), 318
Practical Salinity Unit (PSU), 54
Precipitation Estimation from Remotely Sensed Information using Articial Neural Networks (PERSIANN), 40
  PERSIANN Cloud Classification System (PERSIANN-CCS), **45**, **45**
precipitation estimation on multi-sensor data, 41–46
  global, 44–45
  missions on, 45–46
  NEXRAD, 43–44
  radar principles of, 41–43
Precipitation Radar (PR), 129
Priestly-Taylor Jet Propulsion Laboratory (PT-JPL) model, 327, 331–332, 335
principal component analysis, 70, 222
proximal sensing of wetlands, 223–225
pseudo-invariant features (PIFs), 225
Public Works Department, 71
Pulicat Lake, 100

Q

Quantitative Precipitation Estimates (QPE), 128
Quantitative Precipitation Forecasts (QPF), 128
QuickBird data, 70, 218

R

Radar Imaging Satellite (RISAT), 138
RADAR Satellite (RADARSAT), 70, 126
RADARSAT-1, 217
RADARSAT-2, 217
Radiative Transfer Models (RTM), 182
Radio Detection and Ranging (RADAR), 41, **43**, 70, 138, **141**

radiometric land surface temperature methods, 335–348
  comparison, 348
  instantaneous/daily ET, 347–348
  one-source models, 336–341
  SSEB model, 343–347
  SSEBop, 343–347
  T_R-based methods, 343–347
  two-source models, 341–343
rainfed crops, 406–414
  corn crop, 409–414
  eucalyptus and pasture crops, 406–409
random forest (RF), 221
rapid geomorphic assessments (RGAs), 21
rayleigh scattering, 43
recurrent neural networks (RNN), 221
Reduced Emissions from Deforestation and Degradation (REDD+), 209
regional local/field-scale studies, 47, 50, 52–53
relative radiometric normalization (RRN) method, 225
relative sea level rise (RSLR), 208–209
resultant scattering mechanisms, *116*
Roanoke River floodplain (United States), 19
rock glacier movement, *14*
Root Mean Square Errors (RMSE), 126
rough open water (ROW), 122
runoff (R), 39

S

SAFER (Simple Algorithm for Evapotranspiration Retrieving), 385, 387, *389*, 391
salt marshes
  biophysical characteristics of, 239
  environments, *224*
  habitats, 226, 239, 253
  mapping of, 223, 487
  patches, 242
  species, *224*
  stress in, 243
  tidal, 226
  vegetation, 223, 226
sand dunes, 85
sandstone-shale aquifers, 87
sandstone-shale/carbonate rocks, 87–88
satellite mission, 54
  SMAP, 54–55
  SWOT, 55
Satellite Pour l'Observation de la Terre (SPOT), 69, 197, 218, 287
satellite precipitation products, **48–49**
ScanSAR (SC), 150
scatterplot-based methods, 348–353, *349*
  coupling T_R–VI scatterplots with SVAT models, 352–353
  day–night temperature diference and VI methods, 352
  surface-to-air temperature diference and VI methods, 351
  T_R–albedo methods, 351–352
  T_R–VI methods, 350
Schoenoplectus acutus, 245, 249, 250, 253
seasonal evapotranspiration, estimation, 354–355
seeded region growing (SRG), 181

self-organizing maps (SOMs), 148
semi-automatic object-based flood detection, 150–156, **151**, 154
  methodology, 151–154
  results, 154–156
semi-consolidated/consolidated sedimentary rock, 86–89. *see also under* groundwater
sensible heat flux (H), 46
Sentinel-1, 138, 161, 169, 217, 225, 479, 483
Sentinel-2, 225, 330, 331, 351, 358, 360, 361, 475, 479, 483, 486
Sentinel-3, 386, 360, 361
shear fractures, 97. *see also* fractured hard rocks
shortwave infrared (SWIR), 40
Shuttle Imaging Radar missions (SIR-A, SIR-B, and SIR-C), 217
Shuttle Radar Topography Mission (SRTM), 25, 72, 125–126, 138, 164, 187–188, 197, 206
Simple Remote Sensing Evapotranspiration Model (Sim-ReSET), 342
simple visual interpretation, 222
simplified remote sensing, *79*
Simplied Surface Energy Balance (SSEBop), 343–347
Simplified Surface Energy Balance Index (S-SEBI MAP), 40
Simplified Surface Energy Balance (SSEB) model, 343–347
sink hole, *88*
smooth open water (SOW), 122, 141–143
snake algorithms, 149
snow cover (SC), 425–457
  daily cloud-free (*see* cloud-free snow cover mapping)
  flexible multiday combinations of, 431–437
  livestock disasters in pastoral area, 451–456
    accuracy assessment, **454**
    early warning and risk assessment, 452
    factors of, 452
    flow chart, *453*
    livestock factors, 452
    models, 452–456
    snow and weather factors, 452
    social economy factors, 452
    terrain and grassland factors, 452
  MODIS mapping and standard products, principles of, 427–430
    and accuracy, 429–430
    binary/fractional mapping, 427–429
  overview, 425–426
  parameters, 437–439
  scaling effect in, 446–451
    downscaling methods, 448–451
    upscaling methods, 446–448
  as water resource
    for lake-level and watershed analyses, 439–446
    over lake basin for lake-level analysis, 440–442
    quantitative water resource assessment, 442–446
    snowmelt runoff model (*see* snowmelt runoff model)
snow cover duration (SCD), 437–439
snow-covered area (SCA), 426
snow cover end dates (SCED), 437–439
snow cover index (SCI), 437–439
snow cover onset dates (SCOD), 437–439
snow-mapping algorithm (SNOMAP), 428

Index

snowmelt runoff model, 442–446
   datasets and methodology, 443
   results, 443–446
snow season length (SSL), 439
Snow Telemetry (SNOTEL) stations, 430
snow water equivalent (SWE), 426, 432–436
Soil and the Moisture and Ocean Salinity (SMOS), 41, 52, 54
Soil Conservation Service (SCS), 126–127
Soil Moisture, Active and Passive (SMAP), 52, 54–55, 57, 334
soil moisture estimation, 41, 52–55, **53**
   global, 54
   regional local/ field-scale studies, 52–53
   satellite mission on, 54–55
soils/land use/drainage, 72–73
Solar-Induced Chlorophyll Fluorescence (SIF), 178–180, *180*
Solar Ultraviolet Imager (SUVI), 46
SONAR, 19
Southern Great Plains (SGP), 53
Spaceborne Imaging Radar (SIR), 69–70, 72, 138
Space Environment In Situ Suite (SEISS), 46
Space Shuttle Topographic Mission (SRTM), 283
spatial disaggregation, 357–361
spatial resolution, **137**
Special Sensor Microwave Imager (SSMI/S), 128, 426, 474
species discrimination, 205
Specific Leaf Area (SLA), 178
spectral reflectance ratios, 178
spectral response function (SRF), 225
Spinning Enhanced Visible and Infrared Imager (SEVIRI) radiometer, 353
Sriharikota launch station, 100
SRTM water body mask (SWBD), 164
streambank retreat (SR), 21
StripMap (SM), 150
Submerged Mangrove Recognition Index (SMRI), 184, *185*
support vector machine (SVM), 221
Surface Energy Balance Algorithm for Land (SEBAL), 40, 326, 335, 336–341, **337–339**
surface energy balance (SEB) methods, 46–47, 48, *49*, 51, 335, 336–341
Surface Energy Balance System (SEBS), 40, 336, 473
surface scattering, 116–117
Surface Water and Ocean Topography (SWOT), 55, 57
surface water/groundwater-irrigated fields, *75*
surface water mapping, 222
suspended sediment concentration (SSC), 23
synthetic aperture radars (SARs), 7–9, 14, 28, 54–55, 69–70, 116–123, 137–169, *138*, 140–143, 145–146, 147, 153–154, 168, 217, *284*
   ability of, 18
   civil spaceborne, **139**
   flood studies with, 481–483
   misclassification of flooding in, **140**
   role in landslide analysis, 15, *15*
   SAR/ancillary data, 116–123, **119–121**
   SAR-based water detection, 146–150
      change detection, 148
      contextual classification, 148–149
      integration of auxiliary data, 149–150
      thresholding, 147–148
      visual interpretation, 147
   SAR interaction/water bodies, 139–146
      partially submerged vegetation, 143–145
      smooth open water, 141–143
   SAR interferometry (InSAR), 8, *8*, *14*, 15, 17–18, 28
   scenes, *141*
   simulation of flood extent, 19
   surface-scattered radiation, *140*
   water/flood extent mapping, **146**
   water image, 222
   wetland mapping, 282–283

T

TanDEM-X SAR system, 8
tensile fractures, 97. *see also* fractured hard rocks
tensile joints, 97
TerraSAR-X, 118, 145, 150, *152*, **153**, *155*, 156–157, *160*, 160–162, **162–163**, 167, 218
Terrestrial Laser Scans (TLS), 182
terrestrial water storage (TWS), 55
Thematic Mapper (TM), 69, 74, 89, 102, 126–127
thresholding, 147–148
tidal-driven nutrient exchange, 184–185
TIR Thermal Infrared (TIR), 47
top-of-atmosphere (TOA), 206
top of canopy (TOC), 223
T_R–albedo methods, 351–352
T_R-based methods, 343–347
   Simplified Surface Energy Balance (SSEB) model, 343–347
   SSEBop (operational Simplied Surface Energy Balance), 343–347
tree height (TH), 181
TRMM Microwave Imager (TMI), 40, 129
TRMM Multi-satellite Precipitation Analysis (TMPA), 129
Tropical Rainfall Measuring Mission (TRMM), 40, 45, 128–129, **129**
T_R–VI methods, 350
two-dimensional (2D), 126
two-source energy balance (TSEB) models, 336, 341–343
Typha domingensis, 246, 253

U

unconsolidated rock terrain, 79–86. *see also under* groundwater targeting
uncrewed aerial systems (UAS), 318
U-Net, 221
United Nations Institute for Training (UNOSAT), 136
unmanned aerial vehicles (UAVs), *180–181*, 180–182, 217
upscaling methods, SC, 446–448
urban flooding, 145–146
US Environmental Protection Agency (EPA), 216
US Geological Survey Earth Resources Observation and Science (USGS EROS), 197, 207

V

valley fills, 81–82
variable infiltration capacity (VIC) model, 325, 353
Vegetation Condition Index (VCI), 226–227, 229, 232–233, *233*
vegetation fraction (VF), 239

Vegetation Health Products (VHP), 226
Vegetation Horizontal Occlusion Index (VHOI), 182
vegetation indices (VI), 45, 51–52, 183, 221–222, **241**
    advent of remote sensing estimation of, 51
    application of crop coefficient, 330–331
    -based methods to estimate ET, 327–335
    calibration methods, 328–330
    canopy resistance, 332
    comparison of, 335
    crop coefficients, 328–331
    empirical, 328–331
    MOD16, 332
    from NOAA AVHRR, 73
    from NODIS NDVI, 73
    PT-JPL Model, 331–332
    soil evaporation, 332–334
Vegetation Water Index, 223
virtual machine (VM), 162
visible (VIS), 44
Visible Atmospheric Resistant Index (VARI), 242
visual interpretation, 147
volcanic region groundwater, *90*
volcanic terrain, 89–93. *see also under* groundwater targeting
volcanoes, 89–93, *91*, 476
volumetric backscattering, 116–117
Volumetric Soil Moisture (VSM), 53
VV polarization, 117, 145

W

WaMaPro, **151**, 157–158, 161
water cycle, 39–40
water hyacinth, 144, 223
Water Indication Mask (WAM), 164
water productivity (WP), 383–415
    agricultural growing region scale, 392–401
        Fruit Circuit pole, 399–401
        North of Minas Gerais pole, 396–399
        Petrolina/Juazeiro pole, 392–396
    assessments, **388**
    crop scale, 401–414
        irrigated crops (*see* irrigated crops)
        rainfed crops (*see* rainfed crops)
    effects of climate and land use changes, 384
    modeling of components, 387–392
    overview, 383–386
    policy implications, 414–415
    study regions and data set, *386*, 386–387
wavelet transform method, wetlands vegetation, 229–230
weathered hard rocks, *95*, 95–96. *see also* hard rocks
    exploration parameters in, 96
    groundwater potential in, 95–96
    profile of, 95
weathering, 13, 90, 94–97
Web Processing Service (WPS), 157
West and Central Africa (WCA), IV wetlands. *see* inland valley (IV) wetlands
wet-dry zone, *229*, 231, *231*, **232**
wetland(s), 214–254
    biomass, 221–222
    case studies, 226–254
    challenges and solutions, 223–226
    change detection, 220–221
    classification, 220–221
    defined, 216
    evolution, 217–222
        application, 220–222
        data, 217–220
        techniques, 220–222
    multi-sensor joint observation, 225–226
    new methods in, 226–254
    overview, 216–217
    proximal sensing of, 223–225
    surface water mapping, 222
wetland-related datasets, 218–220
wetlands vegetation, 221–222, 226–238, 253
    climate trend change, 233–234
    complexity of, 237–238
    correlation analysis, 230
    impact of climate change on, 235–237
    indices and data sources, 227–228
    methods, 227–230
    ordinary least squares linear regression method (OLS), 229–230
    overview, 226–227
    responses of, 234–235
    results, 231–238
    selection of, 228
    spatio-temporal trend change
        of NDVI, 231–232
        of VCI, 232–233
    wavelet transform method, 229–230
wetter agroecosystems, 269
White Sands National Monument, *25*
Wide Dynamic Range Vegetation Index (WDRVI), 242
Wide Swath Mode (WSM), 147
WorldCover database, 184
WorldView-2, 182

Y

Yellow River Delta in China, 20